Extended Addition Principle For any events E and F, from a sample space S,

$$P(E \cup F) = P(E) + P(F) - P(E \cap F).$$

Bayes' Formula For any events E and $F_1, F_2, \ldots, F_n$, from a sample space S, where $F_1 \cup F_2 \cup \cdots \cup F_n = S$,

$$P(F_i|E) = \frac{P(F_i) \cdot P(E|F_i)}{P(F_1) \cdot P(E|F_1) + \cdots + P(F_n) \cdot P(E|F_n)}.$$

Probability in a Bernoulli Experiment If p is the probability of success in a single trial of a Bernoulli experiment, the probability of x successes and $n - x$ failures in n independent repeated trials of the experiment is

$$\binom{n}{x} p^x (1 - p)^{n-x}.$$

Mean The mean of the n numbers, $x_1, x_2, x_3, \ldots, x_n$, is

$$\bar{x} = \frac{\sum(x)}{n}.$$

Standard Deviation The standard deviation of the n numbers, $x_1, x_2, x_3, \ldots, x_n$, with mean $\bar{x}$, is

$$s = \sqrt{\frac{\sum (x - \bar{x})^2}{n - 1}}.$$

Binomial Distribution Suppose an experiment is a series of n independent repeated trials, where the probability of a success in a single trial is always p. Let x be the number of successes in the n trials. Then the probability that exactly x successes will occur in n trials is given by

$$\binom{n}{x} p^x (1 - p)^{n-x}.$$

The mean is

$$\mu = np,$$

and the standard deviation is

$$\sigma = \sqrt{np(1 - p)}.$$

Fourth Edition

MATHEMATICS WITH APPLICATIONS

IN THE MANAGEMENT, NATURAL, AND SOCIAL SCIENCES

Margaret L. Lial
American River College

Charles D. Miller
American River College

SCOTT, FORESMAN AND COMPANY
Glenview, Illinois
London, England

Alternative Books

Mathematics and Calculus with Applications, Second Edition, is a longer book, featuring more topics, especially in calculus.

Finite Mathematics, Third Edition, features an introduction to the broad areas of linear mathematics and probability for students of business and life science. *Finite Mathematics* is written at a higher level than this text.

Calculus with Applications, Third Edition, gives an extensive treatment of calculus for those students needing a separate course in calculus.

Computer Applications for Finite Mathematics and Calculus by Donald R. Coscia (with accompanying diskettes for Apple II or IBM PC) is designed for courses in finite mathematics; calculus for the life, management, and social sciences; statistics; or discrete mathematics.

Cover photo: Jerome and Laya Wiesner Building, MIT by I. M. Pei & Partners. Colored wall bands by Kenneth Noland. Photo © 1985 Steve Rosenthal.

Library of Congress Cataloging-in-Publication Data

Lial, Margaret L.
  Mathematics with applications in the management, natural, and social sciences.
  Includes index.
  1. Mathematics—1961–   . I. Miller, Charles David. II. Title.
QA37.2L5   1987   510   86-25998
ISBN 0–673–18464–1

Preface

Mathematics with Applications, Fourth Edition, introduces the mathematical topics needed by students of management, social science, and natural science for a combined finite mathematics and calculus course. It also can be used for a separate course in either finite mathematics or applied calculus as well as for a course in applied college algebra. Previous editions have been used in courses such as Mathematics with Applications, Mathematics for Management and Social Science, Mathematics for Business and Economics, Finite Mathematics, Introduction to Analysis, Algebra with Applications, and College Mathematics.

The only prerequisite assumed is a course in algebra. A review of algebra is given in Chapter 1, and a diagnostic pretest is included in the *Instructor's Guide* to help determine which students need review.

Features new in the fourth edition include the following.

The *algebra review* (Chapter 1) is presented in greater depth and includes more challenging exercises on rational expressions, exponents, and absolute value equations and inequalities.

The chapters on *matrices and systems of linear equations* have been combined into one chapter (Chapter 6) to provide a clearer presentation which uses systems to motivate matrices.

The chapter on *linear programming* (Chapter 7) has been thoroughly and carefully revised.

The discussion of *permutations and combinations* (Chapter 8) now provides a stronger distinction between permutations and combinations.

The *derivatives of exponential and logarithmic functions* are now included in the same chapter (Chapter 11) as all the other derivatives.

The treatment of *multivariate functions* (Chapter 12) has been expanded.

The *fundamental theorem of calculus* is now presented in a separate section (Chapter 13).

The following popular features are retained from previous editions.

Problems at the side help test student understanding. By working the more than 400 marginal problems as topics are encountered, students can quickly locate their source of difficulty.

Examples and exercises are extensive. This book continues to have substantially more examples (more than 400) and exercises (3900 drill and 1300 applications) than other leading books.

Realistic applications help students grasp new concepts and see the many interesting ways in which mathematics is used. The applications in this book are as realistic as possible.

One or more case studies follow most of the chapters of the book. These cases present the topics of the chapters as they apply to real-life situations. We have tried to keep the cases short and simple. The concepts of the course come alive when students can see how mathematics is used at Upjohn or Booz, Allen and Hamilton, for example.

The book can be used for a variety of courses, including the following.

Finite Mathematics and Calculus (one year or less) Use the entire book; cover topics from Chapters 1–4 as needed before proceeding to further topics.

Finite Mathematics (one semester or one or two quarters) Use as much of Chapters 1–4 as needed, and then go into the topics of Chapters 5–10 as time and local needs permit.

Calculus (one semester or quarter) Use Chapters 1–4 as necessary, and then use Chapters 11–13.

College Algebra with Applications (one semester or quarter) Use Chapters 1–8 with the topics of Chapters 7 and 8 being optional.

Chapter interdependence is as follows.

Chapter		**Prerequisite**
1	Fundamentals of Algebra	None
2	Linear Models	None
3	Polynomial and Rational Models	Chapter 2
4	Exponential and Logarithmic Models	Chapter 2
5	Mathematics of Finance	None
6	Systems of Linear Equations and Matrices	None
7	Linear Programming	Chapters 2 and 6
8	Sets, Counting, and Probability	None
9	Further Topics in Probability	Chapter 8
10	Statistics	Chapter 8
11	Differential Calculus	Chapters 2–4
12	Applications of the Derivative	Chapter 11
13	Integral Calculus	Chapters 11–12

Additional materials for this book include the following.

The *Instructor's Guide* contains a complete testing program, including an algebra pretest (with answers), a bank of test items for each chapter (with answers), and answers to even-numbered text exercises. In addition, the manual offers background material for instructors and presents computer programs in BASIC, for Leontief models and for Markov chains, and the program LINPRO, for the simplex method of linear programming.

The *Study Guide and Student's Solutions Manual* contains the solutions to the odd-numbered exercises. Detailed solutions are provided for each major concept presented within each section, with helpful hints, cautions, and suggestions. More condensed solutions are then presented for the remaining odd-numbered problems.

Many instructors helped us prepare this revision. In particular, we would like to thank Garret Etgen, University of Houston; George Evanovich, Iona College; Patricia Hirschy, Delaware Technical and Community College; Alec Ingraham, New Hampshire College; Donald Mason, Elmhurst College; Carol Nessmith, Georgia Southern College; and Daniel Symancyk, Anne Arundel Community College.

The text answers were checked by James Hodge, College of Lake County; Louis F. Hoelzle, Bucks County Community College; and Wing M. Park, College of Lake County.

At Scott Foresman we were very fortunate to be able to work with an extremely talented group of people: Bill Poole, Pam Carlson, and Marge Prullage helped establish the overall framework of the revision, and Adam Bryer helped with the detailed changes.

Margaret L. Lial
Charles D. Miller

TO THE STUDENT

Using the text Side problems, which help reinforce skills and pinpoint sources of difficulty, are referred to in the text by numbers within colored squares: ■ at the ends of examples and ① elsewhere. Squares without numbers (■) mark the ends of examples that have no side problems.

Computer exercises Section, chapter review, and case exercises requiring the use of a computer are highlighted by colored exercise numbers.

Additional help If you would like more help with mathematics, you may want to get a copy of the *Study Guide and Student's Solutions Manual,* containing solutions to the odd-numbered exercises. Detailed hints and cautions are provided for the major concepts of each section, and more condensed solutions are presented for the remaining problems. Your local college bookstore either has this book or can order it for you.

Contents

Chapter 9   Further Topics in Probability   403

Chapter 10   Statistics   445

Appendix: Tables   676

Answers to Selected Exercises   692

Index   737

1 Fundamentals of Algebra

This book is about the application of mathematics to various subjects, mainly business, social science, and biology. Almost all of these applications begin with some real-world problem that is solved by creating a mathematical model. Mathematical models are discussed in more detail later. Basically, a **mathematical model** is an equation (or other mathematical relationship) that represents a given problem.

For example, suppose the real-world problem is to find the area of a floor, 12 feet by 18 feet. Here, a mathematical model would be the formula for the area of a rectangle, Area = Length × Width, or $A = LW$. Substituting the numbers, 18 for L and 12 for W into this formula gives the area as $A = 18 \times 12 = 216$ square feet.

Most mathematical models involve algebra. Algebra is used first to set up a model and then to simplify the resulting equations. Since algebra is so vital to a study of the applications of mathematics, this book begins with a review of some of the fundamental ideas of algebra.

1.1 The Real Numbers

The various types of numbers used in this book can be explained with a diagram called a **number line.** Draw a number line by choosing any point on a horizontal line and labeling it 0. Then choose any point to the right of 0 and label it 1. The distance between 0 and 1 gives a unit of measure that can be used repeatedly to locate points to the right of 1, labeled 2, 3, 4, and so on, and points to the left of 0, labeled $-1, -2, -3, -4$, and so on. A number line with several sample numbers located (or **graphed**) on it is shown in Figure 1.1. ☐

1 Draw a number line and graph the numbers $-4, -1,$ 0, 1, 2.5, 13/4 on it.

Answer:

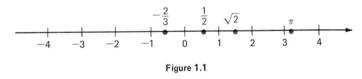

Figure 1.1

Any number that can be associated with a point on the number line is called a **real number.*** All the numbers used in this book are real numbers. The names of the most common types of real numbers are as follows.

---

*Not all numbers are real numbers. An example of a number that is not a real number is $\sqrt{-1}$.

The Real Numbers	
Natural (counting) numbers	1, 2, 3, 4, . . .
Whole numbers	0, 1, 2, 3, 4, . . .
Integers	. . . , $-3, -2, -1, 0, 1, 2, 3, . . .$
Rational numbers	All numbers of the form p/q, where p and q are integers, with $q \neq 0$
Irrational numbers	Real numbers that are not rational

The three dots in this box show that the numbers continue indefinitely in the same way. The relationships among these types of numbers are shown in Figure 1.2. Notice, for example, that the integers are also rational numbers and real numbers, but the integers are not irrational numbers.

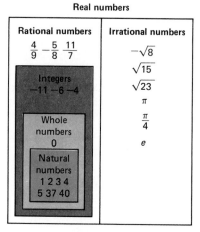

Figure 1.2

One example of an irrational number is π, the ratio of the circumference of a circle to its diameter. The number π can be approximated by writing $\pi \approx 3.14159$ or $\pi \approx 22/7$ ($\approx$ means "is approximately equal to"), but there is no rational number that is exactly equal to π. Another irrational number can be found by constructing a triangle having a 90° angle, with the two shortest sides each 1 unit long, as shown in Figure 1.3. The third side can be shown to have a length which is irrational (the length is $\sqrt{2}$ units).

Figure 1.3

Many whole numbers have square roots which are irrational numbers; in fact, if a whole number is not the square of an integer, then its square root is irrational.

Example 1 List all the names of sets of numbers that apply to the following.
(a) 6

Consult Figure 1.2 to see that 6 is a counting number, whole number, integer, rational number, and real number.
(b) 3/4

This number is rational and real.
(c) $\sqrt{8}$

Since 8 is not the square of an integer, $\sqrt{8}$ is irrational and real.

Some basic properties of the real numbers are given below.

2 Name all the types of numbers that apply to the following.
(a) -2
(b) $-5/8$
(c) π

Answer:
(a) Integer, rational, real
(b) Rational, real
(c) Irrational, real

Properties of the Real Numbers For all real numbers a, b, and c:

Commutative properties
$$a + b = b + a$$
$$ab = ba$$

Associative properties
$$(a + b) + c = a + (b + c)$$
$$(ab)c = a(bc)$$

Identity properties
There exists a unique real number 0 such that
$$a + 0 = a \quad \text{and} \quad 0 + a = a.$$
There exists a unique real number 1 such that
$$a \cdot 1 = a \quad \text{and} \quad 1 \cdot a = a.$$

Inverse properties
There exists a unique real number $-a$ such that
$$a + (-a) = 0 \quad \text{and} \quad (-a) + a = 0.$$
If $a \neq 0$, there exists a unique real number $1/a$ such that
$$a \cdot \frac{1}{a} = 1 \quad \text{and} \quad \frac{1}{a} \cdot a = 1.$$

Distributive property
$$a(b + c) = ab + ac$$

Example 2 The following statements are examples of the commutative property. Notice that the order of the numbers changes from one side of the equals sign to the other.

(a) $6 + x = x + 6$

(b) $(6 + x) + 9 = (x + 6) + 9$

(c) $(6 + x) + 9 = 9 + (6 + x)$

(d) $5 \cdot (9 \cdot 8) = (9 \cdot 8) \cdot 5$

(e) $5 \cdot (9 \cdot 8) = 5 \cdot (8 \cdot 9)$ ■

Example 3 The following statements are examples of the associative properties. Here the order of the numbers does not change, but the placement of the parentheses does change.

(a) $4 + (9 + 8) = (4 + 9) + 8$

(b) $3(9x) = (3 \cdot 9)x$

(c) $(\sqrt{3} + \sqrt{7}) + 2\sqrt{6} = \sqrt{3} + (\sqrt{7} + 2\sqrt{6})$ ▣3

3 Name the property illustrated in each of the following examples.

(a) $(2 + 3) + 9$
$= (3 + 2) + 9$

(b) $(2 + 3) + 9$
$= 2 + (3 + 9)$

(c) $(2 + 3) + 9$
$= 9 + (2 + 3)$

(d) $(4 \cdot 6)p = (6 \cdot 4)p$

(e) $4(6p) = (4 \cdot 6)p$

Answer:

(a) Commutative property

(b) Associative property

(c) Commutative property

(d) Commutative property

(e) Associative property

The identity properties give some special properties of the numbers 0 and 1. Since 0 preserves the identity of a real number under addition, 0 is the **identity element for addition.** In the same way, 1 preserves the identity of a real number under multiplication and is the **identity element for multiplication.**

Example 4 By the identity properties.

(a) $-8 + 0 = -8$,

(b) $(-9)1 = -9$. ■

The number $-a$ is the **additive inverse** of a, and $1/a$ is the **multiplicative inverse** of the nonzero real number a.

Example 5 By the inverse properties,

(a) $9 + (-9) = 0$,

(b) $-15 + 15 = 0$,

(c) $6 \cdot \dfrac{1}{6} = 1$,

(d) $-8 \cdot \left(\dfrac{1}{-8}\right) = 1$,

(e) $\dfrac{1}{\sqrt{5}} \cdot \sqrt{5} = 1$.

(f) There is no real number x such that $0 \cdot x = 1$, so 0 has no inverse for multiplication. **4**

4 Name the property illustrated in each of the following examples.

(a) $2 + 0 = 2$

(b) $-\dfrac{1}{4} \cdot (-4) = 1$

(c) $-\dfrac{1}{4} + \dfrac{1}{4} = 0$

(d) $1 \cdot \dfrac{2}{3} = \dfrac{2}{3}$

Answer:

(a) identity property

(b) inverse property

(c) inverse property

(d) identity property

One of the most important properties of the real numbers, and the only one that involves both addition and multiplication, is the distributive property. The next example shows how this property is applied.

Example 6 By the distributive property,
(a) $9(6 + 4) = 9 \cdot 6 + 9 \cdot 4$,
(b) $3(x + y) = 3x + 3y$,
(c) $-\sqrt{5}(m + 2) = -m\sqrt{5} - 2\sqrt{5}$. (Here the radical is written after the variable, so we won't confuse, for example, $-\sqrt{5}m$ and $-\sqrt{5m}$.)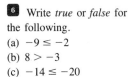

5 Use the distributive property to complete each of the following.
(a) $4(-2 + 5)$
(b) $2(a + b)$
(c) $-3(p + 1)$

Answer:
(a) $4(-2) + 4(5)$
(b) $2a + 2b$
(c) $-3p + (-3)(1)$

Comparing two real numbers requires symbols that indicate their order on the number line. The following symbols are used to indicate that one number is greater than or less than another number.

> $<$ means *is less than* $\leq$ means *is less than or equal to*
> $>$ means *is greater than* $\geq$ means *is greater than or equal to*

The definitions below show how the number line is used to decide which of two given numbers is the greater.

> For real numbers a and b,
> if a is to the left of b on a number line, then $a < b$;
> if a is to the right of b on a number line, then $a > b$.

6 Write *true* or *false* for the following.
(a) $-9 \leq -2$
(b) $8 > -3$
(c) $-14 \leq -20$

Answer:
(a) True
(b) True
(c) False

Example 7 Write *true* or *false* for each of the following.
(a) $8 < 12$
   This statement says that 8 is less than 12, which is true.
(b) $-6 > -3$
   The graph of Figure 1.4 shows both -6 and -3. Since -6 is to the *left* of -3, then $-6 < -3$, and the given statement is false.
(c) $-2 \leq -2$
   Since $-2 = -2$, this statement is true. **6**

Figure 1.4

A number line can be used to draw the graph of a set of numbers as shown in the next few examples.

7 Graph all integers x such that
(a) $-3 < x < 5$
(b) $1 \leq x \leq 5$.

Answer:
(a) graph with points at $-3, -2, -1, 0, 1, 2, 3, 4, 5$
(b) graph with points at $0, 1, 2, 3, 4, 5, 6$

Example 8 Graph all integers x such that $1 < x < 5$.
   The only integers between 1 and 5 are 2, 3, and 4. These integers are graphed on the number line of Figure 1.5. **7**

number line with points marked at 2, 3, 4 on scale $-3, -2, -1, 0, 1, 2, 3, 4, 5$

Figure 1.5

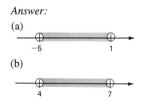

8 Graph all real numbers x such that
(a) $-5 < x < 1$
(b) $4 < x < 7$.

Answer:
(a)

(b)

9 Graph all real numbers x such that
(a) $x \geq 4$
(b) $-2 \leq x \leq 1$.

Answer:
(a)

(b)

10 Simplify the following.
(a) $16 \div 8 + 9 \div 3$
(b) $[-7 + (-9)](-4) - 8(3)$
(c) $\dfrac{-11 - (-12) - 4 \cdot 5}{-8 - (-3)(-5)}$
(d) $\dfrac{36 \div 4 \cdot 3 \div 9 + 1}{9 \div (-6) \cdot 8 - 4}$

Answer:
(a) 5
(b) 40
(c) 19/23
(d) $-1/4$

Example 9 Graph all real numbers x such that $-2 < x < 3$.
   The graph includes all the real numbers between -2 and 3 and not just the integers. Graph these numbers by drawing a heavy line from -2 to 3 on the number line, as in Figure 1.6. Open dots at -2 and 3 show that neither of these points belongs to the graph. **8**

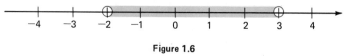

Figure 1.6

Example 10 Graph all real numbers x such that $x \geq -2$.
   Start at -2 and draw a heavy line to the right, as in Figure 1.7. Put a heavy dot at -2 to show that -2 itself is part of the graph. **9**

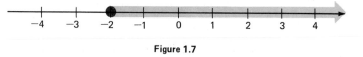

Figure 1.7

Order of Operations Avoid possible ambiguity when working problems with real numbers with the following *order of operations,* which has been agreed on as the most useful. (This order of operations is used by computers and many calculators.)

Order of Operations

If parentheses or square brackets are present:

1. Work separately above and below any fraction bar.
2. Use the rules below within each set of parentheses or square brackets. Start with the innermost and work outward.

If no parentheses are present:

1. Find all powers and roots, working from left to right.
2. Do any multiplications or divisions in the order in which they occur, working from left to right.
3. Do any additions or subtractions in the order in which they occur, working from left to right.

Example 11 Use the order of operations to simplify the following.
(a) $6 \div 3 + 2 \cdot 4 = 2 + 2 \cdot 4 = 2 + 8 = 10$

(b) $\dfrac{-9(-3) + (-5)}{2(-8) - 5(3)} = \dfrac{27 + (-5)}{-16 - 15} = \dfrac{22}{-31} = -\dfrac{22}{31}$

(c) $-(3 - 5) - [2 - (9 - 13)] = -(-2) - [2 - (-4)]$
$= 2 - [6]$
$= -4$ **10**

Absolute Value Distance is always given as a nonnegative number. For example, the distance from 0 to -2 on a number line is 2, the same as the distance from 0 to 2. The *absolute value* of both 2 and -2 is 2. The absolute value of a number a gives the distance on the number line from a to 0. Write the absolute value of the real number a as $|a|$. For example, the distance on the number line from 9 to 0 is 9, as is the distance from -9 to 0. (See Figure 1.8.) By definition, $|9| = 9$ and $|-9| = 9$.

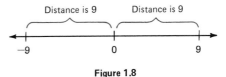

Figure 1.8

The definition of absolute value can be stated more formally as follows.

Absolute Value For any real number a,

$$|a| = a \qquad \text{if } a \geq 0$$
$$|a| = -a \qquad \text{if } a < 0.$$

The second part of this definition requires a little care. If a is a *negative* number, then $-a$ is a *positive* number. Thus, for any value of a, $|a| \geq 0$.

11 Find the following.
(a) $|-6|$
(b) $-|7|$
(c) $-|-2|$
(d) $|-3 - 4|$
(e) $|2 - 7|$

Answer:
(a) 6
(b) -7
(c) -2
(d) 7
(e) 5

Example 12 Find each of the following.
(a) $|5|$
    Since $5 > 0$, then $|5| = 5$.
(b) $|-5| = 5$
(c) $-|-5| = -(5) = -5$
(d) $|0| = 0$
(e) $|8 - 9|$
    First simplify the expression inside the absolute value bars.
$$|8 - 9| = |-1| = 1$$
(f) $|-4 - 7| = |-11| = 11$ **11**

1.1 Exercises

Name all the types of numbers that apply to the following. (See Example 1.)

1. 6

2. -9

3. -7

4. 0

5. $\dfrac{1}{2}$

6. $-\dfrac{5}{11}$

7. $\sqrt{7}$

8. $-\sqrt{11}$

9. π

10. $\dfrac{1}{\pi}$

Label each of the following as true *or* false.

11. Every integer is a rational number.

12. Every integer is a whole number.

13. Every whole number is an integer.

14. Some whole numbers are not natural numbers.

15. There is a natural number that is not a whole number.

16. Every rational number is a natural number.

17. Every natural number is a rational number.

18. No whole numbers are rational.

Identify the properties that are illustrated in each of the following. Some will require more than one property. Assume all variables represent real numbers. (See Examples 2–6.)

19. $8 \cdot 9 = 9 \cdot 8$

20. $3 + (-3) = 0$

21. $3 + (-3) = (-3) + 3$

22. $0 + (-7) = (-7) + 0$

23. $-7 + 0 = -7$

24. $8 + (12 + 6) = (8 + 12) + 6$

25. $[9(-3)] \cdot 2 = 9[(-3) \cdot 2]$

26. $8(m + 4) = 8m + 8 \cdot 4$

27. $x(y + 2) = xy + 2x$

28. $(7 - y) + 0 = 7 - y$

29. $8(4 + 2) = (2 + 4)8$

30. $x \cdot \dfrac{1}{x} + x \cdot \dfrac{1}{x} = x\left(\dfrac{1}{x} + \dfrac{1}{x}\right)$ (if $x \neq 0$)

Graph each of the following on a number line. (See Example 8.)

31. All integers x such that $-5 < x < 5$

32. All integers x such that $-4 < x < 2$

33. All whole numbers x such that $x \leq 3$

34. All whole numbers x such that $1 \leq x \leq 8$

35. All natural numbers x such that $-1 < x < 5$

36. All natural numbers x such that $x \leq 2$

Graph all real numbers x satisfying the following conditions. (See Examples 9 and 10.)

37. $x \leq 4$

38. $x > 5$

39. $6 \leq x$

40. $x \geq 8$

41. $x > -2$

42. $-4 < x$

43. $-5 < x < -3$

44. $3 < x < 5$

45. $-3 \leq x \leq 6$

46. $8 \leq x \leq 14$

47. $1 < x \leq 6$

48. $-4 \leq x < 3$

Simplify each of the following expressions, using the order of operations given in the text. (See Example 11.)

49. $9 \div 3 \cdot 4 \cdot 2$

50. $18 \cdot 3 \div 9 \div 2$

51. $8 + 7 \cdot 2 + (-5)$

52. $-9 + 6 \cdot 5 + (-8)$

53. $(-9 + 4 \cdot 3)(-7)$

54. $-15(-8 - 4 \div 2)$

55. $-(-13 - 8) - [(-4 - 5) - (-8 + 15)]$

56. $[8 - (-7 - 19)] + (-4 - 11) - (-7 - 2)$

57. $\dfrac{-8 + (-4)(-6) \div 12}{4 - (-3)}$

58. $\dfrac{15 \div 5 \cdot 4 \div 6 - 8}{-6 - (-5) - 8 \div 2}$

59. $\dfrac{17 \div (3 \cdot 5 + 2) \div 8}{-6 \cdot 5 - 3 - 3(-11)}$

60. $\dfrac{-12(-3) + (-8)(-5) + (-6)}{17(-3) + 4 \cdot 8 + (-7)(-2) - (-5)}$

61. $\dfrac{-9.23(5.87) + 6.993}{1.225(-8.601) - 148(.0723)}$

62. $\dfrac{189.4(3.221) - 9.447(-8.772)}{4.889[3.177 - 8.291(3.427)]}$

Evaluate each of the following. (See Example 12.)

63. $-|-4|$

64. $-|-2|$

65. $|6 - 4|$

66. $|3 - 17|$

67. $-|12 + (-8)|$

68. $|-6 + (-15)|$

69. $|8 - (-9)|$

70. $|-3 - (-2)|$

71. $|8| - |-4|$

72. $|-9| - |-12|$

73. $-|-4| - |-1 - 14|$

74. $-|6| - |-12 - 4|$

In each of the following problems, fill in the blank with either =, <, or >, so that the resulting statement is true.

75. $|5|$ ____ $|-5|$

76. $|3|$ ____ $|-3|$

77. $-|7|$ ____ $|7|$

78. $-|-4|$ ____ $|4|$

79. $|10 - 3|$ ____ $|3 - 10|$

80. $|6 - (-4)|$ ____ $|-4 - 6|$

81. $|1 - 4|$ ____ $|4 - 1|$

82. $|10 - 8|$ ____ $|8 - 10|$

83. $|-2 + 8|$ ____ $|2 - 8|$

84. $|3 + 1|$ ____ $|-3 - 1|$

85. $|3| \cdot |-5|$ ____ $|3(-5)|$

86. $|3| \cdot |2|$ _____ $|3(2)|$

87. $|3 - 2|$ _____ $|3| - |2|$

88. $|5 - 1|$ _____ $|5| - |1|$

89. In general, if a and b are any real numbers having the same sign (both negative or both positive), is it always true that $|a + b| = |a| + |b|$?

90. If a and b are any real numbers, is it always true that $|a + b| = |a| + |b|$?

91. If a and b are any two real numbers, is it always true that $|a - b| = |b - a|$?

92. For which real numbers b does $|2 - b| = |2 + b|$?

Management *Use inequality symbols to rewrite each of the following statements, which are based on a recent article in* Business Week *magazine.* Let x represent the unknown in each exercise.*

Example: Rewrite ''Sales of Compaq Computer Corporation in a recent year were at least \$500 million'' by letting x represent the sales in millions of dollars. Then $x \geq 500$. ■

93. Compaq Computer Corp. earnings in a recent year were at least \$26 million.

94. Compaq's market share increased to about 15% of the \$9 billion personal computer market.

95. Prices of the Compaq II model of computer start at \$3499.

96. The original Compaq computer now sells for \$1500 to \$2000.

97. Other IBM compatible computers cost between \$1000 and \$1500.

98. The newest Compaq portable weighs less than 25 pounds.

Social Science *Sociologists measure the status of an individual within a society by evaluating for that individual the number x, which gives the percentage of the population with less income than the given person, and the number y, the percentage of the population with less education. The average status is defined as $(x + y)/2$, while the individual's status incongruity is defined by $|(x - y)/2|$. People with high status incongruities would include unemployed Ph.D.'s (low x, high y) and millionaires who didn't make it past the second grade (high x, low y).*

99. What is the highest possible average status for an individual? The lowest?

100. What is the highest possible status incongruity for an individual? The lowest?

101. Jolene Rizzo makes more money than 56% of the population and has more education than 78%. Find her average status and status inconguity.

102. A popular movie star makes more money than 97% of the population and is better educated than 12%. Find the average status and status incongruity for this individual.

1.2 Linear Equations and Applications

One of the main uses of algebra is to solve equations. An **equation** states that two mathematical expressions are equal. Examples of equations include $x + 6 = 9$, $4y + 8 = 12$, and $9z = -36$. The letter in each equation, the unknown, is called the **variable.**

A **solution** of an equation is a number that can be substituted for the variable in the equation to produce a true statement. For example, substituting the number 9 for x in the equation $2x + 1 = 19$ gives

$$2x + 1 = 19$$
$$2(\mathbf{9}) + 1 = 19 \qquad \text{Let } x = 9$$
$$18 + 1 = 19. \qquad \text{True}$$

---

*From ''Compaq Still Confounds Its Skeptics,'' *Business Week*, March 3, 1986, pp. 97, 99.

[1] Is -4 a solution of the following equations?
(a) $3x + 5 = -7$
(b) $2x - 3 = 5$

Answer:
(a) Yes
(b) No

This true statement indicates that 9 is a solution of $2x + 1 = 19$. [1]

Equations that can be written in the form $ax + b = c$, where a, b, and c are real numbers, with $a \neq 0$, are called **linear equations.** Examples of linear equations include $5y + 9 = 16$, $8x = 4$, and $-3p + 5 = -8$. Examples of equations that are *not* linear include $|x| = 4$, $2x^2 = 5x + 6$, and $\sqrt{x + 2} = 4$.

The following properties are used to solve equations.

Properties of Equality   For any real numbers a, b, and c,

(a) if $a = b$, then $a + c = b + c$.   **Addition property of**
(The same number may be   **equality**
added to both sides of an
equation.)

(b) if $a = b$, then $ac = bc$.   **Multiplication property of**
(The same number may be   **equality**
multiplied on both sides of
an equation.)

Recall that subtraction is defined in terms of addition: for all real numbers a and b,

$$a - b = a + (-b);$$

and division is defined in terms of multiplication: for all real numbers a and all nonzero real numbers b,

$$\frac{a}{b} = a \cdot \frac{1}{b}.$$

For this reason, special properties for subtraction and division are not needed, as the following examples show.

Example 1   Solve the linear equation $5x - 3 = 12$.

Using the addition property of equality, add 3 to both sides. This isolates the term containing the variable on one side of the equals sign.

$$5x - 3 = 12$$
$$5x - 3 + 3 = 12 + 3 \quad \text{Add 3 to both sides}$$
$$5x = 15$$

To get $1x$ instead of $5x$ on the left, use the fact that $(1/5) \cdot 5 = 1$, and multiply both sides of the equation by $1/5$.

$$5x = 15$$
$$\frac{1}{5}(5x) = \frac{1}{5}(15) \quad \text{Multiply both sides by } \frac{1}{5}$$
$$1x = 3$$
$$x = 3$$

2 Solve the following.

(a) $3p - 5 = 19$

(b) $4y + 3 = -5$

(c) $-2k + 6 = 2$

Answer:

(a) 8

(b) -2

(c) 2

The solution of the original equation, $5x - 3 = 12$, is 3, which can be checked by substituting 3 for x in the original equation. **2**

Example 2 Solve $2k + 3(k - 4) = 2(k - 3)$.

First simplify this equation using the distributive property. By this property, $3(k - 4)$ is $3k - 3 \cdot 4$, or $3k - 12$. Also, $2(k - 3)$ is $2k - 2 \cdot 3$, or $2k - 6$. The equation can now be written as

$$2k + 3(k - 4) = 2(k - 3)$$
$$2k + 3k - 12 = 2k - 6.$$

On the left, $2k + 3k = (2 + 3)k = 5k$, again by the distributive property, which gives

$$5k - 12 = 2k - 6.$$

One way to proceed is to add $-2k$ to both sides.

$$5k - 12 + (-2k) = 2k - 6 + (-2k) \qquad \text{Add } -2k \text{ to both sides}$$
$$3k - 12 = -6$$

Now add 12 to both sides.

$$3k - 12 + 12 = -6 + 12 \qquad \text{Add 12 to both sides}$$
$$3k = 6$$

Finally, multiply both sides by 1/3.

$$\frac{1}{3}(3k) = \frac{1}{3}(6) \qquad \text{Multiply both sides by } \frac{1}{3}$$
$$k = 2$$

The solution is 2. Check this result by substituting 2 for k in the original equation. **3**

3 Solve the following.

(a) $3(m - 6) + 2(m + 4)$
$= 4m - 2$

(b) $-2(y + 3) + 4y$
$= 3(y + 1) - 6$

Answer:

(a) 8

(b) -3

The next three examples show how to simplify the solution of linear equations involving fractions. Solve these equations by multiplying both sides of the equation by a **common denominator,** a number that can be divided (with remainder 0) by each denominator in the equation.

Example 3 Solve $\dfrac{r}{10} - \dfrac{2}{15} = \dfrac{3r}{20} - \dfrac{1}{5}$.

Here the denominators are 10, 15, 20, and 5. Each of these numbers can be divided into 60; therefore 60 is a common denominator. Multiply both sides of the equation by 60.

$$60\left(\frac{r}{10} - \frac{2}{15}\right) = 60\left(\frac{3r}{20} - \frac{1}{5}\right)$$

Use the distributive property to eliminate the denominators.

$$60\left(\frac{r}{10}\right) - 60\left(\frac{2}{15}\right) = 60\left(\frac{3r}{20}\right) - 60\left(\frac{1}{5}\right)$$

$$6r - 8 = 9r - 12$$

Add $-6r$ and 12 to both sides.

$$6r - 8 + (-6r) + 12 = 9r - 12 + (-6r) + 12$$
$$4 = 3r$$

Multiply both sides by 1/3 to get the solution

$$r = \frac{4}{3}.$$

4 Solve the following.

(a) $\dfrac{x}{2} - \dfrac{x}{4} = 6$

(b) $\dfrac{2x}{3} + \dfrac{1}{2} = \dfrac{x}{4} - \dfrac{9}{2}$

Answer:

(a) 24

(b) -12

Check this solution in the original equation. **4**

Example 4 Solve $\dfrac{4}{3(k + 2)} - \dfrac{k}{3(k + 2)} = \dfrac{5}{3}.$

Multiply both sides of the equation by the common denominator $3(k + 2)$. (Here $k \neq -2$, since $k = -2$ would give a zero denominator, making the fraction meaningless.)

$$3(k + 2) \cdot \frac{4}{3(k + 2)} - 3(k + 2) \cdot \frac{k}{3(k + 2)}$$
$$= 3(k + 2) \cdot \frac{5}{3}$$

$4 - k = 5(k + 2)$	
$4 - k = 5k + 10$	Distributive property
$4 - k + k = 5k + 10 + k$	Add k to both sides
$4 = 6k + 10$	
$4 + (-10) = 6k + 10 + (-10)$	Add -10 to both sides
$-6 = 6k$	
$-1 = k$	Multiply by $\dfrac{1}{6}$

5 Solve the equation

$$\frac{5p + 1}{3(p + 1)}$$
$$= \frac{3p - 3}{3(p + 1)} + \frac{9p - 3}{3(p + 1)}.$$

Answer:

1

The solution is -1. Check this solution in the original equation. **5**

Example 5 Solve $\dfrac{x}{x - 2} = \dfrac{2}{x - 2} + 2.$

Multiply both sides of the equation by $x - 2$, assuming that $x - 2 \neq 0$. This gives

$$x = 2 + 2(x - 2)$$
$$x = 2 + 2x - 4$$
$$x = 2.$$

Recall the assumption that $x - 2 \neq 0$. Since $x = 2$, then $x - 2 = 0$, and the multiplication property of equality does not apply. To see this, substitute 2 for x in the original equation—this substitution produces a zero denominator. Since division by zero is not defined, there is no solution for the given equation.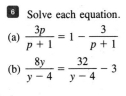

Sometimes an equation must be solved that has more than one letter. Solving for x as in Example 6 is called **solving for a specified variable.** As a general rule, the first few letters of the alphabet, a, b, c, and so on, are used to represent constants, while letters such as x, y, and z are used for variables.

Example 6   Solve for x: $3(2x - 5a) + 4b = 4x - 2$.

Use the distributive property to get

$$6x - 15a + 4b = 4x - 2.$$

Treat x as the variable, the other letters as constants. Get all terms with x on one side of the equals sign, and all terms without x on the other side.

$$6x - 4x = 15a - 4b - 2 \qquad \text{Isolate terms with } x \text{ on the left}$$
$$2x = 15a - 4b - 2 \qquad \text{Combine terms}$$
$$x = \frac{15a - 4b - 2}{2}. \qquad \text{Multiply by } \frac{1}{2}$$

The final equation is solved for x, as required. **7**

Example 7   The formula

$$A = \frac{24f}{b(p + 1)}$$

gives the approximate annual interest rate* for a consumer loan paid off with monthly payments. Here f is the finance charge on the loan, p is the total number of payments, and b is the original balance of the loan. Solve the formula for p.

Treat p as the variable and the other letters as constants. The goal is to isolate p on one side of the equals sign. Begin by multiplying both sides of the formula by $p + 1$, assuming that $p \neq -1$. Since p is the number of payments, p is greater than 0 here.

---

*This formula is not accurate enough for the requirements of federal law.

6 Solve each equation.

(a) $\dfrac{3p}{p + 1} = 1 - \dfrac{3}{p + 1}$

(b) $\dfrac{8y}{y - 4} = \dfrac{32}{y - 4} - 3$

Answer:
Neither equation has a solution.

7 Solve for x.
(a) $2x - 7y = 3xk$
(b) $8(4 - x) + 6p$
   $= -5k - 11yx$

Answer:

(a) $x = \dfrac{7y}{2 - 3k}$

(b) $x = \dfrac{5k + 32 + 6p}{8 - 11y}$

$$A = \frac{24f}{b(p + 1)}$$

$$(p + 1)A = (p + 1)\frac{24f}{b(p + 1)}$$

$$(p + 1)A = \frac{24f}{b}$$

Now think of undoing the operations that have been performed on p. Undo the multiplication by A by multiplying both sides of the equation by $1/A$.

$$\frac{1}{A}(p + 1)A = \frac{1}{A} \cdot \frac{24f}{b}$$

$$p + 1 = \frac{24f}{Ab}.$$

(We must assume $A \neq 0$. Why is this a very safe assumption here?) Finally, undo the addition of 1 to p by adding -1 on each side to get

$$p = \frac{24f}{Ab} - 1,$$

8 Solve $J\left(\dfrac{m}{k} + a\right) = m$

for k.

Answer:

$\dfrac{Jm}{m - Ja}$

a result solved for p. **8**

Word Problems   One of the main reasons for learning mathematics is to be able to use it to solve practical problems. However, for many students, learning how to apply mathematical skills to applications is the most difficult task they face. Hints that may help with applications are given in the rest of this section.

A common difficulty with "word problems" is trying to do everything at once. It is usually best to attack the problem in stages as follows.

Solving Word Problems

1. Decide on the unknown. Name it with some variable that you *write down*. Many students try to skip this step. They are eager to get on with the writing of an equation. But this is an important step. If you don't know what the variable represents, how can you write a meaningful equation or interpret a result?

2. Decide on a variable expression to represent any other unknowns in the problem. For example, if x represents the width of a rectangle, and you know that the length is one more than twice the width, then *write down* that the length is $1 + 2x$.

3. Use the results of Steps 1 and 2 to write an equation.

4. Solve the equation.

5. Check the solution in the words of the original problem.

The following examples illustrate this approach.

Example 8 If the length of a side of a square is increased by 3 cm, the new perimeter is 40 cm more than twice the length of the side of the original square. Find the length of a side of the original square.

> *Step 1* What should the variable represent? To find the length of a side of the original square, let
>
> $$x = \text{length of a side of the original square.}$$
>
> *Step 2* The length of a side of the new square is 3 cm more than the length of a side of the old square, so
>
> $$x + 3 = \text{length of a side of the new square.}$$

Now write a variable expression for the new perimeter. Since the perimeter of a square is four times the length of a side,

$$4(x + 3) = \text{the perimeter of the new square.}$$

> *Step 3* Now write an equation by looking again at the information in the problem. The new perimeter is 40 more than twice the length of a side of the original square, so the equation is

the new perimeter	is	40	more than	twice the side of the original square
$4(x + 3)$	$=$	40	$+$	$2x.$

> *Step 4* Now solve the equation. The result should then be checked by using the wording of the original problem.
>
> $$4(x + 3) = 40 + 2x$$
> $$4x + 12 = 40 + 2x$$
> $$2x = 28$$
> $$x = 14$$

The length of a side of the new square would be $14 + 3 = 17$ cm; its perimeter would be $4(17) = 68$ cm. Twice the length of the side of the original square is $2(14) = 28$ cm. Since $40 + 28 = 68$ cm, the solution checks with the words of the original problem. **9**

9 (a) A triangle has a perimeter of 45 cm. Two of the sides of the triangle are equal in length, with the third side 9 cm longer than either of the two equal sides. Find the lengths of the sides of the triangle.
(b) A rectangle has a perimeter which is five times its width. The length is 4 more than the width. Find the length and width of the rectangle.

Answer:

(a) 12 cm, 12 cm, 21 cm
(b) Length is 12, width is 8

Example 9 Chuck travels 80 km in the same time that Mary travels 180 km. Mary travels 50 km per hour faster than Chuck. Find the speed of each person.
Use the steps given above.

> *Step 1* Use x to represent Chuck's speed, and $x + 50$ to represent Mary's speed, which is 50 km per hour faster than Chuck's.

Step 2   Constant velocity problems of this kind require the distance formula

$$d = rt,$$

where d is the distance traveled in t hours at a constant rate of speed r. The distance traveled by each person is given, along with the fact that the time traveled by each person is the same. Solve the formula $d = rt$ for t.

$$d = rt$$

$$\frac{1}{r} \cdot d = \frac{1}{r} \cdot rt$$

$$\frac{d}{r} = t$$

For Chuck, $d = 80$ and $r = x$, giving $t = 80/x$. For Mary, $d = 180$, $r = x + 50$, and $t = 180/(x + 50)$. Use these facts to complete a chart, which organizes the information given in the problem.

	d	r	t
Chuck	80	x	$\dfrac{80}{x}$
Mary	180	$x + 50$	$\dfrac{180}{x + 50}$

Step 3   Since both people traveled for the same time, the equation is

$$\frac{80}{x} = \frac{180}{x + 50}.$$

Step 4   Multiply both sides of the equation by $x(x + 50)$.

$$x(x + 50)\frac{80}{x} = x(x + 50)\frac{180}{x + 50}$$

$$80(x + 50) = 180x$$

$$80x + 4000 = 180x$$

$$4000 = 100x$$

$$40 = x$$

Since x represents Chuck's speed, Chuck went 40 km per hour. Mary's speed is $x + 50$, or $40 + 50 = 90$ km per hour. Check these results in the words of the original problem.   **10**

10 (a) Tom and Dick are in a foot race. Tom runs at 7 mph and Dick runs at 5 mph. If they start at the same time, how long will it be until they are 1/2 mile apart?
(b) In part (a), suppose the run has a staggered start. If Dick starts first, and Tom starts 10 minutes later, how long will it be until they are neck and neck?

Answer:
(a) 15 min (1/4 hour)
(b) Tom runs 25 min.

Example 10 Rick and Debbi are refinishing an antique table. Working alone, Rick would need 8 days to finish the table, while Debbi would require 10. If they work together, how long will it take to complete the project?

Step 1 Let x represent the number of days it will take Rick and Debbi working together to complete the table.

Step 2 In one day, Rick will do 1/8 of the table, while Debbi will do 1/10. Working together, in one day they will complete $1/x$ of the job.

Step 3 The amount of the job done by each person in one day must equal the amount done working together in one day, or

$$\frac{1}{8} + \frac{1}{10} = \frac{1}{x}.$$

Step 4 Multiply both sides of this equation by $40x$, getting

$$5x + 4x = 40$$
$$9x = 40$$
$$x = \frac{40}{9} \quad \left(\text{or } 4\frac{4}{9}\right).$$

Working together they can finish the table in 4 4/9 days. ▣ 11

Example 11 Mark Webber receives a $14,000 bonus from his company. He invests part of the money in tax-free bonds at 6% per year and the remainder at 16% per year. He earns $1390 after one year in interest from the investments. Find the amount he has invested at each rate.

Let x represent the amount Jones invests at 6%, so that $14,000 - x$ is the amount invested at 16%. Since interest is given by the product of principal, rate, and time,

interest at 6% $= x \cdot 6\% \cdot 1 = .06x$
interest at 16% $= (14,000 - x)(16\%)(1) = .16(14,000 - x).$

Since the total interest is $1390,

$$.06x + .16(14,000 - x) = 1390.$$

Solve this equation.

$$.06x + .16(14,000) - .16x = 1390$$
$$-.10x + 2240 = 1390$$
$$x = 8500$$

Thus, $8500 was invested at 6%, and $14,000 - $8500 = $5500 at 16%. ▣ 12

▣ 11 (a) Wendy can do a job in 7 hours, while her assistant needs 11. How long would it take them if they worked together?
(b) Factory A can produce a batch of parts twice as fast as factory B. Working together, the two factories need 2 days to fill an order. How long would it take factory B working alone?

Answer:
(a) 77/18 hours
(b) 6 days

▣ 12 An investor owns two pieces of property. One, worth twice as much as the other, returns 12% in annual interest, while the other returns 8%. Find the value of each piece of property if the total annual interest earned is $8000.

Answer:
$50,000 and $25,000

1.2 Exercises

Solve each of the following equations. See Examples 1–5.

1. $4x - 1 = 15$

2. $-3y + 2 = 5$

3. $.2m - .5 = .1m + .7$

4. $.01p + 3.1 = 2.03p - 2.96$

5. $\dfrac{5}{6}k - 2k + \dfrac{1}{3} = \dfrac{2}{3}$

6. $\dfrac{3}{4} + \dfrac{1}{5}r - \dfrac{1}{2} = \dfrac{4}{5}r$

7. $3r + 2 - 5(r + 1) = 6r + 4$

8. $5(a + 3) + 4a - 5 = -(2a - 4)$

9. $2[m - (4 + 2m) + 3] = 2m + 2$

10. $4[2p - (3 - p) + 5] = -7p - 2$

11. $\dfrac{3x - 2}{7} = \dfrac{x + 2}{5}$

12. $\dfrac{2p + 5}{5} = \dfrac{p + 2}{3}$

13. $\dfrac{x}{3} - 7 = 6 - \dfrac{3x}{4}$

14. $\dfrac{y}{3} + 1 = \dfrac{2y}{5} - 4$

15. $\dfrac{1}{4p} + \dfrac{2}{p} = 3$

16. $\dfrac{2}{t} + 6 = \dfrac{5}{2t}$

17. $\dfrac{m}{2} - \dfrac{1}{m} = \dfrac{6m + 5}{12}$

18. $-\dfrac{3k}{2} + \dfrac{9k - 5}{6} = \dfrac{11k + 8}{k}$

19. $\dfrac{2r}{r - 1} = 5 + \dfrac{2}{r - 1}$

20. $\dfrac{3x}{x + 2} = \dfrac{1}{x + 2} - 4$

21. $\dfrac{4}{x - 3} - \dfrac{8}{2x + 5} + \dfrac{3}{x - 3} = 0$

22. $\dfrac{5}{2p + 3} - \dfrac{3}{p - 2} = \dfrac{4}{2p + 3}$

23. $\dfrac{3}{2m + 4} = \dfrac{1}{m + 2} - 2$

24. $\dfrac{8}{3k - 9} - \dfrac{5}{k - 3} = 4$

Solve each of the following equations for x. (See Example 6.) (In Exercises 29–32, recall that $a^2 = a \cdot a$.)

25. $2(x - a) + b = 3x + a$

26. $5x - (2a + c) = a(x + 1)$

27. $ax + b = 3(x - a)$

28. $4a - ax = 3b + bx$

29. $x = a^2x - ax + 3a - 3$

30. $2a = ax - a - 6x + 6$

31. $a^2x + 3x = 2a^2$

32. $ax + b^2 = bx - a^2$

Solve each equation for the specified variable. Assume all denominators are nonzero. (See Example 7.)

33. $PV = k$ for V

34. $i = prt$ for p

35. $V = V_0 + gt$ for g

36. $S = S_o + gt^2 + k$ for g

37. $A = \dfrac{1}{2}(B + b)h$ for B

38. $C = \dfrac{5}{9}(F - 32)$ for F

39. $\dfrac{1}{R} = \dfrac{1}{r_1} + \dfrac{1}{r_2}$ for R

40. $m = \dfrac{Ft}{v_1 - v_2}$ for v_2

Solve each of the following equations. Round to the nearest hundredth.

41. $9.06x + 3.59(8x - 5) = 12.07x + .5612$

42. $-5.74(3.1 - 2.7p) = 1.09p + 5.2588$

43. $\dfrac{2.5x - 7.8}{3.2} + \dfrac{1.2x + 11.5}{5.8} = 6$

44. $\dfrac{4.19x + 2.42}{.05} - \dfrac{5.03x - 9.74}{.02} = 1$

45. $\dfrac{2.63r - 8.99}{1.25} - \dfrac{3.90r - 1.77}{2.45} = r$

46. $\dfrac{8.19m + 2.55}{4.34} - \dfrac{8.17m - 9.94}{1.04} = 4m$

Natural Science *In the metric system of weights and measures, temperature is measured in degrees Celsius (°C) instead of degrees Fahrenheit (°F). To convert back and forth between the two systems, use*

$$C = \frac{5(F - 32)}{9} \quad and \quad F = \frac{9}{5}C + 32.$$

In each of the following exercises, convert to the other system. Round answers to the nearest tenth of a degree if necessary.

47. 20°C **48.** 100°C **49.** 59°F **50.** 86°F

51. 100°F **52.** 350°F **53.** 40°C **54.** 85°C

Management *Example 7 introduced the formula for the approximate annual interest rate of a loan paid off with monthly payments:*

$$A = \frac{24f}{b(p + 1)}.$$

Use this formula to find the value of the variables not given in each of the following. Round A to the nearest percent and round other variables to the nearest whole numbers. (This formula is not accurate enough for the requirements of federal law.)

55. $f = \$800,\quad b = \$4000,\quad p = 36;\quad$ find A

56. $f = \$60,\quad b = \$740,\quad p = 12;\quad$ find A

57. $A = 14\%,\quad b = \$2000,\quad p = 36;\quad$ find f

58. $A = 11\%,\quad b = \$1500,\quad p = 24;\quad$ find f

59. $A = 16\%,\quad f = \$370,\quad p = 36;\quad$ find b

60. $A = 10\%,\quad f = \$490,\quad p = 48;\quad$ find b

Management *When a loan is paid off early, a portion of the finance charge must be returned to the borrower. By one method of calculating finance charge (called the rule of 78), the amount of unearned interest (finance charge to be returned) is given by*

$$u = f \cdot \frac{n(n + 1)}{q(q + 1)}$$

where u represents unearned interest, f is the original finance charge, n is the number of payments remaining when the loan is paid off, and q is the original number of payments. Find the amount of the unearned interest in each of the following.

61. Original finance charge = $800, loan scheduled to run 36 months, paid off with 18 payments remaining.

62. Original finance charge = $1400, loan scheduled to run 48 months, paid off with 12 payments remaining.

63. Original finance charge = $950, loan scheduled to run 24 months, paid off with 6 payments remaining.

64. Original finance charge = $175, loan scheduled to run 12 months, paid off with 3 payments remaining.

Solve each word problem. (See Examples 8–11.)

65. A triangle has a perimeter of 27 cm. One side is twice as long as the shortest side. The third side is 7 cm longer than the shortest side. Find the length of the shortest side.

66. The length of a rectangle is three inches less than twice the width. The perimeter is 54 inches. Find the width.

67. Ms. Prullage invests $20,000 received from an insurance settlement in two ways: some at 13%, and some at 16%. Altogether, she makes $2840 per year interest. How much is invested at each rate?

68. Grey Thornton received $52,000 profit from the sale of some land. He invested part at 15% interest, and the rest at 19% interest. He earned a total of $9040 interest per year. How much did he invest at each rate?

69. Matt Whitney won $100,000 in a state lottery. He paid income tax of 40% on the winnings. Of the rest, he invested some at 8 1/2% and some at 16%, making $5,550 interest per year. How much is invested at each rate?

70. Mary Collins earned $48,000 from royalties on her cookbook. She paid a 40% income tax on these royalties. The balance was invested in two ways, at 7 1/2% and at 10 1/2%. The investments produce $2550 interest income per year. Find the amount invested at each rate.

71. Lucy Day bought two plots of land for a total of $120,000. On the first plot, she made a profit of 15%. On the second, she lost 10%. Her total profit was $5500. How much did she pay for each piece of land?

72. Suppose $20,000 is invested at 12%. How much additional money must be invested at 16% to produce a yield of 14.4% on the entire amount invested?

73. The Old Time Goodies Store sells mixed nuts. Cashews sell for $4 per quarter kilogram, hazelnuts for $3 per quarter kilogram, and peanuts for $1 per quarter kilogram. How many kilograms of peanuts should be added to 10 kilograms of cashews and 8 kilograms of hazelnuts to make a mixture which will sell for $2.50 per quarter kilogram?

74. The Old Time Goodies Store also sells candy. They want to prepare 200 kilograms of a mixture for a special Halloween promotion to sell at $4.84 per kilogram. How much $4 per kilogram candy should be mixed with $5.20 per kilogram candy for the required mix?

Natural Science *Exercises 75 and 76 depend on the idea of the* octane rating *of gasoline, a measure of its antiknock qualities. Actual gasoline blends are compared to standard fuels. In one measure of octane, a standard fuel is made with only two ingredients: heptane and isooctane. For this fuel, the octane rating is the percent of isooctane. For example, a gasoline with an octane rating of 98 has the same antiknock properties as a standard fuel that is 98% isooctane.*

75. How many liters of 94 octane gasoline should be mixed with 200 liters of 99 octane gasoline to get a mixture that is 97 octane?

76. A service station has 92 octane and 98 octane gasoline. How many liters of each should be mixed to provide 12 liters of 96 octane gasoline for a chemistry experiment?

77. On a vacation trip, José averaged 50 mph traveling from Amarillo to Flagstaff. Returning by a different route which covered the same number of miles, he averaged 55 mph. What is the distance between the two cities if his total traveling time was 32 hours?

78. Cindy Smith left by plane to visit her mother in Hartford, 420 kilometers away. Fifteen minutes later, her mother left to meet her at the airport. She drove the 20 kilometers to the airport at 40 kph, arriving just as the plane taxied in. What was the speed of the plane?

79. Pat Schmelling took 20 minutes to drive her boat upstream to water ski at her favorite spot. Coming back later in the day, at the same boat speed, took her 15 minutes. If the current in that part of the river is 5 kph, what was her boat speed?

80. Joe traveled against the wind in a small plane at 180 mph for 3 hours. The return trip with the wind took 2.8 hours. What was the speed of the wind?

81. Mark can clean the house in 9 hours and Wendy in 6. How long will it take them if they work together to clean it?

82. Helen can paint a room in 5 hours. Jay can paint the same room in 4 hours. How long will it take them to paint the room together?

83. Two chemical plants are polluting a river. If plant A produces a predetermined maximum amount of pollution twice as fast as plant B, and together they produce the maximum pollution in 26 hours, how long will it take plant B alone?

84. A sewage treatment plant has two inlet pipes to its settling pond. One can fill the pond in 10 hours, the other in 12 hours. If the first pipe is open for 5 hours and then the second pipe is opened, how long will it take to fill the pond?

85. An inlet pipe can fill Dominic's pool in 5 hours, while an outlet pipe can empty it in 8 hours. In his haste to watch television, Dominic left both pipes open. How long would it then take to fill the pool?

86. Suppose Dominic discovered his error (see Exercise 85) after an hour-long program. If he then closed the outlet pipe, how much longer would be needed to fill the pool?

1.3 Linear Inequalities

Linear equations were discussed in the previous section. In this section, **linear inequalities,** such as

$$5x < 10, \qquad 2m + 1 > 7, \qquad \text{and} \qquad 5x - 2 \le 3,$$

are discussed. To solve a linear inequality, simplify it to the form $x < b$. (Throughout this section, definitions and theorems are given only for $<$, but they are equally valid for $>$, $\le$, or $\ge$.) The following properties are used to simplify an inequality.

> **Properties of Inequality** For real numbers a, b, and c:
>
> (a) if $a < b$, then $a + c < b + c$;
>
> (b) if $a < b$, and if $c > 0$, then $ac < bc$;
>
> (c) if $a < b$, and if $c < 0$, then $ac > bc$.

Pay careful attention to part (c): if both sides of an inequality are multiplied by a negative number, the direction of the inequality symbol must be reversed. For example, starting with the true statement $-3 < 5$ and multiplying both sides by the positive number 2 gives

$$-3 \cdot 2 < 5 \cdot 2,$$

or

$$-6 < 10,$$

still a true statement. On the other hand, starting with $-3 < 5$ and multiplying both sides by the negative number -2 gives a true result only if the direction of the inequality symbol is reversed:

$$-3(-2) > 5(-2)$$
$$6 > -10. \quad \boxed{1}$$

1 (a) First multiply both sides of $-6 < -1$ by 4, and then multiply both sides of $-6 < -1$ by -7.
(b) Multiply both sides of $9 \geq -4$ first by 2, and then by -5.
(c) First add 4, and then add -6 to both sides of $-3 < -1$.

Answer:
(a) $-24 < -4$; $42 > 7$
(b) $18 \geq -8$; $-45 \leq 20$
(c) $1 < 3$; $-9 < -7$

Example 1 Solve $3x + 5 > 11$. Graph the solution.
First, add -5 to both sides.

$$3x + 5 + (-5) > 11 + (-5)$$
$$3x > 6$$

Now multiply both sides by 1/3.

$$\frac{1}{3}(3x) > \frac{1}{3}(6)$$
$$x > 2$$

(Why was the direction of the inequality symbol not changed?) A graph of the solution is shown in Figure 1.9. The endpoint is open to show that 2 is not a solution of the inequality. **2**

2 Solve the following. Graph each solution.
(a) $5z - 11 < 14$
(b) $-3k \leq -12$
(c) $-8y \geq 32$

Answer:
(a) $z < 5$

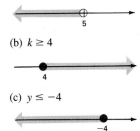

(b) $k \geq 4$

(c) $y \leq -4$

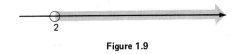

Figure 1.9

Example 2 Solve $4 - 3y \leq 7 + 2y$. Graph the solution.
Add -4 to both sides.

$$4 - 3y + (-4) \leq 7 + 2y + (-4)$$
$$-3y \leq 3 + 2y$$

Add $-2y$ to both sides. Remember that *adding* to both sides never changes the direction of the inequality symbol.

$$-3y + (-2y) \le 3 + 2y + (-2y)$$
$$-5y \le 3$$

Multiply both sides by $-1/5$. Since $-1/5$ is negative, change the direction of the inequality symbol.

$$-\frac{1}{5}(-5y) \ge -\frac{1}{5}(3)$$

$$y \ge -\frac{3}{5}$$

3 Solve the following. Graph each solution.
(a) $8 - 6t \ge 2t + 24$
(b) $-4r + 3(r + 1) < 2r$

Answer:
(a) $t \le -2$

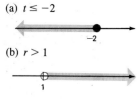

(b) $r > 1$

Figure 1.10 shows a graph of the solution. **3**

Figure 1.10

Example 3 Solve $-2 < 5 + 3m < 20$. Graph the solution.
  The inequality $-2 < 5 + 3m < 20$ says that $5 + 3m$ is *between* -2 and 20. Solve this inequality with an extension of the properties given above. Work as follows, first adding -5 to each part.

$$-2 + (-5) < 5 + 3m + (-5) < 20 + (-5)$$
$$-7 < 3m < 15$$

Now multiply each part by $1/3$.

$$-\frac{7}{3} < m < 5$$

4 Solve each of the following. Graph each solution.
(a) $9 < k + 5 < 13$
(b) $-6 \le 2z + 4 \le 12$

Answer:
(a) $4 < k < 8$

(b) $-5 \le z \le 4$

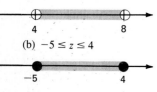

A graph of the solution is given in Figure 1.11. **4**

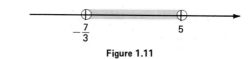

Figure 1.11

  The solutions of some inequalities result in graphs with separate parts as shown in the next example.

Example 4 Solve $3x - 5 < 7$ or $2x - 1 > 13$. Graph the solution.
  Begin by solving each inequality separately, keeping the "or" between the two solutions.

$$\begin{array}{ccc} 3x - 5 < 7 & \text{or} & 2x - 1 > 13 \\ 3x < 12 & \text{or} & 2x > 14 \\ x < 4 & \text{or} & x > 7 \end{array}$$

The final inequality says that x may be any number less than 4 or any number greater than 7. The solution includes the numbers graphed in Figure 1.12. There is no shortcut way to write the solution $x < 4$ or $x > 7$. ▣ 5

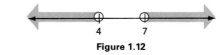

Figure 1.12

5 Solve each inequality. Graph each solution.
(a) $5p - 10 < -20$ or
$2p + 6 > 8$
(b) $1 - 4y \geq 5$ or
$2 - 3y \leq -7$

Answer:
(a) $p < -2$ or $p > 1$

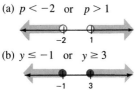

(b) $y \leq -1$ or $y \geq 3$

Example 5 The formula for converting from Celsius to Fahrenheit temperature is

$$F = \frac{9}{5}C + 32.$$

What Celsius temperature range corresponds to 32°F to 77°F?

The Fahrenheit temperature range is $32 < F < 77$. Since $F = (9/5)C + 32$,

$$32 < \frac{9}{5}C + 32 < 77.$$

Solve the inequality for C.

$$32 < \frac{9}{5}C + 32 < 77$$

$$0 < \frac{9}{5}C < 45$$

$$0 < C < \frac{5}{9} \cdot 45$$

$$0 < C < 25$$

6 In Example 5, what Celsius temperatures correspond to 5°F to 95°F?

Answer:
$-15°C$ to $35°C$

The corresponding Celsius temperature range is 0°C to 25°C. ▣ 6

1.3 Exercises

Solve each inequality. Graph each solution in Exercises 1–32. (See Examples 1–4.)

1. $6x \leq -18$

2. $4m > -32$

3. $-3p < 18$

4. $-5z \leq 40$

5. $-9a < 0$

6. $-4k \geq 0$

7. $2x + 1 \leq 9$

8. $3y - 2 < 10$

9. $-3p - 2 \geq 1$

10. $-5t + 3 \geq -2$

11. $6k - 4 < 3k - 1$

12. $2a - 2 > 4a + 2$

13. $m - (4 + 2m) + 3 < 2m + 2$

14. $2p - (3 - p) \leq -7p - 2$

15. $-2(3y - 8) \geq 5(4y - 2)$

16. $5r - (r + 2) \geq 3(r - 1) + 5$

17. $3p - 1 < 6p + 2(p - 1)$

18. $x + 5(x + 1) > 4(2 - x) + x$

19. $-7 < y - 2 < 4$

20. $-3 < m + 6 < 2$

21. $8 \leq 3r + 1 \leq 13$

22. $-6 < 2p - 3 \leq 5$

23. $-4 \leq \dfrac{2k - 1}{3} \leq 2$

24. $-1 \leq \dfrac{5y + 2}{3} \leq 4$

25. $z + 1 \leq 2$ or $z - 5 \geq 1$

26. $2y + 2 \leq 5$ or $3y + 4 \geq 22$

27. $6m + 4 \geq 4 + m$ or $2m + 6 < -2 - 2m$

28. $5 - 2t > 3$ or $8 - 4t < 2 - 2t$

29. $\dfrac{3}{2}b - 2 < 4$ or $\dfrac{3}{4}b + \dfrac{1}{3} > \dfrac{19}{3}$

30. $\dfrac{x}{4} + 3 < \dfrac{9}{4}$ or $\dfrac{x}{2} - 4 > -\dfrac{7}{2}$

31. $\dfrac{3}{5}(2p + 3) \geq \dfrac{1}{10}(5p + 1)$

32. $\dfrac{8}{3}(z - 4) \leq \dfrac{2}{9}(3z + 2)$

33. $7.6092k \geq 2.28276$

34. $1.3075m < -1.569$

35. $8.0413z - 9.7268 < 1.7251z - 0.25250$

36. $3.2579 + 5.0824k > 0.76423k + 6.280619$

37. $-(1.42m + 7.63) + 3(3.7m - 1.12)$
$\leq 4.81m - 8.555$

38. $3(8.14a - 6.32) - (4.31a - 4.84) > 0.342a + 9.499$

Find the unknown numbers in each of the following.

39. Five times a number is between -8 and 6.

40. Half a number is between -4 and -1.

41. When 5 is added to three times a number, the result is greater than or equal to 11.

42. If 4 is subtracted from a number, the result is at least 9.

43. One third of a number is added to 2, giving a result at least 8.

44. Seven times a number, minus 5, is no more than 3.

Solve each of the following word problems. (See Example 5.)

45. A student has a total of 970 points so far in her algebra class. At the end of the course she must have 81% of the 1300 points possible in order to get a B. What is the lowest score she can earn on the 100-point final to get a B in the class?

46. Bill has twice as many dimes as nickels and he has at least 15 coins. At least how many nickels does he have?

47. A nearby business college charges a tuition of $6440 annually. Tom makes no more than $1610 per year in his summer job. What is the least number of summers that he must work in order to make enough for one year's tuition?

48. A nurse must make sure that Ms. Carlson receives at least 30 units of a certain drug each day. This drug comes from red pills or green pills, each of which provides three units of the drug. The patient must have twice as many red pills as green pills. Find the smallest number of green pills that will satisfy the requirement.

Management *A product will* **break even,** *or produce a profit, only if the revenue from selling the product at least equals the cost of producing it. Find all values of x where the following products will at least break even.*

49. The cost to produce x units of wire is $C = 50x + 5000$, while the revenue is $R = 60x$.

50. The cost to produce x units of squash is $C = 100x + 6000$, while the revenue is $R = 500x$.

51. $C = 85x + 900$; $R = 105x$

52. $C = 70x + 500$; $R = 60x$

53. $C = 1000x + 5000$; $R = 900x$

54. Bill and Cheryl Bradkin went to Portland, Maine, for a week. They needed to rent a car, so they checked out two rental firms. Avis wanted $28 per day, with no mileage fee. Downtown Toyota wanted $108 per week and 14¢ per mile. Let x represent the number of miles that the Bradkins would drive in one week. Set up an inequality expressing the rates of the two firms. Then decide how many miles they would have to drive before the Avis car was the better deal.

1.4 Absolute Value Equations and Inequalities

Recall from Section 1.1 that the absolute value of the number a, written $|a|$, gives the distance on a number line from a to 0. For example, $|4| = 4$, and $|-7| = 7$. In this section equations and inequalities involving absolute value are discussed.

Example 1   Solve the equation $|x| = 3$.

There are two numbers whose absolute value is 3, namely 3 and -3. The solutions of the given equation are 3 and -3. ■

Example 2   Solve $|p - 4| = 2$.

This equation will be satisfied if the expression inside the absolute value bars, $p - 4$, equals either 2 or -2:

$$p - 4 = 2 \quad \text{or} \quad p - 4 = -2.$$

Solving these two equations produces

$$p = 6 \quad \text{or} \quad p = 2,$$

so that 6 and 2 are solutions for the original equation. As before, check by substituting in the original equation. ■

1 Solve each equation.
(a) $|y| = 9$
(b) $|r + 3| = 1$
(c) $|2k - 3| = 7$

Answer:
(a) 9, -9
(b) -2, -4
(c) 5, -2

Example 3   Solve $|4m - 3| = |m + 6|$.

The quantities in absolute value bars must either be equal or be negatives of one another to satisfy the equation. That is,

$$4m - 3 = m + 6 \quad \text{or} \quad 4m - 3 = -(m + 6)$$
$$3m = 9 \qquad\qquad\qquad 4m - 3 = -m - 6$$
$$m = 3 \qquad\qquad\qquad\qquad 5m = -3$$
$$m = -\frac{3}{5}.$$

2 Solve each equation.
(a) $|r + 6| = |2r + 1|$
(b) $|5k - 7| = |10k - 2|$

Answer:
(a) 5, $-7/3$
(b) -1, 3/5

Check that the solutions for the original equation are 3 and $-3/5$. ■

The next examples show how to solve inequalities with absolute value.

Example 4   Solve $|x| < 5$, and then solve the inequality $|x| > 5$.

Since absolute value gives the distance from a number to 0, the inequality $|x| < 5$ is true for all real numbers whose distance from 0 is less than 5. This includes all numbers from -5 to 5, or

$$-5 < x < 5.$$

A graph of the solution is shown in Figure 1.13.

$$\begin{array}{c} -5 \qquad\qquad 5 \end{array}$$

Figure 1.13

In a similar way, the solution of $|x| > 5$ is given by all those numbers whose distance from 0 is *greater* than 5. This includes the numbers satisfying $x < -5$ or

3 Solve each inequality. Graph each solution.

(a) $|x| \leq 1$

(b) $|y| \geq 3$

Answer:

(a) $-1 \leq x \leq 1$

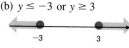

(b) $y \leq -3$ or $y \geq 3$

4 Solve each inequality. Graph each solution.

(a) $|p + 3| < 4$

(b) $|2k - 1| \leq 7$

Answer:

(a) $-7 < p < 1$

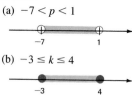

(b) $-3 \leq k \leq 4$

5 Solve each inequality. Graph each solution.

(a) $|y - 2| > 5$

(b) $|3k - 1| \geq 2$

(c) $|2 + 5r| - 4 \geq 1$

Answer:

(a) $y < -3$ or $y > 7$

(b) $k \leq -\dfrac{1}{3}$ or $k \geq 1$

(c) $r \leq -\dfrac{7}{5}$ or $r \geq \dfrac{3}{5}$

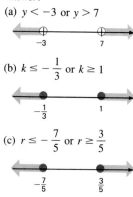

$x > 5$. A graph of the solution, which is written

$$x < -5 \qquad \text{or} \qquad x > 5,$$

is shown in Figure 1.14. **3**

Figure 1.14

Example 4 suggests the following generalizations.

Assume a and b are real numbers with $b > 0$.

1. Solve $|a| < b$, by solving $-b < a < b$.
2. Solve $|a| > b$, by solving $a < -b$ or $a > b$.

Example 5 Solve $|x - 2| < 5$.

Replace a with $x - 2$ and b with 5 in property (1) above. Now solve $|x - 2| < 5$ by solving the inequality

$$-5 < x - 2 < 5.$$

Add 2 to each part, getting the solution

$$-3 < x < 7,$$

which is graphed in Figure 1.15. **4**

Figure 1.15

Example 6 Solve $|2 - 7m| - 1 > 4$.

Use the properties given above, by first adding 1 on both sides.

$$|2 - 7m| > 5$$

Now use property (2) from above to solve $|2 - 7m| > 5$ by solving the inequality

$$2 - 7m < -5 \qquad \text{or} \qquad 2 - 7m > 5.$$

Solve each part separately.

$$-7m < -7 \qquad \text{or} \qquad -7m > 3$$

$$m > 1 \qquad \text{or} \qquad m < -\frac{3}{7}$$

The solution, written $m > 1$ or $m < -3/7$, is graphed in Figure 1.16. **5**

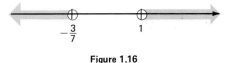

Figure 1.16

Example 7 Solve $|2 - 5x| \geq -4$.

The absolute value of a number is always nonnegative. Therefore, $|2 - 5x| \geq -4$ is always true, so that the solution is the set of all real numbers.   **6**

Absolute value inequalities can be used to indicate how far a number may be from a given number. The next example illustrates this use of absolute value.

Example 8 Write each statement using absolute value.

(a) k is at least 4 units from 1.

Another way of saying "k is at least 4 units from 1" is "the difference between k and 1 is at least 4." See Figure 1.17(a). Since k may be on either side of 1 on the number line, k may be less than -3 or greater than 5. Write this statement using absolute value as follows:

$$|k - 1| \geq 4.$$

(b) p is within 2 units of 5.

This statement means that the difference between p and 5 must be less than or equal to 2. See Figure 1.17(b). Using absolute value notation, the statement is written as

$$|p - 5| \leq 2. \quad \boxed{7}$$

6 Solve each inequality.
(a) $|5m - 2| > -1$
(b) $|2 + 3a| < -3$
(c) $|6 + r| > 0$

Answer:
(a) All real numbers
(b) No solution
(c) All real numbers except -6

7 Write each statement using absolute value.
(a) m is at least 3 units from 5.
(b) t is within .01 of 4.

Answer:
(a) $|m - 5| \geq 3$
(b) $|t - 4| \leq .01$

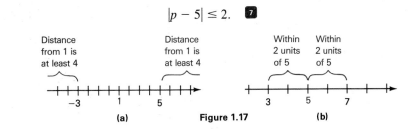

Figure 1.17

1.4 Exercises

Solve each of the following equations. (See Examples 1–3.)

1. $|a - 2| = 1$

2. $|x - 3| = 2$

3. $|3m - 1| = 2$

4. $|4p + 2| = 5$

5. $|5 - 3x| = 3$

6. $|-3a + 7| = 3$

7. $\left|\dfrac{z - 4}{2}\right| = 5$

8. $\left|\dfrac{m + 2}{2}\right| = 7$

9. $\left|\dfrac{5}{r - 3}\right| = 10$

10. $\left|\dfrac{3}{2h - 1}\right| = 4$

11. $\left|\dfrac{6y + 1}{y - 1}\right| = 3$

12. $\left|\dfrac{3a - 4}{2a + 3}\right| = 1$

13. $|2k - 3| = |5k + 4|$

14. $|p + 1| = |3p - 1|$

15. $|4 - 3y| = |7 + 2y|$

16. $|2 + 5a| = |4 - 6a|$

Solve each of the following inequalities. Graph each solution. (See Examples 4–7.)

17. $|x| \le 3$

18. $|y| \le 10$

19. $|m| > 1$

20. $|z| > 5$

21. $|a| < -2$

22. $|b| > -5$

23. $|x| - 3 \le 7$

24. $|r| + 3 \le 10$

25. $|2x + 5| < 3$

26. $\left|x - \dfrac{1}{2}\right| < 2$

27. $|3m - 2| > 4$

28. $|4x - 6| > 10$

29. $|3z + 1| \ge 7$

30. $|8b + 5| \ge 7$

31. $\left|5x + \dfrac{1}{2}\right| - 2 < 5$

32. $\left|x + \dfrac{2}{3}\right| + 1 < 4$

Find all unknown numbers in each of the following.

33. The absolute value of a number is no more than 6.

34. If the absolute value of a number is found, the result is at least 5.

35. If 6 is added to four times a number, the absolute value of the result is no more than 1.

36. Add -2 to twice a number. The absolute value of the result is less than or equal to 5.

37. The absolute value of the sum of a number and 2 is found. Eight is subtracted from the absolute value. The final result is at least 4.

38. Three times a number is added to 8. The absolute value of this sum is found. Two is added to the absolute value. The final result is no more than 7.

Write each of the following statements using absolute value. (See Example 8.)

39. x is within 4 units of 2.

40. m is no more than 8 units from 9.

41. z is no less than 2 units from 12.

42. p is at least 5 units from 9.

43. k is 6 units from 9.

44. r is 5 units from 3.

45. If x is within .0004 units of 2, then y is within .00001 units of 7.

46. y is within .001 units of 10 whenever x is within .02 units of 5.

1.5 Polynomials

A **polynomial** in one variable is an algebraic expression of the form

$$a_n x^n + a_{n-1} x^{n-1} + \cdots + a_1 x + a_0,$$

where $a_0, a_1, a_2, \ldots,$ and a_n are real numbers, n is a natural number, and $a_n \ne 0$. Examples of polynomials include

$$5x^4 + 2x^3 + 6x, \quad 8m^3 + 9m^2 - 6m + 3, \quad 10p, \quad \text{and} \quad -9.$$

Expressions that are *not* polynomials include

$$8x^3 + \frac{6}{x}, \quad \frac{9 + x}{2 - x}, \quad \text{and} \quad \frac{-p^2 + 5p + 3}{2p - 1}.$$

As an example, the entire expression $9p^4$ is called a **term,** the number 9 is called the **coefficient,** p is the **variable,** and 4 is the **exponent.** The expression p^4 means $p \cdot p \cdot p \cdot p$, while p^2 means $p \cdot p$. (Exponents are discussed in more detail later in this chapter.) The **degree of a term** with only one variable is the exponent on the variable. For example, the term $9p^4$ has degree 4. The **degree of a polynomial** is the highest degree of any of its terms.

Polynomials can be added or subtracted by combining terms using the distributive property. For example,

$$12y^4 + 6y^4 = (12 + 6)y^4 = 18y^4$$

and

$$-2m^2 + 8m^2 = (-2 + 8)m^2 = 6m^2.$$

The polynomial $8y^4 + 2y^5$ cannot be further simplified.

Two terms having the same variable and the same exponent are called **like terms;** other terms are called **unlike terms.** Only like terms may be combined. Subtract polynomials using the fact that $a - b = a + (-b)$. The next example shows how to add and subtract polynomials by combining terms.

Example 1 Add or subtract as indicated.
(a) $(8x^3 - 4x^2 + 6x) + (3x^3 + 5x^2 - 9x + 8)$
Combine like terms.

$$\begin{aligned}
(8x^3 &- 4x^2 + 6x) + (3x^3 + 5x^2 - 9x + 8) \\
&= (8x^3 + 3x^3) + (-4x^2 + 5x^2) + (6x - 9x) + 8 \\
&= 11x^3 + x^2 - 3x + 8
\end{aligned}$$

(b) $(-4x^4 + 6x^3 - 9x^2 - 12) + (-3x^3 + 8x^2 - 11x + 7)$
$$= -4x^4 + 3x^3 - x^2 - 11x - 5$$

(c) $(2x^2 - 11x + 8) - (7x^2 - 6x + 2)$
$$\begin{aligned}
&= (2x^2 - 11x + 8) + (-7x^2 + 6x - 2) \\
&= -5x^2 - 5x + 6 \quad \boxed{1}
\end{aligned}$$

The distributive property is also used when multiplying polynomials. For example, the product of $8x$ and $6x - 4$ is found as follows.

$$\begin{aligned}
\mathbf{8x(6x - 4)} &= \mathbf{8x(6x)} - \mathbf{8x(4)} \qquad \text{Distributive property} \\
&= 48x^2 - 32x \qquad\qquad x \cdot x = x^2
\end{aligned}$$

The product of $3p - 2$ and $5p + 1$ can be found by using the distributive property twice:

$$\begin{aligned}
\mathbf{(3p - 2)(5p + 1)} &= \mathbf{(3p - 2)(5p)} + \mathbf{(3p - 2)(1)} \\
&= \mathbf{3p(5p)} - \mathbf{2(5p)} + \mathbf{3p(1)} - \mathbf{2(1)} \\
&= 15p^2 - 10p + 3p - 2 \\
&= 15p^2 - 7p - 2. \quad \boxed{2}
\end{aligned}$$

Factoring As shown above, multiplication of polynomials is performed using the distributive property. The reverse process, where a polynomial is written as a product of other polynomials, is called **factoring.** For example, one way to factor the number 18 is to write it as the product $9 \cdot 2$. Since $18 = 9 \cdot 2$, both 9 and 2 are called **factors** of 18. The integer factors of 18 are 2, 9; -2, -9; 6, 3; -6, -3; 18, 1; -18, -1.

---

1 Add or subtract.
(a) $(-2x^2 + 7x + 9)$
  $+ (3x^2 + 2x - 7)$
(b) $(4x + 6) - (13x - 9)$
(c) $(9x^3 - 8x^2 + 2x)$
  $- (9x^3 - 2x^2 - 10)$

Answer:
(a) $x^2 + 9x + 2$
(b) $-9x + 15$
(c) $-6x^2 + 2x + 10$

2 Find the following products.
(a) $-6r(2r - 5)$
(b) $11m(8m + 3)$
(c) $(5k - 1)(2k + 3)$
(d) $(7z - 3)(2z + 5)$

Answer:
(a) $-12r^2 + 30r$
(b) $88m^2 + 33m$
(c) $10k^2 + 13k - 3$
(d) $14z^2 + 29z - 15$

The algebraic expression $15m + 45$ is made up of two terms, $15m$ and 45. Each of these terms can be divided by 15. In fact, $15m = 15 \cdot m$ and $45 = 15 \cdot 3$. By the distributive property,

$$15m + 45 = \mathbf{15} \cdot m + \mathbf{15} \cdot 3 = \mathbf{15}(m + 3).$$

Both 15 and $m + 3$ are factors of $15m + 45$. Since 15 divides into all terms of $15m + 45$ and is the largest number that will do so, it is called the **greatest common factor** for the polynomial $15m + 45$. The process of writing $15m + 45$ as $15(m + 3)$ is called **factoring out** the greatest common factor.

Example 2   Factor out the greatest common factor.
(a) $12p - 18q$
   Both $12p$ and $18q$ are divisible by 6, and

$$12p - 18q = \mathbf{6} \cdot 2p - \mathbf{6} \cdot 3q = \mathbf{6}(2p - 3q).$$

(b) $8x^3 - 9x^2 + 15x$
   Each of these terms is divisible by x.

$$8x^3 - 9x^2 + 15x = (8x^2) \cdot x - (9x) \cdot x + 15 \cdot x$$
$$= x(8x^2 - 9x + 15) \quad \boxed{3}$$

❸ Factor out the greatest common factor.
(a) $12r + 9k$
(b) $75m^2 + 100n^2$
(c) $6m^4 - 9m^3 + 12m^2$

Answer:
(a) $3(4r + 3k)$
(b) $25(3m^2 + 4n^2)$
(c) $3m^2(2m^2 - 3m + 4)$

A polynomial may not have a greatest common factor (other than 1), and yet may still be factorable. For example, the polynomial $x^2 + 5x + 6$ can be factored as $(x + 2)(x + 3)$. Check this: find the product $(x + 2)(x + 3)$; you should get $x^2 + 5x + 6$.

Starting with a polynomial such as $x^2 + 5x + 6$, how can it be written as the product $(x + 2)(x + 3)$? There are two different ways to factor a polynomial of three terms such as $x^2 + 5x + 6$, depending on whether the coefficient of x^2 is 1, or a number other than 1. If the coefficient is 1, proceed as shown in the following examples.

Example 3   Factor $y^2 + 8y + 15$.
   Since the coefficient of y^2 is 1, factor by finding two numbers whose *product* is 15, and whose *sum* is 8. Use trial and error to find these numbers. Begin by listing all pairs of integers having a product of 15. As you do this, also form the sum of the numbers.

Products	Sums
$15 \cdot 1 = 15$	$15 + 1 = 16$
$\mathbf{5 \cdot 3 = 15}$	$\mathbf{5 + 3 = 8}$
$(-1) \cdot (-15) = 15$	$-1 + (-15) = -16$
$(-5) \cdot (-3) = 15$	$-5 + (-3) = -8$

The numbers 3 and 5 have a product of 15 and a sum of 8, so $y^2 + 8y + 15$ factors as

$$y^2 + 8y + 15 = (y + 3)(y + 5).$$

4 Factor the following.
(a) $m^2 + 11m + 30$
(b) $h^2 + 10h + 9$

Answer:
(a) $(m + 5)(m + 6)$
(b) $(h + 9)(h + 1)$

5 Factor the following.
(a) $a^2 - 7a + 10$
(b) $r^2 - 5r - 14$
(c) $m^2 + 3m - 40$

Answer:
(a) $(a - 5)(a - 2)$
(b) $(r - 7)(r + 2)$
(c) $(m + 8)(m - 5)$

The result can also be written as $(y + 5)(y + 3)$. **4**

Example 4 Factor $p^2 - 8p - 20$.
Find two numbers whose product is -20 and whose sum is -8. Make a list of all pairs of integers whose product is -20. Choose from this list that pair whose sum is -8. You should find the pair -10 and 2; the product of these numbers is -20 and their sum is -8. Therefore,

$$p^2 - 8p - 20 = (p - 10)(p + 2). \quad \boxed{5}$$

For a trinomial such as $kx^2 + mx + n$, where k, m, and n are integers and $k \neq 1$, any factorization will be of the form $(ax + b)(cx + d)$, with a, b, c, and d integers. Multiplying out the product $(ax + b)(cx + d)$ gives

$$(ax + b)(cx + d) = acx^2 + (ad + bc)x + bd,$$

which equals $kx^2 + mx + n$ only if

$$ac = k, \qquad ad + bc = m, \qquad \text{and} \qquad bd = n. \qquad (*)$$

Thus, to factor a trinomial $kx^2 + mx + n$, four integers, a, b, c, and d, must be found satisfying the conditions given in equations (*). If no such numbers exist, the trinomial cannot be factored and is a **prime polynomial.**

Example 5 Factor each of the following polynomials.
(a) $6p^2 - 7p - 5$
Find integers a, b, c, and d so that

$$6p^2 - 7p - 5 = (ap + b)(cp + d).$$

Using the results given in equations (*) above, find integers a, b, c, and d such that $ac = 6$, $ad + bc = -7$, and $bd = -5$. To find these integers, try various possibilities. Since $ac = 6$, let $a = 2$ and $c = 3$. Since $bd = -5$, let $b = -5$ and $d = 1$, giving

$$(2p - 5)(3p + 1) = 6p^2 - 13p - 5. \qquad \text{(incorrect)}$$

Make another attempt; try

$$(3p - 5)(2p + 1) = 6p^2 - 7p - 5. \qquad \text{(correct)}$$

The trinomial $6p^2 - 7p - 5$ factors as $(3p - 5)(2p + 1)$.
(b) $2x^2 + 9xy - 5y^2$
The factors of $2x^2$ are $2x$ and x; the possible factors of -5 are -5 and 1, or 5 and -1. Try various combinations of these factors until one works (if, indeed, any work). For example, the product $(2x + 5y)(x - y)$ is

$$\begin{aligned} (2x + 5y)(x - y) &= (2x + 5y)(x) - (2x + 5y)(y) \\ &= 2x^2 + 5xy - 2xy - 5y^2 \\ &= 2x^2 + 3xy - 5y^2. \end{aligned}$$

This product is not correct. Try another combination:

$$(2x - y)(x + 5y) = (2x - y)(x) + (2x - y)(5y)$$
$$= 2x^2 - xy + 10xy - 5y^2$$
$$= 2x^2 + 9xy - 5y^2.$$

This combination led to the correct polynomial;

$$2x^2 + 9xy - 5y^2 = (2x - y)(x + 5y).$$

(c) $r^2 + 6r + 7$

It is not possible to find integers a, b, c, and d so that

$$r^2 + 6r + 7 = (ar + b)(cr + d);$$

therefore, $r^2 + 6r + 7$ cannot be factored and is a prime polynomial. **6**

The following two special factorizations occur so often that they are listed for future reference.

$x^2 - y^2 = (x + y)(x - y)$	**Difference of two squares**
$x^2 + 2xy + y^2 = (x + y)^2$ $\left.\vphantom{\begin{matrix}a\\b\end{matrix}}\right\}$ $x^2 - 2xy + y^2 = (x - y)^2$	**Perfect square**

Example 6 Factor each of the following.

(a) $4m^2 - 9$

First, $4m^2 - 9$ is the difference of two squares, since $4m^2 = (2m)^2$ and $9 = 3^2$. Use the pattern for the difference of two squares, letting $2m$ replace x and 3 replace y. Then the pattern $x^2 - y^2 = (x + y)(x + y)$ becomes

$$4m^2 - 9 = (2m)^2 - 3^2$$
$$= (2m + 3)(2m - 3).$$

(b) $64p^2 - 49q^2 = (8p)^2 - (7q)^2 = (8p + 7q)(8p - 7q)$

(c) $x^2 + 36$ cannot be factored.

(d) $a^2 + 12a + 36$

Since the first and last terms are perfect squares, this trinomial may be a perfect square. Check the middle term to see: it should be $2xy$. Here a replaces x and 6 replaces y in the pattern, so the middle term should be $2(a)(6) = 12a$. The middle term *is* $12a$, so the trinomial can be factored as

$$a^2 + 12a + 36 = (a + 6)^2.$$

(e) $9p^2 - 24pq + 16q^2$

Since $9p^2 = (3p)^2$ and $16q^2 = (4q)^2$, let $3p$ replace x and $4q$ replace y in the pattern for a perfect square. The middle term is

$$24pq = 2(3p)(4q),$$

so the trinomial is factored as

$$9p^2 - 24pq + 16q^2 = (3p - 4q)^2. \quad \boxed{7}$$

---

6 Factor the following:

(a) $3k^2 + k - 2$

(b) $3m^2 + 5mn - 2n^2$

(c) $6p^2 + 13pq - 5q^2$

Answer:

(a) $(3k - 2)(k + 1)$

(b) $(3m - n)(m + 2n)$

(c) $(2p + 5q)(3p - q)$

---

7 Factor the following.

(a) $r^2 - 81$

(b) $9p^2 - 49$

(c) $y^2 + 100$

(d) $m^2 - 8mn + 16n^2$

(e) $100k^2 - 60kp + 9p^2$

Answer:

(a) $(r + 9)(r - 9)$

(b) $(3p + 7)(3p - 7)$

(c) Prime

(d) $(m - 4n)^2$

(e) $(10k - 3p)^2$

Finally, two more special types of factorizations occur from time to time.

$$x^3 - y^3 = (x - y)(x^2 + xy + y^2) \quad \textbf{Difference of two cubes}$$
$$x^3 + y^3 = (x + y)(x^2 - xy + y^2) \quad \textbf{Sum of two cubes}$$

8 Factor the following.
(a) $a^3 + 1000$
(b) $z^3 - 64$
(c) $1000m^3 - 27z^3$

Answer:
(a) $(a + 10)(a^2 - 10a + 100)$
(b) $(z - 4)(z^2 + 4z + 16)$
(c) $(10m - 3z)(100m^2 + 30mz + 9z^2)$

Example 7 Factor each of the following.
(a) $k^3 - 8$

Use the pattern for the difference of two cubes, since $k^3 = (k)^3$ and $8 = (2)^3$, to get

$$k^3 - 8 = (k - 2)(k^2 + 2k + 4).$$

(b) $m^3 + 125 = m^3 + 5^3 = (m + 5)(m^2 - 5m + 25)$
(c) $8k^3 - 27z^3 = (2k)^3 - (3z)^3 = (2k - 3z)(4k^2 + 6kz + 9z^2)$ **8**

1.5 Exercises

Add or subtract as indicated. (See Example 1.)

1. $(8m + 9) + (6m - 3)$

2. $(-7p - 11) + (8p + 5)$

3. $(-2k - 3) - (7k - 8)$

4. $(12z + 10) - (3z + 9)$

5. $(2x^2 - 6x + 11) + (-3x^2 + 7x - 2)$

6. $(-3a^2 + 2a - 5) + (7a^2 + 2a + 9)$

7. $(-4y^2 - 3y + 8) - (2y^2 - 6y - 2)$

8. $(7b^2 + 2b - 5) - (3b^2 + 2b - 6)$

9. $(2x^3 - 2x^2 + 4x - 3) - (2x^3 + 8x^2 - 1)$

10. $(3y^3 + 9y^2 - 11y + 8) - (-4y^2 + 10y - 6)$

11. $(.613x^2 - 4.215x + 0.892) - .47(2x^2 - 3x + 5)$

12. $.83(5r^2 - 2r + 7) - (7.12r^2 + 6.423r - 2)$

Find each of the following products.

13. $3p(2p - 5)$

14. $4y(8y + 1)$

15. $-9m(2m^2 + 3m - 1)$

16. $2a(4a^2 - 6a + 3)$

17. $(6k - 1)(2k - 3)$

18. $(8r + 3)(r - 1)$

19. $(3y + 5)(2y - 1)$

20. $(2a - 5)(4a + 3)$

21. $(5r - 3s)(5r + 4s)$

22. $(9k + q)(2k - q)$

23. $(.012x - .17)(.3x + .54)$

24. $(6.2m - 3.4)(.7m + 1.3)$

Factor out the greatest common factor in each of the following. (See Example 2.)

25. $25k + 30$

26. $6m - 12$

27. $4z + 4$

28. $9y - 9$

29. $8x + 6y + 4z$

30. $15p + 9q + 12s$

31. $6r^2 + 4r + 8$

32. $15z^2 - 10z + 25$

33. $m^3 - 9m^2 + 6m$

34. $y^3 + 6y^2 + 8y$

35. $8a^3 - 16a^2 + 24a$

36. $3y^3 + 24y^2 + 9y$

37. $25p^4 - 20p^3q + 100p^2q^2$

38. $60m^4 - 120m^3n + 50m^2n^2$

Factor each of the following. If a polynomial cannot be factored, write prime. *Factor out the greatest common factor as necessary. (See Examples 3–5.)*

39. $m^2 + 9m + 14$

40. $p^2 - 2p - 15$

41. $x^2 + 4x - 5$

42. $y^2 + y - 72$

43. $z^2 + 9z + 20$

44. $k^2 + 8k + 15$

45. $b^2 - 8b + 7$

46. $r^2 + r - 20$

47. $a^2 + 4ab + 5b^2$

48. $y^2 - 6yx + 8x^2$

49. $s^2 + 2st - 35t^2$

50. $n^2 - 12np - 35p^2$

51. $y^2 - 4yz - 21z^2$

52. $r^2 + rs - 42s^2$

53. $6a^2 - 48a - 120$

54. $8h^2 - 24h - 320$

55. $3m^3 + 12m^2 + 9m$

56. $3y^4 - 18y^3 + 15y^2$

57. $2x^2 - 5x - 3$

58. $3r^2 - r - 2$

59. $3a^2 + 10a + 7$

60. $4y^2 + y - 3$

61. $2a^2 - 17a + 30$

62. $3k^2 + 2k - 8$

63. $15y^2 + y - 2$

64. $6x^2 + x - 1$

65. $3p^2 - 7p + 10$

66. $8r^2 + r + 6$

67. $5a^2 - 7ab - 6b^2$

68. $12s^2 + 11st - 5t^2$

69. $21m^2 + 13mn + 2n^2$

70. $20y^2 + 39yx - 11x^2$

71. $24a^4 + 10a^3b - 4a^2b^2$

72. $18x^5 + 15x^4z - 75x^3z^2$

73. $32z^5 - 20z^4a - 12z^3a^2$

74. $15x^4 - 7x^3p - 4x^2p^2$

Factor each of the following. (See Examples 6 and 7.)

75. $x^2 - 64$

76. $y^2 - 144$

77. $9m^2 - 25$

78. $4p^2 - 9$

79. $121a^2 - 100$

80. $144m^2 - 169$

81. $9x^2 + 64$

82. $100a^2 + 9$

83. $z^2 + 14zy + 49y^2$

84. $y^2 + 20yx + 100x^2$

85. $m^2 - 6mn + 9n^2$

86. $a^2 - 10ab + 25b^2$

87. $9p^2 - 24p + 16$

88. $16m^2 + 40m + 25$

89. $a^3 - 216$

90. $b^3 + 125$

91. $8r^3 - 27s^3$

92. $1000p^3 + 27q^3$

93. $64m^3 + 125$

94. $216y^3 - 343$

95. $1000y^3 - z^3$

96. $125p^3 + 8q^3$

1.6 Rational Expressions

Later chapters of the book contain work with algebraic fractions. Examples of these fractions (called **rational expressions**) include

$$\frac{8}{x-1}, \qquad \frac{3x^2 + 4x}{5x - 6}, \qquad \text{and} \qquad \frac{2 + \dfrac{1}{y}}{y}.$$

Since rational expressions involve quotients, it is important to keep in mind values of the variables that make denominators zero. For example, 1 cannot be used as a replacement for x in the first rational expression above, and 6/5 cannot be used in the middle one, since these values make the respective denominators equal 0.

The rules for operations with rational expressions are given below.

 What values of the variable make each denominator equal 0?

(a) $\dfrac{5}{x-3}$

(b) $\dfrac{2x-3}{4x-1}$

(c) $\dfrac{x+2}{x}$

Answer:

(a) 3

(b) 1/4

(c) 0

Operations with rational expressions For all mathematical expressions P, $Q \neq 0$, R, and $S \neq 0$,

(a) $\dfrac{P}{Q} = \dfrac{PS}{QS}$ Fundamental property

(b) $\dfrac{P}{Q} \cdot \dfrac{R}{S} = \dfrac{PR}{QS}$ Multiplication

(c) $\dfrac{P}{Q} + \dfrac{R}{Q} = \dfrac{P+R}{Q}$ Addition

(d) $\dfrac{P}{Q} - \dfrac{R}{Q} = \dfrac{P-R}{Q}$ Subtraction

(e) $\dfrac{P}{Q} \div \dfrac{R}{S} = \dfrac{P}{Q} \cdot \dfrac{S}{R}$, $R \neq 0.$ Division

The following examples illustrate these operations.

Example 1 Write each of the following rational expressions in lowest terms (so that the numerator and denominator have no common factor except 1).

(a) $\dfrac{12m}{18}$

Both $12m$ and 18 are divisible by 6. By operation (a) above,

$$\frac{12m}{18} = \frac{2m \cdot 6}{3 \cdot 6} = \frac{2m}{3}.$$

(b) $\dfrac{8x + 16}{4} = \dfrac{8(x + 2)}{4} = \dfrac{4 \cdot 2(x + 2)}{4} = 2(x + 2)$

The numerator, $8x + 16$, was factored so that the common factor could be identified. The answer could also be written as $2x + 4$, if desired.

(c) $\dfrac{k^2 + 7k + 12}{k^2 + 2k - 3} = \dfrac{(k + 4)(k + 3)}{(k - 1)(k + 3)} = \dfrac{k + 4}{k - 1}$ **2**

The values of k in Example 1(c) are restricted to $k \neq 1$ and $k \neq -3$. From now on, such restrictions will be assumed when working with rational expressions.

Example 2 Multiply.

(a) $\dfrac{2}{3} \cdot \dfrac{y}{5}$

Use operation (b) above; multiply the numerators and then the denominators.

$$\frac{2}{3} \cdot \frac{y}{5} = \frac{2 \cdot y}{3 \cdot 5} = \frac{2y}{15}$$

The result, $2y/15$, is in lowest terms.

(b) $\dfrac{3y + 9}{6} \cdot \dfrac{18}{5y + 15}$

Factor where possible.

$$\frac{3y + 9}{6} \cdot \frac{18}{5y + 15} = \frac{3(y + 3)}{6} \cdot \frac{18}{5(y + 3)}$$

$$= \frac{3 \cdot 18(y + 3)}{6 \cdot 5(y + 3)} \qquad \text{Multiply numerators and denominators}$$

$$= \frac{3 \cdot 6 \cdot 3(y + 3)}{6 \cdot 5(y + 3)} \qquad 18 = 6 \cdot 3$$

$$= \frac{3 \cdot 3}{5} \qquad \text{Write in lowest terms}$$

$$= \frac{9}{5}$$

2 Write each of the following in lowest terms.

(a) $\dfrac{12k + 36}{18}$

(b) $\dfrac{15m + 30m^2}{5m}$

(c) $\dfrac{2p^2 + 3p + 1}{p^2 + 3p + 2}$

Answer:

(a) $\dfrac{2(k + 3)}{3}$ or $\dfrac{2k + 6}{3}$

(b) $3(1 + 2m)$ or $3 + 6m$

(c) $\dfrac{2p + 1}{p + 2}$

(c) $\dfrac{m^2 + 5m + 6}{m + 3} \cdot \dfrac{m}{m^2 + 3m + 2}$

$= \dfrac{(m + 2)(m + 3)}{m + 3} \cdot \dfrac{m}{(m + 2)(m + 1)}$ Factor

$= \dfrac{m(m + 2)(m + 3)}{(m + 3)(m + 2)(m + 1)}$ Multiply

$= \dfrac{m}{m + 1}$ Lowest terms   **3**

3 Multiply.

(a) $\dfrac{3r^2}{5} \cdot \dfrac{20}{9r}$

(b) $\dfrac{y - 4}{y^2 - 2y - 8} \cdot \dfrac{y^2 - 4}{3y}$

Answer:

(a) $\dfrac{4r}{3}$

(b) $\dfrac{y - 2}{3y}$

Example 3 Divide.

(a) $\dfrac{8x}{5} \div \dfrac{11x^2}{20}$

As shown in operation (e) above, invert the second expression and multiply.

$\dfrac{8x}{5} \div \dfrac{11x^2}{20} = \dfrac{8x}{5} \cdot \dfrac{20}{11x^2}$ Invert and Multiply

$= \dfrac{8x \cdot 20}{5 \cdot 11x^2}$ Multiply

$= \dfrac{32}{11x}$ Lowest terms

(b) $\dfrac{9p - 36}{12} \div \dfrac{5(p - 4)}{18}$

$= \dfrac{9p - 36}{12} \cdot \dfrac{18}{5(p - 4)}$ Invert and multiply

$= \dfrac{9(p - 4)}{12} \cdot \dfrac{18}{5(p - 4)}$ Factor

$= \dfrac{27}{10}$ Multiply and write in lowest terms   **4**

4 Divide.

(a) $\dfrac{5m}{16} \div \dfrac{m^2}{10}$

(b) $\dfrac{2y - 8}{6} \div \dfrac{5y - 20}{3}$

(c) $\dfrac{m^2 - 2m - 3}{m(m + 1)} \div \dfrac{m + 4}{5m}$

Answer:

(a) $\dfrac{25}{8m}$

(b) $\dfrac{1}{5}$

(c) $\dfrac{5(m - 3)}{m + 4}$

Example 4 Add or subtract as indicated.

(a) $\dfrac{4}{5k} - \dfrac{11}{5k}$

As operation (d) above shows, when two rational expressions have the same denominators, subtract by subtracting the numerators.

$\dfrac{4}{5k} - \dfrac{11}{5k} = \dfrac{4 - 11}{5k} = -\dfrac{7}{5k}$

(b) $\dfrac{7}{p} + \dfrac{9}{2p} + \dfrac{1}{3p}$

These three denominators are different; operation (c) for addition requires the same denominators. Find a common denominator, one which can be divided by p, $2p$, and $3p$. A common denominator here is $6p$. Rewrite each rational expression, using operation (a), with a denominator of $6p$. Then add using operation (c).

$$\frac{7}{p} + \frac{9}{2p} + \frac{1}{3p} = \frac{6 \cdot 7}{6 \cdot p} + \frac{3 \cdot 9}{3 \cdot 2p} + \frac{2 \cdot 1}{2 \cdot 3p} \qquad \text{Operation (a)}$$

$$= \frac{42}{6p} + \frac{27}{6p} + \frac{2}{6p}$$

$$= \frac{42 + 27 + 2}{6p} \qquad \text{Operation (c)}$$

$$= \frac{71}{6p}$$

(c) $\dfrac{3}{k-1} - \dfrac{1}{k}$

The common denominator is $k(k-1)$.

$$\frac{3}{k-1} - \frac{1}{k} = \frac{3 \cdot k}{k(k-1)} - \frac{1(k-1)}{k(k-1)} \qquad \text{Operation (a)}$$

$$= \frac{3k}{k(k-1)} - \frac{k-1}{k(k-1)}$$

$$= \frac{3k - (k-1)}{k(k-1)} \qquad \text{Operation (d)}$$

$$= \frac{3k - k + 1}{k(k-1)} \qquad\qquad -(k-1) = -1(k-1)$$
$$\qquad\qquad\qquad\qquad\qquad\qquad = -k + 1$$

$$= \frac{2k+1}{k(k-1)} \quad \boxed{5}$$

Complex Fractions Any quotient of two rational expressions is called a **complex fraction.** Complex fractions can often be simplified by the methods shown in the following examples.

Example 5 Simplify each complex fraction.

(a) $\dfrac{6 - \dfrac{5}{k}}{1 + \dfrac{5}{k}}$

Multiply both numerator and denominator by the common denominator k.

$$\frac{6 - \dfrac{5}{k}}{1 + \dfrac{5}{k}} = \frac{k\left(6 - \dfrac{5}{k}\right)}{k\left(1 + \dfrac{5}{k}\right)} \qquad \text{Multiply by } \frac{k}{k}$$

---

⑤ Add or subtract.

(a) $\dfrac{1}{m} - \dfrac{7}{m}$

(b) $\dfrac{3}{4r} + \dfrac{8}{3r}$

(c) $\dfrac{1}{m-2} - \dfrac{3}{2(m-2)}$

Answer:

(a) $-\dfrac{6}{m}$

(b) $\dfrac{41}{12r}$

(c) $\dfrac{-1}{2(m-2)}$

$$\frac{k\left(6 - \dfrac{5}{k}\right)}{k\left(1 + \dfrac{5}{k}\right)} = \frac{6k - k\left(\dfrac{5}{k}\right)}{k + k\left(\dfrac{5}{k}\right)} \qquad \text{Distributive property}$$

$$= \frac{6k - 5}{k + 5} \qquad \text{Simplify}$$

(b) $\dfrac{\dfrac{a}{a + 1} + \dfrac{1}{a}}{\dfrac{1}{a} + \dfrac{1}{a + 1}}$

Multiply both numerator and denominator by the common denominator of all the fractions, in this case $a(a + 1)$. Doing so gives

Simplify each complex fraction.

(a) $\dfrac{t - \dfrac{1}{t}}{2t + \dfrac{3}{t}}$

(b) $\dfrac{\dfrac{m}{m + 2} + \dfrac{1}{m}}{\dfrac{1}{m} - \dfrac{1}{m + 2}}$

$$\frac{\dfrac{a}{a + 1} + \dfrac{1}{a}}{\dfrac{1}{a} + \dfrac{1}{a + 1}} = \frac{\left(\dfrac{a}{a + 1} + \dfrac{1}{a}\right)a(a + 1)}{\left(\dfrac{1}{a} + \dfrac{1}{a + 1}\right)a(a + 1)} = \frac{a^2 + (a + 1)}{(a + 1) + a} = \frac{a^2 + a + 1}{2a + 1}.$$

As an alternate method of solution, first perform the indicated additions in the numerator and denominator, and then divide.

$$\frac{\dfrac{a}{a + 1} + \dfrac{1}{a}}{\dfrac{1}{a} + \dfrac{1}{a + 1}} = \frac{\dfrac{a^2 + 1(a + 1)}{a(a + 1)}}{\dfrac{1(a + 1) + 1(a)}{a(a + 1)}} = \frac{\dfrac{a^2 + a + 1}{a(a + 1)}}{\dfrac{2a + 1}{a(a + 1)}}$$

$$= \frac{a^2 + a + 1}{a(a + 1)} \cdot \frac{a(a + 1)}{2a + 1} = \frac{a^2 + a + 1}{2a + 1} \quad \text{\small 6}$$

Answer:

(a) $\dfrac{t^2 - 1}{2t^2 + 3}$

(b) $\dfrac{m^2 + m + 2}{2}$

1.6 Exercises

Write each of the following in lowest terms. Factor as necessary. (See Example 1.)

1. $\dfrac{6m}{24}$

2. $\dfrac{8k}{56}$

3. $\dfrac{7z^2}{14z}$

4. $\dfrac{32y^2}{16y}$

5. $\dfrac{25p^3}{10p^2}$

6. $\dfrac{14z^3}{6z^2}$

7. $\dfrac{8k + 16}{9k + 18}$

8. $\dfrac{20r + 10}{30r + 15}$

9. $\dfrac{3(t + 5)}{(t + 5)(t - 3)}$

10. $\dfrac{-8(y - 4)}{(y + 2)(y - 4)}$

11. $\dfrac{8x^2 + 16x}{4x^2}$

12. $\dfrac{36y^2 + 72y}{9y}$

13. $\dfrac{m^2 - 4m + 4}{m^2 + m - 6}$

14. $\dfrac{r^2 - r - 6}{r^2 + r - 12}$

15. $\dfrac{x^2 + 3x - 4}{x^2 - 1}$

16. $\dfrac{z^2 - 5z + 6}{z^2 - 4}$

17. $\dfrac{8m^2 + 6m - 9}{16m^2 - 9}$

18. $\dfrac{6y^2 + 11y + 4}{3y^2 + 7y + 4}$

Multiply or divide as indicated in each of the following. Write all answers in lowest terms. (See Examples 2 and 3.)

19. $\dfrac{9k^2}{25} \cdot \dfrac{5}{3k}$

20. $\dfrac{21m^3}{9m} \cdot \dfrac{12m^2}{7m}$

21. $\dfrac{15p^3}{9p^2} \div \dfrac{6p}{10p^2}$

22. $\dfrac{3r^2}{9r^3} \div \dfrac{8r^3}{6r}$

23. $\dfrac{a + b}{2p} \cdot \dfrac{12}{5(a + b)}$

24. $\dfrac{3(x - 1)}{y} \cdot \dfrac{2y}{7(x - 1)}$

25. $\dfrac{a - 3}{16} \div \dfrac{a - 3}{32}$

26. $\dfrac{9}{2(4 - y)} \div \dfrac{3}{4 - y}$

27. $\dfrac{2k + 8}{6} \div \dfrac{3k + 12}{2}$

28. $\dfrac{5m + 25}{10} \cdot \dfrac{12}{6m + 30}$

29. $\dfrac{9y - 18}{6y + 12} \cdot \dfrac{3y + 6}{15y - 30}$

30. $\dfrac{12r + 24}{36r - 36} \div \dfrac{6r + 12}{8r - 8}$

31. $\dfrac{4a + 12}{2a - 10} \div \dfrac{a^2 - 9}{a^2 - a - 20}$

32. $\dfrac{6r - 18}{9r^2 + 6r - 24} \cdot \dfrac{12r - 16}{4r - 12}$

33. $\dfrac{k^2 - k - 6}{k^2 + k - 12} \cdot \dfrac{k^2 + 3k - 4}{k^2 + 2k - 3}$

34. $\dfrac{n^2 - n - 6}{n^2 - 2n - 8} \div \dfrac{n^2 - 9}{n^2 + 7n + 12}$

35. $\dfrac{m^2 + 3m + 2}{m^2 + 5m + 4} \div \dfrac{m^2 + 5m + 6}{m^2 + 10m + 24}$

36. $\dfrac{y^2 + y - 2}{y^2 + 3y - 4} \div \dfrac{y^2 + 3y + 2}{y^2 + 4y + 3}$

37. $\dfrac{2m^2 - 5m - 12}{m^2 - 10m + 24} \div \dfrac{4m^2 - 9}{m^2 - 9m + 18}$

38. $\dfrac{6n^2 - 5n - 6}{6n^2 + 5n - 6} \cdot \dfrac{12n^2 - 17n + 6}{12n^2 - n - 6}$

Add or subtract as indicated in each of the following. Write all answers in lowest terms. (See Example 4.)

39. $\dfrac{8}{r} + \dfrac{6}{r}$

40. $\dfrac{7}{m} - \dfrac{4}{m}$

41. $\dfrac{3}{2k} + \dfrac{5}{3k}$

42. $\dfrac{8}{5p} + \dfrac{3}{4p}$

43. $\dfrac{2}{3y} - \dfrac{1}{4y}$

44. $\dfrac{6}{11z} - \dfrac{5}{2z}$

45. $\dfrac{a + 1}{2} - \dfrac{a - 1}{2}$

46. $\dfrac{y + 6}{5} - \dfrac{y - 6}{5}$

47. $\dfrac{3}{p} + \dfrac{1}{2}$

48. $\dfrac{9}{r} - \dfrac{2}{3}$

49. $\dfrac{2}{y} - \dfrac{1}{4}$

50. $\dfrac{6}{11} + \dfrac{3}{a}$

51. $\dfrac{1}{6m} + \dfrac{2}{5m} + \dfrac{4}{m}$

52. $\dfrac{8}{3p} + \dfrac{5}{4p} + \dfrac{9}{2p}$

53. $\dfrac{1}{m - 1} + \dfrac{2}{m}$

54. $\dfrac{8}{y + 2} - \dfrac{3}{y}$

55. $\dfrac{6}{r} - \dfrac{5}{r - 2}$

56. $\dfrac{8}{a - 1} - \dfrac{5}{a}$

57. $\dfrac{8}{3(a - 1)} + \dfrac{2}{a - 1}$

58. $\dfrac{5}{2(k + 3)} + \dfrac{2}{k + 3}$

59. $\dfrac{2}{5(k - 2)} + \dfrac{3}{4(k - 2)}$

60. $\dfrac{11}{3(p + 4)} - \dfrac{5}{6(p + 4)}$

61. $\dfrac{2}{x^2 - 2x - 3} + \dfrac{5}{x^2 - x - 6}$

62. $\dfrac{3}{m^2 - 3m - 10} + \dfrac{5}{m^2 - m - 20}$

63. $\dfrac{2y}{y^2 + 7y + 12} - \dfrac{y}{y^2 + 5y + 6}$

64. $\dfrac{-r}{r^2 - 10r + 16} - \dfrac{3r}{r^2 + 2r - 8}$

65. $\dfrac{3k}{2k^2 + 3k - 2} - \dfrac{2k}{2k^2 - 7k + 3}$

66. $\dfrac{4m}{3m^2 + 7m - 6} - \dfrac{m}{3m^2 - 14m + 8}$

In each of the following exercises, simplify the complex fraction. (See Example 5.)

67. $\dfrac{1 + \dfrac{1}{x}}{1 - \dfrac{1}{x}}$

68. $\dfrac{2 - \dfrac{2}{y}}{2 + \dfrac{2}{y}}$

69. $\dfrac{\dfrac{1}{x+1}-\dfrac{1}{x}}{\dfrac{1}{x}}$

70. $\dfrac{\dfrac{1}{y+3}-\dfrac{1}{y}}{\dfrac{1}{y}}$

73. $\dfrac{1+\dfrac{1}{1-b}}{1-\dfrac{1}{1+b}}$

74. $\dfrac{m-\dfrac{1}{m^2-4}}{\dfrac{1}{m+2}}$

71. $\dfrac{\dfrac{1}{x+h}-\dfrac{1}{x}}{h}$

72. $\dfrac{\dfrac{1}{(x+h)^2}-\dfrac{1}{x^2}}{h}$

1.7 Integer Exponents

As mentioned in Section 1.5,

$$a^2 = a \cdot a, \text{ while } a^3 = a \cdot a \cdot a,$$

and so on. This section gives a more general meaning to the symbol a^n. First, recall that n is the **exponent** in a^n, and a is the **base**.

If n is a natural number, then

$$a^n = a \cdot a \cdot a \cdots a,$$

where a appears as a factor n times.

Example 1 Evaluate the following.
(a) $2^3 = 2 \cdot 2 \cdot 2 = 8$
  Read 2^3 as "2 cubed."
(b) $5^2 = 5 \cdot 5 = 25$
  Read 5^2 as "5 squared."
(c) $4^5 = 4 \cdot 4 \cdot 4 \cdot 4 \cdot 4 = 1024$
  Read 4^5 as "4 to the fifth power."
(d) $\left(\dfrac{3}{4}\right)^2 = \dfrac{3}{4} \cdot \dfrac{3}{4} = \dfrac{9}{16}$ **1**

1 Evaluate the following.
(a) 6^3
(b) 5^4
(c) 1^7
(d) $\left(\dfrac{2}{5}\right)^3$

Answer:
(a) 216
(b) 625
(c) 1
(d) 8/125

A common error in using exponents occurs with expressions such as $4 \cdot 3^2$. The exponent of 2 applies only to the base 3, so that

$$4 \cdot 3^2 = 4 \cdot 3 \cdot 3 = 36.$$

On the other hand,

$$(4 \cdot 3)^2 = (4 \cdot 3)(4 \cdot 3) = 12 \cdot 12 = 144,$$

and so

$$4 \cdot 3^2 \neq (4 \cdot 3)^2.$$

A zero exponent is defined as follows.

If a is any nonzero real number, then
$$a^0 = 1.$$

The symbol 0^0 is not defined.

Example 2 Evaluate the following.
(a) $6^0 = 1$
(b) $(-9)^0 = 1$
(c) $-(4^0) = -(1) = -1$ ◼2

Negative exponents are defined next.

If n is a natural number, and if $a \neq 0$, then
$$a^{-n} = \frac{1}{a^n}.$$

2 Evaluate the following.
(a) 17^0
(b) 30^0
(c) $(-10)^0$
(d) $-(12)^0$

Answer:
(a) 1
(b) 1
(c) 1
(d) -1

Example 3 Evaluate the following.

(a) $3^{-2} = \dfrac{1}{3^2} = \dfrac{1}{9}$

(b) $5^{-4} = \dfrac{1}{5^4} = \dfrac{1}{625}$

(c) $9^{-1} = \dfrac{1}{9^1} = \dfrac{1}{9}$

(d) $-4^{-2} = -\dfrac{1}{4^2} = -\dfrac{1}{16}$

(e) $\left(\dfrac{3}{4}\right)^{-1} = \dfrac{1}{\left(\dfrac{3}{4}\right)^1} = \dfrac{1}{\dfrac{3}{4}} = \dfrac{4}{3}$

(f) $\left(\dfrac{2}{3}\right)^{-3} = \dfrac{1}{\left(\dfrac{2}{3}\right)^3} = \dfrac{1}{\dfrac{2^3}{3^3}} = 1 \cdot \dfrac{3^3}{2^3} = \dfrac{3^3}{2^3} = \dfrac{27}{8}$ ◼3

3 Evaluate the following.
(a) 6^{-2}
(b) -6^{-3}
(c) -3^{-4}
(d) $\left(\dfrac{5}{8}\right)^{-1}$
(e) $\left(\dfrac{1}{2}\right)^{-4}$

Answer:
(a) 1/36
(b) $-1/216$
(c) $-1/81$
(d) 8/5
(e) 16

Parts (e) and (f) of Example 3 involve work with fractions that can lead to error. For a useful shortcut with such fractions, use the properties of division of rational numbers and the definition of a negative exponent to get

$$\left(\frac{a}{b}\right)^{-n} = \frac{1}{\left(\dfrac{a}{b}\right)^n} = \frac{1}{\dfrac{a^n}{b^n}} = 1 \cdot \frac{b^n}{a^n} = \frac{b^n}{a^n} = \left(\frac{b}{a}\right)^n.$$

For all positive integers n and nonzero real numbers a and b,

$$\left(\frac{a}{b}\right)^{-n} = \left(\frac{b}{a}\right)^{n}.$$

Example 4 Evaluate each of the following.

(a) $\left(\frac{1}{3}\right)^{-2} = \left(\frac{3}{1}\right)^{2} = 9$

4 Evaluate the following.

(a) $\left(\frac{7}{3}\right)^{-2}$

(b) $\left(\frac{5}{9}\right)^{-3}$

Answer:
(a) 9/49
(b) 729/125

(b) $\left(\frac{3}{4}\right)^{-3} = \left(\frac{4}{3}\right)^{3} = \frac{64}{27}$ **4**

The definitions of exponents given above can be used to prove the following properties of exponents.

Properties of Exponents For any integers m and n, and any real numbers a and b for which the following exist,

(a) $a^{m} \cdot a^{n} = a^{m+n}$ **Product property**

(b) $\dfrac{a^{m}}{a^{n}} = a^{m-n}$ **Quotient property**

(c) $(a^{m})^{n} = a^{mn}$

(d) $(ab)^{m} = a^{m} \cdot b^{m}$ **Power properties**

(e) $\left(\dfrac{a}{b}\right)^{m} = \dfrac{a^{m}}{b^{m}}.$

5 Simplify the following.

(a) $9^{6} \cdot 9^{4}$

(b) $\dfrac{8^{7}}{8^{3}}$

(c) $\dfrac{14^{9}}{14^{12}}$

(d) $(13^{4})^{3}$
(e) $(6y)^{4}$

(f) $\left(\dfrac{3}{4}\right)^{2}$

Answer:
(a) 9^{10}
(b) 8^{4}
(c) $1/14^{3}$
(d) 13^{12}
(e) $6^{4}y^{4}$
(f) $3^{2}/4^{2}$ or 9/16

Example 5 Use the properties of exponents to simplify each of the following. Leave answers with exponents.

(a) $7^{4} \cdot 7^{6} = 7^{4+6} = 7^{10}$ Property (a)

(b) $\dfrac{9^{14}}{9^{6}} = 9^{14-6} = 9^{8}$ Property (b)

(c) $\dfrac{r^{9}}{r^{17}} = r^{9-17} = r^{-8} = \dfrac{1}{r^{8}}$ $(r \neq 0)$ Property (b)

(d) $(2m^{3})^{4} = 2^{4} \cdot (m^{3})^{4} = 2^{4}m^{12}$ Properties (c) and (d)

(e) $(3x)^{4} = 3^{4}x^{4}$ Property (d)

(f) $\left(\dfrac{9}{7}\right)^{6} = \dfrac{9^{6}}{7^{6}}$ Property (e) **5**

Example 6 Use the definition of negative exponents and the properties of exponents to simplify each of the following expressions. Give answers with only positive exponents. Assume all variables represent nonzero real numbers.

6 Simplify the following. Give answers with only positive exponents. Assume all variables represent nonzero real numbers.

(a) $\dfrac{(t^{-1})^2}{t^{-5}}$ (b) $\dfrac{(3z)^{-1}z^4}{z^2}$

(c) $\dfrac{3^{-5}\cdot 3^{-2}}{3^{-6}\cdot 3^3}$

(d) $4^{-1}+2^{-2}$

(e) $\dfrac{3^{-1}+2^{-2}}{2^{-2}}$

Answer:
(a) t^3 (d) 1/2
(b) $z/3$ (e) 7/3
(c) $1/3^4$

(a) $\dfrac{(m^3)^{-2}}{m^4}=\dfrac{m^{-6}}{m^4}=m^{-6}m^{-4}=m^{-10}=\dfrac{1}{m^{10}}$

(b) $\dfrac{(2y)^3y^{-1}}{y^2}=\dfrac{2^3y^3y^{-1}}{y^2}=\dfrac{2^3y^2}{y^2}=2^3$

(c) $\dfrac{2^{-3}\cdot 2^{-1}}{2^4\cdot 2^{-7}}=\dfrac{2^{-4}}{2^{-3}}=2^{-4-(-3)}=2^{-1}=\dfrac{1}{2}$

(d) $2^{-1}+3^{-1}=\dfrac{1}{2}+\dfrac{1}{3}=\dfrac{5}{6}$

(e) $\dfrac{2^{-1}-3^{-2}}{2^{-1}}=\dfrac{\frac{1}{2}-\frac{1}{3^2}}{\frac{1}{2}}=\dfrac{\frac{1}{2}-\frac{1}{9}}{\frac{1}{2}}=\dfrac{\frac{9}{18}-\frac{2}{18}}{\frac{1}{2}}=\dfrac{7}{18}\cdot\dfrac{2}{1}=\dfrac{7}{9}$ **6**

1.7 Exercises

Evaluate each of the following. Write all answers without exponents. (See Examples 1–4.)

1. 7^3 2. 4^2
3. 8^{-1} 4. 9^{-2}
5. 2^{-3} 6. 3^{-4}
7. 5^{-1} 8. 6^{-3}
9. 8^{-3} 10. 12^{-2}
11. $\left(\dfrac{3}{4}\right)^3$ 12. $\left(\dfrac{2}{3}\right)^4$
13. $\left(\dfrac{5}{8}\right)^2$ 14. $\left(\dfrac{6}{7}\right)^3$
15. $\left(\dfrac{1}{2}\right)^{-3}$ 16. $\left(\dfrac{1}{5}\right)^{-3}$
17. $\left(\dfrac{2}{7}\right)^{-2}$ 18. $\left(\dfrac{4}{5}\right)^{-3}$

Simplify each of the following. Write all answers using only positive exponents. (See Examples 5 and 6.)

19. $\dfrac{3^8}{3^2}$ 20. $\dfrac{4^9}{4^7}$
21. $\dfrac{7^5}{7^9}$ 22. $\dfrac{8^6}{8^{12}}$
23. $\dfrac{3^{-4}}{3^2}$ 24. $\dfrac{9^{-2}}{9^5}$

25. $\dfrac{2^{-5}}{2^{-2}}$ 26. $\dfrac{14^{-3}}{14^{-8}}$
27. $\dfrac{6^{-1}}{6}$ 28. $\dfrac{15}{15^{-1}}$
29. $4^{-3}\cdot 4^6$ 30. $5^{-9}\cdot 5^{10}$
31. $7^{-5}\cdot 7^{-2}$ 32. $9^{-1}\cdot 9^{-3}$
33. $\dfrac{8^9\cdot 8^{-7}}{8^{-3}}$ 34. $\dfrac{5^{-4}\cdot 5^6}{5^{-1}}$
35. $\dfrac{10^8\cdot 10^{-10}}{10^4\cdot 10^2}$ 36. $\dfrac{2^{-4}\cdot 2^{-3}}{2^6\cdot 2^{-5}}$
37. $\left(\dfrac{5^{-6}\cdot 5^3}{5^{-2}}\right)^{-1}$ 38. $\left(\dfrac{8^{-3}\cdot 8^4}{8^{-2}}\right)^{-2}$

Simplify each of the following. Assume all variables represent positive real numbers. Write answers with only positive exponents. (See Examples 5 and 6.)

39. $\dfrac{x^4\cdot x^3}{x^5}$ 40. $\dfrac{y^9\cdot y^7}{y^{13}}$
41. $\dfrac{(4k^{-1})^2}{2k^{-5}}$ 42. $\dfrac{(3z^2)^{-1}}{z^5}$
43. $\dfrac{7^{-1}\cdot 7r^{-3}}{7^2\cdot (r^{-2})^2}$ 44. $\dfrac{12^3\cdot 12^5\cdot y^{-2}}{12^{-1}\cdot (y^{-3})^{-2}}$
45. $\dfrac{6k^{-4}\cdot (3k^{-1})^{-2}}{2^3\cdot k^2}$ 46. $\dfrac{8p^{-3}\cdot (4p^2)^{-2}}{p^{-5}}$

47. $(m^4p^{-2})^2 \cdot (m^{-1}p^2)$ **48.** $(z^3x^5)^{-1} \cdot (z^{-4}x^{-2})$

49. $(5a^2b^{-3})^{-1} \cdot (3^{-1}a^{-2}b^2)^{-2}$

50. $(2p^2q^{-3})^{-2} \cdot (7p^{-1}q^2)^{-1}$

Evaluate each expression, letting $a = 2$, $b = -3$, and $c = 0$.

51. $a^3 + b$ **52.** $b^3 + a^2$

53. $-b^2 + 3(c + 5)$ **54.** $2a^2 - b^4$

55. $a^7b^9c^5$ **56.** $12a^{15}b^6c^8$

57. $a^b + b^a$ **58.** $b^c + a^c$

59. $a^{-1} + b^{-1}$ **60.** $b^{-2} - a^{-1}$

61. $a^{-b} + b^{-c}$ **62.** $a^{2b} + b^b$

Perform all indicated operations and write answers with positive exponents. Assume all variables represent positive real numbers. (See Example 6.)

63. $2^{-1} - 3^{-1}$ **64.** $4^{-1} + 5^{-1}$

65. $(6^{-1} + 2^{-1})^{-1}$ **66.** $(5^{-1} - 10^{-1})^{-1}$

67. $\dfrac{3^{-2} - 4^{-1}}{4^{-1}}$ **68.** $\dfrac{6^{-1} + 5^{-2}}{6^{-1}}$

69. $\dfrac{a^{-1} + b^{-1}}{(ab)^{-1}}$ **70.** $\dfrac{p^{-1} - q^{-1}}{(pq)^{-2}}$

71. $(a + b)^{-1}(a^{-1} + b^{-1})$ **72.** $(m^{-1} + n^{-1})^{-1}$

1.8 Rational Exponents and Radicals

In this section the definition of a^n will be extended to include rational values of n and not just integer values. We begin by defining $a^{1/n}$ for positive integer values of n. Any definition of $a^{1/n}$ should be consistent with the properties of exponents given in the previous section. In particular, $(b^m)^n = b^{mn}$ should still be valid. Replacing b^m with $a^{1/n}$ gives

$$(a^{1/n})^n = a^{(1/n)n} = a^1 = a.$$

This means that $a^{1/n}$ should be defined as an **nth root of a.** For example, $a^{1/2}$ indicates a second root or **square root** of a.

There are two numbers whose square is 16, both 4 and -4. To distinguish between them, the symbol $a^{1/n}$ is reserved for the *positive* root, so

$$16^{1/2} = 4.$$

Use $-16^{1/2} = -4$ to indicate the negative root, because $-16^{1/2}$ means $-(16^{1/2})$ which equals $-(4)$ or -4. There is no real number square root of -16, so $(-16)^{1/2}$ is not a real number.

The symbol $a^{1/3}$ represents the third root or **cube root** of a. For example,

$$216^{1/3} = 6 \qquad \text{since } 6^3 = 216$$

and
$$(-8)^{1/3} = -2 \qquad \text{since } (-2)^3 = -8.$$

This discussion is summarized below.

If a is a real number and n is a positive integer, then the **nth root** of a, written $a^{1/n}$, is defined as follows.		
n is	**$a \geq 0$**	**$a < 0$**
even	$a^{1/n}$ is the positive real number such that $(a^{1/n})^n = a$.	$a^{1/n}$ is not a real number.
odd	$a^{1/n}$ is the real number such that $(a^{1/n})^n = a$.	

Example 1 Evaluate the following roots.
(a) $36^{1/2} = 6$ since $6^2 = 36$.
(b) $-100^{1/2} = -10$
(c) $-(225^{1/2}) = -15$
(d) $625^{1/4} = 5$ since $5^4 = 625$.
(e) $(-1296)^{1/4}$ is not a real number, but $-1296^{1/4} = -6$.
(f) $(-27)^{1/3} = -3$
(g) $-32^{1/5} = -2$ **1**

1 Evaluate the following.
(a) $16^{1/2}$
(b) $16^{1/4}$
(c) $-256^{1/2}$
(d) $(-256)^{1/2}$
(e) $-8^{1/3}$
(f) $243^{1/5}$

Answer:
(a) 4
(b) 2
(c) -16
(d) Not a real number
(e) -2
(f) 3

For more general rational exponents, the symbol $a^{m/n}$ must be defined so that the properties for exponents still hold. For the first power property to hold,

$$(a^{1/n})^m \text{ must equal } a^{m/n}.$$

This suggests the following definition.

> For all integers m and all positive integers n, and for all real numbers a for which $a^{1/n}$ exists,
>
> $$a^{m/n} = (a^{1/n})^m.$$

Example 2 Evaluate the following.
(a) $27^{2/3} = (27^{1/3})^2 = 3^2 = 9$
(b) $32^{2/5} = (32^{1/5})^2 = 2^2 = 4$
(c) $64^{4/3} = (64^{1/3})^4 = 4^4 = 256$
(d) $25^{3/2} = (25^{1/2})^3 = 5^3 = 125$ **2**

2 Evaluate the following.
(a) $16^{3/4}$
(b) $25^{5/2}$
(c) $32^{7/5}$
(d) $100^{3/2}$

Answer:
(a) 8
(b) 3125
(c) 128
(d) 1000

It can be shown that all the properties given above for integer exponents also apply to rational exponents.

Example 3 Simplify the following. Assume all variables represent positive real numbers.

(a) $\dfrac{27^{1/3} \cdot 27^{5/3}}{27^3} = \dfrac{27^{1/3+5/3}}{27^3}$ Product property

$\qquad = \dfrac{27^2}{27^3} = 27^{2-3}$ Quotient property

$\qquad = 27^{-1} = \dfrac{1}{27}$ Definition of negative exponent

(b) $81^{5/4} \cdot 4^{-3/2} = (81^{1/4})^5 \cdot (4^{1/2})^{-3}$ Definition of $a^{m/n}$

$\qquad = 3^5 \cdot 2^{-3}$ Definition of $a^{1/n}$

$\qquad = \dfrac{3^5}{2^3} \quad \text{or} \quad \dfrac{243}{8}$ Definition of negative exponent

(c) $6y^{2/3} \cdot 2y^{1/2} = 12y^{2/3+1/2} = 12y^{7/6}$

(d) $\left(\dfrac{3m^{5/6}}{y^{3/4}}\right)^2 \cdot \left(\dfrac{8y^{-3}}{m^6}\right)^{2/3} = \dfrac{9m^{5/3}}{y^{3/2}} \cdot \dfrac{4}{m^4y^2}$ Power property (e)

$$\frac{9m^{5/3}}{y^{3/2}} \cdot \frac{4}{m^4 y^2} = \frac{36m^{(5/3)-4}}{y^{(3/2)+2}} \qquad \text{Product property}$$

$$= \frac{36m^{-7/3}}{y^{7/2}} = \frac{36}{m^{7/3}y^{7/2}}$$

(e) $m^{2/3}(m^{7/3} + 2m^{1/3}) = (m^{2/3+7/3} + 2m^{2/3+1/3}) = m^3 + 2m$ ▢**3**

3 Simplify. Assume all variables represent positive real numbers.
(a) $6^{2/5} \cdot 6^{3/5}$

(b) $\dfrac{8^{2/3} \cdot 8^{-4/3}}{8^2}$

(c) $3x^{1/4} \cdot 5x^{5/4}$

(d) $\left(\dfrac{2k^{1/3}}{p^{5/4}}\right)^2 \cdot \left(\dfrac{4k^{-2}}{p^5}\right)^{3/2}$

(e) $a^{5/8}(2a^{3/8} + a^{-1/8})$

Answer:
(a) 6
(b) $1/8^{8/3}$ or $1/2^8$
(c) $15x^{3/2}$
(d) $32/(p^{10}k^{7/3})$
(e) $2a + a^{1/2}$

Radicals The symbol $a^{1/n}$ was defined above as the nth root of a, for appropriate values of a and n. An alternative notation for $a^{1/n}$ uses radicals.

If n is an even natural number and $a \ge 0$, or n is an odd natural number,

$$a^{1/n} = \sqrt[n]{a}.$$

The symbol $\sqrt[n]{}$ is a **radical sign,** the number a is the **radicand,** and n is the **index** of the radical. The familiar symbol $\sqrt{a}$ is used instead of $\sqrt[2]{a}$.

Example 4 Simplify the following.
(a) $\sqrt[4]{16} = 16^{1/4} = 2$
(b) $\sqrt[5]{-32} = -2$
(c) $\sqrt[3]{1000} = 10$
(d) $\sqrt[6]{\dfrac{64}{729}} = \dfrac{2}{3}$ ▢**4**

4 Simplify.
(a) $\sqrt[3]{27}$
(b) $\sqrt[4]{625}$
(c) $\sqrt[6]{64}$
(d) $\sqrt[3]{\dfrac{64}{125}}$

Answer:
(a) 3
(b) 5
(c) 2
(d) 4/5

The symbol $a^{m/n}$ also can be written in an alternative notation using radicals.

For all rational numbers m/n and all real numbers a for which $\sqrt[n]{a}$ exists,

$$a^{m/n} = (\sqrt[n]{a})^m \qquad \text{or} \qquad a^{m/n} = \sqrt[n]{a^m}.$$

A word of warning: $\sqrt[n]{x^n}$ cannot be written simply as x. For example, if $x = -5$,

$$\sqrt{x^2} = \sqrt{(-5)^2} = \sqrt{25} = 5 \ne x.$$

Use the following rule to avoid the fact that a negative value of x can produce a positive result for an even root.

For any real number a and any natural number n,

$$\sqrt[n]{a^n} = |a| \text{ if } n \text{ is even}$$

and

$$\sqrt[n]{a^n} = a \text{ if } n \text{ is odd.}$$

To get around this difficulty that $\sqrt[n]{a^n}$ is not necessarily equal to a, we shall assume that all variables in radicands represent only nonnegative numbers.

The properties of exponents can be written with radicals as shown below.

> For all real numbers a and b and positive integers m and n for which all indicated roots exist,
>
> (a) $\sqrt[n]{a} \cdot \sqrt[n]{b} = \sqrt[n]{ab}$
>
> (b) $\dfrac{\sqrt[n]{a}}{\sqrt[n]{b}} = \sqrt[n]{\dfrac{a}{b}}$ $(b \neq 0)$
>
> (c) $\sqrt[m]{\sqrt[n]{a}} = \sqrt[mn]{a}.$

Example 5 Simplify the following.

(a) $\sqrt{6} \cdot \sqrt{54} = \sqrt{6 \cdot 54} = \sqrt{324} = 18$

(b) $\sqrt{\dfrac{7}{64}} = \dfrac{\sqrt{7}}{\sqrt{64}} = \dfrac{\sqrt{7}}{8}$

(c) $\sqrt[7]{\sqrt[3]{2}} = \sqrt[21]{2}$ **5**

Simplifying Radicals When working with numbers, it is customary to write them in their simplest form. For example, $10/2$ is written as 5, $-9/6$ is written as $-3/2$, and $4/20$ is written as $1/5$. Expressions with radicals also should be written in their simplest form. The **simplified form** of a radical is defined as follows.

> **Simplified Radicals** An expression with radicals is simplified when all the following conditions are satisfied.
>
> (a) The radicand has no factors with powers greater than or equal to the index of the radical.
>
> (b) The index of the radical and the power of the radicand have no common factor except 1.
>
> (c) All denominators are rational numbers.
>
> (d) All indicated operations have been performed (if possible).

Example 6 Simplify each of the following. Assume all variables represent positive real numbers.

(a) $\sqrt{175} = \sqrt{25 \cdot 7} = \sqrt{25} \cdot \sqrt{7} = 5\sqrt{7}$

(b) $\sqrt[3]{81x^5y^7z^6} = \sqrt[3]{27 \cdot 3 \cdot x^3 \cdot x^2 \cdot y^3 \cdot y^3 \cdot y \cdot z^3 \cdot z^3}$

$= \sqrt[3]{(27x^3y^3y^3z^3z^3)(3x^2y)}$

$= 3xyyzz\sqrt[3]{3x^2y}$

$= 3xy^2z^2\sqrt[3]{3x^2y}$ **6**

5 Simplify.

(a) $\sqrt{3} \cdot \sqrt{27}$

(b) $\sqrt{\dfrac{3}{49}}$

(c) $\sqrt[5]{\sqrt[4]{3}}$

Answer:

(a) 9

(b) $\dfrac{\sqrt{3}}{7}$

(c) $\sqrt[20]{3}$

6 Simplify.

(a) $\sqrt{80}$

(b) $\sqrt[3]{32m^4p^6q^2}$

Answer:

(a) $4\sqrt{5}$

(b) $2mp^2\sqrt[3]{4mq^2}$

Radicals with the same radicand and the same index, such as $3\sqrt[4]{11pq}$ and $-7\sqrt[4]{11pq}$, are called **like radicals.** Only like radicals can be added or subtracted. As shown in part (b) of the next example, it is sometimes necessary to simplify radicals before adding or subtracting. As before, assume that all variables represent only positive numbers.

Example 7 Simplify the following.

(a) $3\sqrt[4]{11pq} - 7\sqrt[4]{11pq} = (3 - 7)\sqrt[4]{11pq} = -4\sqrt[4]{11pq}$

(b) $\sqrt{98x^3y} + 3x\sqrt{32xy}$

First remove all perfect square factors from under the radical. Then use the distributive property, as follows.

$$\sqrt{98x^3y} + 3x\sqrt{32xy} = \sqrt{49 \cdot 2 \cdot x^2 \cdot x \cdot y} + 3x\sqrt{16 \cdot 2 \cdot x \cdot y}$$
$$= 7x\sqrt{2xy} + (3x)(4)\sqrt{2xy}$$
$$= 7x\sqrt{2xy} + 12x\sqrt{2xy}$$
$$= 19x\sqrt{2xy} \quad \boxed{7}$$

7 Simplify. Assume all variables represent positive real numbers.

(a) $5\sqrt[3]{4x^2z} - 3\sqrt[3]{4x^2z}$

(b) $\sqrt{27m^3n} + 2m\sqrt{12mn}$

Answer:

(a) $2\sqrt[3]{4x^2z}$

(b) $7m\sqrt{3mn}$

Multiplying radical expressions is much like multiplying polynomials.

Example 8 Multiply the following.

(a) $(\sqrt{2} + 3)(\sqrt{8} - 5) = \sqrt{2}(\sqrt{8}) - \sqrt{2}(5) + 3\sqrt{8} - 3(5)$
$$= \sqrt{16} - 5\sqrt{2} + 3(2\sqrt{2}) - 15$$
$$= 4 - 5\sqrt{2} + 6\sqrt{2} - 15$$
$$= -11 + \sqrt{2}$$

(b) $(\sqrt{7} - \sqrt{10})(\sqrt{7} + \sqrt{10}) = (\sqrt{7})^2 - (\sqrt{10})^2$
$$= 7 - 10$$
$$= -3 \quad \boxed{8}$$

8 Multiply.

(a) $(\sqrt{5} - \sqrt{2})(3 + \sqrt{2})$

(b) $(\sqrt{3} + \sqrt{7})(\sqrt{3} - \sqrt{7})$

Answer:

(a) $3\sqrt{5} + \sqrt{10} - 3\sqrt{2} - 2$

(b) -4

Rationalizing the Denominator The process of rationalizing the denominator (writing denominators with no radicals) is explained in the next examples.

Example 9 Rationalize each denominator.

(a) $\dfrac{4}{\sqrt{3}}$

To rationalize the denominator, multiply by $\sqrt{3}/\sqrt{3}$ so that the denominator is $\sqrt{3} \cdot \sqrt{3} = 3$, a rational number.

$$\frac{4}{\sqrt{3}} \cdot \frac{\sqrt{3}}{\sqrt{3}} = \frac{4\sqrt{3}}{3}$$

(b) $\sqrt[4]{\dfrac{3}{5}}$

9 Rationalize the denominator.

(a) $\dfrac{2}{\sqrt{5}}$

(b) $\sqrt[3]{\dfrac{3}{2}}$

Answer:

(a) $\dfrac{2\sqrt{5}}{5}$

(b) $\dfrac{\sqrt[3]{12}}{2}$

Start by using the fact that the radical of a quotient can be written as the quotient of radicals.

$$\sqrt[4]{\frac{3}{5}} = \frac{\sqrt[4]{3}}{\sqrt[4]{5}}$$

Multiply numerator and denominator by $\sqrt[4]{5^3}$ to rationalize the denominator. Use this number so that the denominator will be $\sqrt[4]{5} \cdot \sqrt[4]{5^3} = \sqrt[4]{5^4} = 5$, a rational number. This multiplication gives

$$\frac{\sqrt[4]{3}}{\sqrt[4]{5}} = \frac{\sqrt[4]{3} \cdot \sqrt[4]{5^3}}{\sqrt[4]{5} \cdot \sqrt[4]{5^3}} = \frac{\sqrt[4]{3 \cdot 5^3}}{\sqrt[4]{5^4}} = \frac{\sqrt[4]{375}}{5}. \quad \textbf{9}$$

10 Rationalize the denominator.

(a) $\dfrac{4}{1 + \sqrt{3}}$

(b) $\dfrac{1 - \sqrt{2}}{1 + \sqrt{2}}$

Answer:

(a) $-2 + 2\sqrt{3}$

(b) $-3 + 2\sqrt{2}$

Example 10 Rationalize the denominator of $\dfrac{1}{1 - \sqrt{2}}$.

The best approach here is to multiply both the numerator and denominator by the conjugate* of the denominator, in this case $1 + \sqrt{2}$. As suggested by Example 8(b), the product $(1 - \sqrt{2})(1 + \sqrt{2})$ is rational.

$$\frac{1}{1 - \sqrt{2}} = \frac{1(1 + \sqrt{2})}{(1 - \sqrt{2})(1 + \sqrt{2})} = \frac{1 + \sqrt{2}}{1 - 2}$$

$$= \frac{1 + \sqrt{2}}{-1} = -1 - \sqrt{2} \quad \textbf{10}$$

1.8 Exercises

Evaluate each of the following. Write all answers without exponents. (See Examples 1–2.)

1. $81^{1/2}$

2. $16^{1/4}$

3. $27^{1/3}$

4. $81^{1/4}$

5. $8^{2/3}$

6. $9^{3/2}$

7. $1000^{2/3}$

8. $64^{3/2}$

9. $32^{2/5}$

10. $32^{6/5}$

11. $-125^{2/3}$

12. $-125^{4/3}$

13. $\left(\dfrac{4}{9}\right)^{1/2}$

14. $\left(\dfrac{16}{25}\right)^{1/2}$

15. $\left(\dfrac{64}{27}\right)^{1/3}$

16. $\left(\dfrac{8}{125}\right)^{1/3}$

17. $16^{-5/4}$

18. $625^{-1/4}$

19. $\left(\dfrac{27}{64}\right)^{-1/3}$

20. $\left(\dfrac{121}{100}\right)^{-3/2}$

21. $2^{-1} + 4^{-1}$

22. $2^{-2} + 3^{-2}$

Simplify each of the following. Write all answers using only positive exponents. Assume all variables represent positive real numbers. (See Example 3.)

23. $2^{1/2} \cdot 2^{3/2}$

24. $5^{3/8} \cdot 5^{5/8}$

25. $27^{2/3} \cdot 27^{-1/3}$

26. $9^{-3/4} \cdot 9^{1/4}$

27. $\dfrac{4^{2/3} \cdot 4^{5/3}}{4^{1/3}}$

28. $\dfrac{3^{-5/2} \cdot 3^{3/2}}{3^{7/2} \cdot 3^{-9/2}}$

29. $\dfrac{6^{2/5} \cdot 6^{-4/5}}{6^2 \cdot 6^{-1/5}}$

30. $\dfrac{3^{-5/3} \cdot 3^{2/3}}{3^{-1} \cdot 3^{4/3}}$

---

*The conjugate of $a\sqrt{m} + b\sqrt{n}$ is $a\sqrt{m} - b\sqrt{n}$.

31. $\dfrac{4^{-2/3} \cdot 4^{1/5}}{4^{5/3}}$

32. $\dfrac{8^{1/4} \cdot 8^{-3/4}}{8^{-5/3}}$

33. $(2p)^{1/2} \cdot (2p^3)^{1/3}$

34. $(5k^2)^{3/2} \cdot (5k^{1/3})^{3/4}$

35. $\dfrac{(mn)^{1/5}(m^2n)^{-2/5}}{m^{1/3}n}$

36. $\dfrac{(r^2s)^{3/2}(rs^2)^{1/4}}{r^3s}$

37. $x^{3/4}(x^2 - 3x^3)$

38. $p^{2/3}(2p^{1/3} + 5p)$

39. $2z^{1/2}(3z^{-1/2} + z^{1/2})$

40. $4a^{2/3}(2a^{1/3} - 3a^{3/4})$

Simplify each of the following. Assume that all variables represent positive real numbers. (See Examples 4–6 and 9.)

41. $\sqrt[3]{125}$

42. $\sqrt[4]{81}$

43. $\sqrt[4]{1296}$

44. $\sqrt[3]{343}$

45. $\sqrt[5]{-3125}$

46. $\sqrt[7]{-128}$

47. $\sqrt{50}$

48. $\sqrt{45}$

49. $\sqrt{2000}$

50. $\sqrt{3200}$

51. $-\sqrt[4]{32}$

52. $-\sqrt[4]{243}$

53. $-\sqrt{\dfrac{9}{5}}$

54. $\sqrt{\dfrac{3}{8}}$

55. $-\sqrt[3]{\dfrac{3}{2}}$

56. $-\sqrt[3]{\dfrac{4}{5}}$

57. $\sqrt[4]{\dfrac{3}{2}}$

58. $\sqrt[4]{\dfrac{32}{81}}$

59. $\sqrt{24 \cdot 3^2 \cdot 2^4}$

60. $\sqrt{18 \cdot 2^3 \cdot 4^2}$

61. $\sqrt{8z^5x^8}$

62. $\sqrt{24m^6n^5}$

63. $\sqrt{m^2n^7p^8}$

64. $\sqrt{81p^5x^2y^6}$

65. $\sqrt{\dfrac{2}{3x}}$

66. $\sqrt{\dfrac{5}{3p}}$

67. $\sqrt{\dfrac{x^5y^3}{3^2}}$

68. $\sqrt{\dfrac{g^3h^5}{r^3}}$

69. $\sqrt[3]{\sqrt{4}}$

70. $\sqrt[4]{\sqrt[3]{2}}$

71. $\sqrt[6]{\sqrt[3]{x}}$

72. $\sqrt[8]{\sqrt[4]{x}}$

Simplify each of the following. Assume that all variables represent positive real numbers. (See Examples 7 and 8.)

73. $4\sqrt{3} - 5\sqrt{12} + 3\sqrt{75}$

74. $2\sqrt{5} - 3\sqrt{20} + 2\sqrt{45}$

75. $\sqrt{50} - 8\sqrt{8} + 4\sqrt{18}$

76. $6\sqrt{27} - 3\sqrt{12} + 5\sqrt{48}$

77. $3\sqrt{28p} - 4\sqrt{63p} + \sqrt{112p}$

78. $9\sqrt{8k} + 3\sqrt{18k} - \sqrt{32k}$

79. $3\sqrt[3]{16} - 4\sqrt[3]{2}$

80. $\sqrt[3]{2} - \sqrt[3]{16} + 2\sqrt[3]{54}$

81. $2\sqrt[3]{3} + 4\sqrt[3]{24} - \sqrt[3]{81}$

82. $\sqrt[3]{32} - 5\sqrt[3]{4} + \sqrt[3]{108}$

83. $(\sqrt{2} + 3)(\sqrt{2} - 3)$

84. $(\sqrt{5} + \sqrt{2})(\sqrt{5} - \sqrt{2})$

85. $(\sqrt[3]{11} - 1)(\sqrt[3]{11^2} + \sqrt[3]{11} + 1)$

86. $(\sqrt[3]{7} + 3)(\sqrt[3]{7^2} - 3\sqrt[3]{7} + 9)$

87. $(3\sqrt{2} + \sqrt{3})(2\sqrt{3} - \sqrt{2})$

88. $(4\sqrt{5} - 1)(3\sqrt{5} + 2)$

Rationalize the denominator of each of the following. Assume that all variables represent nonnegative numbers and that no denominators are zero. (See Example 10.)

89. $\dfrac{3}{1 - \sqrt{2}}$

90. $\dfrac{2}{1 + \sqrt{5}}$

91. $\dfrac{\sqrt{3}}{4 + \sqrt{3}}$

92. $\dfrac{2\sqrt{7}}{3 - \sqrt{7}}$

93. $\dfrac{4 - \sqrt{2}}{2 - \sqrt{2}}$

94. $\dfrac{\sqrt{3} - 1}{\sqrt{3} - 2}$

95. $\dfrac{p}{\sqrt{p} + 2}$

96. $\dfrac{\sqrt{r}}{3 - \sqrt{r}}$

97. Management The price of a certain type of solar heater is approximated by p, where

$$p = 2x^{1/2} + 3x^{2/3}$$

and x is the number of units supplied. Find the price when the supply is 64 units.

98. Management The demand for a certain commodity and the price are related by the equation

$$p = 1000 - 200x^{-2/3} \qquad (x > 0)$$

where x is the number of units of the product demanded. Find the price when the demand is 27.

Social Science *In our system of government, the president is elected by the electoral college, and not by individual voters. Because of this, smaller states have a greater voice in the selection of a president than they would otherwise. Two political scientists have studied the problems of cam-*

paigning for president under the current system and have concluded that candidates should allot their money according to the formula

$$\text{amount for large state} = \left(\frac{E_{\text{large}}}{E_{\text{small}}}\right)^{3/2} \times \text{amount for a small state.}$$

Here E_{large} represents the electoral vote of the large state, and E_{small} represents the electoral vote of the small state. Find the amount that should be spent in each of the following larger states if $1,000,000 is spent in the small state and the following statements are true.

99. The large state has 48 electoral votes and the small state has 3.

100. The large state has 36 electoral votes and the small state has 4.

Natural Science *A Delta Airlines map gives a formula for calculating the visible distance from a jet plane to the horizon. On a clear day, this distance is approximated by*

$$D = 1.22x^{1/2},$$

where x is altitude in feet, and D is distance to the horizon in miles. Find D for an altitude of

101. 5000 feet;　　　102. 10,000 feet;

103. 30,000 feet;　　104. 40,000 feet.

Natural Science *(These problems require a calculator with a y^x key.) The Galápagos Islands are a chain of islands ranging in size from .2 to 2249 square miles. A biologist has shown that the number of different land-plant species on an island in this chain is related to the size of the island by*

$$S = 28.6A^{.32},$$

where A is the area of an island in square miles and S is the number of different plant species on that island. Estimate S (rounding to the nearest whole number) for islands of area

105. 1 square mile;　　106. 25 square miles;

107. 300 square miles;　108. 2000 square miles.

1.9 Quadratic Equations

An equation of the form $ax + b = 0$ is a linear equation, while an equation with 2 as the highest exponent is called **quadratic.**

If a, b, and c are real numbers, with $a \neq 0$, then

$$ax^2 + bx + c = 0,$$

is a **quadratic equation.**

(Why is the restriction $a \neq 0$ necessary?) A quadratic equation written in the form $ax^2 + bx + c = 0$ is said to be in **standard form.**

　　The simplest method of solving a quadratic equation, but one that is not always easily applied, is by factoring. This method depends on the **zero-factor property.**

Zero-Factor Property If a and b are real numbers, with $ab = 0$, then $a = 0$ or $b = 0$ or both.

Example 1 Solve the equation $(x - 4)(3x + 7) = 0$.

　　By the zero-factor property, the product $(x - 4)$ $(3x + 7)$ can equal 0 only if one of the factors equals 0. That is, the product equals zero only if $x - 4 = 0$ or

$3x + 7 = 0$. Solving each of these equations separately will give the solutions of the original equation.

$$x - 4 = 0 \quad \text{or} \quad 3x + 7 = 0$$
$$x = 4 \quad \text{or} \quad 3x = -7$$
$$x = -\frac{7}{3}$$

The solutions of the equation $(x - 4)(3x + 7) = 0$ are 4 and $-7/3$. Check these solutions by substitution in the original equation. **1**

1 Solve the following equations.
(a) $(y - 6)(y + 2) = 0$
(b) $(5k - 3)(k + 5) = 0$
(c) $(2r - 9)(3r + 5) = 0$

Answer:
(a) 6, -2
(b) 3/5, -5
(c) 9/2, $-5/3$

Example 2 Solve $6r^2 + 7r = 3$.

Write the equation in standard form as

$$6r^2 + 7r - 3 = 0.$$

Now factor $6r^2 + 7r - 3$ to get

$$(3r - 1)(2r + 3) = 0.$$

By the zero-factor property, the product $(3r - 1)(2r + 3)$ can equal 0 only if

$$3r - 1 = 0 \quad \text{or} \quad 2r + 3 = 0.$$

Solve each of these equations separately to find that the solutions of the original equation are $1/3$ and $-3/2$. Check these solutions by substitution in the original equation. **2**

2 Solve the following.
(a) $y^2 + 3y = 10$
(b) $2r^2 + 9r = 5$
(c) $3k^2 = 2k + 8$

Answer:
(a) 2, -5
(b) 1/2, -5
(c) $-4/3$, 2

Not all quadratic equations can be readily solved by the zero-factor property. Some quadratic equations for which the zero-factor property is not useful can be solved by the **square-root property.**

> **Square-Root Property** If $b > 0$, then the solutions of
>
> $$x^2 = b$$
>
> are $\sqrt{b}$ and $-\sqrt{b}$.

The two solutions are sometimes abbreviated $\pm\sqrt{b}$.

Example 3 Solve each equation.
(a) $m^2 = 17$

By the square root property, the solutions are $\sqrt{17}$ and $-\sqrt{17}$, abbreviated $\pm\sqrt{17}$.

(b) $(y - 4)^2 = 11$

Use a generalization of the square root property, working as follows:

$$(y - 4)^2 = 11$$
$$y - 4 = \sqrt{11} \quad \text{or} \quad y - 4 = -\sqrt{11}$$
$$y = 4 + \sqrt{11} \quad \text{or} \quad y = 4 - \sqrt{11}.$$

3 Solve each equation.

(a) $p^2 = 21$

(b) $(m + 7)^2 = 15$

(c) $(2k - 3)^2 = 5$

Answer:

(a) $\pm\sqrt{21}$

(b) $-7 \pm \sqrt{15}$

(c) $(3 \pm \sqrt{5})/2$

Abbreviate the solutions as $4 \pm \sqrt{11}$. **3**

As suggested by Example 3(b), any quadratic equation can be solved using the square root property if the equation can be written in the form $(x + n)^2 = k$. The next example shows how to write a quadratic equation in this form.

Example 4 Solve $9z^2 - 12z - 1 = 0$.

We need to find constants n and k so that the given equation can be written in the form $(z + n)^2 = k$. Expanding $(z + n)^2$ gives $z^2 + 2zn + n^2$, where 1 is the coefficient of z^2. Get a coefficient of 1 in the given equation by multiplying both sides by 1/9.

$$z^2 - \frac{4}{3}z - \frac{1}{9} = 0$$

Now add 1/9 on both sides:

$$z^2 - \frac{4}{3}z = \frac{1}{9}. \tag{*}$$

Find a number that will make the left side a perfect square. The term $-4z/3$ is twice the product of z and n, the number whose square is needed to make the left side a perfect square. That is, n must satisfy

$$2zn = -\frac{4}{3}z,$$

or

$$n = -\frac{2}{3}.$$

If $n = -2/3$, then $n^2 = 4/9$. Add $\frac{4}{9}$ to both sides of equation (*) to get

$$z^2 - \frac{4}{3}z + \frac{4}{9} = \frac{1}{9} + \frac{4}{9}.$$

Factoring on the left and adding on the right gives

$$\left(z - \frac{2}{3}\right)^2 = \frac{5}{9}.$$

Now use the square root property and simplify the square root to get

$$z - \frac{2}{3} = \sqrt{\frac{5}{9}} \qquad \text{or} \qquad z - \frac{2}{3} = -\sqrt{\frac{5}{9}}$$

$$z - \frac{2}{3} = \frac{\sqrt{5}}{3} \qquad \text{or} \qquad z - \frac{2}{3} = -\frac{\sqrt{5}}{3}$$

$$z = \frac{2}{3} + \frac{\sqrt{5}}{3} \qquad \text{or} \qquad z = \frac{2}{3} - \frac{\sqrt{5}}{3}.$$

These two solutions may be written as

$$z = \frac{2 + \sqrt{5}}{3} \quad \text{or} \quad z = \frac{2 - \sqrt{5}}{3},$$

4 Solve by completing the square.

(a) $m^2 + 2m = 3$

(b) $6r^2 - r = 2$

Answer:

(a) $1, -3$

(b) $-1/2, 2/3$

abbreviated as $(2 \pm \sqrt{5})/3$. **4**

This process of changing $9z^2 - 12z - 1 = 0$ into the equivalent equation $(z - 2/3)^2 = 5/9$ is called **completing the square**.* The method of completing the square can be used on the general quadratic equation,

$$ax^2 + bx + c = 0 \quad (a \neq 0),$$

to convert it to one whose solution can be found by the square root property. This will give a general formula for solving any quadratic equation. Going through the necessary algebra produces the following important result.

Quadratic Formula   The solutions of the quadratic equation $ax^2 + bx + c = 0$, where $a \neq 0$, are given by

$$x = \frac{-b \pm \sqrt{b^2 - 4ac}}{2a}.$$

Example 5   Solve $x^2 - 4x - 5 = 0$ by the quadratic formula.

The equation is already in standard form (it has 0 alone on one side of the equals sign) so that the letters a, b, and c of the quadratic formula can be identified. The coefficient of the squared term gives the value of a; here $a = 1$. Also, $b = -4$ and $c = -5$. (Be careful to get the correct signs.) Substitute these values into the quadratic formula.

$$x = \frac{-(-4) \pm \sqrt{(-4)^2 - 4(1)(-5)}}{2(1)} \qquad \text{Let } a = 1, b = -4, c = -5$$

$$= \frac{4 \pm \sqrt{16 + 20}}{2} \qquad (-4)^2 = (-4)(-4) = 16$$

$$x = \frac{4 \pm 6}{2} \qquad \sqrt{16 + 20} = \sqrt{36} = 6$$

The $\pm$ sign represents the two solutions of the equation. First use $+$ and then use $-$ to find each of the solutions.

5 Use the quadratic formula to solve each of the following equations.

(a) $3x^2 + 11x - 4 = 0$

(b) $2z^2 - 7z - 4 = 0$

Answer:

(a) $1/3, -4$

(b) $-1/2, 4$

$$x = \frac{4 + 6}{2} = \frac{10}{2} = 5 \quad \text{or} \quad x = \frac{4 - 6}{2} = \frac{-2}{2} = -1$$

The two solutions are 5 and -1. Check by substituting each value in the original equation. **5**

---

*Completing the square is discussed further in Section 3.1.

Example 6 Solve $x^2 + 1 = 4x$.

First add $-4x$ to both sides, to get 0 alone on the right side.

$$x^2 - 4x + 1 = 0$$

Now identify the letters a, b, and c. Here $a = 1$, $b = -4$, and $c = 1$. Substitute these numbers into the quadratic formula.

$$x = \frac{-(-4) \pm \sqrt{(-4)^2 - 4(1)(1)}}{2(1)}$$

$$= \frac{4 \pm \sqrt{16 - 4}}{2}$$

$$= \frac{4 \pm \sqrt{12}}{2}$$

Simplify the solutions by writing $\sqrt{12}$ as $\sqrt{4 \cdot 3} = \sqrt{4} \cdot \sqrt{3} = 2\sqrt{3}$. Substituting $2\sqrt{3}$ for $\sqrt{12}$ gives

$$= \frac{4 \pm 2\sqrt{3}}{2}$$

$$= \frac{2(2 \pm \sqrt{3})}{2} \qquad \text{Factor } 4 \pm 2\sqrt{3}$$

$$x = 2 \pm \sqrt{3}.$$

The two solutions are $2 + \sqrt{3}$ and $2 - \sqrt{3}$.

The exact values of the solutions are $2 + \sqrt{3}$ and $2 - \sqrt{3}$. In many cases a decimal approximation of these solutions is needed. Use a calculator or Table 2 in the back of the book to find that $\sqrt{3} \approx 1.732$, so that (to the nearest thousandth) the solutions are

$$2 + \sqrt{3} \approx 2 + 1.732 = 3.732$$

or

$$2 - \sqrt{3} \approx 2 - 1.732 = .268. \quad \blacksquare \; 6$$

6 Find exact and approximate solutions for the following.
(a) $y^2 - 2y = 2$
(b) $x^2 - 6x + 4 = 0$

Answer:
(a) Exact: $1 + \sqrt{3}$, $1 - \sqrt{3}$; approximate: 2.732, $-.732$
(b) Exact: $3 + \sqrt{5}$, $3 - \sqrt{5}$; approximate: 5.236, .764

The above quadratic equations all had two different solutions. This is not always the case, as the following example shows.

Example 7 **(a)** Solve $9x^2 - 30x + 25 = 0$.

Here $a = 9$, $b = -30$, and $c = 25$. By the quadratic formula,

$$x = \frac{-(-30) \pm \sqrt{(-30)^2 - 4(9)(25)}}{2(9)}$$

$$= \frac{30 \pm \sqrt{900 - 900}}{18}$$

$$= \frac{30 \pm 0}{18} = \frac{30}{18} = \frac{5}{3}.$$

The given equation has only one solution.

(b) Solve $x^2 - 6x + 10 = 0$.

Since $a = 1$, $b = -6$, and $c = 10$,

$$x = \frac{-(-6) \pm \sqrt{(-6)^2 - 4(1)(10)}}{2(1)} = \frac{6 \pm \sqrt{36 - 40}}{2} = \frac{6 \pm \sqrt{-4}}{2}.$$

Recall from Section 1.8 that $\sqrt{-4}$ is not a real number, so there are no *real number* solutions to this equation. **7**

7 Solve the following equations.

(a) $9k^2 - 6k + 1 = 0$

(b) $4m^2 + 28m + 49 = 0$

(c) $2x^2 - 5x + 5 = 0$

Answer:

(a) 1/3

(b) −7/2

(c) No real number solutions

Example 8 A landscape architect wants to make an exposed gravel border of uniform width around a small shed behind a company plant. The shed is 10 feet by 6 feet. He has enough gravel to cover 36 square feet. How wide should the border be?

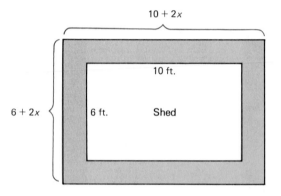

10 + 2x

6 + 2x

10 ft.

6 ft.   Shed

Figure 1.18

A sketch of the shed with border is shown in Figure 1.18. Let x represent the width of the border. Then the width of the large rectangle is $6 + 2x$ and its length is $10 + 2x$. The area of the large rectangle is $(6 + 2x)(10 + 2x)$. The area occupied by the shed is $6 \cdot 10 = 60$. The area of the border is found by subtracting the area of the shed from the area of the large rectangle. This difference should be 36 square feet, giving the equation

$$(6 + 2x)(10 + 2x) - 60 = 36.$$

Solve this equation with the following sequence of steps.

$$60 + 32x + 4x^2 - 60 = 36$$
$$4x^2 + 32x - 36 = 0$$
$$x^2 + 8x - 9 = 0$$
$$(x + 9)(x - 1) = 0$$

The solutions are −9 and 1. The number −9 cannot be the width of the border, so the solution is to make the border one foot wide.

8 Solve each word problem.

(a) A plane traveling at a constant rate goes 810 kilometers with the wind, then turns around and travels for 720 kilometers against the wind. If the speed of the wind is 15 kilometers per hour and the total flight took 6 hours, find the speed of the plane.

(b) The length of a picture is 2 in more than the width. It is mounted on a mat that extends 2 in beyond the picture on all sides. What are the dimensions of the picture if the area of the mat is 99 sq in?

Answer:

(a) 255 kph

(b) 5 in by 7 in

Some equations that are not quadratic can be solved as quadratic equations by making a suitable substitution. Such equations are called **quadratic in form.**

Example 9 Solve $4m^4 - 9m^2 + 2 = 0$.

Use the substitutions

$$x = m^2 \quad \text{and} \quad x^2 = m^4$$

to rewrite the equation as

$$4x^2 - 9x + 2 = 0,$$

a quadratic equation that can be solved by factoring.

$$4x^2 - 9x + 2 = 0$$
$$(x - 2)(4x - 1) = 0$$

$$x - 2 = 0 \quad \text{or} \quad 4x - 1 = 0$$
$$x = 2 \quad \text{or} \quad 4x = 1$$
$$x = \frac{1}{4}$$

Since $x = m^2$,

$$m^2 = 2 \quad \text{or} \quad m^2 = \frac{1}{4}$$

$$m = \pm\sqrt{2} \quad \text{or} \quad m = \pm\frac{1}{2}.$$

There are four solutions: $-\sqrt{2}$, $\sqrt{2}$, $-1/2$, and $1/2$. **9**

9 Solve each equation.
(a) $9x^4 - 23x^2 + 10 = 0$
(b) $4x^4 = 7x^2 - 3$

Answer:
(a) $-\sqrt{5}/3$, $\sqrt{5}/3$, $-\sqrt{2}$, $\sqrt{2}$
(b) $-\sqrt{3}/2$, $\sqrt{3}/2$, -1, 1

Example 10 Solve $4 + \dfrac{1}{z - 2} = \dfrac{3}{2(z - 2)^2}$.

First, substitute u for $z - 2$ to get

$$4 + \frac{1}{z - 2} = \frac{3}{2(z - 2)^2}$$

$$4 + \frac{1}{u} = \frac{3}{2u^2}.$$

Now multiply both sides of the equation by the common denominator $2u^2$; then solve the resulting quadratic equation.

$$2u^2\left(4 + \frac{1}{u}\right) = 2u^2\left(\frac{3}{2u^2}\right)$$

$$8u^2 + 2u = 3$$

$$8u^2 + 2u - 3 = 0$$

$$(4u + 3)(2u - 1) = 0$$

$$4u + 3 = 0 \qquad \text{or} \qquad 2u - 1 = 0$$

$$u = -\frac{3}{4} \qquad \text{or} \qquad u = \frac{1}{2}$$

Since $u = z - 2$,

$$z - 2 = -\frac{3}{4} \qquad \text{or} \qquad z - 2 = \frac{1}{2}$$

$$z = -\frac{3}{4} + 2 \qquad\qquad z = \frac{1}{2} + 2$$

$$z = \frac{5}{4} \qquad \text{or} \qquad z = \frac{5}{2}.$$

10 Solve each equation.

(a) $\dfrac{13}{3(p + 1)} +$

$\dfrac{5}{3(p + 1)^2} = 2$

(b) $1 + \dfrac{1}{a} = \dfrac{5}{a^2}$

Answer:

(a) $-4/3$, $3/2$

(b) $(-1 + \sqrt{21})/2$,
$(-1 - \sqrt{21})/2$

The solutions are 5/4 and 5/2. Check both solutions in the original equation. **10**

The next example shows how to solve an equation for a specified variable, when the equation is quadratic in that variable.

11 Solve each of the following equations for the indicated variable. Assume no variable equals 0.

(a) $k = mp^2 - bp$ for p

(b) $r = \dfrac{APk^2}{3}$ for k

Answer:

(a) $\dfrac{b \pm \sqrt{b^2 + 4mk}}{2m}$

(b) $\pm\sqrt{\dfrac{3r}{AP}}$ or $\dfrac{\pm\sqrt{3rAP}}{AP}$

Example 11 Solve $v = mx^2 + x$ for x. (Assume $m \neq 0$.)

The equation is quadratic in x because of the x^2 term. Use the quadratic formula, first writing the equation in standard form.

$$v = mx^2 + x$$
$$0 = mx^2 + x - v$$

Let $a = m$, $b = 1$, and $c = -v$. Then the quadratic formula gives

$$x = \frac{-1 \pm \sqrt{1^2 - 4(m)(-v)}}{2m}$$

$$x = \frac{-1 \pm \sqrt{1 + 4mv}}{2m}. \quad \textbf{11}$$

1.9 Exercises

Solve each of the following equations. (See Examples 1 and 2.)

1. $(y - 5)(y + 4) = 0$

2. $(m + 3)(m - 1) = 0$

3. $x^2 + 5x + 6 = 0$

4. $y^2 - 3y + 2 = 0$

5. $r^2 - 5r - 6 = 0$

6. $y^2 - y - 12 = 0$

7. $a^2 + 5a = 24$

8. $y^2 = 2y + 15$

9. $x^2 = 3 + 2x$

10. $3m + 4 = m^2$

11. $m^2 + 16 = 8m$

12. $y^2 + 49 = 14y$

13. $2k^2 - k = 10$

14. $6x^2 = 7x + 5$

15. $6x^2 - 5x = 4$

16. $9s^2 + 12s = -4$

17. $m(m - 7) = -10$

18. $z(2z + 7) = 4$

19. $9x^2 - 16 = 0$

20. $25y^2 - 64 = 0$

21. $16x^2 - 16x = 0$

22. $12y^2 - 48y = 0$

Solve the following equations by the square root property or by completing the square. (See Examples 3 and 4.)

23. $x^2 = 29$

24. $p^2 = 37$

25. $(m - 3)^2 = 5$

26. $(p + 2)^2 = 7$

27. $(3k - 1)^2 = 19$

28. $(4t + 5)^2 = 3$

29. $q^2 + 2q = 8$

30. $m^2 + 5m - 6 = 0$

31. $4z^2 - 4z = 1$

32. $11p^2 - 11p + 1 = 0$

Use the quadratic formula to solve each of the following equations. If the solutions involve square roots, give both the exact and approximate solutions. (See Examples 5–7.)

33. $3x^2 - 5x + 1 = 0$

34. $2x^2 - 7x + 1 = 0$

35. $2m^2 = m + 4$

36. $p^2 + p - 1 = 0$

37. $k^2 - 10k = -20$

38. $r^2 = 13 - 12r$

39. $2x^2 + 12x + 5 = 0$

40. $5x^2 + 4x = 1$

41. $2r^2 - 7r + 5 = 0$

42. $8x^2 = 8x - 3$

43. $6k^2 - 11k + 4 = 0$

44. $8m^2 - 10m + 3 = 0$

45. $x^2 + 3x = 10$

46. $2x^2 = 3x + 5$

47. $2x^2 - 7x + 30 = 0$

48. $3k^2 + k = 6$

49. $5m^2 + 5m = 0$

50. $8r^2 + 16r = 0$

Give all real number solutions of the following equations. (See Examples 9 and 10.)

51. $m^4 - 3m^2 - 10 = 0$

52. $2t^4 + 3t^2 - 2 = 0$

53. $6y^4 = y^2 + 15$

54. $4a^4 = 2 - 7a^2$

55. $b^4 - 3b^2 - 5 = 0$

56. $2x^4 - 2x^2 - 1 = 0$

57. $6(p - 3)^2 + 5(p - 3) - 6 = 0$

58. $12(q + 4)^2 - 13(q + 4) - 4 = 0$

59. $2(p + 5)^2 - 3(p + 5) = 20$

60. $4(2y - 3)^2 - (2y - 3) - 3 = 0$

61. $1 + \dfrac{7}{2a} = \dfrac{15}{2a^2}$

62. $5 - \dfrac{4}{k} - \dfrac{1}{k^2} = 0$

63. $-\dfrac{2}{3z^2} + \dfrac{1}{3} + \dfrac{8}{3z} = 0$

64. $2 + \dfrac{5}{x} + \dfrac{1}{x^2} = 0$

Solve the following problems. (See Example 8.)

65. Find two consecutive even integers whose product is 288.

66. The sum of the squares of two consecutive integers is 481. Find the integers.

67. Two integers have a sum of 10. The sum of the squares of the integers is 148. Find the integers.

68. A shopping center has a rectangular area of 40,000 square yards enclosed on three sides for a parking lot.

The length is 200 yards more than twice the width. Find the length and width of the lot.

69. An ecology center wants to set up an experimental garden. It has 300 meters of fencing to enclose a rectangular area of 5000 square meters. Find the length and width of the rectangle.

70. Fred went into a frame-it-yourself shop. He wanted a frame 3 inches longer than wide. The frame he chose extends 1.5 inches beyond the picture on each side. Find the outside dimensions of the frame if the area of the unframed picture is 70 square inches.

71. Joan wants to buy a rug for a room that is 12 feet by 15 feet. She wants to leave a uniform strip of floor around the rug. She can afford 108 square feet of carpeting. What dimensions should the rug have?

72. Max can clean the garage in 9 hours less time than his brother Paul. Working together, they can do the job in 20 hours. How long would it take each one to do the job alone?

73. Paula drives 10 mph faster than Steve. Both start at the same time for Atlanta from Chattanooga, a distance of about 100 miles. It takes Steve 1/3 of an hour longer than Paula to make the trip. What is Steve's average speed?

74. Amy walks 1 mph faster than her friend Lisa. In a walk for charity, both walked the full distance of 24 miles. Lisa took 2 hours longer than Amy. What was Lisa's average speed?

Solve each of the following equations for the indicated variable. Assume all denominators are nonzero, and that all variables represent positive real numbers. (See Example 11.)

75. $S = \dfrac{1}{2} gt^2$ for t

76. $a = \pi r^2$ for r

77. $L = \dfrac{d^4 k}{h^2}$ for h

78. $F = \dfrac{kMv^2}{r}$ for v

79. $S = S_0 + gt^2 + k$ for t

80. $P = \dfrac{E^2 R}{(r + R)^2}$ for R

81. $S = 2\pi rh + 2\pi r^2$ for r

82. $pm^2 - 8qm + \dfrac{1}{r} = 0$ for m

1.10 Quadratic and Rational Inequalities

A **quadratic inequality** is an inequality of the form $ax^2 + bx + c > 0$ (or $<$, or $\leq$, or $\geq$). The highest exponent is always 2. Examples of quadratic inequalities include

$$x^2 - x - 12 < 0, \qquad 3y^2 + 2y \geq 0, \qquad \text{and} \qquad m^2 \leq 4.$$

A method of solving quadratic inequalities is shown in the next few examples.

Example 1 Solve the quadratic inequality $x^2 - x - 12 < 0$.
Since $x^2 - x - 12 = (x - 4)(x + 3)$, the given inequality is the same as

$$(x - 4)(x + 3) < 0.$$

The product of $x - 4$ and $x + 3$ is to be less than 0, or negative. The product will be negative if the two factors $x - 4$ and $x + 3$ have opposite signs. The factor $x - 4$ is positive when $x - 4 > 0$, or $x > 4$. Also, $x - 4$ is negative if $x < 4$. In the same way, $x + 3$ is positive when $x > -3$ and negative when $x < -3$. This information is shown on the *sign graph* of Figure 1.19.

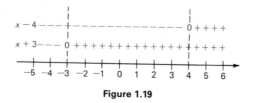

Figure 1.19

The product $(x - 4)(x + 3)$ will be negative when $x - 4$ and $x + 3$ have opposite signs. From the sign graph of Figure 1.19 these expressions have opposite signs whenever x is between -3 and 4. The solution is written $-3 < x < 4$. A graph of this solution is shown in Figure 1.20. **1**

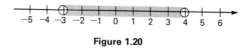

Figure 1.20

1 Solve the following. Graph the solution.
(a) $y^2 + 2y - 3 < 0$
(b) $2p^2 + 3p - 2 < 0$

Answer:
(a) $-3 < y < 1$

(b) $-2 < p < 1/2$

Example 2 Solve the quadratic inequality $r^2 + 3r \geq 4$.
First rewrite the inequality so that one side is 0.

$$r^2 + 3r \geq 4$$
$$r^2 + 3r - 4 \geq 0 \qquad \text{Add } -4 \text{ to both sides}$$

Factor $r^2 + 3r - 4$ as $(r + 4)(r - 1)$. The factor $r + 4$ is positive when $r > -4$ and negative when $r < -4$. Also, the factor $r - 1$ is positive when $r > 1$ and negative when $r < 1$. Use this information to produce the sign graph of Figure 1.21. The product $(r + 4)(r - 1)$ will be positive when $r + 4$ and $r - 1$ have the same signs, either both positive or both negative. From Figure 1.21 the solution is seen to include those numbers less than or equal to -4, together with those numbers greater than or equal to 1. This solution is written

$$r \le -4 \qquad \text{or} \qquad r \ge 1.$$

A graph of the solution is given in Figure 1.22. **2**

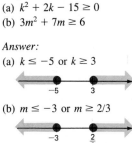

2 Solve each inequality. Graph each solution.
(a) $k^2 + 2k - 15 \ge 0$
(b) $3m^2 + 7m \ge 6$

Answer:
(a) $k \le -5$ or $k \ge 3$

(b) $m \le -3$ or $m \ge 2/3$

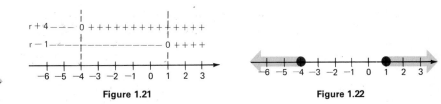

Figure 1.21

Figure 1.22

Although the inequality in the next example is not a quadratic inequality, it can be solved in much the same way as a quadratic inequality.

Example 3 Solve $q^3 - 4q > 0$.

The left side of the inequality can be factored as

$$q^3 - 4q > 0$$
$$q(q^2 - 4) > 0$$
$$q(q + 2)(q - 2) > 0.$$

The factor q is positive when q is positive and negative when q is negative; $q + 2$ is positive when $q > -2$ and negative for $q < -2$; $q - 2$ is positive when $q > 2$ and negative when $q < 2$. This information is shown in the sign graph of Figure 1.23. As the sign graph shows, when $-2 < q < 0$, two factors are negative and one is positive, so the product $q(q + 2)(q - 2)$ is positive. Also, when $q > 2$, all three factors are positive giving a positive product. The solution,

$$-2 < q < 0 \qquad \text{or} \qquad q > 2,$$

is graphed in Figure 1.24. **3**

3 Solve each inequality. Graph each solution.
(a) $m^3 - 9m > 0$
(b) $2k^3 - 50k \le 0$

Answer:
(a) $-3 < m < 0$
    or $m > 3$

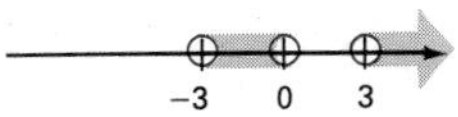

(b) $k \le -5$ or
    $0 \le k \le 5$

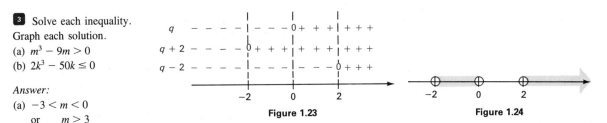

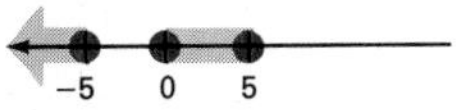

Figure 1.23

Figure 1.24

Rational Inequalities Inequalities with quotients of algebraic expressions are called **rational inequalities**. These inequalities often can be solved with a sign graph.

Example 4 Solve the rational inequality $\dfrac{5}{x + 4} \ge 1$.

It is tempting to begin the solution by multiplying both sides of the inequality by $x + 4$, but to do this it would be necessary to consider whether $x + 4$ is positive or negative. Instead, subtract 1 from both sides of the inequality, getting

$$\frac{5}{x + 4} - 1 \geq 0.$$

Write the left side as a single fraction.

$$\frac{5}{x + 4} - \frac{x + 4}{x + 4} \geq 0 \qquad \text{Get a common denominator}$$

$$\frac{5 - (x + 4)}{x + 4} \geq 0 \qquad \text{Subtract fractions}$$

$$\frac{5 - x - 4}{x + 4} \geq 0 \qquad \text{Distributive property}$$

$$\frac{1 - x}{x + 4} \geq 0.$$

The quotient can change sign only when the denominator is 0 or when the numerator is 0. This happens when

$$1 - x = 0 \qquad \text{or} \qquad x + 4 = 0$$
$$x = 1 \qquad \text{or} \qquad x = -4.$$

Make a sign graph as before. See Figure 1.25.

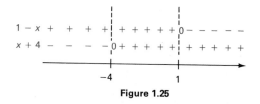

Figure 1.25

This time consider the sign of the quotient of the two quantities rather than their product. The quotient of two numbers is positive if both the numbers are positive or if both the numbers are negative. On the other hand, the quotient is negative if the two numbers have opposite signs. The sign graph in Figure 1.25 suggests that values in the interval $-4 < x < 1$ give a positive quotient and are part of the solution. With a quotient, the endpoints must be considered separately to make sure that no denominator is 0. In this inequality, -4 makes the denominator 0, while 1 satisfies the given inequality. Write the solution as $-4 < x \leq 1$. ■

4 Solve each inequality.

(a) $\dfrac{3}{x - 2} \geq 4$

(b) $\dfrac{p}{1 - p} < 3$

Answer:

(a) $2 < x \leq \dfrac{11}{4}$

(b) $p < \dfrac{3}{4}$ or $p > 1$

As suggested by Example 4, be very careful with the endpoints of the intervals in the solution of rational inequalities.

Example 5 Solve $\dfrac{2x - 1}{3x + 4} < 5$.

Begin by subtracting 5 on both sides and combining the terms on the left into a single fraction. This process gives

$$\frac{2x - 1}{3x + 4} < 5$$

$$\frac{2x - 1}{3x + 4} - 5 < 0$$

$$\frac{2x - 1 - 5(3x + 4)}{3x + 4} < 0$$

$$\frac{-13x - 21}{3x + 4} < 0.$$

Draw a sign graph by first solving the equations

$$-13x - 21 = 0 \qquad \text{and} \qquad 3x + 4 = 0,$$

getting the solutions

$$x = -\frac{21}{13} \qquad \text{and} \qquad x = -\frac{4}{3}.$$

Use the values $-21/13$ and $-4/3$ to divide the number line into three intervals. Now complete a sign graph and find the intervals where the quotient is negative. See Figure 1.26. The quotient is negative when $x < -21/13$ or $x > -4/3$. Neither endpoint satisfies the given inequality.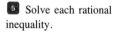

5 Solve each rational inequality.

(a) $\dfrac{3y - 2}{2y + 5} < 1$

(b) $\dfrac{3c - 4}{2 - c} \geq -5$

Answer:

(a) $-5/2 < y < 7$

(b) $c \leq 4/3$ or $c > 2$

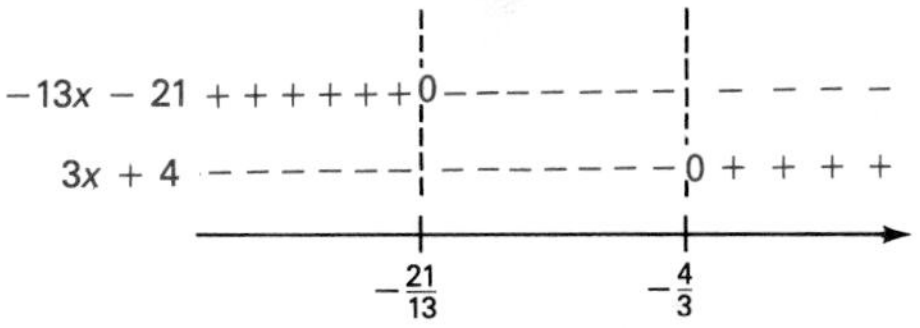

$-13x - 21 + + + + + +\ 0 - - - - - - - \ - \ - \ - \ -$

$3x + 4 \ - - - - - - \ - - - - - - \ 0 + + + +$

$-\frac{21}{13} \qquad -\frac{4}{3}$

Figure 1.26

1.10 Exercises

Solve each of the following quadratic inequalities. Graph each solution. (See Examples 1 and 2.)

1. $(m + 2)(m - 4) < 0$

2. $(k - 1)(k + 2) > 0$

3. $(t + 6)(t - 1) \geq 0$

4. $(y - 2)(y + 3) \leq 0$

5. $y^2 - 3y + 2 < 0$

6. $z^2 - 4z - 5 \leq 0$

7. $2k^2 + 7k - 4 > 0$

8. $6r^2 - 5r - 4 > 0$

9. $q^2 - 7q + 6 \leq 0$

10. $2k^2 - 7k - 15 \leq 0$

11. $6m^2 + m > 1$

12. $10r^2 + r \leq 2$

13. $2y^2 + 5y \leq 3$

14. $3a^2 + a > 10$

15. $x^2 \leq 25$

16. $y^2 \geq 4$

17. $p^2 - 16p > 0$

18. $r^2 - 9r < 0$

Solve the following inequalities. (See Example 3.)

19. $t^3 - 16t \leq 0$

20. $z^3 - 64z \leq 0$

21. $2r^3 - 98r > 0$

22. $3m^3 - 27m > 0$

23. $4m^3 + 7m^2 - 2m > 0$

24. $6p^3 - 11p^2 + 3p > 0$

Solve the following rational inequalities. (See Examples 4 and 5.)

25. $\dfrac{m - 3}{m + 5} \leq 0$

26. $\dfrac{r + 1}{r - 4} > 0$

27. $\dfrac{k - 1}{k + 2} > 1$

28. $\dfrac{a - 5}{a + 2} < -1$

29. $\dfrac{3}{x - 6} \leq 2$

30. $\dfrac{1}{k - 2} < \dfrac{1}{3}$

31. $\dfrac{2y + 3}{y - 5} \leq 1$

32. $\dfrac{a + 2}{3 + 2a} \leq 5$

33. $\dfrac{7}{k + 2} \geq \dfrac{1}{k + 2}$

34. $\dfrac{5}{p + 1} > \dfrac{12}{p + 1}$

Management *The commodity market is very unstable; money can be made or lost quickly when invested in soybeans, wheat, pork bellies, and so on. Suppose that an investor kept track of her total profit, P, at time t, measured in months, after she began investing and found that*

$$P = 4t^2 - 29t + 30.$$

For example, 4 months after she began investing, her total profits were

$$P = 4 \cdot 4^2 - 29 \cdot 4 + 30 \qquad \text{Let } t = 4$$
$$= 4 \cdot 16 - 116 + 30$$
$$= 64 - 116 + 30$$
$$P = -22,$$

so that she was $22 "in the hole" after 4 months.

35. Find her total profit after 8 months.

36. Find her total profit after 10 months.

37. Find when she has just broken even. (Let $P = 0$, and solve the resulting equation.)

38. Find the time intervals where she has been ahead. (Solve the inequality $4t^2 - 29t + 30 > 0$.)

39. **Management** For most items, as the price increases, the demand for the item decreases. For example, suppose that the number, d, of office copiers sold at a price p (in hundreds of dollars) is

$$d = \frac{36{,}000}{p}$$

while the supply of copiers, s, is given by

$$s = 36p + 120.$$

Here the number of units supplied increases as the price increases. Find the value of p at which supply and demand are equal. (This value of p is called the **equilibrium price**.)

40. Find the equilibrium price (see Exercise 39) for an item having

$$d = \frac{9000}{p} \qquad \text{and} \qquad s = 4p - 20.$$

Chapter 1   Review Exercises

Name the numbers from the list -12, -6, $-9/10$, $-\sqrt{7}$, $-\sqrt{4}$, 0, $1/8$, $\pi/4$, 6, $\sqrt{11}$ that are

1. natural numbers;

2. whole numbers;

3. integers;

4. rational numbers;

5. irrational numbers;

6. real numbers.

For Exercises 7–16, choose all words from the following list which apply: natural numbers; whole numbers; integers; rational numbers; irrational numbers; real numbers; meaningless.

7. 22

8. 0

9. $\sqrt{36}$

10. $-\sqrt{25}$

11. -1

12. $\dfrac{5}{8}$

13. $\sqrt{15}$

14. $-\dfrac{6}{0}$

15. $\dfrac{3\pi}{4}$

16. $\pi - \pi$

Write the following numbers in numerical order from smallest to largest.

17. $-7,\ -3,\ 8,\ \pi,\ -2,\ 0$

18. $\dfrac{5}{6},\ \dfrac{1}{2},\ -\dfrac{2}{3},\ -\dfrac{5}{4},\ -\dfrac{3}{8}$

19. $|6 - 4|,\ -|-2|,\ |8 + 1|,\ -|3 - (-2)|$

20. $\sqrt{7},\ -\sqrt{8},\ -|\sqrt{16}|,\ |-\sqrt{12}|$

Write without absolute value bars.

21. $-|-6| + |3|$

22. $|-5| + |-9|$

23. $7 - |-8|$

24. $|-2| - |-7 + 3|$

Graph each of the following on a number line.

25. $x \geq -3$

26. $-4 < x \leq 6$

27. $x < -2$

28. $x \leq 1$

Use the order of operations to simplify.

29. $(-6 + 2 \cdot 5)(-2)$

30. $-4(-7 - 9 \div 3)$

31. $\dfrac{-8 + (-6)(-3) \div 9}{6 - (-2)}$

32. $\dfrac{20 \div 4 \cdot 2 \div 5 - 1}{-9 - (-3) - 12 \div 3}$

Solve each equation.

33. $4k - 11 + 3k = 2k - 1$

34. $2(k + 3) - 4(1 - k) = 6$

35. $2m + 7 = 3m + 1$

36. $4k - 2(k - 1) = 12$

37. $5y - 2(y + 4) = 3(2y + 1)$

38. $\dfrac{x - 3}{2} = \dfrac{2x + 1}{3}$

39. $\dfrac{p}{2} - \dfrac{3p}{4} = 8 + \dfrac{p}{3}$

40. $\dfrac{2r}{5} - \dfrac{r - 3}{10} = \dfrac{3r}{5}$

41. $\dfrac{2z}{5} - \dfrac{4z - 3}{10} = \dfrac{-z + 1}{10}$

42. $\dfrac{p}{p + 2} - \dfrac{3}{4} = \dfrac{2}{p + 2}$

43. $\dfrac{2m}{m - 3} = \dfrac{6}{m - 3} + 4$

44. $\dfrac{15}{k + 5} = 4 - \dfrac{3k}{k + 5}$

Solve for x.

45. $9px - 2 = x$

46. $11x - 5y = 2yx$

47. $3(x + 2b) + a = 2x - 6$

48. $9x - 11(k + p) = x(a - 1)$

49. $\dfrac{x}{m - 2} = kx - 3$

50. $r^2x - 5x = 3r^2$

Solve each of the following problems.

51. A stereo is on sale for 15% off. The sale price is $425. What was the original price?

52. To make a special mix for Valentine's Day, the owner of a candy store wants to combine chocolate hearts which sell for $5 per pound with candy kisses which sell for $3.50 per pound. How many pounds of each kind should be used to get 30 pounds of a mix which can be sold for $4.50 per pound?

53. Ed and Lucy are stuffing envelopes for a political campaign. Working together, they can stuff 5000 envelopes in 4 hours. When Lucy worked at the job alone, it took her 6 hours to stuff the 5000 envelopes. How long would it take Ed working alone to stuff 5000 envelopes?

54. Alison can ride her bike to the university library in 20 minutes. The trip home, which is all uphill, takes her half an hour. If her rate is 8 mph slower on the return trip, how far does she live from the library?

Solve each of the following inequalities.

55. $-9x < 4x + 7$

56. $11y \geq 2y - 8$

57. $-5z - 4 \geq 3(2z - 5)$

58. $-(4a + 6) < 3a - 2$

59. $3r - 4 + r > 2(r - 1)$

60. $7p - 2(p - 3) \leq 5(2 - p)$

61. $5 \leq 2x - 3 \leq 7$

62. $-8 < 3a - 5 < -2$

Solve each equation.

63. $|a + 4| = 7$

64. $|-y + 2| = 3$

65. $\left|\dfrac{r - 5}{3}\right| = 6$

66. $\left|\dfrac{2 - 3a}{7}\right| = 9$

67. $\left|\dfrac{8r - 1}{2}\right| = 7$

68. $\left|\dfrac{3p - 5}{3}\right| = 2$

69. $|5r - 1| = |2r + 3|$

70. $|k + 7| = |3k - 8|$

Solve each inequality.

71. $|m| \leq 7$

72. $|r| < 2$

73. $|p| > 3$

74. $|z| > -1$

75. $|b| \leq -1$

76. $|5m - 8| \leq 2$

77. $|7k - 3| < 5$

78. $|2p - 1| > 2$

79. $|3r + 7| > 5$

80. $|2 - 5y| < -1$

Perform each of the following operations.

81. $(-9m^2 + 11m - 7) + (2m^2 - 12m + 4)$

82. $(3q^3 - 9q^2 + 6) + (4q^3 - 8q + 3)$

83. $(5r^4 - 6r^2 + 2r) - (-3r^4 + 2r^2 - 9r)$

84. $(-7z^3 + 8z^2 - z) - (z^3 + 4z^2 + 10z)$

85. $-(r^5 + 2r^4 + 8r^2) + 3(r^4 + 9r^2)$

86. $2(3y^6 - 9y^2 + 2y) - (5y^6 - 10y^2 - 4y)$

87. $(8y - 7)(2y + 7)$

88. $(9k + 2)(4k - 3)$

89. $(7z + 10y)(3z - 5y)$

90. $(2r + 11s)(4r - 9s)$

91. $(3k - 5m)^2$

92. $(4a - 3b)^2$

93. $(5x - 2)^3$

94. $(3m + 2)^4$

95. $(3w - 2)(5w^2 - 4w + 1)$

96. $(2k + 5)(3k^3 - 4k^2 + 8k - 2)$

Factor as completely as possible.

97. $7z^2 - 9z^3 + z$

98. $6m^3 + 4m^2 - 3m$

99. $12p^5 - 8p^4 + 20p^3$

100. $15y^7 - 25y^4 + 50y^3$

101. $r^2 + rp - 42p^2$

102. $z^2 - 6zk - 16k^2$

103. $6m^2 - 13m - 5$

104. $4k^2 + 11k - 3$

105. $3m^2 - 8m - 35$

106. $16p^2 + 24p + 9$

107. $144p^2 - 169q^2$

108. $81z^2 - 25x^2$

109. $8y^3 - 1$

110. $125a^3 + 216$

Perform each operation.

111. $\dfrac{6p}{7} \cdot \dfrac{28p}{15}$

112. $\dfrac{9r^2}{16} \div \dfrac{27r}{32}$

113. $\dfrac{m^2 - 2m - 3}{m(m - 3)} \div \dfrac{m + 1}{4m}$

114. $\dfrac{3m - 9}{8m} \cdot \dfrac{16m + 24}{15}$

115. $\dfrac{k^2 + k}{8k^3} \cdot \dfrac{4}{k^2 - 1}$

116. $\dfrac{x^2 + x - 2}{x^2 + 5x + 6} \div \dfrac{x^2 + 3x - 4}{x^2 + 4x + 3}$

117. $\dfrac{5}{2r} + \dfrac{6}{r}$

118. $\dfrac{3}{8y} + \dfrac{1}{2y} + \dfrac{7}{9y}$

119. $\dfrac{8}{r - 1} - \dfrac{3}{r}$

120. $\dfrac{1}{4y} + \dfrac{8}{5y}$

121. $\dfrac{\dfrac{1}{p} + \dfrac{1}{q}}{1 - \dfrac{1}{pq}}$

122. $\dfrac{\dfrac{2}{r} - \dfrac{3}{5}}{\dfrac{1}{r} + \dfrac{4}{5}}$

Simplify each of the following. Write all answers without negative exponents. Assume all variables represent positive real numbers.

123. 4^{-2}

124. 6^{-1}

125. 5^{-3}

126. 8^{-2}

127. $\left(\dfrac{3}{4}\right)^{-3}$

128. $\left(\dfrac{5}{8}\right)^{-2}$

129. $6^4 \cdot 6^{-3}$

130. $7^4 \cdot 7^{-6}$

131. $\dfrac{8^{-5}}{8^{-3}}$

132. $\dfrac{6^{-2}}{6^3}$

133. $\dfrac{9^4 \cdot 9^{-5}}{(9^{-2})^2}$

134. $\dfrac{k^4 \cdot k^{-3}}{(k^{-2})^{-3}}$

135. $4^{-1} + 2^{-1}$

136. $3^{-2} + 3^{-1}$

137. $125^{2/3}$

138. $128^{3/7}$

139. $9^{-5/2}$

140. $\left(\dfrac{144}{49}\right)^{-1/2}$

141. $\dfrac{5^{1/3} \cdot 5^{1/2}}{5^{3/2}}$

142. $\dfrac{2^{3/4} \cdot 2^{-1/2}}{2^{1/4}}$

143. $(3a^2)^{1/2} \cdot (3^2a)^{3/2}$

144. $(4p)^{2/3} \cdot (2p^3)^{3/2}$

145. $\sqrt[3]{27}$

146. $\sqrt[4]{625}$

147. $\sqrt[5]{-32}$

148. $\sqrt[6]{-64}$

149. $\sqrt{24}$

150. $\sqrt{63}$

151. $\sqrt[3]{54p^3q^5}$

152. $\sqrt[4]{64a^5b^3}$

153. $\sqrt{\dfrac{5n^2}{6m}}$

154. $\sqrt{\dfrac{3x^3}{2z}}$

155. $\sqrt[3]{\sqrt{5}}$

156. $\sqrt[4]{\sqrt[3]{10}}$

157. $2\sqrt{3} - 5\sqrt{12}$

158. $8\sqrt{7} + 2\sqrt{28}$

159. $5\sqrt[3]{16r^2s^3} - s\sqrt[3]{54r^2}$

160. $2\sqrt[4]{16mn^2} - \sqrt[4]{81mn^2}$

161. $(\sqrt{5} - 1)(\sqrt{5} + 1)$

162. $(\sqrt{7} - \sqrt{3})(\sqrt{7} + \sqrt{3})$

163. $(2\sqrt{5} - \sqrt{3})(\sqrt{5} + 2\sqrt{3})$

164. $(4\sqrt{7} + \sqrt{2})(3\sqrt{7} - \sqrt{2})$

165. $\dfrac{\sqrt{2}}{1 + \sqrt{3}}$

166. $\dfrac{4 + \sqrt{2}}{4 - \sqrt{5}}$

Solve each equation. Give only real number solutions.

167. $x^2 = 7$

168. $p^2 = 23$

169. $(b + 7)^2 = 5$

170. $(2p + 1)^2 = 7$

171. $x^2 - 4x + 3 = 0$

172. $y^2 + 9y + 8 = 0$

173. $2p^2 + 3p = 2$

174. $2y^2 = 15 + y$

175. $x^2 - 2x = 2$

176. $r^2 + 4r = 1$

177. $2m^2 - 12m = 11$

178. $9k^2 + 6k = 2$

179. $2a^2 + a - 15 = 0$

180. $12x^2 = 8x - 1$

181. $2q^2 - 11q = 21$

182. $3x^2 + 2x = 16$

183. $6k^4 + k^2 = 1$

184. $21p^4 = 2 + p^2$

185. $2x^4 = 7x^2 + 15$

186. $3m^4 + 20m^2 = 7$

187. $3 = \dfrac{13}{z} + \dfrac{10}{z^2}$

188. $1 + \dfrac{13}{p} + \dfrac{40}{p^2} = 0$

189. $2 + \dfrac{15}{x - 1} + \dfrac{18}{(x - 1)^2} = 0$

190. $\dfrac{5}{(2t + 1)^2} = 2 - \dfrac{9}{2t + 1}$

Solve each equation for the specified variable.

191. $p = \dfrac{E^2R}{(r + R)^2}$ for r

192. $p = \dfrac{E^2R}{(r + R)^2}$ for E

193. $K = s(s - a)$ for s

194. $kz^2 - hz - t = 0$ for z

Solve each inequality.

195. $r^2 + r - 6 < 0$

196. $y^2 + 4y - 5 \geq 0$

197. $2z^2 + 7z \geq 15$

198. $3k^2 \leq k + 14$

199. $8a^2 + 10a > 3$

200. $9m^2 + 13m > 10$

201. $3r^2 - 5r \leq 0$

202. $2q^2 + q > 0$

203. $\dfrac{m + 2}{m} \leq 0$

204. $\dfrac{q - 4}{q + 3} > 0$

205. $\dfrac{5}{p + 1} > 2$

206. $\dfrac{6}{a - 2} \leq -3$

207. $\dfrac{2}{r + 5} \leq \dfrac{3}{r - 2}$

208. $\dfrac{1}{z - 1} > \dfrac{2}{z + 1}$

Work each word problem.

209. Pam Carlson wants to fence off a rectangular playground beside an apartment building. The building forms one boundary, so she needs to fence only the other three sides. The area of the playground is to be 11,250 square meters. She has enough material to build 325 meters of fence. Find the length and width of the playground.

210. Steve and Paula sell pies. It takes Paula 1 hour longer than Steve to bake a day's supply of pies. Working together, it takes them 1 1/5 hours to bake the pies. How long would it take Steve working alone?

Case 1

Estimating Seed Demands—The Upjohn Company*

The Upjohn Company has a subsidiary which buys seeds from farmers and then resells them. Each spring the firm contracts with farmers to grow the seeds. The firm must decide on the number of acres that it will contract for. The problem faced by the company is that the demand for seeds is not constant, but fluctuates from year to year. Also, the number of tons of seed produced per acre varies, depending on weather and other factors. In an attempt to decide the number of acres that should be planted in order to maximize profits, a company mathematician created a model of the variables involved in determining the number of acres to plant.

The analysis of this model required advanced methods that we will not go into. We can, however, give the conclusion; the number of acres that will maximize profit in the long run is found by solving the equation

$$F(AX + Q) = \frac{(S - C_p)X - C_A}{(S - C_p + C_c)X} \qquad \textbf{(1)}$$

for A. For $z = (AX + Q)$, $F(z)$ represents the chances that z tons of seed will be demanded by the marketplace. The variables in the equation are

A = number of acres of land contracted by the company

X = quantity of seed produced per acre of land (in tons)

Q = quantity of seed in inventory from previous years

S = selling price per ton of seed

C_p = variable cost (production, marketing, etc.) per ton of seed

C_c = cost to carry over one ton of seed from previous year

C_A = variable cost per acre of land.

To advise management of the number of acres of seed to contract for, the mathematician studied past records to find the values of the various variables. From these records and from predictions of future trends, it was concluded that S = \$10,000 per ton, X = .1 ton per acre (on the average), Q = 200 tons, C_p = \$5000 per ton, C_A = \$100 per acre, C_c = \$3000 per ton.

By the same process, $F(z)$ is found to be approximated by

$$F(z) = \frac{z}{1000} - \frac{1}{2}, \quad \text{if} \quad 500 \leq z \leq 1500 \text{ tons.} \qquad \textbf{(2)}$$

Exercises

1. Evaluate $AX + Q$ using the values of X and Q given above.

2. Find $F(AX + Q)$, using equation (2) and your results from Exercise 1.

3. Solve equation (1) for A.

4. How many acres should be planted?

5. How many tons of seed will be produced?

6. Find the total revenue that will be received from the sale of the seeds.

---

*Based on work by David P. Rutten, Senior Mathematician, The Upjohn Company, Kalamazoo, Michigan.

2   Linear Models

When using mathematics to solve a real-world problem, it is usually necessary to set up a *mathematical model*—a mathematical description of the real-world situation. Constructing a mathematical model requires a solid understanding of the situation to be modeled, along with sound knowledge of the possible mathematical ideas that can be used to construct the model. The very richness and diversity of contemporary mathematics too often serves as a barrier between a person in another field and the mathematical tools needed by that person. There are so many useful parts of mathematics that it is often hard to know which to pick for a particular model. One way to avoid this problem is to have a thorough understanding of the most useful basic mathematical tools available for model building.

This chapter looks at the mathematics of *linear* models—those used for data whose graphs can be approximated by a straight line. The first section discusses the basic ideas of functions as a preparation for understanding both linear and nonlinear models.

2.1   Functions

A common problem in many real-life situations is to describe relationships between quantities. For example, assuming that the number of hours a student studies each day is related to the grade received in a course, how can the relationship be expressed? One way is to set up a table showing the hours of study and the corresponding grade that results. Such a table might appear as shown at the side.

Hours of study	Grade
3	A
$2\frac{1}{2}$	B
2	C
1	D
0	F

In other relationships, a formula of some sort is often used to describe how the value of one quantity depends on the value of another. For example, if a certain bank account pays 9% interest per year, then the interest, I, that a deposit of P dollars would earn in one year is given as

$$I = .09 \times P, \quad \text{or} \quad I = .09P.$$

The formula, $I = .09P$, describes the relationship between interest and the amount of money deposited.

In this example, P, which represents the amount of money deposited, is called the **independent variable,** while I is the **dependent variable.** (The amount of interest earned *depends* on the amount of money deposited.) When a specific number, say 2000, is substituted for P, then I takes on *one* specific value—in this case, $.09 \times 2000 = 180$. The variable I is said to be a *function* of P. By definition,

> a **function** is a rule which assigns to each element from one set*
> exactly one element from another set.

*A set is just a collection of objects. See Chapter 8 for more details.

In almost every use of functions in this book, the "rule" mentioned in the box will be given by an equation, such as the equation $I = .09P$ above. As a shortcut, "the function defined by the rule $y = 3x - 5$" is often called just "the function $y = 3x - 5$."

The set of all possible values for the independent variable in a function is called the **domain** of the function; the set of all possible values for the dependent variable is the **range**.

The domain and range may or may not be the same set.

In the following examples, both linear and nonlinear functions are discussed to clarify the function concepts.

Example 1 Do the following represent functions? (Assume x represents the independent variable, an assumption made throughout this book.)
(a) $y = -4x + 11$

For a given value of x, a corresponding value of y is found by multiplying x by -4 and then adding 11 to the result. This process will take a given value of x and produce exactly one value of y. (For example, if $x = -7$, then $y = -4(-7) + 11 = 39$.) Since one value of the independent variable leads to exactly one value of the dependent variable, $y = -4x + 11$ is a function.
(b) $y^2 = x$

Suppose $x = 36$. Then $y^2 = x$ becomes $y^2 = 36$, from which $y = 6$ or $y = -6$. Here one value of the independent variable leads to *two* values of the dependent variable, so that $y^2 = x$ does not represent a function. ∎

Example 2 Find the domain and range for each function.
(a) $y = -4x + 11$

In this equation any real number chosen for x gives a meaningful value for y, so the domain is the set of all real numbers. As x takes on every real number value, the resulting values for y, the range, also include all real numbers, so the range is the set of all real numbers.
(b) $y = 2x + 7$, $x = 1, 2, 3,$ or 4

The independent variable x is restricted to the values 1, 2, 3, or 4, making the domain the set $\{1, 2, 3, 4\}$. If $x = 1$, then $y = 2x + 7$ becomes $y = 2 \cdot 1 + 7 = 9$, while if $x = 2$, then $y = 2 \cdot 2 + 7 = 11$. If $x = 3$, then $y = 13$, while $x = 4$ produces $y = 15$. The range, the set of all possible values of y, is $\{9, 11, 13, 15\}$. ∎

f(x) Notation Letters such as f, g, or h are often used to name functions. For example, f could be used to name the function

$$y = 5 - 3x.$$

To show that this function is named f, and also to show that x is the independent variable, it is common to replace y with $f(x)$ (read "f of x") to get

$$f(x) = 5 - 3x.$$

1 Do the following define functions?
(a) $y = -6x + 1$
(b) $y = x^2$
(c) $x = y^2 - 1$
(d) $y < x + 2$

Answer:
(a) Yes
(b) Yes
(c) No
(d) No

2 Give the domain and range.
(a) $y = 3x + 1$
(b) $y = 2x - 5$,
    $x = 0, 1, 2, 3, 4$
(c) $y = x^2$

Answer:
(a) Both are the set of all real numbers
(b) Domain: $\{0, 1, 2, 3, 4\}$; range: $\{-5, -3, -1, 1, 3\}$
(c) Domain: set of all real numbers; range: $y \geq 0$

(Do not confuse the notation $f(x)$ with the product $f \cdot x$.) If 2 is chosen as a value of x, then $f(x)$ becomes $f(2) = 5 - 3 \cdot 2 = 5 - 6 = -1$, written

$$f(2) = -1.$$

In a similar manner,

$$f(-4) = 5 - 3(-4) = 17, \quad f(0) = 5, \quad f(-6) = 23,$$

3 Let $f(x) = -3x - 8$.
Find
(a) $f(0)$
(b) $f(-1)$
(c) $f(5)$.

Answer:
(a) -8
(b) -5
(c) -23

and so on. **3**

Example 3 Let $g(x) = x^2 - 4x + 5$. Find $g(3)$, $g(0)$, and $g(a)$.
To find $g(3)$, substitute 3 for x.

$$g(3) = 3^2 - 4(3) + 5 = 9 - 12 + 5 = 2.$$

Find $g(0)$ and $g(a)$ in the same way.

$$g(0) = 0^2 - 4(0) + 5 = 5$$
$$g(a) = a^2 - 4a + 5. \quad \blacksquare\ 4$$

4 Let $f(x) = 5x^2 - 2x + 1$. Find
(a) $f(1)$
(b) $f(3)$
(c) $f(m)$.

Answer:
(a) 4
(b) 40
(c) $5m^2 - 2m + 1$

Example 4 Suppose the sales of a small company have been estimated to be

$$S(x) = 125 + 80x,$$

where $S(x)$ represents the total sales in thousands of dollars in year x, with $x = 0$ representing 1986. Estimate the sales in each of the following years.
(a) 1986
Since $x = 0$ corresponds to 1986, the sales for 1986 are given by $S(0)$. Substituting 0 for x gives

$$S(0) = 125 + 80(0) \qquad \text{Let } x = 0$$
$$= 125.$$

Since $S(x)$ represents sales in thousands of dollars, the sales would be estimated as 125×1000, or \$125,000, in 1986.
(b) 1990
To estimate sales in 1990, let $x = 4$.

$$S(4) = 125 + 80(4) = 125 + 320 = 445,$$

so that sales should be about \$445,000 in 1990. **5**

5 A developer estimates that the total cost of building x large apartment complexes in a year is approximated by
$A(x) = x^2 + 80x + 60$,
where $A(x)$ represents the cost in hundred thousands of dollars. Find the cost of building
(a) 4 complexes
(b) 10 complexes.

Answer:
(a) \$39,600,000
(b) \$96,000,000

Graphs Given a function $y = f(x)$, any value of x in the domain of f produces a single value for y. This pair of numbers, one for x and one for y, can be written as an **ordered pair,** (x, y).

For example, let $y = f(x) = 8 + x^2$. If $x = 1$, then $f(1) = 8 + 1^2 = 9$, producing the ordered pair $(1, 9)$. (Always write the value of the independent variable first.) If $x = -3$, then $f(-3) = 8 + (-3)^2 = 17$, giving $(-3, 17)$. Other ordered pairs for this function include $(0, 8)$, $(-1, 9)$, $(2, 12)$, and so on.

A graph of the ordered pairs produced by a function can be drawn with the perpendicular number lines of a **Cartesian coordinate system,** as shown in Figure 2.1. Let the horizontal number line, or **x-axis,** represent the first component of the ordered pairs of the function, and the vertical or **y-axis** represent the second component. The point where the number lines cross is the zero point on both lines; this point is called the **origin.**

 Locate $(-1, 6)$, $(-3, -5)$, $(4, -3)$, $(0, 2)$, and $(-5, 0)$ on a coordinate system.

Answer:

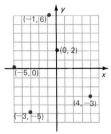

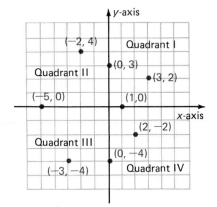

Figure 2.1

Locate the point $(-2, 4)$ on the graph by starting at the origin and counting 2 units to the left on the horizontal axis, and 4 units upward parallel to the vertical axis. This point is shown in Figure 2.1, along with several other sample points. The number -2 is the **x-coordinate** and 4 is the **y-coordinate** of the point $(-2, 4)$.

The x-axis and y-axis divide the graph into four quarters or **quadrants.** For example, Quadrant I includes all those points whose x- and y-coordinates are both positive. The quadrants are numbered as shown in Figure 2.1. The points of the axes themselves belong to no quadrant. The set of points corresponding to the ordered pairs of a function is called the **graph** of the function.

 Let $f(x) = 4x - 7$, with domain $\{-2, -1, 0, 1, 2, 3\}$. Graph the ordered pairs produced by this function.

Answer:

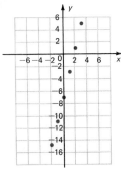

Example 5 Let $f(x) = 7 - 3x$, with domain $\{-2, -1, 0, 1, 2, 3, 4\}$. Graph the ordered pairs produced by this function.

If $x = -2$, then $f(-2) = 7 - 3(-2) = 13$, giving the ordered pair $(-2, 13)$. Additional ordered pairs were found in the same way to get the following table.

x	-2	-1	0	1	2	3	4
y	13	10	7	4	1	-2	-5
Ordered Pair	$(-2, 13)$	$(-1, 10)$	$(0, 7)$	$(1, 4)$	$(2, 1)$	$(3, -2)$	$(4, -5)$

These ordered pairs are graphed in Figure 2.2.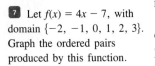

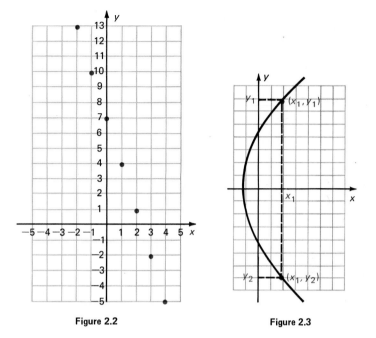

Figure 2.2   **Figure 2.3**

⑧ Does this graph represent a function?

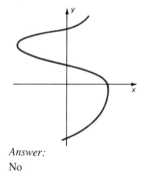

Answer:
No

⑧

For a graph to be the graph of a function, each value of x in the domain of the function must lead to exactly one value of y. In Figure 2.3, for the value $x = x_1$, the graph gives the two y-values, y_1 and y_2. Since the given x-value corresponds to two different y-values, this graph is not the graph of a function. The *vertical line test* for a function is based on this idea.

Vertical Line Test If a vertical line cuts a graph in more than one point, then the graph is not the graph of a function.

2.1   Exercises

List the ordered pairs obtained from each of the following if the domain of x for each exercise is $\{-2, -1, 0, 1, 2, 3\}$. Graph each set of ordered pairs. Give the range. (See Examples 2 and 5.)

1. $y = -4x + 9$ **2.** $y = -6x + 12$

3. $y = -x - 5$ **4.** $y = -2x - 3$

5. $2x + y = 9$ **6.** $3x + y = 16$

7. $y = x(x + 1)$ **8.** $y = (x - 2)(x - 3)$

9. $y = 3 - 4x^2$ **10.** $y = 5 - x^2$

11. $y = \dfrac{1}{x + 3}$ **12.** $y = \dfrac{-2}{x + 4}$

13. $y = \dfrac{3x - 3}{x + 5}$ **14.** $y = \dfrac{2x + 1}{x + 3}$

15. $y = 4$ **16.** $y = -2$

Identify any of the following that represent functions. (See Example 1.)

17. $y = 8x - 3$ **18.** $y = -4 + x$

19. $y = -x^2 + 2x$

20. $y = 8 + x^2$

21. $x = 1 + y^2$

22. $x = 9y - y^2$

23.

24.

25.

26.

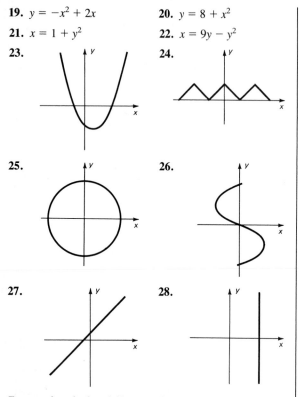

27.

28.

For each of the following functions, find (a) $f(4)$, (b) $f(-3)$, (c) $f(0)$, (d) $f(a)$. (See Example 3.)

29. $f(x) = 3x + 2$

30. $f(x) = 5x - 6$

31. $f(x) = -2x - 4$

32. $f(x) = -3x + 7$

33. $f(x) = 6$

34. $f(x) = 0$

35. $f(x) = 2x^2 + 4x$

36. $f(x) = x^2 - 2x$

37. $f(x) = -x^2 + 5x + 1$

38. $f(x) = -x^2 - x + 5$

39. $f(x) = (x + 1)(x + 2)$

40. $f(x) = (x + 3)(x - 4)$

Let $f(x) = 2x - 3$. Find each of the following. In Exercise 49, first find $f(2)$. (See Example 3.)

41. $f(0)$

42. $f(-1)$

43. $f(-6)$

44. $f(4)$

45. $f(a)$

46. $f(-r)$

47. $f(m + 3)$

48. $f(p - 2)$

49. $f[f(2)]$

50. $f[f(-3)]$

Let $f(x) = -7.61x^3 + 8.94x^2 - 5.03x + 1.87$. Find each of the following. Round to the nearest thousandth.

51. $f(-.408)$

52. $f(1.703)$

53. $f(-3.512)$

54. $f(.0862)$

Management *Work the following exercises.*

55. A chain-saw rental firm charges $7 per day or fraction of a day to rent a saw, plus a fixed fee of $4 for re-sharpening the blade. Let $S(x)$ represent the cost of renting a saw for x days. Find each of the following.

(a) $S\left(\dfrac{1}{2}\right)$

(b) $S(1)$

(c) $S\left(1\dfrac{1}{4}\right)$

(d) $S\left(3\dfrac{1}{2}\right)$

(e) $S(4)$

(f) $S\left(4\dfrac{1}{10}\right)$

(g) $S\left(4\dfrac{9}{10}\right)$

(h) A portion of the graph of $y = S(x)$ is shown here. Explain how it could be continued.

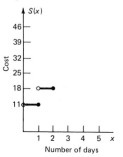

56. To rent a midsized car from Avis costs $40 per day or fraction of a day. If you pick up the car in Boston and drop it off in Utica, there is a fixed $40 charge. Let $C(x)$ represent the cost of renting the car for x days, taking it from Boston to Utica. Find each of the following.

(a) $C\left(\dfrac{3}{4}\right)$

(b) $C\left(\dfrac{9}{10}\right)$

(c) $C(1)$

(d) $C\left(1\dfrac{5}{8}\right)$

(e) $C\left(2\dfrac{1}{9}\right)$

(f) Graph the function $y = C(x)$.

57. Suppose the sales of a small company that sells by mail are approximated by

$$S(t) = 1000 + 50(t + 1),$$

where $S(t)$ represents sales in thousands of dollars. Here t is time in years, with $t = 0$ representing the year 1986. Find the estimated sales in each of the following years. (See Example 4.)

(a) 1986   (b) 1987   (c) 1988   (d) 1989

58. The graph at the side, taken from *Business Week* magazine, December 30, 1985 (page 22), shows the U.S. national debt as a percentage of gross national product since January 1965.*

(a) Is this graph that of a function?

(b) What does the domain represent?

(c) Estimate the range.

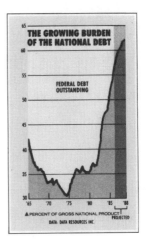

2.2 Linear Models

This section begins a study of *linear functions,* a type of function very important in applications.

A function f is a **linear function** if

$$f(x) = ax + b,$$

where a and b are real numbers.

Examples of linear functions include $y = 2x + 3$, $y = -5$, $y = 3x$, and $2x - 3y = 7$, which can be written as $y = (2/3)x - (7/3)$.

A linear function, such as the function defined by $y = x + 1$, can be graphed by finding several ordered pairs that satisfy the function. For example, if $x = 2$ then

$$y = x + 1$$
$$y = \mathbf{2} + 1 \qquad \text{Let } x = 2$$
$$y = 3.$$

Also, $(0, 1)$, $(4, 5)$, $(-2, -1)$, $(-5, -4)$, $(-3, -2)$, among many others, are ordered pairs that satisfy the equation $y = x + 1$.

---

*"The Growing Burden of the National Debt" reprinted from the December 30, 1985 issue of *Business Week* by special permission. Copyright © 1985 by McGraw-Hill, Inc.

Graph the function by first locating the ordered pairs obtained above, as in Figure 2.4(a). All the points of this graph appear to be on one straight line that can be drawn through the plotted points, as in Figure 2.4(b). This straight line is the graph of $y = x + 1$. Since any vertical line will cut the graph of Figure 2.4(b) in only one point, by the vertical line test, $y = x + 1$ is a function.

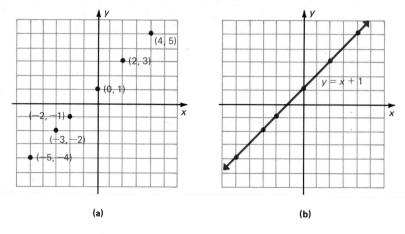

(a)   (b)

Figure 2.4

Example 1 Use the equation $x + 2y = 6$ to complete the ordered pairs $(-6, \ )$, $(-4, \ )$, $(-2, \ )$, $(0, \ )$, $(2, \ )$, $(4, \ )$. Graph these ordered pairs and then draw a straight line through them.

Complete the ordered pair $(-6, \ )$ by letting $x = -6$ in the equation $x + 2y = 6$.

$$x + 2y = 6$$
$$-6 + 2y = 6 \qquad \text{Let } x = -6$$
$$2y = 12 \qquad \text{Add 6 to both sides}$$
$$y = 6$$

The ordered pair is $(-6, 6)$.

In the same way, if $x = -4$, then $y = 5$, giving $(-4, 5)$. Check that the remaining ordered pairs are as graphed in Figure 2.5(a) on the next page. A line is drawn through the points in Figure 2.5(b). ■

Intercepts It can be shown that the graph of any linear function is a straight line. Since a straight line is completely determined from any two distinct points on the line, only two distinct points are needed to draw the graph of a linear function. Two points that are often useful for this purpose are the x-intercept and the y-intercept. The **x-intercept** is the x value (if one exists) where the graph of the equation crosses the x-axis. The **y-intercept** is the y value where the graph crosses the y-axis. The y-intercept of a linear function always exists. At a point where the graph crosses the y-axis, $x = 0$. Also, $y = 0$ at an x-intercept. See Figure 2.6. These facts are used to find the intercepts.

1 Use the equation $2x - 3y = 12$ to complete the ordered pairs $(-6, \ )$, $(-3, \ )$, $(0, \ )$, $(3, \ )$, $(6, \ )$. Graph these points and then draw a straight line through them.

Answer:

$(-6, -8)$, $(-3, -6)$, $(0, -4)$, $(3, -2)$, $(6, 0)$.

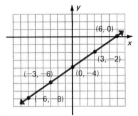

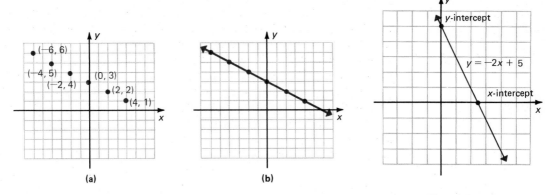

(a)   **(b)**

Figure 2.5   Figure 2.6

Example 2   Use the intercepts to draw the graph of $y = -2x + 5$.

Find the y-intercept, the point where the line crosses the y-axis, by letting $x = 0$ in the equation $y = -2x + 5$.

$$y = -2x + 5$$
$$y = -2(0) + 5 \qquad \text{Let } x = 0$$
$$y = 5$$

2 Find the intercepts of each of the following. Graph the lines.
(a) $3x + 4y = 12$
(b) $5x - 2y = 8$

Answer:
(a) x-intercept 4
   y-intercept 3

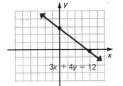

(b) x-intercept 8/5
   y-intercept -4

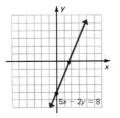

The y-intercept is 5, leading to the ordered pair $(0, 5)$. In the same way, the x-intercept is found by letting $y = 0$.

$$0 = -2x + 5 \qquad \text{Let } y = 0$$
$$2x = 5 \qquad\qquad \text{Add } 2x \text{ to both sides}$$
$$x = \frac{5}{2} = 2\frac{1}{2} \qquad \text{Multiply both sides by } \frac{1}{2}$$

The x-intercept is 5/2, or 2 1/2, so the graph goes through $(5/2, 0)$. These two intercepts lead to the graph of Figure 2.6.

As a check, a third point can be found by choosing another value of x (or y) and finding the corresponding value of the other variable. Check that $(1, 3)$, $(2, 1)$, $(3, -1)$, and $(4, -3)$, among other points, satisfy the equation $y = -2x + 5$ and lie on the line of Figure 2.6.   **2**

In the discussion of intercepts given above, we added the phrase "if one exists" when talking about the place where a graph crosses the x-axis. The next example shows a graph that does not cross the x-axis, and thus has no x-intercept.

Example 3   Graph $y = -3$.

The equation $y = -3$, or equivalently, $y = 0x - 3$, always gives the same y value, -3, for any value of x. Therefore, no value of x will make $y = 0$, so the graph has no x-intercept. Since $y = -3$ is a linear function with a straight-line graph, and since the graph cannot cross the x-axis, the line must be parallel to the x-axis. For any value of x, the value of y is -3, and so the graph is the horizontal line parallel to the x-axis, with y-intercept -3, as shown in Figure 2.7. As the vertical line test shows, the graph is the graph of a function. **3**

Example 3 suggests the following.

> The graph of $y = k$, where k is a real number, is the horizontal line having y-intercept k.

The graph of a function of the form $y = k$ is a horizontal line. Although an equation of the form $x = k$ is *not* the equation of a linear function, it does have a line for a graph, as the next example shows.

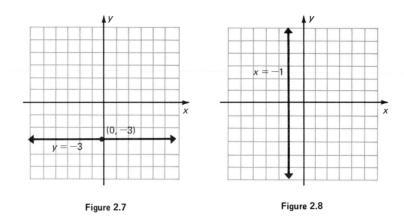

Figure 2.7   Figure 2.8

Example 4   Graph $x = -1$.

Obtain the graph of $x = -1$ by completing some ordered pairs using the equivalent form, $x = 0y - 1$. For example, $(-1, 0), (-1, 1), (-1, 2)$, and $(-1, 4)$ are some ordered pairs satisfying the equation $x = -1$. (The first component of these ordered pairs is always -1, which is what $x = -1$ means.) Here, more than one second component corresponds to the same first component, -1. As the graph of Figure 2.8 shows, a vertical line can cut this graph in more than one point. (In fact, a vertical line cuts the graph in an infinite number of points.) This confirms that $x = -1$ is not a function. **4**

Example 4 suggests the following.

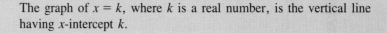

> The graph of $x = k$, where k is a real number, is the vertical line having x-intercept k.

As shown above, $x = -1$ is a linear equation that does not define a linear function. Linear equations of this form, $x = k$, where k is a real number, are the only linear equations that do not define linear functions.

3 Graph $y = -5$. Does this graph represent a function?

Answer:
Yes

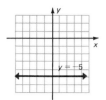

4 Graph $x = 4$. Is this a function?

Answer:
No

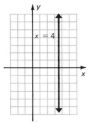

Example 5 Graph $y = -3x$.

Begin by looking for the x-intercept. If $y = 0$, then

$$0 = -3x \qquad \text{Let } y = 0$$
$$0 = x, \qquad \text{Multiply both sides by } -\frac{1}{3}$$

giving the ordered pair $(0, 0)$. Letting $x = 0$ leads to exactly the same ordered pair, $(0, 0)$. Two different points are needed to determine a straight line, and the intercepts have led to only one point. Get a second point by choosing some other value of x (or y). For example, if $x = 2$,

$$y = -3x = -3(2) = -6,$$

giving the ordered pair $(2, -6)$. These two ordered pairs, $(0, 0)$ and $(2, -6)$, were used to get the graph in Figure 2.9. ⑤

⑤ Graph

(a) $x = 5y$;

(b) $5x = y$.

Answer:

(a)

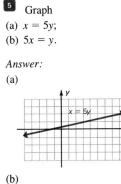

(b)

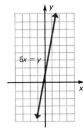

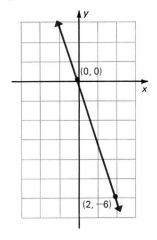

Figure 2.9

Linear functions can be very useful in setting up a mathematical model for a real-life situation. In almost every case, linear (or any other reasonably simple) functions provide only approximations to real-world situations. However, these approximations are often remarkably useful.

Supply and Demand In particular, linear functions are often good choices for **supply and demand curves.** Typically, as the price of an item increases, the demand for the item decreases, while the supply increases. The following example shows this.

Example 6 Suppose that Greg Odjakjian, an economist, has studied the supply and demand for aluminum siding and has determined that price, p, and demand, x, in appropriate units and for an appropriate domain, are related by the linear function

$$p = 60 - \frac{3}{4}x.$$

(a) Find the demand at a price of $40.

Let $p = 40$.

$$p = 60 - \frac{3}{4}x$$

$$40 = 60 - \frac{3}{4}x \qquad \text{Let } p = 40$$

$$-20 = -\frac{3}{4}x \qquad \text{Add } -60 \text{ on both sides}$$

$$\frac{80}{3} = x. \qquad \text{Multiply both sides by } -\frac{4}{3}$$

At a price of $40, 80/3 (or 26 2/3) units will be demanded; this gives the ordered pair (80/3, 40). (It is customary to write the ordered pairs so that price comes second.)

(b) Find the price if the demand is 32 units.

Let $x = 32$.

$$p = 60 - \frac{3}{4}x$$

$$p = 60 - \frac{3}{4}(\mathbf{32}) \qquad \text{Let } x = 32$$

$$p = 60 - 24$$

$$p = 36$$

With a demand of 32 units, the price is $36. This gives the ordered pair (32, 36).

(c) Graph $p = 60 - \frac{3}{4}x$.

Use the ordered pairs (80/3, 40) and (32, 36) to get the demand graph shown in black in Figure 2.10. Only the portion of the graph in Quadrant I is shown, since the supply and demand functions are meaningful only for positive values of p and x. **6**

6 Suppose price and demand are related by $p = 100 - 4x$.

(a) Find the price if the demand is 10 units.

(b) Find the demand if the price is $80.

Answer:

(a) $60

(b) 5 units

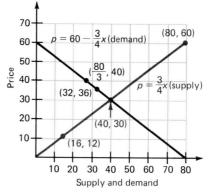

Figure 2.10

Example 7 Suppose that the economist of Example 6 concludes that the price and supply of siding are related by

$$p = \frac{3}{4}x.$$

(a) Find the supply if the price is $60.

$$60 = \frac{3}{4}x \qquad \text{Let } p = 60$$

$$80 = x$$

If the price is $60, then 80 units will be supplied to the marketplace. This gives the ordered pair (80, 60).

(b) Find the price if the supply is 16 units.

$$p = \frac{3}{4}(16) = 12 \qquad \text{Let } x = 16$$

If the supply is 16 units, then the price is $12. This gives the ordered pair (16, 12).

(c) Graph $p = \frac{3}{4}x$.

Use the ordered pairs (80, 60) and (16, 12) to get the supply graph shown in color in Figure 2.10. ■

As shown in the graphs of Figure 2.10, both the supply and the demand functions pass through the point (40, 30). If the price of the siding is more than $30, the supply will exceed the demand. At a price less than $30, the demand will exceed the supply. Only at a price of $30 will demand and supply be equal. For this reason, $30 is called the *equilibrium price*. When the price is $30, demand and supply both equal 40 units, the *equilibrium supply* or *equilibrium demand*. By definition, the **equilibrium price** of a commodity is the price at the point where the supply and demand graphs for that commodity cross. The **equilibrium demand** is the demand at that same point; the **equilibrium supply** is the supply at that point. By definition, the equilibrium supply and the equilibrium demand are equal.

Example 8 Use algebra to find the equilibrium supply for the aluminum siding. (See Examples 6 and 7.)

The equilibrium supply is found when the prices from both supply and demand are equal. From Example 6, $p = 60 - (3/4)x$; in Example 7, $p = (3/4)x$. Set these two expressions for p equal to get the linear equation,

$$60 - \frac{3}{4}x = \frac{3}{4}x$$

$$240 - 3x = 3x \qquad \text{Multiply both sides by 4}$$

$$240 = 6x \qquad \text{Add } 3x \text{ to both sides}$$

$$40 = x.$$

The equilibrium supply is 40 units, the same answer found above. **7**

7 The demand for a certain commodity is related to the price by $p = 80 - (2/3)x$. The supply is related to the price by $p = (4/3)x$. Find

(a) the equilibrium demand;
(b) the equilibrium price.

Answer:
(a) 40
(b) 160/3

2.2 Exercises

Graph each of the following linear equations. Identify any that are not *linear functions. (See Examples 1-5.)*

1. $y = 2x + 1$

2. $y = 3x - 1$

3. $y = 3x + 2$

4. $y = x + 5$

5. $3y + 4x = 12$

6. $4y + 5x = 10$

7. $y = -2$

8. $x = 4$

9. $6x + y = 12$

10. $x + 3y = 9$

11. $x - 5y = 4$

12. $2y + 5x = 20$

13. $x + 5 = 0$

14. $y - 4 = 0$

15. $5y - 3x = 12$

16. $2x + 7y = 14$

17. $8x + 3y = 10$

18. $9y - 4x = 12$

19. $y = 2x$

20. $y = -5x$

21. $y = -4x$

22. $y = 2x$

23. $x - 3y = 0$

24. $x + 4y = 0$

Management *Work the following exercises.*

25. Suppose that the demand and price for a certain model of electric can opener are related by

$$p = 16 - \frac{5}{4}x,$$

where p is price, in dollars, and x is demand. Find the price for a demand of
(a) 0 units;   **(b)** 4 units;   **(c)** 8 units.
Find the demand for the electric can opener at a price of
(d) $6;   **(e)** $11;   **(f)** $16.
(g) Graph $p = 16 - (5/4)x$.
Suppose the price and supply of the can openers are related by

$$p = \frac{3}{4}x,$$

where x represents the supply, and p the price. Find the supply when the price is
(h) $0;   **(i)** $10;   **(j)** $20.
(k) Graph $p = (3/4)x$ on the same axes used for part (g).
(l) Find the equilibrium supply.
(m) Find the equilibrium price.

26. Let the supply and demand functions for strawberry-flavored licorice be

$$\text{supply: } p = \frac{3}{2}x \quad \text{and} \quad \text{demand: } p = 81 - \frac{3}{4}x.$$

(a) Graph these on the same axes.
(b) Find the equilibrium demand.
(c) Find the equilibrium price.

27. Let the supply and demand functions for butter pecan ice cream be given by

$$\text{supply: } p = \frac{2}{5}x \quad \text{and} \quad \text{demand: } p = 100 - \frac{2}{5}x.$$

(a) Graph these on the same axes.
(b) Find the equilibrium demand.
(c) Find the equilibrium price.

28. Let the supply and demand functions for sugar be given by

$$\text{supply: } p = 1.4x - .6$$

and
$$\text{demand: } p = -2x + 3.2.$$

(a) Graph these on the same axes.
(b) Find the equilibrium demand.
(c) Find the equilibrium price.

29. In a recent issue of *Business Week,* the president of Insta-Tune, a chain of franchised automobile tune-up shops, says that people who buy a franchise and open a shop pay a weekly fee of

$$y = .07x + \$135$$

to company headquarters. Here y is the fee and x is the total amount of money taken in during the week by the tune-up center. Find the weekly fee if x is
(a) $0;   **(b)** $1000;   **(c)** $2000;   **(d)** $3000.
(e) Graph y.

30. In a recent issue of *The Wall Street Journal,* we are told that the relationship between the amount of money that an average family spends on food, x, and the amount of money it spends on eating out, y, is approximated by the model $y = .36x$. Find y if x is
(a) $40;   **(b)** $80;   **(c)** $120.
(d) Graph y.

2.3   Slope and the Equations of a Line

As mentioned in the previous section, the graph of a straight line is completely determined by two different points on the line. The graph of a straight line also can be drawn knowing only *one* point on the line *if* the "steepness" of the line is known, too. The number that represents the "steepness" of a line is called the *slope* of that line.

To see how slope is defined, start with Figure 2.11, which shows a line passing through the two different points $(x_1, y_1) = (-3, 5)$ and $(x_2, y_2) = (2, -4)$. The difference in the two x values,

$$x_2 - x_1 = 2 - (-3) = 5$$

in this example, is called the **change in x.** The symbol Δx (read "delta x") is used to represent the change in x. In the same way, Δy represents the **change in y.** In this example,

$$\Delta y = y_2 - y_1 = -4 - 5 = -9.$$

These symbols are used to define the slope of a line.

> The **slope** of the line through the two points (x_1, y_1) and (x_2, y_2), where $x_1 \neq x_2$, is defined as the quotient of the change in y and the change in x, or
>
> $$\textbf{slope} = \frac{\textbf{change in } y}{\textbf{change in } x} = \frac{\Delta y}{\Delta x} = \frac{y_2 - y_1}{x_2 - x_1}.$$

The slope of the line in Figure 2.11 is

$$\text{slope} = \frac{\Delta y}{\Delta x} = \frac{-4 - 5}{2 - (-3)} = -\frac{9}{5}.$$

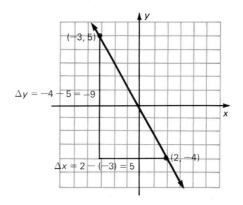

Figure 2.11

Using similar triangles from geometry, it can be shown that the slope is independent of the choice of points on the line. That is, the same value of the slope will be obtained for *any* choice of two different points on the line.

Example 1 Find the slope of the line through the points $(-7, 6)$ and $(4, 5)$.
Let $(x_1, y_1) = (-7, 6)$. Then $(x_2, y_2) = (4, 5)$. Use the definition of slope:

$$\text{slope} = \frac{\Delta y}{\Delta x} = \frac{5 - 6}{4 - (-7)} = -\frac{1}{11}.$$

The slope also could have been found by letting $(x_1, y_1) = (4, 5)$ and $(x_2, y_2) = (-7, 6)$. In that case,

$$\text{slope} = \frac{6 - 5}{-7 - 4} = \frac{1}{-11} = -\frac{1}{11},$$

the same answer. The order in which coordinates are subtracted does not matter, as long as it is done consistently from numerator to denominator. ∎

1 Find the slope of the line through
(a) $(6, 11), (-4, -3)$
(b) $(-3, 5), (-2, 8)$.

Answer:
(a) 7/5
(b) 3

The slope of a line is a measure of the steepness of the line. Figure 2.12 shows examples of lines with different slopes. Lines with positive slopes go up as x goes from left to right along the x-axis, while lines with negative slopes go down as x goes from left to right.

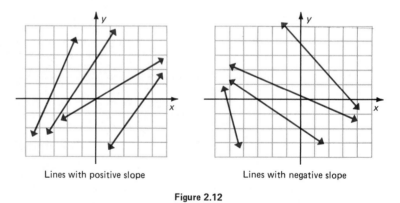

Lines with positive slope   Lines with negative slope

Figure 2.12

Example 2 Find the slope of the line $3x - 4y = 12$.
The slope is found using two different points on the line. The intercepts can be used for the two points. First let $x = 0$ and then let $y = 0$ to find the intercepts.

If $x = 0$,		If $y = 0$	
$3x - 4y = 12$		$3x - 4y = 12$	
$3(0) - 4y = 12$	Let $x = 0$	$3x - 4(0) = 12$	Let $y = 0$
$-4y = 12$		$3x = 12$	
$y = -3$		$x = 4$	

These results give the ordered pairs $(0, -3)$ and $(4, 0)$. Now the slope can be found from the definition.

$$\text{slope} = \frac{0 - (-3)}{4 - 0} = \frac{3}{4} \quad \boxed{2}$$

It can be shown that parallel lines have the same slope and lines with the same slope are parallel. Also, the slopes of perpendicular lines, neither of which is vertical, have a product of -1, and two lines whose slopes have a product of -1 are perpendicular.

Example 3 Tell whether each of the following pairs of lines are *parallel, perpendicular,* or *neither.*

(a) $2x + 3y = 5$ and $4x + 5 = -6y$

Find the slope of each line by first finding two points on each line. The intercepts are a good choice. [3]

From Problem 3 at the side, both slopes are $-2/3$, so the lines are parallel.

(b) $3x = y + 7$ and $x + 3y = 4$

Verify that the slope of $3x = y + 7$ is 3 and the slope of $x + 3y = 4$ is $-1/3$. Since $3(-1/3) = -1$, these lines are perpendicular.

(c) $x + y = 4$ and $x - 2y = 3$

The slope of the first line is -1 and of the second line is $1/2$. Since the slopes are not equal, and their product is $-1/2$, not -1, the lines are neither parallel nor perpendicular. [4]

Slope-Intercept Form A generalization of the method of Example 2 can be used to find the equation of a line given its y-intercept and slope. Assume that a line has y-intercept b, so that it goes through $(0, b)$. Let the slope of the line be m. If (x, y) is any point on the line *other* than $(0, b)$, using the definition of slope with the points $(0, b)$ and (x, y) gives

$$m = \frac{y - b}{x - 0}$$

$$m = \frac{y - b}{x}$$

$$mx = y - b$$

from which

$$y = mx + b.$$

This result is summarized as follows.

> **Slope-Intercept Form** If a line has slope m and y-intercept b, then
>
> $$\boldsymbol{y = mx + b}$$
>
> is the **slope-intercept form** of the equation of the line.

---

[2] Find the slope of
(a) $8x + 5y = 9$
(b) $2x - 7y = 6$
(c) $8y = x$.

Answer:
(a) $-8/5$
(b) $2/7$
(c) $1/8$

[3] Use $x = 0$ and $y = 0$ to get two ordered pairs and then find the slope of each line.
(a) $2x + 3y = 5$
(b) $4x + 5 = -6y$

Answer:
(a) $(0, 5/3)$ and $(5/2, 0)$; $-2/3$
(b) $(0, -5/6)$ and $(-5/4, 0)$; $-2/3$

[4] Tell if the lines in each of the following pairs are *parallel, perpendicular,* or *neither.*
(a) $x - 2y = 6$ and $2x + y = 5$
(b) $3x + 4y = 8$ and $x + 3y = 2$
(c) $2x - y = 7$ and $2y = 4x - 5$

Answer:
(a) Perpendicular
(b) Neither
(c) Parallel

Example 4 Find an equation for the line with y-intercept $7/2$ and slope $-5/2$. Use the slope-intercept form with $b = 7/2$ and $m = -5/2$.

$$y = mx + b$$

$$y = -\frac{5}{2}x + \frac{7}{2} \quad \boxed{5}$$

5 Find an equation for the line with
(a) y-intercept -3 and slope $2/3$;
(b) y-intercept $1/4$ and slope $-3/2$.

Answer:

(a) $y = \dfrac{2}{3}x - 3$

(b) $y = -\dfrac{3}{2}x + \dfrac{1}{4}$

The slope-intercept form of the equation of a line shows that the slope can be found from the equation by solving for y. Then the coefficient of x is the slope and the constant term is the y-intercept.

For example, we found in Example 2 that the slope of the line $3x - 4y = 12$ is $3/4$. This slope can also be found by solving for y.

$$3x - 4y = 12$$
$$-4y = -3x + 12 \qquad \text{Add } -3x$$
$$y = \frac{3}{4}x - 3 \qquad \text{Divide by } -4$$

The coefficient of x gives the slope, $3/4$.

Example 5 Find the slope and y-intercept for each of the following lines.
(a) $5x - 3y = 1$

Solve for y:

$$5x - 3y = 1$$
$$-3y = -5x + 1$$
$$y = \frac{5}{3}x - \frac{1}{3}.$$

The slope is $5/3$ and the y-intercept is $-1/3$.

(b) $-9x + 6y = 2$

Solve for y:

$$-9x + 6y = 2$$
$$6y = 9x + 2$$
$$y = \frac{3}{2}x + \frac{1}{3}.$$

6 Find the slope and y-intercept for
(a) $x + 4y = 6$;
(b) $3x - 2y = 1$.

Answer:

(a) $-1/4$; $3/2$
(b) $3/2$; $-1/2$

The slope is $3/2$ and the y-intercept is $1/3$. $\boxed{6}$

The slope and y-intercept of a line can be used to draw the graph of the line as shown in the next example.

Example 6 Use the slope and y-intercept to graph $3x - 2y = 2$.

Solve for y:

$$3x - 2y = 2$$
$$-2y = -3x + 2$$
$$y = \frac{3}{2}x - 1.$$

The slope is 3/2 and the y-intercept is -1.

To draw the graph, first locate the y-intercept -1, as shown in Figure 2.13. Use the slope to find a second point on the graph. If m represents the slope, then

$$m = \frac{\Delta y}{\Delta x} = \frac{3}{2}.$$

If x changes by 2 units ($\Delta x = 2$), then y will change by 3 units ($\Delta y = 3$). Find the second point by starting at the y-intercept graphed in Figure 2.13 and moving 2 units to the right and 3 units up. Once this second point is located, a line can be drawn through it and the y-intercept.

 Use the slope and y-intercept to graph
(a) $2x + 5y = 10$;
(b) $-x + y = 4$.

Answer:
(a)

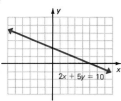

(b)

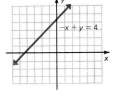

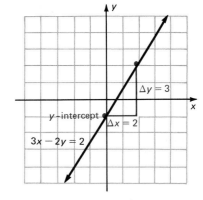

Figure 2.13

Point-Slope Form The slope-intercept form of the equation of a line involves the slope and the y-intercept. Sometimes, however, the slope of a line is known, together with one point (perhaps *not* the y-intercept) that the line goes through. The *point-slope form* of the equation of a line is used to find an equation in this case. Let (x_1, y_1) be any fixed point on the line and let (x, y) represent any other point on the line. If m is the slope of the line, then, by the definition of slope,

$$\frac{y - y_1}{x - x_1} = m,$$

or

$$y - y_1 = m(x - x_1).$$

Point-Slope Form If a line has slope m and passes through the point (x_1, y_1), then

$$y - y_1 = m(x - x_1),$$

is the **point-slope form** of the equation of the line.

Example 7 Find an equation of the line with the given slope and going through the given point.

(a) $(-4, 1)$, $m = -3$

Use the point-slope form since a point the line goes through, together with the slope of the line, is known. Substitute the values $x_1 = -4$, $y_1 = 1$, and $m = -3$ into the point-slope form.

$$y - y_1 = m(x - x_1)$$
$$y - 1 = -3[x - (-4)] \qquad \text{Let } y_1 = 1, m = -3, x_1 = -4$$
$$y - 1 = -3(x + 4)$$
$$y - 1 = -3x - 12 \qquad \text{Distributive property}$$
$$y = -3x - 11$$

(b) $(3, -7)$, $m = 5/4$

$$y - y_1 = m(x - x_1)$$
$$y - (-7) = \frac{5}{4}(x - 3) \qquad \text{Let } y_1 = -7, m = \frac{5}{4}, x_1 = 3$$

$$y + 7 = \frac{5}{4}(x - 3)$$
$$4y + 28 = 5(x - 3) \qquad \text{Multiply both sides by 4}$$
$$4y + 28 = 5x - 15 \qquad \text{Distributive property}$$
$$4y = 5x - 43 \qquad \boxed{8}$$

> **8** Find an equation of the line having the given slope and going through the given point.
>
> (a) $m = -3/5$, $(5, -2)$
> (b) $m = 1/3$, $(6, 8)$
>
> *Answer:*
> (a) $5y = -3x + 5$
> (b) $3y = x + 18$

The point-slope form can also be used to find an equation of a line given two different points that the line goes through. The procedure for doing this is shown in the next example.

Example 8 Find an equation of the line through $(5, 4)$ and $(-10, -2)$.

Begin by using the definition of slope to find the slope of the line that passes through the two points.

$$\text{slope} = m = \frac{-2 - 4}{-10 - 5} = \frac{-6}{-15} = \frac{2}{5}$$

Use $m = 2/5$ and either of the given points in the point-slope form. If $(x_1, y_1) = (5, 4)$, then

$$y - y_1 = m(x - x_1)$$
$$y - 4 = \frac{2}{5}(x - 5) \qquad \text{Let } y_1 = 4, m = \frac{2}{5}, x_1 = 5$$
$$5y - 20 = 2(x - 5) \qquad \text{Multiply both sides by 5}$$
$$5y - 20 = 2x - 10 \qquad \text{Distributive property}$$
$$5y = 2x + 10.$$

> **9** Find an equation of the line through
>
> (a) $(2, 3)$ and $(-4, 6)$;
> (b) $(-8, 2)$ and $(3, -6)$.
>
> *Answer:*
> (a) $2y = -x + 8$
> (b) $11y = -8x - 42$

Check that the result is the same when $(x_1, y_1) = (-10, -2)$. $\boxed{9}$

Example 9 Find an equation of the line through $(8, -4)$ and $(-2, -4)$.

First find the slope.
$$m = \frac{-4 - (-4)}{-2 - 8} = \frac{0}{-10} = 0$$

Choose, say, $(8, -4)$ as (x_1, y_1), and use the point-slope form.

$$y - y_1 = m(x - x_1)$$
$$y - (-4) = 0(x - 8) \qquad \text{Let } y_1 = -4, \; m = 0, \; x_1 = 8$$
$$y + 4 = 0 \qquad\qquad 0(x - 8) = 0$$
$$y = -4 \quad \boxed{10}$$

10 Find an equation of the line through $(-2, 5)$ and $(7, 5)$.

Answer:

$y = 5$

As shown in the previous section, $y = -4$ represents a horizontal line with y-intercept -4. Example 9 suggests the following.

> The slope of every horizontal line is 0.

Example 10 Find an equation of the line through $(4, 3)$ and $(4, -6)$.

Begin by finding the slope.
$$m = \frac{-6 - 3}{4 - 4} = \frac{-9}{0}$$

Division by 0 is not defined, so the slope is undefined. Graphing the given ordered pairs $(4, 3)$ and $(4, -6)$ and drawing a line through them gives a vertical line. From the last section, vertical lines have equations of the form $x = k$, where k can be any real number. Since the x-coordinate of the two ordered pairs given above is 4, the desired equation is $x = 4$. $\quad \boxed{11}$

11 Find an equation of the line through $(-5, 1)$ and $(-5, 7)$.

Answer:

$x = -5$

Example 10 suggests the following.

> The slope of every vertical line is undefined.

A summary of the equations of lines discussed in this section follows.

Equation	Description
$ax + by = c$	If $a \neq 0$ and $b \neq 0$, line has x-intercept c/a and y-intercept c/b.
$x = k$	**Vertical line,** x-intercept k, no y-intercept, undefined slope
$y = k$	**Horizontal line,** y-intercept k, no x-intercept, slope 0
$y = mx + b$	**Slope-intercept form,** slope m, y-intercept b
$y - y_1 = m(x - x_1)$	**Point-slope form,** slope m, line passes through (x_1, y_1)

2.3 Exercises

In each of the following exercises, find the slope, if it exists, of the line through the given pair of points. (See Example 1.)

1. $(-8, 6)$, $(2, 4)$

2. $(-3, 2)$, $(5, 9)$

3. $(-1, 4)$, $(2, 6)$

4. $(3, -8)$, $(4, 1)$

5. The origin and $(-4, 6)$

6. The origin and $(8, -2)$

7. $(-2, 9)$, $(-2, 11)$

8. $(7, 4)$, $(7, 12)$

9. $(3, -6)$, $(-5, -6)$

10. $(5, -11)$, $(-9, -11)$

Tell whether each pair of lines is parallel, perpendicular, or neither. (See Example 3.)

11. $4x - 3y = 6$ and $3x + 4y = 8$

12. $2x - 5y = 7$ and $15y = 5 + 6x$

13. $3x + 2y = 8$ and $6y = 5 - 9x$

14. $x - 3y = 4$ and $y = 1 - 3x$

15. $4x = 2y + 3$ and $2y = 2x + 3$

16. $2x - y = 6$ and $x - 2y = 4$

17. $2x - y = 9$ and $x = 2y$

18. $4x = 2y + 7$ and $y = 2x$

Find the slope and y-intercept of each of the following lines. (See Examples 2 and 5.)

19. $y = 3x + 4$

20. $y = -3x + 2$

21. $y + 4x = 8$

22. $y - x = 3$

23. $3x + 4y = 5$

24. $2x - 5y = 8$

25. $3x + y = 0$

26. $y - 4x = 0$

27. $2x + 5y = 0$

28. $3x - 4y = 0$

29. $y = 8$

30. $y = -4$

31. $y + 2 = 0$

32. $y - 3 = 0$

33. $x = -8$

34. $x = 3$

Graph the line through the given point with the given slope. (See Example 7.)

35. $(-4, 2)$, $m = \dfrac{2}{3}$

36. $(3, -2)$, $m = \dfrac{3}{4}$

37. $(-5, -3)$, $m = -2$

38. $(-1, 4)$, $m = 2$

39. $(8, 2)$, $m = 0$

40. $(2, -4)$, $m = 0$

41. $(6, -5)$, undefined slope

42. $(-8, 9)$, undefined slope

43. $(0, -2)$, $m = \dfrac{3}{4}$

44. $(0, -3)$, $m = \dfrac{2}{5}$

45. $(5, 0)$, $m = \dfrac{1}{4}$

46. $(-9, 0)$, $m = \dfrac{5}{2}$

Find an equation for the line with the given y-intercept and slope. (See Examples 4 and 6.)

47. 4, $m = -\dfrac{3}{4}$

48. -3, $m = \dfrac{2}{3}$

49. -2, $m = -\dfrac{1}{2}$

50. $\dfrac{3}{2}$, $m = \dfrac{1}{4}$

51. $\dfrac{5}{4}$, $m = \dfrac{3}{2}$

52. $-\dfrac{3}{8}$, $m = \dfrac{3}{4}$

Find an equation for each of the following lines. (See Examples 4 and 7–10.)

53. Through $(-4, 1)$, $m = 2$

54. Through $(5, 1)$, $m = -1$

55. Through $(0, 3)$, $m = -3$

56. Through $(-2, 3)$, $m = 3/2$

57. Through $(3, 2)$, $m = 1/4$

58. Through $(0, 1)$, $m = -2/3$

59. Through $(-1, 1)$ and $(2, 5)$

60. Through $(4, -2)$ and $(6, 8)$

61. Through $(9, -6)$ and $(12, -8)$

62. Through $(-5, 2)$ and $(7, 5)$

63. Through $(-8, 4)$ and $(-8, 6)$

64. Through $(2, -5)$ and $(4, -5)$

65. Through $(-1, 3)$ and $(0, 3)$

66. Through $(2, 9)$ and $(2, -9)$

Many real-world situations can be approximately described by a straight-line graph. One way to find the equation of such a straight line is to use two typical data points from the graph and the point-slope form of the equation of a line. In Exercises 67–71, assume that the data in each problem can be approximated fairly closely by a straight line. Use the given information to find the equation of the line. Find the slope of each of the lines.

67. Management A company finds that it can make a total of 20 solar heaters for $13,900, while 10 heaters cost $7500. Let y be the total cost to produce x solar heaters.

68. Management The sales of a small company were $27,000 in its second year of operation and $63,000 in its fifth year. Let y represent sales in year x.

69. Social Science According to research done by the political scientist James March, if the Democrats win 45% of the two-party vote for the House of Representatives, they win 42.5% of the seats. If the Democrats win 55% of the vote, they win 67.5% of the seats. Let y be the percent of seats won, and x the percent of the two-party vote.

70. Social Science If the Republicans win 45% of the two-party vote, they win 32.5% of the seats (see Exercise 69.) If they win 60% of the vote, they get 70% of the seats. Let y represent the percent of the seats, and x the percent of the vote.

71. Natural Science When a certain industrial pollutant is introduced into a river, the reproduction of catfish declines. In a given period of time, 3 tons of the pollutant results in a fish population of 37,000. Also, 12 tons of pollutant produce a fish population of 28,000. Let y be the fish population when x tons of pollutant are introduced into the river.

72. Natural Science A person's tibia bone goes from ankle to knee. A male with a tibia 40 cm in length will have a height of 177 cm, while a male's tibia 43 cm in length corresponds to a height of 185 cm.
 (a) Write a linear equation showing how the height of a male, h, relates to the length of his tibia, t.
 (b) Estimate the height of a male having a tibia of length 38 cm; 45 cm.
 (c) Estimate the length of the tibia of a male for a height of 190 cm.

73. Natural Science The radius bone goes from the wrist to the elbow. A female whose radius bone is 24 cm long would be 167 cm tall, while a radius of 26 cm in a female corresponds to a height of 174 cm.
 (a) Write a linear equation showing how the height of a female, h, corresponds to the length of her radius bone, r.
 (b) Estimate the height of a female having a radius of length 23 cm; 27 cm.
 (c) Estimate the length of a radius bone of a female for a height of 170 cm.

74. Show that b gives the y-intercept in the slope-intercept form $y = mx + b$.

2.4 Linear Mathematical Models

If the principles causing a certain event to happen are well understood, then the mathematical model describing that event can be very accurate. As a rule, the mathematical models constructed in the physical sciences are excellent at predicting events. For example, Hooke's law for an elastic spring states that the distance a spring stretches is given by

$$d = kf,$$

where f is the force applied to stretch the spring and k is a known constant for a particular spring. Using this equation, d can be predicted exactly for a given value of f. ①

The situation is different in the fields of management and social science. Mathematical models in these fields tend to be only approximations, and often gross approximations at that. So many variables are involved that no mathematical model

① Suppose k is 2.1 for a particular spring. Find d for the following forces.
 (a) 15 pounds
 (b) 8 pounds
 (c) 30 pounds

Answer:
 (a) 31.5
 (b) 16.8
 (c) 63

can ever hope to produce results comparable to those in physical science. In spite of the limitations of mathematical models, they have found a large and increasing acceptance in management and economic decision making. There is one main reason for this—mathematical models can produce very useful results.

Let us now look at some mathematical models of real-world situations.

Sales Analysis It is common to compare the change in sales of two companies by comparing the rates at which these sales change. If the sales of the two companies can be approximated by linear functions, the work of the last section can be used to find rates of change.

Example 1 The chart below shows sales in two different years for two different companies.

Company	Sales in 1986	Sales in 1989
A	$10,000	$16,000
B	5000	14,000

The sales of Company A increased from $10,000 to $16,000 over this 3-year period, for a total increase of $6000. The average rate of change of sales is

$$\frac{\$6000}{3} = \$2000.$$

Assuming that the sales of Company A have increased linearly (that is, that the sales can be closely approximated by a linear function), then an equation describing the sales can be found by finding the equation of the line through (0, 10,000) and (3, 16,000), where $x = 0$ represents 1986 and $x = 3$ represents 1989. The slope of the line is

$$\frac{16,000 - 10,000}{3 - 0} = 2000,$$

which is the same as the annual rate of change found above. Using the point-slope form of the equation of a line gives

$$y - 10,000 = 2000(x - 0)$$
$$y = 2000x + 10,000,$$

2 Assume that the sales of Company B from Example 1 have also increased linearly. Find an equation giving its sales and the average rate of change of sales.

Answer:

$y = 3000x + 5,000; 3000$

as an equation describing sales. **2**

As Example 1 and Problem 2 at the side suggest, the average rate of change is the same as the slope of the line. This is always true for data that can be modeled with a linear function.

Example 2 Suppose that a researcher has concluded that a dosage of x grams of a certain stimulant causes a rat to gain

$$y = 2x + 50$$

grams of weight, for appropriate values of x.

(a) If the researcher administers 30 grams of the stimulant, how much weight will the rat gain?

Let $x = 30$. Then the rat will gain

$$y = 2(30) + 50 = 110$$

grams of weight.

(b) What is the average weight gain?

The average rate of change of weight gain with respect to the amount of stimulant is given by the slope of the line. The slope of $y = 2x + 50$ is 2, so that 2 grams is the change in weight when the dose is changed by 1 gram. **3**

Cost Analysis It is common in manufacturing for the cost of making an item to consist of two parts. One part is a **fixed cost** for designing the product, establishing a factory, training workers, and so on. Within broad limits, the fixed cost is constant for a particular product and does not change as more items are made. The second part of the cost is a **variable cost** per item for labor, materials, packing, shipping, and so on. The variable cost may well be the same per item, with total variable cost increasing as the number of items increases.

Example 3 Suppose that the cost of producing clock-radios can be approximated by the linear model

$$C(x) = 12x + 100,$$

where $C(x)$ is the cost in dollars to produce x radios. The cost to produce 0 radios is

$$C(0) = 12(0) + 100 = 100,$$

or $100. This amount, $100, is the fixed cost.

Once the company has invested the fixed cost into the clock-radio project, what is the additional cost per radio? To find out, first find the cost of a total of 5 radios:

$$C(5) = 12(5) + 100 = 160,$$

or $160. The cost of 6 radios is

$$C(6) = 12(6) + 100 = 172,$$

or $172. The sixth radio costs $172 - $160 = $12 to produce. In the same way, the 81st radio costs $C(81) - C(80) = \$1072 - \$1060 = \$12$ to produce. In fact, the $(n + 1)$st radio costs

$$C(n + 1) - C(n) = [12(n + 1) + 100] - [12n + 100] = 12$$

dollars to produce. Since each additional radio costs $12 to produce, $12 is the variable cost per radio. The number 12 is also the slope of the cost function, $C(x) = 12x + 100$. ■

In economics, the cost of producing an additional item is called the **marginal cost** of that item. In the clock-radio example, the marginal cost of each radio is $12. **4**

3 A certain new anticholesterol drug is related to the blood cholesterol level by the linear model

$$y = 280 - 3x,$$

when x is the dosage of the drug (in grams) and y is the blood cholesterol level.
(a) Find the blood cholesterol level if 12 grams of the drug are administered.
(b) In general, an increase of 1 gram in the dose causes what change in the blood cholesterol level?

Answer:
(a) 244
(b) a decrease of 3 units

4 The cost in dollars to produce x kilograms of chocolate candy is given by $C(x)$, where in dollars

$$C(x) = 3.5x + 800.$$

Find each of the following.
(a) The fixed cost
(b) The total cost for 12 kilograms
(c) The variable cost
(d) The marginal cost of the 40th kilogram

Answer:
(a) $800
(b) $842
(c) $3.50 per kilogram
(d) $3.50

A summary of this discussion follows.

If a cost function is given by a linear function of the form $C(x) = mx + b$, then m represents the **variable cost** per item and b the **fixed cost.** Conversely, if the fixed cost of producing an item is b and the variable cost is m, then the **cost function,** $C(x)$, for producing x items, is given by $C(x) = mx + b$.

Example 4 In a certain city, a taxi company charges riders a fixed fee of $1.50 plus $1.80 per mile. Write a cost function, $C(x)$, which is a mathematical model for a ride of x miles.

Here the fixed cost is $b = 1.50$ dollars, with a variable cost of $m = 1.80$ dollars. The cost function, $C(x)$, is

$$C(x) = 1.80x + 1.50.$$

For example, a taxi ride of 4 miles will cost $C(4) = 1.80(4) + 1.50 = 8.70$, or $8.70. For each additional mile, the cost increases by $1.80. **5**

5 Avis charges $29 for a one-day rental of a certain model car, plus 24¢ per mile. Write a cost function, $C(x)$, giving the cost of driving the car x miles in one day.

Answer:

$C(x) = .24x + 29$

Example 5 The variable cost for raising a certain type of frog for laboratory study is $12 per unit of frogs, while the cost to produce 100 units is $1500. Find the cost function, $C(x)$, given that it is linear.

Since the cost function is linear, it can be expressed in the form $C(x) = mx + b$. The variable cost of $12 per unit gives the value for m in the model. The model can be written as $C(x) = 12x + b$. To find b, use the fact that the cost of producing 100 units of frogs is $1500, or $C(100) = 1500$. Substituting $x = 100$ and $C(x) = 1500$ into $C(x) = 12x + b$ gives

$$C(x) = 12x + b$$
$$1500 = 12 \cdot 100 + b$$
$$1500 = 1200 + b$$
$$b = 300.$$

6 The total cost of producing 10 units of a business calculator is $220. The variable cost per calculator is $14. Find the cost function, $C(x)$, if it is linear.

Answer:

$C(x) = 14x + 80$

The desired model is given by $C(x) = 12x + 300$, where the fixed cost is $300. **6**

Depreciation Because machines and equipment wear out or become obsolete over a period of time, business firms must take into account the amount of value that the equipment has lost during each year of its useful life. This lost value, called **depreciation,** may be calculated in several ways. The simplest way is to use **straight-line,** or **linear,** depreciation in which an item having a useful life of n years is assumed to lose $1/n$ of its value each year. For example, a typewriter with a ten-year life would be assumed to lose 1/10 of its value each year, and 4/10 of its value in 4 years.

A machine may have some **scrap value** at the end of its useful life, so depreciation is calculated on **net cost**—the difference between purchase price and scrap value. To find the annual straight-line depreciation on an item having a net cost of x dollars and a useful life of n years, multiply the net cost by the fraction of the value lost each year, $1/n$. The annual straight-line depreciation is then

$$\text{annual straight-line depreciation} = \frac{1}{n}x.$$

Example 6 An asset has a purchase price of $100,000 and a scrap value of $40,000. The useful life of the asset is 10 years. Find each of the following for this asset.

(a) Net cost

Since net cost is the difference between purchase price and scrap value,

$$x = \text{net cost} = \$100,000 - \$40,000 = \$60,000.$$

(b) Annual depreciation

The useful life of the asset is 10 years. Therefore, 1/10 of the net cost is depreciated each year. The annual depreciation by the straight-line method is

$$\frac{1}{n}x = \frac{1}{10} \cdot 60,000 = 6000,$$

or $6000.

(c) Undepreciated balance after 4 years

The total amount that will be depreciated over the life of the asset is $60,000. In four years, the depreciation will be

$$4 \times \$6000 = \$24,000,$$

and the undepreciated balance will be

$$\$60,000 - \$24,000 = \$36,000. \quad \boxed{7}$$

Straight-line depreciation is the easiest method of depreciation to use, but it often does not accurately reflect the rate at which assets actually lose value. Some assets, such as new cars, lose more value annually at the beginning of their useful life than at the end. For this reason, two other methods of depreciation, the sum-of-the-years'-digits method (discussed in Exercises 41 and 42 below) and double declining balance often are used.

Break-even Analysis A company can make a profit only if the revenue it receives from its customers exceeds the cost of producing its goods and services. The point at which revenue just equals cost is called the **break-even point.**

Example 7 A firm producing chicken feed finds that the total cost, $C(x)$, of producing x units is given by

$$C(x) = 20x + 100.$$

7 A backhoe (used for digging basements, swimming pools, and other holes) costs $65,000, has a useful life of 7 years, and a scrap value of $16,000. Find the

(a) net cost;

(b) annual depreciation;

(c) undepreciated balance after 5 years.

Answer:

(a) $49,000

(b) $7000

(c) $14,000

The revenue, $R(x)$, from selling x units is given by the product of the price per unit and the number of units sold. If the feed sells for \$24 per unit, then the revenue from selling x units is

$$R(x) = 24x.$$

The firm will break even (no profit and no loss), as long as revenue just equals cost, or $R(x) = C(x)$. This is true whenever

$$R(x) = C(x)$$
$$24x = 20x + 100 \qquad \text{Substitute for } R(x) \text{ and } C(x)$$
$$4x = 100 \qquad \text{Add } -20x \text{ to both sides}$$
$$x = 25.$$

Here the break-even point is at $x = 25$.

The graphs of $C(x) = 20x + 100$ and $R(x) = 24x$ are shown in Figure 2.14. The break-even point is shown on the graph. If the company produces more than 25 units (if $x > 25$), it makes a profit; if $x < 25$ it loses money. 8

8 For a certain magazine, $C(x) = .20x + 1200$, while $R(x) = .50x$, where x is the number of magazines sold. Find the break-even point.

Answer:
4000 magazines

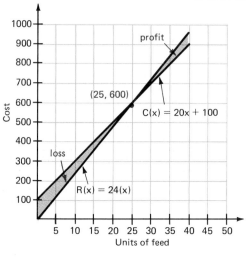

Figure 2.14

2.4 Exercises

1. **Management** Suppose the sales of a particular brand of electric guitar are

$$S(x) = 300x + 2000,$$

where $S(x)$ represents the number of guitars sold in year x, with $x = 0$ corresponding to 1982. Find the sales in each of the following years.
(a) 1984 (b) 1985 (c) 1986 (d) 1982
(e) Find the annual rate of change of the sales.

2. **Natural Science** If the population of ants in an anthill is

$$A(x) = 1000x + 6000,$$

where $A(x)$ represents the number of ants present at the end of month x, find the number of ants present at the end of each of the following months. Let $x = 0$ represent June.

(a) June   (b) July

(c) August   (d) December

(e) What is the monthly rate of change of the number of ants?

3. **Natural Science** Let $N(x) = -5x + 100$ represent the number of bacteria (in thousands) present in a certain tissue culture at time x, measured in hours, after an antibacterial spray is introduced into the environment. Find the number of bacteria present at each of the following times.

(a) $x = 0$ (b) $x = 6$ (c) $x = 20$

(d) What is the hourly rate of change in the number of bacteria? Interpret the negative sign in the answer.

4. Let $R(x) = -8x + 240$ represent the number of students in a large business mathematics class, where x represents the number of hours of study required weekly. Find the number of students present at each of the following levels of required study.

(a) $x = 0$ (b) $x = 5$ (c) $x = 10$

(d) What is the rate of change of the number of students in the class with respect to the number of hours of study. Interpret the negative sign in the answer.

(e) The professor in charge of the class likes to have exactly 16 students. How many hours of study must be required in order to have exactly 16 students? (Hint: Find a value of x such that $R(x) = 16$.)

5. **Management** Assume that the sales of a certain appliance dealer are approximated by a linear function. Suppose that sales were $850,000 in 1981 and $1,262,500 in 1986. Let $x = 0$ represent 1981.

(a) Find the equation giving the dealer's yearly sales.

(b) What were the dealer's sales in 1984?

(c) Estimate sales in 1989.

6. **Management** Assume that the sales of a certain automobile parts company are approximated by a linear function. Suppose that sales were $200,000 in 1979, and $1,000,000 in 1986. Let $x = 0$ represent 1979 and $x = 7$ represent 1986.

(a) Find the equation giving the company's yearly sales.

(b) Find the sales in 1981.

(c) Estimate the sales in 1988.

7. **Management** Suppose the number of bottles of a vitamin, $V(x)$, on hand at the beginning of the day in a health food store is given by

$$V(x) = 600 - 20x$$

where $x = 1$ corresponds to June 1, and x is measured in days. If the store is open every day of the month, find the number of bottles on hand at the beginning of each of the following days.

(a) June 6   (b) June 12   (c) June 24

(d) When will the last bottle from this stock be sold?

(e) What is the daily rate of change of this stock?

8. **Social Science** In psychology, the just-noticeable-difference (JND) for some stimulus is defined as the amount by which the stimulus must be increased so that a person will perceive it as having just barely increased. For example, suppose a research study indicates that a line 40 centimeters in length must be increased to 42 cm before a subject thinks that it is longer. In this case, the JND would be $42 - 40 = 2$ cm. In a particular experiment, the JND (y) is given by

$$y = 0.03x,$$

where x represents the original length of the line. Find the JND for lines having the following lengths.

(a) 10 cm   (b) 20 cm   (c) 50 cm

(d) 100 cm

(e) Find the rate of change in the JND with respect to the original length of the line.

Management *Write a cost function for each of the following. Identify all variables used. (See Example 4.)*

9. A chain-saw rental firm charges $12 plus $1 per hour.

10. A trailer-hauling service charges $45 plus $2 per mile.

11. A parking garage charges 35¢ plus 30¢ per half hour.

12. For a one-day rental, a car rental firm charges $14 plus 6¢ per mile.

Management *Assume that each of the following can be expressed as a linear cost function. Find the appropriate cost function in each case. (See Example 5.)*

13. Fixed cost, $100; 50 items cost $1600 to produce.

14. Fixed cost, $400; 10 items cost $650 to produce.

15. Fixed cost, $1000; 40 items cost $2000 to produce.

16. Fixed cost, $8500; 75 items cost $11,875 to produce.

17. Variable cost, $50; 80 items cost $4500 to produce.

18. Variable cost, $120; 100 items cost $15,800 to produce.

19. Variable cost, $90; 150 items cost $16,000 to produce.

20. Variable cost, $120; 700 items cost $96,500 to produce.

21. Management The manager of a local restaurant told us that his cost function for producing coffee is $C(x) = .097x$, where $C(x)$ is the total cost in dollars of producing x cups. (He is ignoring the cost of the coffee pot and the cost of labor.) Find the total cost of producing the following numbers of cups.
(a) 1000 cups   (b) 1001 cups
(c) Find the marginal cost of the 1001st cup.
(d) What is the marginal cost for *any* cup?

22. Management In deciding whether or not to set up a new manufacturing plant, company analysts have decided that a reasonable function for the total cost to produce x items is

$$C(x) = 500,000 + 4.75x.$$

(a) Find the total cost to produce 100,000 items.
(b) Find the marginal cost of the items to be produced in this plant.

Management *Assume that each of the following can be expressed as a linear cost function. Find (a) the cost function; (b) the revenue function; (c) the break-even point. (See Example 7.)*

	Fixed cost	Variable cost	Item sells for
23.	$500	$10	$35
24.	$180	$11	$20
25.	$250	$18	$28
26.	$1500	$30	$80
27.	$2700	$100	$125
28.	$165	$20	$25

Management *Let $C(x)$ be the total cost to manufacture x items. Then the quotient $(C(x))/x$ is the **average cost per item**. Use this definition in Exercises 29 and 30.*

29. $C(x) = 800 + 20x$; find the average cost per item if x is
(a) 10;   (b) 50;   (c) 200.

30. $C(x) = 500,000 + 4.75x$; find the average cost per item if x is
(a) 1000;   (b) 5000;   (c) 10,000.

31. Management The graph below shows the total sales, y, in thousands of dollars from the distribution of x thousand catalogs. Find the average rate of change of sales with respect to the number of catalogs distributed for the following changes in x. (See Example 1.)
(a) 10 to 20   (b) 10 to 40
(c) 20 to 30   (d) 30 to 40

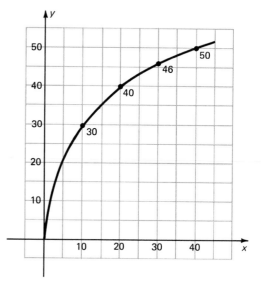

32. Management The graph below shows annual sales (in units) of a typical product. Sales increase slowly at first to some peak, hold steady for a while, and then decline as the product goes out of style. Find the average annual rate of change in sales for the following changes in years.
(a) 1 to 3   (b) 2 to 4   (c) 3 to 6
(d) 5 to 7   (e) 7 to 9   (f) 8 to 11
(g) 9 to 10   (h) 10 to 12

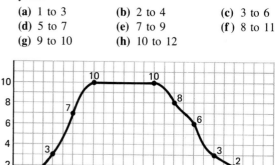

Management *For each of the assets in Exercises 33–38 find the straight-line depreciation in year four, and the undepreciated balance after 4 years. (See Example 6.)*

33. Cost: $50,000; scrap value: $10,000; life: 20 years

34. Cost: $120,000; scrap value: $0; life: 10 years

35. Cost: $80,000; scrap value: $20,000; life: 30 years

36. Cost: $720,000; scrap value: $240,000; life: 12 years

37. Cost: $1,400,000; scrap value: $200,000; life: 8 years

38. Cost: $2,200,000; scrap value: $400,000; life: 12 years

39. Management Suppose an asset has a net cost of $80,000 and a four-year life.
  (a) Find the straight-line depreciation in each of years 1, 2, 3, and 4 of the item's life.
  (b) Find the sum of all depreciation for the four-year life.

40. Management A forklift truck has a net cost of $12,000, with a useful life of 5 years.
  (a) Find the straight-line depreciation in each of years 1, 2, 3, 4, and 5 of the forklift's life.
  (b) Find the sum of all depreciation for the five-year life.

Management *Some assets, such as new cars, lose more value annually at the beginning of their useful life than at the end. By one method of depreciation for such assets, called the* **sum-of-the-years'-digits,** *the depreciation in year j, which is called A_j, is given by*

$$A_j = \frac{n - j + 1}{n(n + 1)} \cdot 2I,$$

where n is the useful life of the item, $1 \le j \le n$, and I is the net cost of the item. (Another approach to this method of depreciation is shown in Exercise 39 of Section 6.4.)

41. For a certain asset, $n = 4$ and $I = \$10,000$.
  (a) Use the sum-of-the-years'-digits method to find the depreciation for each of the four years covering the useful life of the asset.
  (b) What would be the annual depreciation by the straight-line method?

42. A machine tool costs $105,000 and has a scrap value of $25,000, with a useful life of 4 years.
  (a) Use the sum-of-the-years'-digits method to find the depreciation in each of the four years of the machine tool's life.
  (b) Find the total depreciation by this method.

Management *For each of the following assets, use the sum-of-the-years'-digits method to find the depreciation in year 1 and year 4.*

43. Cost: $36,500; scrap value: $8500; life: 10 years

44. Cost: $6250; scrap value: $250; life: 5 years

45. Cost: $18,500; scrap value: $3900; life: 6 years

46. Cost: $275,000; scrap value: $25,000; life: 20 years

47. The cost to produce x units of wire is $C(x) = 50x + 5000$, while the revenue is $R(x) = 60x$. Find the break-even point and the revenue at the break-even point.

48. The cost to produce x units of squash is $C(x) = 100x + 6000$, while the revenue is $R(x) = 500x$. Find the break-even point.

Management *You are the manager of a firm. You are considering the manufacture of a new product, so you ask the accounting department to produce cost estimates and the sales department to produce sales estimates. After you receive the data, you must decide whether or not to go ahead with production of the new product. Analyze the following data (find a break-even point) and then decide what you would do. (See Example 7.)*

49. $C(x) = 105x + 6000$; $R(x) = 250x$; no more than 400 units can be sold.

50. $C(x) = 275x + 5625$; $R(x) = 320x$; no more than 200 units can be sold.

51. $C(x) = 80x + 7000$; $R(x) = 95x$; no more than 400 units can be sold.

52. $C(x) = 65x + 9500$; $R(x) = 80x$; no more than 600 units can be sold.

53. $C(x) = 140x + 3000$; $R(x) = 125x$ (Hint: what does a negative value of x mean?)

54. $C(x) = 1750x + 95,000$; $R(x) = 1750x$

55. The sales of a certain furniture company in thousands of dollars are shown in the chart below.

x (year)	y (sales)
0	48
1	59
2	66
3	75
·4	80
5	90

(a) Graph this data, plotting years on the x-axis and sales on the y-axis. (Note that the data points can be closely approximated by a straight line.)

(b) Draw a line through the points (2, 66) and (5, 90). The other four points should be close to this line. (These two points were selected as "best" representing the line that could be drawn through the data points.)

(c) Use the two points of (b) to find an equation for the line that approximates the data.

(d) Complete the following chart.

Year	Sales (actual)	Sales (predicted from equation of (c))	Difference, actual minus predicted
0			
1			
2			
3			
4			
5			

(e) Use the result of (c) to predict sales in year 7.

(f) Do the same for year 9.

56. Social Science Most people are not very good at estimating the passage of time. Some people's estimations are too fast, and others', too slow. One psychologist has constructed a mathematical model for actual time as a function of estimated time; if y represents actual time and x estimated time, then

$$y = mx + b,$$

where m and b are constants that must be determined experimentally for each person.

Suppose that for a particular person, $m = 1.25$ and $b = -5$. Find y if x is

(a) 30 minutes   (b) 60 minutes;

(c) 120 minutes;   (d) 180 minutes,

Suppose that for another person, $m = .85$ and $b = 1.2$. Find y if x is

(e) 15 minutes;   (f) 30 minutes;

(g) 60 minutes;   (h) 120 minutes.

For this same person, find x if y is

(i) 60 minutes;   (j) 90 minutes.

 Graph each function.

(a) $y = |x + 3|$

(b) $y = 2 - |x|$

Answer:

(a)

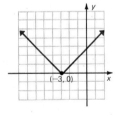

(b)

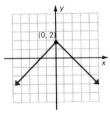

2.5 Other Useful Functions

Many functions used in later work are nonlinear. Several of these useful functions are introduced in the next chapter. This section discusses functions which are not linear, but whose graphs are made up of parts of straight lines.

The **absolute value function,** defined by $y = |x|$, is a common nonlinear function. Since $|x|$ can be found for any real number x, the domain is the set of all real numbers. Also, $|x| \geq 0$ for any real number x, so the range is the set of all nonnegative real numbers. If $x \geq 0$, then $y = |x| = x$, so the graph is the line $y = x$ for nonnegative values of x. On the other hand, if $x < 0$, then $y = |x| = -x$, and the graph is the same as that of $y = -x$ for negative values of x. The final graph is shown in Figure 2.15. By the vertical line test, the graph is that of a function.

Example 1 Graph $y = |x - 2|$.

Again, the domain is the set of all real numbers, with the range the set of all nonnegative real numbers. If $x = 2$, then $y = 0$. This point, (2, 0), is the lowest point on the graph. Graphing the point (2, 0) and a few other selected points, leads to the graph of Figure 2.16. This graph has the same shape as that of $y = |x|$, but the "vertex" point is shifted, or *translated,* 2 units right, from (0, 0) to (2, 0).

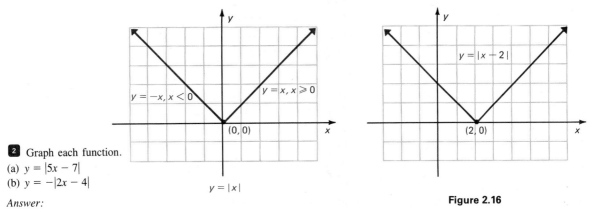

$y = |x|$

Figure 2.15

Figure 2.16

2 Graph each function.
(a) $y = |5x - 7|$
(b) $y = -|2x - 4|$

Answer:

(a)

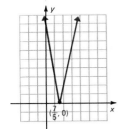

(b)

Example 2 Graph $y = |3x + 4|$.

Plotting a few ordered pairs and connecting them with two rays gives the graph of Figure 2.17. As the graph shows, the "vertex" is translated to $(-4/3, 0)$. The two parts of the graph form a smaller angle than that of $y = |x|$. **2**

The graphs of the absolute value functions of Examples 1 and 2 are made up of portions of two different straight lines. Such functions, called **piece-wise functions,** are often defined with different equations for different parts of the domain, as in the next example.

Example 3 Graph the function

$$y = \begin{cases} x + 1 & \text{if } x \leq 2 \\ -2x + 7 & \text{if } x > 2. \end{cases}$$

For $x \leq 2$, graph $y = x + 1$. For $x > 2$, graph $y = -2x + 7$, as shown in Figure 2.18. In this example, the two parts meet at the point $(2, 3)$. **3**

3 Graph $f(x)$, where

$$f(x) = \begin{cases} -2x + 5 & \text{if } x < 2 \\ x - 4 & \text{if } x \geq 2. \end{cases}$$

Answer:

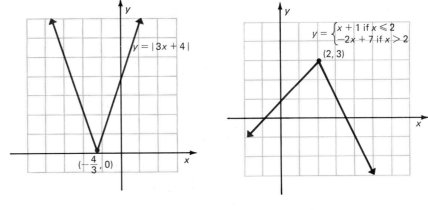

Figure 2.17

Figure 2.18

4 Graph $y = [(1/2)x + 1]$.

Answer:

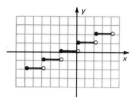

The **greatest integer function,** written $y = [x]$, is defined by saying that $[x]$ is the greatest integer less than or equal to x. For example, $[8] = 8$, $[-5] = -5$, $[\pi] = 3$, $[12\ 1/9] = 12$, $[-2.001] = -3$, and so on.

Example 4 Graph $y = [x]$.

For x in the interval $0 \le x < 1$, the value of $[x] = 0$. For values of x where $1 \le x < 2$, $[x] = 1$, and so on. Thus, the graph, as shown in Figure 2.19, consists of a series of line segments. In each case, the left endpoint of the segment is included, and the right endpoint is excluded. The domain of the function is the set of all real numbers, and the range is the set of integers. **4**

The greatest integer function, graphed in Figure 2.19, is an example of a **step function.**

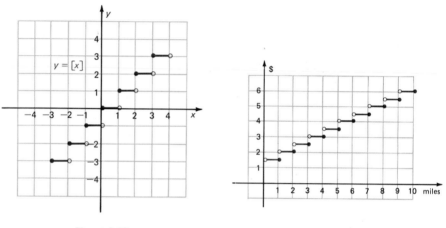

Figure 2.19   Figure 2.20

5 Assume that the post office charges 30¢ per ounce, or fraction of an ounce, to mail a letter. Graph the ordered pairs (ounces, cost).

Answer:

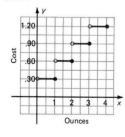

Example 5 The Terrific Taxi Company charges $1 plus 50¢ per mile or part of a mile. Find the cost of a trip of 5 miles; of 4.5 miles; of 9.5 miles. Graph the ordered pairs (miles, cost). Is this graph the graph of a function?

The cost for a 5-mile trip is

$$1 + 5(.50) = 3.50 \qquad \text{or} \qquad \$3.50.$$

For a 4.5-mile trip, the cost will be the same as for a 5-mile trip, $3.50. A 9.5-mile trip costs the same as a 10-mile trip: $1 + 10(.50) = 6$, or $6. The graph, a step function, is shown in Figure 2.20. The vertical line test shows that the graph is that of a function. **5**

Example 6 Business graphs are often made up of portions of straight lines. For example, the graph of Figure 2.21 shows one prediction of interest rates for two kinds of real estate loans for the period of the 1980s. (Interest rates from 1966 are included for comparison.) To get this graph, points are plotted for different years and connected with straight line segments. The graph suggests that mortgages for single family homes (the lower graph) averaged about 14% in 1980 and a little over 11% in 1985. Rates are predicted to be about 10% in 1989.   ■

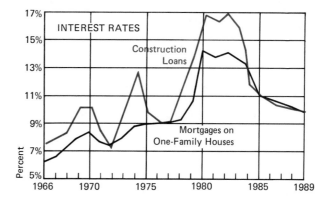

Mortgage interest rates, more than anything else, will control the pace of home-building over the next several years. Last month some rates in Washington, D.C., reached a paralyzing 17 percent, unheard of since Civil War days. By Townsend-Greenspan's estimate for FORTUNE, mortgage rates will fall enough over the remainder of this year to lower the 1980 average to 14 percent. Even so, it will probably be 1989 before 10 percent home loans are seen again.[*]

Figure 2.21

2.5   Exercises

Graph each of the following. Identify any that are not functions. (See Examples 1 and 2.)

1. $y = |x + 1|$

2. $y = |x - 1|$

3. $y = |2 - x|$

4. $y = |-3 - x|$

5. $y = |5x + 4|$

6. $3y = |x + 2|$

7. $y + 1 = |x - 2|$

8. $2y - 1 = |x + 3|$

9. $x = |y|$

10. $x = |y + 1|$

11. $y = |x| + 4$

12. $y = 2|x| - 1$

13. $2x = 3|y| - 1$

14. $x - 1 = |3y + 2|$

15. $y = |x| - x$

16. $x = |y| + y$

Graph each of the following functions. (See Example 3.)

17. $y = \begin{cases} x - 1 & \text{if } x \le 3 \\ 2 & \text{if } x > 3 \end{cases}$

18. $y = \begin{cases} 6 - x & \text{if } x \le 3 \\ 3x - 6 & \text{if } x > 3 \end{cases}$

19. $y = \begin{cases} 4 - x & \text{if } x < 2 \\ 1 + 2x & \text{if } x \ge 2 \end{cases}$

20. $y = \begin{cases} -2 & \text{if } x \ge 1 \\ 2 & \text{if } x < 1 \end{cases}$

21. $y = \begin{cases} 2x + 1 & \text{if } x \ge 0 \\ x & \text{if } x < 0 \end{cases}$

22. $y = \begin{cases} 5x - 4 & \text{if } x \ge 1 \\ x & \text{if } x < 1 \end{cases}$

23. $y = \begin{cases} 2 + x & \text{if } x < -4 \\ -x & \text{if } -4 \le x \le 5 \\ 3x & \text{if } x > 5 \end{cases}$

[*] "Mortgage interest rates" by William Reduto, *Fortune* Magazine, April 7, 1980. Reprinted by permission.

24. $y = \begin{cases} -2x & \text{if} & x < -3 \\ 3x + 1 & \text{if} & -3 \le x < 2 \\ -4x & \text{if} & x \ge 2 \end{cases}$

25. $y = \begin{cases} |x| & \text{if} & x > -2 \\ x & \text{if} & x \le -2 \end{cases}$

26. $y = \begin{cases} |x| - 1 & \text{if} & x > -1 \\ x - 1 & \text{if} & x \le -1 \end{cases}$

Graph each of the following functions. (See Example 4.)

27. $y = [-x]$

28. $y = [2x]$

29. $y = [2x - 1]$

30. $y = [3x + 1]$

31. $y = [3x]$

32. $y = [3x] + 1$

33. $y = [3x] - 1$

34. $y = x - [x]$

Solve the following exercises. (See Example 5.)

35. A delivery company charges $2 plus 25¢ per mile or part of a mile. Find the cost for a trip of
   (a) 3 miles;   (b) 3.4 miles;
   (c) 3.9 miles;   (d) 5 miles.
   (e) Graph the ordered pairs (miles, cost).
   (f) Is this a function?

36. The charge to rent a Haul-It-Yourself Trailer is $25 plus $1 per hour or portion of an hour. Find the cost to rent a trailer for
   (a) 2 hours;   (b) 1.5 hours;
   (c) 4 hours;   (d) 3.7 hours.
   (e) Graph the ordered pairs (hours, cost).
   (f) Is this a function?

37. A college typing service charges $3 plus $7 per hour or fraction of an hour. Graph the ordered pairs (hours, cost).

38. A parking garage charges 35¢ plus 30¢ per hour or fraction of an hour. Graph the ordered pairs (hours, cost).

39. For a lift truck rental of no more than three days, the charge is $300. An additional charge of $75 is made for each day or portion of a day after three. Graph the ordered pairs (days, cost).

40. A car rental costs $37 for one day, which includes 50 free miles. Each additional 25 miles, or portion, costs $10. Graph the ordered pairs (miles, cost).

41. **Natural Science** When a diabetic takes long-acting insulin, the insulin reaches its peak effect on the blood sugar level in about three hours. This effect remains fairly constant for five hours, then declines, and is very low until the next injection. In a typical patient, the level of blood sugar might be given by the following function.

$$i(t) = \begin{cases} 40t + 100 & \text{if} & 0 \le t \le 3 \\ 220 & \text{if} & 3 < t \le 8 \\ -80t + 860 & \text{if} & 8 < t \le 10 \\ 60 & \text{if} & 10 < t \le 24 \end{cases}$$

Here $i(t)$ is the blood sugar level, in appropriate units, at time t measured in hours from the time of the injection. Chuck takes his insulin at 6 A.M. Find the blood sugar level at each of the following times.
   (a) 7 A.M.   (b) 9 A.M.   (c) 10 A.M.
   (d) noon   (e) 2 P.M.   (f) 5 P.M.
   (g) midnight   (h) Graph $y = i(t)$.

42. **Natural Science** A factory begins emitting particulate matter into the atmosphere at 8 A.M. each workday, with the emissions continuing until 4 P.M. The level of pollutants, $P(t)$, measured by a monitoring station 1/2 mile away is approximated as follows, where t represents the number of hours since 8 A.M.

$$P(t) = \begin{cases} 75t + 100 & \text{if} & 0 \le t \le 4 \\ 400 & \text{if} & 4 < t < 8 \\ -100t + 1200 & \text{if} & 8 \le t \le 10 \\ -\dfrac{50}{7}t + \dfrac{1900}{7} & \text{if} & 10 < t < 24 \end{cases}$$

Find the level of pollution at
   (a) 9 A.M.   (b) 11 A.M.
   (c) 5 P.M.   (d) 7 P.M.
   (e) midnight.   (f) Graph $y = P(t)$.

43. **Management** The **elasticity of demand** is the percent by which the demand for a product changes as price changes. For example, an elasticity of -1 means that demand changes at the same rate as price, so that a 10% price increase will cause a 10% drop in demand. An elasticity of $-.4$ means that a 10% price increase will cause a $.4 \times 10\% = 4\%$ drop in demand. Recently, there has been much controversy over the elasticity of demand for gasoline. It is now agreed that the short-term elasticity is small (you still have to get to work tomorrow), but the longer term elasticity is much higher (your next car will be much more fuel efficient). One recent projection of gasoline elasticity is as follows, where $e(t)$ represents elasticity at time t measured in years from some base year.

$$e(t) = \begin{cases} -.10 & \text{for} \quad 0 \le t \le 2 \\ -.25 & \text{for} \quad 2 < t \le 5 \\ -.50 & \text{for} \quad 5 < t \le 8 \\ -.75 & \text{for} \quad 8 < t \le 12 \\ -1.10 & \text{for} \quad 12 < t \le 16 \end{cases}$$

For example, a 10% price increase now will cause a $.75 \times 10\% = 7.5\%$ drop in demand in years 9 through 12.

(a) Graph $y = e(t)$.

(b) Give the domain and range for e.

44. Management Normally, an increase in price will produce an increase in the supply of an item. However, there are a few items where an increase in price produces a *decrease* in supply. An example is labor—in developing countries with few consumer goods available, an increase in wage rates can actually lead to workers putting in fewer hours. Such situations produce *backward-bending* supply curves. Some economists now feel that oil may be on a backward bending supply curve—as the price increases, some exporting nations can get sufficient revenue for their needs with the sale of fewer barrels of oil. To see how these curves work, let x be the number of millions of barrels of oil produced in some fixed time period, and let p be the price per barrel. Based on certain published data, the supply curve must be graphed as follows.

(a) For $10 \le x \le 12$, graph $p = (5/2)x$.

(b) For $12 < x \le 14$, graph $p = x + 18$.

(c) For $32 \le p \le 38$, graph $x = 14$.

(d) For $11 \le x \le 14$, graph $p = -2x + 66$.

Chapter 2   Review Exercises

List the ordered pairs obtained from each of the following. Assume that the domain of x for each exercise is $\{-3, -2, -1, 0, 1, 2, 3\}$. Graph each set of ordered pairs. Give the range.

1. $2x - 5y = 10$

2. $3x + 7y = 21$

3. $y = (2x + 1)(x - 1)$

4. $y = (x + 4)(x + 3)$

5. $y = -2 + x^2$

6. $y = 3x^2 - 7$

7. $y = \dfrac{2}{x^2 + 1}$

8. $y = \dfrac{-3 + x}{x + 10}$

9. $y + 1 = 0$

10. $y = 3$

For each function defined as follows, find (a) $f(6)$, (b) $f(-2)$, (c) $f(-4)$, (d) $f(r + 1)$.

11. $f(x) = 4x - 1$

12. $f(x) = 3 - 4x$

13. $f(x) = -x^2 + 2x - 4$

14. $f(x) = 8 - x - x^2$

15. Let $f(x) = 5x - 3$ and $g(x) = -x^2 + 4x$. Find each of the following.

(a) $f(-2)$

(b) $g(3)$

(c) $g(-4)$

(d) $f(5)$

(e) $g(-k)$

(f) $g(3m)$

(g) $g(k - 5)$

(h) $f(3 - p)$

(i) $f[g(-1)]$

(j) $g[f(2)]$

16. Assume that it costs 30¢ to mail a letter weighing one ounce or less, with each additional ounce, or portion of an ounce, costing 27¢. Let $C(x)$ represent the cost to mail a letter weighing x ounces. Find the cost of mailing a letter of the following weights.

(a) 3.4 ounces

(b) 1.02 ounces

(c) 5.9 ounces

(d) 10 ounces

(e) Graph C.

(f) Give the domain and range for C.

Graph each of the following lines. Find the slope in Exercises 17–24, and find an equation of the line in Exercises 25–28.

17. $y = 4x + 3$

18. $y = 6 - 2x$

19. $3x - 5y = 15$

20. $2x + 7y = 14$

21. $x + 2 = 0$

22. $y = 1$

23. $y = 2x$

24. $x + 3y = 0$

25. Through $(2, -4)$, $m = 3/4$

26. Through $(0, 5)$, $m = -2/3$

27. Through $(-4, 1)$, $m = 3$

28. Through $(-3, -2)$, $m = -1$

29. The supply and demand for a certain commodity are related by

supply: $p = 6x + 3$;   demand: $p = 19 - 2x$

where p represents the price at a supply or demand,

respectively, of x units. Find the supply and the demand when the price is

(a) 10;   (b) 15;   (c) 18.

(d) Graph both the supply and the demand functions on the same axes.

(e) Find the equilibrium price.

(f) Find the equilibrium supply; the equilibrium demand.

30. For a particular product, 72 units will be supplied at a price of 6, while 104 units will be supplied at a price of 10. Write a supply function for this product.

Find the slope of the line through each of the following pairs of points.

31. $(-2, 5)$ and $(4, 7)$

32. $(4, -1)$ and $(3, -3)$

33. $(0, 0)$ and $(11, -2)$

34. $(0, 0)$ and $(0, 7)$

Find the slope of each of the following lines.

35. $2x + 3y = 15$ **36.** $4x - y = 7$

37. $x + 4 = 9$ **38.** $3y - 1 = 14$

39. $y = x$ **40.** $y = -2$

41. $x + 5y = 0$ **42.** $y - 3x = 0$

Find an equation for each of the following lines.

43. Through $(5, -1)$, slope 2/3

44. Through $(8, 0)$, slope $-1/4$

45. Through $(5, -2)$ and $(1, 3)$

46. Through $(2, -3)$ and $(-3, 4)$

47. Undefined slope, through $(-1, 4)$

48. Slope 0, through $(-2, 5)$

49. x-intercept -3, y-intercept 5

50. x-intercept $-2/3$, y-intercept 1/2

Find each of the following linear cost functions.

51. Eight units of paper cost $300; fixed cost is $60.

52. Fixed cost is $2000; 36 units cost $8480.

53. Twelve units cost $445; 50 units cost $1585.

54. Thirty units cost $1500; 120 units cost $5640.

55. Eighty units cost $2735; 175 units cost $5775.

56. The graph below, from *Time* magazine, October 28, 1985 (page 70), shows the sales in millions of vehicles for the last few years.*

Find the average rate of change of sales during the following intervals. (Estimate the sales from the graph as carefully as possible.)

(a) Domestic cars: 1978 to 1979

(b) Domestic cars: 1980 to 1985

(c) Domestic trucks: 1982 to 1984

(d) Imported cars: 1978 to 1985

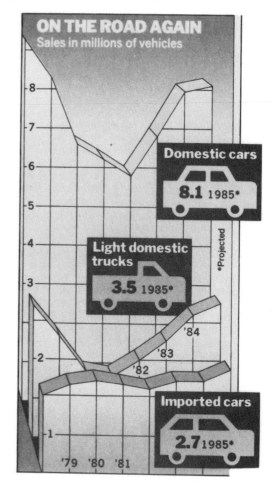

---

*"On the Road Again" from "It's a Jungle Out There." Copyright © 1985 Time Inc. All rights reserved. Reprinted by permission from *Time*.

57. The cost of producing x units of a product is $C(x)$, where

$$C(x) = 20x + 100.$$

The product sells for $40 per unit.
(a) Find the break-even point.
(b) What revenue will the company receive if it sells just the number of units from part (a)?

58. A product can be sold for $25 per unit. The cost of producing x units in a certain plant is $C(x) = 24x + 5000$. Should the item be produced?

Graph each of the following functions.

59. $y = -|x|$

60. $y = |x| - 3$

61. $y = -|x| - 2$

62. $y = -|x + 1| + 3$

63. $y = 2|x - 3| - 4$

64. $y = [x - 3]$

65. $y = \left[\left(\dfrac{1}{2}\right)x - 2\right]$

66. $y = \begin{cases} -4x + 2 & \text{if } x \le 1 \\ 3x - 5 & \text{if } x > 1 \end{cases}$

67. $y = \begin{cases} 3x + 1 & \text{if } x < 2 \\ -x + 4 & \text{if } x \ge 2 \end{cases}$

68. $y = \begin{cases} |x| & \text{if } x < 3 \\ 6 - x & \text{if } x \ge 3 \end{cases}$

69. A trailer hauling service charges $45, plus $2 per mile or part of a mile. Graph the ordered pairs (miles, cost).

70. Let f be a function which gives the cost to rent a floor polisher for x hours. The cost is a flat $3 for cleaning the polisher plus $4 per day or fraction of a day for using the polisher.
(a) Graph f.
(b) Give the domain and range of f.

Case 2

Marginal Cost—Booz, Allen and Hamilton*

Booz, Allen and Hamilton is a large management consulting firm. One of the services it provides to client companies is profitability studies, which show ways in which the client can increase profit levels. The client company requesting the analysis presented in this case is a large producer of a staple food. The company buys from farmers, and then processes the food in its mills, resulting in a finished product. The company sells both at retail under its own brands, and in bulk to other companies who use the product in the manufacture of convenience foods.

The client company has been reasonably profitable in recent years, but the management retained Booz, Allen and Hamilton to see whether its consultants could suggest ways of increasing company profits. The management of the company had long operated with the philosophy of trying to process and sell as much of its product as possible, since they felt this would lower the average processing cost per unit sold. However, the consultants found that the client's fixed mill costs were quite low, and that, in fact, processing extra units made the cost per unit start to increase. (There are several reasons for this: the company must run three shifts, machines break down more often, and so on.)

In this case, we shall discuss the marginal cost of two of the company's products. The marginal cost (cost of producing an extra unit) of production for product A was found by the consultants to be approximated by the linear function

$$y = .133x + 10.09,$$

where x is the number of units produced (in millions) and y is the marginal cost.

For example, at a level of production of 3.1 million units, an additional unit of product A would cost about

$$y = .133(3.1) + 10.09$$
$$\approx \$10.50.\dagger$$

At a level of production of 5.7 million units, an extra unit costs \$10.85. Figure 1 to the right shows a graph of the marginal cost function from $x = 3.1$ to $x = 5.7$, the domain to which the function above was found to apply.

The selling price for product A is \$10.73 per unit, so that, as shown on the graph of Figure 1, the company was losing money on many units of the product that it sold. Since the selling price could not be raised if the company was to remain competitive, the consultants recommended that production of product A be cut.

For product B, the Booz, Allen and Hamilton consultants found a marginal cost function given by

$$y = .0667x + 10.29,$$

with x and y as defined above. Verify that at a production level of 3.1 million units, the marginal cost is \$10.50, while at a production level of 5.7 million units, the marginal cost is \$10.67. Since the selling price of this product is \$9.65, the consultants again recommended a cutback in production.

The consultants ran similar cost analyses of other products made by the company, and then issued their recommendations: the company should reduce total production by 2.1 million units. The analysts predicted that this would raise profits for the products under discussion from \$8.3 million annually to \$9.6 million, which is very close to what actually happened when the client took this advice.

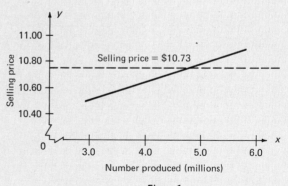

Figure 1

*Case study, "Marginal Cost—Booz, Allen and Hamilton" supplied by John R. Dowdle of Booz, Allen & Hamilton, Inc. Reprinted by permission.

†The symbol "≈" means *is approximately equal to*.

Exercises

1. At what level of production, x, was the marginal cost of a unit of product A equal to the selling price?

2. Graph the marginal cost function for product B from $x = 3.1$ million units to $x = 5.7$ million units.

3. Find the number of units for which marginal cost equals the selling price for product B.

4. For product C, the marginal cost of production is

$$y = .133x + 9.46$$

   (a) Find the marginal cost at a level of production of 3.1 million units; of 5.7 million units.
   (b) Graph the marginal cost function.
   (c) For a selling price of $9.57, find the level of production for which the cost equals the selling price.

3 Polynomial and Rational Models

While the linear functions of Chapter 2 provide the mathematical models for many different types of problems, not all real-world situations can be adequately approximated by these functions. In many cases, a curve is needed, and not a straight line. One important set of curves comes from the **polynomial functions,** those defined by polynomial expressions. The first section of this chapter discusses polynomial functions defined by quadratic expressions; more general polynomial functions are discussed in the second section. Rational functions, defined by quotients of polynomials, are discussed in Section 3.3.

3.1 Quadratic Models

A polynomial function defined by a quadratic expression is a *quadratic function*.

> A function f is a **quadratic function** if
> $$f(x) = ax^2 + bx + c,$$
> where a, b, and c are real numbers with $a \neq 0$.

The restriction $a \neq 0$ is made to guarantee that the function is not linear.

Example 1 Graph the quadratic function defined by
$$y = x^2.$$

The function defined by $y = x^2$ is perhaps the simplest of all quadratic functions. (To see that $y = x^2$ is quadratic, let $a = 1$, $b = 0$, and $c = 0$ in the general quadratic function defined by $y = ax^2 + bx + c$.) Graph $y = x^2$ by choosing some values for x, and then find the corresponding values for y, as in the table with Figure 3.1. The resulting points can then be plotted, with a smooth curve drawn through them, as in Figure 3.1. The domain of the function defined by $y = x^2$ is the set of all real numbers, while the range is the set of all nonnegative real numbers. ■

The curve in Figure 3.1 is called a **parabola.** The lowest point on this parabola, $(0, 0)$, is called the **vertex.** It can be proved that every quadratic function has a graph which is a parabola. If the graph of Figure 3.1 were folded in half along the y-axis, the two halves of the parabola would match exactly. This means that the graph of a quadratic function is *symmetric* about a vertical line through the vertex: this line is the **axis** of the parabola.

1 Graph each of the following parabolas.
(a) $y = x^2 - 4$
(b) $y = x^2 + 5$

Answer:
(a)

(b)

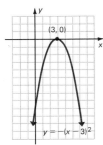

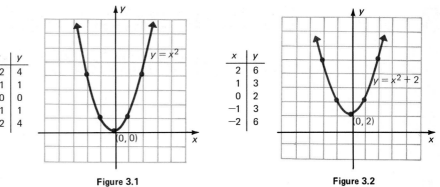

Figure 3.1

Figure 3.2

Parabolas have many useful properties. Cross sections of radar dishes and spotlights form parabolas. Discs often visible on the sidelines of televised football games are microphones having reflectors with parabolic cross sections. These microphones are used by the television networks to pick up the shouted signals of the quarterbacks.

Example 2 Graph the quadratic function defined by $y = x^2 + 2$.

For any value of x that might be chosen, the corresponding value of y will be 2 more than for the parabola graphed in Figure 3.1. This means that the graph of the parabola $y = x^2 + 2$, as shown in Figure 3.2, is shifted 2 units upward compared to the graph of $y = x^2$. For example, (2, 4) and (−1, 1) are on the graph of $y = x^2$, while the corresponding points on the graph of $y = x^2 + 2$ are (2, 6) and (−1, 3). Also, the point (0, 2) is the vertex of the parabola $y = x^2 + 2$, while (0, 0) is the vertex of the parabola $y = x^2$. **1**

Example 3 Graph the function defined by $y = -(x + 2)^2$.

The table in Figure 3.3 shows several ordered pairs that belong to the function. These ordered pairs show that the graph of $y = -(x + 2)^2$ is "upside down" in comparison to $y = x^2$ (because of the negative sign) and also shifted 2 units to the left. Here the vertex, (−2, 0), is the *highest* point on the graph. **2**

2 Graph $y = -(x - 3)^2$.

Answer:

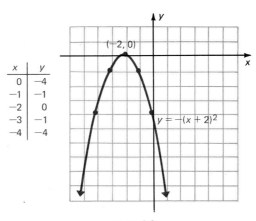

x	y
0	−4
−1	−1
−2	0
−3	−1
−4	−4

Figure 3.3

As shown in Example 2, the graph of $y = x^2 + 2$ was shifted upward 2 units in comparison with the graph of $y = x^2$. In Example 3, the graph of $y = -(x + 2)^2$ is shifted 2 units to the left when compared to $y = x^2$, and opens downward. Both upward (or downward) and side-to-side shifts can occur in the same parabola, as in the next example.

Example 4 Graph $y = 2(x - 3)^2 - 5$.

Using ideas similar to those of the examples above, this parabola has vertex $(3, -5)$ and opens upward. The coefficient 2 causes the values of y to increase more rapidly than in the parabola $y = x^2$, so that the graph in this example is "narrower" than the graph of $y = x^2$. Figure 3.4 shows a table of ordered pairs satisfying $y = 2(x - 3)^2 - 5$, as well as the graph of the function. **3**

3 Graph the following.

(a) $y = (x + 4)^2 - 3$
(b) $y = -2(x - 3)^2 + 1$

Answer:

(a)

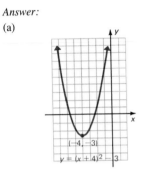

(b)

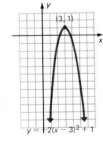

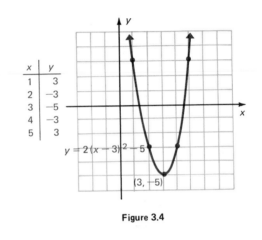

x	y
1	3
2	-3
3	-5
4	-3
5	3

$y = 2(x - 3)^2 - 5$

$(3, -5)$

Figure 3.4

In summary,

> if f is a quadratic function defined by $y = a(x - h)^2 + k$, then the graph of the function f is a parabola having its vertex at (h, k). If $a > 0$, the parabola opens upward; if $a < 0$, it opens downward. If $0 < |a| < 1$, the parabola is "broader" than $y = x^2$, while if $|a| > 1$, the parabola is "narrower" than $y = x^2$.

The vertex and axis of a parabola can be found quickly if the equation of the parabola is in the form

$$y = a(x - h)^2 + k$$

for real numbers $a \neq 0$, h and k. An equation not given in this form can be converted to it by a process called **completing the square,** first discussed in Section 1.9. This process is reviewed in the following example.

Example 5 Graph $y = x^2 - 2x + 3$ by completing the square.

As mentioned, this parabola can be graphed by first completing the square to rewrite the equation in the form $y = a(x - h)^2 + k$. Begin by writing,

$$y = x^2 - 2x + 3$$

as

$$y = (x^2 - 2x \quad) + 3.$$

The expression inside the parentheses must be written as the square of some quantity. Do this by taking half the coefficient of x, $(1/2)(-2) = -1$, and then squaring this result: $(-1)^2 = 1$.* Now add and subtract 1 inside the parentheses. This gives

$$y = (x^2 - 2x + 1 - 1) + 3$$

or

$$y = (x^2 - 2x + 1) + (-1 + 3).$$

Factor $x^2 - 2x + 1$ as $(x - 1)^2$, to get

$$y = (x - 1)^2 + 2.$$

By this result, the graph of the given function is a parabola with the vertex at $(1, 2)$. Since $a = 1$ is positive, the parabola opens upward. Plotting a few additional points gives the graph shown in Figure 3.5. **4**

4 Complete the square for each of the following. Then graph the parabola.
(a) $y = x^2 - 6x + 11$
(b) $y = x^2 + 8x + 18$

Answer:
(a) $y = (x - 3)^2 + 2$

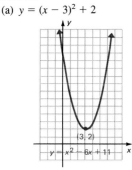

(b) $y = (x + 4)^2 + 2$

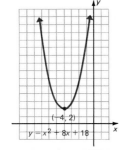

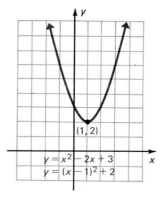

Figure 3.5

Example 6 Graph $y = -2x^2 + 12x - 19$.

Get y in the form $a(x - h)^2 + k$ by first factoring -2 from $-2x^2 + 12x$.

$$y = -2x^2 + 12x - 19 = -2(x^2 - 6x) - 19$$

---

*Take *half* the coefficient of x since $(x + b)^2 = x^2 + 2xb + b^2$. The constant b can be found by setting $2b$ equal to the coefficient of x and solving. Find b^2 by squaring b.

Take half of -6: $(1/2)(-6) = -3$. Square this result: $(-3)^2 = 9$. Add and subtract 9 inside the parentheses.

$$y = -2(x^2 - 6x + 9 - 9) - 19$$
$$y = -2(x^2 - 6x + 9) + (-2)(-9) - 19$$

Simplify and factor to get

$$y = -2(x - 3)^2 - 1.$$

Since $a = -2$ is negative, the parabola opens downward. The vertex is at $(3, -1)$, and the parabola is narrower than $y = x^2$. Use these results and plot additional ordered pairs as needed to get the graph of Figure 3.6. ▪

 Complete the square and graph the following.
(a) $y = 3x^2 - 12x + 14$
(b) $y = -x^2 + 6x - 12$

Answer:
(a) $y = 3(x - 2)^2 + 2$

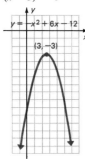

(b) $y = -(x - 3)^2 - 3$

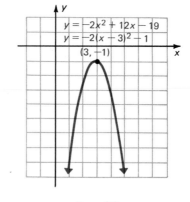

Figure 3.6

The fact that the vertex of a parabola of the form $y = ax^2 + bx + c$ is the highest or lowest point on the graph can be used in applications to find a maximum or a minimum value.

Example 7 Toni Tyson owns and operates Aunt Elmyra's Blueberry Pies. She has hired a consultant to analyze her business operations. The consultant tells Tyson that her profits, $P(x)$, from the sale of x units of pies, are given by

$$P(x) = 120x - x^2.$$

How many units of pies should she make in order to maximize profit? What is the maximum profit?

The expression giving profit can be rewritten as $P(x) = -x^2 + 120x + 0$. Complete the square to rewrite $P(x)$ as follows.

$$P(x) = -x^2 + 120x$$
$$= -(x^2 - 120x)$$
$$= -(x^2 - 120x + 3600 - 3600)$$
$$= -(x^2 - 120x + 3600) + 3600$$
$$P(x) = -(x - 60)^2 + 3600$$

As shown in Figure 3.7, the graph of P is a parabola with vertex (60, 3600) opening downward. Since the parabola opens downward, the vertex leads to *maximum* profit. The maximum profit of $3600 is obtained when 60 units of pies are made. Here the profit increases as more and more pies are made, up to the point $x = 60$, and then decreases as more and more pies are made past this point. **6**

6 When a company sells x units of a product, its profit is $P(x) = -2x^2 + 40x + 280$. Find
(a) the number of units that should be sold so that maximum profit is received;
(b) the maximum profit.

Answer:
(a) 10 units
(b) $480

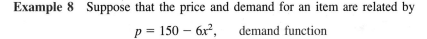

P(x)

3600

(60, 3600)

$P(x) = 120x - x^2$

30   60   90   120   x

Figure 3.7

Example 8 Suppose that the price and demand for an item are related by

$$p = 150 - 6x^2, \quad \text{demand function}$$

where p is the price and x is the number of items demanded (in hundreds). The price and supply are related by

$$p = 10x^2 + 2x, \quad \text{supply function}$$

where x is the supply of the item (in hundreds). Find the equilibrium demand (and supply) and the equilibrium price.

The graphs of both of these are parabolas; these parabolas can be graphed by completing the square, giving the results shown in Figure 3.8 on the next page. Only those portions of the graphs that lie in the first quadrant are included, since neither supply, demand, nor price can be negative.

The equilibrium demand and supply, by definition, occur at the same price p. Setting $150 - 6x^2$, where x is the demand, equal to $10x^2 + 2x$, where x is the supply, gives an equation that leads to the equilibrium supply and demand.

$$150 - 6x^2 = 10x^2 + 2x.$$

Solve this quadratic equation as follows.

$$0 = 16x^2 + 2x - 150 \qquad \text{Add } -150 \text{ and } 6x^2 \text{ to both sides}$$

$$0 = 8x^2 + x - 75 \qquad \text{Multiply both sides by } \frac{1}{2}$$

This equation can be solved by the quadratic formula of Chapter 1. Here $a = 8$, $b = 1$, and $c = -75$.

$$x = \frac{-1 \pm \sqrt{1 - 4(8)(-75)}}{2(8)}$$

$$= \frac{-1 \pm \sqrt{1 + 2400}}{16} \qquad -4(8)(-75) = 2400$$

$$= \frac{-1 \pm 49}{16} \qquad \sqrt{1 + 2400} = \sqrt{2401} = 49$$

$$x = \frac{-1 + 49}{16} = \frac{48}{16} = 3 \quad \text{or} \quad x = \frac{-1 - 49}{16} = -\frac{50}{16} = -\frac{25}{8}$$

It is not possible to make $-25/8$ units, so discard that answer, and use only $x = 3$. Since x represents supply and demand (in hundreds), equilibrium demand (and supply) is $3 \cdot 100$, or 300 units. Find the equilibrium price by substituting 3 for x in either the supply or the demand function. Using the supply function gives

$$p = 10x^2 + 2x$$
$$p = 10 \cdot 3^2 + 2 \cdot 3 \qquad \text{Let } x = 3$$
$$= 10 \cdot 9 + 6$$
$$p = 96. \quad \blacksquare \; 7$$

7 The price and demand for an item are related by $p = 32 - x^2$, while price and supply are related by $p = x^2$. Find
(a) the equilibrium supply;
(b) the equilibrium price.

Answer:
(a) 4
(b) 16

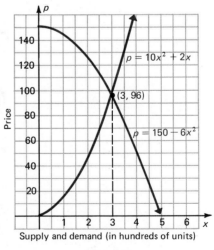

Figure 3.8

3.1 Exercises

1. Graph the functions defined in parts (a)–(d) on the same coordinate system.

(a) $f(x) = 2x^2$ (b) $f(x) = 3x^2$

(c) $f(x) = \dfrac{1}{2}x^2$ (d) $f(x) = \dfrac{1}{3}x^2$

(e) How does the coefficient affect the shape of the graph?

2. Graph the functions defined in parts (a)–(d) on the same coordinate system.

(a) $y = \dfrac{1}{2}x^2$ (b) $y = -\dfrac{1}{2}x^2$

(c) $y = 4x^2$ (d) $y = -4x^2$

(e) What effect does the minus sign have on the graph?

3. Graph the functions defined in parts (a)–(d) on the same coordinate system.

(a) $f(x) = x^2 + 2$ (b) $f(x) = x^2 - 1$

(c) $f(x) = x^2 + 1$ (d) $f(x) = x^2 - 2$

(e) How do these graphs differ from the graph of $f(x) = x^2$?

4. Graph the functions defined in parts (a)–(d) on the same coordinate system.

(a) $f(x) = (x - 2)^2$ (b) $f(x) = (x + 1)^2$

(c) $f(x) = (x + 3)^2$ (d) $f(x) = (x - 4)^2$

(e) How do these graphs differ from the graph of $f(x) = x^2$?

Graph the following parabolas. Find the vertex of each. (See Examples 1–6.)

5. $y = -x^2$ **6.** $y = -\dfrac{1}{2}x^2$

7. $y = -x^2 + 1$ **8.** $y = x^2 - 3$

9. $y = 3x^2 - 2$ **10.** $y = -2x^2 + 4$

11. $y = (x + 2)^2$ **12.** $y = (x - 3)^2$

13. $y = -(x - 4)^2$ **14.** $y = -(x + 5)^2$

15. $y = -2(x - 3)^2$ **16.** $y = -3(x + 4)^2$

17. $y = (x - 1)^2 - 3$ **18.** $y = (x - 2)^2 + 1$

19. $y = -(x + 4)^2 + 2$ **20.** $y = -(x + 1)^2 - 3$

21. $y = x^2 - 4x + 6$ **22.** $y = x^2 + 6x + 3$

23. $y = x^2 + 12x + 1$ **24.** $y = x^2 - 10x + 3$

25. $y = 2x^2 + 4x + 1$ **26.** $y = 3x^2 + 6x + 5$

27. $y = 2x^2 - 4x + 5$ **28.** $y = -3x^2 + 24x - 46$

29. $y = -x^2 + 6x - 6$ **30.** $y = -x^2 + 2x + 5$

Find several points satisfying each of the following and then sketch the graph.

31. $y = .14x^2 + .56x - .3$

32. $y = .82x^2 + 3.28x - .4$

33. $y = -.09x^2 - 1.8x + .5$

34. $y = -.35x^2 + 2.8x - .3$

35. Management George Lobell runs a sandwich shop. By studying data for his past costs, he has found that a mathematical model describing the cost of operating his shop is given by

$$C(x) = x^2 - 10x + 40,$$

where $C(x)$ is the daily cost to make x units of sandwiches.

(a) Complete the square for $C(x)$.

(b) Graph the parabola resulting from part (a).

(c) From the vertex of the parabola, find the number of units of sandwiches George must sell to produce minimum cost.

(d) What is the minimum cost?

36. Management Josette Skelnik runs a taco stand. She has found that her profits are approximated by

$$P(x) = -x^2 + 60x - 400,$$

where $P(x)$ is the profit from selling x units of tacos.

(a) Complete the square on $P(x)$.

(b) Graph the parabola.

(c) Find the number of units of tacos that Josette should make to produce maximum profit.

(d) What is the maximum profit?

37. Management French fries produce a tremendous profit (150% to 300%) for many fast food restaurants. Management, therefore, desires to maximize the number of bags sold. Suppose that a mathematical model connecting p, the profit per day from french fries (in hundreds of dollars), and x, the price per bag (in dimes), is $p = -x^2 + 6x - 1$.

(a) Find the price per bag that leads to maximum profit.

(b) What is the maximum profit?

38. Natural Science A researcher in physiology has decided that a good mathematical model for the number of impulses fired after a nerve has been stimulated is given by $y = -x^2 + 20x - 60$,

where y is the number of responses per millisecond, and x is the number of milliseconds since the nerve was stimulated.
(a) When will the maximum firing rate be reached?
(b) What is the maximum firing rate?

39. Management Suppose the price p, is related to the demand x, where x is measured in hundreds of items, by

$$p = 640 - 5x^2.$$

Find p when the demand is
(a) 0; (b) 5; (c) 10.
(d) Graph $p = 640 - 5x^2$.

Suppose that the price and supply of the item are related by $p = 5x^2$.
(e) Graph $p = 5x^2$ on the axes used in part (d).
(f) Find the equilibrium supply.
(g) Find the equilibrium price.

Management *Suppose that the supply and demand of a certain textbook are related to price by*

$$\text{supply: } p = \frac{1}{5}x^2; \text{ demand: } p = -\frac{1}{5}x^2 + 40.$$

40. Use a calculator to estimate the demand at a price of
(a) 10; (b) 20; (c) 30; (d) 40.
(e) Graph $p = -(1/5)x^2 + 40$.
Approximate the supply at a price of
(f) 5; (g) 10; (h) 20; (i) 30.
(j) Graph $p = (1/5)x^2$ on the axes used in part (e).

41. Find the equilibrium demand in Exercise 39.

42. Find the equilibrium price in Exercise 40.

43. Management The revenue of a charter bus company depends on the number of unsold seats. If the revenue, $R(x)$, is given by

$$R(x) = 5000 + 50x - x^2,$$

where x is the number of unsold seats, find the number of unsold seats which will produce maximum revenue. Find the maximum revenue.

44. Natural Science If an object is thrown upward with an initial velocity of 32 feet per second, then its height, in feet, above the ground after t seconds is given by

$$h = 32t - 16t^2.$$

Find the maximum height attained by the object. Find the number of seconds it takes for the object to hit the ground.

45. Management A charter flight charges a fare of $200 per person, plus $4 per person for each unsold seat on the plane. If the plane holds 100 passengers, and if x represents the number of unsold seats, find the following.
(a) An expression for the total revenue received for the flight. (Hint: Multiply the number of people flying, $100 - x$, by the price per ticket.)
(b) The graph for the expression of part (a).
(c) The number of unsold seats that will produce the maximum revenue.
(d) The maximum revenue.

46. Management The demand for a certain type of cosmetic is given by

$$p = 500 - x,$$

where p is the price when x units are demanded.
(a) Find the revenue, $R(x)$, that would be obtained at a demand of x. (Hint: revenue = demand × price.)
(b) Graph the revenue function, $R(x)$.
(c) From the graph of the revenue function, estimate the price that will produce the maximum revenue.
(d) What is the maximum revenue?

47. Natural Science Between the months of June and October, the percent of maximum possible chlorophyll production in a leaf is approximated by $C(x)$, where

$$C(x) = 10x + 50.$$

Here x is time in months, with $x = 1$ representing June. From October through December, $C(x)$ is approximated by

$$C(x) = -20(x - 5)^2 + 100,$$

where x is as above. Find the percent of maximum possible chlorophyll production in each of the following months.
(a) June (b) July (c) September
(d) October (e) November (f) December

48. Natural Science Use your results in Exercise 47 to sketch a graph of $y = C(x)$ from June through December. In which month is chlorophyll production a maximum?

49. A culvert is shaped like a parabola, 18 cm across the top and 12 cm deep. How wide is the culvert 8 cm from the top?

50. Let x be in the interval $0 \le x \le 1$. Use a graph to suggest that the product $x(1 - x)$ is always less than or equal to 1/4. For what values of x does the product equal 1/4?

Graph $y = x^3 - 5$.

Answer:

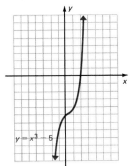

2 Graph the following.
(a) $y = x^5 - 2$
(b) $y = -x^3 + 3$

Answer:
(a)

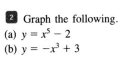

(b)

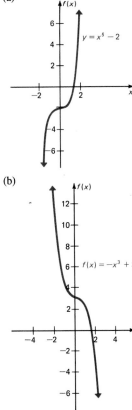

3.2 Polynomial Models

Earlier we discussed linear and quadratic functions and found their graphs. Both these functions are special types of *polynomial functions,* functions defined by a polynomial expression.

A function f is a **polynomial function of degree n,** where n is a nonnegative integer, if

$$f(x) = a_n x^n + a_{n-1} x^{n-1} + \cdots + a_1 x + a_0,$$

where $a_n, a_{n-1}, \ldots, a_1$ and a_0 are real numbers with $a_n \neq 0$.

For $n = 1$, a polynomial function takes the form

$$f(x) = a_1 x + a_0,$$

a linear function. A linear function, therefore, is a polynomial function of degree 1. (An exception: a linear function of the form $f(x) = a_0$ for a real number a_0 is a polynomial function of degree 0.) A polynomial function of degree 2 is a quadratic function that can be written in the form

$$f(x) = a_2 x^2 + a_1 x + a_0.$$

Perhaps the simplest polynomial functions are those defined by an expression of the form $y = x^n$, so some of these functions are graphed in the next examples.

Example 1 Graph $y = x^3$.

First, find several ordered pairs belonging to the graph, as shown in the table beside Figure 3.9 on the next page. Then plot the ordered pairs and draw a smooth curve through them, getting the graph in Figure 3.9. **1**

Example 2 Graph each of the following.
(a) $f(x) = (1/2)x^3$

The graph will be "broader" than that of $f(x) = x^3$, but will have the same general shape. The graph goes through the points $(-2, -4)$, $(-1, -1/2)$, $(0, 0)$, $(1, 1/2)$, and $(2, 4)$. See Figure 3.10.
(b) $f(x) = (3/2)x^4$

The following table gives some ordered pairs.

x	-2	-1	0	1	2
y	24	3/2	0	3/2	24

The graph is shown in Figure 3.11. **2**

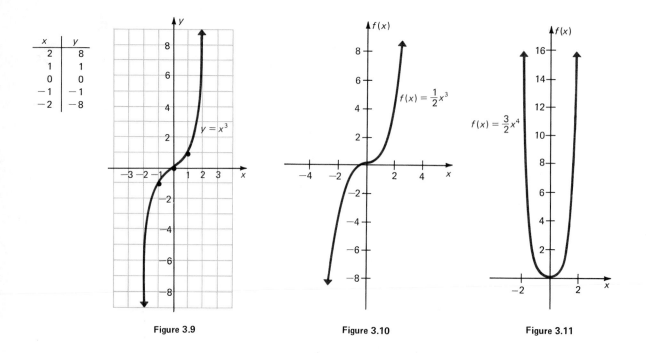

x	y
2	8
1	1
0	0
−1	−1
−2	−8

$y = x^3$

$f(x) = \frac{1}{2}x^3$

$f(x) = \frac{3}{2}x^4$

Figure 3.9   **Figure 3.10**   **Figure 3.11**

The graphs above suggest that the domain of a polynomial function is the set of all real numbers. The range of a polynomial function of odd degree (1, 3, 5, 7, and so on) is also the set of all real numbers. Graphs typical of polynomial functions of odd degree are shown in Figure 3.12.

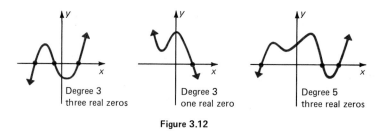

Degree 3
three real zeros

Degree 3
one real zero

Degree 5
three real zeros

Figure 3.12

A polynomial function of even degree has a range that takes the form $y \le k$ or $y \ge k$, for some real number k. Figure 3.13 shows two graphs of typical polynomial functions of even degree.

It is difficult to graph most polynomial functions without the use of calculus. A large number of points must be plotted to get a reasonably accurate graph. How-

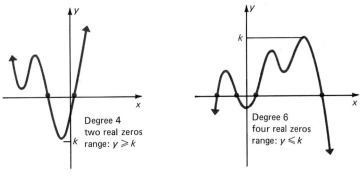

Figure 3.13

ever, if the expression defining a polynomial function can be factored, its graph can be approximated without plotting too many points, as shown in the next examples.

Also, these examples show the **x-intercepts** used in drawing the graph. Recall: an x-intercept is a point where the graph crosses the x-axis. The value of y is 0 when the graph crosses the x-axis, so find any x-intercepts by letting $f(x) = 0$. ③

Example 3   Graph $f(x) = (2x + 3)(x - 1)(x + 2)$.

Multiplying out the expression on the right would show that f is a third degree polynomial, called a *cubic* polynomial. Sketch the graph of f by first finding any x-intercepts; do this by setting $f(x) = 0$.

$$f(x) = 0$$
$$(2x + 3)(x - 1)(x + 2) = 0$$

Solve this equation by placing each of the three factors equal to 0.

$$2x + 3 = 0 \quad \text{or} \quad x - 1 = 0 \quad \text{or} \quad x + 2 = 0$$
$$x = -\frac{3}{2} \qquad\qquad x = 1 \qquad\qquad x = -2$$

The three numbers, $-3/2$, 1, and -2, divide the x-axis into four regions:

$$x < -2, \quad -2 < x < -\frac{3}{2}, \quad -\frac{3}{2} < x < 1, \quad \text{and} \quad 1 < x.$$

These regions are shown in Figure 3.14.

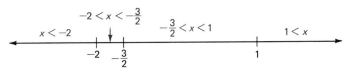

Figure 3.14

③ Identify the x-intercepts of each graph.

(a)

(b)

Answer:
(a) -2, 5/4, 2
(b) -3

In any of these regions, the values of $f(x)$ are either always positive or always negative. Find the sign of $f(x)$ in each region by selecting an x-value in each region and substituting to determine if the values of the function are positive or negative in that region. A typical selection of test points and the results of the tests are shown below.

Region	Test point	Value of f(x)	Sign of f(x)
$x < -2$	-3	-12	negative
$-2 < x < -\dfrac{3}{2}$	$-\dfrac{7}{4}$	$\dfrac{11}{32}$	positive
$-\dfrac{3}{2} < x < 1$	0	-6	negative
$1 < x$	2	28	positive

When the values of $f(x)$ are negative, the graph is below the x-axis, and when the values of $f(x)$ are positive, the graph is above the x-axis, so that the graph looks something like the sketch in Figure 3.15. If necessary, the sketch could be improved by plotting additional points in each region.   **4**

4 Graph
$f(x) = (3x - 4)(x + 2)(x - 3)$.

Answer:

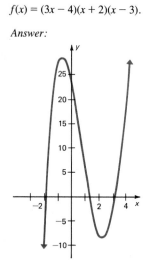

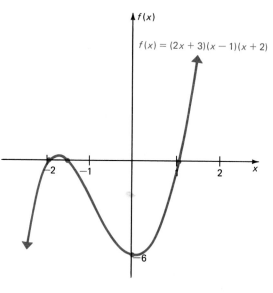

$$f(x) = (2x + 3)(x - 1)(x + 2)$$

Figure 3.15

Example 4   Sketch the graph of $f(x) = 3x^4 + x^3 - 2x^2$.
The polynomial can be factored as follows.

$$3x^4 + x^3 - 2x^2 = x^2(3x^2 + x - 2)$$
$$= x^2(3x - 2)(x + 1)$$

Find the x-intercepts by letting $f(x) = 0$. The polynomial is 0 at $x = 0$, $x = 2/3$, and $x = -1$. These three values divide the x-axis into four regions:

$$x < -1, \qquad -1 < x < 0, \qquad 0 < x < \frac{2}{3}, \qquad \text{and} \qquad \frac{2}{3} < x.$$

Determine the sign of $f(x)$ in each region by substituting an x-value from each region into the function to get the following information.

Region	Test point	Value of $f(x)$	Sign of $f(x)$	Location relative to axis
$x < -1$	-2	32	positive	above
$-1 < x < 0$	$-\dfrac{1}{2}$	$-\dfrac{7}{16}$	negative	below
$0 < x < \dfrac{2}{3}$	$\dfrac{1}{2}$	$-\dfrac{3}{16}$	negative	below
$\dfrac{2}{3} < x$	1	2	positive	above

 5 Graph
$f(x) = 3x^3 + 5x^2 - 2x$.

Answer:

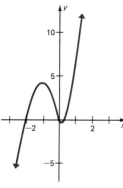

6 Graph
$y = x^3 - 2x^2 - 5x + 6$.

Answer:

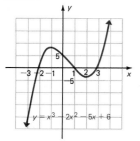

With the values of x used for the test points and the corresponding values of $f(x)$, sketch the graph as shown in Figure 3.16. **5**

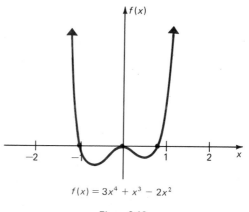

$$f(x) = 3x^4 + x^3 - 2x^2$$

Figure 3.16

If the expression defining a polynomial function cannot be factored readily, the only way that we currently have of sketching the graph of the function is to plot enough points so that the general shape of the graph can be approximated. A calculator is often very helpful in making the necessary calculations.

Example 5 Graph $y = x^3 - 2x^2 - x + 2$.

Start by completing several different ordered pairs, as in the table of Figure 3.17. Plot the points and draw a smooth curve through them to get the result shown in Figure 3.17 on the next page. **6**

x	y
3	8
2	0
1	0
0	2
−1	0
−2	−12

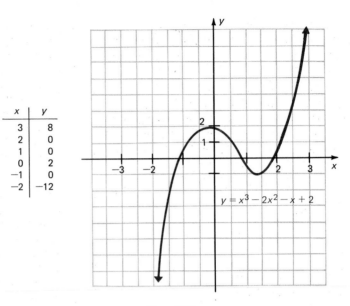

Figure 3.17

Example 6 Graph $y = 2x^4 + 3x^3 - 4x^2 - 3x + 2$.

Again complete several ordered pairs and draw a smooth curve through them, getting the graph shown in Figure 3.18. **7**

7 Graph
$y = x^4 - 4x^3 - x^2 + 16x - 12$.

Answer:

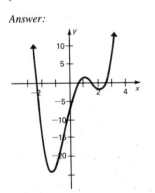

x	y
2	36
1	0
1/2	0
0	2
−1	0
−2	0
−3	56

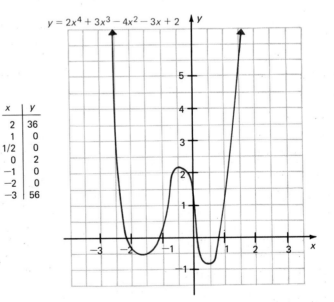

Figure 3.18

3.2 Exercises

Graph each of the following polynomial functions. (See Examples 1–6.)

1. $y = x^3 + 2$ **2.** $y = x^3 - 1$

3. $y = x^4$ **4.** $y = x^4 - 2$

5. $y = 2x(x - 3)(x + 2)$

6. $y = -x(2x + 1)(x - 3)$

7. $y = (x + 2)(x - 3)(x + 4)$

8. $y = (x - 3)(x - 1)(x + 1)$

9. $y = x^2(x - 2)(x + 3)$

10. $y = x^2(x + 1)(x - 1)$

11. $y = (x + 2)(x - 2)(x + 1)(x - 1)$

12. $y = (x - 3)(x + 3)(x - 2)(x + 2)$

13. $y = x^3 - 7x - 6$

14. $y = x^3 + x^2 - 4x - 4$

15. $y = x^3 - 2x^2 - 5x + 6$

16. $y = x^3 + 3x^2 - 4x - 12$

17. $y = x^4 - 5x^2 + 6$

18. $y = x^4 - 2x^2 - 8$

19. $y = x^4 + 2x^3 - 7x^2 - 8x + 12$

20. $y = x^4 - x^3 - 7x^2 + x + 6$

21. $y = 8x^4 - 2x^3 - 47x^2 - 52x - 15$

22. $y = 6x^4 - x^3 - 23x^2 - 4x + 12$

23. $y = 2x^4 - 3x^3 + 4x^2 + 5x - 1$

24. $y = -x^4 - 2x^3 + 3x^2 + 3x + 5$

25. **Natural Science** The polynomial function defined by

$$A(x) = -.015x^3 + 1.058x$$

gives the approximate alcohol concentration (in tenths of a percent) in an average person's bloodstream x hours after drinking about eight ounces of 100 proof whiskey. The function is approximately valid for $0 \le x < 9$. Find the following values.

   (a) $A(1)$ **(b)** $A(2)$ **(c)** $A(4)$

   (d) $A(6)$ **(e)** $A(8)$ **(f)** Graph $A(x)$.

   (g) Using the graph you drew for part (f), estimate the time of maximum alcohol concentration.

(h) In one state, a person is legally drunk if the blood alcohol concentration exceeds .15%. Use the graph of part (f) to estimate the period in which this average person is legally drunk.

26. **Natural Science** A technique for measuring cardiac output depends on the concentration of a dye after a known amount is injected into a vein near the heart. In a normal heart, the concentration of the dye at time x (in seconds) is given by the function defined by

$$g(x) = -.006x^4 + .140x^3 - .053x^2 + 1.79x.$$

Graph $g(x)$.

27. **Natural Science** The pressure of the oil in a reservoir tends to drop with time. By taking sample pressure readings for a particular oil reservoir, petroleum engineers have found that the change in pressure is given by

$$P(t) = t^3 - 25t^2 + 200t,$$

where t is time in years from the date of the first reading.

   (a) Graph $P(t)$.

   (b) For what time period is the change in pressure (drop) increasing? decreasing?

28. **Natural Science** During the early part of the 20th century, the deer population of the Kaibab Plateau in Arizona experienced a rapid increase, because hunters had reduced the number of natural predators. The increase in population depleted the food resources and eventually caused the population to decline. For the period from 1905 to 1930, the deer population was approximated by

$$D(x) = -.125x^5 + 3.125x^4 + 4000,$$

where x is time in years from 1905.

   (a) Use a calculator to find enough points to graph $D(x)$.

   (b) From the graph, over what period of time (from 1905 to 1930) was the population increasing? relatively stable? decreasing?

In many applications, it is necessary to find the maximum or minimum values of a function. (Finding such values is one of the main uses of calculus.) We can find approximate maximum or minimum values of polynomial functions for given intervals by first evaluating the function at the left endpoint of the given interval. Then add .1 to the value of x and reevaluate the polynomial. Keep doing this until the right endpoint of the interval is reached. Then identify the approximate maximum and minimum value for the polynomial on the interval.

29. $y = x^3 + 4x^2 - 8x - 8$, $.3 \le x \le 1$

30. $y = 2x^3 - 5x^2 - x + 1$, $-1 \le x \le 0$

31. $y = x^4 - 7x^3 + 13x^2 + 6x - 28$, $-2 \le x \le -1$

32. $y = x^4 - 7x^3 + 13x^2 + 6x - 28$, $2 \le x \le 3$

Get a table of ordered pairs over the given interval at intervals of .5 for the functions defined as follows. Then graph the function.

33. $f(x) = x^3 + 3x^2 - 2x + 1$,
  for $-3 \le x \le 1.5$

34. $f(x) = -3x^4 - 2x^3 + x^2 + x$,
  for $-3 \le x \le 1.5$

35. $f(x) = -x^4 + 2x^3 - 3x^2 + 4x - 7$,
  for $-1 \le x \le 2$

36. $f(x) = 2x^3 - 3x^2 + 4x - 7$,
  for $-1 \le x \le 2$

3.3   Rational Functions

A *rational function* is defined by an expression that is a quotient of two polynomials, with denominator not zero.

A function f is a **rational function** if

$$f(x) = \frac{P(x)}{Q(x)}, \quad Q(x) \ne 0,$$

where $P(x)$ and $Q(x)$ are polynomials.

Since any values of x which make $Q(x) = 0$ are excluded from the domain of $f(x)$, a rational function usually has a graph with one or more breaks in it. The next few examples show how to graph rational functions.

Example 1   Graph the rational function defined by $y = \dfrac{2}{1 + x}$.

This function is undefined for $x = -1$, since -1 leads to a 0 denominator. For this reason, the graph of this function will not intersect the vertical line $x = -1$. Since x can take on any value except -1, the values of x can approach -1 as closely as desired from either side of -1, as shown in the following table of values.

x approaches -1

↓

x	$-.5$	$-.8$	$-.9$	$-.99$	-1.01	-1.1	-1.2	-1.5
$1 + x$	.5	.2	.1	.01	$-.01$	$-.1$	$-.2$	$-.5$
$\dfrac{2}{1 + x}$	4	10	20	200	-200	-20	-10	-4

↑

$|y|$ gets larger and larger

The table above suggests that as x gets closer and closer to -1, from either side, the sum $1 + x$ gets closer and closer to 0, and $|2/(1 + x)|$ gets larger and larger. The vertical line $x = -1$ that is approached by the curve is called a *vertical asymptote*.

As $|x|$ gets larger and larger, $y = 2/(1 + x)$ gets closer and closer to 0 as shown in the table below.

x	-101	-11	-2	0	9	99
$1 + x$	-100	-10	-1	1	10	100
$y = \dfrac{2}{1 + x}$	$-.02$	$-.2$	-2	2	$.2$	$.02$

The horizontal line $y = 0$ is called a *horizontal asymptote* for this graph. Using the asymptotes and plotting the intercept and a few points (shown with the figure) gives the graph of Figure 3.19.

 Graph the following.

(a) $y = \dfrac{3}{5 - x}$

(b) $y = \dfrac{-4}{x + 4}$

Answer:

(a)

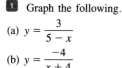

(b)

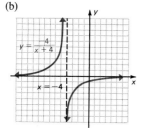

x	y
3	1/2
2	2/3
1	1
0	2
$-1/2$	4
$-3/2$	-4
-2	-2
-3	-1
-4	$-2/3$

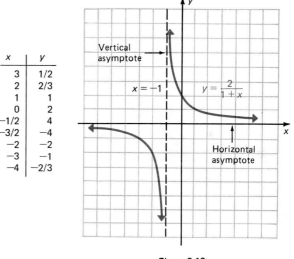

Figure 3.19

As suggested by Example 1,

if a number k makes the denominator equal 0 in the expression defining a rational function, then the line $x = k$ is a **vertical asymptote** for the graph of the function.*

Also, wherever the values of y approach but do not equal some number k as $|x|$ gets larger and larger, the line $y = k$ is a **horizontal asymptote** for the graph.

---

*Actually it is also necessary to make sure that $x = k$ does not also make the numerator equal to 0. If both the numerator and denominator are 0, then there may be no vertical asymptote at k.

Example 2 Graph $y = \dfrac{3x + 2}{2x + 4}$.

The value $x = -2$ makes the denominator equal to 0, so the line $x = -2$ is a vertical asymptote. Find a horizontal asymptote by letting values of x get larger and larger, as in the following chart.

x	$\dfrac{3x + 2}{2x + 4}$	*Ordered pair*
10	$\dfrac{32}{24} = 1.33$	(10, 1.33)
20	$\dfrac{62}{44} = 1.41$	(20, 1.41)
100	$\dfrac{302}{204} = 1.48$	(100, 1.48)
100,000	$\dfrac{300{,}002}{200{,}004} = 1.49998$	(100,000, 1.49998)

The chart suggests that as x gets larger and larger, $(3x + 2)/(2x + 4)$ gets closer and closer to 1.5, or 3/2, making the line $y = 3/2$ a horizontal asymptote. Use a calculator to show that as x gets more negative and takes on the values -10, -100, -1000, $-100,000$, and so on, the graph again approaches the line $y = 3/2$. Using these asymptotes and plotting several points leads to the graph of Figure 3.20. ◼

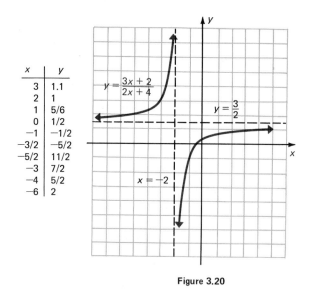

x	y
3	1.1
2	1
1	5/6
0	1/2
−1	−1/2
−3/2	−5/2
−5/2	11/2
−3	7/2
−4	5/2
−6	2

Figure 3.20

In Example 2 above, we found that $y = 3/2$ is a horizontal asymptote for the rational function defined by $y = (3x + 2)/(2x + 4)$. An equation for the horizontal

2 Graph the following.

(a) $y = \dfrac{3x + 5}{x - 3}$

(b) $y = \dfrac{2 - x}{x + 3}$

Answer:

(a)

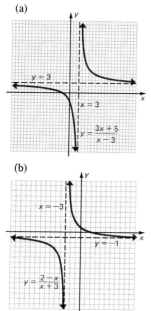

(b)

3 Graph

$$y = \dfrac{-2x}{(x + 4)(2x - 3)}.$$

Answer:

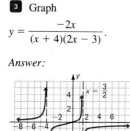

asymptote can also be found by solving $y = (3x + 2)/(2x + 4)$ for x. Do this by multiplying both sides of the equation by $2x + 4$ to get

$$y(2x + 4) = 3x + 2$$

or

$$2xy + 4y = 3x + 2.$$

Collect all terms containing x on one side of the equation:

$$2xy - 3x = 2 - 4y.$$

Factor out x on the left and solve for x.

$$x(2y - 3) = 2 - 4y$$

$$x = \frac{2 - 4y}{2y - 3}$$

This form of the equation shows that y cannot take on the value 3/2. This means that the line $y = 3/2$ is a horizontal asymptote. 2

Example 3 Graph $y = \dfrac{x}{(x - 1)(x + 3)}$.

There are two vertical asymptotes, $x = 1$ and $x = -3$. As $|x|$ gets larger and larger, both numerator and denominator get larger and larger. In this case it is not easy to see what happens to y. One way to decide on the graph is to find several ordered pairs for the function where $|x|$ is relatively large. The table of values given with Figure 3.21 suggests that the values of y approach 0 as $|x|$ gets larger and larger, so the horizontal asymptote is the line $y = 0$. Complete the graph by plotting the ordered pairs and any intercepts from the table and by using the vertical asymptotes. The graph is shown in Figure 3.21. 3

x	y
-8	$-.01$
-6	$-.3$
-4	$-.8$
-1	$.25$
0	$.0$
2	$.4$
4	$.2$
6	$.1$

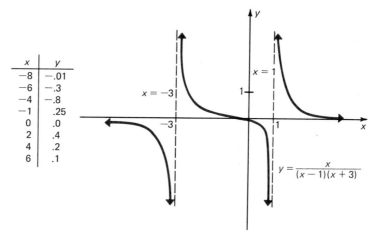

Figure 3.21

The final two examples of this section show mathematical models using rational functions.

Example 4 In a recent year, the U.S. Maritime Administration estimated that the cost of building an oil tanker of 50,000 deadweight tons in the United States was $409 per ton. The cost per ton for a 100,000-ton tanker was $310, while the cost per ton for a 400,000-ton tanker was $178.

Figure 3.22 shows these values plotted on a graph, where x represents tons (in thousands) and y represents the cost per ton. There is a gap in the information presented by the government agency: the data skips from $x = 100$ to $x = 400$, with no intermediate values. If a curve could be fitted through the given data points, then any desired intermediate values could be approximated.

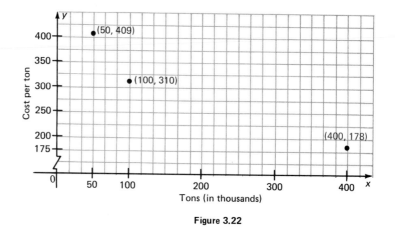

Figure 3.22

Many different functions might give a good mathematical model for the given data. The graph of Figure 3.22 suggests that a rational function might be a good choice. Using methods not discussed here,

$$y = \frac{110,000}{x + 225}$$

was found to be a reasonable mathematical model for approximating the given data. The known values $x = 50$, $x = 100$, and $x = 400$ can be substituted to test the "goodness of fit" of the model. Doing this gives the following results.

x	$y(given)$	$y(from\ model)$
50	409	400
100	310	338
400	178	176

As the chart shows, the rational function is a good mathematical model for the data, so it can be used to approximate y for intermediate values of x. Letting $x = 150$, for example, gives

$$y = \frac{110,000}{150 + 225} = 293.$$

Also, $y = 259$ when $x = 200$, while if $x = 300$, then $y = 210$. Use these points to graph the function, as in Figure 3.23. ■ **4**

4 Using the model of Example 4, find the cost per ton to build a ship of
(a) 75,000 tons;
(b) 350,000 tons.

Answer:
(a) $367
(b) $191

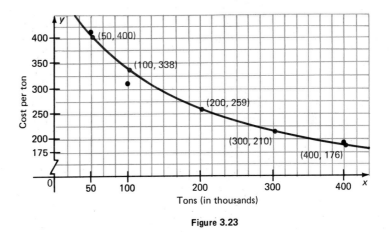

Figure 3.23

Example 5 In many situations involving environmental pollution, much of the pollutant can be removed from the air or water at a fairly reasonable cost, but the last, small part of the pollutant can be very expensive to remove.

Cost as a function of the percentage of pollutant removed from the environment can be calculated for various percentages of removal, with a curve fitted through the resulting data points. This curve then leads to a mathematical model of the situation. Rational functions often are a good choice for these **cost-benefit models.**

For example, suppose a cost-benefit model is given by

$$y = \frac{18x}{106 - x},$$

where y is the cost (in thousands of dollars) of removing x percent of a certain pollutant. The domain of x is the set of all numbers from 0 to 100 inclusive; any amount of pollutant from 0% to 100% can be removed. To remove 100% of the pollutant here would cost

$$y = \frac{18(100)}{106 - 100} = 300,$$

or $300,000. Check that 95% of the pollutant can be removed for $155,000, 90% for $101,000, and 80% for $55,000. Using these points, as well as others that could be obtained from the function above, leads to the graph shown in Figure 3.24. **5**

5 Using the mathematical model of Example 5, find the cost to remove the following percents of pollutants.
(a) 70%
(b) 85%
(c) 90%
(d) 98%

Answer:
(a) $35,000
(b) About $73,000
(c) About $101,000
(d) About $221,000

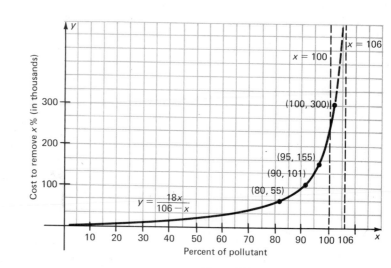

Figure 3.24

3.3 Exercises

Graph each of the following. (See Examples 1–3.)

1. $y = \dfrac{1}{x + 2}$

2. $y = \dfrac{1}{x - 1}$

3. $y = \dfrac{-4}{x - 3}$

4. $y = \dfrac{-1}{x + 3}$

5. $y = \dfrac{2}{x}$

6. $y = \dfrac{-3}{x}$

7. $y = \dfrac{2}{3 + 2x}$

8. $y = \dfrac{4}{5 + 3x}$

9. $y = \dfrac{3x}{x - 1}$

10. $y = \dfrac{4x}{3 - 2x}$

11. $y = \dfrac{x}{x - 9}$

12. $y = \dfrac{3x}{x - 4}$

13. $y = \dfrac{x + 1}{x - 4}$

14. $y = \dfrac{x - 3}{x + 5}$

15. $y = \dfrac{2x - 1}{4x + 2}$

16. $y = \dfrac{3x - 6}{6x - 1}$

17. $y = \dfrac{1 - 2x}{5x + 20}$

18. $y = \dfrac{6 - 3x}{4x + 12}$

19. $y = \dfrac{-x - 4}{3x + 6}$

20. $y = \dfrac{-x + 8}{2x + 5}$

21. $y = \dfrac{1}{(x + 1)(x - 3)}$

22. $y = \dfrac{1}{(x - 2)(x + 4)}$

23. $y = \dfrac{3}{(x + 4)^2}$

24. $y = \dfrac{2}{(x - 3)^2}$

25. $y = \dfrac{3x}{(x + 1)(x - 2)}$

26. $y = \dfrac{2x}{(x + 2)(x + 4)}$

27. $y = \dfrac{5x}{x^2 - 1}$

28. $y = \dfrac{-x}{x^2 - 4}$

29. Management Suppose that the average cost per unit, $C(x)$, of producing x units of margarine is given by

$$C(x) = \frac{500}{x + 30}.$$

Find the average cost per unit of producing each of the following quantities.
(a) 10 units   **(b)** 20 units   **(c)** 50 units
(d) 70 units   **(e)** 100 units
(f) Graph $y = C(x)$.

30. Management Using the information given in Example 4 of the text, estimate the cost per ton for building ships weighing
(a) 25,000 tons;   **(b)** 125,000 tons;
(c) 250,000 tons;   **(d)** 275,000 tons.

31. Natural Science Suppose a cost-benefit model (see Example 5) is given by

$$y = \frac{6.5x}{102 - x},$$

where y is the cost in thousands of dollars of removing x percent of a certain pollutant. Find the cost of removing the following percents of pollutants.
(a) 0%   **(b)** 50%   **(c)** 80%   **(d)** 90%
(e) 95%   **(f)** 99%   **(g)** 100%
(h) Graph the function.

32. Natural Science Suppose a cost-benefit model is given by

$$y = \frac{6.7x}{100 - x},$$

where y is the cost in thousands of dollars of removing x percent of a given pollutant. Find the cost of removing each of the following percents of pollutants.
(a) 50%   **(b)** 70%   **(c)** 80%   **(d)** 90%
(e) 95%   **(f)** 98%   **(g)** 99%
(h) Is it possible, according to this model, to remove *all* the pollutant?
(i) Graph the function.

33. Natural Science In Exercise 32, what percent of pollutant can be removed for
(a) $10,000?   **(b)** $38,000?

34. Antique car fans often enter their cars in a *concours d'elegance* in which a maximum of 100 points can be awarded to a particular car. Points are awarded for the general attractiveness of the car. Based on a recent article in *Business Week,* we constructed the following mathematical model for the cost, in thousands of dollars, of restoring a car so that it will win x points.

$$C(x) = \frac{10x}{49(101 - x)}$$

Find the cost of restoring a car so that it will win
(a) 99 points;   **(b)** 100 points.

35. Natural Science To calculate the drug dosage for a child, pharmacists may use the formula

$$d(x) = \frac{Dx}{x + 12},$$

where x is the child's age in years and D is the adult dosage. The adult dosage of Naldecon is 70 mg.
(a) What is the vertical asymptote for the graph of this function?
(b) What is the horizontal asymptote?
(c) Graph $d(x)$.

36. Natural Science In electronics, the circuit gain is given by

$$G(R) = \frac{R}{r + R},$$

where R is the resistance of a temperature sensor in the circuit and r is a constant. Let $r = 1000$ ohms.
(a) Find any vertical asymptotes of the graph of the function.
(b) Find any horizontal asymptotes of the graph of the function.
(c) Graph $G(R)$.

Management *In management,* **product-exchange functions** *give the relationship between quantities of two items that can be produced by the same machine or factory. For example, an oil refinery can produce gasoline, heating oil, or a combination of the two; a winery can produce red wine, white wine, or a combination of the two. Sketch the Quadrant I portion of the graph of each of the functions defined as follows, and then estimate the maximum quantities of each product that can be produced. (Hint: look at the intercepts.)*

37. The product-exchange function for gasoline, x, and heating oil, y, in hundreds of gallons per day, is

$$y = \frac{125,000 - 25x}{125 + 2x}.$$

38. A drug factory found the product-exchange function for a red tranquilizer, x, and a blue tranquilizer, y, is

$$y = \frac{900,000,000 - 30,000x}{x + 90,000}.$$

Management *In recent years the economist Arthur Laffer has been a center of controversy because of his* **Laffer curve,** *an idealized version of which is shown below. According to this curve,* increasing *a tax rate, say from x_1 percent to x_2 percent on the graph, can actually lead to a* decrease *in government revenue. All economists agree on the endpoints—0 revenue at tax rates of both 0% and 100%, but there is much disagreement on the location of the rate x_1 that produces maximum revenue.*

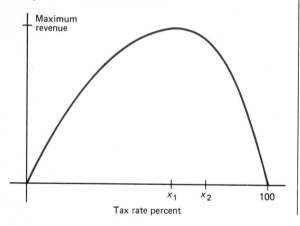

Tax rate percent

39. Suppose an economist studying the Laffer curve produced the rational function

$$y = \frac{60x - 6000}{x - 120}$$

where y is government revenue in millions from a tax rate of x percent, and $50 \le x \le 100$. Find the revenue from a tax rate of

(a) 50%; **(b)** 60%; **(c)** 80%;
(d) 100%. **(e)** Graph the function.

40. Suppose the economist studies a different tax, this time producing

$$y = \frac{80x - 8000}{x - 110},$$

with y giving government revenue in tens of millions of dollars for a tax rate of x percent, with the function valid for $55 \le x \le 100$. Find the revenue from a tax rate of

(a) 55%; **(b)** 60%; **(c)** 70%; **(d)** 90%;
(e) 100%. **(f)** Graph the function.

Use a computer to find several ordered pairs and then sketch the graphs of the functions defined as follows. Identify any horizontal or vertical asymptotes.

41. $f(x) = \dfrac{-2x^2 + x - 1}{2x + 3}$ **42.** $f(x) = \dfrac{3x + 2}{x^2 - 4}$

43. $f(x) = \dfrac{2x^2 - 5}{x^2 - 1}$ **44.** $f(x) = \dfrac{4x^2 - 1}{x^2 + 1}$

Chapter 3 Review Exercises

Graph each of the following.

1. $y = x^2 - 4$ **2.** $y = 6 - x^2$

3. $y = -(x - 1)^2$ **4.** $y = (x + 2)^2$

5. $y = 3(x + 1)^2 - 5$

6. $y = -(1/4)(x - 2)^2 + 3$

7. $y = -(x + 3)^2 - 2$

8. $y = (x - 4)^2 + 3$

9. $y = x^2 - 4x + 2$

10. $y = x^2 + 2x - 3$

11. $y = -x^2 + 6x - 3$

12. $y = -x^2 - 4x + 1$

13. $y = 4x^2 - 8x + 3$

14. $y = 2x^2 + 4x - 3$

15. $y = -3x^2 - 12x - 8$

16. $y = -3x^2 - 6x + 2$

17. $y = -x^2 + 5x - 2$

18. $y = -x^2 - 3x + 4$

Find the maximum or minimum value for the following.

19. $y = x^2 - 4x + 1$ **20.** $y = x^2 + 6x + 3$

21. $y = -3x^2 - 12x - 1$ **22.** $y = -2x^2 + 4x - 3$

23. $y = 4x^2 - 8x + 3$ **24.** $y = -3x^2 - 6x + 2$

Solve each of the following.

25. The commodity market is very unstable; money can be made or lost quickly when investing in soybeans, wheat, pork bellies, and the like. Suppose that an investor kept track of her total profit, P, at time t, measured in months, after she began investing, and found that

$$P = 4t^2 - 29t + 30.$$

Find the time intervals where she has been ahead. (*Hint:* $t > 0$ in this case.)

26. Suppose the velocity of an object is given by

$$v = 2t^2 - 5t - 12,$$

where t is time in seconds. (Here t can be positive or negative.) Find the intervals where the velocity is negative.

27. An analyst has found that his company's profits, in hundreds of thousands of dollars, are given by

$$P = 3x^2 - 35x + 50,$$

where x is the amount, in hundreds of dollars, spent on advertising. Decide for what values of x the company makes a profit.

28. The manager of a large apartment complex has found that the profit is given by

$$P = -x^2 + 250x - 15,000,$$

where x is the number of units rented. Decide for what values of x the complex produces a profit.

29. Find the rectangular region of maximum area that can be enclosed with 200 meters of fencing.

30. Find the rectangular region of maximum area that can be enclosed with 200 meters of fencing if no fencing is needed along one side of the region.

Graph each of the following polynomial functions.

31. $y = x^3 - 2$ **32.** $y = 4 - x^3$

33. $y = -x^4 + 1$ **34.** $y = 2 + x^4$

35. $y = -(x - 3)^3$ **36.** $y = (x + 1)^3$

37. $y = (x + 2)^4$ **38.** $y = -(x - 4)^4$

39. $y = x(2x - 1)(x + 2)$

40. $y = 3x(3x + 2)(x - 1)$

41. $y = 2x^3 - 11x^2 - 2x + 2$

42. $y = x^3 - 3x^2 - 4x - 2$

43. $y = x^4 - 4x^3 - 5x^2 + 14x - 15$

44. $y = x^4 + x^3 - 7x^2 - x + 6$

Graph each of the following rational functions.

45. $y = \dfrac{1}{x - 3}$ **46.** $y = \dfrac{-2}{x + 4}$

47. $y = \dfrac{-3}{2x - 4}$ **48.** $y = \dfrac{5}{3x + 7}$

49. $y = \dfrac{5x + 5}{3x - 5}$ **50.** $y = \dfrac{3 - x}{2x - 9}$

51. $y = \dfrac{-2}{(x - 1)(x + 3)}$ **52.** $y = \dfrac{4}{(x + 2)(x + 5)}$

53. $y = \dfrac{4x}{(2x + 3)(x - 4)}$ **54.** $y = \dfrac{-x}{(3x + 5)(x + 2)}$

55. A cost-benefit curve for pollution control is given by

$$y = \frac{9.2x}{106 - x},$$

where y is the cost in thousands of dollars of removing x percent of a specific industrial pollutant. Find y for each of the following values of x.

(a) $x = 50$ (b) $x = 98$

(c) What percent of the pollutant can be removed for $22,000?

56. The average cost to make x units of a product is given by $A(x)$, where

$$A(x) = \frac{6x}{2x + 5},$$

with x representing the number of units made. Find the average cost to make

(a) 10 units,   (b) 30 units,

(c) 50 units,   (d) 100 units.

4 Exponential and Logarithmic Models

Exponential functions are possibly the single most important type of function used in practical applications. Exponential functions, and the closely related logarithmic functions, are used to describe growth and decay and other important ideas in management, social science, and biology.

4.1 Exponential Functions

Chapter 1 showed how to evaluate 2^x for rational values of x. For example,

$$2^3 = 8, \quad 2^{1/3} = \sqrt[3]{2},$$

$$2^{-1} = \frac{1}{2}, \quad 2^{3/4} = \sqrt[4]{2^3} = \sqrt[4]{8}.$$

What about evaluating 2^x for irrational values of x? For example, what meaning can be given to $2^{\sqrt{2}}$? We know that $\sqrt{2} \approx 1.414$, and we know that we can evaluate 2^1, $2^{1.4}$, $2^{1.41}$, and so on.* In this way, the value of $2^{\sqrt{2}}$ can be approximated as accurately as necessary by replacing $\sqrt{2}$ with a decimal approximation to as many decimal places as required. For example, replacing $\sqrt{2}$ with 1.414 gives $2^{\sqrt{2}} \approx 2.665$. It turns out that these approximations approach closer and closer to one single, fixed number, so it is reasonable to assume that 2^x has a real number value for any value of x, rational or irrational. Therefore, $y = 2^x$ defines a function which has the set of real numbers as domain. This function is an example of an *exponential function*.

A function f is an **exponential function** with base a if, for any real number x,

$$f(x) = a^x,$$

where $a > 0$ and $a \neq 1$.

(If $a = 1$, the function is the constant function $f(x) = 1$.)

Example 1 Graph the following exponential functions.
(a) $y = 2^x$

---

*It is possible to approximate $2^{1.4}$ and $2^{1.41}$ with logarithms or certain calculators.

Begin by making a table of values of x and y, as shown beside Figure 4.1(a). Then plot these points and draw a smooth curve through them, getting the graph shown in Figure 4.1(a). The graph approaches the negative x-axis but will never touch it, since y cannot be 0. This makes the x-axis a horizontal asymptote. The graph suggests that as x gets larger and larger, values of 2^x also get larger and larger. However, there is no vertical asymptote.

(b) $y = 2^{-x}$

By the properties of exponents,

$$2^{-x} = \frac{1}{2^x} = \left(\frac{1}{2}\right)^x.$$

1 Graph $y = \left(\frac{1}{3}\right)^x$.

Construct a table of values and draw a smooth curve through the resulting points. [see Figure 4.1(b)]. **1**

Answer:

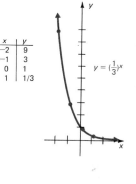

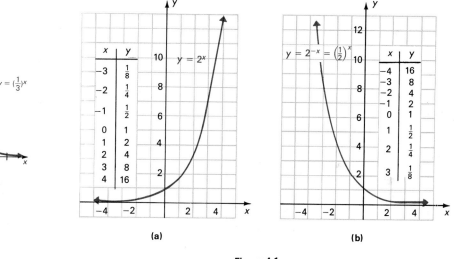

(a)   (b)

Figure 4.1

As suggested by the graphs in Figure 4.1, the domain of an exponential function defined by an expression of the form $y = a^x$ includes all real numbers, and the range includes all positive real numbers. The base a is restricted to positive values, with negative or zero bases not allowed. For example, the domain of $y = (-4)^x$ could not include such numbers as $x = 1/2$ or $x = 1/4$. The resulting graph would be at best a series of separate points having little practical use.

The graph of $y = 2^x$ in Figure 4.1(a) is typical of the graphs of $y = a^x$ where $a > 1$. For larger values of a, the graph rises to the right more steeply, but the general shape is similar to that of $y = 2^x$. When $0 < a < 1$, the graph decreases to the right like the graph of $y = (1/2)^x$ in Figure 4.1(b). The graphs of $y = 2^x$ and $y = 2^{-x}$ are mirror images of each other with respect to the y-axis. The graphs of several typical exponential functions in Figure 4.2 illustrate these statements.

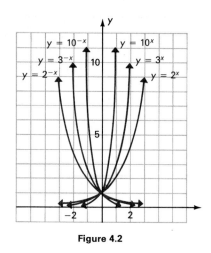

Figure 4.2

Example 2   Graph $y = 2^{-x^2}$.

Plotting several values of x and y gives the graph in Figure 4.3. Both the negative sign and the fact that x is squared affect the shape of the graph, with the final result looking quite different from the typical graphs of exponential functions in Figures 4.1 and 4.2. Graphs such as this are important in probability, where the normal curve has an equation similar to the one in this example. ■

2 Graph $y = \left(\dfrac{1}{2}\right)^{-x^2}$.

Answer:

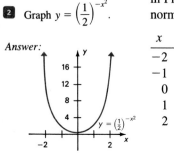

x	y
-2	$1/16$
-1	$1/2$
0	1
1	$1/2$
2	$1/16$

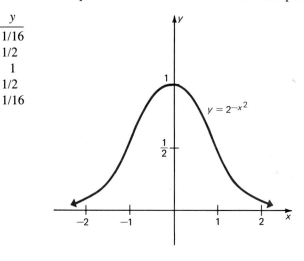

Figure 4.3

The graphs of typical exponential functions in Figure 4.2 suggest that a given value of x leads to exactly one value of a^x and each value of a^x corresponds to exactly one value of x. Because of this, an equation with a variable in the exponent, called an **exponential equation,** can often be solved using the following property.

If $a \neq 1$ and $a^x = a^y$, then $x = y$.

Be careful: both bases must be the same. The value $a = 1$ is excluded since $1^2 = 1^3$, for example, even though $2 \neq 3$. (This property is not contradicted by the graph of $y = 2^{-x^2}$ in Figure 4.3 since that graph is *not* the graph of an exponential function.)

As an example, solve $2^{3x} = 2^7$ using this property, as follows:

$$2^{3x} = 2^7$$
$$3x = 7$$
$$x = \frac{7}{3}. \quad \boxed{3}$$

③ Solve each equation.
(a) $6^x = 6^4$
(b) $3^{2x} = 3^9$
(c) $5^{-4x} = 5^3$

Answer:
(a) 4
(b) 9/2
(c) −3/4

Example 3 Solve $9^x = 27$.

First rewrite both sides of the equation so the bases are the same. Since $9 = 3^2$ and $27 = 3^3$,

$$(3^2)^x = 3^3$$
$$3^{2x} = 3^3$$
$$2x = 3$$
$$x = \frac{3}{2}. \quad \boxed{4}$$

④ Solve each equation.
(a) $8^{2x} = 4$
(b) $5^{3x} = 25^4$
(c) $36^{-2x} = 6$

Answer:
(a) 1/3
(b) 8/3
(c) −1/4

As noted earlier, exponential functions have many practical applications. For example, in situations which involve growth or decay of a population, the size of the population at a given time t often is determined by an exponential function of t.

Example 4 The oxygen consumption of yearling salmon (in appropriate units) increases exponentially with speed of swimming according to

$$f(x) = 100(3)^{.6x},$$

where x is the speed in feet per second. Find each of the following.
(a) $f(0)$

Substitute 0 for x:

$$f(0) = 100(3)^{.6(0)} = 100(3)^0 = 100 \cdot 1 = 100.$$

When the fish are still (their speed is 0) their oxygen consumption is 100 units.
(b) $f(5)$

Replace x with 5.

$$f(5) = 100(3)^{.6(5)} = 100(3)^3 = 2700$$

A speed of 5 feet per second increases the oxygen consumption to 2700 units. **⑤**

⑤ The number of organisms present at time t is given by
$$f(t) = 75(2)^{.5t}$$
Find the number of organisms present at
(a) $t = 0,$
(b) $t = 2,$
(c) $t = 4.$

Answer:
(a) 75
(b) 150
(c) 300

4.1 Exercises

Which of the following define exponential functions? (Hint: Sketch the graph and then decide.)

1. $f(x) = 5^x - 1$

2. $f(x) = 2x^5$

3. $f(x) = 4x^3 - 1$

4. $f(x) = 2 \cdot 3^{5x-1}$

Let $f(x) = (2/3)^x$. Find each of the following values.

5. $f(2)$

6. $f(-1)$

7. $f\left(\dfrac{1}{2}\right)$

8. $f(0)$

Graph each of the following exponential functions. (See Examples 1 and 2.)

9. $y = 3^x$

10. $y = 4^x$

11. $y = 3^{-x}$

12. $y = 4^{-x}$

13. $y = \left(\dfrac{1}{4}\right)^x$

14. $y = \left(\dfrac{1}{3}\right)^x$

15. $y = \left(\dfrac{1}{3}\right)^{-x}$

16. $y = \left(\dfrac{1}{4}\right)^{-x}$

17. $y = 3^{2x}$

18. $y = 4^{x/2}$

19. $y = 2^{-x/2}$

20. $y = 3^{-2x}$

21. $y = 2^{x^2}$

22. $y = 3^{x^2}$

23. $y = 3^{-x^2}$

24. $y = 2^{-x^2}$

25. $y = 2^{x+1}$

26. $y = 3^{x+1}$

27. $y = x \cdot 2^x$

28. $y = x^2 \cdot 2^x$

29. $y = 2^{|x|}$

30. $y = 2^{-|x|}$

Solve each of the following equations. (See Example 3.)

31. $5^x = 25$

32. $3^x = \dfrac{1}{9}$

33. $2^x = \dfrac{1}{8}$

34. $4^x = 64$

35. $4^x = 8$

36. $25^x = 125$

37. $16^x = 64$

38. $\left(\dfrac{1}{8}\right)^x = 8$

39. $\left(\dfrac{3}{4}\right)^x = \dfrac{16}{9}$

40. $5^{-2x} = \dfrac{1}{25}$

41. $3^{x-1} = 9$

42. $16^{-x+1} = 8$

43. $25^{-2x} = 3125$

44. $16^{x+2} = 64^{2x-1}$

45. $81^{-2x} = 3^{x-1}$

46. $7^{-x} = 49^{x+3}$

47. $2^{|x|} = 16$

48. $5^{-|x|} = \dfrac{1}{25}$

49. $\left(\dfrac{1}{3}\right)^{|x-4|} = \dfrac{1}{81}$

50. $\left(\dfrac{3}{2}\right)^{|x+5|} = \dfrac{27}{8}$

51. $2^{x^2-4x} = \dfrac{1}{16}$

52. $5^{x^2+x} = 1$

53. $8^{x^2} = 2^{5x}$

54. $9^x = 3^{x^2}$

Work the following exercises.

55. If \$1 is deposited into an account paying 12% per year compounded annually, then after t years the account will contain

$$y = (1 + .12)^t = (1.12)^t$$

dollars.

(a) Use a calculator to complete the following table.

t	0	1	2	3	4	5	6	7	8	9	10
y	1					1.76					3.11

(b) Graph $y = (1.12)^t$.

56. If money loses value at the rate of 8% per year, the value of \$1 in t years is given by

$$y = (1 - .08)^t = (.92)^t.$$

(a) Use a calculator to complete the following table.

t	0	1	2	3	4	5	6	7	8	9	10
y	1					.66					.43

(b) Graph $y = (.92)^t$.

57. Use the results of Exercise 56(a) to answer the following questions.

(a) Suppose a house is valued at \$65,000 today. Estimate the value of a similar house in 10 years. (Hint: Solve the equation $.43t = \$65,000$.)

(b) Find the value of a \$20 textbook in 8 years.

58. **Social Science** Under certain conditions, the number of individuals of a species that is newly introduced into an area can double every year. That is, if t represents the number of years since the species was introduced into the area, and y represents the number of individuals, then

$$y = 6 \cdot 2^t$$

if 6 animals were introduced into the area originally.

(a) Complete the following table.

t	0	1	2	3	4	5	6	7	8	9	10
y	6					192					6144

(b) Graph $y = 6 \cdot 2^t$.

59. Social Science   Suppose the population of a city is

$$P(t) = 1,000,000(2^{.2t}),$$

where t represents time measured in years. Find each of the following values.

(a) $P(0)$ (b) $P(5/2)$ (c) $P(5)$ (d) $P(10)$
(e) Graph $P(t)$.

60. Natural Science   Suppose the quantity in grams, of a radioactive substance present at time t is

$$Q(t) = 500(3^{-.5t})$$

where t is measured in months. Find the quantity present at each of the following times.

(a) $t = 0$ (b) $t = 2$ (c) $t = 4$ (d) $t = 10$
(e) Graph $Q(t)$.

61. Natural Science   *Escherichia coli* is a strain of bacteria that occurs naturally in many different situations. Under certain conditions, the number of these bacteria present in a colony is given by

$$E(t) = E_0 \cdot 2^{t/30},$$

where $E(t)$ is the number of bacteria present t minutes after the beginning of an experiment, and E_0 is the number present when $t = 0$. Let $E_0 = 1,000,000$, and use a calculator with a y^x key to find the number of bacteria at the following times.

(a) $t = 5$ (b) $t = 10$ (c) $t = 15$
(d) $t = 20$ (e) $t = 30$ (f) $t = 60$
(g) $t = 90$ (h) $t = 120$

62. Social Science   A person learning certain skills involving repetition tends to learn quickly at first. Then learning tapers off and approaches some upper limit. Suppose the number of characters per minute a typesetter can produce is given by

$$p(t) = 250 - 120(2.8)^{-.5t}$$

where t is the number of months the typesetter has been in training. Find each of the following.

(a) $p(2)$ (b) $p(4)$ (c) $p(10)$
(d) Graph $p(t)$.

63. Suppose the domain of $y = 2^x$ is restricted to include only rational values of x (which are the only values discussed prior to this section.) Describe in words the resulting graph.

64. In our definition of exponential function, we ruled out negative values of a. However, in a textbook on mathematical economics, published by a well-known publisher, the author obtained a "graph" of $y = (-2)^x$ by

plotting the following points and drawing a smooth curve through them.

x	-4	-3	-2	-1	0	1	2	3
y	$\dfrac{1}{16}$	$-\dfrac{1}{8}$	$\dfrac{1}{4}$	$-\dfrac{1}{2}$	1	-2	4	-8

The graph oscillates very neatly from positive to negative values of y. Comment on this approach. (This example shows the dangers of point plotting when drawing graphs.)

Natural Science   Generally speaking, the larger an organism, the greater its complexity (as measured by counting the number of different types of cells present). The figure shows an estimate of the maximum number of cells (or the largest volume) in various organisms plotted against the number of types of cells found in those organisms.*

65. Use the graph to estimate the maximum number of cells and the corresponding volume for each of the following organisms.

(a) whale   (b) sponge   (c) green alga

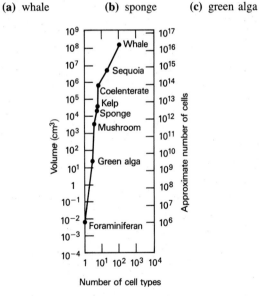

Number of cell types

66. From the graph estimate the number of cell types in each of the following organisms.

(a) mushroom   (b) kelp   (c) sequoia

---

*From *On Size and Life* by Thomas A. McMahon and John Tyler Bonner. Copyright © 1983 by Thomas A. McMahon and John Tyler Bonner. Reprinted by permission of W. H. Freeman and Company.

4.2 Applications of Exponential Functions

Perhaps the single most useful exponential function is the function defined by $y = e^x$, where e is an irrational number that occurs often in practical applications. To see how the number e is used in a practical problem, let us begin with the formula for compound interest (interest paid on both principal and interest). If P dollars is deposited in an account paying an annual rate of interest i compounded (paid) m times per year, the account will contain

$$P\left(1 + \frac{i}{m}\right)^{nm}$$

dollars after n years. (This formula will be derived in the next chapter.)

For example, suppose \$1000 is deposited into an account paying 8% per year compounded quarterly, or four times a year. After 10 years the account will contain

$$P\left(1 + \frac{i}{m}\right)^{nm} = 1000\left(1 + \frac{.08}{4}\right)^{10(4)}$$
$$= 1000(1 + .02)^{40} = 1000(1.02)^{40},$$

in dollars. The number $(1.02)^{40}$ can be found in financial tables, or by using a calculator with a y^x key. To five decimal places, $(1.02)^{40} = 2.20804$. The amount on deposit after 10 years is

$$1000(1.02)^{40} = 1000(2.20804) = 2208.04,$$

or \$2208.04.

Suppose now that a lucky investment produces annual interest of 100%, so that $i = 1.00$, or $i = 1$. Suppose also that only \$1 can be deposited at this rate, and only for one year. Then $P = 1$ and $n = 1$. Substituting into the formula for compound interest gives

$$P\left(1 + \frac{i}{m}\right)^{nm} = 1\left(1 + \frac{1}{m}\right)^{1(m)} = \left(1 + \frac{1}{m}\right)^{m}$$

As interest is compounded more and more often, the value of this expression will increase. If $m = 1$ (interest is compounded annually),

$$\left(1 + \frac{1}{m}\right)^{m} = \left(1 + \frac{1}{1}\right)^{1} = 2^1 = 2,$$

so that \$1 becomes \$2 in one year.

A calculator with a y^x key gives the results for larger and larger values of m shown in the table at the side. These results have been rounded to five decimal places. The table suggests that as m increases, the value of $(1 + 1/m)^m$ gets closer and closer to some fixed number. It turns out that this is indeed the case. This fixed number is called e. To nine decimal places,

m	$\left(1 + \dfrac{1}{m}\right)^{m}$
1	2
2	2.25
5	2.48832
10	2.59374
25	2.66584
50	2.69159
100	2.70481
500	2.71557
1000	2.71692
10,000	2.71815
1,000,000	2.71828

$$e = 2.718281828.$$

① Use Table 3 or a calculator to find each of the following.
(a) $e^{.06}$
(b) $e^{-.06}$
(c) $e^{2.30}$
(d) $e^{-2.30}$

Answer:
(a) 1.06184
(b) .94176
(c) 9.97418
(d) .10026

Table 3 in the back of the book gives various powers of e. Also, some calculators will give values of e^x. In Figure 4.4, the functions defined by $y = 2^x$, $y = e^x$, and $y = 3^x$ are graphed for comparison. ①

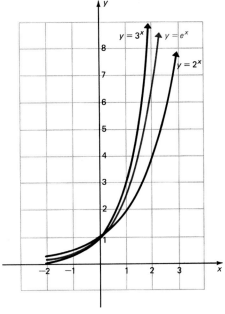

Figure 4.4

In many situations in biology, economics, and the social sciences, a quantity changes at a rate proportional to the quantity present. In such cases, the amount present at time t is a function of t called the *exponential growth function*.

> **Exponential Growth Function**   Let y_0 be the amount or number of a quantity present at time $t = 0$. Then, under certain conditions, the amount present at time t is given by
>
> $$y = y_0 e^{kt},$$
>
> for some constant k.

The next example illustrates exponential growth.

Example 1   Suppose the population of a midwestern city is

$$P(t) = 10,0000e^{.04t},$$

where t represents time measured in years. The population at time $t = 0$ is

$$P(0) = 10,000e^{(.04)0}$$
$$= 10,000e^0$$
$$= 10,000(1)$$
$$= 10,000.$$

2 In Example 1, find the population of the city after
(a) 1 year,
(b) 3 years,
(c) 7 years,
(d) 10 years.

Answer:
(a) about 10,400
(b) about 11,300
(c) about 13,200
(d) about 14,900

3 Suppose the number of bacteria in a culture at time t is
$$y = 500e^{.4t},$$
where t is measured in hours.
(a) How many bacteria are present at $t = 0$?
(b) How many bacteria are present at $t = 10$?

Answer:
(a) 500
(b) about 27,300

The population of the city is 10,000 at time $t = 0$, written $P_0 = 10,000$. The population of the city at year $t = 5$ is
$$P(5) = 10,000e^{(.04)5}$$
$$= 10,000e^{.2}.$$

The number $e^{.2}$ can be found in Table 3 or by using a suitable calculator. By either of these methods, $e^{.2} = 1.22140$ (to five decimal places), so that
$$P(5) = 10,000(1.22140) = 12,214.$$

In five years the population of the city will be about 12,000. 2

Example 2 Suppose the amount, y, of a certain radioactive substance present at time t is given by
$$y = 1000e^{-.1t},$$
where t is measured in days and y is measured in grams.
(a) How much of the substance will be present at time $t = 0$?
When $t = 0$,
$$y = 1000e^{-.1(0)}$$
$$= 1000 \cdot 1$$
$$= 1000,$$
with $y_0 = 1000$ grams.
(b) How much will be present at time $t = 5$?
$$y = 1000e^{-.1(5)}$$
$$= 1000e^{-.5}$$
$$\approx 1000(.60653)$$
$$= 606.53 \text{ grams} \quad 3$$

Example 3 Sales of a new product often grow rapidly at first and then begin to level off with time. For example, suppose the sales, $S(x)$, in some appropriate unit, of a new model typewriter are approximated by
$$S(x) = 1000 - 800e^{-x},$$
where x represents the number of years the typewriter has been on the market. Calculate S_0, $S(1)$, $S(2)$, and $S(4)$. Graph $S(x)$.
Find S_0 by letting $x = 0$:
$$S_0 = S(0) = 1000 - 800 \cdot 1 = 200.$$

Using Table 3,
$$S(1) = 1000 - 800e^{-1}$$
$$\approx 1000 - 800(.36787)$$
$$\approx 1000 - 294$$
$$= 706.$$

4 Suppose that sales of a product are given by $S(x) = 1200 - 800e^{-.5x}$. Find (a) $S(0)$, (b) $S(2)$, (c) $S(4)$, (d) $S(10)$.
(e) Find an equation of the horizontal asymptote for the graph.
(f) Graph $y = S(x)$.

Answer:
(a) 400 (b) 906
(c) 1090 (d) 1190
(e) $S(x) = 1200$
(f)

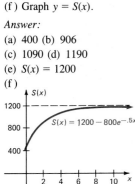

5 Suppose the value of the assets of a certain company at time t are given by

$$y = 100,000 - 75,000e^{-.2t}$$

where t is measured in years. Find y for the following values of t.
(a) $t = 0$ (b) $t = 5$
(c) $t = 10$ (d) $t = 25$
(e) Find the horizontal asymptote.
(f) Graph y.

Answer:
(a) 25,000 (b) 72,410
(c) 89,850 (d) 99,500
(e) $y = 100,000$
(f)

In the same way, verify that $S(2) = 892$ and $S(4) = 985$. Plotting several such points leads to the graph shown in Figure 4.5. The line $y = 1000$ is a horizontal asymptote. As the graph suggests, sales will tend to level off with time and gradually approach a level of 1000 units.   **4**

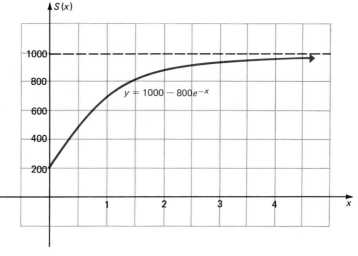

Figure 4.5

Example 4 Assembly line operations tend to have a high turnover of employees, forcing the companies involved to spend much time and effort in training new workers. It has been found that a worker new to the operation of a certain task on the assembly line will produce items according to

$$P(x) = 25 - 25e^{-.3x}$$

where $P(x)$ is the number of items produced by the worker on day x. How many items will be produced by a new worker on day 8?
    Evaluate $P(8)$.

$$\begin{aligned}
P(8) &= 25 - 25e^{-.3(8)} \\
&= 25 - 25e^{-2.4} \\
&\approx 25 - 25(.09071) \\
&\approx 25 - 2.3 \\
&= 22.7.
\end{aligned}$$

On the eighth day, the worker can be expected to produce about 23 items. Plotting several such points leads to the graph of $P(x)$ shown in Figure 4.6.   **5**

    Graphs such as the one in Figure 4.6 are called **learning curves.** According to such a graph, a new worker tends to learn quickly at first; then learning tapers off and approaches some upper limit. This is characteristic of the learning of certain types of skills involving the repetitive performance of the same task.

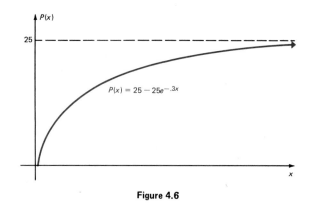

Figure 4.6

Example 5 Under certain conditions the total number of facts of a certain kind that are remembered is approximated by

$$N(t) = y_0\left(\frac{1 + e}{1 + e^{t+1}}\right),$$

where $N(t)$ is the number of facts remembered at time t, measured in days, and y_0 is the number of facts remembered initially. Graph $y = N(t)$.

Plot some points to help draw the graph. For example, if $t = 0$,

$$N(0) = y_0\left(\frac{1 + e}{1 + e^1}\right) = y_0(1) = y_0.$$

If $t = 1$,

$$N(1) = y_0\left(\frac{1 + e}{1 + e^2}\right) \approx y_0\left(\frac{3.718}{8.389}\right) \approx .44y_0.$$

Plotting several such points (see the problem at the side) leads to the graph of Figure 4.7. Graphs such as this are called **forgetting curves.**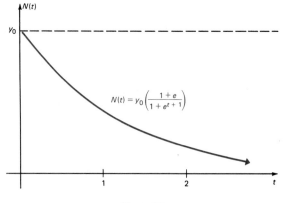

In Example 5, find
(a) $N(2)$,
(b) $N(3)$,
(c) $N(5)$.

Answer:
(a) $.18y_0$
(b) $.07y_0$
(c) $.009y_0$

$N(t)$

y_0

$$N(t) = y_0\left(\frac{1 + e}{1 + e^{t + 1}}\right)$$

Figure 4.7

4.2 Exercises

1. Social Science Suppose the population of a city, $P(t)$, is given by

$$P(t) = 1,000,000e^{.02t},$$

where t represents time measured in years. Find each of the following values.
(a) P_0 **(b)** $P(2)$ **(c)** $P(4)$ **(d)** $P(10)$

2. Natural Science A population of lice $L(t)$ is given by

$$L(t) = 100e^{.1t},$$

where t is measured in months. Find each of the following values.
(a) L_0 **(b)** $L(1)$ **(c)** $L(6)$ **(d)** $L(12)$

3. Natural Science Suppose the quantity, $Q(t)$, measured in grams, of a radioactive substance present at time t is given by

$$Q(t) = 500e^{-.05t},$$

where t is measured in days. Find the quantity present at each of the following times.
(a) $t = 0$ **(b)** $t = 4$ **(c)** $t = 8$ **(d)** $t = 20$

4. Natural Science The amount of a chemical in grams that will dissolve in a solution is given by

$$C(t) = 10e^{.02t},$$

where t is the temperature of the solution. Find each of the following.
(a) $C(0°)$ **(b)** $C(10°)$ **(c)** $C(30°)$
(d) $C(100°)$

5. Natural Science Let the number of bacteria, $B(t)$, present in a certain culture be given by

$$B(t) = 25,000e^{.2t},$$

where t is time measured in hours, and $t = 0$ corresponds to noon. Find the number of bacteria present at each of the following times.
(a) Noon   **(b)** 1 P.M.   **(c)** 2 P.M.   **(d)** 5 P.M.

6. Natural Science When a bactericide is introduced into a certain culture, the number of bacteria present, $D(t)$, is given by

$$D(t) = 50,000e^{-.01t},$$

where t is time measured in hours. Find the number of bacteria present at each of the following times.
(a) $t = 0$ **(b)** $t = 5$ **(c)** $t = 20$ **(d)** $t = 50$

7. In Chapter 5, it is shown that P dollars compounded continuously (every instant) at an annual rate of interest i would amount to

$$A = Pe^{ni}$$

at the end of n years. How much would \$20,000 amount to at 8% compounded continuously for the following number of years?
(a) 1 year   **(b)** 5 years   **(c)** 10 years

8. The number (in hundreds) of fish in a small commercial pond is given by

$$F(t) = 27 - 15e^{-.8t},$$

where t is in years. Find each of the following.
(a) F_0 **(b)** $F(1)$ **(c)** $F(5)$ **(d)** $F(10)$

9. The number of words per minute that an average typist can type is given by

$$W(x) = 60 - 30e^{-.5t},$$

where t is time in months after the beginning of a typing class. Find each of the following.
(a) W_0 **(b)** $W(1)$ **(c)** $W(4)$ **d** $W(6)$

10. Management Sales of a new model can opener are approximated by

$$S(x) = 5000 - 4000e^{-x},$$

where x represents the number of years that the can opener has been on the market, and $S(x)$ represents sales in thousands. Find each of the following.
(a) $S(0)$ **(b)** $S(1)$ **(c)** $S(2)$ **(d)** $S(5)$
(e) $S(10)$
(f) Find the horizontal asymptote for the graph.
(g) Graph $y = S(x)$.

11. Management Assume that a person new to an assembly line will produce

$$P(x) = 500 - 500e^{-x}$$

items per day, where x is time measured in days. Find each of the following.
(a) $P(0)$ **(b)** $P(1)$ **(c)** $P(2)$
(d) $P(5)$ **(e)** $P(10)$
(f) Find the horizontal asymptote for the graph.
(g) Graph $y = P(x)$.

12. Management   Experiments have shown that sales of a product, under relatively stable market conditions, but in the absence of promotional activities such as advertising, tend to decline at a constant yearly rate. This rate of sales decline varies considerably from product to product, but seems to remain the same for any particular product. The sales decline can often be expressed by

$$S(t) = S_0 e^{-at},$$

where $S(t)$ is the rate of sales at time t measured in years, S_0 is the rate of sales at time $t = 0$, and a is the sales decay constant. Suppose the sales decay constant for a particular product is $a = .10$. Let $S_0 = 50,000$. Find

(a) $S(1)$;   (b) $S(3)$.

13. Management   In Exercise 12, suppose $S_0 = 80,000$ and $a = .05$. Find

(a) $S(2)$;   (b) $S(10)$.

14. Social Science   A sociologist has shown that the fraction $y(t)$ of people in a group who have heard a rumor after t days is approximated by

$$y(t) = \frac{y_0 e^{kt}}{1 - y_0(1 - e^{kt})},$$

where y_0 is the fraction of people who have heard the rumor at time $t = 0$, and k is a constant. A graph of $y(t)$ for a particular value of k is shown in the figure.

(a) If $k = .1$ and $y_0 = .05$, find $y(10)$.

(b) If $k = .2$ and $y_0 = .10$, find $y(5)$.

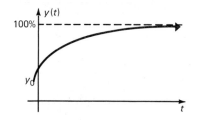

15. The higher a student's grade-point average, the fewer applications the student must send to medical schools (other things being equal). Using information given in a guidebook for prospective medical school students, we constructed the following mathematical model of the number of applications that a student should send out:

$$y = 540 e^{-1.3x},$$

where y is the number of applications that should be sent out by a person whose grade-point average is x. Here $2.0 \leq x \leq 4.0$. Use a calculator with a y^x key to find the number of applications that should be sent out by students having a grade-point average of

(a) 2.0;   (b) 2.5;   (c) 3.0;

(d) 3.5;   (e) 3.9;   (f) 4.0.

16. Natural Science   Many environmental situations place effective limits on the growth of the numbers of an organism in an area. Many such limited growth situations are described by the **logistic function,** defined by

$$G(t) = \frac{m \cdot G_0}{G_0 + (m - G_0)e^{-kmt}},$$

where G_0 is the initial number present, m is the maximum possible size of the population, and k is a positive constant. Assume $G_0 = 1000$, $m = 25,000$, and $k = .04$.

(a) Find $G(5)$.   (b) Find $G(10)$.

Management   *Wal-Mart stores, headquartered in Bentonville, Arkansas, is one of the fastest growing retail chains in the country. Using data from the October 14, 1985 issue of* Business Week *magazine (page 146),* we constructed the mathematical model*

$$R(t) = .3 e^{.3356t}$$

to approximate annual revenues of the chain, where R(t) is revenue in billions in year t, with t = 0 representing 1975.

17. Use this model to estimate sales in each of the following years. Compare the estimates with the numbers on the graph on the following page.

(a) 1975   (b) 1980   (c) 1984   (d) 1985

18. For net income, we constructed the mathematical model

$$P(t) = 9 e^{.3632t},$$

---

*Two graphs, "Big Payoffs from Minding the Store," reprinted from the October 14, 1985 issue of *Business Week* by special permission. Copyright © 1985 by McGraw-Hill, Inc.

where $P(t)$ is net income in millions in year t, with $t = 0$ representing 1975. Use this model to estimate net income in each of the following years.

(a) 1975 **(b)** 1980 **(c)** 1984 **(d)** 1985

Natural Science *Newton's Law of Cooling says that the rate at which a body cools is proportional to the difference in temperature between the body and an environment into which it is introduced. Using calculus, the temperature $F(t)$ of the body at time t after being introduced into an environment having constant temperature T_0 is*

$$F(t) = T_0 + Ce^{-kt},$$

where C and k are constants. Use this result in Exercises 19–24.

19. Find the temperature of an object when $t = 4$ if $T_0 = 125$, $C = .8$, and $k = .2$.

20. Find the temperature of an object when $t = 9$ if $T_0 = 180$, $C = .5$, and $k = .6$.

21. A piece of metal is heated to 300°C and then placed in a cooling liquid at 50°C. After 4 minutes, the metal has cooled to 175°C. Find its temperature after 12 minutes.

22. Boiling water, at 100°C, is placed in a freezer at 0°C. The temperature of the water is 50°C after 24 minutes. Find the temperature of the water after 96 minutes.

23. A volcano discharges lava at 800°C. The surrounding air has a temperature of 20°C. The lava cools to 410°C in five hours. Find its temperature after 15 hours.

24. Paisley refuses to drink coffee cooler than 95°F. She makes coffee with a temperature of 170°F in a room with a temperature of 70°F. The coffee cools to 120°F in 10 minutes. What is the longest time she can let the coffee sit before she drinks the coffee?

Natural Science *The pressure of the atmosphere, $p(h)$, in pounds per square inch, is given by*

$$p(h) = p_0e^{-kh},$$

where h is the height above sea level and p_0 and k are constants. The pressure at sea level is 15 pounds per square inch and the pressure is 9 pounds per square inch at a height of 12,000 feet.

25. Find the pressure at an altitude of 6000 feet.

26. What would be the pressure encountered by a spaceship at an altitude of 150,000 feet?

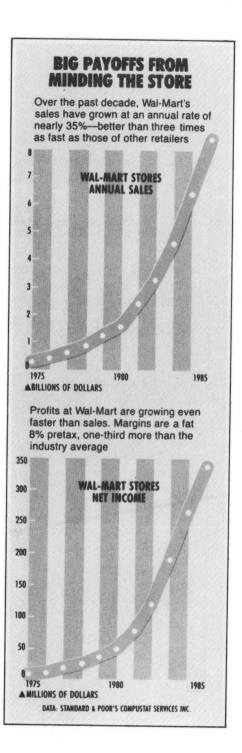

BIG PAYOFFS FROM MINDING THE STORE

Over the past decade, Wal-Mart's sales have grown at an annual rate of nearly 35%—better than three times as fast as those of other retailers

WAL-MART STORES ANNUAL SALES

▲BILLIONS OF DOLLARS

Profits at Wal-Mart are growing even faster than sales. Margins are a fat 8% pretax, one-third more than the industry average

WAL-MART STORES NET INCOME

▲MILLIONS OF DOLLARS

DATA: STANDARD & POOR'S COMPUSTAT SERVICES INC.

4.3 Logarithmic Functions

The previous sections discussed exponential fucntions. Now, this section discusses logarithmic functions, which can be obtained from exponential functions using the following definition.

Definition of Logarithm For $a > 0$ and $a \neq 1$,

$$y = \log_a x \qquad \text{means} \qquad a^y = x.$$

(Read $y = \log_a x$ as "y is the logarithm of x to the base a.") For example, the exponential statement $2^4 = 16$ can be translated into a logarithmic statement using the definition: $4 = \log_2 16$.

This key definition should be memorized. It is important to remember the location of the base and exponent in each part of the definition.

1 Write the logarithmic form of
(a) $5^3 = 125$;
(b) $3^{-4} = 1/81$;
(c) $8^{2/3} = 4$.

Answer:
(a) $\log_5 125 = 3$
(b) $\log_3 (1/81) = -4$
(c) $\log_8 4 = 2/3$

$$\text{exponent}$$
$$\downarrow$$
$$\text{logarithmic form:} \quad y = \log_a x$$
$$\uparrow$$
$$\text{base}$$

$$\text{exponent}$$
$$\downarrow$$
$$\text{exponential form:} \quad a^y = x$$
$$\uparrow$$
$$\text{base}$$

By the definition, a logarithm is an exponent: the exponent on the base a that will yield the number x.

2 Write the exponential form of
(a) $\log_{16} 4 = 1/2$;
(b) $\log_3 (1/9) = -2$;
(c) $\log_{16} 8 = 3/4$.

Answer:
(a) $16^{1/2} = 4$
(b) $3^{-2} = 1/9$
(c) $16^{3/4} = 8$

Example 1 The chart below shows several pairs of equivalent statements. The same statement is written in both exponential and logarithmic form.

Exponential form	Logarithmic form
(a) $3^2 = 9$	$\log_3 9 = 2$
(b) $(1/5)^{-2} = 25$	$\log_{1/5} 25 = -2$
(c) $10^5 = 100{,}000$	$\log_{10} 100{,}000 = 5$
(d) $4^{-3} = 1/64$	$\log_4 1/64 = -3$
(e) $2^{-4} = 1/16$	$\log_2 1/16 = -4$
(f) $e^0 = 1$	$\log_e 1 = 0$ **1** **2**

A logarithmic function can be defined as follows.

A function f is a **logarithmic function** with base a if

$$f(x) = \log_a x,$$

where $a > 0$ and $a \neq 1$.

---

3 (a) Graph $y = 3^x$ and $y = \log_3 x$ on the same axes.

(b) Graph $y = (1/3)^x$ and $y = \log_{1/3} x$ on the same axes.

Answer:

(a)

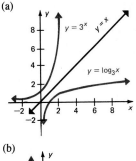

(b)

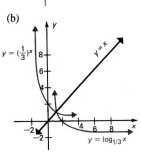

To graph a logarithmic function, rewrite it in exponential form, as in the next example.

Example 2 Graph the following logarithmic functions.

(a) $y = \log_2 x$

Rewrite $y = \log_2 x$ in exponential form as $2^y = x$. Then choose various values of y and find the corresponding values of x. This gives the points used in Figure 4.8(a) to graph $y = \log_2 x$. For comparison, a graph of $y = 2^x$ is also shown on the graph. These graphs are mirror images with respect to the line $y = x$.

(b) $y = \log_{1/2} x$

This graph is found by rewriting $y = \log_{1/2} x$ in exponential form as $(1/2)^y = x$. Then choose several values of y and find the corresponding values of x. Use the results to get the graph of $y = \log_{1/2} x$ in Figure 4.8(b). Again, for comparison, a graph of $y = (1/2)^x$ is included. These two graphs are mirror images with respect to $y = x$. **3**

	$x = 2^y$	
y	x	(x, y)
-1	$2^{-1} = \dfrac{1}{2}$	$\left(\dfrac{1}{2}, -1\right)$
0	$2^0 = 1$	$(1, 0)$
1	$2^1 = 2$	$(2, 1)$
2	$2^2 = 4$	$(4, 2)$
3	$2^3 = 8$	$(8, 3)$

	$x = \left(\dfrac{1}{2}\right)^y$	
y	x	(x, y)
-2	$\left(\dfrac{1}{2}\right)^{-2} = 4$	$(4, -2)$
-1	$\left(\dfrac{1}{2}\right)^{-1} = 2$	$(2, -1)$
0	$\left(\dfrac{1}{2}\right)^0 = 1$	$(1, 0)$
1	$\left(\dfrac{1}{2}\right)^1 = \dfrac{1}{2}$	$\left(\dfrac{1}{2}, 1\right)$
2	$\left(\dfrac{1}{2}\right)^2 = \dfrac{1}{4}$	$\left(\dfrac{1}{4}, 2\right)$

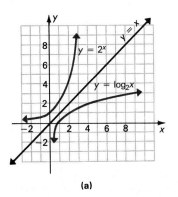

(a)

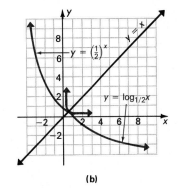

(b)

Figure 4.8

Typical graphs of logarithmic and exponential functions are shown below.

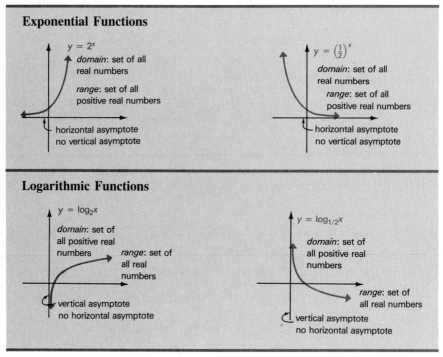

Exponential Functions

$y = 2^x$
domain: set of all real numbers

range: set of all positive real numbers

horizontal asymptote
no vertical asymptote

$y = \left(\frac{1}{2}\right)^x$
domain: set of all real numbers
range: set of all positive real numbers

horizontal asymptote
no vertical asymptote

Logarithmic Functions

$y = \log_2 x$
domain: set of all positive real numbers
range: set of all real numbers

vertical asymptote
no horizontal asymptote

$y = \log_{1/2} x$
domain: set of all positive real numbers
range: set of all real numbers

vertical asymptote
no horizontal asymptote

The exponential function defined by $f(x) = 2^x$ and the logarithmic function $g(x) = \log_2 x$ are closely related. For example, $f(3) = 2^3 = 8$, while $g(8) = \log_2 8 = 3$. Thus, $f(3) = 8$ and $g(8) = 3$. Also, $f(5) = 32$ and $g(32) = 5$. In fact, for any number m, if $f(m) = p$, then $g(p) = m$. Functions related in this way are called **inverses** of each other. Every logarithmic function is the inverse of some exponential function. A more complete discussion of inverse functions is given in most standard intermediate algebra and college algebra books.

The usefulness of logarithmic functions depends in large part on the following *properties of logarithms*.

Properties of Logarithms Let x and y be any positive real numbers and r any real number. Let a be a positive real number, $a \neq 1$. Then

(a) $\log_a xy = \log_a x + \log_a y$;   (b) $\log_a \dfrac{x}{y} = \log_a x - \log_a y$;

(c) $\log_a x^r = r \log_a x$;

(d) $\log_a a = 1$;

(e) $\log_a 1 = 0$.

Prove property (a) as follows. Let

$$m = \log_a x \qquad \text{and} \qquad n = \log_a y.$$

Then, by the definition of logarithm,

$$a^m = x \qquad \text{and} \qquad a^n = y.$$

Multiply to get

$$a^m \cdot a^n = x \cdot y,$$

or, by a property of exponents,

$$a^{m+n} = xy.$$

Use the definition of logarithm to rewrite this last statement as

$$\log_a xy = m + n.$$

Replace m with $\log_a x$ and n with $\log_a y$ to get

$$\log_a xy = \log_a x + \log_a y.$$

Properties (b) and (c) can be proven in a similar way. Since $a^1 = a$ and $a^0 = 1$, properties (d) and (e) come from the definition of logarithm.

4 Use the properties of logarithms to rewrite and evaluate each of the following, given $\log_3 7 \approx 1.77$ and $\log_3 5 \approx 1.46$.
(a) $\log_3 35$
(b) $\log_3 7/5$
(c) $\log_3 25$
(d) $\log_3 3$
(e) $\log_3 1$

Answer:
(a) 3.23
(b) .31
(c) 2.92
(d) 1
(e) 0

Example 3 Using the properties of logarithms, if $\log_6 7 \approx 1.09$ and $\log_6 5 \approx .9$,
(a) $\log_6 35 = \log_6 7 \cdot 5 = \log_6 7 + \log_6 5 \approx 1.09 + .9 = 1.99$;
(b) $\log_6 5/7 = \log_6 5 - \log_6 7 \approx -.19$;
(c) $\log_6 5^3 = 3 \log_6 5 \approx 3(.9) = 2.7$;
(d) $\log_6 6 = 1$;
(e) $\log_6 1 = 0$. **4**

Example 4 If all the following variable expressions represent positive numbers, then for $a > 0$, $a \neq 1$,
(a) $\log_a x + \log_a (x - 1) = \log_a x(x - 1)$
(b) $\log_a \dfrac{x^2 + 4}{x + 6} = \log_a (x^2 + 4) - \log_a (x + 6)$
(c) $\log_a 9x^5 = \log_a 9 + \log_a x^5 = \log_a 9 + 5 \log_a x$. **5**

5 Simplify, using the properties of logarithms.
(a) $\log_a 5x + \log_a 3x^4$
(b) $\log_a 3p - \log_a 5q$
(c) $4 \log_a k - 3 \log_a m$

Answer:
(a) $\log_a 15x^5$
(b) $\log_a 3p/(5q)$
(c) $\log_a k^4/m^3$

Historically, one of the main applications of logarithms has been as an aid to numerical calculation. Many numerical problems can be simplified using the properties above and tables of logarithms. Since our number system has base 10, logarithms to base 10 are most convenient for numerical calculations. Base 10 logarithms are called **common logarithms.** For simplicity, $\log_{10} x$ is abbreviated $\log x$. With this notation,

$$\log 1000 = 3,$$
$$\log 100 = 2,$$
$$\log 1 = 0,$$
$$\log .01 = -2,$$

and so on.

The use of common logarithms to simplify numerical calculations is discussed in the Appendix to this chapter. Common logarithms have few applications other than numerical calculations. In most other practical applications of logarithms, the number e is used as base. (Recall: To seven decimal places, $e = 2.7182818$.) Logarithms to base e are called **natural logarithms,** with the natural logarithm of x written $\ln x$. While common logarithms may seem more ''natural'' than logarithms to base e, there are good reasons for using natural logarithms instead. The most important reason is discussed in Section 11.8. Natural logarithms can be found with some calculators, or in Table 4. ⑥

The next example shows how to use properties of logarithms to find values not given directly in Table 4.

Example 5 Use a calculator or Table 4 to find the following logarithms.
(a) $\ln 83$

With a calculator, press the keys for 83, then press the ln key, and read the result, 4.4188, to four decimal places. Table 4 does not give $\ln 83$. However, the value of $\ln 83$ can be found from the properties of logarithms:

$$\ln 83 = \ln (8.3 \times 10)$$
$$= \ln 8.3 + \ln 10$$
$$\approx 2.1163 + 2.3026$$
$$= 4.4189.$$

This result is slightly different from the answer found above when using a calculator. This difference is due to rounding error.
(b) $\ln 36$

A calculator gives $\ln 36 = 3.5835$ to four decimal places. To use the table, first use properties of logarithms, since 36 is not listed directly in Table 4.

$$\ln 36 = \ln 6^2$$
$$= 2 \ln 6$$
$$\approx 2(1.7918)$$
$$= 3.5836$$

⑥ Use Table 4 or a calculator to find the following.
(a) $\ln 6.1$
(b) $\ln 20$
(c) $\ln .8$
(d) $\ln .1$

Answer:
(a) 1.8083
(b) 2.9957
(c) −.2231
(d) −2.3026

Alternatively, find $\ln 36$ as follows:

7 Use Table 4 or a calculator to find the following to four decimal places.
(a) $\ln 15$
(b) $\ln 28$
(c) $\ln 270$

Answer:
(a) 2.7081
(b) 3.3322
(c) 5.5984

$$\ln 36 = \ln 9 \cdot 4 = \ln 9 + \ln 4 \approx 2.1972 + 1.3863 = 3.5835. \quad \boxed{7}$$

The next example shows how to solve equations involving logarithms.

Example 6 Solve each of the following equations.
(a) $\log_x 8/27 = 3$

First, use the definition of logarithm and write the expression in exponential form.

$$x^3 = \frac{8}{27}$$

$$x^3 = \left(\frac{2}{3}\right)^3$$

$$x = \frac{2}{3}$$

The solution is 2/3.
(b) $\log_4 x = 5/2$

In exponential form, the given statement becomes

$$4^{5/2} = x$$

$$(4^{1/2})^5 = x$$

$$2^5 = x$$

$$32 = x.$$

The solution is 32. **8**

8 Solve each equation.
(a) $\log_x 6 = 1$
(b) $\log_5 25 = m$
(c) $\log_{27} x = 2/3$

Answer:
(a) 6
(b) 2
(c) 9

In the next example, properties of logarithms are needed to solve equations.

Example 7 Solve $\log_2 x - \log_2 (x - 1) = 1$.

By the properties of logarithms, the left-hand side can be simplified as

$$\log_2 x - \log_2 (x - 1) = \log_2 \frac{x}{x - 1}.$$

The original equation then becomes

$$\log_2 \frac{x}{x - 1} = 1.$$

Use the definition of logarithm to write this last result as

$$\frac{x}{x - 1} = 2^1 \qquad \text{or} \qquad \frac{x}{x - 1} = 2.$$

Solve this equation:

$$\frac{x}{x-1} \cdot (x-1) = 2(x-1)$$

$$x = 2(x-1)$$
$$x = 2x - 2$$
$$-x = -2$$
$$x = 2.$$

9 Solve each equation.
(a) $\log_5 x + 2\log_5 x = 3$
(b) $\log_6 (a + 2)$

$$- \log_6 \frac{a-7}{5} = 1$$

Answer:
(a) 5
(b) 52

The domain of logarithmic functions includes only positive real numbers, so it is *necessary* to check this proposed solution in the original equation.

$$\log_2 x - \log_2 (x - 1) = 1$$
$$\log_2 2 - \log_2 (2 - 1) = 1 \qquad \text{Let } x = 2$$
$$1 - 0 = 1 \qquad \text{Definition of logarithm}$$

The solution, 2, checks. **9**

4.3 Exercises

Write each of the following in logarithmic form. (See Example 1.)

1. $2^3 = 8$

2. $5^2 = 25$

3. $3^4 = 81$

4. $6^3 = 216$

5. $\left(\frac{1}{3}\right)^{-2} = 9$

6. $\left(\frac{3}{4}\right)^{-2} = \frac{16}{9}$

Write each of the following in exponential form. (See Example 1.)

7. $\log_2 128 = 7$

8. $\log_3 81 = 4$

9. $\log_5 \frac{1}{25} = -2$

10. $\log_2 \frac{1}{8} = -3$

11. $\log 10,000 = 4$

12. $\log .00001 = -5$

Evaluate each of the following.

13. $\log 1000$

14. $\log 100$

15. $\log .01$

16. $\log .0001$

17. $\log_5 25$

18. $\log_9 81$

19. $\log_4 64$

20. $\log_6 216$

21. $\log_2 \frac{1}{4}$

22. $\log_3 \frac{1}{27}$

23. $\log_9 \frac{1}{81}$

24. $\log_4 \frac{1}{16}$

25. $\log_2 \sqrt[3]{\frac{1}{4}}$

26. $\log_8 \sqrt[4]{\frac{1}{2}}$

27. $\log_6 36^4$

28. $\log_5 125^2$

29. $\log_e \sqrt{e}$

30. $\log_e \frac{1}{e}$

31. Complete the following table of values for $y = \log_3 x$.

x	1/27						9
y	-3	-2	-1	0	1	2	3

Graph $y = \log_3 x$ using the same scale on both axes.

32. Complete the following table of values for $y = 3^x$.

x	-3	-2	-1	0	1	2	3
y		1/9					27

Graph $y = 3^x$ on the same axes you used in Exercise 31. Compare the two graphs. How are they related?

Graph each of the following. (See Example 2.)

33. $y = \log_4 x$

34. $y = \log x$

35. $y = \log_3 (x - 1)$

36. $y = \log_2 (1 + x)$

37. $y = \log_2 x^2$

38. $y = \log_2 |x|$

Given $\log_{10} 2 \approx .3010$ and $\log_{10} 3 \approx .4771$, find each of the following without using a calculator or a table. (See Example 3.)

39. $\log_{10} 6$ **40.** $\log_{10} 12$ **41.** $\log_{10} 9$

42. $\log_{10} 20$ **43.** $\log_{10} 30$ **44.** $\log_{10} 36$

Given $\ln 2 \approx .69315$ and $\ln 3 \approx 1.09861$, evaluate the following without using a calculator or a table. (See Example 3.)

45. $\ln 6$ **46.** $\ln 12$ **47.** $\ln \dfrac{3}{2}$

48. $\ln \dfrac{2}{3}$ **49.** $\ln 3^4$ **50.** $\ln 2^5$

51. $\ln 2^{.1}$ **52.** $\ln 3^{.2}$ **53.** $\ln 3^{-.2}$

54. $\ln 2^{-.1}$ **55.** $\ln 18$ **56.** $\ln 24$

Use the properties of logarithms and write each of the following as a sum, difference, or product. Assume all variable expressions represent positive real numbers. (See Example 4.)

57. $\log_3 \dfrac{2}{5}$ **58.** $\log_4 \dfrac{6}{7}$ **59.** $\log_9 7m$

60. $\log_5 8p$ **61.** $\log_3 \dfrac{3x}{5k}$ **62.** $\log_7 \dfrac{11p}{13y}$

63. $\log_k \dfrac{pq^2}{m}$ **64.** $\log_z \dfrac{x^5 y^3}{3}$

65. $\log_r (5m + 7n)$ **66.** $\log_y (8k + 7z)$

67. $\log_3 \dfrac{5\sqrt{2}}{\sqrt[4]{7}}$ **68.** $\log_2 \dfrac{9\sqrt[3]{5}}{\sqrt[4]{3}}$

Find each of the following natural logarithms. (See Example 5.)

69. $\ln 20$ **70.** $\ln 35$ **71.** $\ln 60$

72. $\ln 50$ **73.** $\ln 800$ **74.** $\ln 920$

75. $\ln 532$ **76.** $\ln 255$ **77.** $\ln 768$

78. $\ln 324$ **79.** $\ln 58{,}500$ **80.** $\ln 12{,}400$

81. $\ln 59.43$ **82.** $\ln 1.074$ **83.** $\ln .428$

84. $\ln .397$

Suppose $\log_b 2 = a$ and $\log_b 3 = c$. Use the properties of logarithms to find the following logarithms.

85. $\log_b 8$ **86.** $\log_b 24$ **87.** $\log_b 54$

88. $\log_b 144$ **89.** $\log_b (72b)$ **90.** $\log_b (4b^2)$

Solve each of the following equations. (See Examples 6 and 7.)

91. $\log_x 25 = -2$ **92.** $\log_x \dfrac{1}{16} = -2$

93. $\log_9 27 = m$ **94.** $\log_8 4 = z$

95. $\log_y 8 = \dfrac{3}{4}$ **96.** $\log_r 7 = \dfrac{1}{2}$

97. $\log_3 (5x + 1) = 2$ **98.** $\log_5 (9x - 4) = 1$

99. $\log_4 x - \log_4 (x + 3) = -1$

100. $\log_9 m - \log_9 (m - 4) = -2$

101. $\log_3 (y + 2) = \log_3 (y - 7) + \log_3 4$

102. $\log_8 (z - 6) = 2 - \log_8 (z + 15)$

103. $\log_7 k - \log_7 (k + 1) = \log_7 5$

104. $\log_2 (5 + 4y) - \log_2 (3 + y) = \log_2 3$

Work the following exercises.

105. Management Suppose the sales of a certain product are approximated by

$$S(t) = 125 + 83 \log(5t + 1),$$

where $S(t)$ is sales in thousands of dollars t years after the product was introduced on the market. Find
(a) $S(0)$ (b) $S(2)$
(c) $S(4)$ (d) $S(31)$.
(e) Graph $y = S(t)$.

106. Natural Science The population of an animal species that is introduced into a certain area may grow rapidly at first but then grow more slowly as time goes on. A logarithmic function can provide an excellent description of such growth. Suppose that the population of foxes, $F(t)$, in an area t months after the foxes were introduced there is

$$F(t) = 500 \log (2t + 3).$$

Find the population of foxes at the following times:
(a) when they are first released into the area (that is, when $t = 0$);
(b) after 3 months;
(c) after 15 months.
(d) Graph $y = F(t)$.

Management In many applications, data is graphed on a logarithmic scale, where differences between successive measurements are not always the same. For example, the *graph at the side from the July 30, 1979, issue of* Fortune *magazine (page 58),* shows the price/performance ratio for various models of IBM computers. Notice on the vertical scale that the distance from* 100 *to* 500 *is the same as the distance from* 1000 *to* 5000. *This is characteristic of a graph drawn with logarithmic scales.*

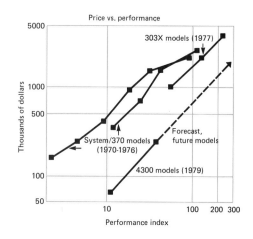

107. Estimate the price for a 4300 model computer producing a performance index of 10. Do the same for a System/370 (use the upper graph).

108. To locate a performance index of 50 on the horizontal scale, first find that log 50 ≈ 1.7, so that 50 would be located about .7, or 70% of the way from 10 to 100. Locate 50 on the horizontal axis, and then estimate the price for a model 4300 computer with a performance index of 50.

4.4 Applications of Logarithmic Functions

This section first shows an additional method of solving exponential and logarithmic equations and then shows several applications using this method. Some of these applications depend on the following result.

Let x and y be positive numbers. Let b be a positive number, $b \neq 1$.

If $x = y$, then $\log_b x = \log_b y$.

If $\log_b x = \log_b y$, then $x = y$.

For convenience, base e is generally used in applications.

Example 1 Solve $3^x = 5$.

Since 3 and 5 cannot easily be written with the same base, the methods of Section 4.1 cannot be used to solve this equation. Instead use the result given above and take natural logarithms of both sides.

$$3^x = 5$$
$$\ln 3^x = \ln 5$$
$$x \ln 3 = \ln 5 \qquad \text{Property of logarithms}$$
$$x = \frac{\ln 5}{\ln 3}$$

---

*Graph adapted from "Price vs. Performance" from How the 4300 Fits I.B.M.'s New Strategy by Bro Uttal from *Fortune,* July 30, 1979. Copyright © 1979 Time, Inc. All rights reserved. Reprinted by permission.

From Table 4 or a calculator, ln 5 ≈ 1.6094 and ln 3 ≈ 1.0986, so

$$x \approx \frac{1.6094}{1.0986}.$$

Use a calculator to divide 1.6094 by 1.0986:

$$x = 1.465,$$

rounded to the nearest thousandth.

A calculator with a y^x key can be used to check this answer. Evaluate $3^{1.465}$; the result should be approximately 5. This step verifies that, to the nearest thousandth, the solution of the given equation is 1.465. **1**

By the definition of logarithm, ln e = 1. This fact, along with properties of logarithms, can be used to solve equations involving powers of e, as the next example shows.

Example 2 Solve each equation, and round solutions to the nearest thousandth.
(a) $e^{.04m} = 13$

Take natural logarithms of both sides.

$$e^{.04m} = 13$$
$$\ln e^{.04m} = \ln 13$$

Use a property of logarithms to rewrite this last equation as

$$.04m \ln e = \ln 13.$$

Since ln e = 1, the equation becomes

$$.04m(1) = \ln 13$$
$$.04m = \ln 13$$
$$m = \frac{\ln 13}{.04}$$
$$m \approx \frac{2.5649}{.04}$$
$$m = 64.123 \quad \text{(rounded)}.$$

(b) $e^{2-x} = .798$

Work as above.

$$\ln e^{2-x} = \ln .798$$
$$(2 - x)\ln e = \ln .798$$
$$2 - x = \ln .798$$
$$-x = -2 + \ln .798$$
$$x = 2 - \ln .798$$
$$x \approx 2 - (-.226)$$
$$x = 2.226 \quad \boxed{2}$$

1 Solve each equation. Round to the nearest thousandth.
(a) $2^x = 7$
(b) $5^m = 50$
(c) $3^y = 17$

Answer:
(a) 2.807
(b) 2.431
(c) 2.579

2 Solve each equation. Round to the nearest thousandth.
(a) $e^{.1x} = 11$
(b) $e^{-x/2} = 5$
(c) $e^{3+x} = .893$

Answer:
(a) 23.979
(b) −3.219
(c) −3.113

Example 3 Solve $3^{2x-1} = 4^{x+2}$.

Taking natural logarithms on both sides gives

$$\ln 3^{2x-1} = \ln 4^{x+2}.$$

Now use a property of logarithms.

$$(2x - 1) \ln 3 = (x + 2) \ln 4$$
$$2x \ln 3 - \ln 3 = x \ln 4 + 2 \ln 4$$
$$2x \ln 3 - x \ln 4 = 2 \ln 4 + \ln 3$$

Factor out x on the left to get

$$x(2 \ln 3 - \ln 4) = 2 \ln 4 + \ln 3$$

or

$$x = \frac{2 \ln 4 + \ln 3}{2 \ln 3 - \ln 4}.$$

Using properties of logarithms,

$$x = \frac{\ln 16 + \ln 3}{\ln 9 - \ln 4}$$

or, finally,

$$x = \frac{\ln 48}{\ln \frac{9}{4}}.$$

This quotient could be approximated by a decimal if desired.

$$x = \frac{\ln 48}{\ln 2.25} \approx \frac{3.8712}{.8109} \approx 4.774$$

To the nearest thousandth, the solution is 4.774. To find $\ln 2.25$ with Table 4, write $\ln 2.25$ as $\ln 1.5^2 = 2 \ln 1.5$. **3**

3 Solve each equation. Round to the nearest thousandth.
(a) $6^m = 3^{2m-1}$
(b) $5^{6a-3} = 2^{4a+1}$

Answer:
(a) 2.710
(b) .802

Example 4 Solve $3e^{x^2} = 600$.

First, divide each side by 3 to get

$$e^{x^2} = 200.$$

Now take natural logarithms on both sides; then use properties of logarithms.

$$e^{x^2} = 200$$
$$\ln e^{x^2} = \ln 200$$
$$x^2 \ln e = \ln 200$$

Since $\ln e = 1$,

$$x^2 = \ln 200$$
$$x = \pm \sqrt{\ln 200}$$
$$x \approx \pm 2.302.$$

The solutions are ± 2.302, rounding to the nearest thousandth. (The symbol $\pm$ is used as a shortcut for writing the two solutions, 2.302 and -2.302.) **4**

4 Solve each equation. Round to the nearest thousandth.
(a) $8e^{k^2} = 16$
(b) $e^{x^2+1} = 35$
(c) $e^{2x^2-3} = 9$

Answer:
(a) $\pm.833$
(b) ±1.599
(c) ±1.612

The result given at the beginning of this section, with the properties of logarithms given in Section 4.3, is useful in solving logarithmic equations, as shown in the next examples.

Example 5 Solve $\log_a (x + 4) - \log_a (x + 2) = \log_a x$.
Using a property of logarithms, rewrite the equation as

$$\log_a \frac{x + 4}{x + 2} = \log_a x.$$

Then

$$\frac{x + 4}{x + 2} = x$$
$$x + 4 = x(x + 2)$$
$$x + 4 = x^2 + 2x$$
$$x^2 + x - 4 = 0.$$

By the quadratic formula,

$$x = \frac{-1 \pm \sqrt{1 + 16}}{2}$$

so that

$$x = \frac{-1 + \sqrt{17}}{2} \qquad \text{or} \qquad x = \frac{-1 - \sqrt{17}}{2}.$$

$\text{Log}_a x$ cannot be evaluated for $x = (-1 - \sqrt{17})/2$, since this number is negative and not in the domain of $\log_a x$. By substitution verify that $(-1 + \sqrt{17})/2$ is a solution. **5**

5 Solve each equation.
(a) $\log_2 (p + 9) - \log_2 p$
     $= \log_2 (p + 1)$
(b) $\log_3 (m + 1) -$
     $\log_3 (m - 1) = \log_3 m$

Answer:
(a) 3
(b) $1 + \sqrt{2} \approx 2.414$

Example 6 Suppose that $A(t)$, the amount of a certain radioactive substance present at time t, is given by

$$A(t) = 1000e^{-.1t},$$

where t is measured in days and $A(t)$ in grams. At time $t = 0$,

$$A(0) = 1000e^{-.1(0)}$$
$$= 1000$$

grams of the substance present. Also,

$$A(5) = 1000e^{-.1(5)}$$
$$= 1000e^{-.5}$$
$$\approx 1000(.60653)$$
$$= 606.53,$$

so that about 607 grams are still present after 5 days. Now let us find the half-life of the substance. (The **half-life** of a radioactive substance is the time it takes for exactly half the sample to decay.)

Find the half-life by finding a value of t such that $A(t) = (1/2)(1000) = 500$ grams. That is, find the half-life by solving the equation

$$500 = 1000e^{-.1t}.$$

First, divide both sides by 1000, obtaining

$$\frac{1}{2} = e^{-.1t}.$$

Now take natural logarithms of both sides. This gives

$$\ln\frac{1}{2} = \ln e^{-.1t}.$$

Using a property of logarithms,

$$\ln\frac{1}{2} = (-.1t)(\ln e),$$

and since $\ln e = 1$,

$$\ln\frac{1}{2} = -.1t,$$

or

$$t = \frac{\ln\frac{1}{2}}{-.1}.$$

Since $\ln 1/2 = \ln .5$, use Table 4 or a calculator to get

$$t \approx \frac{-.6931}{-.1} \approx 6.9.$$

It will take about 6.9 days for half the sample to decay. **6** **7**

Example 7 Carbon 14 is a radioactive isotope of carbon which has a half-life of about 5600 years. The earth's atmosphere contains much carbon, mostly in the form of carbon dioxide gas, with small traces of carbon 14. Most atmospheric carbon is the nonradioactive isotope, carbon 12. The ratio of carbon 14 to carbon 12 is virtually constant in the atmosphere. However, as a plant absorbs carbon dioxide from the air in the process of photosynthesis, the carbon 12 stays in the plant while the carbon 14 is converted into nitrogen. Thus, in a plant, the ratio of carbon 14 to carbon 12 is smaller than the ratio in the atmosphere. Even when the plant dies, this ratio will continue to decrease. Based on this, a method of dating objects called **carbon-14 dating** has been developed.

(a) Suppose a skeleton has been discovered in which the ratio of carbon 14 to carbon 12 is only about half the ratio found in the atmosphere. How long ago did this individual die?

As mentioned above, in 5600 years half the carbon 14 in a specimen will decay, so the individual died about 5600 years ago.

6 The amount of a substance present at time t (in hours) is given by

$$A(t) = 530e^{-.2t},$$

and $A(t)$ is measured in grams. How much of the substance will remain after 5 hours? That is, find $A(5)$.

Answer:

About 195 grams

7 Find the half-life of the substance in Problem 6 above.

Answer:

About 3.5 hours

(b) Let R be the (nearly constant) ratio of carbon 14 to carbon 12 found in the atmosphere, and let r be the ratio as found in an observed specimen. It can then be shown that the relationship between R and r is given by

$$R = r \cdot e^{(t \ln 2)/5600},$$

where t is the number of years since the death of the specimen. Verify the formula for $t = 0$.

To do this, substitute 0 for t in the formula. This gives

$$R = r \cdot e^{(0 \cdot \ln 2)/5600}$$
$$= r \cdot e^0$$
$$= r \cdot 1$$
$$= r.$$

This result is correct; when $t = 0$, the specimen has just died, so that R and r should be the same.

(c) Suppose a specimen is found in which $r = (2/3)R$. Estimate the age of the specimen.

Use the relationship given in (b) above and substitute $(2/3)R$ for r.

$$R = r \cdot e^{(t \ln 2)/5600}$$
$$= \frac{2}{3}R \cdot e^{(t \ln 2)/5600}$$

Dividing both sides by R, and multiplying both sides by 3/2 gives

$$\frac{3}{2} = e^{(t \ln 2)/5600}.$$

Taking natural logarithms of both sides,

$$\ln \frac{3}{2} = \ln e^{(t \ln 2)/5600}.$$

Using properties of logarithms,

$$\ln \frac{3}{2} = \left(\frac{t \ln 2}{5600} \right) \ln e.$$

Since $\ln e = 1$,

$$\ln \frac{3}{2} = \frac{t \ln 2}{5600}.$$

Solve this equation for t, the age of the specimen, by multiplying both sides by 5600/ln 2. This gives

$$\frac{5600 \ln 3/2}{\ln 2} = t.$$

8 What is the age of a specimen in which $r = (1/3)R$?

Answer:
About 8880 years

Using Table 4 or a calculator, $\ln 3/2 = \ln 1.5 \approx .4055$ and $\ln 2 \approx .6931$. Thus,

$$t \approx \frac{5600(.4055)}{.6931} \approx 3280,$$

so that the specimen is about 3280 years old. **8**

4.4 Exercises

Solve each of the following equations. Round to the nearest thousandth. (See Examples 1–5.)

1. $3^x = 6$

2. $4^x = 12$

3. $7^x = 8$

4. $13^p = 55$

5. $3^{a+2} = 5$

6. $5^{2-x} = 12$

7. $6^{1-2k} = 8$

8. $2^{k-3} = 11$

9. $5 \cdot 4^{3m-1} = 5 \cdot 12^{m+2}$

10. $8 \cdot 3^{2m-5} = 8 \cdot 13^{m-1}$

11. $e^{k-1} = 4$

12. $e^{2-y} = 12$

13. $2e^{5a+2} = 8$

14. $10e^{3z-7} = 5$

15. $2^x = -3$

16. $(1/4)^p = -4$

17. $\left(1 + \dfrac{r}{2}\right)^5 = 9$

18. $\left(1 + \dfrac{n}{4}\right)^3 = 12$

19. $100(1 + .02)^{3+n} = 150$

20. $500(1 + .05)^{p/4} = 200$

21. $2^{x^2-1} = 12$

22. $3^{2-x^2} = 4$

23. $2(e^x + 1) = 10$

24. $5(e^{2x} - 2) = 15$

25. $\log (t - 1) = 1$

26. $\log q^2 = 1$

27. $\log (x - 3) = 1 - \log x$

28. $\log (z - 6) = 2 - \log (z + 15)$

29. $\ln (y + 2) = \ln (y - 7) + \ln 4$

30. $\ln p - \ln (p + 1) = \ln 5$

31. $\ln (3x - 1) - \ln (2 + x) = \ln 2$

32. $\ln (8k - 7) - \ln(3 + 4k) = \ln (9/11)$

33. $\ln (5 + 4y) - \ln (3 + y) = \ln 3$

34. $\ln m + \ln (2m + 5) = \ln 7$

35. $\ln x + 1 = \ln (x - 4)$

36. $\ln (4x - 2) = \ln 4 - \ln (x - 2)$

37. $2 \ln (x - 3) = \ln (x + 5) + \ln 4$

38. $\ln (k + 5) + \ln (k + 2) = \ln 14k$

39. $\log_5 (r + 2) + \log_5 (r - 2) = 1$

40. $\log_4 (z + 3) + \log_4 (z - 3) = 1$

41. $\log_3 (a - 3) = 1 + \log_3 (a + 1)$

42. $\log w + \log (3w - 13) = 1$

43. $\log_2 \sqrt{2y^2} - 1 = 1/2$

44. $\log_2 (\log_2 x) = 1$

45. $\log z = \sqrt{\log z}$

46. $\log x^2 = (\log x)^2$

47. $\log_x 5.87 = 2$

48. $\log_x 11.9 = 3$

49. $1.8^{p+4} = 9.31$

50. $3.7^{5z-1} = 5.88$

Work the following exercises. (See Example 6.)

51. Natural Science The amount of a certain radioactive specimen present at time t (measured in seconds) is given by

$$A(t) = 5000e^{-.02t},$$

where $A(t)$ is the amount measured in grams. Find each of the following.

(a) $A(0)$ (b) $A(5)$ (c) $A(20)$

(d) The half-life of the specimen

52. Social Science The population of a small mining town has been decreasing, with the population at time t in years given by

$$P(t) = 8000e^{-.04t}.$$

Find each of the following.

(a) $P(0)$ (b) $P(5)$ (c) $P(10)$

(d) How many years will it take for the town to lose half its population?

(e) When will the population be down to 2000 people?

53. Natural Science Find the half-life for the following radioactive substances, if the amount of the substance present after t days is as follows.
(a) $A(t) = 5000e^{-.03t}$
(b) $A(t) = 2350e^{-.08t}$
(c) $A(t) = 18,000e^{-.0002t}$

54. Natural Science The number of bacteria in a certain culture, $B(t)$, is approximated by

$$B(t) = 250,000e^{-.04t},$$

where t is time measured in hours. Find each of the following.

(a) B_0 (b) $B(5)$ (c) $B(20)$
(d) The time it will take until only 125,000 bacteria are present
(e) The time it will take until only 25,000 bacteria are present

Using Example 7, find the age of a specimen for each of the following.

55. $r = .8R$ **56.** $r = .4R$

57. $r = .1R$ **58.** $r = .01R$

59. Natural Science A large cloud of radioactive debris from a nuclear explosion has floated over the Pacific Northwest, contaminating much of the hay supply. Consequently, farmers in the area are concerned that the cows who eat this hay will give contaminated milk. (The tolerance level for radioactive iodine in milk is 0.) The percent of the initial amount of radioactive iodine still present in the hay after t days is approximated by $P(t)$, which is given by the mathematical model

$$P(t) = 100e^{-.1t}.$$

(a) Find the percent remaining after 4 days.
(b) Find the percent remaining after 10 days.
(c) Some scientists feel that the hay is safe after the percent of radioactive iodine has declined to 10% of the original amount. Solve the equation $10 = 100e^{-.1t}$ to find the number of days before the hay may be used.
(d) Other scientists believe that the hay is not safe until the level of radioactive iodine has declined to only 1% of the original level. Find the number of days that this would take.

60. Social Science The number of years, $N(r)$, since two independently evolving languages split off from a com-

mon ancestral language is approximated by

$$N(r) = -5000 \ln r,$$

where r is the proportion of the words from the ancestral language that are common to both languages now. Find each of the following.

(a) $N(.9)$ (b) $N(.5)$ (c) $N(.3)$
(d) How many years have elapsed since the split if 70% of the words of the ancestral language are common to both languages today?
(e) If two languages split off from a common ancestral language about 1000 years ago, find r.

Natural Science *For Exercises 61–64, recall that* log x *represents the common (or base 10) logarithm of x.*

61. The loudness of sounds is measured in a unit called a decibel. To do this, a very faint sound, called the *threshold sound*, is assigned an intensity I_0. If a particular sound has intensity I, then the decibel rating of this louder sound is

$$10 \cdot \log \frac{I}{I_0}.$$

Find the decibel ratings of sounds having the following intensities.

(a) $100I_0$ (b) $1000I_0$
(c) $100,000I_0$ (d) $1,000,000I_0$

62. Find the decibel ratings of the following sounds, having intensities as given. Round answers to the nearest whole number.

(a) Whisper, $115I_0$
(b) Busy street, $9,500,000I_0$
(c) Heavy truck, 20 m away, $1,200,000,000I_0$
(d) Rock music, $895,000,000,000I_0$
(e) Jetliner at takeoff, $109,000,000,000,000I_0$

63. The intensity of an earthquake, measured on the *Richter Scale*, is given by

$$\log \frac{I}{I_0},$$

where I_0 is the intensity of an earthquake of a certain (small) size. Find the Richter Scale ratings of earthquakes having intensity

(a) $1000I_0$ (b) $1,000,000I_0$ (c) $100,000,000I_0$.

64. The San Francisco earthquake of 1906 had a Richter Scale rating of 8.6. Use a calculator with a y^x key to express the intensity of this earthquake as a multiple of I_0. (See Exercise 63.)

To find the maximum permitted levels of certain pollutants in fresh water, the EPA has established the functions defined in Exercises 65–66, where M(h) is the maximum permitted level of pollutant for a water hardness of h milligrams per liter. Find M(h) in each case. (These results give the maximum permitted average concentration in micrograms per liter for a 24-hour period.)

65. Copper: $M(h) = e^r$, where $r = 0.65 \ln h - 1.94$ and $h = 9.7$.

66. Lead: $M(h) = e^r$, where $r = 1.51 \ln h - 3.37$ and $h = 8.4$.

The graphs below are plotted on a logarithmic scale, *where differences between successive measurements are not always the same. Data that do not plot in a linear pattern on the usual Cartesian axes often form a linear pattern when plotted on a logarithmic scale. Notice that on the vertical scale, the distance from 1 to 2 is not the same as the distance from 2 to 3, and so on. This is characteristic of a graph drawn with logarithmic scales.*

67. The graph below gives the rate of oxygen consumption for resting guinea pigs of various sizes. This rate is proportional to body mass raised to the power .67.* Estimate the oxygen consumption for a guinea pig with body mass of .3 kg. Do the same for one with body mass of .7.

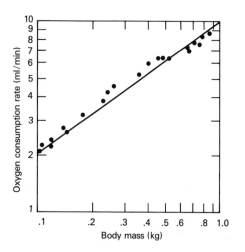

68. The graph below gives the world weight-lifting records, as log W_T, plotted against the logarithm of body weight. Here W_T is the total weight lifted in three lifts: the press, the snatch, and the clean-and-jerk. The numbers beside each point indicate the body weight class, in pounds.*

(a) Find the record for a weight of 150 pounds and for a weight of 165 pounds. (Use base 10 logarithms.)

(b) Find the body weight that corresponds to a record of 750 pounds.

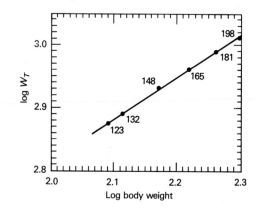

In the central Sierra Nevada Mountains of California, the percent of moisture that falls as snow rather than rain is approximated reasonably well by $p = 86.3 \ln h - 680$, where p is the percent of moisture as snow at an altitude h (in feet). (Assume $h \geq 3000$.)

69. Find the percent of moisture that falls as snow at the following altitudes:

(a) 3000 ft, **(b)** 4000 ft, **(c)** 7000 ft.

70. Graph p.

*From *On Size and Life* by Thomas A. McMahon and John Tyler Bonner. Copyright © 1983 by Thomas A. McMahon and John Tylor Bonner. Reprinted by permission of W. H. Freeman and Company.

Appendix: Common Logarithms

As mentioned earlier, common logarithms are logarithms to base 10. In this appendix we discuss the use of common logarithms as an aid in numerical calculations. Recall from Section 4.3 that the common logarithm of a number x is written log x. Also recall from Section 4.3 that we can use the definition of logarithm to write

$$\log 1000 = 3, \qquad \log 1 = 0,$$
$$\log 100 = 2, \qquad \log .1 = -1,$$
$$\log 10 = 1, \qquad \log .01 = -2,$$

and so on. In order to find the common logarithm of a number that is not a power of 10, special logarithm tables must be used. An excerpt from such a table is given below, and a complete table is given in Table 5 at the back of the book. The table below can be used to find log 4230 by writing 4230 as 4.23×1000, so that

$$\log 4230 = \log (4.23 \times 1000).$$

Using properties of logarithms,

$$\log 4230 = \log 4.23 + \log 1000.$$

Since log 1000 = 3,

$$\log 4230 = \log 4.23 + 3.$$

To find log 4.23 in the table, find 4.2 down the left column, and find 3 across the top row. This leads to the table entry .6263, so that log 4.23 = .6263. Finally,

$$\log 4230 = \log 4.23 + 3$$
$$= .6263 + 3$$
$$= 3.6263.$$

The integer part of the logarithm (3 in the example above) is called the **characteristic,** while the decimal part (.6263 in the example) is called the **mantissa.** The mantissa determines the digits, while the characteristic is used to locate the decimal point.

x	0	1	2	3	4	5	6	7	8	9
4.0	.6021	.6031	.6042	.6053	.6064	.6075	.6085	.6096	.6107	.6117
4.1	.6128	.6138	.6149	.6160	.6170	.6180	.6191	.6201	.6212	.6222
4.2	.6232	.6243	.6253	.6263	.6274	.6284	.6294	.6304	.6314	.6325
4.3	.6335	.6345	.6355	.6365	.6375	.6385	.6395	.6405	.6415	.6425
4.4	.6435	.6444	.6454	.6464	.6474	.6484	.6493	.6503	.6513	.6522

The characteristic can be found by writing the number in scientific notation. For example, to find log 4230, write

$$\log 4230 = \log (4.230 \times 10^3).$$

The exponent on 10 gives the characteristic. The mantissa is found from the table as explained above.

Example 1 Find each of the following common logarithms.
(a) log 418,000

Since $418{,}000 = 4.18 \times 100{,}000 = 4.18 \times 10^5$,

$$\begin{aligned}
\log 418{,}000 &= \log (4.18 \times 10^5) \\
&= \log 4.18 + \log 10^5 \\
&= .6212 + 5 \\
&= 5.6212.
\end{aligned}$$

(b) log 4.47 = .6503

Here the characteristic is 0. ■

Example 2 Find log .00587.

Use scientific notation and the properties of logarithms to get

$$\begin{aligned}
\log .00587 &= \log (5.87 \times 10^{-3}) \\
&= \log 5.87 + \log 10^{-3} \\
&= .7686 + (-3) \\
&= .7686 - 3.
\end{aligned}$$

The logarithm is usually left in this form. A calculator would give the answer as -2.2314, the algebraic sum of .7686 and -3. In the calculator answer the mantissa is a negative number. When using a table of logarithms, this is not the best form for the answer, since it is not clear which number is the mantissa and which is the characteristic. ■

Sometimes the logarithm of a number is known, and the number must be found. The desired number is called the **antilogarithm,** sometimes abbreviated *antilog.* For example, .756 is the antilogarithm of $.8785 - 1$, since

$$\log .756 = .8785 - 1.$$

Find this result by looking for .8785 in the body of the logarithm table. You should find 7.5 at the left and 6 at the top. Since the characteristic of $.8785 - 1$ is -1, the antilogarithm is

$$7.56 \times 10^{-1} = .756.$$

Example 3 Find the antilogarithm of the following logarithms.
(a) 6.6503

---

1 Find each logarithm.
(a) log 23,500
(b) log 81.7
(c) log 5.84

Answer:
(a) 4.3711
(b) 1.9122
(c) .7664

---

2 Find each logarithm.
(a) log .0985
(b) log .359
(c) log .00000804

Answer:
(a) .9934 − 2
(b) .5551 − 1
(c) .9053 − 6

As shown in the logarithm table, .6503 = log 4.47. Then 6 will be the exponent on 10 when the required number is written in scientific notation. Thus, the antilogarithm of 6.6503 is

$$4.47 \times 10^6 = 4{,}470{,}000.$$

(b) .6096 − 2

The characteristic is −2 and the mantissa is .6096. From the logarithm table, .6096 = log 4.07. The antilogarithm of .6096 − 2 is

$$4.07 \times 10^{-2} = .0407. \quad \boxed{3}$$

3 Find the antilogarithm for each of the following logarithms.
(a) 3.8751
(b) .4048 − 1

Answer:
(a) 7500
(b) .254

We can now use properties of logarithms from Section 4.3, along with Table 5, to simplify numerical calculations.

Example 4 Evaluate $\dfrac{4350(5.86)}{28.3}$.

By the properties of logarithms, $\log xy = \log x + \log y$, and $\log x/y = \log x - \log y$, so

$$\log \left[\frac{(4350)(5.86)}{28.3} \right] = \log (4350)(5.86) - \log 28.3$$

$$= \log 4350 + \log 5.86 - \log 28.3.$$

From Table 5, log 4350 = 3.6385, log 5.86 = .7679, and log 28.3 = 1.4518. Thus,

$$\log 4350 + \log 5.86 - \log 28.3 = 3.6385 + .7679 - 1.4518$$
$$= 2.9546.$$

Also from Table 5, we see that 9.01 is the number whose mantissa is closest to .9546. Thus,

$$2.9546 = 2 + .9546$$
$$= \log 100 + \log 9.01$$
$$= \log 901,$$

4 Use logarithms to evaluate
(a) 1.09(155)
(b) $\dfrac{854}{9.72}$
(c) $\dfrac{(647)(.932)}{28.5}$.

Answer:
(a) 169
(b) 87.9
(c) 21.2

with $\dfrac{(4350)(5.86)}{28.3} \approx 901.$ $\quad \boxed{4}$

Example 5 Use logarithms to evaluate $21^{3.4}$.

By the properties of logarithms,

$$\log 21^{3.4} = 3.4 \log 21.$$

From Table 5, log 21 = 1.3222. Thus,

$$\log 21^{3.4} = 3.4 \log 21$$
$$= 3.4(1.3222)$$
$$= 4.4955.$$

5 Use logarithms to evaluate $1.6^{12.5}$.

Answer:
356

From Table 5 again, the antilogarithm of .4955 is 3.13, so

$$21^{3.4} \approx 31{,}300. \quad \boxed{5}$$

Example 6   The scrap value of a machine is the value of the machine at the end of its useful life. By one method of calculating scrap value, where it is assumed a constant percentage of value is lost annually, the scrap value S is given by

$$S = C(1 - r)^n,$$

where C is the original cost, n is the useful life of the machine in years, and r is the constant annual percentage of value lost. Find the scrap value of a machine costing $30,000, having a useful life of 12 years, and a constant annual rate of value loss of 15%.

Here $C = \$30,000$, $n = 12$, and $r = 15\% = .15$. The scrap value is given by

$$S = C(1 - r)^n$$
$$= 30,000(1 - .15)^{12}$$
$$= 30,000(.85)^{12}.$$

Use common logarithms to evaluate this last quantity:

$$\log 30,000(.85)^{12} = \log 30,000 + \log (.85)^{12}$$
$$= \log 30,000 + 12 \log .85.$$

From Table 5, $\log .85 = 0.9294 - 1$, or just $-.0706$. Thus, since $\log 30,000 = 4.4771$,

$$\log 30,000(.85)^{12} = 4.4771 + 12(-.0706)$$
$$= 3.6299.$$

Finally, from Table 5,

$$S = 30,000(.85)^{12} \approx \$4260.$$

The scrap value of the machine is about $4260.   **6**

6 Use logarithms to find the scrap value of a machine after 10 years. The machine cost $42,700 new and loses 20% of its value annually.

Answer:
About $4580

Appendix Exercises

Find each of the following common logarithms. (See Examples 1 and 2.)

1. log 749 **2.** log 8560 **3.** log 9.71

4. log 86.4 **5.** log 7.13 **6.** log 912

7. log .810 **8.** log .0712 **9.** log .00348

10. log .0000426

Find the antilogarithms of the following. (See Example 3.)

11. .3502 **12.** .6675 **13.** 2.9484

14. 3.8470 **15.** 1.4886 **16.** 1.1847

17. .9355 − 2 **18.** .7210 − 3 **19.** .9294 − 4

20. .4771 − 1

Use logarithms to evaluate each of the following. (See Examples 4 and 5.)

21. $\dfrac{(71.4)(23.6)}{48.2}$ **22.** $\dfrac{(7150)(9830)}{4280}$

23. $12.8^{2.2}$ **24.** $3.67^{2.7}$

25. $\sqrt{34,500}$ **26.** $\sqrt{23.7}$

(Hint: $\sqrt{a} = a^{1/2}$)

27. $\sqrt[3]{689}$ **28.** $\sqrt[3]{1.93}$

29. $67.8^{3/2}$ **30.** $26,700^{5/4}$

Management   *Use the formula for scrap value presented in Example 6 to find the scrap value for each of the following machines.*

31. Original cost, $54,000; life, 8 years; annual rate of value loss, 12%

32. Original cost, $178,000; life, 11 years; annual rate of value loss, 14%

33. Management Midwest Creations finds that its total sales, $T(x)$, in thousands, from the distribution of x catalogs, where x is measured in thousands, is approximated by

$$T(x) = 500 \log (x + 1).$$

Find the total sales resulting from the distribution of

(a) 0 catalogs; **(b)** 5000 catalogs;
(c) 24,000 catalogs; **(d)** 49,000 catalogs.

34. Natural Science The number of fish, $F(t)$, present in a tank at time t, measured in days, is given by

$$F(t) = 4750t^{1.3}.$$

Find the number of fish present at each of the following times.

(a) $t = 8$ **(b)** $t = 12$ **(c)** $t = 30$

35. Natural Science Philadelphia Chemicals, Inc. has to store a new gas which it plans to produce. The pressure p (in pounds per cubic foot) and volume v (in cubic feet) of the gas are related by the formula

$$pv^{1.6} = 800.$$

Find v if $p = 31.2$ pounds per cubic foot.

36. Social Science A common problem in archaeology is determining estimates of populations. Several methods have been used to calculate the number of people who once occupied a site. One method relates the total surface area of a site to the number of occupants. If P represents the population of a site which covers an area of a square units, then

$$\log P = k \log a,$$

where k is an appropriate constant which varies for hilly, coastal, or desert environments, or for sites with single family dwellings or multiple family dwellings.* Find the populations of sites with the following areas, in square meters (m²), using $k = .8$.

(a) 230 m² **(b)** 95m² **(c)** 20,000 m²

Natural Science *In chemistry, the pH of a solution is defined as*

$$pH = -\log [H_3O^+]$$

where $[H_3O^+]$ is the hydronium ion concentration in moles per liter. Find the pH of each of the following, whose hydronium ion concentration is given. Round to the nearest tenth.

37. Grapefruit, 6.3×10^{-4}

38. Crackers, 3.9×10^{-9}

39. Limes, 1.6×10^{-2}

40. Sodium hydroxide (lye), 3.2×10^{-14}

Chapter 4 Review Exercises

Solve each of the following equations.

1. $2^{3x} = \dfrac{1}{8}$ **2.** $\left(\dfrac{9}{16}\right)^x = \dfrac{3}{4}$

3. $9^{2y-1} = 27^y$ **4.** $\dfrac{1}{2} = \left(\dfrac{b}{4}\right)^{1/4}$

Graph each of the following.

5. $y = 5^x$ **6.** $y = 5^{-x}$

7. $y = \log_5 x$ **8.** $y = \log_{1/5} x$

9. A company finds that its new workers produce

$$P(x) = 100 - 100e^{-.8x}$$

items per day, after x days on the job. Find each of the following.

(a) $P(0)$ **(b)** $P(1)$ **(c)** $P(5)$
(d) How many items per day would you expect an experienced worker to produce?

10. The amount of a certain radioactive material, in grams, present after t days, is given by

$$A(t) = 800e^{-.04t}.$$

Find $A(t)$ if
(a) $t = 0$, **(b)** $t = 5$.

---

*From "The Quantitative Investigation of Indian Mounds," University of California Publications in American Archaeology and Ethnology, Vol. 40. Published in 1950 by the University of California Press. Reprinted by permission of the University of California Press.

Handwritten notes: $e^{4.} = 82.9$; $\log$ Number = exponent, $\log$ base, $\log_2 64 = 6$

Write each of the following using logarithms.

11. $2^6 = 64$ **12.** $3^{1/2} = \sqrt{3}$

13. $e^{.09} = 1.09417$ **14.** $10^{1.07918} = 12$

Write each of the following using exponents.

15. $\log_2 32 = 5$ **16.** $\log_{10} 100 = 2$

17. $\ln 82.9 = 4.41763$ **18.** $\log 15.46 = 1.18921$

Handwritten note: $\log_e 82.9 =$

Evaluate each of the following.

19. $\log_3 81$ **20.** $\log_8 16$

21. $\log_{32} 16$ **22.** $\log_{25} 5$

23. $\log_{100} 1000$ **24.** $\log_{1/2} 4$

Given $\log_5 3 \approx .68$ and $\log_5 2 \approx .43$, evaluate each of the following.

25. $\log_5 18$ **26.** $\log_5 3/4$

27. $\log_5 \sqrt{8}$ **28.** $\log_5 \sqrt{6}$

Find each of the following natural logarithms.

29. $\ln 6.2$ **30.** $\ln 700$

31. $\ln 483$ **32.** $\ln .504$

Simplify using the properties of logarithms.

33. $\log_5 3k + \log_5 7k^3$ **34.** $\log_3 2y^3 - \log_3 8y^2$

35. $2 \cdot \log_2 x - 3 \cdot \log_2 m$ **36.** $5 \cdot \log_4 r - 3 \cdot \log_4 r^2$

Solve each equation. Round to the nearest thousandth.

37. $8^p = 19$ **38.** $3^z = 11$

39. $5 \cdot 2^{-m} = 35$ **40.** $2 \cdot 15^{-k} = 18$

41. $e^{-5-2x} = 5$ **42.** $e^{3x-1} = 12$

43. $10^{2r-3} = 17$ **44.** $8^{9y-4} = 15$

45. $6^{2-m} = 2^{3m+1}$ **46.** $5^{3r-1} = 6^{2r+5}$

47. $(1 + .003)^k = 1.089$ **48.** $(1 + .094)^z = 2.387$

49. $4 \cdot 3^{x^2} = 15$ **50.** $3 \cdot 5^{p^2} = 28$

51. $\log (m + 2) = 1$ **52.** $\log x^2 = 2$

53. $\log (p - 1) = 1 + \log p$

54. $\ln (m + 3) - \ln m = \ln 2$

55. $2 \ln (y + 1) = \ln (y^2 - 1) + \ln 5$

56. $\log_3 k + \log_3 (k + 2) = \log_3 8$

57. The height, in meters, of the members of a certain tribe is approximated by

$$h = .5 + \log t,$$

where t is the tribe member's age in years, and $1 \le t \le 20$. Find the height of a tribe member of age

(a) 2 years   (b) 5 years

(c) 10 years   (d) 20 years

58. The turnover of legislators is a problem of interest to political scientists. One model of legislative turnover in the U.S. House of Representatives is given by

$$M = 434e^{-.08t},$$

where M is the number of continuously serving members at time t.* This model is based on the 1965 membership of the House. Thus, 1965 corresponds to $t = 0$, 1966 to $t = 1$, and so on. Find the number of continuously serving members in each of the following years.

(a) 1969   (b) 1973   (c) 1979

Find each of the following common logarithms.

59. $\log 246$ **60.** $\log .0387$

61. $\log .000317$ **62.** $\log .978$

Find the antilogarithm of each of the following logarithms.

63. 3.4983 **64.** 5.7243

65. $.6493 - 2$ **66.** $.8142 - 5$

Use common logarithms to approximate each of the following.

67. $\dfrac{(6.49)(28.3)}{(71.5)^2}$ **68.** $2.43^{3.2}$

69. $(8.1)^{.81}$ **70.** $\sqrt[5]{\dfrac{27.1}{4.33}}$

---

*Excerpt from "Exponential Models of Legislative Turnover" by Thomas W. Casstevens. Reprinted by permission of COMAP, Arlington, MA.

Case 3

The Van Meegeren Art Forgeries*

After the liberation of Belgium at the end of World War II, officials began a search for Nazi collaborators. One person arrested as a collaborator was a minor painter, H. A. Van Meegeren; he was charged with selling a valuable painting by the Dutch artist Vermeer (1632–1675) to the Nazi Hermann Goering. He defended himself from the very serious charge of collaboration by claiming that the painting was a fake—he had forged it himself.

He also claimed that the beautiful and famous painting "Disciples at Emmaus," as well as several other supposed Vermeers, was his own work. To prove this, he did another "Vermeer" in his prison cell. An international panel of experts was assembled, which pronounced as forgeries all the "Vermeers" in question.

Many people would not accept the verdict of this panel for the painting "Disciples at Emmaus"; it was felt to be too beautiful to be the work of a minor talent such as Van Meegeren. In fact, the painting was declared genuine by a noted art scholar and sold for $170,000. The question of the authenticity of this painting continued to trouble art historians, who began to insist on conclusive proof one way or the other. This proof was given by a group of scientists at Carnegie-Mellon University, using the idea of radioactive decay.

The dating of objects is based on radioactivity; the atoms of certain radioactive elements are unstable, and within a given time period a fixed fraction of such atoms will spontaneously disintegrate, forming atoms of a new element. If t_0 represents some initial time, N_0 represents the number of atoms present at time t_0, and N represents the number present at some later time t, then it can be shown (using physics and calculus) that

$$t - t_0 = \frac{1}{\lambda} \cdot \ln \frac{N_0}{N}$$

where λ is a "decay constant" that depends on the radioactive substance under consideration.

*From "The Van Meegeren Art Forgeries" by Martin Braun from *Applied Mathematical Sciences*, Vol. 15. Copyright © 1975. Published by Springer-Verlag New York, Inc. Reprinted by permission.

If t_0 is the time that the substance was formed or made, then $t - t_0$ is the age of the item. Thus, the age of an item is given by

$$\frac{1}{\lambda} \cdot \ln \frac{N_0}{N}.$$

The decay constant λ can be readily found, as can N, the number of atoms present now. The problem is N_0—we can't find a value for this variable. However, it is possible to get reasonable ranges for the values of N_0. This is done by studying the white lead in the painting. This pigment has been used by artists for over 2000 years. It contains a small amount of the radioactive substance lead-210 and an even smaller amount of radium-226.

Radium-226 disintegrates through a series of intermediate steps to produce lead-210. The lead-210, in turn, decays to form polonium-210. This last process, lead-210 to polonium-210, has a half-life of 22 years. That is, in 22 years half the initial quantity of lead-210 will decay to polonium-210.

When lead ore is processed to form white lead, most of the radium is removed with other waste products. Thus, most of the supply of lead-210 is cut off, with the remainder beginning to decay very rapidly. This process continues until the lead-210 in the white lead is once more in equilibrium with the small amount of radium then present. Let $y(t)$ be the amount of lead-210 per gram of white lead present at time t since manufacture of the pigment. Let r represent the number of disintegrations of radium-226 per minute per gram of white lead. (Actually, r is a function of time, but the half life of radium-226 is so long in comparison to the time interval in question that we assume it to be a constant.) If λ is the decay constant for lead-210, then it can be shown that

$$y(t) = \frac{r}{\lambda} [1 - e^{-\lambda(t-t_0)}] + y_0 e^{-\lambda(t-t_0)}. \qquad \textbf{(1)}$$

All variables in this result can be evaluated except y_0, the amount of lead-210 present originally. To get around this problem, we use the fact that the original amount of lead-210 was in radioactive equilibrium with the larger amount of radium-226 in the ore from which the metal was extracted. We therefore take samples of different ores and compute the rate of disintegration of radium-226. The results are as shown in the table.

Location of ore	Disintegrations of radium-226 per minute per gram of white lead
Oklahoma	4.5
S.E. Missouri	.7
Idaho	.18
Idaho	2.2
Washington	140
British Columbia	.4

The numbers in the table vary from .18 to 140—quite a range. Since the number of disintegrations is proportional to the amount of lead-210 present originally, we must conclude that y_0 also varies over a tremendous range. Thus, equation (1) cannot be used to obtain even a crude estimate of the age of a painting. However, we want to distinguish only between a modern forgery and a genuine painting that would be 300 years old.

To do this, we observe that if the painting is very old compared to the 22 year half life of lead-210, then the amount of radioactivity from the lead-210 will almost equal the amount of radioactivity from the radium-226. On the other hand, if the painting is modern, then the amount of radioactivity from the lead-210 will be much greater than the amount from the radium-226.

We want to know if the painting is modern or 300 years old. To find out, let $t - t_0 = 300$ in equation (1), getting

$$\lambda y_0 = \lambda \cdot y(t) \cdot e^{300\lambda} - r(e^{300\lambda} - 1) \qquad (2)$$

after some rearrangement of terms. If the painting is modern, then λy_0 should be a very large number; λy_0 represents the number of disintegrations of the lead-210 per minute per gram of white lead at the time of manufacture. By studying samples of white lead, we can conclude that λy_0 should never be anywhere near as large as 30,000. Thus, we use equation (2) to calculate λy_0; if our result is greater than 30,000 we conclude that the painting is a modern forgery. The details of this calculation are left for the exercises.

Exercises

1. To calculate λ, use the formula

$$\lambda = \frac{\ln 2}{\text{half-life}}.$$

Find λ for lead-210, whose half-life is 22 years.

2. For the painting "Disciples at Emmaus," the current rate of disintegration of the lead-210 was measured and found to be $\lambda \cdot y(t) = 8.5$. Also, r was found to be .8. Use this information and equation (2) to calculate λy_0. Based on your results, what do you conclude about the age of the painting?

The table below lists several other possible forgeries. Decide which of them must be modern forgeries.

Title	$\lambda \cdot y(t)$	r
3. "Washing of Feet"	12.6	.26
4. "Lace Maker"	1.5	.4
5. "Laughing Girl"	5.2	.6
6. "Woman Reading Music"	10.3	.3

5

Mathematics of Finance

Not too many years ago, money could be borrowed by the largest and most secure corporations for 3% (and home mortgages could be had for 4 1/2%). Today, however, even the largest corporations must pay at least 10% or 12% for their money, and over 20% at times. This makes it important that both the corporation management and consumers (who pay 21% or more to retailers such as Sears or Wards) have a good understanding of the cost of borrowing money. The charge for borrowing money is called **interest.** The formulas for interest are developed in this chapter.

5.1 Simple Interest and Discount

Simple interest is interest paid only on the amount deposited and not on past interest. The amount deposited is the **principal, P.** The **rate** of interest, r, is given as a percent per year, and t is the **time,** measured in years. Simple interest, I, is the product of the principal, rate, and time. (To use the formula for simple interest, write the rate r in decimal form.)

> The **simple interest I** for t years on an amount of P dollars at a rate of interest r per year for t years is
>
> $$I = Prt.$$

1 Find the simple interest for the following.

(a) $1000 at 8% for 2 years

(b) $5500 at 10.5% for $1\frac{1}{2}$ years

(c) $12,200 at 11.75% for 7 months

Answer:

(a) $160

(b) $866.25

(c) $836.21

Throughout this chapter, whenever a rate of interest is given, it is always given *per year* unless stated otherwise.

Example 1 The simple interest on $15,000 at 12 1/4% for 11 months is found from the formula $I = Prt$, with $P = \$15,000$, $r = 12\ 1/4\% = .1225$, and $t = 11/12$ of a year.

$$I = Prt$$

$$I = \$15,000(.1225)\left(\frac{11}{12}\right)$$

$$I = \$1684.38 \text{ rounded to the nearest cent} \quad \boxed{1}$$

A deposit of P dollars today at a rate of interest r for t years produces interest of $I = Prt$. This interest, added to the original principal P, gives

$$P + Prt = P(1 + rt).$$

This result, called the **future value** of P dollars at an interest rate r for t years, is summarized as follows. (When loans are involved, the future value is often called the *maturity value* of the loan.)

> The **future value** or **maturity value** A of P dollars for t years at a rate of interest r per year is
>
> $$A = P(1 + rt).$$

Example 2 Find the maturity value for each of the following loans.
(a) A loan of $2500 to be repaid in 8 months with interest of 14%

The loan is for 8 months, or $8/12 = 2/3$ of a year. The maturity value is

$$A = P(1 + rt)$$

$$A = 2500\left[1 + .14\left(\frac{2}{3}\right)\right]$$

$$\approx 2500[1 + .09333] = 2733.33 \qquad \text{(rounded)}$$

or $2733.33. (Interest is rounded to the nearest cent, as is customary in financial problems. Since money is usually rounded to the nearest cent, the "is approximately equal to" symbol $\approx$ is not used.) Of this maturity value,

$$\$2733.33 - \$2500 = \$233.33$$

represents interest.
(b) A loan of $11,280 for 85 days at 11% interest

It is common to assume 360 days in a year when working with simple interest. We shall usually make such an assumption in this book. The maturity value in this example is

$$A = 11{,}280\left[1 + .11\left(\frac{85}{360}\right)\right] \approx 11{,}280[1.0259722] = 11{,}572.97,$$

or $11,572.97. **2**

Part (b) of Example 2 assumed 360 days in a year. Interest found using 360 days is called **ordinary interest,** while interest found using 365 days is **exact interest.**

Present Value A sum of money that can be deposited today to yield some larger amount in the future is called the **present value** of that future amount. Let P be the present value of some amount A at some time t (in years) in the future. Assume a rate of interest r. As above, the future value of this sum is

$$A = P(1 + rt).$$

2 Find the maturity value of each loan.
(a) $10,000 at 15% for 6 months
(b) $8970 at 17% for 9 months
(c) $95,106 at 14.8% for 76 days

Answer:

(a) $10,750
(b) $10,113.68
(c) $98,077.53

Since P is the present value, divide both sides of this last result by $1 + rt$ to get

$$P = \frac{A}{1 + rt}.$$

In summary,

the **present value** P of a future amount of A dollars at a rate of simple interest r per year for t years is

$$P = \frac{A}{1 + rt}.$$

Example 3 Find the present value of the following future amounts.
(a) $10,000 in one year, if interest is 13%

Here $A = 10{,}000$, $t = 1$, and $r = .13$. Use the formula given above.

$$P = \frac{10{,}000}{1 + (.13)(1)} = \frac{10{,}000}{1.13} = 8849.56$$

If $8849.56 were deposited today, at 13% interest, a total of $10,000 would be in the account in one year. These two sums, $8849.56 today, and $10,000 in a year, are equivalent (at 13%); one becomes the other in a year. The amount of interest earned is $10{,}000 - \$8849.56 = \$1150.44$.

(b) $32,000 in four months at 9% interest

$$P = \frac{32{,}000}{1 + (.09)\left(\dfrac{4}{12}\right)} = \frac{32{,}000}{1.03} = 31{,}067.96$$

A deposit of $31,067.96 today, at 9% interest, would produce $32,000 in four months. **3**

Example 4 Because of a court settlement, Charlie Dawkins owes $5000 to Arnold Parker. The money must be paid in ten months, with no interest. Suppose Dawkins wishes to pay the money today. What amount should Parker be willing to accept? Assume an interest rate of 11%.

The amount that Parker should be willing to accept is given by the present value:

$$P = \frac{5000}{1 + (.11)\left(\dfrac{10}{12}\right)} \approx \frac{5000}{1.09167} = 4580.14.$$

Parker should be willing to accept $4580.14 today in settlement of the obligation. **4**

3 Find the present value of the following future amounts. Assume 15% interest.
(a) $7500 in 1 year
(b) $89,000 in 5 months
(c) $164,200 in 125 days

Answer:
(a) $6521.74
(b) $83,764.71
(c) $156,071.29

4 Mark McFarland is owed $19,500 by Carl Tyson. The money will be paid in 11 months, with no interest. If the current interest rate is 19%, how much should McFarland be willing to accept today in settlement of the debt?

Answer:
$16,607.52

Simple Discount   It is not an uncommon practice to have interest deducted from the amount of a loan before giving the balance to the borrower. The money that is deducted is called the **discount,** with the amount of money actually received by the borrower called the **proceeds.**

Example 5   Elizabeth Thornton agrees to pay $8500 to her banker in nine months. The banker subtracts a discount of 15% and gives the balance to Thornton. Find the amount of the discount and the proceeds.

The discount is found in the same way that simple interest is found.

$$\text{Discount} = 8500(.15)\left(\frac{9}{12}\right) = 956.25$$

The proceeds are found by subtracting the discount from the original amount.

$$\text{Proceeds} = \$8500 - \$956.25 = \$7543.75 \quad \boxed{5}$$

In Example 5, the borrower was charged a discount of 15%. However, 15% is *not* the interest rate paid, since 15% applies to the $8500, while the borrower actually received only $7543.75. Find the rate of interest actually paid by the borrower as in the next example.

5   Craig Barth signs an agreement at his bank to pay the bank $25,000 in 5 months. The bank charges a 20% discount rate. Find the amount of the discount and the amount Barth actually receives.

Answer:

$2083.33; $22,916.67

Example 6   Find the actual rate of interest paid by Thornton in Example 5.

Find the rate of interest paid by Thornton with the formula for simple interest, $I = Prt$, with r the unknown. Since the borrower received only $7543.75, then $I = 956.25$, $P = 7543.75$, and $t = 9/12$. Substitute these values into $I = Prt$.

$$I = Prt$$

$$956.25 = 7543.75(r)\left(\frac{9}{12}\right)$$

$$\frac{956.25}{7543.75\left(\frac{9}{12}\right)} = r$$

$$.169 \approx r$$

The interest rate paid by the borrower is about 16.9%. (This rate actually paid is called the **effective rate.**)   **6**

6   Refer to Problem 5 at the side above, and find the effective rate of interest paid by Barth.

Answer:

21.8%

Let D represent the amount of discount on a loan. Then $D = Art$, where A is the maturity value of the loan (the amount borrowed plus interest), and r is the stated rate of interest. The amount actually received, the proceeds, can be written as $P = A - D$, or $P = A - Art$, from which $P = A(1 - rt)$.

The formulas for discount interest can be summarized as follows.

If D is the discount on a loan having a maturity value A at a rate of interest r per year for t years, the **proceeds** P are

$$P = A - D \quad \text{or} \quad P = A(1 - rt).$$

One common use of discount interest is in *discounting a note,* a process by which a promissory note due at some time in the future can be converted to cash now.

Example 7 Jim Levy owes $4250 to Don Reynolds. The loan is payable in one year, at 12% interest. Reynolds needs cash to buy a new car, so three months before the loan is payable he goes to his bank to have the loan discounted. The bank charges a 16% discount fee. Find the amount of cash he will receive from the bank.

First find the maturity value of the loan, the amount Levy must pay to Reynolds. By the formula for maturity value,

$$A = P(1 + rt)$$
$$A = 4250[1 + (.12)(1)]$$
$$= 4250 (1.12) = 4760,$$

7 A firm accepts a $21,000 note, due in 7 months with interest of 19.5%. Suppose the firm discounts the note at a bank, 75 days before it is due. Find the amount the firm would receive if the bank charges a 22.4% discount rate.

or $4760.

The bank applies its discount rate to this total:

$$\text{Amount of discount} = 4760(.16)\left(\frac{3}{12}\right) = 190.40.$$

(Remember that the loan was discounted three months before it was due.) Reynolds actually receives

$$\$4760 - \$190.40 = \$4569.60$$

Answer:
$22,297.27

in cash from the bank. Three months later, the bank would get $4760 from Levy. **7**

5.1 Exercises

Find the simple interest. (See Examples 1 and 2.)

1. $1000 at 12% for one year

2. $4500 at 16% for one year

3. $25,000 at 21% for nine months

4. $3850 at 17% for eight months

5. $1974 at 16.2% for seven months

6. $3724 at 14.1% for eleven months

Find the simple interest. Assume a 360-day year. Also, assume 30 days in each month.

7. $12,000 at 14% for 72 days

8. $38,000 at 19% for 216 days

9. $5147.18 at 17.3% for 58 days

10. $2930.42 at 13.9% for 123 days

11. $7980 at 15%; the loan was made May 7 and is due on September 19

12. $5408 at 20%; the loan was made August 16 and is due on December 30

Find the simple interest. Assume 365 days in a year, and use the exact number of days in a month. (Assume 28 days in February.)

13. $7800 at 16%; made on July 7 and due October 25

14. $11,000 at 15%; made on February 19 and due May 31

15. $2579 at 17.6%; made on October 4 and due March 15

16. $37,098 at 19.2%; made on September 12 and due July 30

Find the present value of the future amounts. Assume 360 days in a year. (See Example 3.)

17. $15,000 in 8 months, money earns 16%

18. $48,000 in 9 months, money earns 14%

19. $5276 in 3 months, money earns 17.4%

20. $6892 in 7 months, money earns 18.2%

21. $15,402 in 125 days, money earns 19.3%

22. $29,764 in 310 days, money earns 21.4%

Find the proceeds for the following amounts. Assume 360 days in a year. (See Example 5.)

23. $7150, discount rate 16%, length of loan 11 months

24. $9450, discount rate 18%, length of loan 7 months

25. $358, discount rate 21.6%, length of loan 183 days

26. $509, discount rate 23.2%, length of loan 238 days

Work the following exercises.

27. Donna Sharp borrowed $25,900 from her father to start a flower shop. She repaid him after eleven months, with interest of 18.4%. Find the total amount she repaid.

28. A corporation accountant forgot to pay the firm's income tax of $725,896.15 on time. The government charged a penalty of 17.7% interest for the 34 days the money was late. Find the total amount, tax and penalty, that was paid. (Use a 365-day year.)

29. Tuition of $1769 will be due when the spring term begins, in four months. What amount should a student deposit today, at 6.25%, to have enough to pay the tuition?

30. A firm of attorneys has ordered seven new IBM typewriters, at a cost of $2104 each. The machines will not be delivered for seven months. What amount could the firm deposit in an account paying 15.42% to have enough to pay for the machines?

31. Roy Gerard needs $5196 to pay for remodeling work on his house. He plans to repay the loan in 10 months. His bank loans money at a discount rate of 17%. Find the amount of his loan.

32. Mary Collins decides to go back to college. To get to school she buys a small car for $6100. She decides to borrow the money at the bank, where they charge a 19.8% discount rate. If she will repay the loan in 7 months, find the amount of the loan.

33. Marge Prullage signs a $4200 note at the bank. The bank charges a 17.2% discount rate. Find the net proceeds if the note is for ten months. Find the effective interest rate charged by the bank.

34. A bank charges a 23.1% discount rate on a $1000 note for 90 days. Find the effective rate.

35. Patrick Kelley owes $7000 to the Eastside Music Shop. He has agreed to pay the amount in 7 months, at an interest rate of 21%. Two months before the loan is due to be paid, the store discounts it at the bank. The bank charges a 23.7% discount rate. How much money does the store receive?

36. A building contractor gives a $13,500 note to a plumber. The note is due in nine months, with interest of 19%. Three months after the note is signed, the plumber discounts it at the bank. The bank charges a 21.1% discount rate. How much money does the plumber actually receive?

5.2 Compound Interest

With simple interest, the interest is calculated only on the amount of principal. If the interest is then left on deposit with the original principal for an additional time period, another calculation of interest would have interest paid on the original principal, as well as the previous interest. Interest paid on principal as well as previous interest left on deposit is called **compound interest.** Find a formula for compound interest by first supposing that P dollars is deposited at a rate of interest i per year.

(While r is used with simple interest, it is common to use i with compound interest.) The interest earned during the first year is found with the formula for simple interest:

$$\text{First year interest} = P \cdot i \cdot 1 = Pi.$$

At the end of one year, the amount on deposit will be the sum of the original principal and the interest earned, or

$$P + Pi = P(1 + i). \tag{1}$$

If the deposit earns compound interest, the interest earned during the second year is found from the total amount on deposit at the end of the first year. Thus, the interest earned during the second year (again found by the formula for simple interest), is given by

$$P(1 + i)(i)(1) = P(1 + i)i, \tag{2}$$

so that the total amount on deposit at the end of the second year is given by the sum of the amounts from (1) and (2) above, or

$$P(1 + i) + P(1 + i)i = P(1 + i) \cdot (1 + i)$$
$$= P(1 + i)^2.$$

In the same way, the total amount on deposit at the end of three years is

$$P(1 + i)^3.$$

Generalizing, in n years the total amount on deposit is $P(1 + i)^n$, called the **compound amount.**

Interest can be compounded more than once a year. Common **periods of compounding** include *semiannually* (two periods per year), *quarterly* (four periods per year), *monthly* (twelve periods) and *daily* (usually 365 periods per year). Find the interest rate per period by dividing i, the interest rate per *year,* by m, the number of periods of compounding per *year.*

The formula in the next box can be derived in much the same way as was the formula above.

If P dollars is deposited for n years with m periods of compounding per year at a rate of interest i per year, the **compound amount** A is

$$A = P\left(1 + \frac{i}{m}\right)^{nm}.$$

Example 1 Suppose $1000 is deposited for 6 years in an account paying 8% per year compounded annually.
(a) Find the compound amount.

In the formula above, $P = 1000$, $i = 8\% = .08$, $m = 1$, and $n = 6$. The compound amount is

$$A = P\left(1 + \frac{i}{m}\right)^{nm}$$

$$A = 1000\left(1 + \frac{.08}{1}\right)^{6 \cdot 1} = 1000(1.08)^6.$$

The number $(1.08)^6$ can be found with some calculators, or by using special compound interest tables. Such a table is given at the back of this book. (See Table 6.) For $(1.08)^6$, look for 8% across the top and 6 (for 6 periods) down the side. You should find 1.58687, so $(1.08)^6 \approx 1.58687$, and

$$A = 1000(1.58687) = 1586.87,$$

or $1586.87, which represents the compound amount.
(b) Find the actual amount of interest earned.

Subtract the initial deposit from the compound amount.

$$\text{Amount of interest} = \$1586.87 - \$1000 = \$586.87 \quad \boxed{1}$$

Example 2 Find the amount of interest earned by a deposit of $1000 for 6 years at 16% compounded quarterly.

Interest compounded quarterly is compounded four times a year. In 6 years, there are $4 \cdot 6 = 24$ quarters, or 24 periods. Interest of 16% per year is 16%/4, or 4%, per quarter. The compound amount is

$$1000(1 + .04)^{24} = 1000(1.04)^{24}.$$

The value of $(1.04)^{24}$ can be found with a calculator or in Table 6. Locate 4% across the top of the table and 24 periods at the left. The table gives the number 2.56330 so

$$A = 1000(2.56330) = 2563.30,$$

or $2563.30. The compound amount is $2563.30 and the interest earned is $2563.30 - \$1000 = \$1563.30$. $\boxed{2}$

Example 3 Find the compound amount if $900 is deposited at 16% compounded semiannually for 8 years.

In 8 years there are $8 \cdot 2 = 16$ semiannual periods. If interest is 16% per year, then 16%/2 = 8% is earned per semiannual period. Use a calculator, or look in Table 6 for 8% and 16 periods, finding the number 3.42594. The compound amount is

$$A = 900(1.08)^{16} = 900(3.42594) = 3083.35,$$

or $3083.35. $\boxed{3}$

The more often interest is compounded within a given time period, the more interest will be earned. Using a calculator with a y^x key and the formula above gives the results shown in the following chart.

---

1 Suppose $17,000 is deposited at 8% compounded annually for 11 years.
(a) Find the compound amount.
(b) Find the amount of interest earned.

Answer:
(a) $39,637.87
(b) $22,637.87

2 Find the compound amount.
(a) $10,000 deposited at 12% compounded semiannually for 7 years
(b) $36,000 deposited at 18% compounded monthly for 2 years

Answer:
(a) $22,609.00
(b) $51,462.00

3 (a) Find the final amount on deposit and the interest earned if $2500 is deposited in an account paying 6% compounded quarterly. The money is deposited for 5 years.
(b) Suppose $4000 is deposited at 16% compounded semiannually for 8 years. Find the final amount on deposit.

Answer:
(a) $A = \$3367.15$; $867.15
(b) $A = \$13,703.76$

Interest on $1000 at 12% per Year for 10 Years

Compounded	Number of periods of compounding	Compound amount	Interest
not at all (simple interest)	—	—	$1200.00
annually	10	$1000(1 + .12)^{10} = \$3105.85$	$2105.85
semiannually	20	$1000\left(1 + \dfrac{.12}{2}\right)^{20} = \$3207.14$	$2207.14
quarterly	40	$1000\left(1 + \dfrac{.12}{4}\right)^{40} = \$3262.04$	$2262.04
monthly	120	$1000\left(1 + \dfrac{.12}{12}\right)^{120} = \$3300.39$	$2300.39
daily	3650	$1000\left(1 + \dfrac{.12}{365}\right)^{3650} = \$3319.46$	$2319.46
hourly	87,600	$1000\left(1 + \dfrac{.12}{8760}\right)^{87,600} = \$3320.09$	$2320.09
every minute	5,256,000	$1000\left(1 + \dfrac{.12}{525,600}\right)^{5,256,000} = \$3320.11$	$2320.11

As suggested by the chart, it makes a big difference whether interest is compounded or not. Interest differs by $905.85 when simple interest is compared to interest compounded annually. However, increasing the frequency of compounding makes smaller and smaller differences in the amount of interest earned. In fact, it can be shown that even if interest is compounded at intervals of time as small as one chooses (such as each hour, each minute, or each second), the total amount of interest earned will be only slightly more than for daily compounding. This is true even for a process called **continuous compounding,** which can be loosely described as compounding every instant.

Interest with continuous compounding is found as follows.

Continuous Compounding The compound amount A for a deposit of P dollars at an interest rate i per year compounded continuously for n years is given by

$$A = Pe^{ni}.$$

The number e comes up in various practical applications (to eight decimal places, $e = 2.71828183$). Values of e^x are given in Table 3.

Example 4 Suppose $5000 is deposited in an account paying 8% compounded continuously for five years. Find the compound amount.

Let $P = 5000$, $n = 5$, and $i = .08$. Then

$$A = 5000e^{5(.08)} = 5000e^{.4}.$$

From Table 3, or a calculator, $e^{.4} \approx 1.49182$, and

$$A = 5000(1.49182) = 7459.10,$$

or $7459.10. Check that daily compounding would have produced a compound amount about 30¢ less. **4**

4 Find the compound amount, assuming continuous compounding. A calculator with an e^x key is needed in part (b).
(a) $12,000 at 10% for 5 years
(b) $22,876 at 17.2% for 9 years

Answer:
(a) $19,784.64
(b) $107,564.25

Example 5 Suppose $24,000 is deposited at 16% per year for 9 years. Use a calculator with a y^x key to find the interest earned by (a) daily, (b) hourly, and (c) continuous compounding. Assume 365 days per year.
(a) In 9 years there are $9 \times 365 = 3285$ daily periods. The compound amount with daily compounding is

$$24{,}000\left(1 + \frac{.16}{365}\right)^{3285} = 101{,}264.74,$$

or $101,264.74. The interest earned is

$$\$101{,}264.74 - \$24{,}000 = \$77{,}264.74.$$

(b) In one year, there are $365 \times 24 = 8760$ hourly periods, while in 9 years there are $9 \times 8760 = 78{,}840$ hourly periods. The compound amount is

$$24{,}000\left(1 + \frac{.16}{8760}\right)^{78{,}840} = 101{,}295.37,$$

or $101,295.37. This amount includes interest of

$$\$101{,}295.37 - \$24{,}000 = \$77{,}295.37,$$

only $30.63 more than when interest is compounded daily.
(c) The compound amount for continuous compounding is

$$24{,}000e^{9(.16)} = 24{,}000e^{1.44}$$
$$= 24{,}000(4.2206958) \qquad \text{Use a calculator}$$
$$= 101{,}296.70,$$

or $101,296.70. With continuous compounding, the interest earned, $101,296.70 - $24,000 = $77,296.70, is only $1.33 more than when interest is compounded hourly. **5**

5 Suppose $5980 is deposited at 20% for 5 years. Find the interest earned by (a) daily, (b) hourly, and (c) continuous compounding.

Answer:
(a) $10,270.87
(b) $10,275.14
(c) $10,275.33

Effective Rate If $1 is deposited at 4% compounded quarterly, a calculator or Table 6 shows that at the end of one year, the compound amount is $1.0406, an increase of 4.06% over the original $1. The actual increase of 4.06% in the money is somewhat higher than the stated increase of 4%. To keep these two numbers straight, 4% is called the **nominal** or **stated** rate of interest, while 4.06% is called the **effective** rate.

Example 6 Find the effective rate corresponding to a nominal rate of 6% compounded semiannually.

Look in Table 6 for an interest rate of 3% for 2 periods. Find the number 1.06090. Alternatively, use a calculator to find $(1.03)^2$. Either method shows that $1 will increase to $1.06090, an actual increase of $1.06090 - 1 = .0609$, or 6.09%. The effective rate is 6.09%. **6**

6 Find the effective rate corresponding to a nominal rate of
(a) 12% compounded monthly;
(b) 16% compounded quarterly.

Answer:
(a) 12.683%
(b) 16.986%

Generalizing from this example, the effective rate of interest is given by the following formula.

> The **effective rate** corresponding to a stated rate of interest i per year compounded m times per year is
> $$\left(1 + \frac{i}{m}\right)^m - 1.$$

Example 7 A bank pays interest of 9% compounded monthly. Find the effective rate.

Use the formula in the box, with $i = .09$ and $m = 12$. The effective rate is
$$\left(1 + \frac{.09}{12}\right)^{12} - 1.$$

Use a calculator with a y^x key to get
$$(1.0075)^{12} - 1 = .0938,$$

7 Find the effective rate corresponding to a nominal rate of
(a) 15% compounded monthly,
(b) 10% compounded quarterly.

Answer:
(a) 16.08%
(b) 10.38%

or 9.38%. **7**

Present Value with Compound Interest The formula for compound interest, $A = P(1 + i/m)^{nm}$, has five variables, A, P, i, m, and n. If the values of any four of these variables are known, the value of the fifth then can be found. In particular if A, the amount of money needed in the future, and also i, m, and n, are known, then the value of P can be found. Here P is the amount that should be deposited today to produce A dollars in n years. The next example shows this.

Example 8 Joan Wilson must pay a lump sum of $6000 in 5 years. What amount deposited today at 8% compounded annually will amount to the needed $6000 in 5 years?

Here $A = 6000$, $i = .08$, $m = 1$, $n = 5$, and P is unknown. Substituting these values into the formula for the compound amount gives
$$6000 = P\left(1 + \frac{.08}{1}\right)^{5(1)} = P(1.08)^5.$$

From a calculator or Table 6, $(1.08)^5 \approx 1.46933$, with

$$6000 \approx P(1.46933)$$

and

$$P \approx \frac{6000}{1.46933} = 4083.49,$$

or $4083.49. If Wilson deposits $4083.49 for 5 years in an account paying 8% compounded annually, she will have $6000 when she needs it. **8**

As the last example shows, $6000 in 5 years is the same as $4083.49 today (if money can be deposited at 8% compounded annually). Recall from the first section that an amount that can be deposited today to yield a given sum in the future is called the *present value* of this future sum.

Example 9 Find the present value of $16,000 in 9 years if money can be deposited at 12% compounded semiannually.

In 9 years there are $2 \cdot 9 = 18$ semiannual periods. A rate of 12% per year is 6% in each semiannual period. Use a calculator, or look in Table 6 (6% across the top and 18 periods down the side) to find 2.85434. The present value is

$$\frac{16,000}{2.85434} = 5605.50.$$

A deposit of $5605.50 today, at 12% compounded semiannually, will produce a total of $16,000 in 9 years. **9**

The formula for compound amount also can be solved for n, as the following example shows.

Example 10 Suppose the general level of inflation in the economy averages 8% per year. Find the number of years it would take for the overall level of prices to double.

We want to find the number of years it will take for $1 worth of goods or services to cost $2. That is, we want to find n in the equation

$$2 = 1(1 + .08)^n,$$

where $A = 2$, $P = 1$, and $i = .08$. This equation simplifies to

$$2 = (1.08)^n.$$

While n can be found using logarithms or certain calculators, a reasonable approximation can be found by reading the 8% column of Table 6. Read down this column for the number closest to 2, which is 1.99900. This number corresponds to 9 periods. The general level of prices will double in about 9 years. This time period is called the **years to double.** **10**

5.2 Exercises

Find the compound amount when the following deposits are made. (See Examples 1–3.)

1. $1000 at 6% compounded annually for 8 years

2. $1000 at 8% compounded annually for 10 years

3. $4500 at 8% compounded annually for 20 years

4. $810 at 8% compounded annually for 12 years

5. $470 at 12% compounded semiannually for 12 years

6. $15,000 at 16% compounded semiannually for 11 years

7. $46,000 at 12% compounded semiannually for 5 years

8. $1050 at 16% compounded semiannually for 13 years

9. $7500 at 16% compounded quarterly for 9 years

10. $8000 at 16% compounded quarterly for 4 years

11. $6500 at 12% compounded quarterly for 6 years

12. $9100 at 12% compounded quarterly for 4 years

Find the amount of interest earned by the following deposits. (See Examples 1–2.)

13. $6000 at 8% compounded annually for 8 years

14. $21,000 at 6% compounded annually for 5 years

15. $43,000 at 10% compounded semiannually for 9 years

16. $7500 at 8% compounded semiannually for 5 years

17. $2196.58 at 20.8% compounded quarterly for 4 years

18. $4915.73 at 21.6% compounded quarterly for 3 years

Find the present value of the following sums. (See Examples 8 and 9.)

19. $4500 at 8% compounded annually for 9 years

20. $11,500 at 8% compounded annually for 12 years

21. $15,902.74 at 19.8% compounded annually for 7 years

22. $27,159.68 at 21.3% compounded annually for 11 years

23. $2000 at 16% compounded semiannually for 8 years

24. $2000 at 12% compounded semiannually for 8 years

25. $8800 at 16% compounded quarterly for 5 years

26. $7500 at 12% compounded quarterly for 9 years

27. If money can be invested at 8% compounded quarterly, which is larger, $1000 now or $1210 in 5 years?

28. If money can be invested at 6% compounded annually, which is larger, $10,000 now or $15,000 in 10 years?

Find the compound amount if $20,000 is invested at 8% compounded continuously for the following number of years. Use a calculator with an e^x or y^x key for Exercises 33 and 34. (See Examples 4 and 5.)

29. 1   30. 5   31. 10

32. 15   33. 3   34. 7

Find the effective rate corresponding to the following nominal rates. (See Examples 6 and 7.)

35. 4% compounded semiannually

36. 8% compounded quarterly

37. 8% compounded semiannually

38. 10% compounded semiannually

39. 12% compounded semiannually

40. 12% compounded quarterly

Find the present value of $17,200 for the following number of years, if money can be deposited at 11.4% compounded continuously. (Use a calculator with an e^x or a y^x key.)

41. 2   42. 4   43. 7   44. 10

Use the ideas of Example 10 to answer questions 45–47. Find the time it would take for the general level of prices in the economy to double if the average annual inflation rate is

45. 4%   46. 5%   47. 6%.

48. (a) The consumption of electricity has increased historically at 6% per year. If it continues to increase at this rate indefinitely, find the number of years before the electric utilities would need to double their generating capacity.

    (b) Suppose a conservation campaign coupled with higher rates caused the demand for electricity to increase at only 2% per year, as it has recently. Find the number of years before the utilities would need to double generating capacity.

Under certain conditions, Swiss banks pay negative interest—they charge you. (You didn't think all that secrecy was free?) Suppose a bank "pays" −2.4% interest compounded annually. Use a calculator and find the compound amount for a deposit of $150,000 after

49. 2 years

50. 4 years

51. 8 years

52. 12 years.

Work the following exercises.

53. Bill Cornett deposited $18,000 in an account paying 12% compounded quarterly. How much interest will he earn in 7 years?

54. Ann Fridl-Marshall placed $25,000 in an account paying 10% compounded semiannually. How much interest will she earn in 5 years?

55. Susan Dollman needs $18,000 in 3 years.
   (a) What amount can Dollman deposit today, at 8% compounded annually, so that she will have the needed money?
   (b) How much interest will be earned?

(c) Suppose Dollman can deposit only $12,000 today. How much would she be short of the $18,000 in 3 years?

56. Brent Alderman must pay $27,000 in settlement of an obligation in 3 years.
   (a) What amount can he deposit today, at 12% compounded monthly, to have enough?
   (b) How much interest will he earn?
   (c) Suppose Alderman can deposit only $12,000 today. How much would he be short of the needed $27,000 in 3 years?

Find the compound amount for the following deposits.

57. $40,552 at 9.13% compounded quarterly for 5 years

58. $11,641.10 at 10.9% compounded monthly for 8 years

59. $673.27 at 8.6% compounded semiannually for 3.5 years

60. $2964.93 at 11.4% compounded monthly for 4.25 years

5.3 Sequences

So far, this chapter has discussed only *lump sums*—a lump sum deposit today that produces a lump sum compounded amount in the future, or the present value of a lump sum amount in the future. In practice, however, it is very common to deal with a *sequence* of periodic payments, such as a car payment or house payment. Formulas for these payments are developed in the next sections, using some of the results of this section.

A **sequence** is a function whose domain is the set of positive integers. For example,

$$a(n) = 2n, \quad n = 1, 2, 3, 4, \ldots$$

defines a sequence. The letter n is used as a variable instead of x to emphasize the fact that the domain includes only positive integers. For the same reason, a is used to name the function instead of f.

The range values of a sequence function, such as

$$a(1) = 2, \quad a(2) = 4, \quad a(3) = 6, \ldots$$

from the sequence given above, are called the **terms** of the sequence. Instead of writing $a(5)$ for the fifth term of the sequence, for example, it is customary to write

$$a_5 = 10. \quad \boxed{1}$$

The symbol a_n is often used for the **general** or **nth term** of a sequence. For example, for the sequence 4, 7, 10, 13, 16, . . . the general term is given by

⟁ **1** For the sequence defined by $a(n) = 2n$, find a_1, a_2, a_8, a_{20}, and a_{51}.

Answer:
$a_1 = 2$, $a_2 = 4$, $a_8 = 16$, $a_{20} = 40$, and $a_{51} = 102$.

$a_n = 1 + 3n$. This formula for a_n can be used to find any desired term of the sequence. For example, the first three terms of the sequence having $a_n = 1 + 3n$ are

$$a_1 = 1 + 3(1) = 4, \quad a_2 = 1 + 3(2) = 7, \quad a_3 = 1 + 3(3) = 10.$$

Also, $a_8 = 25$ and $a_{12} = 37$.

Example 1 Find the first four terms for the sequence having the general term $a_n = -4n + 2$.

Replace n, in turn, with 1, 2, 3, and 4. If $n = 1$,

$$a_1 = -4(1) + 2 = -4 + 2 = -2. \qquad \text{Let } n = 1$$

Also, $\qquad a_2 = -4(2) + 2 = -6. \qquad\qquad \text{Let } n = 2$

When $n = 3$, then $a_3 = -10$; finally, $a_4 = -4(4) + 2 = -14$. The first four terms of this sequence are $-2, -6, -10,$ and -14. ②

Arithmetic Sequences A sequence in which each term after the first is found by adding the same number to the preceding term is called an **arithmetic sequence.** The sequence of Example 1 above is an arithmetic sequence; -4 is added to any term to get the next term.

The sequence

$$8, 13, 18, 23, 28, \ldots$$

is an arithmetic sequence since each term after the first is found by adding 5 to the previous term. The number 5, the difference between any two adjacent terms, is called the **common difference.** ③

If a_1 is the first term of an arithmetic sequence and d is the common difference, then the second term can be found by adding the common difference d to the first term: $a_2 = a_1 + d$. The third term is found by adding d to the second term.

$$a_3 = a_2 + d = (a_1 + d) + d = a_1 + 2d$$

In the same way, $a_4 = a_1 + 3d$ and $a_5 = a_1 + 4d$. Generalizing, the nth term of an arithmetic sequence is given by $a_n = a_1 + (n - 1)d$.

> If an arithmetic sequence has first term a_1 and common difference d, then a_n, the **nth term** of the sequence, is given by
> $$a_n = a_1 + (n - 1)d.$$

Example 2 A company had sales of \$50,000 during its first year of operation. If the sales increase by \$6000 per year, find its sales in the eleventh year.

Since the sales for each year after the first are found by adding \$6000 to the sales of the previous year, the sales form an arithmetic sequence with $a_1 = 50{,}000$ and $d = 6000$. Using the formula for the nth term of an arithmetic sequence, sales

---

② Find the first four terms for each of the following sequences.

(a) $a_n = -9 + 7n$

(b) $a_n = (2n + 3)(n - 1)$

(c) $a_n = \dfrac{n + 1}{n + 2}$

Answer:

(a) $-2, 5, 12, 19$

(b) $0, 7, 18, 33$

(c) $\dfrac{2}{3}, \dfrac{3}{4}, \dfrac{4}{5}, \dfrac{5}{6}$

---

③ Identify any arithmetic sequences.

(a) $7, 15, 23, 31, \ldots$

(b) $-9, -11, -13,$ $-15, \ldots$

(c) $2, 4, 8, 16, 32, \ldots$

Answer:

(a) and (b) are arithmetic sequences; (c) is not.

during the eleventh year are given by

$$a_{11} = 50{,}000 + (11 - 1)6000 = 50{,}000 + 60{,}000 = 110{,}000,$$

or $110,000. ▇4▇

4 Suppose a field has 10 rabbits in it at the end of the first month of an experiment, with the number of rabbits then increasing by 8 per month. Find the number of rabbits present after
(a) 3 months;
(b) 8 months;
(c) 12 months.

Answer:
(a) 26
(b) 66
(c) 98

The formula above gives the nth term of an arithmetic sequence. The next formula shows how to find the sum of the first n terms of an arithmetic sequence. Develop this formula by starting with an arithmetic sequence having first term a_1 and common difference d. Let S_n represent the sum of the first n terms of the sequence. Start by writing S_n as follows:

$$S_n = a_1 + [a_1 + d] + [a_1 + 2d] + \cdots + [a_1 + (n - 1)d].$$

Next, write this same sum in reverse order.

$$S_n = [a_1 + (n - 1)d] + [a_1 + (n - 2)d] + \cdots + [a_1 + d] + a_1$$

Now add corresponding terms of the respective sides of these last two equations.

$$S_n + S_n = (a_1 + [a_1 + (n - 1)d]) + ([a_1 + d] + [a_1 + (n - 2)d]) + \cdots + ([a_1 + (n - 1)d] + a_1)$$

From this,

$$2S_n = [2a_1 + (n - 1)d] + [2a_1 + (n - 1)d] + \cdots + [2a_1 + (n - 1)d].$$

There are n of the $[2a_1 + (n - 1)d]$ terms on the right, making

$$2S_n = n[2a_1 + (n - 1)d],$$

5 Evaluate
$\frac{n}{2}[2a_1 + (n - 1)d]$ if
(a) $n = 20$, $a_1 = 6$, $d = 4$;
(b) $n = 15$, $a_1 = -9$, $d = 7$;
(c) $n = 22$, $a_1 = -12$, $d = -3$.

Answer:
(a) 880
(b) 600
(c) −957

$$S_n = \frac{n}{2}[2a_1 + (n - 1)d]. \quad \boxed{5} \qquad\qquad (1)$$

Since $a_n = a_1 + (n - 1)d$, equation (1) can also be written as

$$S_n = \frac{n}{2}[a_1 + a_1 + (n - 1)d]$$

or

$$S_n = \frac{n}{2}(a_1 + a_n).$$

The following box summarizes this work with arithmetic sequences.

Suppose an arithmetic sequence has first term a_1, common difference d, and nth term a_n. Then the **sum** S_n of the first n terms of the sequence is given by

$$S_n = \frac{n}{2}[2a_1 + (n - 1)d]$$

or

$$S_n = \frac{n}{2}(a_1 + a_n).$$

6 For the sequence 12, 15, 18, 21, . . . , find the sum of

(a) the first 10 terms;

(b) the first 31 terms.

Answer:

(a) 255

(b) 1767

7 Find the sum of the first 500 counting numbers.

Answer:

125,250

8 A new firm sells 60 solar heating systems during its first year in business. Each year after that, it sells 40 more systems than in the preceding year. Find the number of systems that will be sold during the firm's first 10 years in business.

Answer:

2400

9 Identify any geometric sequences.

(a) 10, 5, $2\frac{1}{2}$,

$1\frac{1}{4}$, $\frac{5}{8}$, . . .

(b) −9, 6, −3, 0, −3, 6, −9, . . .

(c) $\frac{3}{4}$, $\frac{3}{2}$, 3, 6, 12, 24, . . .

Answer:

(a) and (c) are geometric; (b) is not.

Either of these two formulas can be used to find the sum of the first n terms of an arithmetic sequence.

Example 3 Find the sum of the first 25 terms of the arithmetic sequence

$$3, \ 8, \ 13, \ 18, \ 23, \ . \ . \ . \ .$$

In this sequence, $a_1 = 3$ and $d = 5$. Using the first formula from above gives

$$S_{25} = \frac{25}{2}[2(3) + (25 - 1)5] \qquad \text{Let } n = 25, \quad a_1 = 3, \quad d = 5$$

$$= \frac{25}{2}[6 + 120]$$

$$= 1575. \quad \boxed{6} \ \boxed{7}$$

Example 4 Find the total sales of the company of Example 2 during its first 11 years.

From Example 2, $a_1 = 50,000$, $a_{11} = 110,000$, and $n = 11$. By the second formula above,

$$S_{11} = \frac{11}{2}(50,000 + 110,000) = 880,000.$$

The total sales for the 11 years are \$880,000. $\boxed{8}$

Geometric Sequences In an arithmetic sequence, each term after the first is found by adding the same number to the preceding term. In a **geometric sequence,** each term after the first is found by *multiplying* the preceding term by the same number. For example,

$$3, \ -6, \ 12, \ -24, \ 48, \ -96, \ . \ . \ .$$

is a geometric sequence with each term after the first found by multiplying the preceding term by −2. The number −2 is called the **common ratio.** $\boxed{9}$

If a_1 is the first term of a geometric sequence and r is the common ratio, then the second term is given by $a_2 = a_1 r$ and the third term by $a_3 = a_2 r = a_1 r^2$. Also, $a_4 = a_1 r^3$, and $a_5 = a_1 r^4$. The next box generalizes from these results.

If a geometric sequence has first term a_1 and common ratio r, then a_n, the **nth term** of the sequence, is given by

$$a_n = a_1 r^{n-1}.$$

Example 5 Find the indicated term for each of the following geometric sequences.

(a) 6, 24, 96, 384, . . . ; find a_7.

Here $a_1 = 6$. Find r by choosing any term except the first and dividing it by the preceding term. Choosing 96 gives

$$r = \frac{96}{24} = 4.$$

Now use the formula for the nth term.

$$
\begin{aligned}
a_7 &= 6(4)^{7-1} \qquad \text{Let } n = 7, \, a_1 = 6, \, r = 4 \\
&= 6(4)^6 \\
&= 6(4096) \qquad 4^6 = 4096 \\
&= 24{,}576
\end{aligned}
$$

(b) $8, \, -16, \, 32, \, -64, \, 128, \, \ldots$; find a_6.

In this sequence $a_1 = 8$ and $r = -2$.

$$
\begin{aligned}
a_6 &= 8(-2)^{6-1} \qquad \text{Let } n = 6, \, a_1 = 8, \, r = -2 \\
&= 8(-2)^5 \\
&= 8(-32) \\
&= -256 \quad \boxed{10} \; \boxed{11}
\end{aligned}
$$

10 Find a_4 and a_7 for the geometric sequence having $a_1 = 5$ and $r = -2$.

Answer:
$a_4 = -40$; $a_7 = 320$

11 The number of bacteria in a culture triples every four hours. If 1000 bacteria are present initially, how many bacteria will be present at the end of
(a) 8 hours?
(b) 12 hours?
(c) 24 hours?

Answer:
(a) 9000
(b) 27,000
(c) 729,000

Often the sum of the first n terms of a geometric sequence is needed. A formula for this sum can be found just as with arithmetic sequences. To find this formula, let a geometric sequence have first term a_1 and common ratio r. Write the sum S_n of the first n terms as

$$S_n = a_1 + a_2 + a_3 + \cdots + a_n.$$

This sum can also be written as

$$S_n = a_1 + a_1 r + a_1 r^2 + \cdots + a_1 r^{n-1}. \tag{2}$$

If $r = 1$, then $S_n = na_1$, the correct result for this case. If $r \neq 1$, multiply both sides of equation (2) by r, obtaining

$$rS_n = a_1 r + a_1 r^2 + a_1 r^3 + \cdots + a_1 r^n. \tag{3}$$

Now subtract corresponding sides of equation (2) from equation (3);

$$
\begin{aligned}
rS_n - S_n &= a_1 r^n - a_1 \\
S_n(r - 1) &= a_1(r^n - 1),
\end{aligned}
$$

or finally

$$S_n = \frac{a_1(r^n - 1)}{r - 1}.$$

The next box summarizes this result.

> If a geometric sequence has first term a_1 and common ratio r, then the **sum** S_n of the first n terms is given by
>
> $$S_n = \frac{a_1(r^n - 1)}{r - 1}, \qquad r \neq 1.$$

12 (a) Find S_4 and S_7 for the geometric sequence 5, 15, 45, 135, ...
(b) Find S_5 for the geometric sequence having $a_1 = -3$ and $r = -5$.

Answer:
(a) $S_4 = 200$, $S_7 = 5465$
(b) $S_5 = -1563$

Example 6 Find the sum of the first six terms of the geometric sequence 3, 12, 48, ...

Here $a_1 = 3$ and $r = 4$. Find S_6 by the result above.

$$S_6 = \frac{3(4^6 - 1)}{4 - 1} \qquad \text{Let } n = 6, \ a_1 = 3, \ r = 4$$

$$= \frac{3(4096 - 1)}{3}$$

$$= 4095 \quad \boxed{12}$$

5.3 Exercises

In Exercises 1–16, a formula for the general term of a sequence is given. Use the formula to find the first five terms of the sequence. Identify each sequence as arithmetic, geometric, *or* neither. *(See Example 1.)*

1. $a_n = 6n + 5$
2. $a_n = 12n - 3$
3. $a_n = 3n - 7$
4. $a_n = 5n - 12$
5. $a_n = -6n + 4$
6. $a_n = -11n + 10$
7. $a_n = 2^n$
8. $a_n = 3^n$
9. $a_n = (-2)^n$
10. $a_n = (-3)^n$
11. $a_n = 3(2^n)$
12. $a_n = -4(2^n)$
13. $a_n = \dfrac{n + 1}{n + 5}$
14. $a_n = \dfrac{2n}{n + 1}$
15. $a_n = \dfrac{1}{n + 1}$
16. $a_n = \dfrac{1}{n + 8}$

Identify each of the following sequences as arithmetic, geometric, *or* neither. *For an arithmetic sequence, give the common difference. For a geometric sequence, give the common ratio.*

17. 6, 14, 22, 30, 38, 46, . . .
18. 40, 46, 52, 58, 64, . . .
19. 5, 8, 11, 14, 17, 20, 23, . . .
20. 23, 34, 45, 54, 63, 72, . . .
21. 4, 12, 36, 108, . . .
22. 7, 14, 28, 56, 112, . . .
23. 2, 5, 9, 14, 20, 27, . . .
24. 1, 4, 9, 16, 25, 36, . . .
25. 12, 9, 6, 3, 0, −3, −6, . . .
26. 37, 31, 25, 19, 13, 7, . . .

27. −18, −15, −12, −9, −6, . . .
28. −21, −17, −13, −9, −5, . . .
29. 3, −6, 12, −24, 48, −96, . . .
30. −5, 10, −20, 40, −80, . . .
31. −5, 6, −7, 8, −9, 10, −11, . . .
32. −12, 9, −6, 3, . . .

Find the indicated term for each of the following arithmetic sequences.

33. $a_1 = 10$, $d = 5$; find a_{13}
34. $a_1 = 6$, $d = 9$; find a_8
35. $a_1 = 8$, $d = 3$; find a_{20}
36. $a_1 = 13$, $d = 7$; find a_{11}
37. 6, 9, 12, 15, 18, . . . ; find a_{25}
38. 14, 17, 20, 23, 26, 29, . . . ; find a_{13}
39. −9, −13, −17, −21, −25, . . . ; find a_{15}
40. −4, −11, −18, −25, −32, . . . ; find a_{18}

Find the sum of the first six terms for each of the following arithmetic sequences. (See Example 3.)

41. 3, 6, 9, 12, . . .
42. 11, 13, 15, 17, . . .
43. 88, 98, 108, . . .
44. 92, 95, 98, . . .
45. $a_1 = 8$, $d = 9$
46. $a_1 = 12$, $d = 6$
47. $a_1 = 7$, $d = -4$
48. $a_1 = 13$, $d = -5$

Find a_5 for each of the following geometric sequences. (See Example 5.)

49. $a_1 = 3$, $r = 2$
50. $a_1 = 5$, $r = 3$
51. $a_1 = -8$, $r = 3$
52. $a_1 = -6$, $r = 2$

53. $a_1 = 1, r = -3$

54. $a_1 = 12, r = -2$

55. $a_1 = 1024, r = \dfrac{1}{2}$

56. $a_1 = 729, r = \dfrac{1}{3}$

Find the sum of the first four terms for each of the following geometric sequences. (See Example 6.)

57. $a_1 = 1, r = 2$

58. $a_1 = 3, r = 3$

59. $a_1 = 5, r = \dfrac{1}{5}$

60. $a_1 = 6, r = \dfrac{1}{2}$

61. $a_1 = 128, r = -\dfrac{3}{2}$

62. $a_1 = 81, r = -\dfrac{2}{3}$

Sums of the terms of a sequence often are written in sigma *notation (also used in statistics). For example, to evaluate*

$$\sum_{i=1}^{4} (3i + 7)$$

(where Σ is the capital Greek letter sigma*), first evaluate $3i + 7$ for $i = 1$, then for $i = 2$, $i = 3$, and finally, $i = 4$. Then add the results to get*

$$\sum_{i=1}^{4} (3i + 7) =$$

$$[3(1) + 7] + [3(2) + 7] + [3(3) + 7] + [3(4) + 7]$$
$$= 10 + 13 + 16 + 19 = 58.$$

Evaluate each of the following sums.

63. $\displaystyle\sum_{i=1}^{5} (3 - 2i)$

64. $\displaystyle\sum_{i=1}^{4} (2i - 5)$

65. $\displaystyle\sum_{i=1}^{8} (2i + 1)$

66. $\displaystyle\sum_{i=1}^{4} (-6i + 8)$

67. $\displaystyle\sum_{i=1}^{5} i(2i + 1)$

68. $\displaystyle\sum_{i=1}^{4} (2i - 1)(i + 2)$

69. $\displaystyle\sum_{i=1}^{4} (3i + 1)(i + 1)$

70. $\displaystyle\sum_{i=1}^{5} (i - 3)(i + 5)$

A sum in the form

$$\sum_{i=1}^{n} (ai + b),$$

where a and b are real numbers, represents the sum of the first n terms of an arithmetic sequence. The first term is

$a_1 = a(1) + b = a + b$ *and the common difference is a. The nth term is $a_n = an + b$. Using these numbers, the sum can be found by using the formula for the sum of the first n terms. Use this method to find the following sums.*

71. $\displaystyle\sum_{i=1}^{5} (2i + 8)$

72. $\displaystyle\sum_{i=1}^{6} (4i - 5)$

73. $\displaystyle\sum_{i=1}^{4} (-8i + 6)$

74. $\displaystyle\sum_{i=1}^{9} (2i - 3)$

75. $\displaystyle\sum_{i=1}^{500} i$

76. $\displaystyle\sum_{i=1}^{1000} i$

Sigma notation also can be used for geometric sequences. A sum in the form

$$\sum_{i=1}^{n} a(b^i)$$

represents the sum of the first n terms of a geometric sequence having first term $a_1 = ab$ and common ratio b. These sums can be found by applying the formula for the sum of the first n terms of a geometric sequence. Use this method to find the following sums.

77. $\displaystyle\sum_{i=1}^{4} 3(2^i)$

78. $\displaystyle\sum_{i=1}^{3} 2(3^i)$

79. $\displaystyle\sum_{i=1}^{4} \dfrac{1}{2}(4^i)$

80. $\displaystyle\sum_{i=1}^{4} \dfrac{3}{2}(2^i)$

81. $\displaystyle\sum_{i=1}^{4} \dfrac{4}{3}(3^i)$

82. $\displaystyle\sum_{i=1}^{4} \dfrac{5}{3}(3^i)$

Work the following exercises. (See Example 2.)

83. Joy Watt is hired for \$11,400 per year, with annual raises of \$600. What will she earn during her eighth year with the company?

84. What will be the total income received by Watt (see Exercise 83) during her first eight years with the company?

85. A certain machine annually loses 30% of the value it had at the beginning of that year. If its initial value is \$10,000, use a calculator to find its value
   (a) at the end of the fifth year,
   (b) at the end of the eighth year.

86. A certain colony of bacteria increases in number by 10% per hour. After five hours, what is the percentage increase in the population over the initial population?

5.4 Annuities

The first two sections of this chapter discussed *lump sum* payments; now this section discusses a *sequence* of equal payments. For example, suppose $1500 is deposited at the end of each year for the next six years, in an account paying 8% per year, compounded annually. How much would be in the account after the six years?

A sequence of equal payments made at equal periods of time is called an **annuity.** If the payments are made at the end of the time period, and if the frequency of payments is the same as the frequency of compounding, the annuity is called an **ordinary annuity.** The time between payments is the **payment period,** with the time from the beginning of the first payment period to the end of the last called the **term** of the annuity. The **future value** of the annuity, the final sum on deposit, is defined as the sum of the compound amounts of all the payments, compounded to the end of the term.

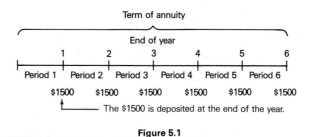

Figure 5.1

Figure 5.1 illustrates an annuity of $1500 at the end of each year for six years. Find the future value of this annuity by looking separately at each of the $1500 payments. The first of these payments will produce a compound amount of

$$1500(1 + .08)^5 = 1500(1.08)^5.$$

Use 5 as the exponent instead of 6 since the money is deposited at the *end* of the first year, and earns interest for only five years. The second payment of $1500 will produce a compound amount of $1500(1.08)^4$. As shown in Figure 5.2, the future value of the annuity is

$$1500(1.08)^5 + 1500(1.08)^4 + 1500(1.08)^3 + 1500(1.08)^2$$
$$+ 1500(1.08)^1 + 1500.$$

(The last payment earns no interest at all.) From Table 6, this sum is

$$1500(1.46933) + 1500(1.36049) + 1500(1.25971) + 1500(1.16640)$$
$$+ 1500(1.08000) + 1500$$
$$\approx \$2204.00 + \$2040.74 + \$1889.57 + \$1749.60 + \$1620 + \$1500$$
$$= \$11,003.91. \quad \boxed{1}$$

$\boxed{1}$ Repeat these steps for an annuity of $2000 at the end of each year for 3 years: assume interest of 8% compounded annually.

(a) The first deposit of $2000 produces a total of _____.

(b) The second deposit becomes _____.

(c) No interest is earned on the third deposit, so the total in the account is _____.

Answer:

(a) $2332.80

(b) $2160.00

(c) $6492.80

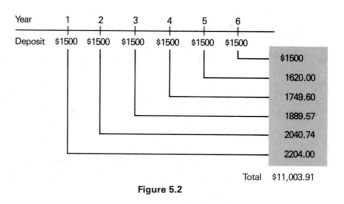

Figure 5.2

To generalize this result, suppose that a payment of R dollars is paid into an account at the end of each period for n periods, at a rate of interest i per period. The first payment of R dollars will produce a compound amount of $R(1 + i)^{n-1}$ dollars, the second payment produces $R(1 + i)^{n-2}$ dollars, and so on; the final payment earns no interest and contributes just R dollars to the total. If A represents the future value of the annuity, then (as shown in Figure 5.3),

$$A = R(1 + i)^{n-1} + R(1 + i)^{n-2} + R(1 + i)^{n-3} + \cdots + R(1 + i) + R$$

or, written in reverse order,

$$A = R + R(1 + i)^1 + R(1 + i)^2 + \cdots + R(1 + i)^{n-1}.$$

This sum is the sum of the first n terms of the geometric sequence having first term R and common ratio $1 + i$. Using the formula for the sum of the first n terms of a geometric sequence from the previous section,

$$A = \frac{R[(1 + i)^n - 1]}{(1 + i) - 1} = \frac{R[(1 + i)^n - 1]}{i} = R\left[\frac{(1 + i)^n - 1}{i}\right].$$

The quantity in brackets is commonly written $s_{\overline{n}|i}$ (read "s-angle-n at i"), so that

$$A = R \cdot s_{\overline{n}|i}.$$

Values of $s_{\overline{n}|i}$ can be found by a calculator or from Table 7, "Amount of an Annuity," at the back of this book.

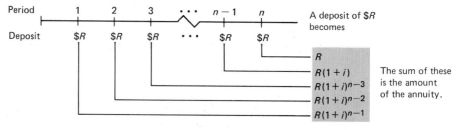

Figure 5.3

As a check, go back to the annuity of $1500 at the end of each year for 6 years; interest was 8% per year compounded annually. In Table 7, these values give the number 7.33593. Multiply this number and 1500:

$$A = 1500(7.33593) = 11,003.90,$$

or $11,003.90. The results differ by 1¢ due to rounding error.

The next box summarizes this work.

> The **future value** A of an annuity of n payments of R dollars each at the end of consecutive interest periods with interest compounded at a rate of interest i per period is
>
> $$A = R\left[\frac{(1 + i)^n - 1}{i}\right] \qquad \text{or} \qquad A = R \cdot s_{\overline{n}|i}.$$

Example 1 Tom Bleser is an athlete who feels that his playing career will last 7 years. To prepare for his future, he deposits $22,000 at the end of each year for 7 years in an account paying 6% compounded annually. How much will he have on deposit after 7 years?

His payments form an ordinary annuity with $R = 22,000$, $n = 7$, and $i = .06$. The future value of this annuity is (by the formula above)

$$A = 22,000\left[\frac{(1.06)^7 - 1}{.06}\right].$$

From Table 7, or a calculator, the number in brackets, $s_{\overline{7}|.06}$, is 8.39384, so that

$$A = 22,000(8.39384) = 184,664.48,$$

or $184,664.48. **2**

Example 2 Suppose $1000 is deposited at the end of each six-month period for 5 years in an account paying 16% compounded semiannually. Find the future value of the annuity.

Interest of 16%/2 = 8% is earned semiannually. In 5 years there are $5 \cdot 2 = 10$ semiannual periods. Find $s_{\overline{10}|.08}$ from Table 7 or a calculator. Looking in the 8% column and down 10 periods gives $s_{\overline{10}|.08} \approx 14.48656$, with a future value of

$$A = 1000(14.48656) = 14,486.56,$$

or $14,486.56. **3**

The results so far have been developed for *ordinary annuities*—those with payments made at the *end* of each time period. These results can be modified slightly to apply to **annuities due**—an annuity where payments are made at the *beginning* of each time period. Find the future value of an annuity due by treating each payment as if it were made at the *end* of the *preceding* period. That is, use Table 7, or a calculator to find $s_{\overline{n}|i}$ for *one additional period;* compensate for this by subtracting the amount of one payment.

2 Johnson Building Materials deposits $2500 at the end of each year in an account paying 8% per year compounded annually. Find the total amount on deposit after

(a) 6 years;
(b) 10 years.

Answer:
(a) $18,339.83
(b) $36,216.40

3 Helen's Dry Goods deposits $5800 at the end of each quarter for 4 years. Find the final amount on deposit if the deposits earn

(a) 8% compounded quarterly;
(b) 16% compounded quarterly.

Answer:
(a) $108,107.88
(b) $126,582.27

Example 3 Find the future value of an annuity due if payments of $500 are made at the beginning of each quarter for 7 years, in an account paying 12% compounded quarterly.

In 7 years, there are 28 quarterly periods. Look in row 29 (28 + 1) of Table 7. Use the 12%/4 = 3% column of the table. You should find the number 45.21885. Multiply this number by 500, the amount of each payment.

$$500(45.21885) = 22,609.43$$

Subtract the amount of one payment from this result.

$$\$22,609.43 - \$500 = \$22,109.43$$

4 Ms. Black deposits $800 at the beginning of each six-month period for 5 years. Find the final amount if the account pays
(a) 8% compounded semiannually;
(b) 16% compounded semiannually.

Answer:
(a) $9989.08
(b) $12,516.39

The account will contain a total of $22,109.43 after 7 years. **4**

Just as the formula $A = P(1 + i)^n$ can be solved for any of its variables, the formula for the future value of an annuity can be used to find the values of variables other than A. In Example 4 below, the amount of money wanted at the end, A, is given, and R, the amount of each payment, must be found.

Example 4 Betsy Martens wants to buy an expensive video camera three years from now. She wants to deposit an equal amount at the end of each quarter for three years in order to accumulate enough money to pay for the camera. The camera costs $2400, and the bank pays 12% interest compounded quarterly. Find the amount of each of the twelve deposits she will make.

This example describes an ordinary annuity with $A = 2400$, $i = .03$ (12%/4 = 3%) and $n = 3 \cdot 4 = 12$ periods. The unknown is the amount of each payment, R. By the formula for the amount of an annuity from above,

$$2400 = R \cdot s_{\overline{12}|.03}.$$

From Table 7, or a calculator,

5 Francisco Arce needs $8000 in 6 years so that he can go on an archeological dig. He wants to deposit equal payments at the end of each quarter so that he will have enough to go on the dig. Find the amount of each payment if the bank pays
(a) 12% compounded quarterly;
(b) 16% compounded quarterly.

Answer:
(a) $232.38
(b) $204.69

$$2400 \approx R(14.19203)$$
$$R = 169.11,$$

or $169.11 (dividing both sides by 14.19203). **5**

The annuity in Example 4 is a **sinking fund:** a fund set up to receive periodic payments. These periodic payments ($169.11 in the example) together with the interest earned by the payments, are designed to produce a given sum at some time in the future. As another example, a sinking fund might be set up to receive money that will be needed to pay off the principal on a loan at some time in the future.

Example 5 The Stockdales are close to retirement. They agree to sell an antique urn to the local museum for $17,000. Their tax adviser suggests that they defer receipt of this money until they retire, 5 years in the future. (At that time, they might well be in a lower tax bracket.) Find the amount of each payment the museum must make into a sinking fund so that it will have the necessary $17,000 in 5 years.

Assume that the museum can earn 8% compounded annually on its money. Also, assume that the payments are made annually.

These payments are the periodic payments into an ordinary annuity. The annuity will amount to $17,000 in 5 years at 8% compounded annually. Using the formula,

$$17,000 = R \cdot s_{\overline{5}|.08}$$

or

$$R = \frac{17,000}{s_{\overline{5}|.08}}$$

gives

$$R \approx \frac{17,000}{5.86660} \qquad \text{From Table 7}$$

$$R = 2897.76,$$

or $2897.76. If the museum deposits $2897.76 at the end of each year for 5 years in an account paying 8% compounded annually, it will have the total amount that it needs. This result is shown in the following table. In these tables, the last payment might well differ slightly from the others because of rounding errors.

Payment number	Amount of deposit	Interest earned	Total in account
1	$2897.76	$ 0	$ 2897.76
2	2897.76	231.82	6027.34
3	2897.76	482.19	9407.29
4	2897.76	752.58	13,057.63
5	2897.76	1044.61	17,000.00

6 A firm must pay off $40,000 worth of bonds in 7 years. Find the amount of each annual payment to be made into a sinking fund if money earns 6% compounded annually.

Answer:
(a) $4765.40

5.4 Exercises

Find each of the following values.

1. $s_{\overline{12}|.05}$ **2.** $s_{\overline{20}|.06}$ **3.** $s_{\overline{16}|.04}$

4. $s_{\overline{40}|.02}$ **5.** $s_{\overline{20}|.01}$ **6.** $s_{\overline{18}|.015}$

7. $s_{\overline{15}|.04}$ **8.** $s_{\overline{30}|.015}$

Find the future value of the following ordinary annuities. Interest is compounded annually. (See Example 1.)

9. $R = 100$, $i = .06$, $n = 4$

10. $R = 1000$, $i = .06$, $n = 5$

11. $R = 10,000$, $i = .05$, $n = 19$

12. $R = 100,000$, $i = .08$, $n = 23$

13. $R = 8500$, $i = .06$, $n = 30$

14. $R = 11,200$, $i = .08$, $n = 25$

15. $R = 46,000$, $i = .06$, $n = 32$

16. $R = 29,500$, $i = .05$, $n = 15$

Find the future value of each of the following ordinary annuities. Payments are made and interest is compounded as given. (See Example 2.)

17. $R = 9200$, 16% interest compounded semiannually for 7 years

18. $R = 3700$, 12% interest compounded semiannually for 11 years

19. $R = 800$, 12% interest compounded semiannually for 12 years

20. $R = 4600$, 16% interest compounded quarterly for 9 years

21. $R = 15,000$, 12% interest compounded quarterly for 6 years

22. $R = 42,000$, 16% interest compounded semiannually for 12 years

Find the future value of each annuity due. Assume that interest is compounded annually. (See Example 3.)

23. $R = 600$, $i = .06$, $n = 8$

24. $R = 1400$, $i = .08$, $n = 10$

25. $R = 20,000$, $i = .08$, $n = 6$

26. $R = 4000$, $i = .06$, $n = 11$

Find the future value of each annuity due. (See Example 3.)

27. Payments of $1000 made at the beginning of each year for 9 years at 8% compounded annually

28. $750 deposited at the beginning of each year for 15 years at 6% compounded annually

29. $100 deposited at the beginning of each quarter for 9 years at 16% compounded quarterly

30. $1500 deposited at the beginning of each semiannual period for 11 years at 16% compounded semiannually

Find the periodic payment that will amount to the given sums under the given conditions. (See Example 4.)

31. $A = \$10,000$, interest is 8% compounded annually, payments made at the end of each year for 12 years

32. $A = \$100,000$, interest is 16% compounded semiannually, payments made at the end of each semiannual period for 9 years

33. $A = \$50,000$, interest is 12% compounded quarterly, payments made at the end of each quarter for 8 years

34. $A = \$147,200$, interest is 12% compounded monthly, payments made at the end of each month for 2 years

Work the following exercises.

35. Pat Dillon deposits $12,000 at the end of each year for 9 years in an account paying 8% interest compounded annually.
   (a) Find the final amount she will have on deposit.
   (b) Pat's brother-in-law works in a bank which pays 6% compounded annually. If she deposits her money in this bank, instead of the one above, how much will she have in her account?
   (c) How much would Dillon lose over 9 years by using her brother-in-law's bank?

36. One bank in town pays 8% compounded semiannually, while another pays 8% compounded quarterly. Find the difference in the interest earned in 5 years on a deposit of $24,000.

37. Pam Parker deposits $2435 at the beginning of each semiannual period for 8 years in an account paying 12% compounded semiannually. She then leaves that money alone, with no further deposits, for an additional 5 years. Find the final amount on deposit after the entire 13-year period.

38. Chuck deposits $10,000 at the beginning of each year for 12 years in an account paying 8% compounded annually. He then puts the total amount on deposit in another account paying 12% compounded semiannually for another 9 years. Find the final amount on deposit after the entire 21-year period.

39. Ray Berkowitz needs $10,000 in 8 years.
   (a) What amount should he deposit at the end of each quarter at 16% compounded quarterly so that he will have his $10,000?
   (b) Find Berkowitz's quarterly deposit if the money is deposited at 12% compounded quarterly.

40. Harv's Meats knows that it must buy a new deboner machine in 4 years. The machine costs $12,000. In order to accumulate enough money to pay for the machine, Harv decides to deposit a sum of money at the end of each six months in an account paying 16% compounded semiannually. How much should each payment be?

41. Barb Silverman wants to buy an $18,000 car in 6 years. How much money must she deposit at the end of each quarter in an account paying 12% compounded quarterly, so that she will have enough to pay for her car?

Recent tax law changes have made Individual Retirement Accounts (IRA's) available to certain workers. By these plans, a person can currently deposit $2000 annually, with taxes deferred on the principal and interest. To attract these deposits, banks have been advertising the amount that will accumulate at retirement. Suppose a 40-year-old deposits $2000 per year until age 65. Find the total in the account with the following assumptions of interest rates. (Assume semiannual compounding with payment made at the end of each semiannual period.)

42. 4%   **43.** 6%   **44.** 8%

45. 10%   **46.** 12%

Find the amount of each payment to be made into a sinking fund so that enough will be present to pay off the indicated loans. (See Example 5.)

47. Loan $2000, money earns 6% compounded annually, 5 annual payments

48. Loan $8500, money earns 8% compounded annually, 7 annual payments

49. Loan $11,000, money earns 16% compounded semiannually, for 6 years

50. Loan $75,000, money earns 12% compounded semiannually, for 4 1/2 years

51. Loan $50,000, money earns 16% compounded quarterly, for 2 1/2 years

52. Loan $25,000, money earns 12% compounded quarterly, for 3 1/2 years

53. Loan $6000, money earns 18% compounded monthly, for 3 years

54. Loan $9000, money earns 18% compounded monthly, for 2 1/2 years

Find the final amount of the annuities in Exercises 55 and 56.

55. $892.17 each month for 2 years in an account paying 7.5% interest compounded monthly

56. $2476.32 each quarter for 5.5 years at 7.81% interest compounded quarterly.

57. Jill Streitsel sells some land in Nevada. She will be paid a lump sum of $60,000 in 7 years. Until then, the buyer pays 8% interest, paid quarterly.
(a) Find the amount of each quarterly interest payment.
(b) The buyer sets up a sinking fund so that enough money will be present to pay off the $60,000. The buyer wants to make semiannual payments into the sinking fund; the account pays 6% compounded semiannually. Find the amount of each payment into the fund.
(c) Prepare a table showing the amount in the sinking fund after each deposit.

58. Jeff Reschke bought a rare stamp for his collection. He agreed to pay a lump sum of $4000 after 5 years. Until then, he pays 6% interest.
(a) Find the amount of each semiannual interest payment.
(b) Reschke sets up a sinking fund so that enough money will be present to pay off the $4000. He wants to make annual payments into the fund. The account pays 8% compounded annually. Find the amount of each payment into the fund.
(c) Prepare a table showing the amount in the sinking fund after each deposit.

5.5 Present Value of an Annuity; Amortization

The previous section discussed how to find the amount of an annuity after a series of equal, periodic payments. This section considers the present value of such an annuity. There are two ways to think of the *present value of an annuity*.

1. The **present value of an annuity** is a lump sum that can be deposited today that will amount to the same final total as would the periodic *payments* of an annuity, or

2. The **present value of an annuity** is a lump sum that could be deposited today so that equal periodic *withdrawals* could be made.

As an example of this second way of looking at the present value of an annuity, find the amount that must be deposited today at 10% compounded annually so that $1500 could be removed at the end of each year for 6 years.

The amount that must be deposited today is the sum of the present values of each of the separate withdrawals. In other words, today's deposit must include the present value of the $1500 to be withdrawn at the end of the first year. Use Table 6 and the formulas for compound interest to first find the present value of $1500 at 10% compounded annually for 1 year.

$$A = P\left(1 + \frac{i}{m}\right)^{nm}$$

$$P = \frac{A}{\left(1 + \dfrac{i}{m}\right)^{nm}}$$

$$P \approx \frac{\$1500}{1.10000} = \$1363.64$$

It is also necessary to deposit today the present value of the $1500 to be withdrawn at the end of the second year. Again use Table 6, this time with 10% and 2 years, finding 1.21000. The present value for the withdrawal at the end of the second year is

$$\frac{\$1500}{1.21000} = \$1239.67.$$

Continuing in this way for all six withdrawals gives the result shown in Figure 5.4.

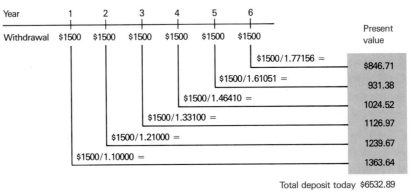

Total deposit today $6532.89

Figure 5.4

A lump sum deposit today of $6532.89 at 10% compounded annually would permit withdrawals of $1500 at the end of each year for 6 years. Also, a lump sum deposit today of $6532.89 left in an account for 6 years would produce the same final total as deposits of $1500 at the end of each year for 6 years, with all deposits at 10% compounded annually.

Find a formula for the present value of an annuity by first recalling that if deposits of R dollars are made at the end of each period for n periods, at a rate of interest i per period, then the account will contain

$$A = R \cdot s_{\overline{n}|i} = R\left[\frac{(1 + i)^n - 1}{i}\right]$$

dollars after n periods. The present value of this annuity is the lump sum P that can be deposited today at a rate of interest i per period which will amount to the same A dollars in n periods. Also, P dollars deposited today will amount to $P(1 + i)^n$ dollars after n periods at a rate of interest i per period. This amount, $P(1 + i)^n$, should be the same as A, the future value of the annuity. Substituting $P(1 + i)^n$ for A in the formula above gives

$$P(1 + i)^n = R\left[\frac{(1 + i)^n - 1}{i}\right].$$

Multiply both sides by $(1 + i)^{-n}$, to solve for P.

$$P = R(1 + i)^{-n}\left[\frac{(1 + i)^n - 1}{i}\right]$$

Use the distributive property; also recall $(1 + i)^{-n}(1 + i)^n = 1$.

$$P = R\left[\frac{(1 + i)^{-n}(1 + i)^n - (1 + i)^{-n}}{i}\right]$$

$$P = R\left[\frac{1 - (1 + i)^{-n}}{i}\right]$$

The amount P is called the **present value of the annuity.** The quantity in brackets is abbreviated as $a_{\overline{n}|i}$, so

$$a_{\overline{n}|i} = \frac{1 - (1 + i)^{-n}}{i}.$$

Values of $a_{\overline{n}|i}$ are given in Table 8, "Present Value of an Annuity," at the back of this text. The next box summarizes the formula for the present value of an annuity. Notice again that i represents rate of interest per period, not rate of interest per year.

The **present value** P of an annuity of n payments of R dollars each at the end of consecutive interest periods with interest compounded at a rate of interest i per period is

$$P = R\left[\frac{1 - (1 + i)^{-n}}{i}\right] \quad \text{or} \quad P = R \cdot a_{\overline{n}|i}.$$

Example 1 What lump sum deposited today at 8% interest compounded annually will yield the same total amount as payments of $1500 at the end of each year for 12 years, also at 8% interest compounded annually?

We want to find the present value of an annuity of $1500 per year for 12 years at 8% compounded annually. Using the table or a calculator, $a_{\overline{12}|.08} \approx 7.53608$, so

$$P = 1500(7.53608) = 11{,}304.12$$

or $11,304.12. A lump sum deposit of $11,304.12 today at 8% compounded annually will yield the same total after 12 years as deposits of $1500 at the end of each year for 12 years at 8% compounded annually.

Let's check this result. The compound amount in 12 years of a deposit of $11,304.12 today at 8% compounded annually can be found by the formula $A = P(1 + i)^n$. From the compound interest table, Table 6, $11,304.12 will produce a total of

$$(11{,}304.12)(2.51817) = 28{,}465.70,$$

or $28,465.70. On the other hand, from the table for $s_{\overline{n}|i}$, Table 7, deposits of $1500 at the end of each year for 12 years, at 8% compounded annually, give

$$1500(18.97713) = 28{,}465.70,$$

or $28,465.70, the same amount found above.

In summary, there are two ways to have $28,465.70 in 12 years at 8% compounded annually—a single deposit of $11,304.12 today, or payments of $1500 at the end of each year for 12 years. **1**

Example 2 Ms. Gonsalez agrees to pay $500 at the end of each year for 9 years to settle a business obligation. If money can be deposited at 10% compounded annually, find the lump sum she could deposit today to have enough money, with principal and interest, to make the payments.

The necessary amount is the present value of an annuity of $500 per year for 9 years, at 10% compounded annually. Use the formula

$$A = R \cdot a_{\overline{n}|i},$$

with $R = 500$, $n = 9$, and $i = 10\%$. From Table 8, $a_{\overline{9}|.10} \approx 5.75902$, so

$$A = \$500(5.75902) = \$2879.51.$$

A deposit of $2879.51 today, at 10% compounded annually, will permit withdrawals of $500 at the end of each year for 9 years. Since the withdrawals total $9 \times \$500 = \$4500$, the interest earned is

$$\$4500 - \$2879.51 = \$1620.49. \quad \textbf{2}$$

The present value formula also can be used if the lump sum is known and the periodic payment of the annuity must be found. The next example shows how to do this.

Example 3 A car costs $6000. After a down payment of $1000, the balance will be paid off in 36 monthly payments with interest of 12% per year on the unpaid balance. Find the amount of each payment.

1 What lump sum deposited today at 8% compounded annually will yield the same total amount as payments of
(a) $650 at the end of each year for 9 years?
(b) $17,580 at the end of each year for 14 years?

Answer:
(a) $4060.48
(b) $144,933.74

2 What lump sum deposited today would permit withdrawals of $1000 at the end of each quarter for 4 years, if the money is deposited at 8% compounded quarterly?

Answer:
$13,577.71

A single lump sum payment of $5000 today would pay off the loan. So, $5000 is the present value of an annuity of 36 monthly payments with interest of $12\%/12 = 1\%$ per month. Find R, the amount of each payment, by starting with $P = R \cdot a_{\overline{n}|i}$; replace P with 5000, n with 36, and i with .01. From Table 8, or a calculator, $a_{\overline{36}|.01} \approx 30.10751$, so

$$5000 = R(30.10751)$$
$$R = 166.07,$$

3 Find the monthly payment in Example 3 if the loan is for 48 months.

Answer:

$131.67

or $166.07. A monthly payment of $166.07 will be needed. **3** **4**

Amortization A loan is **amortized** if both the principal and interest are paid by a sequence of equal periodic payments. In Example 3 above, a loan of $5000 at 12% interest on the unpaid balance could be amortized by paying $166.07 per month for 36 months.

4 Roberta Maring buys a small business for $174,000. She agrees to pay off the cost in payments at the end of each semiannual period for 7 years, with interest of 16% on the unpaid balance. Find the amount of each payment.

Answer:

$21,105.64

Example 4 A speculator agrees to pay $15,000 for a parcel of land; this amount, with interest, will be paid over 4 years in semiannual payments at an interest rate of 12%.

(a) Find the amount of each payment.

If the speculator were to pay $15,000 immediately, there would be no need for any payments at all, making $15,000 the present value of an annuity of R dollars, with $n = 2 \cdot 4 = 8$ periods, and $i = 12\%/2 = 6\% = .06$ per period. Using $P = R \cdot a_{\overline{n}|i}$ with $P = 15,000$ gives

$$15,000 = R \cdot a_{\overline{8}|.06}$$

or

$$R = \frac{15,000}{a_{\overline{8}|.06}}.$$

From Table 8, or a calculator, $a_{\overline{8}|.06} \approx 6.20979$, and

$$R = \frac{15,000}{6.20979} = 2415.54,$$

or $2415.54. Each payment is $2415.54.

(b) Find the portion of the first payment that is applied to the reduction of the debt.

Interest is 12% per year, or 6% per semiannual period. During the first period, the entire $15,000 is owed. Interest on this amount for 6 months (1/2 year) is found by the formula for simple interest:

$$I = Prt = 15,000(.12)\left(\frac{1}{2}\right) = 900,$$

or $900. At the end of 6 months, the speculator makes a payment of $2415.54; since $900 of this represents interest, a total of

$$\$2415.54 - \$900 = \$1515.54$$

is applied to the reduction of the original debt.

(c) Find the balance due after 6 months.

The original balance due is $15,000. After 6 months, $1515.54 is applied to reduction of the debt. The debt owed after 6 months is

5 In Example 4, find the interest owed for the second six-month period and the amount applied to the reduction of the debt.

$$15,000 - \$1515.54 = \$13,484.46. \quad \boxed{5}$$

Continuing this process gives the **amortization schedule** shown below. As the schedule shows, the payment is always the same, except perhaps for a small adjustment in the final payment. Payment 0 is the original amount of the loan.

Answer:
$809.07; $1606.47

Amortization Schedule

Payment number	Amount of payment	Interest for period	Portion to principal	Principal at end of period
0	—	—	—	$15,000.00
1	$2415.54	$900.00	$1515.54	13,484.46
2	2415.54	809.07	1606.47	11,877.99
3	2415.54	712.68	1702.86	10,175.13
4	2415.54	610.51	1805.03	8370.10
5	2415.54	502.21	1913.33	6456.77
6	2415.54	387.41	2028.13	4428.64
7	2415.54	265.72	2149.82	2278.82
8	2415.54	136.73	2278.82	0

6 A new car is to be financed with a $6000, 48-month loan, at 18% on the unpaid balance.
(a) Find the monthly payment necessary to amortize the loan.
(b) Find the principal balance after the first payment.
(c) Find the principal balance after the second payment.

Answer:
(a) $176.25
(b) $5913.75
(c) $5826.21

Example 5 A house is bought for $74,000, with a down payment of $16,000. Interest is charged at 10.25% per year for 30 years. Find the amount of each monthly payment to amortize the loan.

Here, the present value P is 58,000 (74,000 − 16,000). Also, $i = .1025/12 = .0085416667$, and $n = 12 \cdot 30 = 360$. Find R from the formula for the present value of an annuity.

$$58,000 = R \cdot a_{\overline{360}|.0085416667}$$

or

$$58,000 = R\left[\frac{1 - (1 + .0085416667)^{-360}}{.0085416667}\right]$$

Use a financial calculator, or a calculator with a y^x key to get

7 Find the monthly payment in Example 5 for interest rates of
(a) 14%;
(b) 17%.

Answer:
(a) $687.23
(b) $826.89

$$58,000 \approx R\left[\frac{1 - .0467967624}{.0085416667}\right]$$

$$58,000 = R\left[\frac{.9532032376}{.0085416667}\right]$$

$$58,000 \approx R[111.5945249]$$

or

$$R = 519.74.$$

Monthly payments of $519.74 will be required to amortize the loan. **7**

5.5 Exercises

Find each of the following values.

1. $a_{\overline{15}|.06}$ **2.** $a_{\overline{10}|.03}$ **3.** $a_{\overline{18}|.04}$ **4.** $a_{\overline{30}|.01}$

5. $a_{\overline{16}|.01}$ **6.** $a_{\overline{32}|.02}$ **7.** $a_{\overline{6}|.015}$ **8.** $a_{\overline{18}|.015}$

Find the present value of each of the following ordinary annuities. (See Examples 1 and 2.)

9. Payments of $1000 are made annually for 9 years at 8% compounded annually.

10. Payments of $5000 are made annually for 11 years at 6% compounded annually.

11. Payments of $890 are made annually for 16 years at 8% compounded annually.

12. Payments of $1400 are made annually for 8 years at 8% compounded annually.

13. Payments of $10,000 are made semiannually for 15 years at 10% compounded semiannually.

14. Payments of $50,000 are made quarterly for 10 years at 8% compounded quarterly.

15. Payments of $15,806 are made quarterly for 3 years at 15.8% compounded quarterly.

16. Payments of $18,579 are made every six months for 8 years at 19.4% compounded semiannually.

Find the lump sum deposited today that will yield the same total amount as payments of $10,000 at the end of each year for 15 years, at each of the given interest rates. (See Examples 1 and 2.)

17. 4%, compounded annually

18. 5%, compounded annually

19. 6%, compounded annually

20. 8%, compounded annually

21. 12%, compounded annually

22. 16%, compounded annually

Work the following exercises.

23. In his will the late Mr. Hudspeth said that each child in his family could have an annuity of $2000 at the end of each year for 9 years, or the equivalent present value. If money can be deposited at 8% compounded annually, what is the present value?

24. In the ''Million Dollar Lottery,'' a winner is paid a million dollars at the rate of $50,000 per year for 20 years. Assume that these payments form an ordinary annuity, and that the lottery managers can invest money at 6% compounded annually. Find the lump sum that the management must put away to pay off the ''millon dollar'' winner.

25. Lynn Meyers buys a new car costing $6000. She agrees to make payments at the end of each monthly period for 4 years.
    (a) If she pays 12% interest on the unpaid balance, what is the amount of each payment?
    (b) Find the total amount of interest Meyers will pay.

26. Tom Frederickson wants to buy a piece of property for a service station. The seller wants to receive $4000 at the end of each year for the next 12 years. If Frederickson can invest money at 10% compounded annually, find the lump sum that he should deposit today so he will be able to make the annual payments.

27. What sum deposited today at 5% compounded annually for 8 years will provide the same amount as $1000 deposited at the end of each year for 8 years at 6% compounded annually?

28. What lump sum deposited today at 8% compounded quarterly for 10 years will yield the same final amount as deposits of $4000 at the end of each six-month period for 10 years at 6% compounded semiannually?

Find the payment necessary to amortize each of the following loans. Interest is calculated on the unpaid balance. (See Examples 3–4.)

29. $1000, 9 annual payments at 8%

30. $2500, 6 quarterly payments at 16%

31. $41,000, 10 semiannual payments at 12%

32. $90,000, 12 annual payments at 8%

33. $140,000, 15 quarterly payments at 12%

34. $7400, 18 semiannual payments at 16%

35. $5500, 24 monthly payments at 18%

36. $45,000, 36 monthly payments at 18%

Find the monthly house payment necessary to amortize the following loans. Interest is calculated on the unpaid balance. You will need a financial calculator or one with a y^x key. (See Example 5.)

37. $49,560 at 15.75% for 25 years

38. $70,892 at 14.11% for 30 years

39. $53,762 at 16.45% for 30 years

40. $96,511 at 15.57% for 25 years

Work the following exercises.

41. An insurance firm pays $4000 for a new printer for its computer. It amortizes the loan for the printer in 4 annual payments at 8% interest on the unpaid balance. Prepare an amortization schedule for this machine.

42. Large semitrailer trucks cost $72,000 each. Ace Trucking buys such a truck and agrees to pay for it by a loan which will be amortized with 9 semiannual payments at 16% interest on the unpaid balance. Prepare an amortization schedule for this truck.

43. One retailer charges $1048 for a correcting electric typewriter. A firm of tax accountants buys 8 of these machines. They make a down payment of $1200 and agree to amortize the balance with monthly payments at 18% interest on the unpaid balance for 4 years. Prepare an amortization schedule showing the first six payments.

44. When Denise Sullivan opened her law office, she bought $14,000 worth of law books and $7200 worth of office furniture. She paid $1200 down and agreed to amortize the balance with semiannual payments for 5 years at 16% interest on the unpaid balance. Prepare an amortization schedule for this purchase.

45. When Ms. Thompson died, she left $25,000 for her husband. He deposits the money at 6% compounded annually. He wants to make annual withdrawals from the account so that the money (principal and interest) is gone in exactly 8 years.
(a) Find the amount of each withdrawal.
(b) Find the amount of each withdrawal if the money must last 12 years.

46. The trustees of a college have accepted a gift of $150,000. The donor has directed the trustees to deposit the money in an account paying 12% per year, compounded semiannually. The trustees may withdraw money at the end of each six-month period; the money must last 5 years.
(a) Find the amount of each withdrawal.
(b) Find the amount of each withdrawal if the money must last 6 years.

Prepare an amortization schedule for each of the following loans.

47. $37,947.50 with 8.5% interest on the unpaid balance, to be paid with equal annual payments over 10 years

48. $4835.80 at 9.25% interest on the unpaid balance, to be repaid in 5 years in equal semiannual payments

Supplementary Exercises

We have discussed many different types of interest problems. A summary of the various formulas that are used with interest appears below, followed by several exercises using the various formulas.

Simple Interest   The **simple interest** I on an amount of P dollars for t years at a rate of interest r per year is

$$I = Prt.$$

Discounts and Proceeds   If D is the discount on a loan having a maturity value A at a rate of interest r per year for t years, the **proceeds** P are

$$P = A - D \quad \text{or} \quad P = A(1 - rt).$$

Compound Interest   If P dollars is deposited for n years at a rate of interest i per year, compounded m times per year, the **compound amount** A is

$$A = P\left(1 + \frac{i}{m}\right)^{nm}.$$

Future Value of an Annuity   The **future value** A of an annuity of n payments of R dollars each at the end of consecutive interest periods with interest compounded at a rate of interest i per period is

$$A = R\left[\frac{(1 + i)^n - 1}{i}\right] \quad \text{or} \quad A = R \cdot s_{\overline{n}|i}.$$

Present Value of an Annuity   The **present value** P of an annuity of n payments of R dollars each at the end of consecutive interest periods with interest compounded at a rate of interest i per period is

$$P = R\left[\frac{1 - (1 + i)^{-n}}{i}\right] \quad \text{or} \quad P = R \cdot a_{\overline{n}|i}.$$

Supplementary Exercises

Work each of the following interest problems. Round money amounts to the nearest cent, time to the nearest day, and rates to the nearest percent.

Find the present value of $82,000 for the following number of years, if the money can be deposited at 12% compounded quarterly.

1. 3 years

2. 5 years

3. 7 years

4. 10 years

Find the amount of the payment necessary to amortize the following amounts.

5. $72,000 loan, 11 annual payments at 12%

6. $4,250 loan, 13 quarterly payments at 12%

7. $58,000 loan, 23 quarterly payments at 16%

8. $112,000 loan, 32 monthly payments at 12%

Find the compound amount for each of the following. Some of these exercises require a calculator with a y^x key.

9. $1,000 at 10% compounded annually for 17 years

10. $3,520 at 8% compounded annually for 10 years

11. $4,792.35 at 12% compounded semiannually for 5 1/2 years

12. $2,500 at 16% compounded quarterly for 3 3/4 years

13. $115,700 at 8% compounded quarterly for 5 1/4 years

14. $23,500 at 18% compounded monthly for 30 months

15. $567.88 at 5.25% compounded daily for 82 days

16. $12,500 at 9% compounded daily for 3 years

Find the simple interest for each of the following. Assume 360 days in a year.

17. $2500 at 8.5% for 2 years

18. $42,500 at 11.75% for 10 months

19. $32,662 at 8.882% for 225 days

20. $56,009 at 13.221% for 179 days

Find the amount of interest earned by each of the following deposits.

21. $22,500 at 12% compounded quarterly for 5 1/4 years

22. $53,142 at 18% compounded monthly for 32 months

Find the compound amount and the amount of interest earned by a deposit of $32,750 at 10% compounded continuously for the following number of years.

23. 2 years   24. 5 years

25. 7 1/2 years   26. 9.2 years

Find the present value of the following future amounts. Assume 360 days in a year and use simple interest.

27. $50,000 for 11 months, money earns 14%

28. $17,320 for 9 months, money earns 13.5%

29. $122,300 for 138 days, money earns 11.75%

30. $51,800 for 312 days, money earns 12.801%

Find the proceeds for the following. Assume 360 days in a year and use simple interest.

31. $72,113 for 11 months, discount rate 15%

32. $23,561 for 112 days, discount rate 14.33%

33. $267,100 for 271 days, discount rate 15.72%

Work the following exercises. Assume simple interest and 360 days in a year.

34. Tom Davis owes $7850 to a relative. He has agreed to pay the money in 5 months, at an interest rate of 13%.

One month before the loan is due, the relative discounts the loan at the bank. The bank charges a 15.82% discount rate. How much money does the relative receive?

35. Find the principal that must be invested at 11.25% simple interest to earn $937.50 interest in 8 months.

36. A small business invests some spare cash for 3 months at 12.2% interest and earns $244 in interest. Find the amount of the investment.

37. Suppose $7200 is invested at 14.5% interest and earns $435 in interest. Find the time of the loan (in months).

38. A person loans $28,000 to her business. The loan is at 15.5% and earns $3255 of interest. Find the time of the loan in months.

39. A 10-month loan of $42,000 produces $5320 of interest. Find the rate of interest.

40. Suppose $84,720 is deposited for 7 months and earns $8055.46 in interest. Find the rate of interest.

41. A loan of $5600 was made on May 7 for 180 days, at 12%. Find the present value of the loan on the day it was made. Use an interest rate of 10%.

42. Suppose a loan of $52,300 is made on December 17, for 85 days, at 11%. Find the present value of the loan on February 1, assuming an interest rate of 13.5%.

Work the following exercises.

43. In 3 years Ms. Thompson must pay a pledge of $7500 to her college's building fund. What lump sum can she deposit today, at 10% compounded semiannually, so that she will have enough to pay the pledge?

44. According to the terms of a divorce settlement, one spouse must pay the other a lump sum of $2800 in 17 months. What lump sum can be invested today, at 18% compounded monthly, so that enough will be available for the payment?

Find the amount of each of the following annuities.

45. $1000 is deposited at the end of each year for 12 years, money earns 10% compounded annually

46. $2500 is deposited at the end of each semiannual period for 5 1/2 years, money earns 12% compounded semiannually

47. $250 is deposited at the end of each quarter for 7 1/4 years, money earns 12% compounded quarterly

48. $800 is deposited at the end of each month for 1 1/4 years, money earns 18% compounded monthly

49. $100 is deposited at the beginning of each year for 5 years, money earns 8% compounded yearly

50. $5200 is deposited at the beginning of each quarter for 3 3/4 years, money earns 16% compounded quarterly

51. Willa Burke deposits $803.47 at the end of each quarter for 3 3/4 years, in an account paying 12% compounded quarterly. Find the final amount in the account. Find the amount of interest earned.

52. A firm of attorneys deposits $5000 of profit sharing money at the end of each semiannual period for 7 1/2 years. Find the final amount in the account if the deposit earns 10% compounded semiannually. Find the amount of interest earned.

Find the amount of each payment to be made to a sinking fund so that enough money will be available to pay off the indicated loan.

53. $10,000 loan, money earns 10% compounded annually, 7 annual payments

54. $42,000 loan, money earns 12% compounded quarterly, 13 quarterly payments

55. $100,000 loan, money earns 12% compounded semiannually, 9 semiannual payments

56. $53,000 loan, money earns 18% compounded monthly, 35 monthly payments

In Exercises 57–61, find the present value of each ordinary annuity.

57. Payments of $1200 are made annually for 7 years at 10% compounded annually

58. Payments of $500 are made semiannually at 10% compounded semiannually for 3 1/2 years

59. Payments of $1500 are made quarterly for 5 1/4 years at 12% compounded quarterly

60. Payments of $905.43 are made monthly for 2 11/12 years at 18% compounded monthly

61. The owner of Eastside Hallmark borrows $48,000 to expand the business. The money will be repaid in equal payments at the end of each year for 7 years. Interest is 10%. Find the amount of each payment.

62. A small resort must add a swimming pool to compete with a new resort built nearby. The pool will cost $28,000. The resort borrows the money and agrees to repay it with equal payments at the end of each quarter for 6 1/2 years, at an interest rate of 12%. Find the amount of each payment.

Prepare an amortization schedule for each of the following loans. Interest is calculated on the unpaid balance.

63. $8500 loan repaid in semiannual payments for 3 1/2 years at 12%

64. $40,000 loan repaid in quarterly payments for 1 3/4 years at 16%

Chapter 5   Review Exercises

Many of these exercises will require a calculator with a y^x key.
Find the simple interest.

1. $15,903 at 18% for 8 months

2. $4902 at 19.5% for 11 months

3. $42,368 at 15.22% for 5 months

4. $3478 at 17.4% for 88 days (assume a 360-day year)

5. $2390 at 18.7% from May 3 to July 28 (assume 365 days in a year)

6. $69,056.12 at 15.5% from September 13 to March 25 of the following year (assume a 365-day year)

Find the present value of the future amounts. Assume 360 days in a year; use simple interest.

7. $25,000 in 10 months, money earns 17%

8. $459.57 in 7 months, money earns 18.5%

9. $80,612 in 128 days, money earns 16.77%

Find the proceeds. Assume 360 days in a year.

10. $56,882, discount rate 19%, length of loan 5 months

11. $802.34, discount rate 18.6%, length of loan 11 months

12. $12,000, discount rate 17.09%, length of loan 145 days

Work the following exercises.

13. Tom Wilson owes $5800 to his mother. He has agreed to pay the money in 10 months, at an interest rate of 14%. Three months before the loan is due, his mother discounts the loan at the bank. The bank charges a 17.45% discount rate. How much money does his mother receive?

14. Larry DiCenso needs $9812 to buy new equipment for his business. The bank charges a discount of 14%. Find the amount of DiCenso's loan, if he borrows the money for 7 months.

Find the effective annual rate for the following bank discount rates. (Assume a one-year loan of $1000.)

15. 14%   **16.** 17.5%

Find the following compound amounts.

17. $1000 at 8% compounded annually for 9 years

18. $2800 at 6% compounded annually for 10 years

19. $19,456.11 at 12% compounded semiannually for 7 years

20. $312.45 at 16% compounded semiannually for 16 years

21. $1900 at 16% compounded quarterly for 9 years

22. $57,809.34 at 12% compounded quarterly for 5 years

23. $2500 at 18% compounded monthly for 3 years

24. $11,702.55 at 18% compounded monthly for 4 years

Find the amount of interest earned by each of the following deposits.

25. $3954 at 8% compounded annually for 12 years

26. $12,699.36 at 16% compounded semiannually for 7 years

27. $7801.72 at 12% compounded quarterly for 5 years

28. $48,121.91 at 18% compounded monthly for 2 years

29. $12,903.45 at 12.37% compounded quarterly for 29 quarters

30. $34,677.23 at 14.72% compounded monthly for 32 months

Find the present value of the following amounts if money can be invested at the given rate.

31. $5000 in 9 years, 8% compounded annually

32. $12,250 in 5 years, 12% compounded semiannually

33. $42,000 in 7 years, 18% compounded monthly

34. $17,650 in 4 years, 16% compounded quarterly

35. $1347.89 in 3.5 years, 13.77% compounded semiannually

36. $2388.90 in 44 months, 12.93% compounded monthly

Work the following exercises.

37. In four years, Mr. Heeren must pay a pledge of $5000 to his church's building fund. What lump sum can he invest today, at 12% compounded semiannually, so that he will have enough to pay his pledge?

38. Joann Hudspeth must make an alimoney payment of $1500 in 15 months. What lump sum can she invest today, at 18% compounded monthly, so that she will have enough to make the payment?

Write the first five terms for each of the following sequences. Identify any which are arithmetic or geometric.

39. $a_n = -4n + 2$ **40.** $a_n = (-2)^n$

41. $a_n = \dfrac{n + 2}{n + 5}$ **42.** $a_n = (n - 7)(n + 5)$

43. Find a_{12} for the arithmetic sequence having $a_1 = 6$ and $d = 5$.

44. Find the sum of the first 20 terms for the arithmetic sequence having $a_1 = -6$ and $d = 8$.

45. Find a_4 for the geometric sequence with $a_1 = -3$ and $r = 2$.

46. Find the sum of the first 6 terms for the geometric sequence with $a_1 = 8000$ and $r = -1/2$.

In Exercises 47–52, find the value of each annuity.

47. $500 is deposited at the end of each six-month period for 8 years; money earns 16% compounded semiannually

48. $1288 is deposited at the end of each year for 14 years; money earns 8% compounded annually

49. $4000 is deposited at the end of each quarter for 7 years; money earns 16% compounded quarterly

50. $233 is deposited at the end of each month for 4 years; money earns 18% compounded monthly

51. $672 is deposited at the beginning of each quarter for 7 years; money earns 20% compounded quarterly

52. $11,900 is deposited at the beginning of each month for 13 months; money earns 18% compounded monthly

53. Georgette Dahl deposits $491 at the end of each quarter for 9 years. If the account pays 19.4% compounded quarterly, find the final amount in the account.

54. J. Euclid deposits $1526.38 at the beginning of each six-month period in an account paying 20.6% compounded semiannually. How much will be in the account after five years?

Find the amount of each payment to be made into a sinking fund so that enough money will be available to pay off the indicated loan.

55. $6500 loan, money earns 8% compounded annually, 6 annual payments

56. $57,000 loan, money earns 16% compounded semiannually, for 8 1/2 years

57. $233,188 loan, money earns 19.7% compounded quarterly, for 7 3/4 years

58. $1,056,788 loan, money earns 18.12% compounded monthly, for 4 1/2 years

Find the present value of each of the following ordinary annuities.

59. Payments of $850 are made annually for 4 years at 8% compounded annually.

60. Payments of $1500 are made quarterly for 7 years, at 16% compounded quarterly.

61. Payments of $4210 are made semiannually for 8 years, at 18.6% compounded semiannually.

62. Payments of $877.34 are made monthly for 17 months, at 22.4% compounded monthly.

Work the following exercises.

63. Vicki Manchester borrows $20,000 from the bank to help her expand her business. She agrees to repay the money in equal payments at the end of each year for 9 years. Interest is at 18.9% per year on the unpaid balance. Find the amount of each payment.

64. Ken Murrill wants to expand his pharmacy. He does this by taking out a loan of $49,275 from the bank, at an interest rate of 24.2% on the unpaid balance. He repays the loan in 48 monthly payments. Find the amount of each payment needed to amortize the loan.

Find the amount of the payment needed to amortize the following loans. Assume that interest is calculated on the unpaid balance.

65. $80,000 loan, 8% interest, 9 annual payments

66. $3200 loan, 16% interest, 10 quarterly payments

67. $32,000 loan, 19.4% interest, 17 quarterly payments

68. $51,607 loan, 23.6% interest, 32 monthly payments

Find the monthly house payments for the following mortgages. Interest is calculated on the unpaid balance.

69. $56,890 at 14.74% for 25 years

70. $77,110 at 16.45% for 30 years

For many years, the most popular mortgages were those for 25 or 30 years. Recently, however, accelerated mortgages, with 15-year payoffs, have become popular. With these mortgages, a slightly higher monthly payment produces a paid-off loan after 15 years.

71. Assume an $80,000 mortgage at 12% compounded monthly. Find the monthly payment and the total amount of interest paid over the following number of years:
    (a) 30,   **(b)** 25,   **(c)** 15.

72. Assume a $125,000 mortgage at 15% compounded monthly. Find the monthly payment and the total amount of interest paid over the following number of years:
    (a) 30,   **(b)** 25,   **(c)** 15.

Prepare amortization schedules for the following loans. Interest is calculated on the unpaid balance.

73. $5000 at 10% semiannually for 3 years, semiannual payments at the end of each period

74. $12,500 at 12% quarterly for 2 years, quarterly payments at the end of each period

Case 4

A New Look at Athletes' Contracts*

There has been a lot of controversy lately about big-buck contracts with athletes and how this affects the nature of the sports business.

Indeed, the recent headlines of a $40 million contract with Steve Young from Brigham Young University to play football for the USFL has opened Pandora's box. Club owners in all sports are wondering when all of this is going to stop.

To put the issue in its proper perspective, however, it seems to me that the $40 million contract is misleading. Just mentioning dollars misses several points. For one, the USFL has got to be around long enough to pay off. Two, the present value of million dollar payments that stretch 43 years into the future is reduced considerably.

In order to assess the effective value of these kinds of contracts, two important questions must be answered. First, what is the risk of the company to pay off on the periodic principal payments; and second, what is the present value of the return when parlayed into future cash benefits?

Recently, I conducted a survey among a number of sports-minded educators and students here on the campus. The survey was similar to the rating survey conducted by Wall Street rating firms when they evaluate risk of a corporation.

Specifically, the question was: "On a scale of 1 to 10, give me your estimate of the team's ability to continue in business and to meet the commitments of its contracts."

I was surprised to find that the baseball teams were rated with the lowest risk; the National Football League teams next; then the National Basketball Association teams; and then the United States Football League teams coming in last.

The cash flows which these athletes receive in the future have a present value today in total dollars based upon about an 8 percent discount factor each year for expected inflation. By determining the expected present value, the average per annum return of Athlete A who receives megabucks in the future can be compared with Athlete B who receives megabucks not so far into the future.

When risk of paying these megabucks is considered, we arrive at a completely different picture than just the dollar signs on the contract portray.

The following chart of the highly paid athletes, recently taken from a newspaper article, shows the athletes' names in sequence according to the best deal when risk and expected yearly return are assessed. A lesser coefficient of variation provides a statistical ratio of lesser risk to the per annum return for each athlete for each dollar of present value.

Notice that these risk assumptions show George Foster of the New York Mets with the best deal. Dave Winfield and Gary Carter are close behind. Steve Young is last, yet it is claimed that he received the most attractive sports contract ever signed.

It is true that risk is only our perception of the uncertainties of future events, and many times our judgment is wrong. But let's face it. It is the only game in town, unless you are gifted with extrasensory perception or have a crystal ball. In that case, I would appreciate your contacting me because I have work for you.

Anytime we purchase a bond or a like security, we make judgments about the future ability of the firm or municipality to meet its interest and principal payments. We employ expert rating agencies like Moody's or Standard & Poor's to professionally rate these securities. Indeed, the market price of the bond depends upon the judgments of these agencies.

It would seem to me that the time is long overdue for a similar rating system to be implemented to determine the risk of sports organizations—if for no other reason than to help the athlete decide where he would like to hang his hat. But more importantly, to provide a truer financial picture to the fans and broadcasting companies who are asked to pick up the tab. What do you think?

---

*From "A New Look at Athletes' Contracts" by Richard F. Kaufman in *The Sacramento Bee*, April 30, 1984. Reprinted by permission of the author.

| | Present value | | Per annum | |
Some highly-paid athletes	of contract (in millions)	Expected value	Risk assessment	Coefficient of variation
George Foster NL (Mets) $10 million, 5 years	$ 7.985	$1.597	2.3	1.44
Dave Winfield AL (Yankees) $21 million, 10 years	$14.091	$1.409	2.6	1.85
Gary Carter NL (Expos) $15 million, 8 years	$10.775	$1.347	3.0	2.23
Moses Malone NBA (76ers) $13.2 million, 6 years	$10.170	$1.695	5.4	3.19
Larry Bird NBA (Celtics) $15 million, 7 years	$11.157	$1.594	5.2	3.26
Wayne Gretzky NHL (Edmonton Oilers) $21 million, 21 years	$10.170	$ .477	5.8	12.16
Magic Johnson NBA (Lakers) $25 million, 25 years	$10.675	$ .427	5.6	13.12
Steve Young USFL (Express) $40 million, 43 years	$11.267	$ .262	7.2	27.48

Case 5

Present Value*

The Southern Pacific Railroad, with lines running from Oregon to Louisiana, is one of the country's most profitable railroads. The railroad has vast landholdings (granted by the government in the last half of the nineteenth century) and is diversified into trucking, pipelines, and data transmission.

The railroad was recently faced with a decision on the fate of an old bridge which crosses the Kings River in central California. The bridge is on a minor line which carries very little traffic. Just north of the Southern Pacific bridge is another bridge, owned by the Santa Fe Railroad; it too carries little traffic. The Southern Pacific had two alternatives: it could replace its own bridge or it could negotiate with the Sante Fe for the rights to run trains over its bridge. In the second alternative a yearly fee would be paid to the Santa Fe, and new connecting tracks would be built. The situation is shown in the figure.

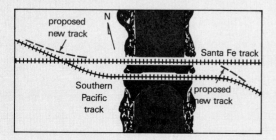

To find the better of these two alternatives, the railroad used the following approach.

1. Calculate estimated expenses for each alternative.
2. Calculate annual cash flows in after tax dollars. At a 48% corporate tax rate, $1 of expenses actually costs the railroad $.52, and $1 of revenue can bring a maximum of $.52 in profit. Cash flow for a given year is found by the following formula.

Cash flow = − .52 (operating and maintenance expenses)
+ .52 (savings and revenue)
+ .48 (depreciation)

3. Calculate the net present values of all cash flows for future years. The formula used is

Net present value = Σ(cash flow in year i)$(1 + k)^{1-i}$

where i is a particular year in the life of the project and k is the assumed annual rate of interest. (Recall: Σ indicates a sum.) The net present value is found for interest rates from 0% to 20%.

4. The interest rate that leads to a net present value of $0 is called the **rate of return** on the project.

Let us now see how these steps worked out in practice.

Alternative 1: Operate over the Sante Fe bridge First, estimated expenses were calculated.

1976	Work done by Southern Pacific on Santa Fe track	$27,000
1976	Work by Southern Pacific on its own track	11,600
1976	Undepreciated portion of cost of current bridge	97,410
1976	Salvage value of bridge	12,690
1977	Annual maintenance of Santa Fe track	16,717
1977	Annual rental fee to Santa Fe	7,382

From these figures and others not given here, annual cash flows and net present values were calculated. The following table was then prepared.

Interest rate, %	Net present value
0	$85,731
4	67,566
8	53,332
12	42,008
16	32,875
20	25,414

Although the table does not show a net present value of $0, the interest rate that leads to that value is 44%. This is the rate of return for this alternative.

*Based on information supplied by The Southern Pacific Transportation Company, San Francisco.

Alternative 2: Build a new bridge Again, estimated expenses were calculated.

1976	Annual maintenance	$ 2,870
1976	Annual inspection	120
1976	Repair trestle	17,920
1977	Install bridge	189,943
1977	Install walks and handrails	15,060
1978	Repaint steel (every 10 years)	10,000
1978	Repair trusses (every 10 years, increases by $200)	2,000
1981	Replace ties	31,000
1988	Repair concrete (every 10 years)	400
2021	Replace ties	31,000

After cash flows and net present values were calculated, the following table was prepared.

Interest rate, %	Net present value
0	$399,577
4	96,784
7.8	0
8	−3,151
12	−43,688
16	−62,615
20	−72,126

In this alternative the net present value is $0 at 7.8%, the rate of return.

Based on this analysis, the first alternative, renting from the Santa Fe, is clearly preferable.

Exercises

Find the cash flow in each of the following years. Use the formula given above.

1. Alternative 1, year 1977, operating expenses $6228, maintennce expenses $2976, savings $26,251, depreciation $10,778

2. Alternative 1, year 1984, same as Exercise 1, but depreciation is only $1347

3. Alternative 2, year 1976, maintenance $2870, operating expenses $6386, savings $26,251, no depreciation

4. Alternative 2, year 1980, operating expenses $6228, maintenance expenses $2976, savings $10,618, depreciation $6736

6

Systems of Linear Equations and Matrices

Many mathematical models involve more than one equation that must be satisfied. Such a set of equations is called a **system of equations.** Any solutions that satisfy all the equations in the system are **solutions of the system.** After discussing systems, this chapter introduces the idea of a **matrix,** and shows how matrices are used to solve systems of linear equations.

6.1 Systems of Linear Equations

An animal feed is made from three ingredients: corn, soybeans, and cottonseed. One unit of each ingredient provides the number of units of protein, fat, and fiber shown in the table. For example, the entries in the first row, .25, .4, and .3, mean that one unit of corn provides twenty-five hundredths (one fourth) of a unit of protein, four tenths of a unit of fat, and three tenths of a unit of fiber.

[1] (a) Check that $x = 40$, $y = 15$, and $z = 30$ is a solution of the first equation.

$$.25x + .4y + .2z = 22$$
$$.25(\underline{}) + .4(\underline{}) + .2(\underline{}) = 22$$
$$\underline{} + \underline{} + \underline{} = 22$$
$$\underline{} = 22$$

Do these three numbers satisfy the equation?

(b) Do these numbers satisfy all of the equations?

Answer:

(a) 40; 15; 30; 10; 6; 6; 22; yes

(b) Yes

	Protein	Fat	Fibers
Corn x	.25	.4	.3
Soybeans y	.4	.2	.2
Cottonseed z	.2	.3	.1

Suppose we need to know the number of units of each ingredient that should be used to make a feed that contains 22 units of protein, 28 units of fat, and 18 units of fiber. Find out by letting x represent the number of units of corn; y, the number of units of soybeans; and z, the number of units of cottonseed required. Since the total amount of protein is to be 22 units,

$$.25x + .4y + .2z = 22.$$

Also, for the 28 units of fat,

$$.4x + .2y + .3z = 28,$$

and, for the 18 units of fiber,

$$.3x + .2y + .1z = 18.$$

These three equations form a *system of equations.*

Solving the problem requires values of x, y, and z that satisfy this system of equations. Verify that $x = 40$, $y = 15$, and $z = 30$ is a solution of the system, since these numbers satisfy all three equations. [1]

A **linear equation in n unknowns** is an equation that can be written in the form

$$a_1x_1 + a_2x_2 + \cdots + a_nx_n = k,$$

where $a_1, a_2, \ldots,$ and k are all real numbers, with at least one of the a_is not zero. For example,

$$\begin{array}{ll} 2x + 3y = 6 & \text{is a linear equation in two unknowns,} \\ 4x - 3y + 2z = 8 & \text{is a linear equation in three unknowns,} \\ 6x - y + 7z + 2w = 1 & \text{is a linear equation in four unknowns,} \end{array}$$

and so on.

A solution of the linear equation

$$a_1x_1 + a_2x_2 + \cdots + a_nx_n = k$$

is a sequence of numbers $s_1, s_2, \ldots, s_n,$ such that

$$a_1s_1 + a_2s_2 + \cdots + a_ns_n = k.$$

The solution may be written between parentheses as $(s_1, s_2, \ldots, s_n)$. For example, $(1, 6, 2)$ is a solution of the equation $3x_1 + 2x_2 - 4x_3 = 7$, since $3(1) + 2(6) - 4(2) = 7$. ②

Because the graph of a linear equation in two variables is a straight line, there are three possibilities for the solution of a system of two linear equations in two variables.

1. The two graphs are lines intersecting at a single point. The coordinates of this point give the solution of the system. [See Figure 6.1(a).]
2. The graphs are distinct parallel lines. When this is the case, the system is **inconsistent;** that is, there is no solution common to both equations. [See Figure 6.1(b).]
3. The graphs are the same line. Here the equations are said to be **dependent,** since any solution of one equation is also a solution of the other. There are an infinite number of solutions. [See Figure 6.1(c).]

② Decide whether $(-1, 4, 3)$ is a solution for each of the following equations.
(a) $x + y + z = 6$
(b) $3x - y + 4z = 5$
(c) $4x + 3y + 9z = 36$

Answer:
(a) Yes
(b) Yes
(c) No

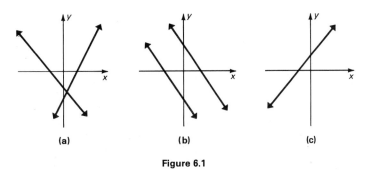

(a) (b) (c)

Figure 6.1

In larger systems, with more equations and more variables, there may also be exactly one solution, an infinite number of solutions, or no solution.

Transformations A linear system of equations can be solved by using properties of algebra to change the system until a simpler equivalent system is found. An **equivalent system** is one that has the same solutions as the given system. There are three transformations that can be applied to a system to get an equivalent system.

> For any system of equations, the following **transformations** produce an equivalent system:
>
> 1. interchange any two equations;
> 2. multiply both sides of an equation by any nonzero real number;
> 3. add to an equation a multiple of any other equation in the system.

Use of these transformations leads to an equivalent system because each transformation is an algebraic manipulation that can be reversed or "undone," allowing a return to the original system. ③

Example 1 Solve the system

$$3x - 4y = 1 \qquad \text{(1)}$$
$$2x + 3y = 12. \qquad \text{(2)}$$

We want to get a system of equations that is equivalent to the given system, but simpler. If the equations are transformed so that the variable x disappears from one of them, the value of y can be found. Start by using the second transformation to multiply both sides of equation (1) by 2. This gives the equivalent system

$$6x - 8y = 2 \qquad \text{(3)}$$
$$2x + 3y = 12. \qquad \text{(2)}$$

Now, using the third transformation, multiply both sides of equation (2) by -3, and add the result to equation (3). Then simplify.

$$(6x - 8y) + (-3)(2x + 3y) = 2 + (-3)(12)$$
$$6x - 8y - 6x - 9y = 2 - 36$$
$$-17y = -34$$
$$y = 2 \qquad \text{(4)}$$

The result of these steps is the equivalent system

$$3x - 4y = 1 \qquad \text{(1)}$$
$$y = 2. \qquad \text{(4)}$$

(Pairing $y = 2$ with either equations (2) or (3) would also give an equivalent system.)

③ Perform the indicated transformation on the system

$$x - 3y = -1$$
$$3x + 2y = 8.$$

(a) Multiply both sides of the first equation by -3.
(b) Add to the second equation the result of multiplying the first equation by -3.

Answer:
(a) The system becomes

$$-3x + 9y = 3$$
$$3x + 2y = 8.$$

(b) The system becomes

$$x - 3y = -1$$
$$11y = 11.$$

Find the value of x by substituting 2 for y in equation (1) to get

$$3x - 4(2) = 1$$
$$3x - 8 = 1$$
$$x = 3.$$

The solution of the given system is (3, 2). The graphs of both equations of the system are shown in Figure 6.2. The graph suggests that (3, 2) satisfies both equations of the system.

 Solve the system of equations

$$3x + 2y = -1$$
$$5x - 3y = 11.$$

Draw the graph of each equation on the same axes.

Answer:
(1, −2)

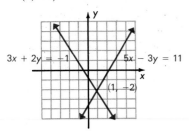

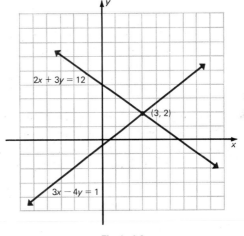

Figure 6.2

Since the method of solution in Example 1 results in the elimination of one variable from an equation of the system, it is called the **elimination method** for solving a system.

Example 2 Solve the system

$$3x - 2y = 4$$
$$-6x + 4y = 7.$$

Eliminate x by multiplying both sides of the first equation by 2 and adding the results to the second equation.

$$2(3x - 2y) + (-6x + 4y) = 2(4) + 7$$
$$6x - 4y - 6x + 4y = 8 + 7$$
$$0 = 15$$

An equivalent system is

$$3x - 2y = 4$$
$$0 = 15.$$

In the second equation of this system, both variables are eliminated with the result a false statement, a signal that these two equations have no common solutions. The system is inconsistent and has no solution. As Figure 6.3 shows, the graphs of the equations of the system produce two distinct parallel lines. **5**

5 Solve the system
$$x - y = 4$$
$$2x - 2y = 3.$$
Draw the graph of each equation on the same axes.

Answer:

No solution, inconsistent system

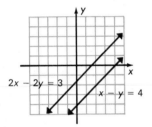

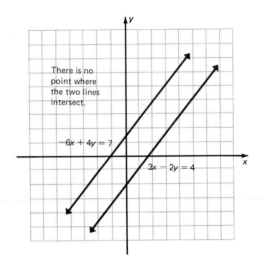

Figure 6.3

Example 3 Solve the system

$$-4x + y = 2$$
$$8x - 2y = -4.$$

Eliminate x by multiplying both sides of the first equation by 2 and adding the results to the second equation.

$$2(-4x + y) + (8x - 2y) = 2(2) + (-4)$$
$$-8x + 2y + 8x - 2y = 4 - 4$$
$$0 = 0$$

An equivalent system is

$$-4x + y = 2$$
$$0 = 0.$$

Both variables are eliminated from the second equation. The result is a true statement, indicating that the two equations have the same graph. When this happens, there is an infinite number of solutions for the system. All the ordered pairs that satisfy the equation $-4x + y = 2$ (or $8x - 2y = -4$) are solutions. See Figure 6.4. **6**

6 Solve each of the following systems.
(a) $3x - 4y = 13$
  $12x - 16y = 52$
(b) $2x + 7y = 8$
  $-4x - 14y = 12$

Answer:

(a) All ordered pairs that satisfy the equation $3x - 4y = 13$ (or $12x - 16y = 52$), dependent system
(b) No solution, inconsistent system

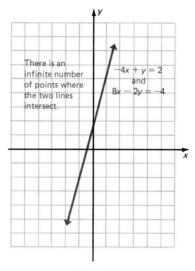

There is an infinite number of points where the two lines intersect.

$-4x + y = 2$
and
$8x - 2y = -4$

Figure 6.4

Several steps are needed to use the elimination method to solve a system of three equations with three variables, as shown in Example 4.

Example 4 Solve the system

$$2x + y - z = 2 \qquad \textbf{(1)}$$
$$x + 3y + 2z = 1 \qquad \textbf{(2)}$$
$$x + y + z = 2. \qquad \textbf{(3)}$$

Begin by eliminating one variable from one pair of equations. Suppose we choose to eliminate x from equations (2) and (3) by multiplying both sides of equation (3) by -1 and then adding the result to equation 2.

$$\begin{aligned} x + 3y + 2z &= 1 \qquad \textbf{(2)}\\ \underline{-x - y - z} &= \underline{-2}\\ 2y + z &= -1 \qquad \textbf{(4)} \end{aligned}$$

The result is an equation in two, instead of three, variables. The equivalent system is

$$2x + y - z = 2 \qquad \textbf{(1)}$$
$$2y + z = -1 \qquad \textbf{(4)}$$
$$x + y + z = 2. \qquad \textbf{(3)}$$

Now eliminate x from another pair of equations, say (1) and (2). This can be done by multiplying both sides of equation (2) by -2 and adding the result to equation (1).

$$2x + y - z = 2 \tag{1}$$
$$\underline{-2x - 6y - 4z = -2}$$
$$-5y - 5z = 0 \tag{5}$$

Again, the result is an equation in only two variables, giving the equivalent system

$$-5y - 5z = 0 \tag{5}$$
$$2y + z = -1 \tag{4}$$
$$x + y + z = 2. \tag{3}$$

Equations (4) and (5) give a system of two equations in two variables that can be solved to find values for y and z. Multiplying both sides of equation (5) by 1/5 and adding the result to equation (4) eliminates z.

$$2y + z = -1 \tag{4}$$
$$\underline{-y - z = 0} \tag{6}$$
$$y = -1$$

An equivalent system now is

$$-5y - 5z = 0 \tag{5}$$
$$y = -1$$
$$x + y + z = 2. \tag{3}$$

Substitute for y in equation (5) to get

$$-5y - 5z = 0 \tag{5}$$
$$-5(-1) - 5z = 0$$
$$-5z = -5$$
$$z = 1.$$

Finally, substituting $y = -1$ and $z = 1$ into one of the original equations, say equation (3), gives

$$x + y + z = 2 \tag{3}$$
$$x + (-1) + 1 = 2$$
$$x + 0 = 2$$
$$x = 2.$$

The solution is $(2, -1, 1)$. **7**

7 Solve each system.

(a) $x + y + z = 2$
$2x - y + 3z = 3$
$3x + y - 4z = -10$

(b) $2x + 4y + z = 2$
$3x - 2y + 3z = -3$
$4x + 5y - z = 16$

Answer:

(a) $(-1, 1, 2)$

(b) $(3, 0, -4)$

Parameters The systems of equations discussed so far have had the same number of equations as variables. When this is the case, there is either *one* solution, *no* solution, or an *infinite number* of solutions. (Refer back to Figure 6.1.) However, mathematical models sometimes lead to a system of equations with fewer equations than variables. Such systems always have either an infinite number of solutions or no solution.

Example 5 Solve the system

$$2x + 3y + 4z = 6 \qquad (1)$$
$$x - 2y + z = 9. \qquad (2)$$

Replace equation (1) by an equation without x by multiplying both sides of equation (2) by -2 and adding to equation (1). The result is

$$7y + 2z = -12, \qquad (3)$$

and the equivalent system is

$$7y + 2z = -12 \qquad (3)$$
$$x - 2y + z = 9. \qquad (2)$$

Multiply both sides of equation (3) by $1/7$. The new system is

$$y + \frac{2}{7}z = -\frac{12}{7} \qquad (4)$$
$$x - 2y + z = 9. \qquad (2)$$

Since there are only two equations, but three variables, we can go no further. Complete the solution by solving equation (4) for y.

$$y = -\frac{2}{7}z - \frac{12}{7}$$

Now substitute the result for y in equation (2), and solve for x.

$$x - 2y + z = 9 \qquad (2)$$
$$x - 2\left(-\frac{2}{7}z - \frac{12}{7}\right) + z = 9$$
$$x + \frac{4}{7}z + \frac{24}{7} + z = 9$$
$$7x + 4z + 24 + 7z = 63$$
$$7x + 11z = 39$$
$$7x = -11z + 39$$
$$x = -\frac{11}{7}z + \frac{39}{7}$$

Both x and y are now given in terms of z. Each choice of a value for z leads to values for x and y. (For this reason, z is called a **parameter.**) For example,

if $z = 1$, then $x = 4$ and $y = -2$,

giving the solution $(4, -2, 1)$. Also,

if $z = -6$, then $x = 15$ and $y = 0$, and

if $z = 0$, then $x = \frac{39}{7}$ and $y = -\frac{12}{7}$,

leading to solutions $(15, 0, -6)$ and $(39/7, -12/7, 0)$. ⑧

⑧ Use the following values of z to find additional solutions for the system of Example 5.

(a) $z = 7$
(b) $z = -14$
(c) $z = 5$

Answer:

(a) $x = -38/7$, $y = -26/7$
(b) $x = 193/7$, $y = 16/7$
(c) $x = -16/7$, $y = -22/7$

There are an infinite number of solutions, since z can take an infinite number of values. This set of solutions is sometimes written as follows.

$$z \text{ arbitrary}$$
$$x = -\frac{11}{7}z + \frac{39}{7}$$
$$y = -\frac{2}{7}z - \frac{12}{7}$$

The solution can also be written as the ordered triple $(-11z/7 + 39/7, -2z/7 - 12/7, z)$. Solving this same system in a different way could make x or y the parameter, instead of z. **9**

9 Solve the system of Example 5 by eliminating z instead of x. Then let x be the arbitrary variable. Verify that the solution is the same by checking the ordered triples found in Example 5.

Answer:

x arbitrary,
$y = 2x/11 - 30/11$,
$z = -7x/11 + 39/11$

The solution of a system with dependent equations, such as the one in Example 3, also can be written with one variable arbitrary. In Example 3, the system was

$$-4x + y = 2 \qquad \qquad (1)$$
$$8x - 2y = -4. \qquad \qquad (2)$$

The solution can be written with x arbitrary by solving equation (1) for y to get

$$-4x + y = 2$$
$$y = 4x + 2.$$

Then the solution is $(x, 4x + 2)$.

Example 5 discussed a system with one more variable than equations. If there are two more variables than equations, there usually will be two parameters, and so on. It is also possible for a system to have *fewer* variables than equations, as in the system

$$2x + 3y = 8$$
$$x - y = 4$$
$$5x + y = 7$$

10 Solve each system.

(a) $2x + y = 16$
$x - y = 2$
$3x + 2y = 18$

(b) $3x + y = 12$
$x - 4y = 17$
$2x + 5y = -5$

Answer:

(a) No solution

(b) $(5, -3)$

that has three equations and two variables. Since each equation has a line as its graph, the possibilities are three lines which intersect at a common point, three lines which cross at three different points, three lines of which two are the same line so that the intersection would be a point, three lines of which two are parallel so that the intersection would be two different points, three lines which are all parallel so that there would be no intersection, and so on. It can be shown that, as in the case of n equations with n variables, the possibilities are a unique solution, no solution, or an infinite number of solutions.

Solve the system above by solving the system made up of the first two equations. The solution, $(4, 0)$, does not satisfy the third equation so that the given system of three equations in two variables has no solution. **10**

Applications Many applications require finding more than one variable. Linear systems can be used to solve these problems. Always begin by reading the problem carefully. Then identify what must be found. Each unknown quantity should be represented by a variable. It is a good idea to *write down* exactly what each variable represents. Then reread the problem, looking for all necessary data. Write that down, too. Finally, look for one or more sentences which lead to equations or inequalities. The next example illustrates these steps.

Example 6 The Busy Bee Bakery sells two special cakes: Chocolate Dream Cake which requires 1/2 hour of the baker's time and costs $3 for ingredients, and Coconut Surprise Cake which requires 1/2 hour to prepare and costs $2 for ingredients. How many of each type can the bakery turn out each day if the baker spends 4 hours per day making these cakes and the bakery spends $18 per day on their ingredients?

Let x represent the number of Chocolate Dream Cakes and y the number of Coconut Surprise Cakes the bakery can make each day. Use the information of the problem to complete a table.

	Number	Baker's time	Cost of ingredients
Chocolate cakes	x	$\frac{1}{2}x$	$3x$
Coconut cakes	y	$\frac{1}{2}y$	$2y$
Totals	—	4	18

Use the information in the table to write two equations.

$$\frac{1}{2}x + \frac{1}{2}y = 4 \qquad \text{(time restriction)}$$

$$3x + 2y = 18 \qquad \text{(cost restriction)}$$

This system of equations can be solved using the methods of this section to get $x = 2$ and $y = 6$. The bakery can make 2 Chocolate Dream Cakes and 6 Coconut Surprise Cakes each day under the conditions of the problem. ■ 11

Example 7 Find the equation of the parabola $y = ax^2 + bx + c$ that goes through the points (0, 3), (2, 15), and (−3, 45).

Since the given points satisfy the equation $y = ax^2 + bx + c$, find suitable values for a, b, and c with the equations

$3 = a(0)^2 + b(0) + c$	or	$3 = c$
$15 = a(2)^2 + b(2) + c$	or	$15 = 4a + 2b + c$
$45 = a(-3)^2 + b(-3) + c$	or	$45 = 9a - 3b + c.$

This system of equations can be solved by the methods of this section to give $a = 4$, $b = -2$, and $c = 3$. The equation of the parabola through the given points is

$$y = 4x^2 - 2x + 3. \quad \blacksquare \ 12$$

11 Write a system of equations to solve Example 6 if the baker spends 1/3 hour per cake on the Dream Cakes and 1/2 hour per cake on the Surprise Cakes, and the costs are changed to $2.50 and $1.50 respectively.

Answer:

$$\frac{1}{3}x + \frac{1}{2}y = 4$$

$$2.5x + 1.5y = 18$$

12 Find the equation of the parabola $y = ax^2 + bx + c$ that goes through the points (0, 4), (2, 0) and (1, 1).

Answer:

$y = x^2 - 4x + 4$

6.1 Exercises

Use the elimination method to solve each of the following systems of two equations in two unknowns. (See Examples 1–3.)

1. $x + y = 9$
$2x - y = 0$

2. $4x + y = 9$
$3x - y = 5$

3. $5x + 3y = 7$
$7x - 3y = -19$

4. $2x + 7y = -8$
$-2x + 3y = -12$

5. $3x + 2y = -6$
$5x - 2y = -10$

6. $-6x + 2y = 8$
$5x - 2y = -8$

7. $2x - 3y = -7$
$5x + 4y = 17$

8. $4m + 3n = -1$
$2m + 5n = 3$

9. $5p + 7q = 6$
$10p - 3q = 46$

10. $12s - 5t = 9$
$3s - 8t = -18$

11. $6x + 7y = -2$
$7x - 6y = 26$

12. $2a + 9b = 3$
$5a + 7b = -8$

13. $3x + 2y = 5$
$6x + 4y = 8$

14. $9x - 5y = 1$
$-18x + 10y = 1$

15. $4x - y = 9$
$-8x + 2y = -18$

16. $3x + 5y + 2 = 0$
$9x + 15y + 6 = 0$

In Exercises 17–20, first multiply both sides of each equation by its common denominator to eliminate the fractions. Then use the elimination method to solve.

17. $\dfrac{x}{2} + \dfrac{y}{3} = 8$

$\dfrac{2x}{3} + \dfrac{3y}{2} = 17$

18. $\dfrac{x}{5} + 3y = 31$

$2x - \dfrac{y}{5} = 8$

19. $\dfrac{x}{2} + y = \dfrac{3}{2}$

$\dfrac{x}{3} + y = \dfrac{1}{3}$

20. $x + \dfrac{y}{3} = -6$

$\dfrac{x}{5} + \dfrac{y}{4} = -\dfrac{7}{4}$

Use the elimination method to solve each of the following systems of three equations in three unknowns. Check your answers. (See Example 4.)

21. $x + y + z = 2$
$2x + y - z = 5$
$x - y + z = -2$

22. $2x + y + z = 9$
$-x - y + z = 1$
$3x - y + z = 9$

23. $x + 3y + 4z = 14$
$2x - 3y + 2z = 10$
$3x - y + z = 9$

24. $4x - y + 3z = -2$
$3x + 5y - z = 15$
$-2x + y + 4z = 14$

25. $x + 2y + 3z = 8$
$3x - y + 2z = 5$
$-2x - 4y - 6z = 5$

26. $3x - 2y - 8z = 1$
$9x - 6y - 24z = -2$
$x - y + z = 1$

27. $2x - 4y + z = -4$
$x + 2y - z = 0$
$-x + y + z = 6$

28. $4x - 3y + z = 9$
$3x + 2y - 2z = 4$
$x - y + 3z = 5$

29. $x + 4y - z = 6$
$2x - y + z = 3$
$3x + 2y + 3z = 16$

30. $3x + y - z = 7$
$2x - 3y + z = -7$
$x - 4y + 3z = -6$

31. $5m + n - 3p = -6$
$2m + 3n + p = 5$
$-3m - 2n + 4p = 3$

32. $2r - 5s + 4t = -35$
$5r + 3s - t = 1$
$r + s + t = 1$

33. $a - 3b - 2c = -3$
$3a + 2b - c = 12$
$-a - b + 4c = 3$

34. $2x + 2y + 2z = 6$
$3x - 3y - 4z = -1$
$x + y + 3z = 11$

Solve each of the following systems of equations. Let x be the arbitrary variable. (See Example 5.)

35. $x + y = 2$
$3x + 3y = 6$

36. $x - y = 1$
$-5x + 5y = -5$

37. $3x + 4y = 12$
$-6x - 8y = -24$

38. $2x - 7y = 3$
$-6x + 21y = -9$

39. $5x + 3y + 4z = 19$
$3x - y + z = -4$

40. $3x + y - z = 0$
$2x - y + 3z = -7$

41. $x + 2y + 3z = 11$
$2x - y + z = 2$

42. $-x + y - z = -7$
$2x + 3y + z = 7$

43. $x + y - z + 2w = -20$
$2x - y + z + w = 11$
$3x - 2y + z - 2w = 27$

44. $4x + 3y + z + 2w = 1$
$-2x - y + 2z + 3w = 0$
$x + 4y + z - w = 12$

Solve each of the following systems of equations.

45. $5x + 2y = 7$
$-2x + y = -10$
$x - 3y = 15$

46. $9x - 2y = -14$
$3x + y = -4$
$-6x - 2y = 8$

47. $\begin{aligned} x + 7y &= 5 \\ 4x - 3y &= 2 \\ -x + 2y &= 10 \end{aligned}$

48. $\begin{aligned} -3x - 2y &= 11 \\ x + 2y &= -14 \\ 5x + y &= -9 \end{aligned}$

49. $\begin{aligned} x + y &= 2 \\ y + z &= 4 \\ x + z &= 3 \\ y - z &= 8 \end{aligned}$

50. $\begin{aligned} 2x + y &= 7 \\ x + 3z &= 5 \\ y - 2z &= 6 \\ x + 4y &= 10 \end{aligned}$

Write each of the following using a system of equations, and then solve. (See Example 7.)

51. Find a and b so that the line $ax + by = 5$ contains the points $(-2, 1)$ and $(-1, -2)$.

52. Find m and b so that the line $y = mx + b$ contains the points $(4, 6)$ and $(-5, -3)$.

53. Find a, b, and c so that the graph of $y = ax^2 + bx + c$ contains the points $(2, 3)$, $(-1, 0)$, and $(-2, 2)$.

54. Find a, b, and c so that the points $(2, 14)$, $(0, 0)$, and $(-1, -1)$ lie on the graph of $y = ax^2 + bx + c$.

Write a system of equations for each of the following problems; then solve the system. (See Example 6.)

55. A working couple earned a total of $4352. The wife earned $64 per day; the husband earned $8 per day less. Find the number of days each worked if the total number of days worked by both was 72.

56. Management Midtown Manufacturing Company makes two products, plastic plates and plastic cups. Both require time on two machines: plates—one hour on machine A and two hours on machine B; cups—three hours on machine A and one hour on machine B. Both machines operate 15 hours a day. How many of each product can be produced in a day under these conditions?

57. Management A company produces two models of bicycles, model 201 and model 301. Model 201 requires 2 hours of assembly time and model 301 requires 3 hours of assembly time. The parts for model 201 cost $25 per bike and the parts for model 301 cost $30 per bike. If the company has a total of 34 hours of assembly time and $365 available per day for these two models, how many of each can be made in a day?

58. Juanita invests $10,000, received from her grandmother, in three ways. With one part, she buys mutual funds which offer a return of 8% per year. The second

part, which amounts to twice the first, is used to buy government bonds at 9% per year. She puts the rest in the bank at 5% annual interest. The first year her investments bring a return of $830. How much did she invest in each way?

59. Management To get the necessary funds for a planned expansion, a small company took out three loans totaling $25,000. The company was able to borrow some of the money at 16%. They borrowed $2000 more than one-half the amount of the 16% loan at 20%, and the rest at 18%. The total annual interest was $4440. How much did they borrow at each rate?

60. Management The business analyst for Midtown Manufacturing wants to find an equation which can be used to project sales of a relatively new product. For the years 1985, 1986, and 1987 sales were $15,000, $32,000, and $123,000 respectively.

(a) Graph the sales for the years 1985, 1986, and 1987, letting the year 1985 equal 0 on the x-axis. Let the values on the vertical axis be in thousands. [For example, the point (1986, 32,000) will be graphed as (1, 32).]

(b) Find the equation of the straight line $ax + by = c$ through the points for 1985 and 1987.

(c) Find the equation of the parabola $y = ax^2 + bx + c$ through the three given points.

(d) Find the projected sales for 1990 first by using the equation from part (b) and then by using the equation from part (c). If you were to estimate sales of the product in 1990 which result would you choose? Why?

In Exercises 61 and 62, find the value of k for which each system has a single solution, and then find the solution of the system.

61. $\begin{aligned} 4x + 8z &= 12 \\ 2y - z &= -2 \\ 3x + y + z &= -1 \\ x + y - kz &= -7 \end{aligned}$

62. $\begin{aligned} 2x + y + z &= 4 \\ 3y + z &= -2 \\ x + y - z &= -3 \\ 4x + kz &= 8 \end{aligned}$

6.2 Solution of Linear Systems by the Gauss-Jordan Method

The Echelon Method Although the elimination method can be used to solve systems with more than two equations, it is a complicated process even for systems of three variables. The **echelon method** is a systematic approach that transforms a system into an equivalent system having the form

$$x + by + cz = d$$
$$y + ez = f$$
$$z = g,$$

where b, c, d, e, f, and g are real numbers. Since the last equation gives the value of z, the complete solution can be found by substitution. The advantage of the echelon method is that it can be carried out with matrix notation (discussed later) making it possible to solve systems of equations by computer.

Example 1 Solve the system

$$2x + y - z = 2$$
$$x + 3y + 2z = 1$$
$$x + y + z = 2.$$

Use transformations to get an equivalent system that has the form shown above. In this new system, the coefficient of x in the first equation should be 1. Use the first transformation to interchange the first two equations. (As an alternate approach, a coefficient of 1 could be found by multiplying both sides of the first equation by 1/2.) The new system follows. (A shorthand notation in color shows the steps involved. The letter R represents the rows of the system.)

$$x + 3y + 2z = 1 \qquad \text{Interchange } R_1, R_2$$
$$2x + y - z = 2$$
$$x + y + z = 2$$

Replace the second equation above with an equation without x by multiplying both sides of the first equation by -2 and adding the results to the second equation. This process gives the new system

$$x + 3y + 2z = 1$$
$$-5y - 5z = 0 \qquad (-2)R_1 + R_2$$
$$x + y + z = 2.$$

Simplify the second equation by multiplying both sides by $-1/5$.

$$x + 3y + 2z = 1$$
$$y + z = 0 \qquad -\frac{1}{5}R_2$$
$$x + y + z = 2$$

Replace the third equation with an equation not containing x by multiplying both sides of the first equation by -1 and adding the results to the third equation to get the new system

$$\begin{aligned} x + 3y + 2z &= 1 \\ y + z &= 0 \\ -2y - z &= 1. \qquad -1R_1 + R_3 \end{aligned}$$

Replace the third equation with an equation not containing y by multiplying both sides of the second equation by 2 and adding the results to the third equation.

$$\begin{aligned} x + 3y + 2z &= 1 \\ y + z &= 0 \\ z &= 1 \qquad 2R_2 + R_3 \end{aligned}$$

The third equation of this system gives the value of z: $z = 1$. Substitute 1 for z in the second equation to get $y = -1$. Finally, substitute 1 for z and -1 for y in the first equation to get $x = 2$. The solution for the given system is $(2, -1, 1)$. **1**

1 Use the echelon method to solve each system.

(a) $\begin{aligned} 2x + y &= -1 \\ x + 3y &= 2 \end{aligned}$

(b) $\begin{aligned} 2x - y + 3z &= 2 \\ x + 2y - z &= 6 \\ -x - y + z &= -5 \end{aligned}$

Answer:

(a) $(-1, 1)$

(b) $(3, 1, -1)$

Solve a linear system in n variables with the **echelon method** by using transformations to perform the following steps.

1. Make the coefficient of the first variable equal to 1 in the first equation and 0 in all other equations.
2. Make the coefficient of the second variable equal to 1 in the second equation and 0 in all the following equations.
3. Make the coefficient of the third variable equal to 1 in the third equation and 0 in all the following equations.
4. Continue in this way until the last equation is of the form $x_n = k$, where k is a constant.

Example 1 showed the echelon method for solving a linear system of equations. The procedure can be simplified using a *matrix*. The variables in a given system are always the same, so only the coefficients and the constants are really needed. As an example, we use a matrix to solve the system of Example 1:

$$\begin{aligned} 2x + y - z &= 2 \\ x + 3y + 2z &= 1 \\ x + y + z &= 2. \end{aligned}$$

This system can be written in an abbreviated form as

$$\begin{bmatrix} 2 & 1 & -1 & 2 \\ 1 & 3 & 2 & 1 \\ 1 & 1 & 1 & 2 \end{bmatrix}.$$

Such a rectangular array of numbers enclosed by brackets is called a **matrix** (plural: **matrices**). Each number in the array is an **element** or **entry.** To separate the constants in the last column of the matrix from the coefficients of the variables, use a vertical line, producing the following **augmented matrix.**

$$\begin{bmatrix} 2 & 1 & -1 & 2 \\ 1 & 3 & 2 & 1 \\ 1 & 1 & 1 & 2 \end{bmatrix}$$

The rows of the augmented matrix can be transformed in the same way as the equations of the system, since the matrix is just a shortened form of the system. The *row operations* on the augmented matrix, which correspond to the transformations of systems of equations, are given in the following theorem.

For any augmented matrix of a system of equations, the following **row operations** produce an augmented matrix of an equivalent system:

1. interchanging any two rows;
2. multiplying the elements of a row by any nonzero real number;
3. adding a multiple of the elements of one row to the corresponding elements of another row.

Row operations, like the transformations of systems of equations, are reversible. If they are used to go from matrix A to matrix B, then it is possible to use row operations to transform B back into A. In addition to their use in solving equations, row operations are very important in the simplex method of Chapter 7. The row operations will be abbreviated with the same notation used earlier for equations.

By the first row operation, the matrix

$$\begin{bmatrix} 1 & 3 & 5 & 6 \\ 0 & 1 & 2 & 3 \\ 2 & 1 & -2 & -5 \end{bmatrix} \quad \text{becomes} \quad \begin{bmatrix} 0 & 1 & 2 & 3 \\ 1 & 3 & 5 & 6 \\ 2 & 1 & -2 & -5 \end{bmatrix} \quad \text{Interchange } R_1, R_2$$

by exchanging the first two rows. Row three is left unchanged.

The second row operation can be used to change

$$\begin{bmatrix} 1 & 3 & 5 & 6 \\ 0 & 1 & 2 & 3 \\ 2 & 1 & -2 & -5 \end{bmatrix} \quad \text{to} \quad \begin{bmatrix} -2 & -6 & -10 & -12 \\ 0 & 1 & 2 & 3 \\ 2 & 1 & -2 & -5 \end{bmatrix} \quad -2R_1$$

by multiplying the elements of the first row of the original matrix by -2, with rows two and three left unchanged.

Using the third row operation,

$$\begin{bmatrix} 1 & 3 & 5 & 6 \\ 0 & 1 & 2 & 3 \\ 2 & 1 & -2 & -5 \end{bmatrix} \quad \text{becomes} \quad \begin{bmatrix} -1 & 2 & 7 & 11 \\ 0 & 1 & 2 & 3 \\ 2 & 1 & -2 & -5 \end{bmatrix} \quad -1R_3 + R_1$$

after first multiplying each element in the third row of the original matrix by -1 and then adding the results to the corresponding elements in the first row of that matrix. Work as follows.

$$\begin{bmatrix} 1 + 2(-1) & 3 + 1(-1) & 5 + (-2)(-1) & 6 + (-5)(-1) \\ 0 & 1 & 2 & 3 \\ 2 & 1 & -2 & -5 \end{bmatrix}$$

$$= \begin{bmatrix} -1 & 2 & 7 & 11 \\ 0 & 1 & 2 & 3 \\ 2 & 1 & -2 & -5 \end{bmatrix}$$

Again rows two and three are left unchanged, *even though the elements of row three were used to transform row one.* ②

2 Perform the following row operations on the matrix

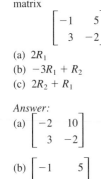

(a) $2R_1$
(b) $-3R_1 + R_2$
(c) $2R_2 + R_1$

Answer:

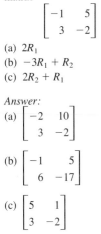

(a) $\begin{bmatrix} -2 & 10 \\ 3 & -2 \end{bmatrix}$

(b) $\begin{bmatrix} -1 & 5 \\ 6 & -17 \end{bmatrix}$

(c) $\begin{bmatrix} 5 & 1 \\ 3 & -2 \end{bmatrix}$

The Gauss-Jordan Method The *Gauss-Jordan method* is an extension of the echelon method of solving systems. Before the Gauss-Jordan method can be used, the system must be in proper form: the terms with variables should be on the left and the constants on the right in each equation, with the variables in the same order in each equation. The following example illustrates the use of the Gauss-Jordan method to solve a system of equations.

Example 2 Solve the system

$$3x - 4y = 1$$
$$5x + 2y = 19.$$

The system is already in the proper form. The procedure for solving the system is parallel to the echelon method except for the last step. We show the echelon method on the left below and the Gauss-Jordan method on the right.

Echelon method	*Gauss-Jordan method*
	Rewrite the system as an augmented matrix.
$3x - 4y = 1$	$\begin{bmatrix} 3 & -4 & 1 \\ 5 & 2 & 19 \end{bmatrix}$
$5x + 2y = 19$	
Multiply both sides of the first equation by $1/3$ so that x has a coefficient of 1.	Using row operation (2) multiply each element of row 1 by $1/3$.
$x - \frac{4}{3}y = \frac{1}{3} \quad \frac{1}{3}R_1$	$\begin{bmatrix} 1 & -\frac{4}{3} & \frac{1}{3} \\ 5 & 2 & 19 \end{bmatrix} \quad \frac{1}{3}R_1$
$5x + 2y = 19$	

Echelon method	*Gauss-Jordan method*

Eliminate x from the second equation by adding -5 times the first equation to the second equation.

$$x - \tfrac{4}{3}y = \tfrac{1}{3}$$
$$\tfrac{26}{3}y = \tfrac{52}{3} \qquad -5R_1 + R_2$$

Using row operation (3) add -5 times the elements of row 1 to the elements of row 2.

$$\begin{bmatrix} 1 & -\tfrac{4}{3} & | & \tfrac{1}{3} \\ 0 & \tfrac{26}{3} & | & \tfrac{52}{3} \end{bmatrix} \qquad -5R_1 + R_2$$

Multiply both sides of the second equation by 3/26 to get $y = 2$.

$$x - \tfrac{4}{3}y = \tfrac{1}{3}$$
$$y = 2 \qquad \frac{3}{26}R_2$$

Multiply the elements of row 2 by 3/26, using row operation (2).

$$\begin{bmatrix} 1 & -\tfrac{4}{3} & | & \tfrac{1}{3} \\ 0 & 1 & | & 2 \end{bmatrix} \qquad \frac{3}{26}R_2$$

Substitute $y = 2$ into the first equation and solve for x to get $x = 3$.

$$x = 3$$
$$y = 2$$

Multiply the elements of row 2 by 4/3 and add to the elements of row 1 (row operation (3)).

$$\begin{bmatrix} 1 & 0 & | & 3 \\ 0 & 1 & | & 2 \end{bmatrix} \qquad \frac{4}{3}R_2 + R_1$$

The solution of the system, (3, 2), can be read directly from the last column of the final matrix.

In both cases, check this solution by substitution in both equations of the original system. ■

When using the Gauss-Jordan method to solve a system, the final matrix will always have zeros above the main diagonal of 1s as well as below. When using row operations to transform a matrix, it is usually best to work column by column from left to right. For each column, the first change should produce a 1 in the main diagonal. Next, perform the steps that give zeros in the remainder of the column. Then proceed to the next column. ③

③ Use the Gauss-Jordan method to solve the system
$$x + 2y = 11$$
$$-4x + y = -8.$$
Give the shorthand notation and the new matrix in (b)–(d).

(a) Set up the augmented matrix.

(b) Get 0 in row 2, column 1.

(c) Get 1 in row 2, column 2.

(d) Finally, get 0 in row 1, column 2.

(e) The solution for the system is _____.

Answer:

(a) $\begin{bmatrix} 1 & 2 & | & 11 \\ -4 & 1 & | & -8 \end{bmatrix}$

(b) $4R_1 + R_2$; $\begin{bmatrix} 1 & 2 & | & 11 \\ 0 & 9 & | & 36 \end{bmatrix}$

(c) $\frac{1}{9}R_2$; $\begin{bmatrix} 1 & 2 & | & 11 \\ 0 & 1 & | & 4 \end{bmatrix}$

(d) $-2R_2 + R_1$; $\begin{bmatrix} 1 & 0 & | & 3 \\ 0 & 1 & | & 4 \end{bmatrix}$

(e) (3, 4)

Example 3 Use the Gauss-Jordan method to solve the system

$$x \qquad\quad + 5z = -6 + y$$
$$3x + 3y \qquad = 10 + z$$
$$x + 3y + 2z = 5.$$

The system must first be rewritten in proper form as follows.

$$x - y + 5z = -6$$
$$3x + 3y - z = 10$$
$$x + 3y + 2z = 5$$

Begin the solution by writing the augmented matrix of the linear system.

$$\begin{bmatrix} 1 & -1 & 5 & | & -6 \\ 3 & 3 & -1 & | & 10 \\ 1 & 3 & 2 & | & 5 \end{bmatrix}$$

This matrix must be transformed into a final matrix of the form

$$\begin{bmatrix} 1 & 0 & 0 & | & m \\ 0 & 1 & 0 & | & n \\ 0 & 0 & 1 & | & p \end{bmatrix},$$

where m, n, and p are real numbers. From this final form of the matrix, read the solution: $x = m$, $y = n$, and $z = p$.

The first element in column one is already 1. Get 0 for the second element in column one by multiplying each element in the first row by -3 and adding the results to the corresponding elements in row two [using row operation (3)].

$$\begin{bmatrix} 1 & -1 & 5 & | & -6 \\ 0 & 6 & -16 & | & 28 \\ 1 & 3 & 2 & | & 5 \end{bmatrix} \quad -3R_1 + R_2$$

Now, change the first element in row three to 0 by multiplying each element of the first row by -1 and adding the results to the corresponding elements of the third row [again using row operation (3)].

$$\begin{bmatrix} 1 & -1 & 5 & | & -6 \\ 0 & 6 & -16 & | & 28 \\ 0 & 4 & -3 & | & 11 \end{bmatrix} \quad -1R_1 + R_3$$

This transforms the first column. Transform the second column in a similar manner. ④

Complete the solution by transforming the third column.

$$\begin{bmatrix} 1 & 0 & \frac{7}{3} & | & -\frac{4}{3} \\ 0 & 1 & -\frac{8}{3} & | & \frac{14}{3} \\ 0 & 0 & 1 & | & -1 \end{bmatrix} \quad \frac{3}{23}R_3$$

$$\begin{bmatrix} 1 & 0 & 0 & | & 1 \\ 0 & 1 & -\frac{8}{3} & | & \frac{14}{3} \\ 0 & 0 & 1 & | & -1 \end{bmatrix} \quad -\frac{7}{3}R_3 + R_1$$

$$\begin{bmatrix} 1 & 0 & 0 & | & 1 \\ 0 & 1 & 0 & | & 2 \\ 0 & 0 & 1 & | & -1 \end{bmatrix} \quad \frac{8}{3}R_3 + R_2$$

④ Continue the solution of the system in Example 3 as follows. Give the shorthand notation and the matrix for each step.

(a) Get 1 in row 2, column 2.

(b) Get 0 in row 1, column 2.

(c) Now get 0 in row 3, column 2.

Answer:

(a) $\frac{1}{6}R_2$;

$$\begin{bmatrix} 1 & -1 & 5 & | & -6 \\ 0 & 1 & -\frac{8}{3} & | & \frac{14}{3} \\ 0 & 4 & -3 & | & 11 \end{bmatrix}$$

(b) $1R_2 + R_1$;

$$\begin{bmatrix} 1 & 0 & \frac{7}{3} & | & -\frac{4}{3} \\ 0 & 1 & -\frac{8}{3} & | & \frac{14}{3} \\ 0 & 4 & -3 & | & 11 \end{bmatrix}$$

(c) $-4R_2 + R_3$;

$$\begin{bmatrix} 1 & 0 & \frac{7}{3} & | & -\frac{4}{3} \\ 0 & 1 & -\frac{8}{3} & | & \frac{14}{3} \\ 0 & 0 & \frac{23}{3} & | & -\frac{23}{3} \end{bmatrix}$$

(Solution continued in the text.)

The linear system associated with this last augmented matrix is

$$x \qquad = \quad 1$$
$$\quad y \quad = \quad 2$$
$$\qquad z = -1,$$

5 Use the Gauss-Jordan method to solve

$$x + y - z = \quad 6$$
$$2x - y + z = \quad 3$$
$$-x + y + z = -4.$$

Answer:

$(3, 1, -2)$

and the solution is $(1, 2, -1)$. **5**

Use the **Gauss-Jordan method** to solve a linear system by performing the following steps.

1. Write all equations with variable terms on the left and constants on the right. Be sure the variables are in the same order in all equations.
2. Write the augmented matrix that corresponds to the system.
3. Use row operations to transform the first column so that the first element is 1 and the remaining elements are 0.
4. Use row operations to transform the second column so that the second element is 1 and the remaining elements are 0.
5. Use row operations to transform the third column so that the third element is 1 and the remaining elements are 0.
6. Continue in this way until the last row is in the form $[0 \ \ 0 \ \ 0 \ldots 0 \ \ 1 \ | \ k]$, where k is a constant.
7. Read the solution to the right of the vertical bar.

Example 4 Use the Gauss-Jordan method to solve the system

$$x + \quad y = 2$$
$$2x + 2y = 5.$$

Begin by writing the augmented matrix.

$$\begin{bmatrix} 1 & 1 & | & 2 \\ 2 & 2 & | & 5 \end{bmatrix}$$

Get 0 for the first element in row two by multiplying the numbers in row one by -2 and adding the results to the corresponding elements in row two.

$$\begin{bmatrix} 1 & 1 & | & 2 \\ 0 & 0 & | & 1 \end{bmatrix} \quad -2R_1 + R_2$$

The next step is to get a 1 for the second element in row two. Since this is impossible, we cannot go further. This matrix gives the system

$$x + \quad y = 2$$
$$0x + 0y = 1.$$

6 Solve each system

(a) $x - y = 4$
 $-2x + 2y = 1$

(b) $3x - 4y = 0$
 $2x + y = 0$

Answer:

(a) No solution

(b) $(0, 0)$

---

7 Continue the solution of the system of Example 5 as follows.

(a) Get 1 in row 2, column 2.

(b) Get 0 in row 1, column 2.

Answer:

(a) $\begin{bmatrix} 1 & 2 & -1 & 0 \\ 0 & 1 & -\frac{4}{7} & -\frac{6}{7} \\ 0 & 0 & 0 & 0 \end{bmatrix}$

(b) $\begin{bmatrix} 1 & 0 & \frac{1}{7} & \frac{12}{7} \\ 0 & 1 & -\frac{4}{7} & -\frac{6}{7} \\ 0 & 0 & 0 & 0 \end{bmatrix}$

---

8 Use the Gauss-Jordan method to solve the following.

(a) $3x + 9y = -6$
 $-x - 3y = 2$

(b) $2x + 9y = 12$
 $4x + 18y = 5$

Answer:

(a) y arbitrary,
$x = -3y - 2$
or $(-3y - 2, y)$

(b) No solution

The second equation simplifies to $0 = 1$, so the system is inconsistent and has no solution. This conclusion can be reached by noting that the row $[\,0\;\;\;0 \mid 1\,]$ has all zeros to the left of the bar. **6**

Example 5 Use the Gauss-Jordan method to solve the system

$$\begin{aligned} x + 2y - z &= 0 \\ 3x - y + z &= 6 \\ -2x - 4y + 2z &= 0. \end{aligned}$$

Start with the augmented matrix.

$$\begin{bmatrix} 1 & 2 & -1 & 0 \\ 3 & -1 & 1 & 6 \\ -2 & -4 & 2 & 0 \end{bmatrix}.$$

The first element in row one is 1. Use row operations to get zeros in the rest of column one.

$$\begin{bmatrix} 1 & 2 & -1 & 0 \\ 0 & -7 & 4 & 6 \\ -2 & -4 & 2 & 0 \end{bmatrix} \quad -3R_1 + R_2$$

$$\begin{bmatrix} 1 & 2 & -1 & 0 \\ 0 & -7 & 4 & 6 \\ 0 & 0 & 0 & 0 \end{bmatrix} \quad 2R_1 + R_3$$

In this example, the row of all zeros in the last matrix indicates that two of the equations (the first and last) are dependent. **7**

Use the result of part (b) at the side to write the system

$$x + \frac{1}{7}z = \frac{12}{7}$$

$$y - \frac{4}{7}z = -\frac{6}{7}.$$

Solving the first equation for x and the second for y gives the solution

$$z \text{ arbitrary}$$

$$y = -\frac{6}{7} + \frac{4}{7}z$$

$$x = \frac{12}{7} - \frac{1}{7}z,$$

or $(12/7 - z/7, -6/7 + 4z/7, z)$. **8**

Although the examples have used only systems with two equations and two variables or three equations and three variables, the Gauss-Jordan method can be used for any system with n equations and n variables. In fact, the method can be used with n equations and m variables, $n \le m$. Notice that the system in Example 5 is actually equivalent to a system of just two equations and three variables. While the method does become tedious, even with three equations and three variables, it is very suitable for use by computers. A computer can produce the solution to a fairly large system very quickly.*

6.2 Exercises

Write the augmented matrix for each of the following systems. **Do not solve.**

1. $2x + 3y = 11$
$\quad x + 2y = 8$

2. $3x + 5y = -13$
$\quad 2x + 3y = -9$

3. $x + 5y = 6$
$\quad\quad\quad y = 1$

4. $2x + 7y = 1$
$\quad 5x = -15$

5. $2x + y + z = 3$
$\quad 3x - 4y + 2z = -7$
$\quad x + y + z = 2$

6. $4x - 2y + 3z = 4$
$\quad 3x + 5y + z = 7$
$\quad 5x - y + 4z = 7$

7. $x + y = 2$
$\quad 2y + z = -4$
$\quad\quad\quad z = 2$

8. $x = 6$
$\quad y + 2z = 2$
$\quad x - 3z = 6$

9. $x = 5$
$\quad y = -2$
$\quad z = 3$

10. $x = 8$
$\quad y + z = 6$
$\quad z = 2$

Write the system of equations associated with each of the following augmented matrices. **Do not solve.**

11. $\begin{bmatrix} 1 & 0 & | & 2 \\ 0 & 1 & | & 3 \end{bmatrix}$

12. $\begin{bmatrix} 1 & 0 & | & 5 \\ 0 & 1 & | & -3 \end{bmatrix}$

13. $\begin{bmatrix} 2 & 1 & | & 1 \\ 3 & -2 & | & -9 \end{bmatrix}$

14. $\begin{bmatrix} 1 & -5 & | & -18 \\ 6 & 2 & | & 20 \end{bmatrix}$

15. $\begin{bmatrix} 1 & 0 & 0 & | & 2 \\ 0 & 1 & 0 & | & 3 \\ 0 & 0 & 1 & | & -2 \end{bmatrix}$

16. $\begin{bmatrix} 1 & 0 & 1 & | & 4 \\ 0 & 1 & 0 & | & 2 \\ 0 & 0 & 1 & | & 3 \end{bmatrix}$

Use the echelon method to solve each of the following systems of equations. (See Example 1.)

17. $x + y = 5$
$\quad x - y = -1$

18. $x + y = -3$
$\quad 2x - 5y = -6$

19. $2x = 10 + 3y$
$\quad 2y = 5 - 2x$

20. $2x - 5y = 10$
$\quad 4x - 5y = 15$

21. $2x - 3y = 2$
$\quad 4x - 6y = 1$

22. $6x - 3y = 1$
$\quad -12x + 6y = -2$

23. $x + y = -1$
$\quad y + z = 4$
$\quad x + z = 1$

24. $x + y - z = 6$
$\quad 2x - y + z = -9$
$\quad x - 2y + 3z = 1$

25. $y = x - 1$
$\quad y = 6 + z$
$\quad z = -1 - x$

26. $x - 2y + z = 5$
$\quad 2x + y - z = 2$
$\quad -2x + 4y - 2z = 2$

27. $2x + 3y + z = 9$
$\quad 4x + y - 3z = -7$
$\quad 6x + 2y - 4z = -8$

28. $5x - 4y + 2z = 4$
$\quad 10x + 3y - z = 27$
$\quad 15x - 5y + 3z = 25$

29. $x + 2y - w = 3$
$\quad 2x + 4z + 2w = -6$
$\quad x + 2y - z = 6$
$\quad 2x - y + z + w = -3$

---

*See D. R. Coscia, *Computer Applications in Finite Mathematics and Calculus,* Scott, Foresman and Company, 1985.

Use the Gauss-Jordan method to solve each of the following systems of equations. (See Examples 2–5.)

30. $x + 2y = 5$
$2x + y = -2$

31. $3x - 2y = 4$
$3x + y = -2$

32. $y = 5 - 4x$
$2x = 3 - y$

33. $4x - 2y = 3$
$-2x + 3y = 1$

34. $x + 2y = 1$
$2x + 4y = 3$

35. $x - y = 1$
$-x + y = -1$

36. $x - z = -3$
$y + z = 9$
$x + z = 7$

37. $x + 3y - 6z = 7$
$2x - y + 2z = 0$
$x + y + 2z = -1$

38. $x = 1 - y$
$2x = z$
$2z = -2 - y$

39. $3x + 5y - z = 0$
$4x - y + 2z = 1$
$-6x - 10y + 2z = 0$

40. $3x + 2y - z = -16$
$6x - 4y + 3z = 12$
$3x + 3y + z = -11$

41. $4x - 2y - 3z = -23$
$-4x + 3y + z = 11$
$8x - 5y + 4z = 6$

42.
$x + 3y - 2z - w = 9$
$2x + 4y + 2w = 10$
$-3x - 5y + 2z - w = -15$
$x - y - 3z + 2w = 6$

43. At rush hours, substantial traffic congestion is encountered at the traffic intersections shown in the figure. (*The streets are all one way.*)

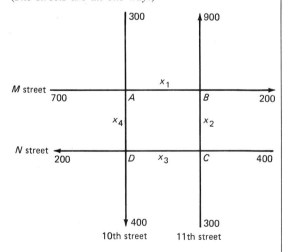

The city wishes to improve the signals at these corners so as to speed the flow of traffic. The traffic engineers first gather data. As the figure shows, 700 cars per hour come down M Street to intersection A; 300 cars per hour come to intersection A on 10th

Street. A total of x_1 of these cars leave A on M Street, while x_4 cars leave A on 10th Street. The number of cars entering A must equal the number leaving, so that

$$x_1 + x_4 = 700 + 300$$
or $$x_1 + x_4 = 1000.$$

For intersection B, x_1 cars enter B on M Street, and x_2 cars enter B on 11th Street. The figure shows that 900 cars leave B on 11th while 200 leave on M. We have

$$x_1 + x_2 = 900 + 200$$
$$x_1 + x_2 = 1100.$$

At intersection C, 400 cars enter on N Street, 300 on 11th Street, while x_2 leave on 11th Street and x_3 leave on N Street. This gives

$$x_2 + x_3 = 400 + 300$$
$$x_2 + x_3 = 700.$$

Finally, intersection D has x_3 cars entering on N and x_4 entering on 10th. There are 400 leaving D on 10th and 200 leaving on N, so that

$$x_3 + x_4 = 400 + 200$$
$$x_3 + x_4 = 600.$$

(a) Use the four equations to set up an augmented matrix, and then use the Gauss-Jordan method to solve it. (Hint: Keep going until you get a row of all zeros.)

(b) Since you got a row of all zeros, the system of equations does not have a unique solution. Write three equations, corresponding to the three nonzero rows of the matrix.

(c) Solve each of the equations for x_4.

(d) One of your equations should have been $x_4 = 1000 - x_1$. What is the largest possible value of x_1 so that x_4 is not negative? What is the largest value of x_4 so that x_1 is not negative?

(e) Your second equation should have been $x_4 = x_2 - 100$. Find the smallest possible value of x_2 so that x_4 is not negative.

(f) For the third equation, $x_4 = 600 - x_3$, find the largest possible values of x_3 and x_4 so that neither variable is negative.

(g) Look at your answers for parts (d)-(f). What is the maximum value of x_4 so that all the equations are satisfied and all variables are nonnegative? of x_3? of x_2? of x_1?

44. Management A manufacturer purchases a part for use at both of its two plants—one in Roseville, California, the other at Akron, Ohio. The part is available in limited quantities from two suppliers. Each supplier has 75 units available. The Roseville plant needs 40 units and the Akron plant requires 75 units. The first supplier charges $70 per unit delivered to Roseville and $90 per unit delivered to Akron. Corresponding costs from the second supplier are $80 and $120. The manufacturer wants to order a total of 75 units from the first, less expensive, supplier, with the remaining 40 units to come from the second supplier. If the company spends $10,750 to purchase the required number of units for the two plants, find the number of units that should be purchased from each supplier for each plant as follows:
(a) Assign variables to the four unknowns.
(b) Write a system of five equations with the four variables. (Not all equations will involve all four variables.)
(c) Use the Gauss-Jordan method to solve the system of equations.

Solve the linear systems in Exercises 45–48.

45. $2.1x + 3.5y + 9.4z = 15.6$
$6.8x - 1.5y + 7.5z = 26.4$
$3.7x + 2.5y - 6.1z = 18.7$

46. $9.03x - 5.91y + 2.68z = 29.5$
$3.94x + 6.82y + 1.53z = 35.4$
$2.79x + 1.68y - 6.23z = 12.1$

47. $10.47x + 3.52y + 2.58z - 6.42w = 218.65$
$8.62x - 4.93y - 1.75z + 2.83w = 157.03$
$4.92x + 6.83y - 2.97z + 2.65w = 462.3$
$2.86x + 19.1\ y - 6.24z - 8.73w = 398.4$

48. $28.6x + 94.5y + 16.0z - 2.94w = 198.3$
$16.7x + 44.3y - 27.3z + 8.9w = 254.7$
$12.5x - 38.7y + 92.5z + 22.4w = 562.7$
$40.1x - 28.3y + 17.5z - 10.2w = 375.4$

In Exercises 49–52, write a system of equations and then solve it.

49. Natural Science Natural Brand plant food is made from three chemicals. The mix must include 10.8% of the first chemical and the other two chemicals must be in the ratio of 4 to 3 as measured by weight. How much of each chemical is required to make 750 kilograms of the plant food?

50. Natural Science Three species of bacteria are fed three foods, I, II, and III. A bacterium of the first species consumes 1.3 units each of foods I and II and 2.3 units of food III each day. A bacterium of the second species consumes 1.1 units of food I, 2.4 units of food II, and 3.7 units of food III each day. A bacterium of the third species consumes 8.1 units of I, 2.9 units of II, and 5.1 units of III each day. If 16,000 units of I, 28,000 units of II, and 44,000 units of III are supplied each day, how many of each species can be maintained in this environment?

51. Natural Science A lake is stocked each spring with three species of fish, A, B, and C. Three foods, I, II, and III, are available in the lake. Each fish of species A requires 1.32 units of food I, 2.9 units of food II, and 1.75 units of food III on the average each day. Species B fish each require 2.1 units of food I, .95 units of food II, and .6 units of food III daily. Species C fish require .86, 1.52, and 2.01 units of I, II, and III per day, respectively. If 490 units of food I, 897 units of food II, and 653 units of food III are available daily, how many of each species should be stocked?

52. Management A company produces three combinations of mixed vegetables which sell in one kilogram packages. Italian style combines .3 kilograms of zucchini, .3 of broccoli, and .4 of carrots. French style combines .6 kilograms of broccoli and .4 of carrots. Oriental style combines .2 kilograms of zucchini, .5 of broccoli, and .3 of carrots. The company has a stock of 16,200 kilograms of zucchini, 41,400 kilograms of broccoli, and 29,400 kilograms of carrots. How many packages of each style should they prepare to use up their supplies?

6.3 Basic Matrix Operations

The study of matrices has been of interest to mathematicians for some time. Recently, however, the use of matrices has assumed greater importance in the fields of management, natural science, and social science because it provides a natural way to organize data. Matrices are yet another type of mathematical model which can be used to solve many of the problems that arise in these fields, from inventory control to models of a nation's economy.

The augmented matrix of the previous section is an example of a **matrix,** a rectangular array of numbers, usually enclosed in brackets or parentheses. Each number is called an element or entry of the matrix. A matrix is often referred to with a capital letter. For example,

$$A = \begin{bmatrix} 2 & -4 & 1 \\ 7 & 8 & 6 \end{bmatrix}$$

is a matrix with two (horizontal) rows and three (vertical) columns.

Example 1 shows the use of matrices to organize data.

Example 1 The EZ Life Company manufactures sofas and armchairs in three models, A, B, and C. The company has regional warehouses in New York, Chicago, and San Francisco. In its August shipment, the company sends 10 model A sofas, 12 model B sofas, 5 model C sofas, 15 model A chairs, 20 model B chairs, and 8 model C chairs to each warehouse.

This data might be organized by first listing it as follows.

| Sofas | 10 model A | 12 model B | 5 model C |
| Chairs | 15 model A | 20 model B | 8 model C |

Alternatively, tabulate the data in a chart.

		Model		
		A	B	C
Furniture	Sofa	10	12	5
	Chair	15	20	8

With the understanding that the numbers in each row refer to the furniture type (sofa, chair) and the numbers in each column refer to the model (A, B, C), the same information can be given by a matrix, as follows.

$$M = \begin{bmatrix} 10 & 12 & 5 \\ 15 & 20 & 8 \end{bmatrix} \quad \blacksquare$$

1 Rewrite this information in a matrix with three rows and two columns.

Answer:

$$\begin{bmatrix} 10 & 15 \\ 12 & 20 \\ 5 & 8 \end{bmatrix}$$

Matrices are classified by their **dimension** or **order,** that is, by the number of rows and columns that they contain. For example, the matrix

$$\begin{bmatrix} 2 & 7 & 5 \\ 4 & 6 & 9 \end{bmatrix}$$

has two rows and three columns. This matrix is said to be of **dimension** 2×3 (read "2 by 3") or **order** 2×3. A matrix with m rows and n columns is of order $m \times n$. The number of rows is always given first.

Example 2

(a) The matrix $\begin{bmatrix} 6 & 5 \\ 3 & 4 \\ 5 & -1 \end{bmatrix}$ is of order 3×2.

(b) $\begin{bmatrix} 5 & 8 & 9 \\ 0 & 5 & -3 \\ -4 & 0 & 5 \end{bmatrix}$ is of order 3×3.

(c) $[1 \quad 6 \quad 5 \quad -2 \quad 5]$ is of order 1×5.

(d) $\begin{bmatrix} 3 \\ -5 \\ 0 \\ 2 \end{bmatrix}$ is of order 4×1.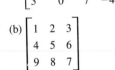

2 Give the order of each of the following matrices.

(a) $\begin{bmatrix} 2 & 1 & -5 & 6 \\ 3 & 0 & 7 & -4 \end{bmatrix}$

(b) $\begin{bmatrix} 1 & 2 & 3 \\ 4 & 5 & 6 \\ 9 & 8 & 7 \end{bmatrix}$

Answer:
(a) 2×4
(b) 3×3

A matrix having the same number of rows as columns is called a **square matrix.** The matrix given in Example 2(b) above is a square matrix, as are

$$\begin{bmatrix} -5 & 6 \\ 8 & 3 \end{bmatrix} \quad \text{and} \quad \begin{bmatrix} 0 & 0 & 0 & 0 \\ -2 & 4 & 1 & 3 \\ 0 & 0 & 0 & 0 \\ -5 & -4 & 1 & 8 \end{bmatrix}.$$

3 Use the numbers 2, 5, −8, 4 to write

(a) a row matrix;
(b) a column matrix;
(c) a square matrix.

Answer:
(a) $[2 \quad 5 \quad -8 \quad 4]$

(b) $\begin{bmatrix} 2 \\ 5 \\ -8 \\ 4 \end{bmatrix}$

(c) $\begin{bmatrix} 2 & 5 \\ -8 & 4 \end{bmatrix}$

or $\begin{bmatrix} 2 & -8 \\ 5 & 4 \end{bmatrix}$

(Other answers are possible.)

A matrix containing only one row is called a **row matrix** or **row vector.** The matrix in Example 2(c) is a row matrix, as are

$$[5 \quad 8], \qquad [6 \quad -9 \quad 2], \qquad \text{and} \qquad [-4 \ 0 \ 0 \ 0].$$

A matrix of only one column, such as in Example 2(d), is a **column matrix** or **column vector.** 3

Equality for matrices is defined as follows.

> Two matrices are **equal** if they are of the same order and if corresponding elements are equal.

Using this definition, the matrices

$$\begin{bmatrix} 2 & 1 \\ 3 & -5 \end{bmatrix} \quad \text{and} \quad \begin{bmatrix} 1 & 2 \\ -5 & 3 \end{bmatrix}$$

are not equal (even though they contain the same elements and are of the same order) since corresponding elements differ.

Example 3 **(a)** From the definition of equality given above, the only way that the statement

$$\begin{bmatrix} 2 & 1 \\ p & q \end{bmatrix} = \begin{bmatrix} x & y \\ -1 & 0 \end{bmatrix}$$

can be true is if $2 = x$, $1 = y$, $p = -1$, and $q = 0$.
(b) The statement

$$\begin{bmatrix} x \\ y \end{bmatrix} = \begin{bmatrix} 1 \\ 4 \\ 0 \end{bmatrix}$$

can never be true, since the two matrices are of different order. (One is 2×1 and the other is 3×1.)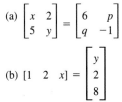

4 Give the values of the variables that make each of the following statements true.

(a) $\begin{bmatrix} x & 2 \\ 5 & y \end{bmatrix} = \begin{bmatrix} 6 & p \\ q & -1 \end{bmatrix}$

(b) $[1 \quad 2 \quad x] = \begin{bmatrix} y \\ 2 \\ 8 \end{bmatrix}$

Answer:
(a) $x = 6$, $y = -1$, $p = 2$, $q = 5$
(b) Can never be true

Addition The matrix given in Example 1,

$$M = \begin{bmatrix} 10 & 12 & 5 \\ 15 & 20 & 8 \end{bmatrix},$$

shows the August shipment from the EZ Life plant to its New York warehouse. If matrix N below gives the September shipment to the same warehouse, what is the total shipment for each item of furniture for these two months?

$$N = \begin{bmatrix} 45 & 35 & 20 \\ 65 & 40 & 35 \end{bmatrix}$$

If 10 model A sofas were shipped in August and 45 in September, then altogether 55 model A sofas were shipped in the two months. Adding the other corresponding entries gives a new matrix, call it Q, which represents the total shipment for the two months.

$$Q = \begin{bmatrix} 55 & 47 & 25 \\ 80 & 60 & 43 \end{bmatrix}$$

It is convenient to refer to Q as the "sum" of M and N.
   The way these two matrices were added illustrates the following definition of addition of matrices.

The **sum** of two $m \times n$ matrices X and Y is the $m \times n$ matrix $X + Y$ in which each element is the sum of the corresponding elements of X and Y.

It is important to remember that only matrices with the same dimensions can be added.

Example 4 Find each sum when possible.

(a) $\begin{bmatrix} 5 & -6 \\ 8 & 9 \end{bmatrix} + \begin{bmatrix} -4 & 6 \\ 8 & -3 \end{bmatrix} = \begin{bmatrix} 5 + (-4) & -6 + 6 \\ 8 + 8 & 9 + (-3) \end{bmatrix} = \begin{bmatrix} 1 & 0 \\ 16 & 6 \end{bmatrix}$

(b) The matrices

$$A = \begin{bmatrix} 5 & 8 \\ 6 & 2 \end{bmatrix} \quad \text{and} \quad B = \begin{bmatrix} 3 & 9 & 1 \\ 4 & 2 & 5 \end{bmatrix}$$

are of different orders, so it is not possible to find the sum $A + B$. **5**

Example 5 The September shipments from the EZ Life Company to the San Francisco and Chicago warehouses are given in matrices S and C below.

$$S = \begin{bmatrix} 30 & 32 & 28 \\ 43 & 47 & 30 \end{bmatrix}, \quad C = \begin{bmatrix} 22 & 25 & 38 \\ 31 & 34 & 35 \end{bmatrix}$$

What was the total amount shipped to the three warehouses in September? (See matrix N above for New York.)

The total of the September shipments is represented by the sum of the three matrices N, S, and C.

$$N + S + C = \begin{bmatrix} 45 & 35 & 20 \\ 65 & 40 & 35 \end{bmatrix} + \begin{bmatrix} 30 & 32 & 28 \\ 43 & 47 & 30 \end{bmatrix} + \begin{bmatrix} 22 & 25 & 38 \\ 31 & 34 & 35 \end{bmatrix}$$

$$= \begin{bmatrix} 97 & 92 & 86 \\ 139 & 121 & 100 \end{bmatrix}$$

For example, from this sum the total number of model C sofas shipped to the three warehouses in September was 86. **6**

The additive inverse of the real number a is $-a$; a similar definition is given for the additive inverse of a matrix.

> The **additive inverse** (or *negative*) of a matrix X is the matrix $-X$ in which each element is the additive inverse of the corresponding element of X.

If

$$A = \begin{bmatrix} 1 & 2 & 3 \\ 0 & -1 & 5 \end{bmatrix} \quad \text{and} \quad B = \begin{bmatrix} -2 & 3 & 0 \\ 1 & -7 & 2 \end{bmatrix},$$

then by the definition of the additive inverse of a matrix,

$$-A = \begin{bmatrix} -1 & -2 & -3 \\ 0 & 1 & -5 \end{bmatrix} \quad \text{and} \quad -B = \begin{bmatrix} 2 & -3 & 0 \\ -1 & 7 & -2 \end{bmatrix}.$$

By the definition of matrix addition, for each matrix X, the sum $X + (-X)$ is a **zero matrix**, O, whose elements are all zeros. There is an $m \times n$ zero matrix for each pair of values of m and n. Zero matrices have the following **identity property:**

---

5 Find each sum when possible.

(a) $\begin{bmatrix} 2 & 5 & 7 \\ 3 & -1 & 4 \end{bmatrix}$

$+ \begin{bmatrix} -1 & 2 & 0 \\ 10 & -4 & 5 \end{bmatrix}$

(b) $\begin{bmatrix} 1 \\ 2 \\ 3 \end{bmatrix} + \begin{bmatrix} 2 & -1 \\ 4 & 5 \\ 6 & 0 \end{bmatrix}$

(c) $[5 \quad 4 \quad -1]$
$\quad + [-5 \quad 2 \quad 3]$

Answer:

(a) $\begin{bmatrix} 1 & 7 & 7 \\ 13 & -5 & 9 \end{bmatrix}$

(b) Not possible

(c) $[0 \quad 6 \quad 2]$

---

6 From the result of Example 5, find the total number of the following shipped to the three warehouses.

(a) Model A chairs
(b) Model B sofas
(c) Model C chairs

Answer:

(a) 139
(b) 92
(c) 100

> If O is the $m \times n$ zero matrix, and A is any $m \times n$ matrix, then
>
> $$A + O = O + A = A.$$

Subtraction The **subtraction** of matrices can be defined in a manner comparable to subtraction for real numbers.

> For two $m \times n$ matrices X and Y, the **difference** of X and Y is the $m \times n$ matrix $X - Y$ in which each element is the difference of the corresponding elements of X and Y, or, equivalently,
>
> $$X - Y = X + (-Y).$$

For example, using A and B from above,

$$A - B = \begin{bmatrix} 1 & 2 & 3 \\ 0 & -1 & 5 \end{bmatrix} - \begin{bmatrix} -2 & 3 & 0 \\ 1 & -7 & 2 \end{bmatrix}$$

$$= \begin{bmatrix} 1 - (-2) & 2 - 3 & 3 - 0 \\ 0 - 1 & -1 - (-7) & 5 - 2 \end{bmatrix}$$

$$= \begin{bmatrix} 3 & -1 & 3 \\ -1 & 6 & 3 \end{bmatrix}.$$

Example 6 **(a)** $[8 \quad 6 \quad -4] - [3 \quad 5 \quad -8] = [5 \quad 1 \quad 4]$
(b) The matrices

$$\begin{bmatrix} -2 & 5 \\ 0 & 1 \end{bmatrix} \quad \text{and} \quad \begin{bmatrix} 3 \\ 5 \end{bmatrix}$$

have different orders and cannot be subtracted. ▨

7 Find each of the following differences when possible.

(a) $\begin{bmatrix} 2 & 5 \\ -1 & 0 \end{bmatrix} - \begin{bmatrix} 6 & 4 \\ 3 & -2 \end{bmatrix}$

(b) $\begin{bmatrix} 1 & 5 & 6 \\ 2 & 4 & 8 \end{bmatrix} - \begin{bmatrix} 2 & 1 \\ 10 & 3 \end{bmatrix}$

(c) $[5 \quad -4 \quad 1]$
$\quad - [6 \quad 0 \quad -3]$

Answer:

(a) $\begin{bmatrix} -4 & 1 \\ -4 & 2 \end{bmatrix}$

(b) Not possible

(c) $[-1 \quad -4 \quad 4]$

Example 7 During September the Chicago warehouse of the EZ Life Company shipped out the following numbers of each model.

$$K = \begin{bmatrix} 5 & 10 & 8 \\ 11 & 14 & 15 \end{bmatrix}$$

What was the Chicago warehouse inventory on October 1, taking into account only the number of items received and sent out during the month?

The number of each kind of item received during September is given by matrix C from Example 5; the number of each model sent out during September is given by matrix K. The October 1 inventory will be represented by the matrix $C - K$ as shown below.

$$\begin{bmatrix} 22 & 25 & 38 \\ 31 & 34 & 35 \end{bmatrix} - \begin{bmatrix} 5 & 10 & 8 \\ 11 & 14 & 15 \end{bmatrix} = \begin{bmatrix} 17 & 15 & 30 \\ 20 & 20 & 20 \end{bmatrix}$$ ▨

6.3 Exercises

Mark each of the following statements as true or false. If false, tell why.

1. $\begin{bmatrix} 1 & 3 \\ 5 & 7 \end{bmatrix} = \begin{bmatrix} 1 & 5 \\ 3 & 7 \end{bmatrix}$

2. $\begin{bmatrix} 1 \\ 2 \\ 3 \end{bmatrix} = \begin{bmatrix} 1 & 2 & 3 \end{bmatrix}$

3. $\begin{bmatrix} x \\ y \end{bmatrix} = \begin{bmatrix} 3 \\ 5 \end{bmatrix}$ if $x = 3$ and $y = 5$.

4. $\begin{bmatrix} 3 & 5 & 2 & 8 \\ 1 & -1 & 4 & 0 \end{bmatrix}$ is a 4 × 2 matrix.

5. $\begin{bmatrix} 1 & 9 & -4 \\ 3 & 7 & 2 \\ -1 & 1 & 0 \end{bmatrix}$ is a square matrix.

6. $\begin{bmatrix} 2 & 4 & -1 \\ 3 & 7 & 5 \\ 0 & 0 & 0 \end{bmatrix} = \begin{bmatrix} 2 & 4 & -1 \\ 3 & 7 & 5 \end{bmatrix}$

Find the order of each of the following. Identify any square, column, or row matrices. (See Example 2.) Give the additive inverse (or negative) of each matrix.

7. $\begin{bmatrix} -4 & 8 \\ 2 & 3 \end{bmatrix}$

8. $\begin{bmatrix} -9 & 6 & 2 \\ 4 & 1 & 8 \end{bmatrix}$

9. $\begin{bmatrix} -6 & 8 & 0 & 0 \\ 4 & 1 & 9 & 2 \\ 3 & -5 & 7 & 1 \end{bmatrix}$

10. $\begin{bmatrix} 8 & -2 & 4 & 6 & 3 \end{bmatrix}$

11. $\begin{bmatrix} 2 \\ 4 \end{bmatrix}$

12. $\begin{bmatrix} -9 \end{bmatrix}$

Find the values of the variables in each of the following. (See Example 3.)

13. $\begin{bmatrix} 2 & 1 \\ 4 & 8 \end{bmatrix} = \begin{bmatrix} x & 1 \\ y & z \end{bmatrix}$

14. $\begin{bmatrix} -5 \\ y \end{bmatrix} = \begin{bmatrix} -5 \\ 8 \end{bmatrix}$

15. $\begin{bmatrix} x + 6 & y + 2 \\ 8 & 3 \end{bmatrix} = \begin{bmatrix} -9 & 7 \\ 8 & k \end{bmatrix}$

16. $\begin{bmatrix} 9 & 7 \\ r & 0 \end{bmatrix} = \begin{bmatrix} m - 3 & n + 5 \\ 8 & 0 \end{bmatrix}$

17. $\begin{bmatrix} -7 + z & 4r & 8s \\ 6p & 2 & 5 \end{bmatrix} + \begin{bmatrix} -9 & 8r & 3 \\ 2 & 5 & 4 \end{bmatrix}$
$= \begin{bmatrix} 2 & 36 & 27 \\ 20 & 7 & 12a \end{bmatrix}$

18. $\begin{bmatrix} a + 2 & 3z + 1 & 5m \\ 4k & 0 & 3 \end{bmatrix} + \begin{bmatrix} 3a & 2z & 5m \\ 2k & 5 & 6 \end{bmatrix}$
$= \begin{bmatrix} 10 & -14 & 80 \\ 10 & 5 & 9 \end{bmatrix}$

Perform the indicated operations where possible. (See Examples 4 and 6.)

19. $\begin{bmatrix} 1 & 2 & 5 & -1 \\ 3 & 0 & 2 & -4 \end{bmatrix} + \begin{bmatrix} 8 & 10 & -5 & 3 \\ -2 & -1 & 0 & 0 \end{bmatrix}$

20. $\begin{bmatrix} 1 & 5 \\ 2 & -3 \\ 3 & 7 \end{bmatrix} + \begin{bmatrix} 2 & 3 \\ 8 & 5 \\ -1 & 9 \end{bmatrix}$

21. $\begin{bmatrix} 1 & 5 & 7 \\ 2 & 2 & 3 \end{bmatrix} + \begin{bmatrix} 4 & 8 & -7 \\ 1 & -1 & 5 \end{bmatrix}$

22. $\begin{bmatrix} 2 & 4 \\ -8 & 1 \end{bmatrix} + \begin{bmatrix} 9 & -3 \\ 8 & 5 \end{bmatrix}$

23. $\begin{bmatrix} 1 & 3 & -2 \\ 4 & 7 & 1 \end{bmatrix} + \begin{bmatrix} 3 & 0 \\ 6 & 4 \\ -5 & 2 \end{bmatrix}$

24. $\begin{bmatrix} 1 & 3 & -2 \\ 4 & 7 & 1 \end{bmatrix} - \begin{bmatrix} 3 & 6 & -5 \\ 0 & 4 & 2 \end{bmatrix}$

25. $\begin{bmatrix} 2 & 8 & 12 & 0 \\ 7 & 4 & -1 & 5 \\ 1 & 2 & 0 & 10 \end{bmatrix} - \begin{bmatrix} 1 & 3 & 6 & 9 \\ 2 & -3 & -3 & 4 \\ 8 & 0 & -2 & 17 \end{bmatrix}$

26. $\begin{bmatrix} 2 & 1 \\ 5 & -3 \\ -7 & 2 \\ 9 & 0 \end{bmatrix} + \begin{bmatrix} 1 & -8 & 0 \\ 5 & 3 & 2 \\ -6 & 7 & -5 \\ 2 & -1 & 0 \end{bmatrix}$

27. $\begin{bmatrix} -4x + 2y & -3x + y \\ 6x - 3y & 2x - 5y \end{bmatrix} + \begin{bmatrix} -8x + 6y & 2x \\ 3y - 5x & 6x + 4y \end{bmatrix}$

28. $\begin{bmatrix} 4k - 8y \\ 6z - 3x \\ 2k + 5a \\ -4m + 2n \end{bmatrix} - \begin{bmatrix} 5k + 6y \\ 2z + 5x \\ 4k + 6a \\ 4m + 2n \end{bmatrix}$

Using matrices

$$O = \begin{bmatrix} 0 & 0 \\ 0 & 0 \end{bmatrix}, \ P = \begin{bmatrix} m & n \\ p & q \end{bmatrix}, \ T = \begin{bmatrix} r & s \\ t & u \end{bmatrix}, \text{ and } X = \begin{bmatrix} x & y \\ z & w \end{bmatrix},$$

verify that the statements in Exercises 29–33 are true.

29. $X + T$ is a 2×2 matrix

30. $X + T = T + X$ (Commutative property of addition of matrices)

31. $X + (T + P) = (X + T) + P$ (Associative property of addition of matrices)

32. $X + (-X) = O$ (Inverse property of addition of matrices)

33. $P + O = P$ (Identity property of addition of matrices)

34. Which of the above properties are valid for matrices that are not square?

35. Management An investment group planning a shopping center decided to include a market, a barber shop, a variety store, a drug store, and a bakery. They estimated the initial cost and the guaranteed rent (both in dollars per square foot) for each type of store, respectively, as follows. Initial cost: 18, 10, 8, 10, and 10; guaranteed rent: 2.7, 1.5, 1.0, 2.0, and 1.7. Write this information first as a 5×2 matrix and then as a 2×5 matrix.

36. Management In reviewing its circulation and advertising, a weekly magazine found that in 1970 its circulation was 770,000, it had 2,817 pages of advertising, and the gross advertising revenue was \$9,718,180. Comparable figures for 1975 were 1,300,000; 3,693;

and \$18,537,057. The figures for 1980 were 1,700,000; 3,809; and \$29,950,700; and for 1985 they were 2,000,000; 3,450; and \$42,598,770. Write this information first as a 4×3 matrix and then as a 3×4 matrix.

37. Natural Science A dietician prepares a diet specifying the amounts a patient should eat of four basic food groups: group I, meats; group II, fruits and vegetables; group III, breads and starches; group IV, milk products. Amounts are given in "exchanges" which represent 1 ounce (meat), 1/2 cup (fruits and vegetables), 1 slice (bread), 8 ounces (milk), or other suitable measurements.

(a) The number of "exchanges" for breakfast for each of the four food groups respectively are 2, 1, 2, and 1; for lunch, 3, 2, 2, and 1; and for dinner, 4, 3, 2, and 1. Write a 3×4 matrix using this information.

(b) The amounts of fat, carbohydrates, and protein in each food group respectively are as follows.
Fat: 5, 0, 0, 10
Carbohydrates: 0, 10, 15, 12
Protein: 7, 1, 2, 8
Use this information to write a 4×3 matrix.

(c) There are 8 calories per unit of fat, 4 calories per unit of carbohydrates, and 5 calories per unit of protein; summarize this data in a 3×1 matrix.

38. Natural Science At the beginning of a laboratory experiment, five baby rats measured 5.6, 6.4, 6.9, 7.6, and 6.1 centimeters in length, and weighed 144, 138, 149, 152, and 146 grams respectively.

(a) Write a 2×5 matrix using this information.

(b) At the end of two weeks, their lengths were 10.2, 11.4, 11.4, 12.7, and 10.8 centimeters, and they weighed 196, 196, 225, 250, and 230 grams. Write a 2×5 matrix with this information.

(c) Use matrix subtraction and the matrices found in (a) and (b) to write a matrix which gives the amount of change in length and weight for each rat.

(d) The following week the rats gained as shown in the matrix below.

$$\begin{matrix} \text{Length} \\ \text{Weight} \end{matrix} \begin{bmatrix} 1.8 & 1.5 & 2.3 & 1.8 & 2.0 \\ 25 & 22 & 29 & 33 & 20 \end{bmatrix}$$

What were their lengths and weights at the end of this week?

6.4 Multiplication of Matrices

Suppose one of the EZ Life Company warehouses receives the following order, written in matrix form, where the entries have the same meaning as in the previous section.

$$\begin{bmatrix} 5 & 4 & 1 \\ 3 & 2 & 3 \end{bmatrix}$$

Later, the store that sent the order asks the company to send five more of the same order. The five new orders can be written as one matrix by multiplying each element in the matrix by 5, giving the product

$$5 \begin{bmatrix} 5 & 4 & 1 \\ 3 & 2 & 3 \end{bmatrix} = \begin{bmatrix} 25 & 20 & 5 \\ 15 & 10 & 15 \end{bmatrix}.$$

In work with matrices, a real number, like the 5 in the product above, is called a **scalar.**

> The **product** of a scalar k and a matrix X is the matrix kX, each of whose elements is k times the corresponding element of X.

For example,

$$(-3) \begin{bmatrix} 2 & -5 \\ 1 & 7 \end{bmatrix} = \begin{bmatrix} -6 & 15 \\ -3 & -21 \end{bmatrix}.$$

While finding the product of two matrices is more involved, such multiplication is important in applications to practical problems. To understand the reasoning behind matrix multiplication, look again at the EZ Life Company. Suppose sofas and chairs of the same model are often sold as sets with matrix W showing the number of each model set in each warehouse.

$$\begin{matrix} & \begin{matrix} A & B & C \end{matrix} \\ \begin{matrix} \text{New York} \\ \text{Chicago} \\ \text{San Francisco} \end{matrix} & \begin{bmatrix} 10 & 7 & 3 \\ 5 & 9 & 6 \\ 4 & 8 & 2 \end{bmatrix} = W \end{matrix}$$

If the selling price of a model A set is $800, of a model B set $1000, and of a model C set $1200, find the total value of the sets in the New York warehouse as follows.

Type	Number of sets		Price of set		Total
A	10	×	$ 800	=	$ 8000
B	7	×	1000	=	7000
C	3	×	1200	=	3600
					$18,600

(Total for New York)

① In this example of the EZ Life Company find the total value of the New York sets if model A sofas sell for $1200, model B for $1600, and model C for $1300.

Answer:
$27,100

The total value of the three kinds of sets in New York is $18,600. ①

The work done in the table above is summarized as follows:

$$10(\$800) + 7(\$1000) + 3(\$1200) = \$18,600.$$

In the same way, the Chicago sets have a total value of

$$5(\$800) + 9(\$1000) + 6(\$1200) = \$20,200,$$

and in San Francisco, the total value of the sets is

$$4(\$800) + 8(\$1000) + 2(\$1200) = \$13,600.$$

Write the selling prices as a column matrix, P, and the total value in each location as a column matrix V.

$$\begin{bmatrix} 800 \\ 1000 \\ 1200 \end{bmatrix} = P \qquad \begin{bmatrix} 18,600 \\ 20,200 \\ 13,600 \end{bmatrix} = V$$

Look back at the elements of W; multiplying the first, second, and third elements of the first row of W by the first, second, and third elements respectively of the column matrix P and then adding these products gives the first element in V. Doing the same thing with the second row of W gives the second element of V; the third row of W leads to the third element of V, suggesting that it is reasonable to write the product of matrices

$$W = \begin{bmatrix} 10 & 7 & 3 \\ 5 & 9 & 6 \\ 4 & 8 & 2 \end{bmatrix} \qquad \text{and} \qquad P = \begin{bmatrix} 800 \\ 1000 \\ 1200 \end{bmatrix}$$

as

$$WP = \begin{bmatrix} 10 & 7 & 3 \\ 5 & 9 & 6 \\ 4 & 8 & 2 \end{bmatrix} \begin{bmatrix} 800 \\ 1000 \\ 1200 \end{bmatrix} = \begin{bmatrix} 18,600 \\ 20,200 \\ 13,600 \end{bmatrix} = V.$$

The product was found by multiplying the elements of the *rows* of the matrix on the left and the corresponding elements of the *column* of the matrix on the right, and then finding the sum of these separate products. Notice that the product of a 3×3 matrix and a 3×1 matrix is a 3×1 matrix.

The **product** AB of an $m \times n$ matrix A and an $n \times k$ matrix B is found as follows. Multiply each element of the *first row* of A by the corresponding element of the *first column* of B. The sum of these n products is the *first row, first column* element of AB. Also, the sum of the products found by multiplying the elements of the *first row* of A times the corresponding elements of the *second column* of B gives the *first row, second column* element of AB, and so on.

Let A be an $m \times n$ matrix and let B be an $n \times k$ matrix. To find the ith row, jth column element of the **product matrix** AB, multiply each element in the ith row of A by the corresponding element in the jth column of B. The sum of these products will give the row i, column j element of AB. The product matrix AB is of order $m \times k$. Be careful: the number of *columns* of A must equal the number of *rows* of B.

Example 1   Find the product AB given

$$A = \begin{bmatrix} 2 & 3 & -1 \\ 4 & 2 & 2 \end{bmatrix} \quad \text{and} \quad B = \begin{bmatrix} 1 \\ 8 \\ 6 \end{bmatrix}.$$

Step 1   Multiply the elements of the first row of A and the corresponding elements of the column of B.

$$\begin{bmatrix} \mathbf{2} & \mathbf{3} & \mathbf{-1} \\ 4 & 2 & 2 \end{bmatrix} \begin{bmatrix} \mathbf{1} \\ \mathbf{8} \\ \mathbf{6} \end{bmatrix} \quad \mathbf{2} \cdot \mathbf{1} + \mathbf{3} \cdot \mathbf{8} + (\mathbf{-1}) \cdot \mathbf{6} = 20$$

Therefore, 20 is the first row entry of the product matrix AB.

Step 2   Multiply the elements of the second row of A and the corresponding elements of B.   ②

Step 3   Write the product using the two entries we found above.

$$AB = \begin{bmatrix} 2 & 3 & -1 \\ 4 & 2 & 2 \end{bmatrix} \begin{bmatrix} 1 \\ 8 \\ 6 \end{bmatrix} = \begin{bmatrix} 20 \\ 32 \end{bmatrix}$$

③

Example 2   Find the product CD given

$$C = \begin{bmatrix} -3 & 4 & 2 \\ 5 & 0 & 4 \end{bmatrix} \quad \text{and} \quad D = \begin{bmatrix} -6 & 4 \\ 2 & 3 \\ 3 & -2 \end{bmatrix}.$$

Step 1

$$\begin{bmatrix} \mathbf{-3} & \mathbf{4} & \mathbf{2} \\ 5 & 0 & 4 \end{bmatrix} \begin{bmatrix} \mathbf{-6} & 4 \\ \mathbf{2} & 3 \\ \mathbf{3} & -2 \end{bmatrix} \quad (\mathbf{-3}) \cdot (\mathbf{-6}) + \mathbf{4} \cdot \mathbf{2} + \mathbf{2} \cdot \mathbf{3} = 32$$

Step 2

$$\begin{bmatrix} \mathbf{-3} & \mathbf{4} & \mathbf{2} \\ 5 & 0 & 4 \end{bmatrix} \begin{bmatrix} -6 & \mathbf{4} \\ 2 & \mathbf{3} \\ 3 & \mathbf{-2} \end{bmatrix} \quad (\mathbf{-3}) \cdot \mathbf{4} + \mathbf{4} \cdot \mathbf{3} + \mathbf{2} \cdot (\mathbf{-2}) = -4$$

---

② Complete this step by filling in the blanks below.

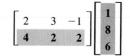

$4(\_\_) + \_\_ (\_\_) + 2(6) = \_\_$
The second row entry of the product is $\_\_$.

Answer:

1; 2; 8; 32; 32

---

③ Find the product AB given

$$A = \begin{bmatrix} 2 & 4 \\ 5 & 6 \end{bmatrix} \quad \text{and}$$

$$B = \begin{bmatrix} -3 \\ 4 \end{bmatrix}$$

Answer:

$$AB = \begin{bmatrix} 10 \\ 9 \end{bmatrix}$$

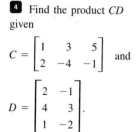 Find the product CD given

$$C = \begin{bmatrix} 1 & 3 & 5 \\ 2 & -4 & -1 \end{bmatrix} \quad \text{and}$$

$$D = \begin{bmatrix} 2 & -1 \\ 4 & 3 \\ 1 & -2 \end{bmatrix}.$$

Answer:

$$CD = \begin{bmatrix} 19 & -2 \\ -13 & -12 \end{bmatrix}$$

5 Matrix A is 4×6 and matrix B is 2×4.
(a) Can AB be found? If so, give its order.
(b) Can BA be found? If so, give its order.

Answer:
(a) No
(b) Yes; 2×6

6 Give the order of each of the following products that can be found.

(a) $\begin{bmatrix} 2 & 4 \\ 6 & 8 \end{bmatrix} \begin{bmatrix} 1 & 2 & 3 \\ 0 & -1 & 2 \end{bmatrix}$

(b) $\begin{bmatrix} 1 & 2 \\ 5 & 10 \\ 12 & 7 \end{bmatrix} \begin{bmatrix} 2 & 4 \\ 3 & 6 \\ 9 & 1 \end{bmatrix}$

(c) $\begin{bmatrix} 5 \\ 2 \\ 4 \end{bmatrix} [1 \quad 0 \quad 6]$

Answer:
(a) 2×3
(b) Not possible
(c) 3×3

Step 3

$$\begin{bmatrix} -3 & 4 & 2 \\ 5 & 0 & 4 \end{bmatrix} \begin{bmatrix} -6 & 4 \\ 2 & 3 \\ 3 & -2 \end{bmatrix} \qquad 5 \cdot (-6) + 0 \cdot 2 + 4 \cdot 3 = -18$$

Step 4

$$\begin{bmatrix} -3 & 4 & 2 \\ 5 & 0 & 4 \end{bmatrix} \begin{bmatrix} -6 & 4 \\ 2 & 3 \\ 3 & -2 \end{bmatrix} \qquad 5 \cdot 4 + 0 \cdot 3 + 4 \cdot (-2) = 12$$

Step 5 The product is

$$CD = \begin{bmatrix} -3 & 4 & 2 \\ 5 & 0 & 4 \end{bmatrix} \begin{bmatrix} -6 & 4 \\ 2 & 3 \\ 3 & -2 \end{bmatrix} = \begin{bmatrix} 32 & -4 \\ -18 & 12 \end{bmatrix}. \quad \blacksquare$$

In Example 2, the product of a 2×3 matrix and a 3×2 matrix was a 2×2 matrix. As mentioned earlier, the product AB of two matrices A and B can be found only if the number of columns of A is the same as the number of rows of B. The final product will have as many rows as A and as many columns as B.

Example 3 Suppose matrix A is 2×2 and matrix B is 2×4. Can the product AB be calculated? What is the order of the product? The following diagram helps decide the answers to these questions.

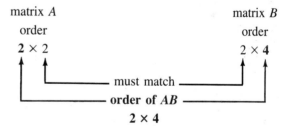

The product of A and B can be calculated because A has two columns and B has two rows. The order of the product is 2×4. **5**

Example 4 Find BA given

$$A = \begin{bmatrix} 1 & -3 \\ 7 & 2 \end{bmatrix} \quad \text{and} \quad B = \begin{bmatrix} 1 & 0 & -1 \\ 3 & 1 & 4 \end{bmatrix}.$$

Since B is a 2×3 matrix and A is a 2×2 matrix, the product BA cannot be found. **6**

Example 5 A contractor builds three kinds of houses, models A, B, and C, with a choice of two styles, Spanish or contemporary. Matrix P shows the number of each kind of house planned for a new 100-home subdivision. The amounts for each of the exterior materials used depend primarily on the style of the house. These amounts are shown in matrix Q. (Concrete is in cubic yards, lumber in units of 1000 board feet, brick in 1000's, and shingles in units of 100 square feet.) Matrix R gives the cost for each kind of material.

$$\begin{array}{c} \\ \text{Model A} \\ \text{Model B} \\ \text{Model C} \end{array} \begin{array}{cc} \text{Spanish} & \text{Contemporary} \\ \left[\begin{array}{cc} 0 & 30 \\ 10 & 20 \\ 20 & 20 \end{array}\right] & = P \end{array}$$

$$\begin{array}{c} \\ \text{Spanish} \\ \text{Contemporary} \end{array} \begin{array}{cccc} \text{Concrete} & \text{Lumber} & \text{Brick} & \text{Shingles} \\ \left[\begin{array}{cccc} 10 & 2 & 0 & 2 \\ 50 & 1 & 20 & 2 \end{array}\right] & & & = Q \end{array}$$

$$\begin{array}{c} \text{Concrete} \\ \text{Lumber} \\ \text{Brick} \\ \text{Shingles} \end{array} \begin{array}{c} \text{Cost per unit} \\ \left[\begin{array}{c} 20 \\ 180 \\ 60 \\ 25 \end{array}\right] = R \end{array}$$

(a) What is the total cost for each model house?

First find PQ to find the cost for each model. The product PQ shows the amount of each material needed for each model house.

$$PQ = \begin{bmatrix} 0 & 30 \\ 10 & 20 \\ 20 & 20 \end{bmatrix} \begin{bmatrix} 10 & 2 & 0 & 2 \\ 50 & 1 & 20 & 2 \end{bmatrix}$$

$$= \begin{array}{cccc} \text{Concrete} & \text{Lumber} & \text{Brick} & \text{Shingles} \\ \left[\begin{array}{cccc} 1500 & 30 & 600 & 60 \\ 1100 & 40 & 400 & 60 \\ 1200 & 60 & 400 & 80 \end{array}\right] & & & \begin{array}{c} \text{Model A} \\ \text{Model B} \\ \text{Model C} \end{array} \end{array}$$

Now multiply PQ and R, the cost matrix, to get the total cost for each model house.

$$\begin{bmatrix} 1500 & 30 & 600 & 60 \\ 1100 & 40 & 400 & 60 \\ 1200 & 60 & 400 & 80 \end{bmatrix} \begin{bmatrix} 20 \\ 180 \\ 60 \\ 25 \end{bmatrix} = \begin{array}{c} \text{Cost} \\ \begin{bmatrix} 72,900 \\ 54,700 \\ 60,800 \end{bmatrix} \begin{array}{c} \text{Model A} \\ \text{Model B} \\ \text{Model C} \end{array} \end{array}$$

(b) How much of each of the four kinds of material must be ordered?

The totals of the columns of matrix PQ will give a matrix whose elements represent the total amounts of each material needed for the subdivision. Call this matrix T, and write it as a row matrix.

$$T = [3800 \quad 130 \quad 1400 \quad 200]$$

(c) What is the total cost for material?

Find the total cost of all the materials by taking the product of matrix T, the matrix showing the total amounts of each material, and matrix R, the cost matrix. (To multiply these and get a 1×1 matrix, representing total cost, we must multiply a 1×4 matrix by a 4×1 matrix. This is why T was written as a row matrix in (b) above.)

$$TR = [3800 \quad 130 \quad 1400 \quad 200] \begin{bmatrix} 20 \\ 180 \\ 60 \\ 25 \end{bmatrix} = [188,400]$$

(d) Suppose the contractor builds the same number of homes in five subdivisions. Calculate the total amount of each material for each model for all five subdivisions.

Multiply PQ by the scalar 5, as follows.

$$5 \begin{bmatrix} 1500 & 30 & 600 & 60 \\ 1100 & 40 & 400 & 60 \\ 1200 & 60 & 400 & 80 \end{bmatrix} = \begin{bmatrix} 7500 & 150 & 3000 & 300 \\ 5500 & 200 & 2000 & 300 \\ 6000 & 300 & 2000 & 400 \end{bmatrix} \blacksquare$$

We can introduce a notation to help keep track of the quantities a matrix represents. For example, we can say that matrix P, from Example 5, represents models/styles, matrix Q represents styles/materials, and matrix R represents materials/cost. In each case the meaning of the rows is written first and the columns second. When we found the product PQ in Example 5, the rows of the matrix represented models and the columns represented materials. Therefore, we can say the matrix product PQ represents models/materials. The common quantity, styles, in both P and Q was eliminated in the produce PQ. Do you see that the product $(PQ)R$ represents models/cost?

In practical problems this notation helps decide in what order to multiply two matrices so that the results are meaningful. In Example 5(c) we could have found either product RT or product TR. However, since T represents subdivisions/materials and R represents materials/cost, the product TR gives subdivisions/cost. ⑦

Exercises 31 and 32 below suggest the truth of the following properties of matrices.

> For any matrices A, B, and C, such that all indicated products exist,
>
> **(a)** AB does not necessarily equal BA (matrix multiplication is *not* commutative), and
>
> **(b)** $A(BC) = (AB)C$ (matrix multiplication is associative).

⑦ Let matrix A be

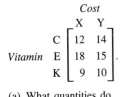

$$\begin{array}{c} \\ Brand \end{array} \begin{array}{c} Vitamin \\ \begin{array}{ccc} C & E & K \end{array} \\ \begin{array}{c} X \\ Y \end{array} \begin{bmatrix} 2 & 7 & 5 \\ 4 & 6 & 9 \end{bmatrix} \end{array}$$

and matrix B be

$$\begin{array}{c} \\ Vitamin \end{array} \begin{array}{c} Cost \\ \begin{array}{cc} X & Y \end{array} \\ \begin{array}{c} C \\ E \\ K \end{array} \begin{bmatrix} 12 & 14 \\ 18 & 15 \\ 9 & 10 \end{bmatrix} \end{array}.$$

(a) What quantities do matrices A and B represent?
(b) What quantities does the product AB represent?
(c) What quantities does the product BA represent?

Answer:
(a) A = brand/vitamin, B = vitamin/cost
(b) AB = brand/cost
(c) Not meaningful, although the product BA can be found

6.4 Exercises

In each of the following exercises, the dimensions of two matrices A and B are given. Find the dimensions of the product AB and the product BA, whenever these products exist. (See Example 3.)

1. A is 2×2, B is 2×2

2. A is 3×3, B is 3×3

3. A is 4×2, B is 2×4

4. A is 3×1, B is 1×3

5. A is 3×5, B is 5×2

6. A is 4×3, B is 3×6

7. A is 4×2, B is 3×4

8. A is 7×3, B is 2×7

Let

$$A = \begin{bmatrix} -2 & 4 \\ 0 & 3 \end{bmatrix} \quad and \quad B = \begin{bmatrix} -6 & 2 \\ 4 & 0 \end{bmatrix}.$$

Find each of the following.

9. $2A$

10. $-3B$

11. $-4B$

12. $5A$

13. $-4A + 5B$

14. $3A - 10B$

Find each of the following matrix products where possible. (See Examples 1, 2, and 4.)

15. $\begin{bmatrix} 1 & 2 \\ 3 & 4 \end{bmatrix} \begin{bmatrix} -1 \\ 7 \end{bmatrix}$

16. $\begin{bmatrix} -1 & 5 \\ 7 & 0 \end{bmatrix} \begin{bmatrix} 6 \\ 2 \end{bmatrix}$

17. $\begin{bmatrix} 2 & 2 & -1 \\ 3 & 0 & 1 \end{bmatrix} \begin{bmatrix} 0 & 2 \\ -1 & 4 \\ 0 & 2 \end{bmatrix}$

18. $\begin{bmatrix} -9 & 2 & 1 \\ 3 & 0 & 0 \end{bmatrix} \begin{bmatrix} 2 \\ -1 \\ 4 \end{bmatrix}$

19. $\begin{bmatrix} -4 & 1 \\ 2 & -3 \end{bmatrix} \begin{bmatrix} 1 & 0 \\ 0 & 1 \end{bmatrix}$

20. $\begin{bmatrix} 1 & 0 \\ 0 & 1 \end{bmatrix} \begin{bmatrix} 3 & -2 \\ 1 & -5 \end{bmatrix}$

21. $\begin{bmatrix} 1 & 0 & 0 \\ 0 & 1 & 0 \\ 0 & 0 & 1 \end{bmatrix} \begin{bmatrix} 3 & -5 & 7 \\ -2 & 1 & 6 \\ 0 & -3 & 4 \end{bmatrix}$

22. $\begin{bmatrix} -8 & 9 & 2 \\ 3 & -4 & 7 \\ -1 & 6 & 3 \end{bmatrix} \begin{bmatrix} 1 & 0 & 0 \\ 0 & 1 & 0 \\ 0 & 0 & 1 \end{bmatrix}$

23. $\begin{bmatrix} 1 & 2 \\ 3 & 4 \end{bmatrix} \begin{bmatrix} -1 & 5 \\ 7 & 0 \end{bmatrix}$

24. $\begin{bmatrix} -1 & 5 \\ 7 & 0 \end{bmatrix} \begin{bmatrix} 1 & 2 \\ 3 & 4 \end{bmatrix}$

25. $\begin{bmatrix} -2 & -3 & 7 \\ 1 & 5 & 6 \end{bmatrix} \begin{bmatrix} 1 \\ 2 \\ 3 \end{bmatrix}$

26. $\begin{bmatrix} 6 \\ 5 \\ 4 \end{bmatrix} [-1 \quad 1 \quad 1]$

27. $\left(\begin{bmatrix} 4 & 3 \\ 1 & 2 \\ 0 & -5 \end{bmatrix} \begin{bmatrix} 2 & -2 \\ 1 & -1 \end{bmatrix} \right) \begin{bmatrix} 10 \\ 0 \end{bmatrix}$

28. $\begin{bmatrix} 4 & 3 \\ 1 & 2 \\ 0 & -5 \end{bmatrix} \left(\begin{bmatrix} 2 & -2 \\ 1 & -1 \end{bmatrix} \begin{bmatrix} 10 \\ 0 \end{bmatrix} \right)$

29. $\begin{bmatrix} 2 & -2 \\ 1 & -1 \end{bmatrix} \left(\begin{bmatrix} 4 & 3 \\ 1 & 2 \end{bmatrix} + \begin{bmatrix} 7 & 0 \\ -1 & 5 \end{bmatrix} \right)$

30. $\begin{bmatrix} 2 & -2 \\ 1 & -1 \end{bmatrix} \begin{bmatrix} 4 & 3 \\ 1 & 2 \end{bmatrix} + \begin{bmatrix} 2 & -2 \\ 1 & -1 \end{bmatrix} \begin{bmatrix} 7 & 0 \\ -1 & 5 \end{bmatrix}$

31. Let

$$A = \begin{bmatrix} -2 & 4 \\ 1 & 3 \end{bmatrix} \quad and \quad B = \begin{bmatrix} -2 & 1 \\ 3 & 6 \end{bmatrix}.$$

(a) Find AB.   **(b)** Find BA.

(c) Did you get the same answer in (a) and (b)? Do you think that matrix multiplication is commutative?

(d) In general, for matrices A and B such that AB and BA both exist, does AB always equal BA?

Given matrices

$$P = \begin{bmatrix} m & n \\ p & q \end{bmatrix}, \quad X = \begin{bmatrix} x & y \\ z & w \end{bmatrix}, \quad T = \begin{bmatrix} r & s \\ t & u \end{bmatrix},$$

verify that the statements in Exercises 32–36 are true.

32. $(PX)T = P(XT)$ (Associative property: see Exercises 27 and 28.)

33. $P(X + T) = PX + PT$ (Distributive property: see Exercises 29 and 30.)

34. PX is a 2×2 matrix

35. $k(X + T) = kX + kT$ for any real number k

36. $(k + h)P = kP + hP$ for any real numbers k and h

37. Management The Bread Box, a small neighborhood bakery, sells four main items: sweet rolls, bread, cake, and pie. The amount of certain major ingredients required to make these items is given in matrix A.

$$A = \begin{array}{c} \\ \\ \\ \\ \end{array}\begin{array}{c} \text{Eggs} \quad \text{Flour*} \quad \text{Sugar*} \quad \text{Butter*} \quad \text{Milk*} \\ \begin{bmatrix} 1 & 4 & \frac{1}{4} & \frac{1}{4} & 1 \\ 0 & 3 & 0 & \frac{1}{4} & 0 \\ 4 & 3 & 2 & 1 & 0 \\ 0 & 1 & 0 & \frac{1}{3} & 0 \end{bmatrix} \end{array} \begin{array}{l} \text{Sweet rolls} \\ \text{(dozen)} \\ \text{Bread (loaves)} \\ \text{Cake (1)} \\ \text{Pie (1)} \end{array}$$

The cost (in cents) for each ingredient when purchased in large lots and in small lots is given by matrix B.

$$\begin{array}{c} \textit{Cost (in cents)} \\ \text{Large lot} \quad \text{Small lot} \end{array}$$

$$B = \begin{bmatrix} 5 & 5 \\ 8 & 10 \\ 10 & 12 \\ 12 & 15 \\ 5 & 6 \end{bmatrix} \begin{array}{l} \text{Eggs} \\ \text{Flour†} \\ \text{Sugar†} \\ \text{Butter†} \\ \text{Milk†} \end{array}$$

(a) Use matrix multiplication to find a matrix representing the comparative costs per item under the two purchase options.

Suppose a day's orders consist of 20 dozen sweet rolls, 200 loaves of bread, 50 cakes, and 60 pies.

(b) Represent these orders as a 1×4 matrix and use matrix multiplication to write as a matrix the amount of each ingredient required to fill the day's orders.

(c) Use matrix multiplication to find a matrix representing the costs under the two purchase options to fill the day's orders.

38. Let I be the matrix $I = \begin{bmatrix} 1 & 0 \\ 0 & 1 \end{bmatrix}$, and let matrices P, X, and T be defined as in Exercises 32–36.

(a) Find IP, PI, IX.

(b) Without calculating, guess what the matrix IT might be.

(c) Suggest a reason for naming a matrix such as I an identity matrix.

39. Management The EZ Life Company buys three machines costing $90,000, $60,000, and $120,000 respectively. They plan to depreciate the machines over a three-year period using the sum-of-the-years'-digits method, where the company assumes that the machines will have 20% of their value left at the end of three years. Thus they will compute the depreciation of 80% of the cost using rates of 1/2, 1/3, and 1/6 respectively for the three-year period.

(a) Write a column matrix C to represent the cost of these three assets.

(b) To get a matrix which represents 80% of the costs, let

$$P = \begin{bmatrix} .8 & 0 & 0 \\ 0 & .8 & 0 \\ 0 & 0 & .8 \end{bmatrix}.$$

Then PC is the desired matrix. Calculate PC.

(c) Write a row matrix R representing the three rates given above.

(d) Find $(PC)R$. This product matrix (which should be 3×3) represents the depreciation charge for each asset for each year of the three-year period.

40. In Exercise 37, Section 6.3, label the matrices found in parts (a), (b), and (c) respectively X, Y, and Z.

(a) Find the product matrix XY. What do the entries of this matrix represent?

(b) Find the product matrix YZ. What do the entries represent?

In Exercises 41 and 42 show that the equation $AX = B$ represents a linear system of two equations in two unknowns. Solve the system and substitute in the matrix equation to check your results.

41. $A = \begin{bmatrix} 2 & 4 \\ 1 & 3 \end{bmatrix}$, $X = \begin{bmatrix} x_1 \\ x_2 \end{bmatrix}$, $B = \begin{bmatrix} 2 \\ -1 \end{bmatrix}$

42. $A = \begin{bmatrix} 1 & 2 \\ -3 & 5 \end{bmatrix}$, $X = \begin{bmatrix} x_1 \\ x_2 \end{bmatrix}$, $B = \begin{bmatrix} -4 \\ 12 \end{bmatrix}$

---

*Measured in cups.   †Cost per cup.

Use the following matrices to find the matrix products in Exercises 43–49.

$$A = \begin{bmatrix} 2 & 3 & -1 & 5 & 10 \\ 2 & 8 & 7 & 4 & 3 \\ -1 & -4 & -12 & 6 & 8 \\ 2 & 5 & 7 & 1 & 4 \end{bmatrix}$$

$$B = \begin{bmatrix} 9 & 3 & 7 & -6 \\ -1 & 0 & 4 & 2 \\ -10 & -7 & 6 & 9 \\ 8 & 4 & 2 & -1 \\ 2 & -5 & 3 & 7 \end{bmatrix}$$

$$C = \begin{bmatrix} -6 & 8 & 2 & 4 & -3 \\ 1 & 9 & 7 & -12 & 5 \\ 15 & 2 & -8 & 10 & 11 \\ 4 & 7 & 9 & 6 & -2 \\ 1 & 3 & 8 & 23 & 4 \end{bmatrix}$$

$$D = \begin{bmatrix} 5 & -3 & 7 & 9 & 2 \\ 6 & 8 & -5 & 2 & 1 \\ 3 & 7 & -4 & 2 & 11 \\ 5 & -3 & 9 & 4 & -1 \\ 0 & 3 & 2 & 5 & 1 \end{bmatrix}$$

43. CD

44. AC

45. CA

46. DC

47. Is $AC = CA$?

48. Is $CD = DC$?

49. Find $C + D$, $(C + D)B$, CB, DB, and $CB + DB$. Does $(C + D)B = CB + DB$?

6.5 Matrix Inverses

In Section 6.3, a zero matrix was defined with properties similar to those of the real number zero, the identity for addition. The real number 1 is the identity element for multiplication; for any real number a, $a \cdot 1 = 1 \cdot a = a$. In this section, an **identity matrix** I is defined that has properties similar to those of the number 1. This identity matrix is then used to find the multiplicative inverse of any square matrix that has an inverse.

If I is to be the identity matrix, the products AI and IA must both equal A. The 2×2 identity matrix that satisfies these conditions is

$$I = \begin{bmatrix} 1 & 0 \\ 0 & 1 \end{bmatrix}. \quad \boxed{1}$$

① Let $A = \begin{bmatrix} 3 & -2 \\ 4 & -1 \end{bmatrix}$

and $I = \begin{bmatrix} 1 & 0 \\ 0 & 1 \end{bmatrix}$.

Find IA and AI.

Answer:

$IA = \begin{bmatrix} 3 & -2 \\ 4 & -1 \end{bmatrix} = A$

and $AI = \begin{bmatrix} 3 & -2 \\ 4 & -1 \end{bmatrix} = A$.

To check that I, as defined above, is really the 2×2 identity matrix, let

$$A = \begin{bmatrix} a & b \\ c & d \end{bmatrix}.$$

Then AI and IA should both equal A.

$$AI = \begin{bmatrix} a & b \\ c & d \end{bmatrix}\begin{bmatrix} 1 & 0 \\ 0 & 1 \end{bmatrix} = \begin{bmatrix} a(1) + b(0) & a(0) + b(1) \\ c(1) + d(0) & c(0) + d(1) \end{bmatrix} = \begin{bmatrix} a & b \\ c & d \end{bmatrix} = A$$

$$IA = \begin{bmatrix} 1 & 0 \\ 0 & 1 \end{bmatrix}\begin{bmatrix} a & b \\ c & d \end{bmatrix} = \begin{bmatrix} 1(a) + 0(c) & 1(b) + 0(d) \\ 0(a) + 1(c) & 0(b) + 1(d) \end{bmatrix} = \begin{bmatrix} a & b \\ c & d \end{bmatrix} = A$$

This verifies that I has been defined correctly. (It can also be shown that I is the only 2×2 identity matrix.)

The identity matrices for 3×3 matrices and 4×4 matrices respectively are

$$I = \begin{bmatrix} 1 & 0 & 0 \\ 0 & 1 & 0 \\ 0 & 0 & 1 \end{bmatrix} \quad \text{and} \quad I = \begin{bmatrix} 1 & 0 & 0 & 0 \\ 0 & 1 & 0 & 0 \\ 0 & 0 & 1 & 0 \\ 0 & 0 & 0 & 1 \end{bmatrix}.$$

By generalizing, an identity matrix can be found for any n by n matrix: this identity matrix will have 1s on the main diagonal from upper left to lower right, with all other entries equal to 0.

Recall that the multiplicative inverse of the nonzero real number a is $1/a$. The product of a and its multiplicative inverse $1/a$ is 1. Now try a similar thing with matrices: given a matrix A, is there a matrix A^{-1} (read "A-inverse") satisfying both

$$AA^{-1} = I \quad \text{and} \quad A^{-1}A = I?$$

It turns out that this matrix inverse A^{-1} can often be found using the row operations of Section 6.2. Note that the symbol A^{-1} does not mean $1/A$; the symbol A^{-1} is just the notation for the inverse of matrix A. Also, *only square matrices can have inverses*. If an inverse exists, it is unique. That is, any given square matrix has no more than one inverse.

Example 1 Given matrices A and B below, decide if they are inverses.

$$A = \begin{bmatrix} 2 & 3 \\ 1 & 8 \end{bmatrix} \quad B = \begin{bmatrix} -1 & 3 \\ 1 & -2 \end{bmatrix}$$

The matrices are inverses if AB and BA both equal I.

$$AB = \begin{bmatrix} 2 & 3 \\ 1 & 8 \end{bmatrix} \begin{bmatrix} -1 & 3 \\ 1 & -2 \end{bmatrix} = \begin{bmatrix} 1 & 0 \\ 7 & -13 \end{bmatrix} \neq I$$

Since $AB \neq I$, the two matrices are not inverses of each other. **2**

2 Given

$$A = \begin{bmatrix} 1 & 2 \\ 4 & 6 \end{bmatrix}$$

and

$$B = \begin{bmatrix} -3 & 1 \\ 2 & -\frac{1}{2} \end{bmatrix},$$

decide if they are inverses.

Answer:
Yes, since $AB = BA = I$.

The rest of this section shows a method for finding the multiplicative inverse of any $n \times n$ matrix that has an inverse. The method depends on three fundamental row operations which may be performed on any row of a matrix. These row operations are the same ones used on augmented matrices in Section 6.2 to obtain an equivalent system.

The **matrix row operations** are:

1. interchanging any two rows of a matrix;

2. multiplying the elements of any row of a matrix by the same nonzero scalar;

3. adding a multiple of the elements of one row of a matrix to the corresponding elements of another row.

As shown earlier, by the first row operation, the matrix

$$\begin{bmatrix} 1 & 3 & 5 \\ 0 & 1 & 2 \\ 2 & 1 & -2 \end{bmatrix} \quad \text{becomes} \quad \begin{bmatrix} 0 & 1 & 2 \\ 1 & 3 & 5 \\ 2 & 1 & -2 \end{bmatrix} \quad \text{Interchange } R_1, R_2$$

by interchanging the first two rows. Row three is left unchanged.
The second row operation allows us to change

$$\begin{bmatrix} 1 & 3 & 5 \\ 0 & 1 & 2 \\ 2 & 1 & -2 \end{bmatrix} \quad \text{to} \quad \begin{bmatrix} -2 & -6 & -10 \\ 0 & 1 & 2 \\ 2 & 1 & -2 \end{bmatrix} \quad -2R_1$$

by multiplying the elements of the first row of the original matrix by -2. Here rows two and three are left unchanged. Using the third row operation,

$$\begin{bmatrix} 1 & 3 & 5 \\ 0 & 1 & 2 \\ 2 & 1 & -2 \end{bmatrix} \quad \text{becomes} \quad \begin{bmatrix} 1 & 3 & 5 \\ 0 & 1 & 2 \\ 0 & -5 & -12 \end{bmatrix} \quad -2R_1 + R_3$$

by first multiplying each element in the first row of the original matrix by -2 and then adding the results to the corresponding elements in the third row of that matrix. Rows one and two are left unchanged even though the elements of row one were used to transform row three. ③

③ Let

$$A = \begin{bmatrix} 6 & 2 & 1 \\ -1 & 3 & 4 \end{bmatrix}$$

(a) Interchange rows one and two.
(b) Multiply row one by $-1/2$.
(c) Add 3 times row one to row two.
(d) Find $3R_2$
(e) Find $2R_1 + R_2$

Answer:

(a) $\begin{bmatrix} -1 & 3 & 4 \\ 6 & 2 & 1 \end{bmatrix}$

(b) $\begin{bmatrix} -3 & -1 & -\frac{1}{2} \\ -1 & 3 & 4 \end{bmatrix}$

(c) $\begin{bmatrix} 6 & 2 & 1 \\ 17 & 9 & 7 \end{bmatrix}$

(d) $\begin{bmatrix} 6 & 2 & 1 \\ -3 & 9 & 12 \end{bmatrix}$

(e) $\begin{bmatrix} 6 & 2 & 1 \\ 11 & 7 & 6 \end{bmatrix}$

The augmented matrices written earlier, such as

$$\begin{bmatrix} 1 & 2 & | & 2 \\ 3 & 4 & | & 5 \end{bmatrix},$$

are really a combination of two matrices written as one. The vertical bar separates the two original matrices

$$\begin{bmatrix} 1 & 2 \\ 3 & 4 \end{bmatrix} \quad \text{and} \quad \begin{bmatrix} 2 \\ 5 \end{bmatrix}.$$

As an example of finding the multiplicative inverse of a matrix, let us look for the inverse of

$$A = \begin{bmatrix} 2 & 4 \\ 1 & -1 \end{bmatrix}.$$

Let the unknown inverse matrix be

$$A^{-1} = \begin{bmatrix} x & y \\ z & w \end{bmatrix}.$$

By the definition of matrix inverse, $AA^{-1} = I$, or

$$AA^{-1} = \begin{bmatrix} 2 & 4 \\ 1 & -1 \end{bmatrix} \begin{bmatrix} x & y \\ z & w \end{bmatrix} = \begin{bmatrix} 1 & 0 \\ 0 & 1 \end{bmatrix}.$$

Use matrix multiplication to get

$$\begin{bmatrix} 2x + 4z & 2y + 4w \\ x - z & y - w \end{bmatrix} = \begin{bmatrix} 1 & 0 \\ 0 & 1 \end{bmatrix}.$$

Setting corresponding elements equal gives the system of equations

$$2x + 4z = 1 \tag{1}$$
$$2y + 4w = 0 \tag{2}$$
$$x - z = 0 \tag{3}$$
$$y - w = 1. \tag{4}$$

Since equations (1) and (3) involve only x and z, while equations (2) and (4) involve only y and w, these four equations lead to two systems of equations,

$$\begin{array}{ccc} 2x + 4z = 1 & & 2y + 4w = 0 \\ x - z = 0 & \text{and} & y - w = 1. \end{array}$$

Writing the two systems as augmented matrices gives

$$\begin{bmatrix} 2 & 4 & | & 1 \\ 1 & -1 & | & 0 \end{bmatrix} \quad \text{and} \quad \begin{bmatrix} 2 & 4 & | & 0 \\ 1 & -1 & | & 1 \end{bmatrix}.$$

Each of these systems can be solved by the Gauss-Jordan method. However, since the elements to the left of the vertical bar are identical, the two systems can be combined into one matrix

$$\begin{bmatrix} 2 & 4 & | & 1 & 0 \\ 1 & -1 & | & 0 & 1 \end{bmatrix}$$

and solved simultaneously as follows. Interchange the two rows to get a 1 in the upper left corner.

$$\begin{bmatrix} 1 & -1 & | & 0 & 1 \\ 2 & 4 & | & 1 & 0 \end{bmatrix} \qquad \text{Interchange } R_1, R_2$$

Multiply row one by -2 and add the results to row two to get

$$\begin{bmatrix} 1 & -1 & | & 0 & 1 \\ 0 & 6 & | & 1 & -2 \end{bmatrix}. \qquad -2R_1 + R_2$$

Now, to get a 1 in the second row, second column position, multiply row two by 1/6.

$$\begin{bmatrix} 1 & -1 & | & 0 & 1 \\ 0 & 1 & | & \frac{1}{6} & -\frac{1}{3} \end{bmatrix} \qquad \frac{1}{6}R_2$$

Finally, add row two to row one to get a 0 in the second column above the 1.

$$\begin{bmatrix} 1 & 0 & | & \frac{1}{6} & \frac{2}{3} \\ 0 & 1 & | & \frac{1}{6} & -\frac{1}{3} \end{bmatrix} \qquad R_2 + R_1$$

The numbers in the first column to the right of the vertical bar give the values of x and z. The second column gives the values of y and w. That is,

$$\left[\begin{array}{cc|cc} 1 & 0 & x & y \\ 0 & 1 & z & w \end{array}\right] = \left[\begin{array}{cc|cc} 1 & 0 & \frac{1}{6} & \frac{2}{3} \\ 0 & 1 & \frac{1}{6} & -\frac{1}{3} \end{array}\right]$$

so that

$$A^{-1} = \begin{bmatrix} x & y \\ z & w \end{bmatrix} = \begin{bmatrix} \frac{1}{6} & \frac{2}{3} \\ \frac{1}{6} & -\frac{1}{3} \end{bmatrix}.$$

Check by multiplying A and A^{-1}. The result should be I.

$$AA^{-1} = \begin{bmatrix} 2 & 4 \\ 1 & -1 \end{bmatrix}\begin{bmatrix} \frac{1}{6} & \frac{2}{3} \\ \frac{1}{6} & -\frac{1}{3} \end{bmatrix} = \begin{bmatrix} \frac{1}{3}+\frac{2}{3} & \frac{4}{3}-\frac{4}{3} \\ \frac{1}{6}-\frac{1}{6} & \frac{2}{3}+\frac{1}{3} \end{bmatrix} = \begin{bmatrix} 1 & 0 \\ 0 & 1 \end{bmatrix} = I.$$

Verify that $A^{-1}A = I$, also. [4]

This procedure for finding the inverse of a matrix can be generalized as follows.

[4] (a) Find A^{-1} if

$$A = \begin{bmatrix} 2 & 2 \\ 4 & 1 \end{bmatrix}.$$

(b) Check your answer by finding AA^{-1} and $A^{-1}A$.

Answer:

(a) $\begin{bmatrix} -\frac{1}{6} & \frac{1}{3} \\ \frac{2}{3} & -\frac{1}{3} \end{bmatrix}$

(b) Both equal $\begin{bmatrix} 1 & 0 \\ 0 & 1 \end{bmatrix}.$

To obtain an **inverse matrix** A^{-1} for any $n \times n$ matrix A for which A^{-1} exists, follow these steps.

1. Form the augmented matrix $[A\,|\,I]$, where I is the $n \times n$ identity matrix.

2. Perform row operations on $[A\,|\,I]$ to get a matrix of the form $[I\,|\,B]$.

3. Matrix B is A^{-1}.

Example 2 Find A^{-1} if $A = \begin{bmatrix} 1 & 0 & 1 \\ 2 & -2 & -1 \\ 3 & 0 & 0 \end{bmatrix}$.

First write the augmented matrix $[A\,|\,I]$.

$$[A\,|\,I] = \left[\begin{array}{ccc|ccc} 1 & 0 & 1 & 1 & 0 & 0 \\ 2 & -2 & -1 & 0 & 1 & 0 \\ 3 & 0 & 0 & 0 & 0 & 1 \end{array}\right]$$

The augmented matrix already has 1 in the upper left-hand corner as needed, so begin by selecting the row operation which will result in a 0 for the first element in row two. Multiply row one by -2 and add the result to row two. This gives

$$\left[\begin{array}{ccc|ccc} 1 & 0 & 1 & 1 & 0 & 0 \\ 0 & -2 & -3 & -2 & 1 & 0 \\ 3 & 0 & 0 & 0 & 0 & 1 \end{array}\right]. \qquad -2R_1 + R_2$$

Get 0 for the first element in row three by multiplying row one by -3 and adding to row three. ⑤

⑤ (a) Complete this step.
(b) Write this row transformation as __.

Answer:
(a)
$$\begin{bmatrix} 1 & 0 & 1 & | & 1 & 0 & 0 \\ 0 & -2 & -3 & | & -2 & 1 & 0 \\ 0 & 0 & -3 & | & -3 & 0 & 1 \end{bmatrix}$$

(b) $-3R_1 + R_3$

Get 1 for the second element in row two by multiplying row two by $-1/2$, obtaining the new matrix

$$\begin{bmatrix} 1 & 0 & 1 & | & 1 & 0 & 0 \\ 0 & 1 & \frac{3}{2} & | & 1 & -\frac{1}{2} & 0 \\ 0 & 0 & -3 & | & -3 & 0 & 1 \end{bmatrix}. \qquad -\frac{1}{2}R_2$$

Get 1 for the third element in row three by multiplying row three by $-1/3$, with the result

$$\begin{bmatrix} 1 & 0 & 1 & | & 1 & 0 & 0 \\ 0 & 1 & \frac{3}{2} & | & 1 & -\frac{1}{2} & 0 \\ 0 & 0 & 1 & | & 1 & 0 & -\frac{1}{3} \end{bmatrix}. \qquad -\frac{1}{3}R_3$$

Now get 0 for the third element in row one by multiplying row three by -1 and adding to row one. ⑥

⑥ (a) Complete this step.
(b) Write this row transformation as __.

Answer:
(a)
$$\begin{bmatrix} 1 & 0 & 0 & | & 0 & 0 & \frac{1}{3} \\ 0 & 1 & \frac{3}{2} & | & 1 & -\frac{1}{2} & 0 \\ 0 & 0 & 1 & | & 1 & 0 & -\frac{1}{3} \end{bmatrix}$$

(b) $-1R_3 + R_1$

Get 0 for the third element in row two by multiplying row three by $-3/2$ and adding to row two.

$$\begin{bmatrix} 1 & 0 & 0 & | & 0 & 0 & \frac{1}{3} \\ 0 & 1 & 0 & | & -\frac{1}{2} & -\frac{1}{2} & \frac{1}{2} \\ 0 & 0 & 1 & | & 1 & 0 & -\frac{1}{3} \end{bmatrix} \qquad -\frac{3}{2}R_3 + R_2$$

This last transformation gives the desired inverse:

$$A^{-1} = \begin{bmatrix} 0 & 0 & \frac{1}{3} \\ -\frac{1}{2} & -\frac{1}{2} & \frac{1}{2} \\ 1 & 0 & -\frac{1}{3} \end{bmatrix}.$$

Confirm this by forming the products $A^{-1}A$ and AA^{-1}. Each should equal I. ⑦

⑦ (a) Find A^{-1} if
$$A = \begin{bmatrix} 2 & 8 \\ -1 & -5 \end{bmatrix}.$$

(b) Check this answer by finding AA^{-1} and $A^{-1}A$.

Answer:
(a) $\begin{bmatrix} \frac{5}{2} & 4 \\ -\frac{1}{2} & -1 \end{bmatrix}$

(b) Each product equals
$$\begin{bmatrix} 1 & 0 \\ 0 & 1 \end{bmatrix}.$$

Example 3 Find A^{-1} if $A = \begin{bmatrix} 2 & -4 \\ 1 & -2 \end{bmatrix}$.

Using row operations to transform the first column of the augmented matrix

$$\begin{bmatrix} 2 & -4 & | & 1 & 0 \\ 1 & -2 & | & 0 & 1 \end{bmatrix}$$

results in the following matrices.

$$\begin{bmatrix} 1 & -2 & | & \frac{1}{2} & 0 \\ 1 & -2 & | & 0 & 1 \end{bmatrix} \qquad \frac{1}{2}R_1$$

$$\begin{bmatrix} 1 & -2 & | & \frac{1}{2} & 0 \\ 0 & 0 & | & -\frac{1}{2} & 1 \end{bmatrix} \qquad -1R_1 + R_2 \cdot$$

8 Find the inverse, if it exists, of each matrix below.

(a) $\begin{bmatrix} 8 & 4 \\ 6 & 3 \end{bmatrix}$ (b) $\begin{bmatrix} 0 & 1 \\ 1 & 0 \end{bmatrix}$

(c) $\begin{bmatrix} 6 & -1 & 2 \\ 4 & 1 & 3 \\ -3 & \frac{1}{2} & -1 \end{bmatrix}$

Answer:

(a) and (c) have no inverse;

(b) is its own inverse.

At this point, the matrix should be changed so that the second element of row two will be 1. Since that element is now 0, there is no way to complete the desired transformation. What is wrong? Remember, near the beginning of this section we mentioned that some matrices do not have inverses. Matrix A is an example of a matrix that has no inverse. In this case there is no matrix A^{-1} such that $AA^{-1} = A^{-1}A = I$. **8**

6.5 Exercises

Use row operation 3 to change each of the following matrices as indicated.

1. $\begin{bmatrix} 2 & 4 \\ 4 & 7 \end{bmatrix}$; $-2R_1 + R_2$

2. $\begin{bmatrix} -1 & 4 \\ 7 & 0 \end{bmatrix}$; $7R_1 + R_2$

3. $\begin{bmatrix} 1 & \frac{1}{5} \\ 5 & 2 \end{bmatrix}$; $-5R_1 + R_2$

4. $\begin{bmatrix} 5 & 7 \\ 2 & -4 \end{bmatrix}$; $\frac{4}{7}R_1 + R_2$

5. $\begin{bmatrix} 1 & 5 & 6 \\ -2 & 3 & -1 \\ 4 & 7 & 0 \end{bmatrix}$; $2R_1 + R_2$

6. $\begin{bmatrix} 2 & 5 & 6 \\ 4 & -1 & 2 \\ 3 & 7 & 1 \end{bmatrix}$; $-6R_3 + R_1$

7. $\begin{bmatrix} -3 & 1 & -4 \\ 2 & 1 & 3 \\ -7 & 5 & 2 \end{bmatrix}$; $-5R_2 + R_3$

8. $\begin{bmatrix} 4 & 10 & -8 \\ 7 & 4 & 3 \\ -1 & 1 & 0 \end{bmatrix}$; $-4R_3 + R_2$

Decide whether or not the given matrices are inverses of each other. (Check to see if their product is I. See Example 1.)

9. $\begin{bmatrix} 2 & 3 \\ 1 & 1 \end{bmatrix}$ and $\begin{bmatrix} -1 & 3 \\ 1 & -2 \end{bmatrix}$

10. $\begin{bmatrix} 5 & 7 \\ 2 & 3 \end{bmatrix}$ and $\begin{bmatrix} 3 & -7 \\ -2 & 5 \end{bmatrix}$

11. $\begin{bmatrix} 2 & 1 \\ 3 & 2 \end{bmatrix}$ and $\begin{bmatrix} 2 & 1 \\ -3 & 2 \end{bmatrix}$

12. $\begin{bmatrix} -1 & 2 \\ 3 & -5 \end{bmatrix}$ and $\begin{bmatrix} -5 & -2 \\ -3 & -1 \end{bmatrix}$

13. $\begin{bmatrix} 1 & 2 & 0 \\ 0 & 1 & 0 \\ 0 & 1 & 0 \end{bmatrix}$ and $\begin{bmatrix} 1 & -2 & 0 \\ 0 & 1 & 0 \\ 0 & -1 & 1 \end{bmatrix}$

14. $\begin{bmatrix} 0 & 1 & 0 \\ 0 & 0 & -2 \\ 1 & -1 & 0 \end{bmatrix}$ and $\begin{bmatrix} 1 & 0 & 1 \\ 1 & 0 & 0 \\ 0 & -1 & 0 \end{bmatrix}$

15. $\begin{bmatrix} 1 & 3 & 3 \\ 1 & 4 & 3 \\ 1 & 3 & 4 \end{bmatrix}$ and $\begin{bmatrix} 7 & -3 & -3 \\ -1 & 1 & 0 \\ -1 & 0 & 1 \end{bmatrix}$

16. $\begin{bmatrix} -1 & 0 & 2 \\ 3 & 1 & 0 \\ 0 & 2 & -3 \end{bmatrix}$ and $\begin{bmatrix} -\frac{1}{5} & \frac{4}{15} & -\frac{2}{15} \\ \frac{3}{5} & \frac{1}{5} & \frac{2}{5} \\ \frac{2}{5} & \frac{2}{15} & -\frac{1}{15} \end{bmatrix}$

Find the inverse, if it exists, for each of the following matrices. (See Examples 2–3.)

17. $\begin{bmatrix} 1 & -1 \\ 2 & 0 \end{bmatrix}$

18. $\begin{bmatrix} -1 & 2 \\ -2 & -1 \end{bmatrix}$

19. $\begin{bmatrix} 3 & -1 \\ -5 & 2 \end{bmatrix}$

20. $\begin{bmatrix} -1 & -2 \\ 3 & 4 \end{bmatrix}$

21. $\begin{bmatrix} -6 & 4 \\ -3 & 2 \end{bmatrix}$

22. $\begin{bmatrix} 5 & 10 \\ -3 & -6 \end{bmatrix}$

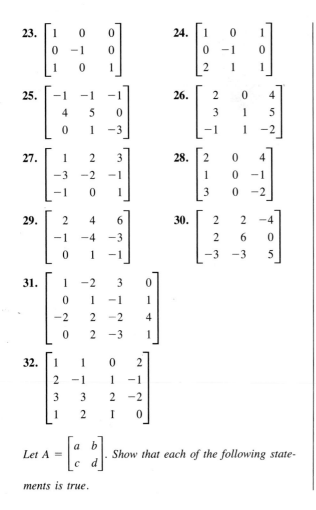

23. $\begin{bmatrix} 1 & 0 & 0 \\ 0 & -1 & 0 \\ 1 & 0 & 1 \end{bmatrix}$

24. $\begin{bmatrix} 1 & 0 & 1 \\ 0 & -1 & 0 \\ 2 & 1 & 1 \end{bmatrix}$

25. $\begin{bmatrix} -1 & -1 & -1 \\ 4 & 5 & 0 \\ 0 & 1 & -3 \end{bmatrix}$

26. $\begin{bmatrix} 2 & 0 & 4 \\ 3 & 1 & 5 \\ -1 & 1 & -2 \end{bmatrix}$

27. $\begin{bmatrix} 1 & 2 & 3 \\ -3 & -2 & -1 \\ -1 & 0 & 1 \end{bmatrix}$

28. $\begin{bmatrix} 2 & 0 & 4 \\ 1 & 0 & -1 \\ 3 & 0 & -2 \end{bmatrix}$

29. $\begin{bmatrix} 2 & 4 & 6 \\ -1 & -4 & -3 \\ 0 & 1 & -1 \end{bmatrix}$

30. $\begin{bmatrix} 2 & 2 & -4 \\ 2 & 6 & 0 \\ -3 & -3 & 5 \end{bmatrix}$

31. $\begin{bmatrix} 1 & -2 & 3 & 0 \\ 0 & 1 & -1 & 1 \\ -2 & 2 & -2 & 4 \\ 0 & 2 & -3 & 1 \end{bmatrix}$

32. $\begin{bmatrix} 1 & 1 & 0 & 2 \\ 2 & -1 & 1 & -1 \\ 3 & 3 & 2 & -2 \\ 1 & 2 & 1 & 0 \end{bmatrix}$

Let $A = \begin{bmatrix} a & b \\ c & d \end{bmatrix}$. *Show that each of the following statements is true.*

33. $IA = A$ **34.** $AI = A$ **35.** $A \cdot O = O$

36. Find A^{-1}. (Assume $ad - bc \neq 0$.) Show that $AA^{-1} = I$.

37. Show that $A^{-1}A = I$.

38. Using the definitions and properties listed in this section, show that for square matrices A and B of the same order, if $AB = O$ and if A^{-1} exists, then $B = O$.

Use matrices C and D to find the inverses in Exercises 39–42.

$$C = \begin{bmatrix} -6 & 8 & 2 & 4 & -3 \\ 1 & 9 & 7 & -12 & 5 \\ 15 & 2 & -8 & 10 & 11 \\ 4 & 7 & 9 & 6 & -2 \\ 1 & 3 & 8 & 23 & 4 \end{bmatrix}$$

$$D = \begin{bmatrix} 5 & -3 & 7 & 9 & 2 \\ 6 & 8 & -5 & 2 & 1 \\ 3 & 7 & -4 & 2 & 11 \\ 5 & -3 & 9 & 4 & -1 \\ 0 & 3 & 2 & 5 & 1 \end{bmatrix}$$

39. C^{-1} **40.** $(CD)^{-1}$

41. D^{-1} **42.** Is $C^{-1}D^{-1} = (CD)^{-1}$?

43. Find $C^{-1}C$. This product *should* equal the identity matrix I, but doesn't *quite* equal I, because of **round-off error**.

44. Find DD^{-1}. Does this product exactly equal I?

6.6 Applications of Matrices

This section gives a variety of applications of matrices.

Solving Systems with Matrices Matrices were used to solve systems of linear equations by the Gauss-Jordan method in Section 6.2. Another way to use matrices to solve linear systems is by writing the system as a matrix equation $AX = B$, where A is the matrix of the coefficients of the variables of the system, X is the matrix of the variables, and B is the matrix of the constants. Matrix A is called the **coefficient matrix**.

To solve the matrix equation $AX = B$, first see if A^{-1} exists. Assuming A^{-1} exists and using the facts that $A^{-1}A = I$ and $IX = X$ along with the associative property of multiplication of matrices gives

$$AX = B$$
$$A^{-1}(AX) = A^{-1}B \qquad \text{Multiply both sides by } A^{-1}$$
$$(A^{-1}A)X = A^{-1}B$$
$$IX = A^{-1}B$$
$$X = A^{-1}B.$$

When multiplying by matrices on both sides of a matrix equation, be careful to multiply in the same order on both sides of the equation, since multiplication of matrices is not commutative (unlike multiplication of real numbers). In summary,

> solve a system of equations $AX = B$, where A is the matrix of coefficients, X is the matrix of variables, and B is the matrix of constants, by first finding A^{-1}. Then, if A^{-1} exists, $X = A^{-1}B$.

This method is most practical in cases where several systems with the same coefficient matrix, but different constants, are to be solved. Then just one inverse matrix must be found.

Example 1 Use the inverse of the coefficient matrix to solve the linear system

$$2x - 3y = 4$$
$$x + 5y = 2.$$

Use the coefficient matrix of the system, together with the matrix of variables and the matrix of constants, to represent the system as a matrix equation.

$$A = \begin{bmatrix} 2 & -3 \\ 1 & 5 \end{bmatrix}, \qquad X = \begin{bmatrix} x \\ y \end{bmatrix}, \qquad \text{and} \qquad B = \begin{bmatrix} 4 \\ 2 \end{bmatrix}$$

The system can then be written in matrix form as the equation $AX = B$ since

$$AX = \begin{bmatrix} 2 & -3 \\ 1 & 5 \end{bmatrix} \begin{bmatrix} x \\ y \end{bmatrix} = \begin{bmatrix} 2x - 3y \\ x + 5y \end{bmatrix} = \begin{bmatrix} 4 \\ 2 \end{bmatrix} = B.$$

Solve the system by first finding A^{-1}. [1]

From the work at the side,

$$A^{-1} = \begin{bmatrix} \frac{5}{13} & \frac{3}{13} \\ -\frac{1}{13} & \frac{2}{13} \end{bmatrix}.$$

Next, find the product $A^{-1}B$.

$$A^{-1}B = \begin{bmatrix} \frac{5}{13} & \frac{3}{13} \\ -\frac{1}{13} & \frac{2}{13} \end{bmatrix} \begin{bmatrix} 4 \\ 2 \end{bmatrix} = \begin{bmatrix} 2 \\ 0 \end{bmatrix}$$

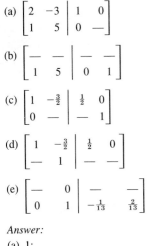

[1] Find A^{-1}, using the following steps.

(a) $\begin{bmatrix} 2 & -3 & | & 1 & 0 \\ 1 & 5 & | & 0 & - \end{bmatrix}$

(b) $\begin{bmatrix} - & - & | & - & - \\ 1 & 5 & | & 0 & 1 \end{bmatrix}$

(c) $\begin{bmatrix} 1 & -\frac{3}{2} & | & \frac{1}{2} & 0 \\ 0 & - & | & - & 1 \end{bmatrix}$

(d) $\begin{bmatrix} 1 & -\frac{3}{2} & | & \frac{1}{2} & 0 \\ - & 1 & | & - & - \end{bmatrix}$

(e) $\begin{bmatrix} - & 0 & | & - & - \\ 0 & 1 & | & -\frac{1}{13} & \frac{2}{13} \end{bmatrix}$

Answer:

(a) 1;

(b) 1, $-3/2$, 1/2, 0;

(c) $\frac{13}{2}$, $-\frac{1}{2}$;

(d) 0, $-\frac{1}{13}$, $\frac{2}{13}$;

(e) 1, 5/13, 3/13

Since $X = A^{-1}B$,

$$X = \begin{bmatrix} x \\ y \end{bmatrix} = \begin{bmatrix} 2 \\ 0 \end{bmatrix}.$$

The solution of the system is $(2, 0)$.

2 (a) Write the matrix of coefficients, the matrix of variables, and the matrix of constants for the system

$$2x + 6y = -14$$
$$-x - 2y = 3.$$

(b) Solve the system in part (a) by using the inverse of the coefficient matrix.

Answer:

(a) $A = \begin{bmatrix} 2 & 6 \\ -1 & -2 \end{bmatrix}$,

$X = \begin{bmatrix} x \\ y \end{bmatrix}$,

$B = \begin{bmatrix} -14 \\ 3 \end{bmatrix}$

(b) $(5, -4)$

Input-Output Models An interesting application of matrix theory to economics was developed by Nobel prize winner Wassily Leontief. His matrix models for studying the interdependencies in an economy are called *input-output* models. In practice these models are very complicated with many variables. We shall discuss only simple examples with a few variables.

Input-output models are concerned with the production and flow of goods (and perhaps services). In an economy with n basic commodities, or sectors, the production of each commodity uses some (perhaps all) of the commodities in the economy as inputs. The amounts of each commodity used in the production of one unit of each commodity can be written as an $n \times n$ matrix A, called the **technological** or **input-output matrix** of the economy.

Example 2 Suppose a simplified economy involves just three commodity categories: agriculture, manufacturing, and transportation, all in appropriate units. Production of one unit of agriculture requires 1/2 unit of manufacturing and 1/4 unit of transportation. Production of one unit of manufacturing requires 1/4 unit of agriculture and 1/4 unit of transportation; while production of one unit of transportation requires 1/3 unit of agriculture and 1/4 unit of manufacturing. Write the input-output matrix of this economy.

The matrix is shown below.

		Output		
		Agriculture	Manufacturing	Transportation
Input	Agriculture	0	$\frac{1}{4}$	$\frac{1}{3}$
	Manufacturing	$\frac{1}{2}$	0	$\frac{1}{4}$
	Transportation	$\frac{1}{4}$	$\frac{1}{4}$	0

$= A$

3 Write a 2 × 2 technological matrix in which one unit of electricity requires 1/2 unit of water and 1/3 unit of electricity, while one unit of water requires no water but 1/4 unit of electricity.

Answer:

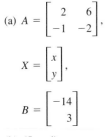

The first column of the input-output matrix represents the amount of each of the three commodities consumed in the production of one unit of agriculture. The second column gives the corresponding amounts required to produce one unit of manufacturing, and the last column gives the amounts needed to produce one unit of transportation. (Although it is unrealistic perhaps that production of one unit of a commodity requires none of that commodity, the simpler matrix involved is useful for our purposes.) **3**

Another matrix used with the input-output matrix is a matrix giving the amount of each commodity produced, called the **production matrix,** or the **vector of gross output.** In an economy producing n commodities, the production matrix can be represented by a column matrix X with entries $x_1, x_2, x_3, \ldots, x_n$.

Example 3 In Example 2, suppose the production matrix is

$$X = \begin{bmatrix} 60 \\ 52 \\ 48 \end{bmatrix}.$$

Then 60 units of agriculture, 52 units of manufacturing and 48 units of transportation are produced. As 1/4 unit of agriculture is used for each unit of manufacturing produced, 1/4 × 52 units of agriculture must be used up in the "production" of manufacturing. Similarly, 1/3 × 48 units of agriculture will be used up in the "production" of transportation. Thus 13 + 16 = 29 units of agriculture are used for production in the economy. Look again at the matrices A and X. Since X gives the number of units of each commodity produced and A gives the amount (in units) of each commodity used to produce one unit of the various commodities, the matrix product AX gives the amount of each commodity used up in the production process.

$$AX = \begin{bmatrix} 0 & \frac{1}{4} & \frac{1}{3} \\ \frac{1}{2} & 0 & \frac{1}{4} \\ \frac{1}{4} & \frac{1}{4} & 0 \end{bmatrix} \begin{bmatrix} 60 \\ 52 \\ 48 \end{bmatrix} = \begin{bmatrix} 29 \\ 42 \\ 28 \end{bmatrix}$$

This product shows that 29 units of agriculture, 42 units of manufacturing, and 28 units of transportation are used to produce 60 units of agriculture, 52 units of manufacturing, and 48 units of transportation. ■

We have seen that the matrix product AX represents the amount of each commodity used in the production process. The remainder (if any) must be enough to satisfy the demand for the various commodities from outside the production system. In an n-commodity economy, this demand can be represented by a **demand matrix** D with entries $d_1, d_2, \ldots, d_n$. The difference between the production matrix, X, and the amount, AX, used in the production process must equal the demand, D, or

$$D = X - AX.$$

In Example 3,

$$D = \begin{bmatrix} 60 \\ 52 \\ 48 \end{bmatrix} - \begin{bmatrix} 29 \\ 42 \\ 28 \end{bmatrix} = \begin{bmatrix} 31 \\ 10 \\ 20 \end{bmatrix}.$$

This result shows that production of 60 units of agriculture, 52 units of manufacturing, and 48 units of transportation would satisfy a demand of 31, 10, and 20 units of each, respectively. ④

In practice, A and D usually are known and X must be found. That is, we need to decide what amounts of production are necessary to satisfy the required demands. Matrix algebra can be used to solve the equation $D = X - AX$ for X.

$$D = X - AX$$
$$D = IX - AX$$
$$D = (I - A)X$$

④ (a) Write a 2 × 1 matrix X to represent gross production of 9000 units of electricity and 12,000 units of water.
(b) Find AX using A from the last problem.
(c) Find D using $D = X - AX$.

Answer:

(a) $\begin{bmatrix} 9000 \\ 12,000 \end{bmatrix}$

(b) $\begin{bmatrix} 6000 \\ 4500 \end{bmatrix}$

(c) $\begin{bmatrix} 3000 \\ 7500 \end{bmatrix}$

If the matrix $I - A$ has an inverse, then

$$X = (I - A)^{-1}D.$$

Example 4 Suppose, in the 3-commodity economy of Examples 2 and 3, there is a demand for 516 units of agriculture, 258 units of manufacturing, and 129 units of transportation. What should production of each commodity be?

The demand matrix is

$$D = \begin{bmatrix} 516 \\ 258 \\ 129 \end{bmatrix}.$$

Find the production matrix by first calculating $I - A$.

$$I - A = \begin{bmatrix} 1 & 0 & 0 \\ 0 & 1 & 0 \\ 0 & 0 & 1 \end{bmatrix} - \begin{bmatrix} 0 & \frac{1}{4} & \frac{1}{3} \\ \frac{1}{2} & 0 & \frac{1}{4} \\ \frac{1}{4} & \frac{1}{4} & 0 \end{bmatrix} = \begin{bmatrix} 1 & -\frac{1}{4} & -\frac{1}{3} \\ -\frac{1}{2} & 1 & -\frac{1}{4} \\ -\frac{1}{4} & -\frac{1}{4} & 1 \end{bmatrix}$$

Using row operations, find the inverse of $I - A$.

$$(I - A)^{-1} = \begin{bmatrix} 1.40 & .50 & .59 \\ .84 & 1.36 & .62 \\ .56 & .47 & 1.30 \end{bmatrix}$$

(The entries are rounded to two decimal places.) Since $X = (I - A)^{-1}D$,

$$X = \begin{bmatrix} 1.40 & .50 & .59 \\ .84 & 1.36 & .62 \\ .56 & .47 & 1.30 \end{bmatrix} \begin{bmatrix} 516 \\ 258 \\ 129 \end{bmatrix} = \begin{bmatrix} 928 \\ 864 \\ 578 \end{bmatrix}$$

(rounded to the nearest whole numbers).

From the last result, we see that production of 928 units of agriculture, 864 of manufacturing, and 578 of transportation is required to satisfy demands of 516, 258, and 129 units respectively. ■

Example 5 An economy depends on two basic products, wheat and oil. To produce one metric ton of wheat requires .25 metric tons of wheat and .33 metric tons of oil. Production of one metric ton of oil consumes .08 metric tons of wheat and .11 metric tons of oil. Find the production which will satisfy a demand of 500 metric tons of wheat and 1000 metric tons of oil.

The input-output matrix, A, and the matrix $I - A$, are

$$A = \begin{bmatrix} .25 & .08 \\ .33 & .11 \end{bmatrix} \quad \text{and} \quad I - A = \begin{bmatrix} .75 & -.08 \\ -.33 & .89 \end{bmatrix}.$$

Next, calculate $(I - A)^{-1}$.

$$(I - A)^{-1} = \begin{bmatrix} 1.39 & .13 \\ .51 & 1.17 \end{bmatrix} \quad \text{(rounded)}$$

5 A simple economy depends on just two products, beer and pretzels.
(a) Suppose 1/2 units of beer and 1/2 unit of pretzels are needed to make 1 unit of beer, and 3/4 unit of beer is needed to make 1 unit of pretzels. Write the technological matrix A for the economy.
(b) Find $I - A$.
(c) Find $(I - A)^{-1}$.
(d) Find the gross production X that will be needed to get a net production of

$$Y = \begin{bmatrix} 100 \\ 1000 \end{bmatrix}.$$

Answer:

(a) $\begin{bmatrix} \frac{1}{2} & \frac{3}{4} \\ \frac{1}{2} & 0 \end{bmatrix}$

(b) $\begin{bmatrix} \frac{1}{2} & -\frac{3}{4} \\ -\frac{1}{2} & 1 \end{bmatrix}$

(c) $\begin{bmatrix} 8 & 6 \\ 4 & 4 \end{bmatrix}$

(d) $\begin{bmatrix} 6800 \\ 4400 \end{bmatrix}$

6 Write the message "when" using 2×1 matrices.

Answer:

$\begin{bmatrix} 23 \\ 8 \end{bmatrix}, \begin{bmatrix} 5 \\ 14 \end{bmatrix}$

To find the production matrix X, use the equation $X = (I - A)^{-1}D$, with

$$D = \begin{bmatrix} 500 \\ 1000 \end{bmatrix}.$$

The production matrix is

$$X = \begin{bmatrix} 1.39 & .13 \\ .51 & 1.17 \end{bmatrix} \begin{bmatrix} 500 \\ 1000 \end{bmatrix} = \begin{bmatrix} 815 \\ 1425 \end{bmatrix}.$$

Production of 815 metric tons of wheat and 1425 metric tons of oil is required to satisfy the indicated demand. 5

Code Theory Governments need sophisticated methods of coding and decoding messages. One example of such an advanced code uses matrix theory. Such a code takes the letters in the words and divides them into groups. (Each space between words is treated as a letter; punctuation is disregarded.) Then, numbers are assigned to the letters of the alphabet. For our purposes, let the letter a correspond to 1, b to 2, and so on. Let the number 27 correspond to a space between words.

For example, the message

mathematics is for the birds

can be divided into groups of three letters each.

    mat   hem   ati   cs–   is–   for   –th   e–b   ird   s––

(We used – to represent a space between words.) We now write a column matrix for each group of three symbols using the corresponding numbers, as determined above, instead of letters. For example, the letters *mat* can be encoded as

$$\begin{bmatrix} 13 \\ 1 \\ 20 \end{bmatrix}.$$

The coded message then consists of the 3×1 column matrices:

$$\begin{bmatrix} 13 \\ 1 \\ 20 \end{bmatrix}, \begin{bmatrix} 8 \\ 5 \\ 13 \end{bmatrix}, \begin{bmatrix} 1 \\ 20 \\ 9 \end{bmatrix}, \begin{bmatrix} 3 \\ 19 \\ 27 \end{bmatrix}, \begin{bmatrix} 9 \\ 19 \\ 27 \end{bmatrix}, \begin{bmatrix} 6 \\ 15 \\ 18 \end{bmatrix}, \begin{bmatrix} 27 \\ 20 \\ 8 \end{bmatrix}, \begin{bmatrix} 5 \\ 27 \\ 2 \end{bmatrix}, \begin{bmatrix} 9 \\ 18 \\ 4 \end{bmatrix}, \begin{bmatrix} 19 \\ 27 \\ 27 \end{bmatrix}.$$ 6

We can further complicate the code by choosing a matrix which has an inverse (in this case a 3×3 matrix, call it M) and finding the products of this matrix and each of the above column matrices. The size of each group, the assignment of numbers to letters, and the choice of matrix M must all be predetermined.

Suppose we choose

$$M = \begin{bmatrix} 1 & 3 & 3 \\ 1 & 4 & 3 \\ 1 & 3 & 4 \end{bmatrix}.$$

If we find the products of M and the column matrices above, we have a new set of column matrices,

$$\begin{bmatrix} 76 \\ 77 \\ 96 \end{bmatrix}, \begin{bmatrix} 62 \\ 67 \\ 75 \end{bmatrix}, \qquad \text{and so on.}$$

The entries of these matrices can then be transmitted to an agent as the message 76, 77, 96, 62, 67, 75, and so on. [7]

[7] Use the matrix given below to find the 2 × 1 matrices to be transmitted for the message you encoded in problem 6 at the side.

$$\begin{bmatrix} 2 & 1 \\ 5 & 0 \end{bmatrix}$$

Answer:

$$\begin{bmatrix} 54 \\ 115 \end{bmatrix}, \begin{bmatrix} 24 \\ 25 \end{bmatrix}$$

When the agent receives the message, it is divided into groups of numbers with each group formed into a column matrix. After multiplying each column matrix by the matrix M^{-1}, the message can be read.

Although this type of code is relatively simple, it is actually difficult to break. Many complications are possible. For example, a long message might be placed in groups of 20, thus requiring a 20 × 20 matrix for coding and decoding. Finding the inverse of such a matrix would require an impractical amount of time if calculated by hand. For this reason some of the largest computers are used by government agencies involved in coding.

Routing The diagram in Figure 6.5 shows the roads connecting four cities. Another way of representing this information is shown in matrix A, where the entries represent the number of roads connecting two cities without passing through another city.* For example, from the diagram we see that there are two roads connecting city 1 to city 4 without passing through either city 2 or 3. This information is entered in row 1, column 4 and again in row 4, column 1 of matrix A.

$$A = \begin{bmatrix} 0 & 1 & 2 & 2 \\ 1 & 0 & 1 & 0 \\ 2 & 1 & 0 & 1 \\ 2 & 0 & 1 & 0 \end{bmatrix}$$

Note that there are 0 roads connecting each city to itself. Also, there is one road connecting cities 3 and 2.

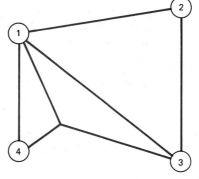

Figure 6.5

---

*From *Matrices with Applications*, section 3.2, example 5, by Hugh G. Campbell. Copyright © 1968, pp. 50–51. Adapted by permission of Prentice-Hall, Englewood Cliffs, New Jersey.

How many ways are there to go from city 1 to city 2, for example, by going through exactly one other city? Since we must go through one other city, we must go through either city 3 or city 4. On the diagram in Figure 6.5, we see that we can go from city 1 to city 2 through city 3 in 2 ways. We can go from city 1 to city 3 in 2 ways and then from city 3 to city 2 in one way, so there are $2 \cdot 1 = 2$ ways to get from city 1 to city 2 through city 3. It is not possible to go from city 1 to city 2 through city 4, because there is no direct route between cities 4 and 2.

The matrix A^2 gives the number of ways to travel between any two cities by passing through exactly one other city. Multiply matrix A by itself, to get A^2. Let the first row, second column entry of A^2 be b_{12}. (We use a_{ij} to denote the entry in the i-th row and j-th column of matrix A.) The entry b_{12} is found as follows.

$$b_{12} = a_{11}a_{12} + a_{12}a_{22} + a_{13}a_{32} + a_{14}a_{42}$$
$$= 0 \cdot 1 + 1 \cdot 0 + 2 \cdot 1 + 2 \cdot 0$$
$$= 2$$

The first product $0 \cdot 1$ in the calculations above represents the number of ways to go from city 1 to city 1, (0), and then from city 1 to city 2, (1). The 0 result indicates that such a trip does not involve a third city. The only nonzero product $(2 \cdot 1)$ represents the two routes from city 1 to city 3 and the one route from city 3 to city 2 which result in the $2 \cdot 1$ or 2 routes from city 1 to city 2 by going through city 3.

Similarly, A^3 gives the number of ways to travel between any two cities by passing through exactly two cities. Also, $A + A^2$ represents the total number of ways to travel between two cities with at most one intermediate city.

The diagram can be given many other interpretations. For example, the lines could represent lines of mutual influence between people or nations, or they could represent communication lines such as telephone lines.

6.6 Exercises

For each of the following, solve the matrix equation $AX = B$ for X. (See Example 1.)

1. $A = \begin{bmatrix} 1 & 3 \\ -2 & 4 \end{bmatrix}$, $B = \begin{bmatrix} 15 \\ 10 \end{bmatrix}$

2. $A = \begin{bmatrix} 2 & -2 \\ 1 & 0 \end{bmatrix}$, $B = \begin{bmatrix} -4 \\ 3 \end{bmatrix}$

3. $A = \begin{bmatrix} -2 & 4 \\ 3 & -1 \end{bmatrix}$, $B = \begin{bmatrix} 40 & -20 \\ 80 & 20 \end{bmatrix}$

4. $A = \begin{bmatrix} 5 & 10 \\ 8 & 14 \end{bmatrix}$, $B = \begin{bmatrix} -15 & 10 \\ 20 & 5 \end{bmatrix}$

5. $A = \begin{bmatrix} 1 & 0 & 2 \\ -1 & 1 & 0 \\ 3 & 0 & 4 \end{bmatrix}$, $B = \begin{bmatrix} 8 \\ 4 \\ -6 \end{bmatrix}$

6. $A = \begin{bmatrix} 2 & 4 & 0 \\ 1 & -2 & 0 \\ 0 & 0 & 3 \end{bmatrix}$, $B = \begin{bmatrix} 72 \\ -24 \\ 48 \end{bmatrix}$

7. $A = \begin{bmatrix} -3 & 0 & 6 \\ 1 & 1 & 0 \\ 0 & 2 & 5 \end{bmatrix}$, $B = \begin{bmatrix} 12 & 3 \\ -6 & 0 \\ 0 & -3 \end{bmatrix}$

8.
$$A = \begin{bmatrix} -2 & 2 & 0 \\ 3 & 1 & 0 \\ 0 & 0 & 1 \end{bmatrix}, \quad B = \begin{bmatrix} 8 & -16 \\ 0 & 32 \\ -11 & 5 \end{bmatrix}$$

Use matrix algebra to solve the following matrix equations for X. Then use the given matrices to find X and check your work.

9. $N = X - MX$, $\quad N = \begin{bmatrix} 8 \\ -12 \end{bmatrix}$, $\quad M = \begin{bmatrix} 0 & 1 \\ -2 & 1 \end{bmatrix}$

10. $A = BX + X$, $\quad A = \begin{bmatrix} 4 & 6 \\ -2 & 2 \end{bmatrix}$, $\quad B = \begin{bmatrix} -2 & -2 \\ 3 & 3 \end{bmatrix}$

Solve each of the systems of equations in Exercises 11–22 using the inverse of the coefficient matrix. The inverses for several of the systems were found in Exercises 23–32 of the previous section. (See Example 1.)

11.
$$\begin{aligned} -x - y - z &= 1 \\ 4x + 5y &= -2 \\ y - 3z &= 3 \end{aligned}$$

12.
$$\begin{aligned} 2x + 4z &= -8 \\ 3x + y + 5z &= 2 \\ -x + y - 2z &= 4 \end{aligned}$$

13.
$$\begin{aligned} 2x + 4y + 6z &= 4 \\ -x - 4y - 3z &= 8 \\ y - z &= -4 \end{aligned}$$

14.
$$\begin{aligned} 2x + 2y - 4z &= 12 \\ 2x + 6y &= 16 \\ -3x - 3y + 5z &= -20 \end{aligned}$$

15.
$$\begin{aligned} x + 2y + 3z &= 5 \\ 2x + 3y + 2z &= 2 \\ -x - 2y - 4z &= -1 \end{aligned}$$

16.
$$\begin{aligned} x + y - 3z &= 4 \\ 2x + 4y - 4z &= 8 \\ -x + y + 4z &= -3 \end{aligned}$$

17.
$$\begin{aligned} x + 2y &= -10 \\ -x + 4z &= 8 \\ - y + z &= 4 \end{aligned}$$

18.
$$\begin{aligned} x + z &= 3 \\ y + 2z &= 8 \\ -x + y &= 4 \end{aligned}$$

19.
$$\begin{aligned} 2x - 2y &= 5 \\ 4y + 8z &= 7 \\ x + 2z &= 1 \end{aligned}$$

20.
$$\begin{aligned} 6x + 5y &= 10 \\ x - z &= 3 \\ -12x - 10y &= -20 \end{aligned}$$

21.
$$\begin{aligned} x - 2y + 3z &= 4 \\ y - z + w &= -8 \\ -2x + 2y - 2z + 4w &= 12 \\ 2y - 3z + w &= -4 \end{aligned}$$

22.
$$\begin{aligned} x + y + 2w &= 3 \\ 2x - y + z - w &= 3 \\ 3x + 3y + 2z - 2w &= 5 \\ x + 2y + z &= 3 \end{aligned}$$

Find the production matrix given the following input-output and demand matrices. (See Examples 4 and 5.)

23. $A = \begin{bmatrix} \frac{1}{2} & \frac{2}{5} \\ \frac{1}{4} & \frac{1}{5} \end{bmatrix}$, $\quad D = \begin{bmatrix} 2 \\ 4 \end{bmatrix}$

24. $A = \begin{bmatrix} \frac{1}{5} & \frac{1}{25} \\ \frac{3}{5} & \frac{1}{20} \end{bmatrix}$, $\quad D = \begin{bmatrix} 3 \\ 10 \end{bmatrix}$

25. $A = \begin{bmatrix} .1 & .03 \\ .07 & .6 \end{bmatrix}$, $\quad D = \begin{bmatrix} 5 \\ 10 \end{bmatrix}$

26. $A = \begin{bmatrix} .01 & .03 \\ .05 & .05 \end{bmatrix}$, $\quad D = \begin{bmatrix} 100 \\ 200 \end{bmatrix}$

27. $A = \begin{bmatrix} .4 & 0 & .3 \\ 0 & .8 & .1 \\ 0 & .2 & .4 \end{bmatrix}$, $\quad D = \begin{bmatrix} 1 \\ 3 \\ 2 \end{bmatrix}$

28. $A = \begin{bmatrix} .1 & .5 & 0 \\ 0 & .3 & .4 \\ .1 & .2 & .1 \end{bmatrix}$, $\quad D = \begin{bmatrix} 10 \\ 4 \\ 2 \end{bmatrix}$

In Exercises 29 and 30, refer to Example 5.

29. Management If the demand is changed to 690 metric tons of wheat and 920 metric tons of oil, how many units of each commodity should be produced?

30. Management Change the technological matrix so that production of one ton of wheat requires 1/5 metric ton of oil (and no wheat), and the production of one metric ton of oil requires 1/3 metric ton of wheat (and no oil). To satisfy the same demand matrix, how many units of each commodity should be produced?

In Exercises 31 and 32, refer to Examples 2 and 3.

31. Management Suppose 1/4 unit of manufacturing and 1/2 unit of transportation are required to produce one unit of agriculture; 1/2 unit of agriculture and 1/4 unit of transportation to produce one unit of manufacturing; and 1/4 unit of agriculture and 1/4 unit of manufacturing to produce one unit of transportation. How many units of each commodity should be produced to satisfy a demand of 1000 units for each commodity?

32. Management In Exercise 31, find the number of units of each commodity which should be produced to satisfy a demand for 500 units of each commodity.

33. Management A primitive economy depends on two basic goods, yams and pork. Production of one bushel of yams requires 1/4 bushel of yams and 1/2 of a pig. To produce one pig requires 1/6 bushel of yams. Find the amount of each commodity which should be produced to get
(a) 1 bushel of yams and 1 pig;
(b) 100 bushels of yams and 70 pigs.

34. Use the methods of the text to encode the message

Arthur is here.

Break the message into groups of two letters and use the matrix

$$M = \begin{bmatrix} -1 & 2 \\ 2 & 5 \end{bmatrix}.$$

35. Encode the following message by breaking it into groups of two letters and using the matrix of Exercise 34.

Attack at dawn unless too cold.

36. Finish encoding the message given in the text.

37. Use matrix A in the discussion on routing in the text to find A^2. Then answer the following questions. How many ways are there to travel from
(a) city 1 to city 3 by passing through exactly one city?
(b) city 2 to city 4 by passing through exactly one city?
(c) city 1 to city 3 by passing through at most one city?
(d) city 2 to city 4 by passing through at most one city?

38. Find A^3. (See Exercise 37.) Then answer the following questions.
(a) How many ways are there to travel between cities 1 and 4 by passing through exactly two cities?
(b) How many ways are there to travel between cities 1 and 4 by passing through at most two cities?

39. Management A small telephone system connects three cities. There are four lines between cities 3 and 2, three lines connecting city 3 with city 1 and two lines between cities 1 and 2.
(a) Write a matrix B to represent this information.
(b) Find B^2.
(c) How many lines which connect cities 1 and 2 go through exactly one other city (city 3)?
(d) How many lines which connect cities 1 and 2 go through at most one other city?

40. Management The figure shows four southern cities served by Delta Airlines.

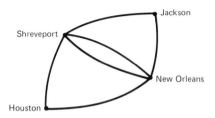

(a) Write a matrix to represent the number of non-stop routes between cities.
(b) Find the number of one-stop flights between Houston and Jackson.
(c) Find the number of flights between Houston and Shreveport which require at most one stop.
(d) Find the number of one-stop flights between New Orleans and Houston.

41. Natural Science The figure shows a food web. The arrows indicate the food sources of each population. For example, cats feed on rats and on mice.

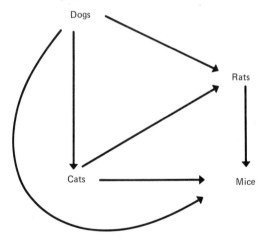

(a) Write a matrix C in which each row and corresponding column represents a population in the food chain. Enter a one when the population in a given row feeds on the population in the given column.
(b) Calculate and interpret C^2.

Chapter 6   Review Exercises

Use the echelon method to solve each of the following linear systems. Identify any dependent or inconsistent systems.

1. $3x - 5y = -18$
  $2x + 7y = 19$

2. $6x + 5y = 53$
  $4x - 3y = 29$

3. $\dfrac{2}{3}x - \dfrac{3}{4}y = 13$
  $\dfrac{1}{2}x + \dfrac{2}{3}y = -5$

4. $3x + 7y = 10$
  $18x + 42y = 50$

5. $4x + 8y = 15$
  $\dfrac{4}{3}x + \dfrac{8}{3}y = 5$

6. $.9x - .2y = .8$
  $.3x + .7y = 4.1$

7. $2x - 3y + z = -5$
  $x + 4y + 2z = 13$
  $5x + 5y + 3z = 14$

8. $x - 3y = 12$
  $2y + 5z = 1$
  $4x + z = 25$

9. A student bought some candy bars, paying 25¢ each for some and 50¢ each for others. The student bought a total of 22 bars, paying a total of $8.50. How many of each kind of bar did he buy?

10. Ink worth $25 a bottle is to be mixed with ink worth $18 per bottle to get 12 bottles of ink worth $20 each. How much of each type of ink should be used?

11. Donna Sharp wins $50,000 in a lottery. She invests part of the money at 6%, twice as much at 7%, and the balance at 9%. Total annual interest is $3800. How much is invested at each rate?

12. The sum of three numbers is 23. The second number is 3 more than the first. The sum of the first and twice the third is 4. Find the three numbers.

Find solutions for the following systems in terms of the arbitrary variable z.

13. $3x - 4y + z = 2$
  $2x + y - 4z = 1$

14. $2x + 3y - z = 5$
  $-x + 2y + 4z = 8$

Use the Gauss-Jordan method to solve each of the following systems.

15. $2x + 3y = 10$
  $-3x + y = 18$

16. $5x + 2y = -10$
  $3x - 5y = -6$

17. $3x + y = -7$
  $x - y = -5$

18. $x - z = -3$
  $y + z = 6$
  $2x - 3z = -9$

19. $2x - y + 4z = -1$
  $-3x + 5y - z = 5$
  $2x + 3y + 2z = 3$

20. $5x - 8y + z = 1$
  $3x - 2y + 4z = 3$
  $10x - 16y + 2z = 3$

For each of the following, find the order of the matrices, find the values of any variables and identify any square, row, or column matrices.

21. $\begin{bmatrix} 2 & 3 \\ 5 & q \end{bmatrix} = \begin{bmatrix} a & b \\ c & 9 \end{bmatrix}$

22. $\begin{bmatrix} 2 & x \\ y & 6 \\ 5 & z \end{bmatrix} = \begin{bmatrix} a & -1 \\ 4 & 6 \\ p & 7 \end{bmatrix}$

23. $[m \quad 4 \quad z \quad -1] = [12 \quad k \quad -8 \quad r]$

24. $\begin{bmatrix} a+5 & 3b & 6 \\ 4c & 2+d & -3 \\ -1 & 4p & q-1 \end{bmatrix} = \begin{bmatrix} -7 & b+2 & 2k-3 \\ 3 & 2d-1 & 4l \\ m & 12 & 8 \end{bmatrix}$

25. $\begin{bmatrix} 6 & m \\ k & 5-x \end{bmatrix} + \begin{bmatrix} r & m+3 \\ k+1 & 2x \end{bmatrix} = \begin{bmatrix} 8 & 10 \\ 12 & -2 \end{bmatrix}$

26. $\begin{bmatrix} 3p & 1 & y \\ q & 5 & z \end{bmatrix} - \begin{bmatrix} 3+p & r & 2y+1 \\ q+2 & m & 10 \end{bmatrix}$
$= \begin{bmatrix} 4p & 2r & 5y \\ 3q & 2m & -z \end{bmatrix}$

27. The activities of a grazing animal can be classified roughly into three categories: grazing, moving, and resting. Suppose horses spend 8 hours grazing, 8 moving, and 8 resting; cattle spend 10 grazing, 5 moving, and 9 resting; sheep spend 7 grazing, 10 moving, and 7 resting; and goats spend 8 grazing, 9 moving, and 7 resting. Write this information as a 4 × 3 matrix.

28. The New York Stock Exchange reports in the daily newspapers give the dividend, price-to-earnings ratio, sales (in hundreds of shares), last price, and change in price for each company. Write the following stock reports as a 4 × 5 matrix. American Telephone & Telegraph: 5, 7, 2532, 52 3/8, −1/4, General Electric: 3, 9, 1464, 56, +1/8. Gulf Oil: 2.50, 5, 4974, 41, −1 1/2. Sears: 1.36, 10, 1754, 18, +1/2.

Given the matrices

$$A = \begin{bmatrix} 4 & 10 \\ -2 & -3 \\ 6 & 9 \end{bmatrix}, \quad B = \begin{bmatrix} 2 & 3 & -2 \\ 2 & 4 & 0 \\ 0 & 1 & 2 \end{bmatrix},$$

$$C = \begin{bmatrix} 5 & 0 \\ -1 & 3 \\ 4 & 7 \end{bmatrix}, \quad D = \begin{bmatrix} 6 \\ 1 \\ 0 \end{bmatrix}, \quad E = [1 \quad 3 \quad -4],$$

$$F = \begin{bmatrix} -1 & 4 \\ 3 & 7 \end{bmatrix}, \quad G = \begin{bmatrix} 2 & 5 \\ 1 & 6 \end{bmatrix},$$

find each of the following that exists.

29. $-A$ **30.** $-E$ **31.** $A + C$

32. $C + 2A$ **33.** $2G - 4F$ **34.** $B - A$

35. $2A - 5C$ **36.** $3B + C$

37. Refer to Exercise 28. Write a 4×2 matrix using the sales and price changes for the four companies. The next day's sales and price changes for the same four companies were 2310, 1258, 5061, 1812 and $-1/4$, $-1/4$, $+1/2$, $+1/2$ respectively. Write a 4×2 matrix using these new sales and price change figures. Use matrix addition to find the total sales and price changes for the two days.

38. An oil refinery in Tulsa sent 110,000 gallons of oil to a Chicago distributor, 73,000 to a Dallas distributor, and 95,000 to an Atlanta distributor. Another refinery in New Orleans sent the following amounts to the same three distributors: 85,000, 108,000, 69,000. The next month the two refineries sent the same distributors new shipments of oil as follows: from Tulsa, 58,000 to Chicago, 33,000 to Dallas, and 80,000 to Atlanta; from New Orleans, 40,000, 52,000, and 30,000 respectively.
(a) Write the monthly shipments from the two distributors to the three refineries as 3×2 matrices.
(b) Use matrix addition to find the total amounts sent to the refineries from each distributor.

Use the matrices given above Exercise 29 to find each of the following that exists.

39. AF **40.** AC **41.** DE

42. ED **43.** BD **44.** EA

45. AFG **46.** EBC

47. An office supply manufacturer makes two kinds of paper clips, standard and extra large. To make a unit of standard paper clips requires 1/4 hour on a cutting ma-

chine and 1/2 hour on a machine which shapes the clips. A unit of extra large paper clips requires 1/3 hour on each machine.
(a) Write this information as a 2×2 matrix (machine/size)
(b) If the cutting machine is operated for 12 hours and the shaping machine for 6 hours, use matrix multiplication to find out how many of each type of clip can be made. (Hint: write the hours as a 1×2 matrix.)

48. Jane Cunningham buys shares of three stocks. Their cost per share and dividend earnings per share are $32, $23, $54, and $1.20, $1.49, and $2.10 respectively. She buys 50 shares of the first stock, 20 shares of the second, and 15 shares of the third.
(a) Write the cost per share and earnings per share of the stocks as a 3×2 matrix.
(b) Write the number of shares of each stock as a 1×3 matrix.
(c) Use matrix multiplication to find the total cost and total dividend earnings of these stocks.

Find the inverse of each of the following matrices that has an inverse.

49. $\begin{bmatrix} -4 & 2 \\ 0 & 3 \end{bmatrix}$ **50.** $\begin{bmatrix} 2 & 1 \\ 5 & 3 \end{bmatrix}$

51. $\begin{bmatrix} 6 & 4 \\ 3 & 2 \end{bmatrix}$ **52.** $\begin{bmatrix} 2 & 0 \\ -1 & 5 \end{bmatrix}$

53. $\begin{bmatrix} 2 & 0 & 4 \\ 1 & -1 & 0 \\ 0 & 1 & -2 \end{bmatrix}$ **54.** $\begin{bmatrix} 2 & -1 & 0 \\ 1 & 0 & 1 \\ 1 & -2 & 0 \end{bmatrix}$

55. $\begin{bmatrix} 2 & 3 & 5 \\ -2 & -3 & -5 \\ 1 & 4 & 2 \end{bmatrix}$ **56.** $\begin{bmatrix} 1 & 3 & 6 \\ 4 & 0 & 9 \\ 5 & 15 & 30 \end{bmatrix}$

Refer again to the matrices given above Exercise 29 to find each of the following that exists.

57. G^{-1} **58.** F^{-1}

59. B^{-1} **60.** $(A + C)^{-1}$

61. $(F - G)^{-1}$ **62.** $F^{-1} - G^{-1}$

Solve each of the following matrix equations $AX = B$ for X.

63. $A = \begin{bmatrix} 2 & 4 \\ -1 & -3 \end{bmatrix}, \quad B = \begin{bmatrix} 8 \\ 3 \end{bmatrix}$

64. $A = \begin{bmatrix} 1 & 3 \\ -2 & 4 \end{bmatrix}$, $\qquad B = \begin{bmatrix} 15 \\ 10 \end{bmatrix}$

65. $A = \begin{bmatrix} 1 & 0 & 2 \\ -1 & 1 & 0 \\ 3 & 0 & 4 \end{bmatrix}$, $\qquad B = \begin{bmatrix} 8 \\ 4 \\ -6 \end{bmatrix}$

66. $A = \begin{bmatrix} 2 & 4 & 0 \\ 1 & -2 & 0 \\ 0 & 0 & 3 \end{bmatrix}$, $\qquad B = \begin{bmatrix} 72 \\ -24 \\ 48 \end{bmatrix}$

Use the method of matrix inverses to solve each of the following equations.

67. $\begin{aligned} x + y &= 4 \\ 2x + 3y &= 10 \end{aligned}$
68. $\begin{aligned} 5x - 3y &= -2 \\ 2x + 7y &= -9 \end{aligned}$

69. $\begin{aligned} 2x + y &= 5 \\ 3x - 2y &= 4 \end{aligned}$
70. $\begin{aligned} x - 2y &= 7 \\ 3x + y &= 7 \end{aligned}$

71. $\begin{aligned} x + y + z &= 1 \\ 2x - y &= -2 \\ 3y + z &= 2 \end{aligned}$
72. $\begin{aligned} x &= -3 \\ y + z &= 6 \\ 2x - 3z &= -9 \end{aligned}$

73. $\begin{aligned} 3x - 2y + 4z &= 4 \\ 4x + y - 5z &= 2 \\ -6x + 4y - 8z &= -2 \end{aligned}$
74. $\begin{aligned} x + 2y &= -1 \\ 3y - z &= -5 \\ x + 2y - z &= -3 \end{aligned}$

Solve each of the following problems by any method.

75. A gold merchant has some 12 carat gold (12/24 pure gold), and some 22 carat gold (22/24 pure). How many grams of each should be mixed to get 25 grams of 15 carat gold?

76. A chemist has some 40% acid solution and some 60% solution. How many liters of each should be used to get 40 liters of a 45% solution?

77. How many pounds of tea worth $4.60 a pound should be mixed with tea worth $6.50 a pound to get 10 pounds of a mixture worth $5.74 a pound?

78. A solution of a drug with a strength of 5% is to be mixed with a 15% solution to get 50 milliliters of an 8% solution. How many milliliters of the 5% solution should be used?

79. The cashier at an amusement park has a total of $2480, made up of fives, tens, and twenties. The total number of bills is 290, and the value of the tens is $60 more than the value of the twenties. How many of each type of bill does the cashier have?

80. Ms. Levy invests $50,000 three ways—at 8%, 8 1/2%, and 11%. In total, she receives $4436.25 per year in interest. The interest from the 11% investment is $80 more than the interest on the 8% investment. Find the amount she has invested at each rate.

Find the production matrix given the following input-output and demand matrices.

81. $A = \begin{bmatrix} .01 & .05 \\ .04 & .03 \end{bmatrix}$ $\qquad D = \begin{bmatrix} 200 \\ 300 \end{bmatrix}$

82. $A = \begin{bmatrix} .2 & .1 & .3 \\ .1 & 0 & .2 \\ 0 & 0 & .4 \end{bmatrix}$ $\qquad D = \begin{bmatrix} 500 \\ 200 \\ 100 \end{bmatrix}$

83. An economy depends on two commodities, goats and cheese. It takes 2/3 of a unit of goats to produce one unit of cheese and 1/2 unit of cheese to produce one unit of goats.
(a) Write the input-output matrix for this economy.
(b) Find the production required to satisfy a demand of 400 units of cheese and 800 units of goats.

84. The matrix below represents the number of direct flights between four cities.

	A	B	C	D
A	0	1	0	1
B	1	0	0	1
C	0	0	0	1
D	1	1	1	0

(a) Find the number of one-stop flights between cities A and C.
(b) Find the total number of flights between cities B and C that are either direct or one-stop.
(c) Find the matrix which gives the number of two-stop flights between these cities.

85. Given the input-output matrix $A = \begin{bmatrix} 0 & \frac{1}{4} \\ \frac{1}{2} & 0 \end{bmatrix}$ and the demand matrix $D = \begin{bmatrix} 2100 \\ 1400 \end{bmatrix}$, find each of the following.
(a) $I - A$ (b) $(I - A)^{-1}$
(c) the production matrix X

Let $M = \begin{bmatrix} 2 & 5 \\ -3 & 1 \end{bmatrix}$.

86. (a) Use M to encode the following message: It is too late.
(b) What matrix should be used to decode this message?

Case 6

Contagion

Suppose that three people have contracted a contagious disease.* A second group of five people may have been in contact with the three infected persons. A third group of six people may have been in contact with the second group. We can form a 3×5 matrix P with rows representing the first group of three and columns representing the second group of five. We enter a one in the corresponding position if a person in the first group has contact with a person in the second group. These direct contacts are called *first-order contacts*. Similarly we form a 5×6 matrix Q representing the first-order contacts between the second and third group. For example, suppose

$$P = \begin{bmatrix} 1 & 0 & 0 & 1 & 0 \\ 0 & 0 & 1 & 1 & 0 \\ 1 & 1 & 0 & 0 & 0 \end{bmatrix} \quad \text{and}$$

$$Q = \begin{bmatrix} 1 & 1 & 0 & 1 & 1 & 1 \\ 0 & 0 & 0 & 0 & 1 & 0 \\ 0 & 0 & 0 & 0 & 0 & 0 \\ 0 & 1 & 0 & 1 & 0 & 0 \\ 1 & 0 & 0 & 0 & 1 & 0 \end{bmatrix}$$

From matrix P we see that the first person in the first group had contact with the first and fourth persons in the second group. Also, none of the first group had contact with the last person in the second group.

A *second-order contact* is an indirect contact between persons in the first and third group through some person in the second group. The product matrix PQ indicates these contacts. Verify that the second row, fourth column entry of PQ is 1. That is, there is one second-order contact between the second person in group 1 and the fourth person in group 3. Let a_{ij} denote the element in the i-th row and j-th column of the matrix PQ. By looking at the products which form a_{24} below, we see that the common contact was with the fourth individual in group 2. (The p_{ij} are entries in P, and the q_{ij} are entries in Q.)

$$a_{24} = p_{21}q_{14} + p_{22}q_{24} + p_{23}q_{34} + p_{24}q_{44} + p_{25}q_{54}$$
$$= 0 \cdot 1 + 0 \cdot 0 + 1 \cdot 0 + 1 \cdot 1 + 0 \cdot 1$$
$$= 1$$

The second person in group 1 and the fourth person in group 3 both had contact with the fourth person in group 2.

This idea could be extended to third-, fourth-, and larger order contacts. It indicates a way to use matrices to trace the spread of a contagious disease. It could also pertain to the dispersal of ideas or anything that might pass from one individual to another.

Exercises

1. Find the second-order contact matrix PQ mentioned in the text.

2. How many second-order contacts were there between the second contagious person and the third person in the third group?

3. Is there anyone in the third group who has had no contacts at all with the first group?

4. The totals of the columns in PQ give the total number of second-order contacts per person, while the column totals in P and Q give the total number of first-order contacts per person. Which person has the most contacts, counting both first- and second-order contacts?

5. Find the matrix product in Exercise 1.

*"First and Second Order Contact to a Contagious Disease" from *Finite Mathematics with Applications to Business, Life Sciences, and Social Sciences* by Stanley Grossman.

Case 7

Leontief's Model of the American Economy

In the April 1965 issue of *Scientific American,* Wassily Leontief explained his input-output system using the 1958 American economy as an example.* He divided the economy into 81 sectors, grouped into six families of related sectors. In order to keep the discussion reasonably simple, we will treat each family of sectors as a single sector and so, in effect, work with a six sector model. The sectors are listed in Table 1.

Table 1

Sector	Examples
Final Nonmetal (FN)	Furniture, processed food
Final Metal (FM)	Household appliances, motor vehicles
Basic Metal (BM)	Machine-shop products, mining
Basic Nonmetal (BN)	Agriculture, printing
Energy (E)	Petroleum, coal
Services (S)	Amusements, real estate

The workings of the American economy in 1958 are described in the input-output table (Table 2) based on Leontief's figures. We will demonstrate the meaning of Table 2 by considering the first left-hand column of numbers. The numbers in this column mean that 1 unit of final nonmetal production requires the consumption of .170 unit of (other) final nonmetal production, .003 unit of final metal output, .025 unit of basic metal products, and so on down the column. Since the unit of measurement that Leontief used for this table is millions of dollars, we conclude that the production of $1 million worth of final nonmetal production consumes $.170 million, or $170,000, worth of other final nonmetal products, $3000 of final metal products, $25,000 of basic metal products, and so on. Similarly, the entry in the column headed FM and opposite S of .074 means that $74,000 worth of input from the service industries goes into the production of $1 million worth of final metal products, and the number .358 in the column headed E and opposite E means that $358,000 worth of energy must be consumed to produce $1 million worth of energy.

* Adapted from *Applied Finite Mathematics* by Robert F. Brown and Brenda W. Brown. Copyright © 1977 by Robert F. Brown and Brenda W. Brown. Reprinted by permission.

Table 2

	FN	FM	BM	BN	E	S
FN	.170	.004	0	.029	0	.008
FM	.003	.295	.018	.002	.004	.016
BM	.025	.173	.460	.007	.011	.007
BN	.348	.037	.021	.403	.011	.048
E	.007	.001	.039	.025	.358	.025
S	.120	.074	.104	.123	.173	.234

By the underlying assumption of Leontief's model, the production of n units (n = any number) of final nonmetal production consumes $.170n$ units of final nonmetal output, $.003n$ units of final metal output, $.025n$ units of basic metal production, and so on. Thus, production of $50 million worth of products from the final nonmetal section of the 1958 American economy required $(.170)(50) = 8.5$ units ($8.5 million) worth of final nonmetal input, $(.003)(50) = .15$ units of final metal input, $(.025)(50) = 1.25$ units of basic metal production, and so on.

Example 1: According to the simplified input-output table for the 1958 American economy, how many dollars worth of final metal products, basic nonmetal products, and services are required to produce $120 million worth of basic metal products?

Each unit ($1 million worth) of basic metal products requires .018 units of final metal products because the number in the BM column of the table opposite FM is .018. Thus, $120 million, or 120 units, requires $(.018)(120) = 2.16$ units, or $2.16 million worth of final metal products. Similarly, 120 units of basic metal production uses $(.021)(120) = 2.52$ units of basic nonmetal production and $(.104)(120) = 12.48$ units of services, or $2.52 million and $12.48 million worth of basic nonmetal output and services respectively. ■

The Leontief model also involves a **bill of demands,** that is, a list of requirements for units of output beyond that required for its inner workings as described in the input-output table. These demands represent exports, surpluses, government and individual consumption, and the like. The bill of demands (in millions) for the simplified version of the 1958 American economy we have been using is shown below.

FN	$99,640
FM	$75,548
BM	$14,444
BN	$33,501
E	$23,527
S	$263,985

We can now use the methods developed in Section 6.5 to answer this question: how many units of output from each sector are needed in order to run the economy and fill the bill of demands? The units of output from each sector required to run the economy and fill the bill of demands is unknown, so we denote them by variables. In our example, there are six quantities which are, at the moment, unknown. The number of units of final nonmetal production required to solve the problem will be our first unknown, because this sector is represented by the first row of the input-output matrix. The unknown quantity of final nonmetal units will be represented by the symbol x_1. Following the same pattern, we represent the unknown quantities in the following manner:

x_1 = units of final nonmetal production required,

x_2 = units of final metal production required,

x_3 = units of basic metal production required,

x_4 = units of basic nonmetal production required,

x_5 = units of energy required,

x_6 = units of services required.

These six numbers are the quantities we are attempting to calculate, and the variables will make up the entries in our production matrix.

To find these numbers, first let A be the 6×6 matrix corresponding to the input-output table.

$$A = \begin{bmatrix} .170 & .004 & 0 & .029 & 0 & .008 \\ .003 & .295 & .018 & .002 & .004 & .016 \\ .025 & .173 & .460 & .007 & .011 & .007 \\ .348 & .037 & .021 & .403 & .011 & .048 \\ .007 & .001 & .039 & .025 & .358 & .025 \\ .120 & .074 & .104 & .123 & .173 & .234 \end{bmatrix}$$

A is the input-output matrix. The bill of demands leads to a 6×1 demand matrix D, and X is the matrix of unknowns.

$$D = \begin{bmatrix} 99,640 \\ 75,548 \\ 14,444 \\ 33,501 \\ 23,527 \\ 263,985 \end{bmatrix} \quad \text{and} \quad X = \begin{bmatrix} x_1 \\ x_2 \\ x_3 \\ x_4 \\ x_5 \\ x_6 \end{bmatrix}$$

To use the formula $D = (I - A)X$, or $X = (I - A)^{-1}D$, we need to find $I - A$.

$$I - A = \begin{bmatrix} 1 & 0 & 0 & 0 & 0 & 0 \\ 0 & 1 & 0 & 0 & 0 & 0 \\ 0 & 0 & 1 & 0 & 0 & 0 \\ 0 & 0 & 0 & 1 & 0 & 0 \\ 0 & 0 & 0 & 0 & 1 & 0 \\ 0 & 0 & 0 & 0 & 0 & 1 \end{bmatrix}$$

$$- \begin{bmatrix} .170 & .004 & 0 & .029 & 0 & .008 \\ .003 & .295 & .018 & .002 & .004 & .016 \\ .025 & .173 & .460 & .007 & .011 & .007 \\ .348 & .037 & .021 & .403 & .011 & .048 \\ .007 & .001 & .039 & .025 & .358 & .025 \\ .120 & .074 & .104 & .123 & .173 & .234 \end{bmatrix}$$

$$= \begin{bmatrix} .830 & -.004 & 0 & -.029 & 0 & -.008 \\ -.003 & .705 & -.018 & -.002 & -.004 & -.016 \\ -.025 & -.173 & .540 & -.007 & -.011 & -.007 \\ -.348 & -.037 & -.021 & .597 & -.011 & -.048 \\ -.007 & -.001 & -.039 & -.025 & .642 & -.025 \\ -.120 & -.074 & -.104 & -.123 & -.173 & .766 \end{bmatrix}$$

By the method given in Section 6.5, we find the inverse (actually an approximation).

$$(I - A)^{-1} = \begin{bmatrix} 1.234 & .014 & .006 & .064 & .007 & .018 \\ .017 & 1.436 & .057 & .012 & .020 & .032 \\ .071 & .465 & 1.877 & .019 & .045 & .031 \\ .751 & .134 & .100 & 1.740 & .066 & .124 \\ .060 & .045 & .130 & .082 & 1.578 & .059 \\ .339 & .236 & .307 & .312 & .376 & 1.349 \end{bmatrix}$$

Therefore, $X = (I - A)^{-1}D =$

$$\begin{bmatrix} 1.234 & .014 & .006 & .064 & .007 & .018 \\ .017 & 1.436 & .057 & .012 & .020 & .032 \\ .071 & .465 & 1.877 & .019 & .045 & .031 \\ .751 & .134 & .100 & 1.740 & .066 & .124 \\ .060 & .045 & .130 & .082 & 1.578 & .059 \\ .339 & .236 & .307 & .312 & .376 & 1.349 \end{bmatrix} \begin{bmatrix} 99,640 \\ 75,548 \\ 14,444 \\ 33,501 \\ 23,527 \\ 263,985 \end{bmatrix}$$

$$= \begin{bmatrix} 131,161 \\ 120,324 \\ 79,194 \\ 178,936 \\ 66,703 \\ 426,542 \end{bmatrix}.$$

From this result,

$$x_1 = 131,161$$
$$x_2 = 120,324$$
$$x_3 = 79,194$$
$$x_4 = 178,936$$
$$x_5 = 66,703$$
$$x_6 = 426,542.$$

In other words, it would require 131,161 units ($131,161 million worth) of final nonmetal production, 120,324 units of final metal output, 79,194 units of basic metal products, and so on to run the 1958 American economy and completely fill the stated bill of demands.

Exercises

1. A much simplified version of Leontief's 42 sector analysis of the 1947 American economy divides the economy into just three sectors: agriculture, manufacturing, and the household (i.e., the sector of the economy which produces labor). It is summarized in the following input-output table.

	Agriculture	Manufacturing	Household
Agriculture	.245	.102	.051
Manufacturing	.099	.291	.279
Household	.433	.372	.011

The bill of demands (in billions of dollars) is shown below.

Agriculture	2.88
Manufacturing	31.45
Household	30.91

(a) Write the input-output matrix A, the demand matrix D, and the matrix X.
(b) Compute $I - A$.
(c) Check that

$$(I - A)^{-1} = \begin{bmatrix} 1.454 & .291 & .157 \\ .533 & 1.763 & .525 \\ .837 & .791 & 1.278 \end{bmatrix}$$

is an approximation to the inverse of $I - A$ by calculating $(I - A)^{-1}(I - A)$.
(d) Use the matrix of part (c) to compute X.
(e) Explain the meaning of the numbers in X in dollars.

2. An analysis of the 1958 Israeli economy* is here simplified by grouping the economy into three sectors: agriculture, manufacturing, and energy. The input-output table is given below.

	Agriculture	Manufacturing	Energy
Agriculture	.293	0	0
Manufacturing	.014	.207	.017
Energy	.044	.010	.216

Exports (in thousands of Israeli pounds) were as follows.

Agriculture	138,213
Manufacturing	17,597
Energy	1,786

(a) Write the input-output matrix A and the demand (export) matrix D.
(b) Compute $I - A$.
(c) Check that

$$(I - A)^{-1} = \begin{bmatrix} 1.414 & 0 & 0 \\ .027 & 1.261 & .027 \\ .080 & .016 & 1.276 \end{bmatrix}$$

is an approximation to the inverse of $I - A$ by calculating $(I - A)^{-1}(I - A)$.
(d) Use the matrix of part (c) to determine the number of Israeli pounds worth of agricultural products, manufactured goods, and energy required to run this model of the Israeli economy and export the stated value of products.

---

*Wassily Leontief, Input-Output Economics (New York: Oxford University Press, 1966), pp. 54–57.

7 Linear Programming

Many realistic problems involve inequalities—a factory can manufacture *no more than* 12 items on a shift, or a medical researcher must interview *at least* one hundred patients to be sure that a new treatment is better than the old treatment.

Linear inequalities, those involving only first degree polynomials in x and y, are used in a process called *linear programming* to *optimize* (find the best solution for) a given situation.

In this chapter we first show how some linear programming problems can be solved by graphical methods. This is followed by a discussion of the *simplex method*, a general method for solving linear programming problems with many variables.

7.1   Graphing Linear Inequalities in Two Variables

Chapter 6 showed examples of linear equations in two variables. Examples of linear *inequalities* in two variables include

$$x + 2y < 4, \qquad 3x + 2y > 6, \qquad \text{and} \qquad 2x - 5y \geq 10.$$

The solution of a linear inequality is an ordered pair that satisfies the inequality. For example $(4, 4)$ is a solution of

$$3x - 2y \leq 6.$$

(Check by substituting 4 for x and 4 for y.) Since a linear inequality has an infinite number of solutions for every choice of a value for x, the best way to show the solution is with a graph.

Example 1   Graph the linear inequality $3x - 2y \leq 6$.

Because of the "$=$" portion of $\leq$, the points of the line $3x - 2y = 6$ satisfy the linear inequality $3x - 2y \leq 6$ and are part of its graph. As in Chapter 1, find the intercepts by first letting $x = 0$ and then letting $y = 0$; use these points to get the graph of $3x - 2y = 6$ shown in Figure 7.1.

The points on the line satisfy "$3x - 2y$ *equals* 6." Locate the points satisfying "$3x - 2y$ *is less than* 6," by first solving $3x - 2y \leq 6$ for y.

$$3x - 2y \leq 6$$
$$-2y \leq -3x + 6$$
$$y \geq \frac{3}{2}x - 3$$

(Recall that multiplying both sides of an inequality by a negative number reverses the direction of the inequality symbol.) As shown in Figure 7.2, the points *above*

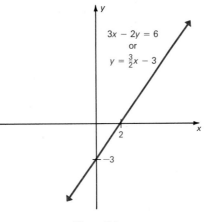

Figure 7.1

1 Graph
(a) $2x + 5y \le 10$,
(b) $x - y \ge 4$.

Answer:
(a)

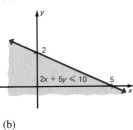

(b)

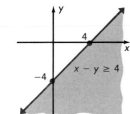

the line $3x - 2y = 6$ satisfy

$$y > \frac{3}{2}x - 3,$$

while those below the line satisfy

$$y < \frac{3}{2}x - 3.$$

The line itself is the **boundary.** In summary, the inequality $3x - 2y \le 6$ is satisfied by all points *on or above* the line $3x - 2y = 6$. Indicate the points above the line by shading, as in Figure 7.3. The line and shaded region of Figure 7.3 make up the graph of the linear inequality $3x - 2y \le 6$. **1**

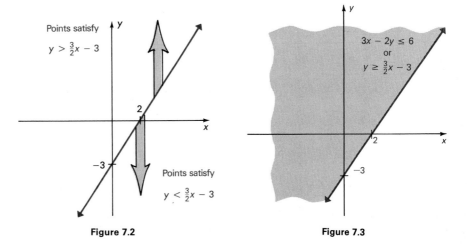

Figure 7.2

Figure 7.3

Example 2 Graph $x + 4y < 4$.

The boundary here is the line $x + 4y = 4$. Since the points on this line do not satisfy $x + 4y < 4$, the line is drawn dashed, as in Figure 7.4. Decide whether to shade the region above the line or the region below the line by solving for y.

$$x + 4y < 4$$
$$4y < -x + 4$$
$$y < -\frac{1}{4}x + 1$$

Since y is less than $(-1/4)x + 1$, the solution is the region below the boundary, as shown by the shaded portion of Figure 7.4. ■

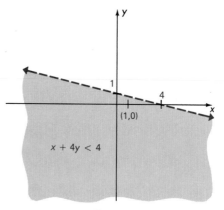

Figure 7.4

2 Graph
(a) $2x + 3y > 12$,
(b) $3x - 2y < 6$.

Answer:
(a)

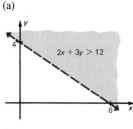

(b)

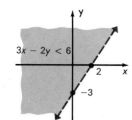

There is an alternate way to find the correct region to shade, or to check the method shown above: choose as a test point any point not on the boundary line. For example, in Example 2 we could choose the point $(1, 0)$, which is not on the line $x + 4y = 4$. Substitute 1 for x and 0 for y in the given inequality.

$$x + 4y < 4$$
$$1 + 4(0) < 4 \qquad \text{Let } x = 1, y = 0$$
$$1 < 4 \qquad \text{True}$$

Since the result $1 < 4$ is true, the test point $(1, 0)$ is on the side of the line where all the points satisfy $x + 4y < 4$. For this reason, shade the side containing $(1, 0)$, as in Figure 7.4. Choosing a different test point, such as $(1, 5)$, might produce a false result when substituted into the given inequality. In this case, shade the side of the line *not including* the test point. 2

As these examples suggest, the graph of a linear inequality is a region in the plane, perhaps including the line which is the boundary of the region. Each of the shaded regions is an example of a **half-plane,** a region on one side of a line. For example, in Figure 7.5 line r divides the plane into half-planes P and Q. The points of line r belong to neither P nor Q. Line r is the boundary of each half-plane.

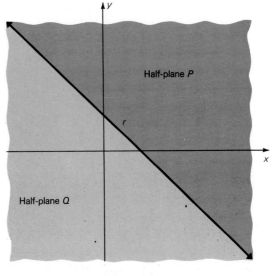

Figure 7.5

Example 3   Graph $x \leq -1$.

Recall that the graph of $x = -1$ is the vertical line through $(-1, 0)$. The points where $x < -1$ lie in the half-plane to the left of the boundary line, as shown in Figure 7.6.

 Graph each of the following.

(a) $x \geq 3$

(b) $y - 3 \leq 0$

Answer:

(a)
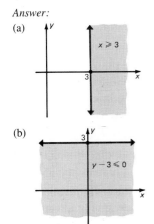

(b)

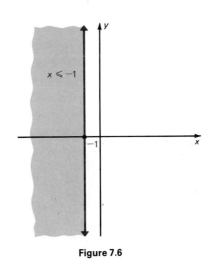

Figure 7.6

Example 4   Graph $x \geq 3y$.

Start by graphing the boundary, $x = 3y$. If $x = 0$, then $y = 0$, giving the point $(0, 0)$. Setting $y = 0$ would produce $(0, 0)$ again. To get a second point on the line, choose another number for x. Choosing $x = 3$ gives $y = 1$, so that $(3, 1)$ is a point on the line $x = 3y$. The points $(0, 0)$ and $(3, 1)$ lead to the line graphed in Figure 7.7. Let us choose $(6, 1)$ as a test point to decide which half-plane to shade. (Any point that is not on the line $x = 3y$ may be used.) Replacing x with 6 and y with 1 in the original inequality gives a true statement, so the half-plane containing $(6, 1)$ is shaded, as shown in Figure 7.7.

4 Graph $2x < y$.

Answer:

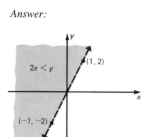

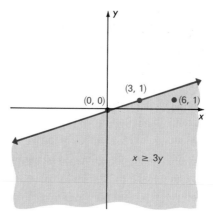

Figure 7.7

The steps used to graph a linear inequality are summarized below.

Graphing a Linear Inequality

1. Draw the graph of the boundary line. Make the line solid if the inequality involves $\leq$ or $\geq$; make the line dashed if the inequality involves $<$ or $>$.
2. Decide which half-plane to shade: either
   (a) solve the inequality for y, getting $y \geq mx + b$ or $y \leq mx + b$; shade the region above the line $y = mx + b$ if $y \geq mx + b$; shade the region below the line if $y \leq mx + b$; or
   (b) choose any point not on the line as a test point; shade the half-plane that includes the test point if the test point satisfies the original inequality; otherwise, shade the half-plane on the other side of the boundary line.

Systems of Inequalities Realistic problems often involve many inequalities. For example, a manufacturing problem might produce inequalities resulting from production requirements, as well as inequalities about cost requirements. A set of at least two inequalities is called a **system of inequalities.** The **solution** of a system of inequalities is made up of all those points that satisfy all the inequalities of the system at the same time. Graph the solution of a system of inequalities by graphing all the inequalities on the same axes and identifying, by heavier shading, the region common to all graphs. The next example shows how this is done.

5 Graph the system
$x + y \leq 6$, $2x + y \geq 4$.

Answer:

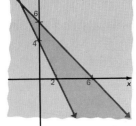

Example 5 Graph the system

$$y < -3x + 12$$
$$x < 2y.$$

The heavily shaded region in Figure 7.8 shows all the points that satisfy both inequalities of the system. Since the points on the boundary lines are not in the solution, the boundary lines are dashed. **5**

The heavily shaded region of Figure 7.8 is sometimes called the **region of feasible solutions,** or just the **feasible region,** since it is made up of all the points that satisfy (are feasible for) each inequality of the system.

Example 6 Graph the feasible region for the system

$$2x - 5y \leq 10$$
$$x + 2y \leq 8$$
$$x \geq 0$$
$$y \geq 0.$$

On the same axes, graph each inequality by graphing the boundary and choosing the appropriate half-plane. Then find the feasible region by locating the intersection (overlap) of all the half-planes. This feasible region is shaded in Figure 7.9. The inequalities $x \geq 0$ and $y \geq 0$ restrict the feasible region to the first quadrant. **6**

6 Graph the feasible region of the system
$x + 4y \leq 8$, $x - y \geq 3$
$x \geq 0$, $y \geq 0$.

Answer:

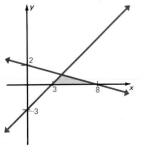

Applications As we shall see in the rest of this chapter, many realistic problems lead to systems of linear inequalities. The next example is typical of such problems.

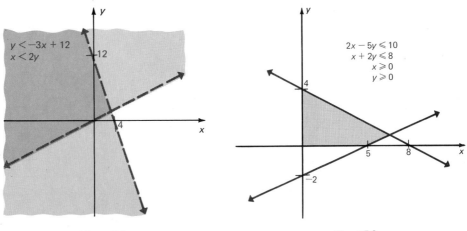

Figure 7.8

Figure 7.9

Example 7 Midtown Manufacturing Company makes plastic plates and cups, both of which require time on two machines. A unit of plates requires one hour on machine A and two on machine B, while a unit of cups requires three hours on machine A and one on machine B. Each machine is operated for at most 15 hours per day. Write a system of inequalities expressing these conditions and graph the feasible region.

Let x represent the number of units of plates to be made, and y represent the number of units of cups. Then make a chart that summarizes the given information.

	Number made	Time on machine A	Time on machine B
Plates	x	1	2
Cups	y	3	1
Maximum time available		15	15

On machine A, x units of plates require a total of $1 \cdot x = x$ hours while y units of cups require $3 \cdot y = 3y$ hours. Since machine A is available no more than 15 hours a day,

$$x + 3y \leq 15.$$

The requirement that machine B be used no more than 15 hours a day gives

$$2x + y \leq 15.$$

It is not possible to produce a negative number of cups or plates, so that

$$x \geq 0 \qquad \text{and} \qquad y \geq 0.$$

The feasible region for this system of inequalities is shown in Figure 7.10. ■

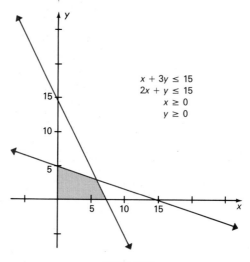

$$x + 3y \leq 15$$
$$2x + y \leq 15$$
$$x \geq 0$$
$$y \geq 0$$

Figure 7.10

7.1 Exercises

Graph each of the following linear inequalities. (See Examples 1–4.)

1. $x + y \le 2$ **2.** $y \le x + 1$

3. $x \ge 3 + y$ **4.** $y \ge x - 3$

5. $4x - y < 6$ **6.** $3y + x > 4$

7. $3x + y < 6$ **8.** $2x - y > 2$

9. $x + 3y \ge -2$ **10.** $2x + 3y \le 6$

11. $4x + 3y > -3$ **12.** $5x + 3y > 15$

13. $2x - 4y < 3$ **14.** $4x - 3y < 12$

15. $x \le 5y$ **16.** $2x \ge y$

17. $-3x < y$ **18.** $-x > 6y$

19. $x + y \le 0$ **20.** $3x + 2y \ge 0$

21. $y < x$ **22.** $y > -2x$

23. $x < 4$ **24.** $y > 5$

25. $y \le -2$ **26.** $x \ge 3$

Graph the feasible region for the following systems of inequalities. (See Examples 5 and 6.)

27. $x + y \le 1$
    $x - y \ge 2$

28. $3x - 4y < 6$
    $2x + 5y > 15$

29. $2x - y < 1$
    $3x + y < 6$

30. $x + 3y \le 6$
    $2x + 4y \ge 7$

31. $-x - y < 5$
    $2x - y < 4$

32. $6x - 4y > 8$
    $3x + 2y > 4$

33. $x + y \le \ \ 4$
    $x - y \le \ \ 5$
    $4x + y \le -4$

34. $3x - 2y \ge \ \ 6$
    $x + \ \ y \le -5$
    $y \le \ \ 4$

35. $-2 < x < 3$
    $-1 \le y \le 5$
    $2x + y < 6$

36. $-2 < x < 2$
    $y > 1$
    $x - y > 0$

37. $2y + x \ge -5$
    $y \le 3 + x$
    $x \ge 0$
    $y \ge 0$

38. $2x + 3y \le \ \ 12$
    $2x + 3y > -6$
    $3x + \ \ y < \ \ 4$
    $x \ge \ \ 0$
    $y \ge \ \ 0$

39. $3x + 4y > 12$
    $2x - 3y < \ \ 6$
    $0 \le y \le \ \ 2$
    $x \ge \ \ 0$

40. $0 \le \ \ x \le \ \ 9$
    $x - 2y \ge \ \ 4$
    $3x + 5y \le 30$
    $y \ge \ \ 0$

Write a system of inequalities and graph the feasible region for each of the following problems. In Exercises 41 and 42, first complete the chart. (See Example 7.)

41. A small pottery shop makes two kinds of planters, glazed and unglazed. The glazed type requires 1/2 hour to throw on the wheel and 1 hour in the kiln. The unglazed type takes 1 hour to throw on the wheel and 6 hours in the kiln. The wheel is available for at most 8 hours a day and the kiln is available for at most 20 hours per day.

	Number	Wheel	Kiln
Glazed	x		
Unglazed	y		
Maximum number of hours available			

42. Carmella and Walt produce handmade shawls and afghans. They spin the yarn, dye it, and then weave it. A shawl requires 1 hour of spinning, 1 hour of dyeing, and 1 hour of weaving. An afghan needs 2 hours of spinning, 1 of dyeing, and 4 of weaving. Together, they spend at most 8 hours spinning, 6 hours dyeing, and 14 hours weaving.

	Number	Hours spinning	Hours weaving	Hours dyeing
Shawls	x			
Afghans	y			
Maximum number of hours available		8	6	14

43. Management   The California Almond Growers have 2400 boxes of almonds to be shipped from their plant in Sacramento to Des Moines and San Antonio. The Des Moines market needs at least 1000 boxes, while the San Antonio market must have at least 800 boxes. Let x = the number of boxes to be shipped to Des Moines and y = the number of boxes to be shipped to San Antonio.

44. Management   A cement manufacturer produces at least 3.2 million barrels of cement annually. He is told by the Environmental Protection Agency that his operation emits 2.5 pounds of dust for each barrel produced.

The EPA has ruled that annual emissions must be reduced to 1.8 million pounds. To do this the manufacturer plans to replace the present dust collectors with two types of electronic precipitators. One type would reduce emissions to .5 pounds per barrel and would cost 16¢ per barrel. The other would reduce the dust to .3 pounds per barrel and would cost 20¢ per barrel. The manufacturer does not want to spend more than .8 million dollars on the precipitators. He needs to know how many barrels he should produce with each type. Let x = the number of barrels in millions produced with the first type and y = the number of barrels in millions produced with the second type.

7.2 Mathematical Models for Linear Programming

Many mathematical models designed to solve problems in business, biology, and economics involve finding the optimum value (either the maximum or the minimum) of a function, subject to certain restrictions. The function to be optimized is called the **objective function.** In a **linear programming** problem, the maximum or minimum value of the objective function is found subject to a set of restrictions, or **constraints,** given by linear inequalities.

This section shows how to set up linear programming problems in two variables. The next section shows how to actually solve such problems.

Example 1 A farmer raises only geese and pigs. She wants to raise no more than 16 animals including no more than 10 geese. She spends $5 to raise a goose and $15 to raise a pig, and has $180 available for this project. Find the maximum profit she can make if each goose produces a profit of $6 and each pig a profit of $20.

In this example, we show how to set up the problem for solution by linear programming, but do not complete the solution. Let x represent the number of geese to be produced, and let y represent the number of pigs. Start by summarizing the information of the problem in a table.

	Number	Cost to raise	Profit each
Geese	x	$ 5	$ 6
Pigs	y	15	20
Maximum available	16	$180	

Use this table to write the necessary constraints. Since the total number of animals cannot exceed 16, the first constraint is

$$x + y \le 16.$$

"No more than 10 geese" leads to

$$x \le 10.$$

The cost to raise x geese at $5 per goose is $5x$ dollars, while the cost for y pigs at $15 each is $15y$ dollars. Since only $180 is available,

$$5x + 15y \leq 180.$$

The number of geese and pigs cannot be negative:

$$x \geq 0, \qquad y \geq 0.$$

The farmer wants to know the number of geese and the number of pigs that should be raised for maximum profit. Each goose produces a profit of $6, and each pig, $20. If z represents total profit, then

$$z = 6x + 20y$$

is the objective function, which is to be maximized.

In summary, the mathematical model for the given linear programming problem is as follows.

Maximize	$z = 6x + 20y$	**(1)**
subject to:	$x + y \leq 16$	**(2)**
	$x \leq 10$	**(3)**
	$5x + 15y \leq 180$	**(4)**
	$x \geq 0, \quad y \geq 0.$	**(5)**

As mentioned before, $z = 6x + 20y$ is the objective function, while inequalities (2)–(5) are the constraints.

Using the methods of the previous section, graph, as in Figure 7.11, the feasible region for the system of inequalities (2)–(5). 1

1 Graph the region of feasible solutions given the constraints

$$x + y \leq 10$$
$$2x - y \leq 20$$
$$x \geq 0$$
$$y \geq 0.$$

Answer:

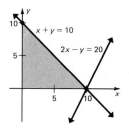

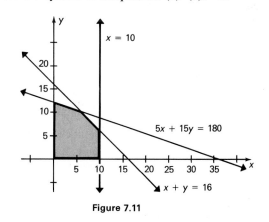

Figure 7.11

Any value in the feasible region of Figure 7.11 will satisfy all the constraints of the problem. However, in most problems only one point in the feasible region will lead to maximum profit. In the next section, we shall see that maximum profit is found by looking at all the *corner points* in the feasible region; a **corner point** is a point in the feasible region where the boundary lines of two constraints cross. The feasible region of Figure 7.11 has been redrawn in Figure 7.12 with all the corner points identified.

The coordinates (0, 12), (0, 0), and (10, 0) can be read directly from the graph. Since the corner points occur where two boundary lines cross, the coordinates of each of the other corner points can be found by solving a system of linear equations. For example, the corner point (6, 10) of Figure 7.12 is found by solving the system

$$x + y = 16$$
$$5x + 15y = 180,$$

using the methods of Chapter 6. Also, the solution of the system

$$x + y = 16$$
$$x = 10$$

gives the corner point (10, 6). ■

[2] Summarize the information from the problem in a table.

Answer:

	Number	Cost of each	Space required	Storage capacity
Ace	x	$40	6 sq ft	8 cu ft
Excello	y	80	8 sq ft	12 cu ft
Maximum available		$560	72 sq ft	

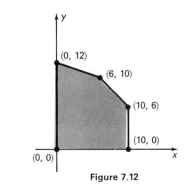

Figure 7.12

Example 2 Write a mathematical model, graph the region of feasible solutions, and find all corner points for the following problem:

An office manager needs to purchase new filing cabinets. He knows that Ace cabinets cost $40 each, require 6 square feet of floor space, and hold 8 cubic feet of files. On the other hand, each Excello cabinet costs $80, requires 8 square feet of floor space, and holds 12 cubic feet. His budget permits him to spend no more than $560 on files, while the office has room for no more than 72 square feet of cabinets. The manager desires the greatest storage capacity within the limitations imposed by funds and space. How many of each type cabinet should he buy?

Let x represent the number of Ace cabinets to be bought and let y represent the number of Excello cabinets. [2]

Start with the constraints imposed by cost and space.

$$40x + 80y \leq 560 \quad \text{Cost}$$
$$6x + 8y \leq 72 \quad \text{Floor space}$$

Since the number of cabinets cannot be negative, $x \geq 0$ and $y \geq 0$. The objective function to be maximized gives the amount of storage capacity provided by some combination of Ace and Excello cabinets. From the information in the table in Problem 2 at the side, the objective function is

$$\text{Storage space} = z = 8x + 12y.$$

In summary, the given problem has produced the following mathematical model.

$$\begin{aligned} \text{Maximize} \quad & z = 8x + 12y \\ \text{subject to:} \quad & 40x + 80y \leq 560 \\ & 6x + 8y \leq 72 \\ & x \geq 0, \quad y \geq 0. \end{aligned}$$

A graph of the feasible region is shown in Figure 7.13. Three of the corner points can be identified from the graph as $(0, 0)$, $(0, 7)$, and $(12, 0)$. The fourth corner point, labeled Q in the figure, can be found by solving the system of equations

$$40x + 80y = 560$$
$$6x + 8y = 72.$$

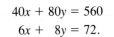 Solve this system to find that Q is the point $(8, 3)$.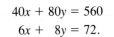

3 Find the corner point labeled Q on the region of feasible solutions given below.

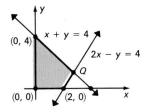

Answer:

$\left(\dfrac{8}{3}, \dfrac{4}{3} \right)$

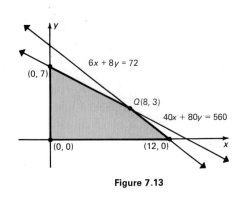

Figure 7.13

Example 3 Write a mathematical model, graph the feasible region, and find all corner points for the following problem:

Certain laboratory animals must have at least 30 grams of protein and at least 20 grams of fat per feeding period. These nutrients come from food A, which costs 18¢ per unit and supplies 2 grams of protein and 4 of fat, and food B, with 6 grams of protein and 2 of fat, costing 12¢ per unit. Food B is bought under a long term contract requiring that at least 2 units of B be used per serving. How much of each food must be bought to produce minimum cost per serving?

Let x represent the amount of food A needed, and y the amount of food B. Use the given information to produce the following table.

Food	Number of units	Grams of protein	Grams of fat	Cost
A	x	2	4	18¢
B	y	6	2	12¢
Minimum required		30	20	

The mathematical model is

$$\text{Minimize} \quad z = .18x + .12y$$
$$\text{subject to:} \quad 2x + 6y \geq 30$$
$$4x + 2y \geq 20$$
$$y \geq 2$$
$$x \geq 0, \quad y \geq 0.$$

(The constraint $y \geq 0$ is redundant because of the constraint $y \geq 2$.) A graph of the feasible region with the corner points identified is shown in Figure 7.14. **4**

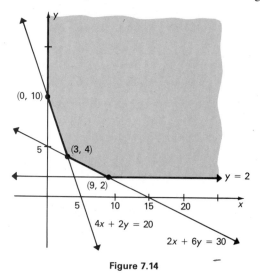

(0, 10)

(3, 4)

(9, 2)

$y = 2$

$4x + 2y = 20$

$2x + 6y = 30$

Figure 7.14

---

4 A popular cereal combines oats and corn. At least 27 tons of the cereal is to be made. For the best flavor, the amount of corn should be at least twice the amount of oats. Corn costs $200 per ton and oats cost $300 per ton. How much of each grain should be used to minimize the cost?
(a) Make a chart to organize the information given in the problem.
(b) Write an equation for the objective function.
(c) Write four inequalities for the constraints.

Answer:
(a)

	Number of tons	Cost/ton
Oats	x	$300
Corn	y	200
	27	

(b) $z = 300x + 200y$
(c) $x + y \geq 27$
$y \geq 2x$
$x \geq 0$
$y \geq 0$

The feasible region in Figure 7.14 is an **unbounded** feasible region—the region extends indefinitely to the upper right. With this region it would not be possible to *maximize* the objective function, because the total cost of the food could always be increased by encouraging the animals to eat more.

7.2 Exercises

Write the constraints in Exercises 1–6 as linear inequalities. Identify all variables used. (Not all the information is used in Exercises 5 and 6.) (See Examples 1–3.)

1. Product A requires 2 hours on machine I, while product B needs 3 hours on the same machine. The machine is available at most 45 hours per week.

2. A cow requires a third of an acre of pasture and a sheep needs a quarter acre. A rancher wants to use at least 120 acres of pasture.

3. John needs at least 25 units of vitamin A per day. Green pills provide 4 units and red pills provide 1.

4. Sandra spends 3 hours selling a small computer and 5 hours selling a larger model. She works no more than 45 hours per week.

5. Coffee costing $6 a pound is to be mixed with coffee costing $5 a pound to get at least 50 pounds of a mixture.

6. A tank in an oil refinery holds 120 gallons. The tank contains a mixture of light oil worth $1.25 per gallon and heavy oil worth $.80 per gallon.

In Exercises 7–16, set up a mathematical model, graph the feasible region, and identify all corner points. Do not try to solve the problem. (See Examples 1–3.)

7. **Management** A manufacturer of refrigerators must ship at least 100 refrigerators to its two West Coast warehouses. Each warehouse holds a maximum of 100 refrigerators. Warehouse A holds 25 refrigerators already, while warehouse B has 20 on hand. It costs $12 to ship a refrigerator to warehouse A and $10 to ship one to warehouse B. How many refrigerators should be shipped to each warehouse to minimize cost? What is the minimum cost?

8. Mark, who is ill, takes vitamin pills. Each day he must have at least 16 units of vitamin A, 5 units of vitamin B_1, and 20 units of vitamin C. He can choose between pill #1 which contains 8 units of A, 1 of B_1, and 2 of C, and pill #2 which contains 2 units of A, 1 of B_1, and 7 of C. Pill #1 costs 15¢ and pill #2 costs 30¢. How many of each pill should he buy in order to minimize his cost? What is the minimum cost?

9. **Management** A machine shop manufactures two types of bolts. Each can be made on any of three groups of machines, but the time required on each group differs, as shown in the table below.

| | Machine groups | | |
	I	II	III
Bolts Type 1	.1 minute	.1 minute	.1 minute
Type 2	.1 minute	.4 minute	.5 minute

Production schedules are made up one day at a time. In a day, there are 240, 720, and 160 minutes available, respectively, on these machines. Type 1 bolts sell for 10¢ and type 2 bolts for 12¢. How many of each type of bolt should be manufactured per day to maximize revenue? What is the maximum revenue?

10. **Management** Seall Manufacturing Company makes color television sets. It produces a bargain set that sells for $100 profit and a deluxe set that sells for $150 profit. On the assembly line the bargain set requires 3 hours, while the deluxe set takes 5 hours. The cabinet shop spends one hour on the cabinet for the bargain set and 3 hours on the cabinet for the deluxe set. Both sets require 2 hours of time for testing and packing. On a particular production run the Seall Company has available 3900 work hours on the assembly line, 2100 work hours in the cabinet shop, and 2200 work hours in the testing and packing department. How many sets of each type should it produce to make maximum profit? What is the maximum profit?

11. **Management** The manufacturing process requires that oil refineries must manufacture at least two gallons of gasoline for every one of fuel oil. To meet the winter demand for fuel oil, at least 3 million gallons a day

must be produced. The demand for gasoline is no more than 6.4 million gallons per day. If the refinery sells gasoline for $1.25 per gallon, and fuel oil for $1 per gallon, how much of each should be produced to maximize revenue? Find the maximum revenue.

12. In a small town in South Carolina, zoning rules require that the window space (in square feet) in a house be at least one-sixth of the space used up by solid walls. The cost to heat the house is 2¢ for each square foot of solid walls and 8¢ for each square foot of windows. Find the maximum total area (windows plus walls) if $16 is available to pay for heat.

13. **Management** A candy company has 100 kilograms of chocolate-covered nuts and 125 kilograms of chocolate-covered raisins to be sold as two different mixes. One mix will contain half nuts and half raisins and will sell for $6 per kilogram. The other mix will contain 1/3 nuts and 2/3 raisins and will sell for $4.80 per kilogram. How many kilograms of each mix should the company prepare for maximum revenue?

14. Ms. Oliveras was given the following advice. She should supplement her daily diet with at least 6000 USP units of vitamin A, at least 195 milligrams of vitamin C, and at least 600 USP units of vitamin D. Ms. Oliveras finds that Mason's Pharmacy carries Brand X vitamin pills at 5¢ each and Brand Y vitamins at 4¢ each. Each Brand X pill contains 3000 USP units of A, 45 milligrams of C, and 75 USP units of D, while the Brand Y pills contain 1000 USP units of A, 50 milligrams of C, and 200 USP units of D. What combination of vitamin pills should she buy to obtain the least possible cost? What is the least possible cost per day?

15. Sam, who is dieting, requires two food supplements, I and II. He can get these supplements from two different products, A and B, as shown in the following table.

		Supplement (grams per serving)	
		I	II
Product	A	3	2
	B	2	4

Sam's physician has recommended that he include at least 15 grams of each supplement in his daily diet. If product A costs 25¢ per serving and product B costs 40¢ per serving, how can he satisfy his requirements most economically? Find the minimum cost.

16. **Management** A small country can grow only two crops for export, coffee and cocoa. The country has 500,000 hectares of land available for the crops. Long-term contracts require that at least 100,000 hectares be devoted to coffee and at least 200,000 hectares to cocoa. Cocoa must be processed locally, and production bottlenecks limit cocoa to 270,000 hectares. Coffee requires two workers per hectare, with cocoa requiring five. No more than 1,750,000 people are available for these crops. Coffee produces a profit of $220 per hectare, and cocoa a profit of $310 per hectare. How many hectares should the country devote to each crop in order to maximize the profit? Find the maximum profit.

7.3 Solving Linear Programming Problems Graphically

The last section showed how to set up a linear programming problem by writing an objective function and the necessary constraints. Then the feasible region was graphed and all corner points identified. In this section, this process is completed. The method of solving these problems from the graph of the feasible region is explained in the next example.

Example 1   Solve the following linear programming problem.

$$\text{Maximize} \qquad z = 2x + 5y$$
$$\text{subject to:} \qquad 3x + 2y \leq 6$$
$$-2x + 4y \leq 8$$
$$x \geq 0, \quad y \geq 0.$$

The feasible region is graphed in Figure 7.15. The coordinates of point A, (1/2, 9/4), can be found by solving the system

$$3x + 2y = 6$$
$$-2x + 4y = 8.$$

Every point in the feasible region satisfies all the constraints. However, we want to find those points that produce the maximum possible value of the objective function. To see how to find this maximum value, let us add to the graph of Figure 7.15 lines which represent the objective function $z \doteq 2x + 5y$ for various sample values of z. By choosing the values 0, 5, 10, and 15 for z, the objective function becomes (in turn)

$$0 = 2x + 5y, \qquad 5 = 2x + 5y, \qquad 10 = 2x + 5y, \qquad 15 = 2x + 5y.$$

These four lines are graphed in Figure 7.16. (Why are all the lines parallel?) The figure shows that z cannot take on the value 15 because the graph for $z = 15$ is entirely outside the feasible region. The maximum possible value of z will be obtained from a line parallel to the others and between the lines representing the objective function when $z = 10$ and $z = 15$. The value of z will be as large as possible and all constraints will be satisfied if this line just touches the feasible region. This occurs at point A. We found above that A has coordinates (1/2, 9/4).

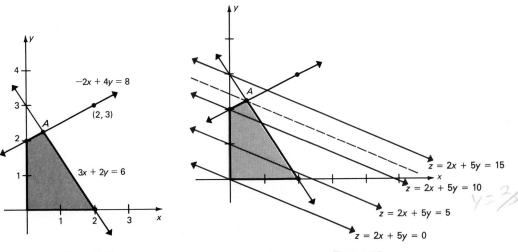

Figure 7.15   Figure 7.16

1 Suppose the objective function in Example 1 is changed to $z = 5x + 2y$.
(a) Sketch the graphs of the objective function when $z = 0$, $z = 5$, and $z = 10$ on the region of feasible solutions given in Figure 8.15.
(b) From the graph, decide what values of x and y will maximize the objective function.

Answer:
(a)

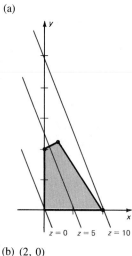

$z = 0$ $z = 5$ $z = 10$

(b) (2, 0)

2 The sketch shows a feasible region with a line representing the graph of the objective function for a particular value of z. Use the sketch to find the values of x and y that (a) minimize z; (b) maximize z.

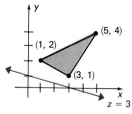

$z = 3$

Answer: (a) (3, 1) (b) (5, 4)

The value of z at this point is

$$z = 2x + 5y$$
$$= 2\left(\frac{1}{2}\right) + 5\left(\frac{9}{4}\right)$$
$$z = 12\frac{1}{4}.$$

The maximum possible value of z is 12 1/4. Of all the points in the feasible region, A leads to the largest possible value of z. **1**

Example 2 Solve the following linear programming problem.

$$\text{Minimize} \quad z = x + 2y$$
$$\text{subject to:} \quad x + y \le 10$$
$$3x + 2y \ge 6$$
$$x \ge 0, \quad y \ge 0.$$

The feasible region and the lines that correspond to $z = 0$, 6, 10, and 14 are shown in Figure 7.17. The minimum value of z will be found where a line (parallel to the other lines for z) just touches the feasible region—that is, at the point (2, 0). The value of z at that point is

$$z = x + 2y = 2 + 2(0) = 2. \quad \text{\textbf{2}}$$

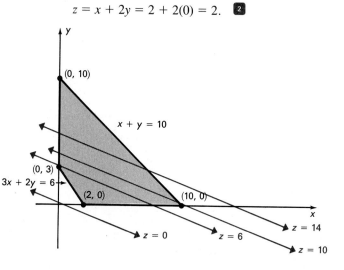

Figure 7.17

Example 3 Solve the following linear programming problem.

$$\text{Minimize} \quad z = 2x + 4y$$
$$\text{subject to:} \quad x + 2y \ge 10$$
$$3x + y \ge 10$$
$$x \ge 0, \quad y \ge 0.$$

Figure 7.18 shows the feasible region and the lines that result when z in the objective function is replaced by 0, 10, 20, 40, and 50. The line representing the objective function touches the region of feasible solutions when $z = 20$. Two corner points, (2, 4) and (10, 0), lie on this line. In this case, both (2, 4) and (10, 0) as well as all the points on the boundary line between them give the same optimum value of z. There is an infinite number of equally "good" values of x and y which give the same minimum value of the objective function $z = 2x + 4y$. This minimum value is 20. **3**

3 The sketch below shows a region of feasible solutions with a line representing the graph of the objective function for a particular value of z. From the sketch decide what ordered pair would minimize z.

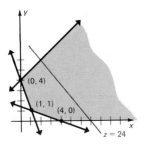

Answer:
(1, 1)

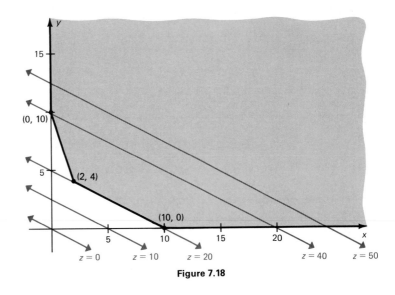

Figure 7.18

The feasible regions in Examples 1 and 2 above are *bounded,* since the regions are enclosed by boundary lines on all sides. Linear programming problems with bounded regions always have solutions. On the other hand, the feasible region in Example 3 is *unbounded* with no way to *maximize* the value of the objective function.

Some general conclusions can be drawn from the method of solution used in Examples 1–3. Figure 7.19 on the next page shows various feasible regions and the lines that result from various values of z. (We assume the lines are in order from left to right as z increases.) In part (a) of the figure, the objective function takes on its minimum value at corner point Q and its maximum value at P. The minimum is again at Q in part (b), but the maximum occurs at P_1 or P_2, or any point on the line segment connecting them. Finally, in part (c), the minimum value occurs at Q, but the objective function has no maximum value because the feasible region is unbounded.

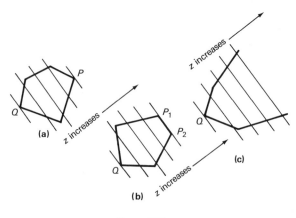

Figure 7.19

The preceding discussion suggests the truth of the **corner point theorem.**

> **Corner Point Theorem** If an optimum value (either a maximum or a minimum) of the objective function exists, it will occur at one or more of the corner points of the feasible region.

This theorem simplifies the job of finding an optimum value: First, graph the feasible region and find all corner points. Then test each corner point in the objective function. Finally, identify the corner point producing the optimum solution. For unbounded regions, first decide whether or not the required optimum can be found; see Example 3.

With the theorem, the problem in Example 1 could have been solved by identifying the four corner points of Figure 7.15: $(0, 0)$, $(0, 2)$, $(1/2, 9/4)$, and $(2, 0)$. Then substituting each of the four points into the objective function $z = 2x + 5y$ would identify the corner point that produces the maximum value of z.

Corner point	Value of $z = 2x + 5y$
$(0, 0)$	$2(0) + 5(0) = 0$
$(0, 2)$	$2(0) + 5(2) = 10$
$(\frac{1}{2}, \frac{9}{4})$	$2(\frac{1}{2}) + 5(\frac{9}{4}) = 12\frac{1}{4}$ (**maximum**)
$(2, 0)$	$2(2) + 5(0) = 4$

From these results, the corner point $(1/2, 9/4)$ yields the maximum value of 12 1/4. This is the same result found earlier. ④

[4] (a) Identify the corner points in the graph from problem 3 at the side.
(b) Which corner point would minimize $z = 2x + 3y$?

Answer:
(a) $(0, 4)$, $(1, 1)$, $(4, 0)$
(b) $(1, 1)$

Example 4 Sketch the feasible region for the following set of constraints:

$$3y - 2x \geq 0$$
$$y + 8x \leq 52$$
$$y - 2x \leq 2$$
$$x \geq 3.$$

Then find the maximum and minimum values of the objective function

$$z = 5x + 2y.$$

The graph in Figure 7.20 shows that the feasible region is bounded. Use the corner points from the graph to find the maximum and minimum values of the objective function.

Corner point	Value of $z = 5x + 2y$	
(3, 2)	$5(3) + 2(2) = 19$	**(minimum)**
(6, 4)	$5(6) + 2(4) = 38$	
(5, 12)	$5(5) + 2(12) = 49$	**(maximum)**
(3, 8)	$5(3) + 2(8) = 31$	

The minimum value of $z = 5x + 2y$ is 19 at the corner point (3, 2). The maximum value is 49 at (5, 12). ▣

5 Use the region of feasible solutions in the sketch to find the following.

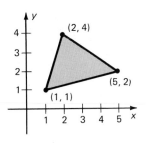

(a) The values of x and y that maximize $z = 2x - y$.
(b) The maximum value of $z = 2x - y$.
(c) The values of x and y that minimize $z = 4x + 3y$.
(d) The minimum value of $z = 4x + 3y$.

Answer:
(a) (5, 2)
(b) 8
(c) (1, 1)
(d) 7

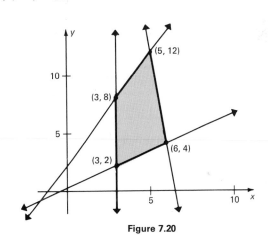

Figure 7.20

The next example completes the solution of Example 2 from the last section.

Example 5 An office manager needs to purchase new filing cabinets. He knows that Ace cabinets cost $40 each, require 6 square feet of floor space, and hold 8 cubic feet of files. On the other hand, each Excello cabinet costs $80, requires 8 square feet of floor space, and holds 12 cubic feet. His budget permits him to spend no more than $560 on files, and the office has room for no more than 72 square feet of cabinets. The manager desires the greatest storage capacity within the limitations imposed by funds and space. How many of each type of cabinet should he buy?

Figure 7.21 on the next page shows the feasible region and corner points found in Example 2 of the previous section. Test these four corner points in the objective function to determine the maximum value of z. The results are shown in the table.

Corner point	Value of $z = 8x + 12y$
(0, 0)	0
(0, 7)	84
(12, 0)	96
(8, 3)	100 (maximum)

The objective function, which represents storage space, is maximized when $x = 8$ and $y = 3$. The manager should buy 8 Ace cabinets and 3 Excello cabinets.   **6**

6 A popular cereal combines oats and corn. At least 27 tons of the cereal is to be made. For the best flavor, the amount of corn should be at least twice the amount of oats. Corn costs $200 per ton and oats cost $300 per ton. How much of each grain should be used to minimize the cost? This problem was set up in problem 4 at the side in Section 8.2.
(a) Graph the feasible region and find the corner points.
(b) Determine the minimum value of the objective function and the point where it occurs.

Answer:
(a)

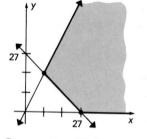

Corner points: (27, 0), (9, 18)
(b) $6300 at (9, 18)

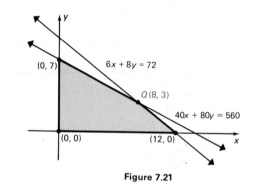

Figure 7.21

A summary of the steps in solving a linear programming problem by the graphical method is given here.

Solving a Linear Programming Problem Graphically

1. Write the objective function and all necessary constraints.
2. Graph the feasible region.
3. Determine the coordinates of each of the corner points.
4. Find the value of the objective function at each corner point.
5. For a bounded region, the solution is given by the corner point producing the optimum value of the objective function.
6. For an unbounded region, check that a solution actually exists. If it does, it will occur at a corner point.

7.3 Exercises

Exercises 1–6 show regions of feasible solutions. Use these regions to find maximum and minimum values of each given objective function. (See Example 4.)

1. $z = 3x + 5y$

2. $z = 6x + y$

3. $z = .40x + .75y$

4. $z = .35x + 1.25y$

5. $z = 2x + 3y$

6. $z = 5x + 6y$

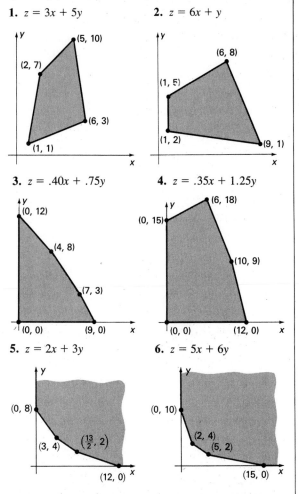

Use graphical methods to solve problems 7–14. (See Example 4.)

7. Maximize $z = 5x + 2y$
subject to: $2x + 3y \le 6$
$4x + y \le 6$
$x \ge 0, \quad y \ge 0.$

8. Minimize $z = x + 3y$
subject to: $x + y \le 10$
$5x + 2y \ge 20$
$-x + 2y \ge 0$
$x \ge 0, \quad y \ge 0.$

9. Maximize $z = 2x + y$
subject to: $3x - y \ge 12$
$x + y \le 15$
$x \ge 2, \quad y \ge 5.$

10. Maximize $z = x + 3y$
subject to: $2x + 3y \le 100$
$5x + 4y \le 200$
$x \ge 10, \quad y \ge 20.$

11. Maximize $z = 4x + 2y$
subject to: $x - y \le 10$
$5x + 3y \le 75$
$x \ge 0, \quad y \ge 0.$

12. Maximize $z = 4x + 5y$
subject to: $10x - 5y \le 100$
$20x + 10y \ge 150$
$x \ge 0, \quad y \ge 0.$

13. Find values of $x \ge 0$ and $y \ge 0$ which maximize $z = 10x + 12y$ subject to each of the following sets of constraints.
(a) $x + y \le 20$
$x + 3y \le 24$
(b) $3x + y \le 15$
$x + 2y \le 18$
(c) $2x + 5y \ge 22$
$4x + 3y \le 28$
$2x + 2y \le 17$

14. Find values of $x \ge 0$ and $y \ge 0$ which minimize $z = 3x + 2y$ subject to each of the following sets of constraints.
(a) $10x + 7y \le 42$
$4x + 10y \ge 35$
(b) $6x + 5y \ge 25$
$2x + 6y \ge 15$
(c) $x + 2y \ge 10$
$2x + y \ge 12$
$x - y \le 8$

In Exercises 15–24, complete the solution of the problems that were set up in Exercises 7–16 of the previous section. (See Example 5.)

15. **Management** A manufacturer of refrigerators must ship at least 100 refrigerators to its two West coast warehouses. Each warehouse holds a maximum of 100 refrigerators. Warehouse A holds 25 refrigerators already, while warehouse B has 20 on hand. It costs $12 to ship a refrigerator to warehouse A and $10 to ship one to warehouse B. How many refrigerators should be shipped to each warehouse to minimize cost? What is the minimum cost?

16. Mark, who is ill, takes vitamin pills. Each day he must have at least 16 units of vitamin A, 5 units of vitamin B_1, and 20 units of vitamin C. He can choose between pill #1 which contains 8 units of A, 1 of B_1, and 2 of C, and pill #2 which contains 2 units of A, 1 of B_1, and 7 of C. Pill #1 costs 15¢ and pill #2 costs 30¢. How many of each pill should he buy in order to minimize his cost? What is the minimum cost?

17. **Management** A machine shop manufactures two types of bolts. Each can be made on any of three groups of machines, but the time required on each group differs, as shown in the table below.

	Machine groups		
	I	II	III
Bolts Type 1	.1 minute	.1 minute	.1 minute
Type 2	.1 minute	.4 minute	.5 minute

Production schedules are made up one day at a time. In a day there are 240, 720, and 160 minutes available, respectively, on these machines. Type 1 bolts sell for 10¢ and type 2 bolts for 12¢. How many of each type of bolt should be manufactured per week to maximize revenue? What is the maximum revenue?

18. **Management** Seall Manufacturing Company makes color television sets. It produces a bargain set that sells for $100 profit and a deluxe set that sells for $150 profit. On the assembly line the bargain set requires 3 hours, while the deluxe set takes 5 hours. The cabinet shop spends one hour on the cabinet for the bargain set and 3 hours on the cabinet for the deluxe set. Both sets require 2 hours of time for testing and packing. On a

particular production run the Seall Company has available 3900 work hours on the assembly line, 2100 work hours in the cabinet shop, and 2200 work hours in the testing and packing department. How many sets of each type should it produce to make maximum profit? What is the maximum profit?

19. **Management** The manufacturing process requires that oil refineries must manufacture at least two gallons of gasoline for every one of fuel oil. To meet the winter demand for fuel oil, at least 3 million gallons a day must be produced. The demand for gasoline is no more than 6.4 million gallons per day. If the refinery sells gasoline for $1.25 per gallon, and fuel oil for $1 per gallon, how much of each should be produced to maximize revenue? Find the maximum revenue.

20. In a small town in South Carolina, zoning rules require that the window space (in square feet) in a house be at least one-sixth of the space used up by solid walls. The cost to heat the house is 2¢ for each square foot of solid walls and 8¢ for each square foot of windows. Find the maximum total area (windows plus walls) if $16 is available to pay for heat.

21. **Management** A candy company has 100 kilograms of chocolate-covered nuts and 125 kilograms of chocolate-covered raisins to be sold as two different mixes. One mix will contain half nuts and half raisins and will sell for $6 per kilogram. The other mix will contain 1/3 nuts and 2/3 raisins and will sell for $4.80 per kilogram. How many kilograms of each mix should the company prepare for maximum revenue? Find the maximum revenue.

22. Ms. Oliveras was given the following advice. She should supplement her daily diet with at least 6000 USP units of vitamin A, at least 195 milligrams of vitamin C, and at least 600 USP units of vitamin D. Ms. Oliveras finds that Mason's Pharmacy carries Brand X vitamin pills at 5¢ each and Brand Y vitamins at 4¢ each. Each Brand X pill contains 3000 USP units of A, 45 milligrams of C, and 75 USP units of D, while Brand Y pills contain 1000 USP units of A, 50 milligrams of C, and 200 USP units of D. What combination of vitamin pills should she buy to obtain the least possible cost? What is the least possible cost per day?

23. Sam, who is dieting, requires two food supplements, I and II. He can get these supplements from two different products, A and B, as shown in the following table.

| | | *Supplement* (grams per serving) | |
		I	II
Product	A	3	2
	B	2	4

Sam's physician has recommended that he include at least 15 grams of each supplement in his daily diet. If product A costs 25¢ per serving and product B costs 40¢ per serving, how can he satisfy his requirements most economically? Find the minimum cost.

24. Management A small country can grow only two crops for export, coffee and cocoa. The country has 500,000 hectares of land available for the crops. Long-term contracts require that at least 100,000 hectares be devoted to coffee and at least 200,000 hectares to cocoa. Cocoa must be processed locally, and production bottlenecks limit cocoa to 270,000 hectares. Coffee requires two workers per hectare, with cocoa requiring five. No more than 1,750,000 people are available for these crops. Coffee produces a profit of $220 per hectare, and cocoa a profit of $310 per hectare. How many hectares should the country devote to each crop in order to maximize the profit? Find the maximum profit.

Management *The importance of linear programming is shown by the inclusion of linear programming problems on most examinations for Certified Public Accountant. Answer the following questions from one such examination.* *

The Random Company manufactures two products, Zeta and Beta. Each product must pass through two processing operations. All materials are introduced at the start of Process No. 1. Random may produce either one product

exclusively or various combinations of both products subject to the following constraints:

	Process no. 1	*Process no. 2*	*Contribution margin per unit*
Hours required to produce one unit of:			
Zeta	1 hour	1 hour	$4.00
Beta	2 hours	3 hours	5.25
Total capacity in hours per day	1,000 hours	1,275 hours	

A shortage of technical labor has limited Beta production to 400 units per day. There are no constraints on the production of Zeta other than the hour constraints in the above schedule. Assume that all relationships between capacity and production are linear.

25. Given the objective to maximize total contribution margin, what is the production constraint for Process No. 1?
(a) Zeta + Beta $\leq$ 1,000
(b) Zeta + 2 Beta $\leq$ 1,000
(c) Zeta + Beta $\geq$ 1,000
(d) Zeta + 2 Beta $\geq$ 1,000

26. Given the objective to maximize total contribution margin, what is the labor constraint for production of Beta?
(a) Beta $\leq$ 400   (b) Beta $\geq$ 400
(c) Beta $\leq$ 425   (d) Beta $\geq$ 425

27. What is the objective function of the data presented?
(a) Zeta + 2 Beta = $9.25
(b) $4.00 Zeta + 3($5.25) Beta = total contribution margin
(c) $4.00 Zeta + $5.25 Beta = total contribution margin
(d) 2($4.00) Zeta + 3($5.25) Beta = total contribution margin

*Material from *Uniform CPA Examinations and Unofficial Answers,* copyright © 1973, 1974, 1975 by the American Institute of Certified Public Accountants, Inc., is reprinted with permission.

7.4 The Simplex Method: Slack Variables and the Pivot

In the previous section, linear programming problems were solved by the graphical method. This method illustrates the basic ideas of linear programming, but it is practical only for problems with two variables. For problems with more than two variables, or with two variables and many constraints, the simplex method is used.*

The simplex method starts with the selection of one corner point (often the origin) from the feasible region. Then, in a systematic way, another corner point is found which improves the value of the objective function. Finally, an optimum solution is reached or it can be seen that none exists.

The simplex method requires a number of steps. We have divided the presentation of these steps into two parts. First, in this section, the problem is set up and the method begun; and then, in the next section the method is completed.

Because the simplex method is used for problems with many variables, it usually is not convenient to use letters such as x, y, z, or w as variable names. Instead, the symbols x_1 (read "x-sub-one"), x_2, x_3, and so on, are used. In the simplex method, all constraints must be expressed in the linear form

$$a_1x_1 + a_2x_2 + a_3x_3 + \cdots \leq b,$$

where x_1, x_2, x_3, . . . are variables and a_1, a_2, . . . , b are constants.

We first discuss the simplex method for linear programming problems in *standard maximum form*.

> **Standard Maximum Form** A linear programming problem is in **standard maximum form** if
>
> 1. the objective function is to be maximized;
> 2. all variables are nonnegative ($x_i \geq 0$);
> 3. all constraints involve $\leq$;
> 4. the constants in the constraints are all nonnegative ($b \geq 0$).

(In Sections 7.6 and 7.7 we discuss problems which do not meet all of these conditions.)

Begin the simplex method by converting the constraints, which are linear inequalities, into linear equations. Do this by adding a nonnegative variable, called a **slack variable,** to each constraint. For example, convert the inequality $x_1 + x_2 \leq 10$ into an equation by adding the slack variable x_3, to get

$$x_1 + x_2 + x_3 = 10, \qquad \text{where } x_3 \geq 0.$$

---

*A computer diskette with the simplex method is included in D. R. Coscia, *Computer Applications in Finite Mathematics and Calculus,* Scott, Foresman and Co., 1985.

The inequality $x_1 + x_2 \leq 10$ says that the sum $x_1 + x_2$ is less than or perhaps equal to 10. The variable x_3 "takes up any slack" and represents the amount by which $x_1 + x_2$ fails to equal 10. For example, if $x_1 + x_2$ equals 8, then x_3 is 2. If $x_1 + x_2 = 10$, the value of x_3 is 0.

Example 1 Restate the following linear programming problem by introducing slack variables.

$$\text{Maximize} \qquad z = 3x_1 + 2x_2 + x_3$$
$$\text{subject to:} \qquad 2x_1 + x_2 + x_3 \leq 150$$
$$2x_1 + 2x_2 + 8x_3 \leq 200$$
$$2x_1 + 3x_2 + x_3 \leq 320$$

with $x_1 \geq 0$, $x_2 \geq 0$, $x_3 \geq 0$.

Rewrite the three constraints as equations by adding slack variables x_4, x_5, and x_6, one for each constraint. By this process, the problem is restated as

$$\text{Maximize} \qquad z = 3x_1 + 2x_2 + x_3$$
$$\text{subject to:} \qquad 2x_1 + x_2 + x_3 + x_4 \qquad\qquad = 150$$
$$2x_1 + 2x_2 + 8x_3 \qquad + x_5 \qquad = 200$$
$$2x_1 + 3x_2 + x_3 \qquad\qquad + x_6 = 320$$

with $x_1 \geq 0$, $x_2 \geq 0$, $x_3 \geq 0$, $x_4 \geq 0$, $x_5 \geq 0$, $x_6 \geq 0$. ◾

1 Rewrite the following set of constraints as equations by adding slack variables.

$$x_1 + x_2 + x_3 \leq 12$$
$$2x_1 + 4x_2 \qquad \leq 15$$
$$x_2 + 3x_3 \leq 10$$

Answer:

$$x_1 + x_2 + x_3 + x_4 = 12$$
$$2x_1 + 4x_2 \qquad + x_5 = 15$$
$$x_2 + 3x_3 + x_6 = 10$$

Adding slack variables to the constraints converts a linear programming problem into a system of linear equations. These equations should have all variables on the left of the equals sign and all constants on the right. All the equations of Example 1 satisfy this condition except for the objective function, $z = 3x_1 + 2x_2 + x_3$, which may be written with all variables on the left as

$$-3x_1 - 2x_2 - x_3 + z = 0.$$

Now the equations of Example 1 can be written as the following augmented matrix.

$$
\begin{array}{ccccccc}
x_1 & x_2 & x_3 & x_4 & x_5 & x_6 & z \\
\end{array}
$$

$$
\left[
\begin{array}{ccccccc|c}
2 & 1 & 1 & 1 & 0 & 0 & 0 & 150 \\
2 & 2 & 8 & 0 & 1 & 0 & 0 & 200 \\
2 & 3 & 1 & 0 & 0 & 1 & 0 & 320 \\
\hline
-3 & -2 & -1 & 0 & 0 & 0 & 1 & 0
\end{array}
\right]
$$

indicators

This matrix is called the **initial simplex tableau.** The numbers in the bottom row, which are from the objective function, are called **indicators** (except for the 0 at the far right). The column headed z will never change in all future work, and will be omitted from now on.

Example 2 Set up the initial simplex tableau for the following problem:

A farmer has 100 acres of available land which he wishes to plant with a mixture of potatoes, corn, and cabbage. It costs him $400 to produce an acre of potatoes, $160 to produce an acre of corn, and $280 to produce an acre of cabbage. He has a maximum of $20,000 to spend. He makes a profit of $120 per acre of potatoes, $40 per acre of corn, and $60 per acre of cabbage. How many acres of each crop should he plant to maximize his profit?

Summarize the given information as follows.

Crop	Number of acres	Cost per acre	Profit per acre
Potatoes	x_1	$400	$120
Corn	x_2	160	40
Cabbage	x_3	280	60
Maximum available	100	$20,000	

If the number of acres allotted to each of the three crops is represented by x_1, x_2, and x_3, respectively, then the constraints of the example can be expressed as

$$x_1 + x_2 + x_3 \leq 100 \quad \text{(number of acres),}$$
$$400x_1 + 160x_2 + 280x_3 \leq 20,000 \quad \text{(production costs),}$$

where x_1, x_2, and x_3 are all nonnegative. The first of these constraints says that $x_1 + x_2 + x_3$ is less than or perhaps equal to 100. Use x_4 as the slack variable, giving the equation

$$x_1 + x_2 + x_3 + x_4 = 100.$$

Here x_4 represents the amount of the farmer's 100 acres that will not be used. (x_4 may be 0 or any value up to 100.)

In the same way, the constraint $400x_1 + 160x_2 + 280x_3 \leq 20,000$ can be converted into an equation by adding a slack variable, x_5:

$$400x_1 + 160x_2 + 280x_3 + x_5 = 20,000.$$

The slack variable x_5 represents any unused portion of the farmer's $20,000 capital. (Again, x_5 may have any value from 0 to 20,000.)

The objective function represents the profit. The farmer wants to maximize

$$z = 120x_1 + 40x_2 + 60x_3.$$

We have now set up the following linear programming problem.

$$\text{Maximize} \quad z = 120x_1 + 40x_2 + 60x_3$$
$$\text{subject to:} \quad x_1 + x_2 + x_3 + x_4 = 100$$
$$400x_1 + 160x_2 + 280x_3 + x_5 = 20,000$$

with $x_1 \geq 0$, $x_2 \geq 0$, $x_3 \geq 0$, $x_4 \geq 0$, $x_5 \geq 0$. Rewrite the objective function as $-120x_1 - 40x_2 - 60x_3 + z = 0$, to get the following initial simplex tableau.

$$
\begin{array}{ccccc}
x_1 & x_2 & x_3 & x_4 & x_5 \\
\end{array}
$$

$$
\left[\begin{array}{ccccc|c}
1 & 1 & 1 & 1 & 0 & 100 \\
400 & 160 & 280 & 0 & 1 & 20{,}000 \\
\hline
-120 & -40 & -60 & 0 & 0 & 0
\end{array}\right] \quad \boxed{2}
$$

2 Set up the initial simplex tableau for the following linear programming problem:

Since the x_4 and x_5 columns in this tableau make up the identity matrix, these variables are called **basic variables.** If we let $x_4 = 100$ (the entry to the far right in the row where the x_4 column has a 1), and let $x_5 = 20{,}000$, the system of equations

Maximize $z = 2x_1 + 3x_2$

subject to: $x_1 + 2x_2 \le 85$

$2x_1 + x_2 \le 92$

$x_1 + 4x_2 \le 104$

with $x_1 \ge 0$, $x_2 \ge 0$.

becomes

$$
\begin{aligned}
x_1 + x_2 + x_3 + x_4 &= 100 \\
400x_1 + 160x_2 + 280x_3 \qquad + x_5 &= 20{,}000
\end{aligned}
$$

Answer:

$$
\left[\begin{array}{ccccc|c}
1 & 2 & 1 & 0 & 0 & 85 \\
2 & 1 & 0 & 1 & 0 & 92 \\
1 & 4 & 0 & 0 & 1 & 104 \\
\hline
-2 & -3 & 0 & 0 & 0 & 0
\end{array}\right]
$$

or

$$
\begin{aligned}
x_1 + x_2 + x_3 + 100 &= 100 \\
400x_1 + 160x_2 + 280x_3 \qquad + 20{,}000 &= 20{,}000
\end{aligned}
$$

$$
\begin{aligned}
x_1 + x_2 + x_3 &= 0 \\
400x_1 + 160x_2 + 280x_3 &= 0.
\end{aligned}
$$

Since $x_1 \ge 0$, $x_2 \ge 0$, and $x_3 \ge 0$, the only possible solution for this last system of equations is $x_1 = 0$, $x_2 = 0$, and $x_3 = 0$. This solution, which corresponds to the origin of the region of feasible solutions, is hardly optimal—it produces a profit of $0 for the farmer. In the next section the simplex method will be used to start with this solution and improve it to find the maximum possible profit. ■

Example 3 Read a solution from the matrix below.

$$
\begin{array}{ccccc}
x_1 & x_2 & x_3 & x_4 & x_5 \\
\end{array}
$$

$$
\left[\begin{array}{ccccc|c}
2 & 1 & 8 & 5 & 0 & 27 \\
9 & 0 & 3 & 12 & 1 & 45 \\
\hline
-2 & 0 & -4 & 0 & 0 & 0
\end{array}\right]
$$

The variables x_2 and x_5 are basic variables. The 1 in the x_2 column is in the first row. This means that $x_2 = 27$ (the far right number in the first row). Also, $x_5 = 45$. Setting $x_2 = 27$ and $x_5 = 45$ forces x_1, x_3, and x_4 to be zero. The solution is thus $x_1 = 0$, $x_2 = 27$, $x_3 = 0$, $x_4 = 0$, and $x_5 = 45$. ■

3 Read a solution from the matrix shown below.

$$
\begin{array}{ccccc}
x_1 & x_2 & x_3 & x_4 & x_5 \\
\end{array}
$$

$$
\left[\begin{array}{ccccc|c}
1 & 2 & 0 & 1 & 10 & 25 \\
0 & 6 & 1 & 3 & 1 & 50
\end{array}\right]
$$

Answer:

$x_1 = 25$, $x_2 = 0$, $x_3 = 50$, $x_4 = 0$, $x_5 = 0$

Solutions read directly from the initial simplex tableau are seldom, if ever, optimal. It is necessary to proceed to other solutions (corresponding to other corner points of the feasible region) until an optimum solution is found. Get these other solutions by using the row transformations of Chapter 6 to change the tableau by "pivoting" about one of the nonzero entries of the tableau. **Pivoting,** explained in the next example, produces a new tableau leading to another solution of the system of equations obtained from the original problem.

Example 4 Pivot about the indicated 2 of the initial simplex tableau

$$
\begin{array}{cccccc}
x_1 & x_2 & x_3 & x_4 & x_5 & x_6 \\
\end{array}
$$

$$
\left[\begin{array}{cccccc|c}
2 & 1 & 1 & 1 & 0 & 0 & 150 \\
2 & 2 & 8 & 0 & 1 & 0 & 200 \\
2 & 3 & 1 & 0 & 0 & 1 & 320 \\
\hline
-3 & -2 & -1 & 0 & 0 & 0 & 0
\end{array}\right].
$$

To pivot about the indicated 2, change x_1 into a basic variable by getting a 1 where the 2 is now and changing all other entries in the x_1 column to 0. Start by multiplying each entry of row 1 by 1/2.

$$
\begin{array}{cccccc}
x_1 & x_2 & x_3 & x_4 & x_5 & x_6 \\
\end{array}
$$

$$
\left[\begin{array}{cccccc|c}
1 & \frac{1}{2} & \frac{1}{2} & \frac{1}{2} & 0 & 0 & 75 \\
2 & 2 & 8 & 0 & 1 & 0 & 200 \\
2 & 3 & 1 & 0 & 0 & 1 & 320 \\
\hline
-3 & -2 & -1 & 0 & 0 & 0 & 0
\end{array}\right] \quad \tfrac{1}{2}R_1
$$

Now get 0 in row 2, column 1 by multiplying each entry in row 1 by -2 and adding the result to the corresponding entry in row 2.

$$
\begin{array}{cccccc}
x_1 & x_2 & x_3 & x_4 & x_5 & x_6 \\
\end{array}
$$

$$
\left[\begin{array}{cccccc|c}
1 & \frac{1}{2} & \frac{1}{2} & \frac{1}{2} & 0 & 0 & 75 \\
0 & 1 & 7 & -1 & 1 & 0 & 50 \\
2 & 3 & 1 & 0 & 0 & 1 & 320 \\
\hline
-3 & -2 & -1 & 0 & 0 & 0 & 0
\end{array}\right] \quad -2R_1 + R_2
$$

Change the 2 in row 3, column 1 to a 0 by a similar process.

$$
\begin{array}{cccccc}
x_1 & x_2 & x_3 & x_4 & x_5 & x_6 \\
\end{array}
$$

$$
\left[\begin{array}{cccccc|c}
1 & \frac{1}{2} & \frac{1}{2} & \frac{1}{2} & 0 & 0 & 75 \\
0 & 1 & 7 & -1 & 1 & 0 & 50 \\
0 & 2 & 0 & -1 & 0 & 1 & 170 \\
\hline
-3 & -2 & -1 & 0 & 0 & 0 & 0
\end{array}\right] \quad -2R_1 + R_3
$$

Finally, change the indicator -3 to 0 by multiplying each entry in row 1 by 3 and adding the result to the corresponding entry in row 4.

$$
\begin{array}{cccccc}
x_1 & x_2 & x_3 & x_4 & x_5 & x_6 \\
\end{array}
$$

$$
\left[\begin{array}{cccccc|c}
1 & \frac{1}{2} & \frac{1}{2} & \frac{1}{2} & 0 & 0 & 75 \\
0 & 1 & 7 & -1 & 1 & 0 & 50 \\
0 & 2 & 0 & -1 & 0 & 1 & 170 \\
\hline
0 & -\frac{1}{2} & \frac{1}{2} & \frac{3}{2} & 0 & 0 & 225
\end{array}\right] \quad 3R_1 + R_4
$$

This simplex tableau gives the solution $x_1 = 75$, $x_2 = 0$, $x_3 = 0$, $x_4 = 0$, $x_5 = 50$, and $x_6 = 170$. Substituting these results into the objective function gives $z = 225$. (The value of z is always the number in the lower right-hand corner.) **4**

In the simplex method, this process is repeated until an optimum solution is found, if one exists. In the next section, we discuss how to decide where to pivot to improve the value of the objective function. We also show how to tell when an optimum solution has been reached or does not exist.

4 Pivot about the indicated 2 in each simplex tableau.

(a)

$$\begin{array}{ccccc} x_1 & x_2 & x_3 & x_4 & x_5 \end{array}$$

$$\left[\begin{array}{ccccc|c} 4 & 2 & 0 & 1 & 0 & 200 \\ 3 & 1 & 1 & 0 & 1 & 300 \end{array}\right]$$

(b)

$$\begin{array}{cccc} x_1 & x_2 & x_3 & x_4 \end{array}$$

$$\left[\begin{array}{cccc|c} 1 & 3 & 1 & 0 & 20 \\ 4 & 2 & 0 & 1 & 50 \end{array}\right]$$

Answer:

(a)

$$\begin{array}{ccccc} x_1 & x_2 & x_3 & x_4 & x_5 \end{array}$$

$$\left[\begin{array}{ccccc|c} 2 & 1 & 0 & \frac{1}{2} & 0 & 100 \\ 1 & 0 & 1 & -\frac{1}{2} & 1 & 200 \end{array}\right]$$

(b)

$$\begin{array}{cccc} x_1 & x_2 & x_3 & x_4 \end{array}$$

$$\left[\begin{array}{cccc|c} -5 & 0 & 1 & -\frac{3}{2} & -55 \\ 2 & 1 & 0 & \frac{1}{2} & 25 \end{array}\right]$$

7.4 Exercises

Convert each of the inequalities in Exercises 1–4 into equations by adding a slack variable. (See Example 1.)

1. $x_1 + 2x_2 \leq 6$

2. $3x_1 + 5x_2 \leq 100$

3. $2x_1 + 4x_2 + 3x_3 \leq 100$

4. $8x_1 + 6x_2 + 5x_3 \leq 250$

For Exercises 5–8, (a) determine the number of slack variables needed; (b) name them; (c) use slack variables to convert each constraint into a linear equation. (See Example 1.)

5. Maximize $z = 10x_1 + 12x_2$
subject to: $4x_1 + 2x_2 \leq 20$
$5x_1 + x_2 \leq 50$
$2x_1 + 3x_2 \leq 25$
and $x_1 \geq 0$, $x_2 \geq 0$.

6. Maximize $z = 1.2x_1 + 3.5x_2$
subject to: $2.4x_1 + 1.5x_2 \leq 10$
$1.7x_1 + 1.9x_2 \leq 15$
and $x_1 \geq 0$, $x_2 \geq 0$.

7. Maximize $z = 8x_1 + 3x_2 + x_3$
subject to: $7x_1 + 6x_2 + 8x_3 \leq 118$
$4x_1 + 5x_2 + 10x_3 \leq 220$
and $x_1 \geq 0$, $x_2 \geq 0$, $x_3 \geq 0$.

8. Maximize $z = 12x_1 + 15x_2 + 10x_3$
subject to: $2x_1 + 2x_2 + x_3 \leq 8$
$x_1 + 4x_2 + 3x_3 \leq 12$
and $x_1 \geq 0$, $x_2 \geq 0$, $x_3 \geq 0$.

Write the solution that can be read from Exercises 9–12. (See Example 3.)

9.

$$\begin{array}{ccccc} x_1 & x_2 & x_3 & x_4 & x_5 \end{array}$$

$$\left[\begin{array}{ccccc|c} 2 & 2 & 0 & 3 & 1 & 15 \\ 3 & 4 & 1 & 6 & 0 & 20 \\ \hline -2 & -1 & 0 & 1 & 0 & 10 \end{array}\right]$$

10.

$$\begin{array}{ccccc} x_1 & x_2 & x_3 & x_4 & x_5 \end{array}$$

$$\left[\begin{array}{ccccc|c} 0 & 2 & 1 & 1 & 3 & 5 \\ 1 & 5 & 0 & 1 & 2 & 8 \\ \hline 0 & -2 & 0 & 1 & 1 & 10 \end{array}\right]$$

11.

$$\begin{array}{cccccc} x_1 & x_2 & x_3 & x_4 & x_5 & x_6 \end{array}$$

$$\left[\begin{array}{cccccc|c} 6 & 2 & 1 & 3 & 0 & 0 & 8 \\ 2 & 2 & 0 & 1 & 0 & 1 & 7 \\ 2 & 1 & 0 & 3 & 1 & 0 & 6 \\ \hline -3 & -2 & 0 & 2 & 0 & 0 & 12 \end{array}\right]$$

12.

$$\begin{array}{cccccc} x_1 & x_2 & x_3 & x_4 & x_5 & x_6 \end{array}$$

$$\left[\begin{array}{cccccc|c} 0 & 2 & 0 & 1 & 2 & 2 & 3 \\ 0 & 3 & 1 & 0 & 1 & 2 & 2 \\ 1 & 4 & 0 & 0 & 3 & 5 & 5 \\ \hline 0 & -4 & 0 & 0 & 4 & 3 & 20 \end{array}\right]$$

Pivot as indicated in the simplex tableau in Exercises 13–18. Read the solution from the final result. (See Example 4.)

13.

$$\begin{array}{ccccc} x_1 & x_2 & x_3 & x_4 & x_5 \end{array}$$

$$\left[\begin{array}{ccccc|c} 1 & 2 & 4 & 1 & 0 & 56 \\ 2 & \boxed{2} & 1 & 0 & 1 & 40 \\ \hline -1 & -3 & -2 & 0 & 0 & 0 \end{array}\right]$$

14.

$$\begin{array}{ccccc} x_1 & x_2 & x_3 & x_4 & x_5 \end{array}$$

$$\left[\begin{array}{ccccc|c} 5 & 4 & 1 & 1 & 0 & 50 \\ 3 & 3 & \boxed{2} & 0 & 1 & 40 \\ \hline -1 & -2 & -4 & 0 & 0 & 0 \end{array}\right]$$

15.

$$\begin{array}{cccccc} x_1 & x_2 & x_3 & x_4 & x_5 & x_6 \end{array}$$

$$\left[\begin{array}{cccccc|c} 2 & 2 & \boxed{1} & 1 & 0 & 0 & 12 \\ 1 & 2 & 3 & 0 & 1 & 0 & 45 \\ 3 & 1 & 1 & 0 & 0 & 1 & 20 \\ \hline -2 & -1 & -3 & 0 & 0 & 0 & 0 \end{array}\right]$$

16.

$$\begin{array}{cccccc} x_1 & x_2 & x_3 & x_4 & x_5 & x_6 \end{array}$$

$$\left[\begin{array}{cccccc|c} 4 & 2 & 3 & 1 & 0 & 0 & 22 \\ 2 & 2 & \boxed{5} & 0 & 1 & 0 & 28 \\ 1 & 3 & 2 & 0 & 0 & 1 & 45 \\ \hline -3 & -2 & -4 & 0 & 0 & 0 & 0 \end{array}\right]$$

17.

$$\begin{array}{cccccc} x_1 & x_2 & x_3 & x_4 & x_5 & x_6 \end{array}$$

$$\left[\begin{array}{cccccc|c} 1 & 1 & 1 & 1 & 0 & 0 & 60 \\ 3 & 1 & \boxed{2} & 0 & 1 & 0 & 100 \\ 1 & 2 & 3 & 0 & 0 & 1 & 200 \\ \hline -1 & -1 & -2 & 0 & 0 & 0 & 0 \end{array}\right]$$

18.

$$\begin{array}{ccccccc} x_1 & x_2 & x_3 & x_4 & x_5 & x_6 & x_7 \end{array}$$

$$\left[\begin{array}{ccccccc|c} 1 & 2 & 3 & 1 & 1 & 0 & 0 & 115 \\ 2 & 1 & 8 & 5 & 0 & 1 & 0 & 200 \\ \boxed{1} & 0 & 1 & 0 & 0 & 0 & 1 & 50 \\ \hline -2 & -1 & -1 & -1 & 0 & 0 & 0 & 0 \end{array}\right]$$

Introduce slack variables as necessary and then write the initial simplex tableau for the linear programming problems in Exercises 19–24. (See Example 2.)

19. Maximize $\quad z = 5x_1 + x_2$
subject to: $\quad 2x_1 + 3x_2 \le 6$
$\qquad\qquad\quad 4x_1 + x_2 \le 6$
with $x_1 \ge 0$ and $x_2 \ge 0$.

20. Maximize $\quad z = x_1 + 3x_2$
subject to: $\quad 2x_1 + 3x_2 \le 100$
$\qquad\qquad\quad 5x_1 + 4x_2 \le 200$
with $x_1 \ge 50$ and $x_2 \ge 50$.

21. Find $x_1 \ge 0$ and $x_2 \ge 0$ such that

$$x_1 + x_2 \le 10$$
$$5x_1 + 2x_2 \le 20$$
$$x_1 + 2x_2 \le 36$$

and $z = x_1 + 3x_2$ is maximized.

22. Find $x_1 \ge 0$ and $x_2 \ge 0$ such that

$$x_1 + x_2 \le 10$$
$$5x_1 + 3x_2 \le 75$$

and $z = 4x_1 + 2x_2$ is maximized.

23. Find $x_1 \ge 0$ and $x_2 \ge 0$ such that

$$3x_1 + x_2 \le 12$$
$$x_1 + x_2 \le 15$$

and $z = 2x_1 + x_2$ is maximized.

24. Find $x_1 \ge 0$ and $x_2 \ge 0$ such that

$$10x_1 + 4x_2 \le 100$$
$$20x_1 + 10x_2 \le 150$$

and $z = 4x_1 + 5x_2$ is maximized.

Management *Set up Exercises 25–30 for solution by the simplex method; that is, express the linear constraints and objective function, add slack variables, and set up the initial simplex tableau. (See Example 2.)*

25. A candy company has 100 kilograms of chocolate-covered nuts and 125 kilograms of chocolate-covered raisins to be sold as two different mixtures. One mix will contain half nuts and half raisins and will sell for $6 per kilogram. The other mix will contain 1/3 nuts and 2/3 raisins, and will sell for $4.80 per kilogram. How many kilograms of each mix should the company prepare for maximum revenue? (This is Exercise 21, Section 7.3.)

26. Seall Manufacturing Company makes color television sets. It produces a bargain set that sells for $100 profit and a deluxe set that sells for $150 profit. On the assembly line the bargain set requires 3 hour's work, while the deluxe set takes 5 hours. The cabinet shop spends one hour on the cabinet for the bargain set and 3 hours on the cabinet for the deluxe set. Both sets require 2 hours of time for testing and packing. On a particular production run the Seall Company has available 3900 work hours on the assembly line, 2100 work hours in the cabinet shop, and 2200 work hours in the testing and packing department. How many sets of each type should it produce to make maximum profit? What is the maximum profit? (See Exercise 18, Section 7.3.)

27. A small boat manufacturer builds three types of fiberglass boats: prams, runabouts, and trimarans. The pram sells at a profit of $75, the runabout at a profit of $90, and the trimaran at a profit of $100. The factory is divided into two sections. Section A does the molding and construction work, while section B does the painting, finishing and equipping. The pram takes 1 hour in section A and 2 hours in section B. The runabout takes 2 hours in A and 5 hours in B. The trimaran takes 3 hours in A and 4 hours in B. Section A has a total of 6240 hours available and section B has 10,800 hours available for the year. The manufacturer has ordered a supply of fiberglass that will build at most 3000 boats, figuring the average amount used per boat. How many of each type of boat should be made to produce maximum profit? What is the maximum profit?

28. Caroline's Quality Candy Confectionery is famous for fudge, chocolate cremes, and pralines. Its candy-making equipment is set up to make 100-pound batches at a time. Currently there is a chocolate shortage and the company can get only 120 pounds of chocolate in the next shipment. On a week's run, the confectionery's cooking and processing equipment is available for a total of 42 machine hours. During the same period the employees have a total of 56 work hours available for packaging. A batch of fudge requires 20 pounds of chocolate while a batch of cremes uses 25 pounds of chocolate. The cooking and processing take 120 minutes for fudge, 150 minutes for chocolate cremes, and 200 minutes for pralines. The packaging times measured in minutes per one pound box are 1, 2, and 3 respectively, for fudge, cremes, and pralines. Determine how many batches of each type of candy the confectionery should make, assuming that the profit per pound box is 50¢ on fudge, 40¢ on chocolate cremes, and 45¢ on pralines. What is the maximum profit?

29. A cat breeder has the following amounts of cat food: 90 units of tuna, 80 units of liver, and 50 units of chicken. To raise a Siamese cat, the breeder must use 2 units of tuna, 1 of liver, and 1 of chicken per day, while raising a Persian cat requires 1, 2, and 1 units respectively per day. If a Siamese cat sells for $12, while a Persian cat sells for $10, how many of each should be raised in order to obtain maximum gross income? What is the maximum gross income?

30. Banal, Inc. produces art for motel rooms. Its painters can turn out mountain scenes, seascapes, and pictures of clowns. Each painting is worked on by three different artists, T, D, and H. Artist T works only 25 hours per week, while D and H work 45 and 40 hours per week, respectively. Artist T spends 1 hour on a mountain scene, 2 hours on a seascape, and 1 hour on a clown. Corresponding times for D and H are 3, 2, and 2 hours, and 2, 1, and 4 hours, respectively. Banal makes $20 on a mountain scene, $18 on a seascape, and $22 from a clown. The head painting packer can't stand clowns, so that no more than 4 clown paintings may be done in a week. Find the number of each type of painting that should be made weekly in order to maximize profit. Find the maximum possible profit.

7.5 Solving Maximization Problems

Linear programming problems were prepared for solution by first converting the constraints to linear equations with slack variables. Then the coefficients of the variables from the linear equations were written as an augmented matrix. Finally, the pivot was used to go from one vertex of the region of feasible solutions to another.

Now we are ready to put all this together and produce an optimum value for the objective function. The procedure is illustrated by completing the example about the farmer. (Recall Example 2 from Section 7.4.)

In the previous section we set up the following simplex tableau.

$$
\begin{array}{ccccc}
x_1 & x_2 & x_3 & x_4 & x_5
\end{array}
$$

$$
\left[
\begin{array}{ccccc|c}
1 & 1 & 1 & 1 & 0 & 100 \\
400 & 160 & 280 & 0 & 1 & 20{,}000 \\
\hline
-120 & -40 & -60 & 0 & 0 & 0
\end{array}
\right]
$$

This tableau leads to the solution $x_1 = 0$, $x_2 = 0$, $x_3 = 0$, $x_4 = 100$, and $x_5 = 20{,}000$. These values produce a value of 0 for z. Since a value of 0 for the farmer's profit is not an optimum, we try to improve this value.

The coefficients of x_1, x_2, and x_3 in the objective function are nonzero, so the profit could be improved by making any one of these variables take on a nonzero value in a solution. Decide which variable to use by looking at the indicators in the last row of the initial simplex tableau above. The coefficient of x_1, -120, is the "most negative" of the indicators. This means that x_1 has the largest coefficient in the objective function, and profit is increased the most by increasing x_1.

If x_1 is nonzero in the solution, then x_1 will be a basic variable. This means that either x_4 or x_5 no longer will be a basic variable. Decide which variable will no longer be basic by starting with the equations of the system,

$$
\begin{aligned}
x_1 + \quad x_2 + \quad x_3 + x_4 \quad\quad &= 100 \\
400x_1 + 160x_2 + 280x_3 \quad\quad + x_5 &= 20{,}000,
\end{aligned}
$$

and solving for x_4 and x_5 respectively.

$$
\begin{aligned}
x_4 &= 100 - x_1 - x_2 - x_3 \\
x_5 &= 20{,}000 - 400x_1 - 160x_2 - 280x_3
\end{aligned}
$$

Only x_1 is being changed to a nonzero value; both x_2 and x_3 keep the value 0. Replacing x_2 and x_3 with 0 gives

$$
\begin{aligned}
x_4 &= 100 - x_1 \\
x_5 &= 20{,}000 - 400x_1.
\end{aligned}
$$

Since both x_4 and x_5 must remain nonnegative, there is a limit to how much the value of x_1 can be increased. The equation $x_4 = 100 - x_1$ (or $x_4 = 100 - 1x_1$) shows that x_1 cannot exceed 100/1, or 100. The second equation, $x_5 = 20{,}000 - 400x_1$, shows that x_1 cannot exceed 20,000/400, or 50. To satisfy both these conditions, x_1 cannot exceed 50, the smaller of 50 and 100. If we let x_1 take the value 50, then $x_1 = 50$, $x_2 = 0$, $x_3 = 0$, and $x_5 = 0$. Since $x_4 = 100 - x_1$, then

$$x_4 = 100 - 50 = 50. \quad \boxed{1}$$

The same result could have been found from the initial simplex tableau given above. Use the tableau by selecting the most negative indicator. (If no indicator is negative, then the value of the objective function cannot be improved.)

$$
\begin{array}{ccccc}
x_1 & x_2 & x_3 & x_4 & x_5 \\
\end{array}
$$

$$
\left[
\begin{array}{ccccc|c}
1 & 1 & 1 & 1 & 0 & 100 \\
400 & 160 & 280 & 0 & 1 & 20{,}000 \\
\hline
-120 & -40 & -60 & 0 & 0 & 0 \\
\end{array}
\right]
$$

↑
———— most negative indicator

The most negative indicator identifies the variable whose value is to be made nonzero. Find the variable which is now basic and which will become nonbasic by calculating the quotients that were found above. Do this by dividing each number from the far right column of the tableau by the corresponding number from the column with the most negative indicator.

Quotients

$$
\begin{array}{ccccc}
& & x_1 & x_2 & x_3 & x_4 & x_5 \\
\end{array}
$$

$$100/1 = 100$$

smaller → $20{,}000/400 = 50$

$$
\left[
\begin{array}{ccccc|c}
1 & 1 & 1 & 1 & 0 & 100 \\
400 & 160 & 280 & 0 & 1 & 20{,}000 \\
\hline
-120 & -40 & -60 & 0 & 0 & 0 \\
\end{array}
\right]
$$

The smaller quotient is 50, from the second row. This identifies 400 as the pivot. Use 400 as pivot, and the appropriate row transformations, to get the second simplex tableau as follows: get 1 in the pivot position by multiplying each element of the second row by 1/400. Then multiply each of the entries in the second row by −1 and add the results to the corresponding entries in the first row, to get a 0 above the pivot. Get a 0 below the pivot, as the first indicator, in a similar way. The new tableau is

$$
\begin{array}{ccccc}
x_1 & x_2 & x_3 & x_4 & x_5 \\
\end{array}
$$

$$
\left[
\begin{array}{ccccc|c}
0 & .6 & .3 & 1 & -.0025 & 50 \\
1 & .4 & .7 & 0 & .0025 & 50 \\
0 & 8 & 24 & 0 & .3 & 6000 \\
\end{array}
\right]
\begin{array}{l}
-1R_2 + R_1 \\
\frac{1}{400}R_2 \\
120R_2 + R_3 \\
\end{array}
$$

and the solution read from this tableau is

$$x_1 = 50, \quad x_2 = 0, \quad x_3 = 0, \quad x_4 = 50, \quad x_5 = 0,$$

the same result found above. The entry 6000 (in color) in the lower right corner of the tableau gives the value of the objective function for this solution:

$$z = \$6000.$$

1 Find the value of $z = 120x_1 + 40x_2 + 60x_3 + 0x_4 + 0x_5$ if $x_1 = 50$, $x_2 = 0$, $x_3 = 0$, $x_4 = 50$, $x_5 = 0$.

Answer:
6000

None of the indicators in the final simplex tableau is negative, which means that the value of z cannot be improved beyond $6000. To see why, recall that the last row gives the coefficients of the objective function. Including the coefficient of 1 for the z column which was dropped gives

$$0x_1 + 8x_2 + 24x_3 + 0x_4 + .3x_5 + z = 6000,$$

or

$$z = 6000 - 0x_1 - 8x_2 - 24x_3 - 0x_4 - .3x_5.$$

Since x_2, x_3, and x_5 are zero, $z = 6000$, but if any of these three variables were to increase, z would decrease.

This result suggests that the optimal solution has been found as soon as no indicators are negative. As long as an indicator is negative, the value of the objective function can be improved. Just find a new pivot and repeat the process until no negative indicators remain.

We can finally state the solution to the problem about the farmer: the optimum value of z is 6000, where $x_1 = 50$, $x_2 = 0$, $x_3 = 0$, $x_4 = 50$, and $x_5 = 0$. That is, the farmer will make a maximum profit of $6000 by planting 50 acres of potatoes. Another 50 acres should be left unplanted. It may seem strange that leaving assets unused can produce a maximum profit, but such results occur fairly often. ②

Example 1 To compare the simplex method with the graphical method, the simplex method is used to solve the problem of Example 2, Section 7.2. The graph is shown again in Figure 7.22. The objective function to be maximized was

$$z = 8x_1 + 12x_2. \qquad \text{storage space}$$

(Since we are using the simplex method, x_1 and x_2 are used as variables instead of x and y.) The constraints were as follows:

$$40x_1 + 80x_2 \leq 560 \qquad \text{cost}$$
$$6x_1 + 8x_2 \leq 72 \qquad \text{floor space}$$
$$x_1 \geq 0, \ x_2 \geq 0.$$

Add a slack variable to each constraint.

$$40x_1 + 80x_2 + x_3 \qquad = 560$$
$$6x_1 + \ 8x_2 \qquad + x_4 = 72.$$

Write the initial simplex tableau.

	x_1	x_2	x_3	x_4	
	40	80	1	0	560
	6	8	0	1	72
	-8	-12	0	0	0

② For the simplex tableau below, find the following.

(a) The pivot

(b) The second tableau

(c) The solution shown in the second tableau

(d) Can the solution be improved?

$$
\begin{array}{ccccc}
x_1 & x_2 & x_3 & x_4 & x_5
\end{array}
$$
$$
\left[\begin{array}{ccccc|c}
1 & 2 & 6 & 1 & 0 & 16 \\
1 & 3 & 0 & 0 & 1 & 25 \\
\hline
-1 & -4 & -3 & 0 & 0 & 0
\end{array}\right]
$$

Answer:

(a) 2

(b)

$$
\begin{array}{ccccc}
x_1 & x_2 & x_3 & x_4 & x_5
\end{array}
$$
$$
\left[\begin{array}{ccccc|c}
\frac{1}{2} & 1 & 3 & \frac{1}{2} & 0 & 8 \\
-\frac{1}{2} & 0 & -9 & -\frac{3}{2} & 1 & 1 \\
\hline
1 & 0 & 9 & 2 & 0 & 32
\end{array}\right]
$$

(c) $(0, 8, 0, 0, 1)$

(d) No

This tableau leads to the solution $x_1 = 0$, $x_2 = 0$, $x_3 = 560$, and $x_4 = 72$, with $z = 0$, which corresponds to the origin in Figure 7.22. The most negative indicator is -12. The necessary quotients are

$$\frac{560}{80} = 7 \quad \text{and} \quad \frac{72}{8} = 9.$$

The smaller quotient is 7, giving 80 as the pivot. Use row transformations to get the new tableau.

$$\begin{bmatrix} \frac{1}{2} & 1 & \frac{1}{80} & 0 & 7 \\ 2 & 0 & -\frac{1}{10} & 1 & 16 \\ -2 & 0 & \frac{3}{20} & 0 & 84 \end{bmatrix} \quad \begin{matrix} \frac{1}{80}R_1 \\ -8R_1 + R_2 \\ 12R_1 + R_3 \end{matrix}$$

The solution from this tableau is $x_1 = 0$, $x_2 = 7$, $x_3 = 0$, and $x_4 = 16$, with $z = 84$, which corresponds to the corner point $(0, 7)$ in Figure 7.22. Because of the indicator -2, the value of z can be improved. The new quotients are

$$\frac{7}{\frac{1}{2}} = 14 \quad \text{and} \quad \frac{16}{2} = 8.$$

Since 8 is smaller, use the 2 in row 2, column 1 as pivot to get the final tableau.

$$\begin{bmatrix} 0 & 1 & \frac{3}{80} & -\frac{1}{4} & 3 \\ 1 & 0 & -\frac{1}{20} & \frac{1}{2} & 8 \\ 0 & 0 & \frac{1}{20} & 1 & 100 \end{bmatrix} \quad \begin{matrix} -\frac{1}{2}R_2 + R_1 \\ \frac{1}{2}R_2 \\ 2R_2 + R_3 \end{matrix}$$

Here the solution is $x_1 = 8$, $x_2 = 3$, $x_3 = 0$, and $x_4 = 0$, with $z = 100$. This solution, which corresponds to the corner point $(8, 3)$ in Figure 7.22, is the same as the solution found earlier. ■

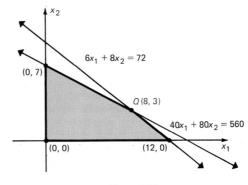

Figure 7.22

As we have seen, each simplex tableau above gave a solution corresponding to one of the corner points of the feasible region. As shown in Figure 7.23, the first solution corresponded to the origin, with $z = 0$. By choosing the appropriate pivot, we moved systematically to a new corner point, $(0, 7)$, which improved the value of z to 84. The next tableau took us to $(8, 3)$, producing the optimum value of $z = 100$. There was no reason to test the last corner point, $(12, 0)$, since the optimum value of z was found before that point was reached. ③

③ A linear programming problem has the initial tableau given below. Use the simplex method to solve the problem.

$$\begin{array}{cccc} x_1 & x_2 & x_3 & x_4 \end{array}$$
$$\begin{bmatrix} 1 & 1 & 1 & 0 & | & 40 \\ 2 & 1 & 0 & 1 & | & 24 \\ \hline -300 & -200 & 0 & 0 & | & 0 \end{bmatrix}$$

Answer:
$x_1 = 0$, $x_2 = 24$, $x_3 = 16$, $x_4 = 0$, $z = 4800$

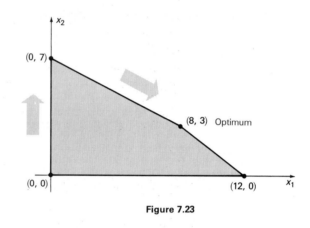

Figure 7.23

The next example shows how to handle zero or a negative number in the column containing the pivot.

Example 2 Find the pivot for the following initial simplex tableau.

$$\begin{array}{ccccc} x_1 & x_2 & x_3 & x_4 & x_5 \end{array}$$
$$\begin{bmatrix} 1 & -2 & 1 & 0 & 0 & | & 100 \\ 3 & 4 & 0 & 1 & 0 & | & 200 \\ 5 & 0 & 0 & 0 & 1 & | & 150 \\ \hline -10 & -25 & 0 & 0 & 0 & | & 0 \end{bmatrix}$$

The most negative indicator is -25. To find the pivot, find the quotients formed by the entries in the right-most column and in the x_2 column: $100/(-2)$, $200/4$, and $150/0$. Since division by 0 is meaningless, disregard the quotient $150/0$, which comes from the equation

$$5x_1 + 0x_2 + 0x_3 + 0x_4 + x_5 = 150.$$

Pivoting in the x_2 column means that x_1 is still 0, and the equation reduces to

$$0x_2 + x_5 = 150 \qquad \text{or} \qquad x_5 = 150 - 0x_2,$$

and so $x_5 = 150$. Then x_5 cannot be zero, as required. In general, disregard quotients with a 0 denominator.

The quotients predict the value of a variable in the solution, and thus cannot be negative. In this example, the quotient $100/(-2) = -50$ comes from the equation

$$x_1 - 2x_2 + x_3 + 0x_4 + 0x_5 = 100.$$

For $x_1 = 0$, this becomes

$$x_3 = 100 + 2x_2.$$

If x_3 is to be nonnegative, then

$$100 + 2x_2 \geq 0$$
$$x_2 \geq -50.$$

Since x_2 must be nonnegative anyway, this equation tells us nothing new. For this reason, disregard negative quotients also.

The only usable quotient is $200/4 = 50$, making 4 the pivot. If all the quotients are either negative or have zero denominators, no optimum solution will be found. The quotients, then, tell whether or not an optimum solution exists. **4**

In summary, the following steps are involved in solving a standard maximum linear programming problem by the simplex method.

4 Find the pivot for the following tableau.

$$
\begin{array}{ccccc}
x_1 & x_2 & x_3 & x_4 & x_5 \\
\end{array}
$$

$$
\left[
\begin{array}{ccccc|c}
0 & 1 & 1 & 0 & 0 & 50 \\
-2 & 3 & 0 & 1 & 0 & 78 \\
2 & 4 & 0 & 0 & 1 & 65 \\
\hline
-5 & -3 & 0 & 0 & 0 & 0 \\
\end{array}
\right]
$$

Answer:
2 (in first column)

Simplex Method

1. Determine the objective function.
2. Write all necessary constraints.
3. Convert each constraint into an equation by adding slack variables.
4. Set up the initial simplex tableau.
5. Locate the most negative indicator. If there are two such indicators, choose either.
6. Form the necessary quotients to find the pivot. Disregard any negative quotients or quotients with a 0 denominator. The smallest nonnegative quotient gives the location of the pivot. If all quotients must be disregarded, no maximum solution exists.* If two quotients are equally the smallest, let either determine the pivot.†
7. Transform the tableau so that the pivot becomes 1 and all other numbers in that column become 0.
8. If the indicators are all positive or 0, this is the final tableau. If not, go back to Step 5 above and repeat the process until a tableau with no negative indicators is obtained.
9. Read the solution from this final tableau. The maximum value of the objective function is the number in the lower right corner of the final tableau.

---

*Some special circumstances are noted at the end of Section 7.6.
†It may be that the first choice of a pivot does not produce a solution. In that case, try the other choice.

Although linear programming problems with more than a few variables would seem to be very complex, in practical applications many entries in the simplex tableau are zeros.

7.5 Exercises

In Exercises 1–6, the initial tableau of a linear programming problem is given. Use the simplex method to solve each problem. (See Examples 1 and 2.)

1.

x_1	x_2	x_3	x_4	x_5	
1	2	4	1	0	8
2	2	1	0	1	10
-2	-5	-1	0	0	0

2.

x_1	x_2	x_3	x_4	x_5	
2	2	1	1	0	10
1	2	3	0	1	15
-3	-2	-1	0	0	0

3.

x_1	x_2	x_3	x_4	x_5	
1	3	1	0	0	12
2	1	0	1	0	10
1	1	0	0	1	4
-2	-1	0	0	0	0

4.

x_1	x_2	x_3	x_4	x_5	x_6	
2	2	1	1	0	0	50
1	1	3	0	1	0	40
4	2	5	0	0	1	80
-2	-3	-5	0	0	0	0

5.

x_1	x_2	x_3	x_4	x_5	x_6	
2	2	8	1	0	0	40
4	-5	6	0	1	0	60
2	-2	6	0	0	1	24
-14	-10	-12	0	0	0	0

6.

x_1	x_2	x_3	x_4	x_5	
3	2	4	1	0	18
2	1	5	0	1	8
-1	-4	-2	0	0	0

Use the simplex method to solve Exercises 7–14.

7. Maximize $z = 4x_1 + 3x_2$
subject to: $2x_1 + 3x_2 \le 11$
$x_1 + 2x_2 \le 6$
and $x_1 \ge 0$, $x_2 \ge 0$.

8. Maximize $z = 2x_1 + 3x_2$
subject to: $3x_1 + 5x_2 \le 29$
$2x_1 + x_2 \le 10$
and $x_1 \ge 0$, $x_2 \ge 0$.

9. Maximize $z = 10x_1 + 12x_2$
subject to: $4x_1 + 2x_2 \le 20$
$5x_1 + x_2 \le 50$
$2x_1 + 2x_2 \le 24$
and $x_1 \ge 0$, $x_2 \ge 0$.

10. Maximize $z = 1.2x_1 + 3.5x_2$
subject to: $2.4x_1 + 1.5x_2 \le 10$
$1.7x_1 + 1.9x_2 \le 15$
and $x_1 \ge 0$, $x_2 \ge 0$.

11. Maximize $z = 8x_1 + 3x_2 + x_3$
subject to: $x_1 + 6x_2 + 8x_3 \le 118$
$x_1 + 5x_2 + 10x_3 \le 220$
and $x_1 \ge 0$, $x_2 \ge 0$, $x_3 \ge 0$.

12. Maximize $z = 12x_1 + 15x_2 + 5x_3$
subject to: $2x_1 + 2x_2 + x_3 \le 8$
$x_1 + 4x_2 + 3x_3 \le 12$
and $x_1 \ge 0$, $x_2 \ge 0$, $x_3 \ge 0$.

13. Maximize $z = x_1 + 2x_2 + x_3 + 5x_4$
subject to: $x_1 + 2x_2 + x_3 + x_4 \le 50$
$3x_1 + x_2 + 2x_3 + x_4 \le 100$
and $x_1 \ge 0$, $x_2 \ge 0$, $x_3 \ge 0$, $x_4 \ge 0$.

14. Maximize $z = x_1 + x_2 + 4x_3 + 5x_4$
subject to: $x_1 + 2x_2 + 3x_3 + x_4 \le 115$
$2x_1 + x_2 + 8x_3 + 5x_4 \le 200$
$x_1 + x_3 \le 50$
and $x_1 \ge 0$, $x_2 \ge 0$, $x_3 \ge 0$, $x_4 \ge 0$.

Set up and solve Exercises 15–20 by the simplex method.

15. **Natural Science** A biologist has 500 kilograms of nutrient A, 600 kilograms of nutrient B, and 300 kilograms of nutrient C. These nutrients will be used to make 4 types of food, whose contents (in percent of nutrient per kilogram of food) and whose "growth values" are as shown below.

Food	Nutrient (%)			Growth value
	A	B	C	
P	0	0	100	90
Q	0	75	25	70
R	37.5	50	12.5	60
S	62.5	37.5	0	50

How many kilograms of each food should be produced in order to maximize total growth value? Find the maximum growth value.

16. **Management** A baker has 150 units of flour, 90 units of sugar, and 150 of raisins. A loaf of raisin bread requires 1 unit of flour, 1 of sugar and 2 of raisins, while a raisin cake needs 5, 2, and 1 units, respectively. If raisin bread sells for $1.05 a loaf and raisin cake for $2.40 each, how many of each should be baked so that gross income is maximized? What is the maximum gross income?

17. **Management** A candy company has 100 kilograms of chocolate-covered nuts and 125 kilograms of chocolate-covered raisins to be sold as two different mixtures. One mix will contain half nuts and half raisins and will sell for $6 per kilogram. The other mix will contain 1/3 nuts and 2/3 raisins, and will sell for $4.80 per kilogram. How many kilograms of each mix should the company prepare for maximum revenue? (See Exercise 13, Section 7.2.) Find the maximum revenue.

18. **Management** Caroline's Quality Candy Confectionery is famous for fudge, chocolate cremes, and pralines. Its candy-making equipment is set up to make 100-pound batches at a time. Currently there is a chocolate shortage and the company can get only 120 pounds of chocolate in the next shipment. On a week's run, the confectionery's cooking and processing equipment is available for a total of 42 machine hours. During the same period the employees have a total of 56 work hours available for packaging. A batch of fudge requires 20 pounds of chocolate while a batch of cremes uses 25 pounds of chocolate. The cooking and

processing take 120 minutes for fudge, 150 minutes for chocolate cremes, and 200 minutes for pralines. The packaging times measured in minutes per one pound box are 1, 2, and 3, respectively for fudge, cremes and pralines. Determine how many batches of each type of candy the confectionery should make, assuming that the profit per pound box is 50¢ on fudge, 40¢ on chocolate cremes, and 45¢ on pralines. Also, find the maximum profit for the week. (See Exercise 28, Section 7.4.)

19. **Management** A manufacturer of bicycles builds one-, three-, and ten-speed models. The bicycles need both aluminum and steel. The company has available 91,800 units of steel and 42,000 units of aluminum. The one-, three-, and ten-speed models need respectively 17, 27, and 34 units of steel, and 12, 21, and 15 units of aluminum. How many of each type of bicycle should be made in order to maximize profit if the company makes $8 per one-speed bike, $12 per three-speed, and $22 per ten-speed? What is the maximum possible profit?

20. **Social Science** A political party is planning a half-hour television show. The show will have 3 minutes of direct requests for money from viewers. Three of the party's politicians will be on the show—a senator, a congresswoman, and a governor. The senator, a party "elder statesman," demands that he be on at least twice as long as the governor. The total time taken by the senator and the governor must be at least twice the time taken by the congresswoman. Based on a pre-show survey, it is believed that 40, 60, and 50 (in thousands) viewers will watch the program for each minute the senator, congresswoman, and governor, respectively are on the air. Find the time that should be allotted to each politician in order to get the maximum number of viewers. Find the maximum number of viewers.

Management *The next two problems come from past CPA examinations.* Select the appropriate answer for each question.*

21. The Ball Company manufactures three types of lamps, labeled A, B, and C. Each lamp is processed in two

---

*Material from *Uniform CPA Examination Questions and Unofficial Answers,* copyright © 1973, 1974, 1975 by the American Institute of Certified Public Accountants, Inc., is reprinted with permission.

departments, I and II. Total available man-hours per day for departments I and II are 400 and 600, respectively. No additional labor is available. Time requirements and profit per unit for each lamp type are as follows.

	A	B	C
Man-hours in I	2	3	1
Man-hours in II	4	2	3
Profit per unit	$5	$4	$3

The company has assigned you as the accounting member of its profit planning committee to determine the numbers of types of A, B, and C lamps that it should produce in order to maximize its total profit from the sale of lamps. The following questions relate to a linear programming model that your group has developed.

(a) The coefficients of the objective function would be
   (1) 4, 2, 3.   (2) 2, 3, 1.
   (3) 5, 4, 3.   (4) 400, 600.

(b) The constraints in the model would be
   (1) 2, 3, 1.   (2) 5, 4, 3.
   (3) 4, 2, 3.   (4) 400, 600.

(c) The constraint imposed by the available man-hours in department I could be expressed as
   (1) $4X_1 + 2X_2 + 3X_3 \leq 400$.
   (2) $4X_1 + 2X_2 + 3X_3 \geq 400$.
   (3) $2X_1 + 3X_2 + 1X_3 \leq 400$.
   (4) $2X_1 + 3X_2 + 1X_3 \geq 400$.

22. The Golden Hawk Manufacturing Company wants to maximize the profits on products A, B, and C. The contribution margin for each product follows.

Product	Contribution margin
A	$2
B	5
C	4

   The production requirements and departmental capacities, by departments, are as follows.

Department	Production requirements by product (hours)			Departmental capacity (Total hours)
	A	B	C	
Assembling	2	3	2	30,000
Painting	1	2	2	38,000
Finishing	2	3	1	28,000

(a) What is the profit-maximization formula for the Golden Hawk Company?
   (1) $2A + $5B + $4C = X$ (where X = profit)
   (2) $5A + 8B + 5C \leq 96,000$
   (3) $2A + $5B + $4C \leq X$
   (4) $2A + $5B + $4C = 96,000$

(b) What is the constraint for the Painting Department of the Golden Hawk Company?
   (1) $1A + 2B + 2C \geq 38,000$
   (2) $2A + $5B + $4C \geq 38,000$
   (3) $1A + 2B + 2C \leq 38,000$
   (4) $2A + 3B + 2C \leq 30,000$

Determine the constraints and the objective function for Exercises 23 and 24, then solve each problem.

23. **Management**   A manufacturer makes two products, toy trucks and toy fire engines. Both are processed in four different departments, each of which has a limited capacity. The sheet metal department can handle at least 1 1/2 times as many trucks as fire engines. The truck assembly department can handle at most 6700 trucks per week, while the fire engine assembly department assembles at most 5500 fire engines weekly. The painting department, which finishes both toys, has a maximum capacity of 12,000 per week. If the profit is $8.50 for a toy truck and $12.10 for a toy fire engine, how many of each item should the company produce to maximize profit?

24. The average weights of the three species stocked in the lake referred to in Section 6.2, Exercise 51, are 1.62, 2.14, and 3.01 kilograms for species A, B, and C, respectively. If the largest amounts of food that can be supplied each day are given as in Exercise 51, how should the lake be stocked to maximize the weight of the fish supported by the lake?

7.6 Mixed Constraints; Minimization Problems

So far the simplex method has been used only to solve standard maximum linear programming problems. In this section this work is extended to include linear programming problems with mixed $\leq$ and $\geq$ constraints. Then we see how solving these problems gives a method of solving minimum problems. (An alternate approach to minimum problems is given in the next section.)

Problems with $\leq$ and $\geq$ Constraints   Suppose a new constraint is added to the farmer problem from Example 2 of Section 7.4: to satisfy orders from regular buyers, the farmer must plant a total of at least 60 acres of the three crops. This constraint introduces the new inequality

$$x_1 + x_2 + x_3 \geq 60.$$

As before, this inequality must be rewritten as an equation in which the variables all represent nonnegative numbers. The inequality $x_1 + x_2 + x_3 \geq 60$ means that

$$x_1 + x_2 + x_3 - x_6 = 60$$

for some nonnegative variable x_6. (Remember that x_4 and x_5 are the slack variables in the problem.)

The new variable, x_6, is called a **surplus variable.** The value of this variable represents the excess number of acres (over 60) which may be planted. Since the total number of acres planted is to be no more than 100 but at least 60, the value of x_6 can vary from 0 to 40.

We must now solve the system of equations

$$
\begin{aligned}
x_1 + \quad x_2 + \quad x_3 + x_4 \qquad\qquad\qquad &= \quad 100 \\
400x_1 + 160x_2 + 280x_3 \qquad + x_5 \qquad\qquad &= 20{,}000 \\
x_1 + \quad x_2 + \quad x_3 \qquad\qquad - x_6 \qquad &= \quad 60 \\
-120x_1 - \ 40x_2 - \ 60x_3 \qquad\qquad\qquad + z &= \quad\ \ 0
\end{aligned}
$$

with x_1, x_2, x_3, x_4, x_5, and x_6 all nonnegative.

Set up the initial simplex tableau. (The z column is omitted as before.)

x_1	x_2	x_3	x_4	x_5	x_6	
1	1	1	1	0	0	100
400	160	280	0	1	0	20,000
1	1	1	0	0	−1	60
−120	−40	−60	0	0	0	0

This tableau gives the solution

$$x_1 = 0, \quad x_2 = 0, \quad x_3 = 0, \quad x_4 = 100, \quad x_5 = 20{,}000, \quad x_6 = -60.$$

But this is not a feasible solution, since x_6 is negative. All the variables in any feasible solution must be nonnegative.

When a negative value of a variable appears, use row operations to transform the matrix until a solution is found in which all variables are nonnegative. The difficulty is caused by the -1 in row three of the matrix. The third column of the usual 3×3 identity matrix does not appear. Correct this by using row transformations to change a column that has nonzero entries (such as the $x_1, x_2,$ or x_3 columns) to one in which the third row entry is 1 and the other entries are 0. The choice of a column is arbitrary. Let's choose the x_2 column and, if this choice does not lead to a feasible solution, try one of the other columns.

The third row entry in the x_2 column is already 1. Using row transformations to get 0's in the rest of the column gives the following tableau.

$$
\begin{array}{cccccc}
x_1 & x_2 & x_3 & x_4 & x_5 & x_6
\end{array}
$$

$$
\left[
\begin{array}{cccccc|c}
0 & 0 & 0 & 1 & 0 & 1 & 40 \\
240 & 0 & 120 & 0 & 1 & 160 & 10{,}400 \\
1 & 1 & 1 & 0 & 0 & -1 & 60 \\
\hline
-80 & 0 & -20 & 0 & 0 & -40 & 2400
\end{array}
\right]
\begin{array}{l}
-1R_3 + R_1 \\
-160R_3 + R_2 \\
\\
40R_3 + R_4
\end{array}
$$

This tableau gives the solution

$$x_1 = 0, \quad x_2 = 60, \quad x_3 = 0, \quad x_4 = 40, \quad x_5 = 10{,}400, \quad \text{and} \quad x_6 = 0,$$

which is feasible. The process of applying row transformations to get a feasible solution is called *phase I* of the solution. In *phase II,* the simplex method is applied as usual. Verify that the pivot is 240.

$$
\begin{array}{cccccc}
x_1 & x_2 & x_3 & x_4 & x_5 & x_6
\end{array}
$$

$$
\left[
\begin{array}{cccccc|c}
0 & 0 & 0 & 1 & 0 & 1 & 40 \\
\boxed{240} & 0 & 120 & 0 & 1 & 160 & 10{,}400 \\
1 & 1 & 1 & 0 & 0 & -1 & 60 \\
\hline
-80 & 0 & -20 & 0 & 0 & -40 & 2400
\end{array}
\right]
$$

$$
\begin{array}{cccccc}
x_1 & x_2 & x_3 & x_4 & x_5 & x_6
\end{array}
$$

$$
\left[
\begin{array}{cccccc|c}
0 & 0 & 0 & 1 & 0 & 1 & 40 \\
1 & 0 & .5 & 0 & .004 & .667 & 43.3 \\
0 & 1 & .5 & 0 & -.004 & -1.667 & 16.7 \\
\hline
0 & 0 & 20 & 0 & .32 & 13.4 & 5864
\end{array}
\right]
\begin{array}{l}
\\
\frac{1}{240}R_2 \\
-1R_2 + R_3 \\
80R_2 + R_4
\end{array}
$$

The second matrix above has been obtained from the first by standard row operations; some numbers in it have been rounded.

This final tableau gives

$$x_1 = 43.3, \quad x_2 = 16.7, \quad x_3 = 0, \quad x_4 = 40, \quad x_5 = 0, \quad \text{and} \quad x_6 = 0.$$

For maximum profit with this new constraint, the farmer should plant 43.3 acres of potatoes, 16.7 acres of corn, and no cabbage. Forty acres of the 100 available should not be planted. The profit will be $5864, less than the $6000 profit if he planted only 50 acres of potatoes. Because of the additional constraint that at least 60 acres must be planted, the profit is reduced.

Example 1 Maximize $\qquad z = 10x_1 + 8x_2$

subject to: $\qquad 4x_1 + 4x_2 \geq 60$

$\qquad\qquad\qquad 2x_1 + 5x_2 \leq 120$

and $x_1 \geq 0$, $x_2 \geq 0$.

Add slack or surplus variables to the constraints as needed to get the system

$$4x_1 + 4x_2 - x_3 \qquad\qquad = 60$$
$$2x_1 + 5x_2 \qquad + x_4 \qquad = 120$$
$$-10x_1 - 8x_2 \qquad\qquad + z = 0.$$

Now write the first simplex tableau.

$$
\begin{array}{cccc}
x_1 & x_2 & x_3 & x_4 \\
\end{array}
$$

$$
\left[
\begin{array}{cccc|c}
4 & 4 & -1 & 0 & 60 \\
2 & 5 & 0 & 1 & 120 \\
\hline
-10 & -8 & 0 & 0 & 0
\end{array}
\right] \;\boxed{1}
$$

① Give the solution from the first tableau.

Answer:

$x_1 = 0$, $x_2 = 0$, $x_3 = -60$, $x_4 = 120$.

The solution, from problem 1 at the side, is not feasible because of the -60. In phase I, use row transformations to modify the tableau to produce a feasible solution. We need a column with 1 in the first row and 0's in the rest of the rows. Let us choose the x_1 column. Multiply the entries in the first row by 1/4 to get 1 in the top row of the column. Then use row transformations to get 0's in the other rows of that column.

$$
\begin{array}{cccc}
x_1 & x_2 & x_3 & x_4 \\
\end{array}
$$

$$
\left[
\begin{array}{cccc|c}
1 & 1 & -\frac{1}{4} & 0 & 15 \\
0 & 3 & \frac{1}{2} & 1 & 90 \\
\hline
0 & 2 & -\frac{5}{2} & 0 & 150
\end{array}
\right]
\begin{array}{l}
\frac{1}{4}R_1 \\
-2R_1 + R_2 \\
10R_1 + R_3
\end{array}
$$

This solution,

$$x_1 = 15, \quad x_2 = 0, \quad x_3 = 0, \quad \text{and} \quad x_4 = 90,$$

is feasible, so phase I is complete. Perform phase II by completing the solution in the usual way. The pivot is 1/2. The next tableau is

$$
\begin{array}{cccc}
x_1 & x_2 & x_3 & x_4 \\
\end{array}
$$

$$
\left[
\begin{array}{cccc|c}
1 & \frac{5}{2} & 0 & \frac{1}{2} & 60 \\
0 & 6 & 1 & 2 & 180 \\
\hline
0 & 17 & 0 & 5 & 600
\end{array}
\right].
$$

No indicators are negative, so the optimum value of the objective function is

$$z = 600 \quad \text{when} \quad x_1 = 60 \quad \text{and} \quad x_2 = 0. \;\blacksquare \;\boxed{2}$$

2 The first tableau of a linear programming problem is given below. Use the second column to complete phase I of the solution and give the feasible solution that results.

$$
\left[
\begin{array}{cccc|c}
3 & 1 & 1 & 0 & 50 \\
4 & 2 & 0 & -1 & 70 \\
\hline
-8 & -10 & 0 & 0 & 0
\end{array}
\right]
$$

Answer:

$$
\left[
\begin{array}{cccc|c}
1 & 0 & 1 & \frac{1}{2} & 15 \\
2 & 1 & 0 & -\frac{1}{2} & 35 \\
\hline
12 & 0 & 0 & -5 & 350
\end{array}
\right]
$$

Solution: $x_1 = 0$, $x_2 = 35$, $x_3 = 15$, $x_4 = 0$

The approach discussed above is used to solve problems where the constraints are mixed $\leq$ and $\geq$ inequalities. The method also can be used to solve a problem where the constant term is negative. (This is not likely to happen in an application, however.)

Minimization Problems   We defined a standard maximum linear programming problem earlier in this chapter. Now we define a **standard minimum linear programming problem.**

Standard Minimum Form   A linear programming problem is in standard minimum form if

1. the objective function is to be minimized;
2. all variables are nonnegative;
3. all constraints involve $\geq$;
4. the constants in the constraints are all nonnegative.

The difference between maximum and minimum problems is in conditions 1 and 3: in standard minimum problems the objective function is to be *minimized,* and all constraints must have $\geq$ instead of $\leq$.

Standard minimum problems can be solved with the method of surplus variables presented above. To solve a minimum problem, first observe that the minimum of an objective function is the same number as the *maximum* of the *negative* of the function.

The next example illustrates the process. We use y_1 and y_2 as variables, and w for the objective function as a reminder that this is a minimum problem.

Example 2   Minimize $w = 3y_1 + 2y_2$
            subject to: $y_1 + 3y_2 \geq 6$
                      $2y_1 + y_2 \geq 3$

and $y_1 \geq 0, \quad y_2 \geq 0$.

Change this to a maximization problem by letting z equal the *negative* of the objective function: $z = -w$. Then find the *maximum* value of z.

$$z = -w = -3y_1 - 2y_2$$

The problem can now be stated as follows.

Maximize $z = -3y_1 - 2y_2$
subject to: $y_1 + 3y_2 \geq 6$
                $2y_1 + y_2 \geq 3$

and $y_1 \geq 0, \quad y_2 \geq 0$.

For phase I of the solution, subtract surplus variables and set up the first tableau.

$$
\begin{array}{c}
\begin{array}{cccc} y_1 & y_2 & y_3 & y_4 \end{array} \\
\begin{array}{c} 1 \\ 2 \\ 3 \end{array}\left[\begin{array}{cccc|c}
1 & 3 & -1 & 0 & 6 \\
2 & 1 & 0 & -1 & 3 \\
3 & 2 & 0 & 0 & 0
\end{array}\right]
\end{array}
$$

The solution, $y_1 = 0$, $y_2 = 0$, $y_3 = -6$, and $y_4 = -3$, contains negative numbers. Row transformations must be used to get a tableau with a feasible solution. Let us use the y_1 column since it already has a 1 in the first row. Begin by getting a 0 in the second row, and then a 0 in the third row.

$$
\begin{array}{c}
\begin{array}{cccc} y_1 & y_2 & y_3 & y_4 \end{array} \\
\left[\begin{array}{cccc|c}
1 & 3 & -1 & 0 & 6 \\
0 & -5 & 2 & -1 & -9 \\
0 & -7 & 3 & 0 & -18
\end{array}\right]
\begin{array}{l} \\ -2R_1 + R_2 \\ -3R_1 + R_3 \end{array}
\end{array}
$$

This tableau has a feasible solution. Now, in phase II, complete the solution as usual by the simplex method. The pivot is -5.

$$
\begin{array}{c}
\begin{array}{cccc} y_1 & y_2 & y_3 & y_4 \end{array} \\
\left[\begin{array}{cccc|c}
1 & 0 & \frac{1}{5} & -\frac{3}{5} & \frac{3}{5} \\
0 & 1 & -\frac{2}{5} & \frac{1}{5} & \frac{9}{5} \\
0 & 0 & \frac{1}{5} & \frac{7}{5} & -\frac{27}{5}
\end{array}\right]
\begin{array}{l} -3R_2 + R_1 \\ -\frac{1}{5}R_2 \\ 7R_2 + R_3 \end{array}
\end{array}
$$

The solution is

$$
y_1 = \frac{3}{5}, \quad y_2 = \frac{9}{5}, \quad y_3 = 0, \quad \text{and} \quad y_4 = 0.
$$

This solution is feasible, and the tableau has no negative indicators. Since $z = -27/5$ and $z = -w$, then $w = 27/5$ is the minimum value, which is obtained when $y_1 = 3/5$ and $y_2 = 9/5$. **3**

Let us summarize the steps involved in the phase I and phase II method used to solve nonstandard problems in this section.

3 Minimize $w = 2y_1 + 3y_2$
subject to: $y_1 + y_2 \geq 10$
$2y_1 + y_2 \geq 16$
and $y_1 \geq 0$, $y_2 \geq 0$.

Answer:
$y_1 = 10$, $y_2 = 0$, $y_3 = 0$,
$y_4 = 4$; $w = 20$

Solving Nonstandard Problems

1. If necessary, convert the problem to a maximum problem.
2. Add slack variables and subtract surplus variables as needed.
3. Write the initial simplex tableau.
4. If the solution from this tableau is not feasible, use row transformations to get a feasible solution (phase I).
5. After a feasible solution is reached, solve by the simplex method (phase II).

We certainly have not covered all the possible complications that can arise in using the simplex method. Some of the difficulties include the following:

1. Some of the constraints may be *equations* instead of inequalities. In this case, *artificial variables* must be used.
2. Occasionally, a transformation will cycle—that is, produce a "new" solution which was an earlier solution in the process. These situations are known as *degeneracies* and special methods are available for handling them.
3. It may not be possible to convert a nonfeasible basic solution to a feasible basic solution. In that case, no solution can satisfy all the constraints. Graphically, this means there is no region of feasible solutions.

(These difficulties are covered in more detail in advanced texts. One example of a text that you might find helpful is *An Introduction to Management Science,* Third Edition, by David R. Anderson, Dennis J. Sweeney, and Thomas A. Williams, 1982, West Publishing Company.)

Two linear programming models in actual use, one on making ice cream, the other on merit pay, are presented at the end of this chapter. These models illustrate the usefulness of linear programming. In most real applications, the number of variables is so large that these problems could not be solved without the use of a method, like the simplex method, that can be adapted to a computer.

7.6 Exercises

Rewrite each system of inequalities, adding slack variables or subtracting surplus variables as necessary. (See Example 1.)

1. $2x_1 + 3x_2 \le 8$
$x_1 + 4x_2 \ge 7$

2. $5x_1 + 8x_2 \le 10$
$6x_1 + 2x_2 \ge 7$

3. $x_1 + x_2 + x_3 \le 100$
$x_1 + x_2 + x_3 \ge 75$
$x_1 + x_2 \quad \ge 27$

4. $2x_1 \quad + x_3 \le 40$
$x_1 + x_2 \quad \ge 18$
$x_1 \quad + x_3 \ge 20$

Convert Exercises 5–8 into maximization problems. (See Example 2.)

5. Minimize $w = 4x_1 + 3x_2 + 2x_3$
subject to: $x_1 + x_2 + x_3 \ge 5$
$x_1 + x_2 \quad \ge 4$
$2x_1 + x_2 + 3x_3 \ge 15$
and $x_1 \ge 0, \quad x_2 \ge 0, \quad x_3 \ge 0.$

6. Minimize $w = 8x_1 + 3x_2 + x_3$
subject to: $7x_1 + 6x_2 + 8x_3 \ge 18$
$4x_1 + 5x_2 + 10x_3 \ge 20$
and $x_1 \ge 0, \quad x_2 \ge 0, \quad x_3 \ge 0.$

7. Minimize $w = x_1 + 2x_2 + x_3 + 5x_4$
subject to: $x_1 + x_2 + x_3 + x_4 \ge 50$
$3x_1 + x_2 + 2x_3 + x_4 \ge 100$
and $x_1 \ge 0, \quad x_2 \ge 0, \quad x_3 \ge 0, \quad x_4 \ge 0.$

8. Minimize $w = x_1 + x_2 + 4x_3$
subject to: $x_1 + 2x_2 + 3x_3 \ge 115$
$2x_1 + x_2 + x_3 \le 200$
$x_1 \quad + x_3 \ge 50$
and $x_1 \ge 0, \quad x_2 \ge 0, \quad x_3 \ge 0.$

Use the simplex method to solve Exercises 9–14. (See Examples 1 and 2.)

9. Maximize $z = 12x_1 + 10x_2$
subject to: $x_1 + 2x_2 \ge 24$
$x_1 + x_2 \le 40$
with $x_1 \ge 0$ and $x_2 \ge 0.$

10. Maximize $z = 6x_1 + 8x_2$
subject to: $3x_1 + 4x_2 \ge 48$
$2x_1 + 4x_2 \le 60$
with $x_1 \ge 0$ and $x_2 \ge 0.$

11. Find $x_1 \geq 0$, $x_2 \geq 0$, and $x_3 \geq 0$ such that

$$x_1 + x_2 + x_3 \leq 150$$
$$x_1 + x_2 + x_3 \geq 100$$

and $z = 2x_1 + 5x_2 + 3x_3$ is maximized.

12. Find $x_1 \geq 0$, $x_2 \geq 0$, and $x_3 \geq 0$ such that

$$x_1 + x_2 + 2x_3 \leq 38$$
$$2x_1 + x_2 + x_3 \geq 24$$

and $z = 3x_1 + 2x_2 + 2x_3$ is maximized.

13. Find $x_1 \geq 0$ and $x_2 \geq 0$ such that

$$x_1 + x_2 \leq 100$$
$$x_1 + x_2 \geq 50$$
$$2x_1 + x_2 \leq 110$$

and $z = 2x_1 + 3x_2$ is maximized.

14. Find $x_1 \geq 0$ and $x_2 \geq 0$ such that

$$x_1 + 2x_2 \leq 18$$
$$x_1 + 3x_2 \geq 12$$
$$2x_1 + 2x_2 \leq 24$$

and $z = 5x_1 + 10x_2$ is maximized.

Solve each of the following by the two-phase method.

15. Find $y_1 \geq 0$, $y_2 \geq 0$ such that

$$10y_1 + 5y_2 \geq 100$$
$$20y_1 + 10y_2 \geq 150$$

and $w = 4y_1 + 5y_2$ is minimized.

16. Minimize $w = 3y_1 + 2y_2$ subject to

$$2y_1 + 3y_2 \geq 60$$
$$y_1 + 4y_2 \geq 40$$

and $y_1 \geq 0$, $y_2 \geq 0$.

17. Minimize $w = 2y_1 + y_2 + 3y_3$ subject to

$$y_1 + y_2 + y_3 \geq 100$$
$$2y_1 + y_2 \quad\;\; \geq 50$$

and $y_1 \geq 0$, $y_2 \geq 0$, $y_3 \geq 0$.

18. Minimize $w = 3y_1 + 2y_2$ subject to

$$y_1 + 2y_2 \geq 10$$
$$y_1 + y_2 \geq 8$$
$$2y_1 + y_2 \geq 12$$

and $y_1 \geq 0$, $y_2 \geq 0$.

Use the simplex method to solve Exercises 19–25.

19. Mark, who is ill, takes vitamin pills. Each day he must have at least 16 units of vitamin A, 5 units of vitamin B_1, and 20 units of vitamin C. He can choose between

pill #1 which costs 10 cents and contains 8 units of A, 1 of B_1, and 2 of C, and pill #2 which costs 20 cents and contains 2 units of A, 1 of B_1, and 7 of C. How many of each pill should he buy in order to minimize his cost?

20. Sam, who is dieting, requires two food supplements, I and II. He can get these supplements from two different products, A and B, as shown in the table.

	Supplement	
	(grams per serving)	
	I	II
Product A	3	2
B	2	4

Sam's physician has recommended that he include at least 15 grams of supplement I but no more than 12 grams of II in his daily diet. If product A costs 25¢ per serving and product B costs 40¢ per serving, how can he satisfy his requirements most economically?

21. Management Brand X Canners produce canned whole tomatoes and tomato sauce. This season, they have available 3,000,000 kilograms of tomatoes for these two products. To meet the demands of regular customers, they must produce at least 80,000 kilograms of sauce and 800,000 kilograms of whole tomatoes. The cost per kilogram is $4 to produce canned whole tomatoes and $3.25 to produce tomato sauce. How many kilograms of tomatoes should they use for each product to minimize cost?

22. Management A brewery produces regular beer and a lower-carbohydrate "light" beer. Steady customers of the brewery buy 12 units of regular beer and 10 units of light beer. While setting up the brewery to produce the beers, the management decides to produce extra beer, beyond that needed to satisfy the steady customers. The cost per unit of regular beer is $36,000 and the cost per unit of light beer is $48,000. The number of units of light beer should not exceed twice the number of units of regular beer. At least twenty additional units of beer can be sold. How much of each type beer should be made so as to minimize total production costs?

23. The chemistry department at a local college decides to stock at least 800 small test tubes and 500 large test tubes. It wants to buy at least 1500 test tubes to take advantage of a special price. Since the small tubes are broken twice as often as the larger, the department will order at least twice as many small tubes as large. If the small test tubes cost 15¢ each and the large ones, made of a cheaper glass, cost 12¢ each, how many of each size should they order to minimize cost?

24. **Management** Topgrade Turf lawn seed mixtures contain three types of seeds: bluegrass, rye, and bermuda. The costs per pound of the three types of seed are 20¢, 15¢, and 5¢. In each mixture there must be at least 20% bluegrass seed and the amount of bermuda must be no more than the amount of rye. To fill current orders, the company must make at least 5000 pounds of the mixture. How much of each kind of seed should be used to minimize cost?

25. **Natural Science** A biologist must make a nutrient for her algae. The nutrient must contain the three basic elements D, E, and F, and must contain at least 10 kilograms of D, 12 kilograms of E, and 20 kilograms of F. The nutrient is made from three ingredients, I, II, and III. The quantity of D, E, and F in one unit of each of the ingredients is as given in the following chart.

One unit of ingredient	Contains the following elements in kilograms			Cost of one unit of ingredient
	D	E	F	
I	4	3	0	4
II	1	2	4	7
III	10	1	5	5

How many units of each ingredient are required to meet her needs at minimum cost?

Determine the constraints and the objective function for Exercises 26 and 27, then solve each problem.

26. **Management** Natural Brand plant food is made from three chemicals. (See Section 6.2, Exercise 49.) In a batch of the plant food there must be at least 81 kilograms of the first chemical and the other two chemicals

must be in the ratio of 4 to 3. If the three chemicals cost $1.09, $.87, and $.65 per kilogram, respectively, how much of each should be used to minimize the cost of producing at least 750 kilograms of the plant food?

27. **Management** A company is developing a new additive for gasoline. The additive is a mixture of three liquid ingredients, I, II, and III. For proper performance, the total amount of additive must be at least 10 ounces per gallon of gasoline. However, for safety reasons, the amount of additive should not exceed 15 ounces per gallon of gasoline. At least 1/4 ounce of ingredient I must be used for every ounce of ingredient II and at least 1 ounce of ingredient III must be used for every ounce of ingredient I. If the cost of I, II, and III is $.30, $.09, and $.27 per ounce, respectively, find the mixture of the three ingredients which produces the minimum cost of the additive. How much of the additive should be used per gallon of gasoline?

Solve the following linear programming problem, which has both "greater than" and "less than" constraints.

28. **Management** A popular soft drink called Sugarlo, which is advertised as having a sugar content of no more than 10%, is blended from five ingredients, each of which has some sugar content. Water may also be added to dilute the mixture. The sugar content of the ingredients and their costs per gallon are given below.

	Ingredient					
	1	2	3	4	5	Water
Sugar content (%)	.28	.19	.43	.57	.22	0
Cost ($/gal.)	.48	.32	.53	.28	.43	.04

At least .01 of the content of Sugarlo must come from ingredients 3 or 4, .01 must come from ingredients 2 or 5, and .01 from ingredients 1 or 4. How much of each ingredient should be used in preparing 15,000 gallons of Sugarlo to minimize the cost?

7.7 Duality

An interesting connection exists between standard maximum and standard minimum problems. It turns out that any solution of a standard maximum problem produces the solution of an associated standard minimum problem, and vice-versa. Each of these associated problems is called the **dual** of the other. One advantage of duals is that standard minimum problems can be solved by the simplex methods already discussed. The idea of a dual is explained with an example.

Example 1 Minimize $\qquad w = 8y_1 + 16y_2$
$\qquad$ subject to: $\qquad y_1 + 5y_2 \geq 9$
$\qquad\qquad\qquad\qquad 2y_1 + 2y_2 \geq 10$

and $y_1 \geq 0, \quad y_2 \geq 0$.

(As mentioned earlier, y_1 and y_2 are used as variables and w as the objective function as a reminder that this is a minimum problem.) Without considering slack variables just yet, write the augmented matrix of the system of inequalities, and include the coefficients of the objective function (not their negatives) as the last row in the matrix.

$$\begin{bmatrix} 1 & 5 & | & 9 \\ 2 & 2 & | & 10 \\ 8 & 16 & | & 0 \end{bmatrix}$$

Look now at the following new matrix, obtained from the one above by interchanging rows and columns.

$$\begin{bmatrix} 1 & 2 & | & 8 \\ 5 & 2 & | & 16 \\ 9 & 10 & | & 0 \end{bmatrix}$$

The *rows* of the first matrix (for the minimizing problem) are the *columns* of the second matrix.

The entries in this second matrix could be used to write the following standard maximum linear programming problem (again ignoring the fact that the numbers in the last row are not negative).

$$\text{Maximize} \qquad z = 9x_1 + 10x_2$$
$$\text{subject to:} \qquad x_1 + 2x_2 \leq 8$$
$$5x_1 + 2x_2 \leq 16$$

with all variables nonnegative.

Figure 7.24 (a) shows the graphical solution to the minimum problem given above, while Figure 7.24 (b) shows the solution of the maximum problem produced by exchanging rows and columns. **1** **2**

1 Use the corner points in Figure 7.24(a) to find the minimum value of $w = 8y_1 + 16y_2$ and where it occurs.

Answer:

48 when $y_1 = 4,\ y_2 = 1$

2 Use Figure 7.24(b) to find the maximum value of $z = 9x_1 + 10x_2$ and where it occurs.

Answer:

48 when $x_1 = 2,\ x_2 = 3$

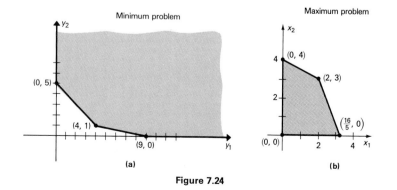

(a)

(b)

Figure 7.24

The two feasible regions in Figure 7.24 are different and the corner points are different, but the values of the objective functions found in Problems 1 and 2 at the side are equal—both are 48. An even closer connection between the two problems is shown by using the simplex method to solve the maximum problem given above.

Maximum problem

$$
\begin{array}{cccc}
x_1 & x_2 & x_3 & x_4 \\
\end{array}
$$

$$
\left[\begin{array}{cccc|c}
1 & 2 & 1 & 0 & 8 \\
5 & 2 & 0 & 1 & 16 \\
\hline
-9 & -10 & 0 & 0 & 0 \\
\end{array}\right]
$$

$$
\begin{array}{cccc}
x_1 & x_2 & x_3 & x_4 \\
\end{array}
$$

$$
\left[\begin{array}{cccc|c}
\frac{1}{2} & 1 & \frac{1}{2} & 0 & 4 \\
4 & 0 & -1 & 1 & 8 \\
\hline
-4 & 0 & 5 & 0 & 40 \\
\end{array}\right]
$$

$$
\begin{array}{cccc}
x_1 & x_2 & x_3 & x_4 \\
\end{array}
$$

$$
\left[\begin{array}{cccc|c}
0 & 1 & \frac{5}{8} & -\frac{1}{8} & 3 \\
1 & 0 & -\frac{1}{4} & \frac{1}{4} & 2 \\
\hline
0 & 0 & 4 & 1 & 48 \\
\end{array}\right]
$$

The maximum is 48 when $x_1 = 2$, $x_2 = 3$.

Notice that the solution to the *minimum problem* is found in the bottom row of the final simplex tableau for the maximum problem. This result suggests that standard minimum problems can be solved by forming the dual standard maximum problem, solving it by the simplex method, and then reading the solution for the minimum problem from the bottom row of the final simplex tableau.

Before using this method to actually solve a minimum problem, let us find the duals of some typical linear programming problems. The process of exchanging the rows and columns of a matrix, which is used to find the dual, is called **transposing** the matrix, and each of the two matrices is the **transpose** of the other.

Example 2 Find the transpose of each matrix.

(a)
$$
A = \left[\begin{array}{ccc}
2 & -1 & 5 \\
6 & 8 & 0 \\
-3 & 7 & -1 \\
\end{array}\right].
$$

Write the rows of matrix A as the columns of the transpose.

$$
\text{Transpose of } A = \left[\begin{array}{ccc}
2 & 6 & -3 \\
-1 & 8 & 7 \\
5 & 0 & -1 \\
\end{array}\right]
$$

3 Give the transpose of each matrix.

(a) $\begin{bmatrix} 2 & 4 \\ 6 & 3 \\ 1 & 5 \end{bmatrix}$

(b) $\begin{bmatrix} 4 & 7 & 10 \\ 3 & 2 & 6 \\ 5 & 8 & 12 \end{bmatrix}$

Answer:

(a) $\begin{bmatrix} 2 & 6 & 1 \\ 4 & 3 & 5 \end{bmatrix}$

(b) $\begin{bmatrix} 4 & 3 & 5 \\ 7 & 2 & 8 \\ 10 & 6 & 12 \end{bmatrix}$

(b) The transpose of $\begin{bmatrix} 1 & 2 & 4 & 0 \\ 2 & 1 & 7 & 6 \end{bmatrix}$ is $\begin{bmatrix} 1 & 2 \\ 2 & 1 \\ 4 & 7 \\ 0 & 6 \end{bmatrix}$. **3**

Example 3 Write the dual of the following standard maximum linear programming problems.

(a) Maximize $z = 2x_1 + 5x_2$
subject to: $x_1 + x_2 \le 10$
$2x_1 + x_2 \le 8$

and $x_1 \ge 0, \quad x_2 \ge 0$.

Begin by writing the augmented matrix for the given problem.

$$\left[\begin{array}{cc|c} 1 & 1 & 10 \\ 2 & 1 & 8 \\ \hline 2 & 5 & 0 \end{array}\right]$$

Form the transpose of this matrix to get

$$\left[\begin{array}{cc|c} 1 & 2 & 2 \\ 1 & 1 & 5 \\ \hline 10 & 8 & 0 \end{array}\right].$$

The dual problem is stated from this second matrix as follows (using y instead of x).

Minimize $w = 10y_1 + 8y_2$
subject to: $y_1 + 2y_2 \ge 2$
$y_1 + y_2 \ge 5$

and $y_1 \ge 0, \quad y_2 \ge 0$.

4 Write the dual of the following linear programming problem.

Minimize
$w = 2y_1 + 5y_2 + 6y_3$
subject to:

$2y_1 + 3y_2 + y_3 \ge 15$
$y_1 + y_2 + 2y_3 \ge 12$
$5y_1 + 3y_2 \qquad \ge 10$
and $y_1 \ge 0, y_2 \ge 0, y_3 \ge 0$.

Answer:

Maximize
$z = 15x_1 + 12x_2 + 10x_3$
subject to:

$2x_1 + x_2 + 5x_3 \le 2$
$3x_1 + x_2 + 3x_3 \le 5$
$x_1 + 2x_2 \qquad \le 6$
and $x_1 \ge 0, x_2 \ge 0, x_3 \ge 0$.

(b) *Given problem*

Minimize $w = 7y_1 + 5y_2 + 8y_3$
subject to: $3y_1 + 2y_2 + y_3 \ge 10$
$y_1 + y_2 + y_3 \ge 8$
$4y_1 + 5y_2 \qquad \ge 25$

and $y_1 \ge 0, \quad y_2 \ge 0, \quad y_3 \ge 0$.

Dual

Maximize $z = 10x_1 + 8x_2 + 25x_3$
subject to: $3x_1 + x_2 + 4x_3 \le 7$
$2x_1 + x_2 + 5x_3 \le 5$
$x_1 + x_2 \qquad \le 8$

and $x_1 \ge 0, \quad x_2 \ge 0, \quad x_3 \ge 0$. **4**

In Example 3, all the constraints of the standard maximum problems were $\le$ inequalities, while all those in the dual minimum problems were $\ge$ inequalities. This is generally the case; inequalities are reversed when the dual problem is stated.

The following table shows the close connection between a problem and its dual.

Given problem	Dual problem
m variables	n variables
n constraints	m constraints
coefficients from objective function	constants
constants	coefficients from objective function

The next theorem, whose proof requires advanced methods, guarantees that a standard minimum problem can be solved by forming a dual standard maximum problem.

Theorem of Duality The objective function w of a minimizing linear programming problem takes on a minimum value if and only if the objective function z of the corresponding dual maximizing problem takes on a maximum value. The maximum value of z equals the minimum value of w.

This method is illustrated in the following example. (This is Example 2 of the previous section; compare this solution with the one given there.)

Example 4 Minimize $\quad w = 3y_1 + 2y_2$
subject to: $\quad y_1 + 3y_2 \geq 6$
$\qquad\qquad 2y_1 + \ y_2 \geq 3$

and $y_1 \geq 0, \quad y_2 \geq 0$.
Use the given information to write the matrix

$$\begin{bmatrix} 1 & 3 & 6 \\ 2 & 1 & 3 \\ \hline 3 & 2 & 0 \end{bmatrix}.$$

Transpose to get the following matrix for the dual problem.

$$\begin{bmatrix} 1 & 2 & 3 \\ 3 & 1 & 2 \\ \hline 6 & 3 & 0 \end{bmatrix}$$

Write the dual problem from this matrix, as follows.

$$\text{Maximize} \qquad z = 6x_1 + 3x_2$$
$$\text{subject to:} \qquad x_1 + 2x_2 \leq 3$$
$$3x_1 + \ x_2 \leq 2$$

and $x_1 \geq 0, \quad x_2 \geq 0$.

Solve this standard maximum problem using the simplex method. Start by introducing slack variables to give the system

$$x_1 + 2x_2 + x_3 \qquad\qquad = 3$$
$$3x_1 + x_2 \qquad + x_4 \qquad = 2$$
$$-6x_1 - 3x_2 - 0x_3 - 0x_4 + z = 0$$

with $x_1 \geq 0$, $x_2 \geq 0$, $x_3 \geq 0$, $x_4 \geq 0$.

The first tableau for this system is given below, with the pivot as indicated.

Quotients	x_1	x_2	x_3	x_4	
$3/1 = 3$	1	2	1	0	3
2/3	3	1	0	1	2
	-6	-3	0	0	0

The simplex method gives the following final tableau.

x_1	x_2	x_3	x_4	
0	1	$\frac{3}{5}$	$-\frac{1}{5}$	$\frac{7}{5}$
1	0	$-\frac{1}{5}$	$\frac{2}{5}$	$\frac{1}{5}$
0	0	$\frac{3}{5}$	$\frac{9}{5}$	$\frac{27}{5}$

The last row of this final tableau shows that the solution of the given *standard minimum problem* is as follows:

The minimum value of $w = 3y_1 + 2y_2$, subject to the given constraints, is 27/5 and occurs when $y_1 = 3/5$ and $y_2 = 9/5$.

The minimum value of w, 27/5, is the same as the maximum value of z. **5**

Let us summarize the steps in solving a standard minimum linear programming problem by the method of duals.

5 Minimize
$w = 10y_1 + 8y_2$
subject to:

$\qquad y_1 + 2y_2 \geq 2$
$\qquad y_1 + y_2 \geq 5$
and $y_1 \geq 0$, $y_2 \geq 0$.

Answer:
$y_1 = 0$, $y_2 = 5$, for a minimum of 40

Solving Minimum Problems with Duals

1. Find the dual standard maximum problem.
2. Solve the maximum problem using the simplex method.
3. The minimum value of the objective function w is the maximum value of the objective function z.
4. The optimum solution is given by the entries in the bottom row of the columns corresponding to the slack variables.

The dual is useful not only in solving minimum problems, but also in seeing how small changes in one variable will affect the value of the objective function. For example, suppose an animal breeder needs at least 6 units per day of nutrient A and at least 3 units of nutrient B and that the breeder can choose between two different feeds, feed 1 and feed 2. Find the minimum cost for the breeder if each bag of feed 1 costs $3 and provides 1 unit of nutrient A and 2 units of B, while each bag of feed 2 costs $2 and provides 3 units of nutrient A and 1 of B.

If y_1 represents the number of bags of feed 1 and y_2 represents the number of bags of feed 2, the given information leads to

$$\text{Minimize} \quad w = 3y_1 + 2y_2$$
$$\text{subject to:} \quad y_1 + 3y_2 \geq 6$$
$$2y_1 + y_2 \geq 3$$

and $y_1 \geq 0, \quad y_2 \geq 0$.

This standard minimum linear programming problem is the one we solved in Example 4 of this section. In that example, we formed the dual and reached the following final tableau:

$$
\begin{array}{cccc}
x_1 & x_2 & x_3 & x_4
\end{array}
$$
$$
\left[
\begin{array}{cccc|c}
0 & 1 & \frac{3}{5} & -\frac{1}{5} & \frac{7}{5} \\
1 & 0 & -\frac{1}{5} & \frac{2}{5} & \frac{1}{5} \\
0 & 0 & \frac{3}{5} & \frac{9}{5} & \frac{27}{5}
\end{array}
\right].
$$

This final tableau shows that the breeder will obtain minimum feed costs by using 3/5 bag of feed 1 and 9/5 bag of feed 2 per day, for a daily cost of 27/5 = 5.40 dollars.

The top two numbers in the right-most column give the **imputed costs** in terms of the necessary nutrients. From the final tableau, $x_2 = 7/5$ and $x_1 = 1/5$, which means that a unit of nutrient A costs $1/5 = .20$ dollars, while a unit of nutrient B costs $7/5 = 1.40$ dollars. The minimum daily cost, $5.40, is found by the following procedure.

$$(\$.20 \text{ per unit of A}) \times (6 \text{ units of A}) = \$1.20$$
$$+ (\$1.40 \text{ per unit of B}) \times (3 \text{ units of B}) = \underline{\$4.20}$$
$$\$5.40 \text{ total}$$

These two numbers from the dual, $.20 and $1.40, also allow the breeder to estimate feed costs for "small" changes in nutrient requirements. For example, an increase of one unit in the requirement for each nutrient would produce a total cost of

$5.40	(6 units of A, 3 of B)
.20	(an extra unit of A)
1.40	(extra of B)
$7.00	total per day.

The numbers .20 and 1.40 are called the **shadow values** of the nutrients. ⑥

⑥ The final tableau of the dual of the problem about filing cabinets in Example 2, Section 7.2 and Example 1, Section 7.5 is given below.

(a) What are the imputed amounts of storage for each unit of cost and floor space?

(b) What are the shadow values of the cost and the floor space?

$$
\begin{array}{cccc}
x_1 & x_2 & x_3 & x_4
\end{array}
$$
$$
\left[
\begin{array}{cccc|c}
0 & 1 & \frac{1}{2} & -\frac{1}{4} & 1 \\
1 & 0 & -\frac{1}{20} & -\frac{1}{80} & \frac{1}{20} \\
0 & 0 & 8 & 3 & 100
\end{array}
\right]
$$

Answer:

(a) cost: 28 sq ft
    floor space: 72 sq ft

(b) $\dfrac{1}{20}, 1$

7.7 Exercises

Find the transpose of each matrix. (See Example 2.)

1. $\begin{bmatrix} 1 & 2 & 3 \\ 3 & 2 & 1 \\ 1 & 10 & 0 \end{bmatrix}$

2. $\begin{bmatrix} 2 & 5 & 8 & 6 & 0 \\ 1 & -1 & 0 & 12 & 14 \end{bmatrix}$

3. $\begin{bmatrix} -1 & 4 & 6 & 12 \\ 13 & 25 & 0 & 4 \\ -2 & -1 & 11 & 3 \end{bmatrix}$

4. $\begin{bmatrix} 1 & 11 & 15 \\ 0 & 10 & -6 \\ 4 & 12 & -2 \\ 1 & -1 & 13 \\ 2 & 25 & -1 \end{bmatrix}$

State the dual problem. (See Example 3.)

5. Maximize $z = 4x_1 + 3x_2 + 2x_3$
   subject to: $x_1 + x_2 + x_3 \le 5$
   $x_1 + x_2 \qquad \le 4$
   $2x_1 + x_2 + 3x_3 \le 15$
   and $x_1 \ge 0, \quad x_2 \ge 0, \quad x_3 \ge 0.$

6. Maximize $z = 8x_1 + 3x_2 + x_3$
   subject to: $7x_1 + 6x_2 + 8x_3 \le 18$
   $4x_1 + 5x_2 + 10x_3 \le 20$
   and $x_1 \ge 0, \quad x_2 \ge 0, \quad x_3 \ge 0.$

7. Minimize $w = y_1 + 2y_2 + y_3 + 5y_4$
   subject to: $y_1 + y_2 + y_3 + y_4 \ge 50$
   $3y_1 + y_2 + 2y_3 + y_4 \ge 100$
   and $y_1 \ge 0, \quad y_2 \ge 0, \quad y_3 \ge 0, \quad y_4 \ge 0.$

8. Minimize $w = y_1 + y_2 + 4y_3$
   subject to: $y_1 + 2y_2 + 3y_3 \ge 115$
   $2y_1 + y_2 + 8y_3 \ge 200$
   $y_1 \qquad + y_3 \ge 50$
   and $y_1 \ge 0, \quad y_2 \ge 0, \quad y_3 \ge 0.$

Use the simplex method to solve. (See Example 4.)

9. Find $y_1 \ge 0$ and $y_2 \ge 0$ such that
   $$2y_1 + 3y_2 \ge 6$$
   $$2y_1 + y_2 \ge 7$$
   and $w = 5y_1 + 2y_2$ is minimized.

10. Find $y_1 \ge 0$ and $y_2 \ge 0$ such that
    $$3y_1 + y_2 \ge 12$$
    $$y_1 + 4y_2 \ge 16$$
    and $w = 2y_1 + y_2$ is minimized.

11. Find $y_1 \ge 0$ and $y_2 \ge 0$ such that
    $$10y_1 + 5y_2 \ge 100$$
    $$20y_1 + 10y_2 \ge 150$$
    and $w = 4y_1 + 5y_2$ is minimized.

12. Minimize $w = 3y_1 + 2y_2$
    subject to: $2y_1 + 3y_2 \ge 60$
    $y_1 + 4y_2 \ge 40$
    and $y_1 \ge 0, \quad y_2 \ge 0.$

13. Minimize $w = 2y_1 + y_2 + 3y_3$
    subject to: $y_1 + y_2 + y_3 \ge 100$
    $2y_1 + y_2 \qquad \ge 50$
    and $y_1 \ge 0, \quad y_2 \ge 0, \quad y_3 \ge 0.$

14. Minimize $w = 3y_1 + 2y_2$
    subject to: $y_1 + 2y_2 \ge 10$
    $y_1 + y_2 \ge 8$
    $2y_1 + y_2 \ge 12$
    and $y_1 \ge 0, \quad y_2 \ge 0.$

Exercises 15 and 17 are repeated from the previous section. Solve them with the method of duals and compare the solutions with Exercises 19 and 21 in Section 7.6.

15. Mark, who is ill, takes vitamin pills. Each day he must have at least 16 units of vitamin A, 5 units of vitamin B_1, and 20 units of vitamin C. He can choose between pill #1 which costs 10 cents and contains 8 units of A, 1 of B_1, and 2 of C, and pill #2 which costs 20 cents and contains 2 units of A, 1 of B_1, and 7 of C. How many of each pill should he buy in order to minimize his cost?

16. Sam, who is dieting, requires two food supplements, I and II. He can get these supplements from two different products, A and B, as shown in the following table.

	Supplement (grams per serving)	
	I	II
Product A	3	2
Product B	2	4

Sam's physician has recommended that he include at least 15 grams of each supplement in his daily diet. If product A costs 25¢ per serving and product B costs 40¢ per serving, how can he satisfy his requirements most economically?

17. **Management** Brand X Canners produce canned whole tomatoes and tomato sauce. This season, they have available 3,000,000 kilograms of tomatoes for these two products. To meet the demands of regular customers, they must produce at least 80,000 kilograms of sauce and 800,000 kilograms of whole tomatoes. The cost per kilogram is $4 to produce canned whole tomatoes and $3.25 to produce tomato sauce. How many kilograms of tomatoes should they use for each product to minimize cost?

18. Refer to the end of Section 7.7, to the example on minimizing the daily cost of feeds.
    (a) Find a combination of feeds that will cost $7.00 and give 7 units of A and 4 units of B.
    (b) Use the dual variables to predict the daily cost of feed if the requirements change to 5 units of A and 4 units of B. Find a combination of feeds to meet these requirements at the predicted price.

19. **Management** A small toy manufacturing firm has 200 squares of felt, 600 ounces of stuffing, and 90 feet of trim available to make two types of toys, a small bear and a monkey. The bear requires 1 square of felt and 4 ounces of stuffing. The monkey requires 2 squares of felt, 3 ounces of stuffing, and 1 foot of trim. The firm makes $1 profit on each bear and $1.50 profit on each monkey. The linear program to maximize profit is

$$\text{Maximize} \quad x_1 + 1.5x_2 = z$$
$$\text{subject to:} \quad x_1 + 2x_2 \leq 200$$
$$4x_1 + 3x_2 \leq 600$$
$$x_2 \leq 90.$$

The final simplex tableau is

$$\begin{bmatrix} 0 & 1 & .8 & -.2 & 0 & 40 \\ 1 & 0 & -.6 & .4 & 0 & 120 \\ 0 & 0 & -.8 & .2 & 1 & 50 \\ \hline 0 & 0 & .6 & .1 & 0 & 180 \end{bmatrix}$$

(a) What is the corresponding dual problem?
(b) What is the optimal solution to the dual problem?
(c) Use the shadow values to estimate the profit the firm will make if their supply of felt increases to 210 squares.
(d) How much profit will the firm make if their supply of stuffing is cut to 590 ounces and their supply of trim is cut to 80 feet?

20. Refer to the problem about the farmer, solved at the beginning of Section 7.5.
    (a) Give the dual problem.
    (b) Use the shadow values to estimate the farmer's profit if land is cut to 90 acres but capital increases to $21,000.
    (c) Suppose the farmer has 110 acres but only $19,000. Find the optimum profit and the planting strategy that will produce this profit.

Determine the constraints and the objective function for Exercises 21 and 22, then solve each problem by the simplex method. Compare the solutions with those for Exercises 26 and 27 of Section 7.6.

21. **Natural Science** Natural Brand plant food is made from three chemicals. In a batch of the plant food there must be at least 81 kilograms of the first chemical and the other two chemicals must be in the ratio of 4 to 3. If the three chemicals cost $1.09, $.87, and $.65 per kilogram, respectively, how much of each should be used to minimize the cost of producing at least 750 kilograms of the plant food?

22. **Management** A company is developing a new additive for gasoline. The additive is a mixture of three liquid ingredients, I, II, and III. For proper performance, the total amount of additive must be at least 10 ounces per gallon of gasoline. However, for safety reasons, the amount of additive should not exceed 15 ounces per gallon of gasoline. At least 1/4 ounce of ingredient I must be used for every ounce of ingredient II and at least 1 ounce of ingredient III must be used for every ounce of ingredient I. If the cost of I, II, and III is $.30, $.09, and $.27 per ounce, respectively, find the mixture of the three ingredients which produces the minimum cost of the additive. How much of the additive should be used per gallon of gasoline?

Chapter 7   Review Exercises

Graph each of the following linear inequalities.

1. $y \geq 2x + 3$

2. $3x - y \leq 5$

3. $3x + 4y \leq 12$

4. $2x - 6y \geq 18$

5. $y \geq x$

6. $y \leq 3$

Graph the solution of each of the following systems of inequalities.

7. $x + y \leq 6$
  $2x - y \geq 3$

8. $4x + y \geq 8$
  $2x - 3y \leq 6$

9. $-4 \leq x \leq 2$
  $-1 \leq y \leq 3$
  $x + y \leq 4$

10. $2 \leq x \leq 5$
  $1 \leq y \leq 7$
  $x - y \leq 3$

11. $x + 3y \geq 6$
  $4x - 3y \leq 12$
  $x \geq 0$
  $y \geq 0$

12. $x + 2y \leq 4$
  $2x - 3y \leq 6$
  $x \geq 0$
  $y \geq 0$

Set up a system of inequalities for each of the following problems; then graph the region of feasible solutions.

13. A bakery makes both cakes and cookies. Each batch of cakes requires two hours in the oven and three hours in the decorating room. Each batch of cookies needs one and a half hours in the oven and two thirds of an hour in the decorating room. The oven is available no more than 15 hours a day, while the decorating room can be used no more than 13 hours a day.

14. A company makes two kinds of pizza, basic and plain. Basic contains cheese and beef, while plain contains onions and beef. The company sells at least three units a day of basic, and at least two units of plain. The beef costs $5 per unit for basic, and $4 per unit for plain. They can spend no more than $50 per day on beef. Dough for basic is $2 per unit, while dough for plain is $1 per unit. The company can spend no more than $16 per day on dough.

Use the given regions to find the maximum and minimum values of the objective function $z = 2x + 4y$.

15.

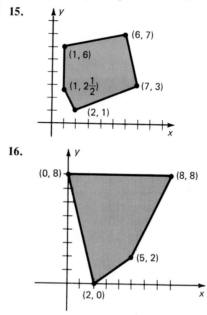

16.

Use the graphical method to solve Exercises 17–22.

17. Maximize $z = 2x + 4y$
  subject to: $3x + 2y \leq 12$
    $5x + y \geq 5$
  with $x \geq 0$ and $y \geq 0$.

18. Find $x \geq 0$ and $y \geq 0$ such that
$$8x + 9y \geq 72$$
$$6x + 8y \geq 72$$
  and $w = 3x + 2y$ is minimized.

19. Find $x \geq 0$ and $y \geq 0$ such that
$$x + y \leq 50$$
$$2x + y \geq 20$$
$$x + 2y \geq 30$$
  and $w = 4x + 2y$ is minimized.

20. Minimize $z = 3x + 2y$
  subject to: $8x + 9y \geq 72$
    $6x + 8y \geq 72$
  and $x \geq 0$, $y \geq 0$.

21. How many batches of cakes and cookies should the bakery of Exercise 13 make in order to maximize profits if cookies produce a profit of $20 per batch and cakes produce a profit of $30 per batch?

22. How many units of each kind of pizza should the company of Exercise 14 make in order to maximize profits if basic sells for $20 per unit and plain for $15 per unit?

*For Exercises 23–26, (**a**) select appropriate variables, (**b**) write the objective function, (**c**) write the constraints as inequalities.*

23. Roberta Hernandez sells three items, A, B, and C, in her gift shop. Each unit of A costs her $5 to buy, $1 to sell and $2 to deliver. For each unit of B, the costs are $3, $2, and $1 respectively, and for each unit of C the costs are $6, $2, and $5 respectively. The profit on A is $4, on B it is $3, and on C, $3. How many of each should she get to maximize her profit if she can spend $1200 to buy, $800 on selling costs, and $500 on delivery costs?

24. An investor is considering three types of investment: a high risk venture into oil leases with a potential return of 15%, a medium risk investment in bonds with a 9% return, and a relatively safe stock investment with a 5% return. He has $50,000 to invest. Because of the risk, he will limit his investment in oil leases and bonds to 30% and his investment in oil leases and stock to 50%. How much should he invest in each to maximize his return assuming investment returns are as expected?

25. The Aged Wood Winery makes two white wines, Fruity and Crystal, from two kinds of grapes and sugar. The wines require the following amounts of each ingredient per gallon and produce a profit per gallon as shown below.

	Grape A (bushels)	Grape B (bushels)	Sugar (pounds)	Profit (dollars)
Fruity	2	2	2	12
Crystal	1	3	1	15

The winery has available 110 bushels of grape A, 125 bushels of grape B, and 90 pounds of sugar. How much of each wine should be made to maximize profit?

26. A company makes three sizes of plastic bags: 5 gallon, 10 gallon and 20 gallon. The production time in hours for cutting, sealing, and packaging a unit of each size is shown below.

Size	Cutting	Sealing	Packaging
5 Gallon	1	1	2
10 Gallon	1.1	1.2	3
20 Gallon	1.5	1.3	4

There are at most 8 hours available each day for each of the three operations. If the profit on a unit of 5-gallon bags is $1, 10-gallon bags is $.90, and 20-gallon bags is $.95, how many of each size should be made per day to maximize the profit?

*For each of the following problems, (**a**) add slack variables and (**b**) set up the first simplex tableau.*

27. Maximize $z = 5x_1 + 3x_2$ subject to:

$$2x_1 + 5x_2 \leq 50$$
$$x_1 + 3x_2 \leq 25$$
$$4x_1 + x_2 \leq 18$$
$$x_1 + x_2 \leq 12$$

and $x_1 \geq 0, \quad x_2 \geq 0.$

28. Maximize $z = 25x_1 + 30x_2$ subject to:

$$3x_1 + 5x_2 \leq 47$$
$$x_1 + x_2 \leq 25$$
$$5x_1 + 2x_2 \leq 35$$
$$2x_1 + x_2 \leq 30$$

and $x_1 \geq 0, \quad x_2 \geq 0.$

29. Maximize $z = 5x_1 + 8x_2 + 6x_3$ subject to:

$$x_1 + x_2 + x_3 \leq 90$$
$$2x_1 + 5x_2 + x_3 \leq 120$$
$$x_1 + 3x_2 \leq 80$$

and $x_1 \geq 0, \quad x_2 \geq 0, \quad x_3 \geq 0.$

30. Maximize $z = 2x_1 + 3x_2 + 4x_3$ subject to:

$$x_1 + x_2 + x_3 \leq 100$$
$$2x_1 + 3x_2 \leq 500$$
$$x_1 + 2x_3 \leq 350$$

and $x_1 \geq 0, \quad x_2 \geq 0, \quad x_3 \geq 0.$

For each of the following, use the simplex method to solve the maximum linear programming problems with initial tableaus as given.

31.

x_1	x_2	x_3	x_4	x_5	
1	2	3	1	0	28
2	4	1	0	1	32
−5	−2	−3	0	0	0

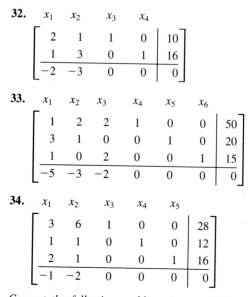

32.

$$\begin{array}{cccc} x_1 & x_2 & x_3 & x_4 \end{array}$$

$$\left[\begin{array}{cccc|c} 2 & 1 & 1 & 0 & 10 \\ 1 & 3 & 0 & 1 & 16 \\ \hline -2 & -3 & 0 & 0 & 0 \end{array}\right]$$

33.

$$\begin{array}{cccccc} x_1 & x_2 & x_3 & x_4 & x_5 & x_6 \end{array}$$

$$\left[\begin{array}{cccccc|c} 1 & 2 & 2 & 1 & 0 & 0 & 50 \\ 3 & 1 & 0 & 0 & 1 & 0 & 20 \\ 1 & 0 & 2 & 0 & 0 & 1 & 15 \\ \hline -5 & -3 & -2 & 0 & 0 & 0 & 0 \end{array}\right]$$

34.

$$\begin{array}{ccccc} x_1 & x_2 & x_3 & x_4 & x_5 \end{array}$$

$$\left[\begin{array}{ccccc|c} 3 & 6 & 1 & 0 & 0 & 28 \\ 1 & 1 & 0 & 1 & 0 & 12 \\ 2 & 1 & 0 & 0 & 1 & 16 \\ \hline -1 & -2 & 0 & 0 & 0 & 0 \end{array}\right]$$

Convert the following problems into maximization problems.

35. Minimize $w = 10y_1 + 15y_2$ subject to:

$$y_1 + y_2 \ge 17$$
$$5y_1 + 8y_2 \ge 42$$

and $y_1 \ge 0, \quad y_2 \ge 0$.

36. Minimize $w = 20y_1 + 15y_2 + 18y_3$ subject to:

$$2y_1 + y_2 + y_3 \ge 112$$
$$y_1 + y_2 + y_3 \ge 80$$
$$y_1 + y_2 \qquad \ge 45$$

and $y_1 \ge 0, \quad y_2 \ge 0, \quad y_3 \ge 0$.

37. Minimize $w = 7y_1 + 2y_2 + 3y_3$ subject to:

$$y_1 + y_2 + 2y_3 \ge 48$$
$$y_1 + y_2 \qquad \ge 12$$
$$y_3 \ge 10$$
$$3y_1 \qquad + y_3 \ge 30$$

and $y_1 \ge 0, \quad y_2 \ge 0, \quad y_3 \ge 0$.

Use the simplex method to solve the following mixed constraint problems.

38. Maximize $z = 2x_1 + 4x_2$ subject to:

$$3x_1 + 2x_2 \le 12$$
$$5x_1 + x_2 \ge 5$$

and $x_1 \ge 0, x_2 \ge 0$.

39. Minimize $w = 4y_1 + 2y_2$ subject to:

$$y_1 + y_2 \le 50$$
$$2y_1 + y_2 \ge 20$$
$$y_1 + 2y_2 \ge 30$$

and $y_1 \ge 0, \ y_2 \ge 0$.

The following tableaus are the final tableaus of minimizing problems solved by letting $w = -z$. Give the solution and the minimum value of the objective function for each problem.

40.

$$\left[\begin{array}{cccccc|c} 0 & 1 & 0 & 2 & 5 & 0 & 17 \\ 0 & 0 & 1 & 3 & 1 & 1 & 25 \\ 1 & 0 & 0 & 4 & 2 & \frac{1}{2} & 8 \\ \hline 0 & 0 & 0 & 2 & 5 & 0 & -427 \end{array}\right]$$

41.

$$\left[\begin{array}{ccccccc|c} 0 & 0 & 2 & 1 & 0 & 6 & 6 & 92 \\ 1 & 0 & 3 & 0 & 0 & 0 & 2 & 47 \\ 0 & 1 & 0 & 0 & 0 & 1 & 0 & 68 \\ 0 & 0 & 4 & 0 & 1 & 0 & 3 & 35 \\ \hline 0 & 0 & 5 & 0 & 0 & 2 & 9 & -1957 \end{array}\right]$$

The following tableaus are the final tableaus of minimizing problems solved by the method of duals. State the solution and the minimum value of the objective function for each problem.

42.

$$\left[\begin{array}{cccccc|c} 1 & 0 & 0 & 3 & 1 & 2 & 12 \\ 0 & 0 & 1 & 4 & 5 & 3 & 5 \\ 0 & 1 & 0 & -2 & 7 & -6 & 8 \\ \hline 0 & 0 & 0 & 5 & 7 & 3 & 172 \end{array}\right]$$

43.

$$\left[\begin{array}{cccccc|c} 0 & 0 & 1 & 6 & 3 & 1 & 2 \\ 1 & 0 & 0 & 4 & -2 & 2 & 8 \\ 0 & 1 & 0 & 10 & 7 & 0 & 12 \\ \hline 0 & 0 & 0 & 9 & 5 & 8 & 62 \end{array}\right]$$

44.

$$\left[\begin{array}{cccc|c} 1 & 0 & 7 & -1 & 100 \\ 0 & 1 & 1 & 3 & 27 \\ \hline 0 & 0 & 7 & 2 & 640 \end{array}\right]$$

45. Solve Exercise 23.

46. Solve Exercise 24.

47. Solve Exercise 25.

48. Solve Exercise 26.

Case 8

Merit Pay—The Upjohn Company*

Individuals doing the same job within the management of a company often receive different salaries. These salaries may differ because of the length of service of an employee, the productivity of an individual worker, and so on. However, for each job there is usually an established minimum and maximum salary.

Many companies make annual reviews of the salary of each of their management employees. At these reviews, an employee may receive a general cost of living increase, an increase based on merit, both, or neither.

In this case, we look at a mathematical model for distributing merit increases in an optimum way. An individual who is due for salary review may be described as shown in Figure 1 below. Here i represents the number of the employee whose salary is being reviewed.

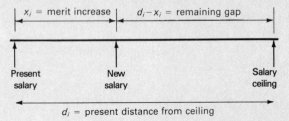

Figure 1

Here the salary ceiling is the maximum salary for the job classification, x_i is the merit increase to be awarded to the individual ($x_i \geq 0$), d_i is the present distance of the current salary from the salary ceiling, and the difference, $d_i - x_i$, is the remaining gap.

We let w_i be a measure of the relative worth of the individual to the company. This is the most difficult variable of the model to actually calculate. One way to evaluate w_i is to give a rating sheet to a number of coworkers and supervisors of employee i. An average rating can be obtained and then divided by the highest rating received by any employee with that same job. This number then gives the worth of employee i relative to all other employees with that job.

*Based on part of a paper by Jack Northam, Head, Mathematical Services Department, The Upjohn Company, Kalamazoo, Michigan.

The best way to allocate money available for merit pay increases is to minimize

$$\sum_{i=1}^{n} w_i(d_i - x_i).$$

This sum is found by multiplying the relative worth of employee i and the salary gap of employee i after any merit increase. Here n represents the total number of employees who have this job. One constraint here is that the total of all merit increases cannot exceed P, the total amount available for merit increases. That is,

$$\sum_{i=1}^{n} x_i \leq P.$$

Also, the increases for an employee must not put that employee over the maximum salary. That is, for employee i,

$$x_i \leq d_i.$$

We can simplify the objective function, using rules from algebra.

$$\sum_{i=1}^{n} w_i(d_i - x_i) = \sum_{i=1}^{n} (w_i d_i - w_i x_i)$$

$$= \sum_{i=1}^{n} w_i d_i - \sum_{i-1}^{n} w_i x_i$$

For a given individual, w_i and d_i are constant. Therefore, the sum of the $w_i d_i$ is some constant, say Z, and

$$\sum_{i=1}^{n} w_i(d_i - x_i) = Z - \sum_{i=1}^{n} w_i x_i.$$

We want to minimize the sum on the left; we can do so by *maximizing* the sum on the right. (Why?) Thus, the original model simplifies to maximizing

$$\sum_{i=1}^{n} w_i x_i,$$

subject to

$$\sum_{i=1}^{n} x_i \leq P \quad \text{and} \quad x_i \leq d_i$$

for each i.

Exercises

Here are current salary information and job evaluation averages for six employees who have the same job. The salary ceiling is $1700 per month.

Employee number	Evaluation average	Current salary
1	570	$1600
2	500	1550
3	450	1500
4	600	1610
5	520	1530
6	565	1420

1. Find w_i for each employee by dividing that employee's evaluation average by the highest evaluation average.

2. Use the simplex method to find the merit increase for each employee. Assume that P is 400.

Case 9

Making Ice Cream*

The first step in the commercial manufacture of ice cream is to blend several ingredients (such as dairy products, eggs, and sugar) to obtain a mix which meets the necessary minimum quality restrictions regarding butterfat content, serum solids, and so on.

Usually, many different combinations of ingredients may be blended to obtain a mix of the necessary quality. Within this range of possible substitutes, the firm desires the combination that produces minimum total cost. This problem is quite suitable for solution by linear programming methods.

A "mid-quality" line of ice cream requires the following minimum percentages of constituents by weight.

Fat	16	%
Serum solids	8	
Sugar solids	16	
Egg solids	.35	
Stabilizer	.25	
Emulsifier	.15	
Total	40.75%	

The balance of the mix is water. A batch of mix is made by blending a number of ingredients, each of which contains one or more of the necessary constituents, and nothing else. The chart above shows the possible ingredients and their costs.

In setting up the mathematical model, use c_1, $c_2, \ldots, c_{14}$ as the cost per unit of the ingredients, and x_1, $x_2, \ldots, x_{14}$ for the quantities of ingredients. The table shows the composition of each ingredient. For example, one pound of ingredient 1 (the 40% cream) contains .400 pounds of fat and .054 pounds of serum solids, with the balance being water.

The ice cream mix is made up in batches of 100 pounds at a time. For a batch to contain 16% fat, at least $100(.16) = 16$ pounds of fat is necessary. Fat is contained in ingredients 1, 2, 3, 4, 5, 6, 10, and 11. The requirement of at least 16 pounds of fat produces the constraint

$$.400x_1 + .230x_2 + .805x_3 + .800x_4 + .998x_5 + .040x_6$$
$$+ .500x_{10} + .625x_{11} \geq 16.$$

Similar constraints can be obtained for the other constituents. Because of the minimum requirements, the total of the ingredients will be at least 40.75 pounds, or

$$x_1 + x_2 + \cdots + x_{13} \geq 40.75.$$

The balance of the 100 pounds is water, making $x_{14} \leq 59.25$.

The table shows that ingredients 12 and 13 must be used—there is no alternate way of getting these constituents into the final mix. For a 100 pound batch of ice cream, $x_{12} = .25$ pound and $x_{13} = .15$ pound. Removing x_{12} and x_{13} as variables permits a substantial simplification of the model; the problem is now reduced to the following system of four constraints:

$$.400x_1 + .230x_2 + .805x_3 + .800x_4 + .998x_5 + .040x_6$$
$$+ .500x_{10} + .625x_{11} \geq 16$$
$$.054x_1 + .069x_2 + .025x_4 + .078x_6 + .280x_7 + .970x_8 \geq 8$$
$$.707x_9 + .100x_{10} \geq 16$$
$$.350x_{10} + .315x_{11} \geq .35.$$

The objective function, which is to be minimized, is

$$c = .298x_1 + .174x_2 + \cdots + 1.090x_{11}.$$

As usual, $x_1 \geq 0$, $x_2 \geq 0, \ldots, x_{14} \geq 0$.

Solving this linear programming problem with the simplex method gives

$$x_2 = 67.394, \quad x_8 = 3.459, \quad x_9 = 22.483, \quad x_{10} = 1.000.$$

The total cost of these ingredients is $14.094. To this, we must add the cost of ingredients 12 and 13, producing total cost of

$$14.094 + (.600)(.25) + (.420)(.15) = 14.307,$$

or $14.307.

*From *Linear Programming in Industry: Theory and Application, An Introduction* by Sven Danø, Fourth revised and enlarged edition. Copyright © 1974 by Springer-Verlag/Wein. Reprinted by permission of Springer-Verlag New York, Inc.

		Ingredients								
		1	2	3	4	5	6	7	8	9
		40% cream	23% cream	Butter	Plastic cream	Butter oil	4% milk	Skim condensed milk	Skim milk powder	Liquid sugar
		Cost ($/lb)								
		.298	**.174**	**.580**	**.576**	**.718**	**.045**	**.052**	**.165**	**.061**
Constituents	1. Fat	.400	.230	.805	.800	.998	.040			
	2. Serum solids	.054	.069		.025		.078	.280	.970	
	3. Sugar solids									.707
	4. Egg solids									
	5. Stabilizer									
	6. Emulsifer									

		Ingredients					
		10	11	12	13	14	
		Sugared egg yolk	Powdered egg yolk	Stabilizer	Emulsifier	Water	
		Cost ($/lb)					
		.425	**1.090**	**.600**	**.420**	**0**	*Requirements*
Constituents	1. Fat	.500	.625				16
	2. Serum solids						8
	3. Sugar solids	.100					16
	4. Egg solids	.350	.315				.35
	5. Stabilizer			1			.25
	6. Emulsifer				1		.15

Exercises

1. Find the mix of ingredients needed for a 100-pound batch of the following grades of ice cream.
   a. ''Generic,'' sold in a white box, with at least 14% fat and at least 6% serum solids (all other ingredients the same)
   b. ''Premium,'' sold with a funny foreign name, with at least 17% fat and at least 16.5% sugar solids (all other ingredients the same)

2. Solve the problem given in this case by using a computer.

8 Sets, Counting, and Probability

A personnel director knows that of 20 applicants for a job, 9 have previous related experience, 5 have MBA degrees and experience, and 12 have MBA degrees or experience. She may really want to know how many have just MBA degrees or what the probability is that a particular applicant has a degree. The terminology and ideas of sets have proved to be very useful in solving problems of this type. Sets help to clarify and classify many mathematical ideas, making them easier to understand. The principles of counting are also useful in finding probabilities. The basics of probability are discussed in this chapter. Further topics in probability and some special applications of probability are discussed in the next chapter.

8.1 Sets

A **set** is a collection of objects. For example, a particular set might contain one of each type of coin now put out by the government. Another set might include all the students in a class. The set containing the numbers 3, 4, and 5 is written

$$\{3, 4, 5\},$$

where set braces, { }, are used to enclose the numbers belonging to the set. The numbers 3, 4, and 5 are called the **elements** or **members** of this set. Show that 4 is an element of the set $\{3, 4, 5\}$ by using the symbol $\in$:

$$4 \in \{3, 4, 5\}.$$

Also, $5 \in \{3, 4, 5\}$. Place a slash through the symbol $\in$ to show that 8 is *not* an element of this set.

$$8 \notin \{3, 4, 5\}$$

Sets are often named with capital letters, so that if

$$B = \{5, 6, 7\},$$

then, for example, $6 \in B$ and $10 \notin B$. ①

Two sets are **equal** if they contain exactly the same elements. The sets $\{5, 6, 7\}$, $\{7, 6, 5\}$, and $\{6, 5, 7\}$ all contain exactly the same elements and are equal. In symbols,

$$\{5, 6, 7\} = \{7, 6, 5\} = \{6, 5, 7\}.$$

① Write *true* or *false*.
(a) $9 \in \{8, 4, -3, -9, 6\}$
(b) $4 \notin \{3, 9, 7\}$
(c) If $M = \{0, 1, 2, 3, 4\}$, then $0 \in M$.

Answer:
(a) False
(b) True
(c) True

344

Sets which do not contain exactly the same elements are *not equal*. For example, the sets {5, 6, 7} and {5, 6, 7, 8} do not contain exactly the same elements and are not equal.

$${5, 6, 7} \neq {5, 6, 7, 8}$$

Sometimes a common property of the elements of a set is more important than a list of the elements in the set. In fact, sometimes this is the only way a set can be specified. This common property can be expressed with **set-builder notation**; the set

$${x \mid x \text{ has property } P}$$

(read "the set of all elements x such that x has property P") represents the set of all elements x having some property P.

Example 1   Write the elements belonging to each of the following sets.
(a) {$x \mid x$ is a counting number less than 5}
   The counting numbers less than 5 make up the set {1, 2, 3, 4}.
(b) {$x \mid x$ is a state that touches Florida} = {Alabama, Georgia}   **2**

When discussing a particular situation or problem, we can identify a **universal set** containing all the elements appearing in any set used in that particular problem. The letter U is used to represent the universal set. The universal set varies depending on the requirements of the problem and is usually stated with the problem. (If it is not stated, it is assumed to include all the elements of every set mentioned.)

For example, the set of numbers between 0 and 3 inclusive differs, depending on the given universal set. If the universal set is the set of integers, then the set of numbers between 0 and 3 inclusive is

$${0, 1, 2, 3}.$$

However, the set of numbers between 0 and 3 inclusive, with the set of real numbers as the universal set, includes the infinite set of numbers

$${x \mid x \text{ is a real number between 0 and 3 inclusive}}.$$

When discussing the set of company employees who favor a certain pension proposal, the universal set might be the set of all company employees. In discussing the species found by Charles Darwin on the Galápagos Islands, the universal set might be the set of all species on all Pacific islands.

Sometimes every element of one set also belongs to another set. For example, if

$$A = {3, 4, 5, 6}$$

and

$$B = {2, 3, 4, 5, 6, 7, 8},$$

2 List the elements in the following sets.
(a) {$x \mid x$ is a counting number more than 5 and less than 8}
(b) {$x \mid x$ is an integer, $-3 < x \leq 1$}

Answer:
(a) {6, 7}
(b) {−2, −1, 0, 1}

then every element of A is also an element of B. This means that A is a **subset** of B, written $A \subset B$. As another example, the set of all presidents of corporations is a subset of the set of all executives of corporations.

Example 2 Write *true* or *false* for each of the following statements.
(a) {3, 4, 5, 6} = {4, 6, 3, 5}
  Both sets contain exactly the same elements; the sets are equal. The statement is true. (The fact that the elements are in a different order doesn't matter.)
(b) {5, 6, 9, 10} $\subset$ {5, 6, 7, 8, 9, 10, 11}
  Every element of the first set is also an element of the second, so the statement is true.
(c) The universal set for parts (a) and (b) could be

$$U = \{0, 1, 2, 3, 4, 5\}.$$

The numbers 6, 7, 8, 9, 10, and 11 must be in the universal set, since they are in the sets of parts (a) and (b), so the statement is false.
(d) The universal set for this example might be

$$U = \{0, 1, 2, 3, 4, 5, 6, 7, 8, 9, 10, 11, 12\}.$$

This statement is true. All the elements listed in the sets of parts (a) and (b) are included in set U. **3**

Figure 8.1 shows a set A which is a subset of a set B, since A is entirely in B. The rectangle represents the universal set, U. Such diagrams, called **Venn diagrams,** are used to clarify relationships among sets.

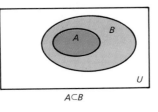

$A \subset B$

Figure 8.1

Sometimes a set has no elements. Some examples are the set of women presidents of the United States, the set of counting numbers less than 1, and the set of men more than 10 feet tall. A set with no elements is called the **empty set.** The symbol $\emptyset$ is used to represent the empty set.

Since it contains no elements, the empty set is a subset of every set: if A is any set then $\emptyset \subset A$. Also, by the definition of subset, every set is a subset of itself: that is, if A is any set, then $A \subset A$. **4**

For any set A,

$$\emptyset \subset A \quad \text{and} \quad A \subset A.$$

---

3 Write *true* or *false*.
(a) {3, 4, 5} $\subset$ {2, 3, 4, 6}
(b) {1, 2, 5, 8} $\subset$
  {1, 2, 5, 8, 10, 11}
(c) {3, 6, 9, 10} $\subset$
  {3, 9, 11, 13}
(d) For this problem U could be
  {1, 2, 3, 4, 5, 6, 7, 8, 9, 10, 11, 12, 13}.

Answer:
(a) False
(b) True
(c) False
(d) True

4
Refer to sets A, B, C, and U in the diagram.

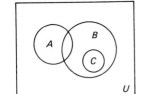

(a) Is $A \subset B$?
(b) Is $C \subset B$?
(c) Is $C \subset U$?
(d) Is $\emptyset \subset A$?

Answer:
(a) No
(b) Yes
(c) Yes
(d) Yes

Example 3 List all possible subsets for each of the following sets.

(a) {7, 8}

There are four subsets of {7, 8}:

$$\emptyset, \quad \{7\}, \quad \{8\}, \quad \{7, 8\}.$$

(b) {a, b, c}

There are eight subsets of {a, b, c}:

$$\emptyset, \quad \{a\}, \quad \{b\}, \quad \{c\}, \quad \{a, b\}, \quad \{a, c\}, \quad \{b, c\}, \quad \{a, b, c\}. \quad \blacksquare$$

In Example 3, the subsets of {7, 8} and the subsets of {a, b, c} were found by trial and error. An alternate method uses a **tree diagram,** a systematic way of listing all the subsets of a given set. Figures 8.2(a) and (b) show tree diagrams for finding the subsets of {7, 8} and {a, b, c}. ⑤

⑤ List all subsets of {w, x, y, z}.

Answer:

∅, {w}, {x}, {y}, {z}, {w, x}, {w, y}, {w, z}, {x, y}, {x, z}, {y, z}, {w, x, y}, {w, x, z}, {w, y, z}, {x, y, z}, {w, x, y, z}

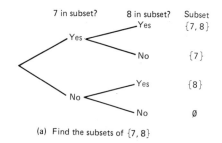

(a) Find the subsets of {7, 8}

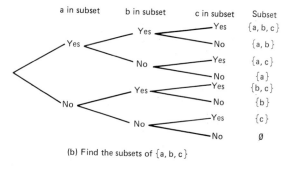

(b) Find the subsets of {a, b, c}

Figure 8.2

Examples and tree diagrams similar to the ones above suggest the following rule.

For a nonnegative integer n, a set containing n distinct elements has 2^n subsets.

Example 4 Find the number of subsets for each of the following sets.
(a) {3, 4, 5, 6, 7}

Since this set has five elements, it has 2^5 or 32 subsets.

(b) {−1, 2, 3, 4, 5, 6, 12, 14}

This set has 8 elements and therefore has 2^8 or 256 subsets.

(c) Ø

The empty set has 0 elements and $2^0 = 1$ subset, Ø itself.

6 Find the number of subsets for each of· the following sets.
(a) {1, 2, 3}
(b) {−6, −5, −4, −3, −2, −1, 0}
(c) {6}

Answer:
(a) 8
(b) 128
(c) 2

Given a set A and a universal set U, the set of all elements of U which do *not* belong to A is called the **complement** of set A. For example, if set A is the set of all the female students in your class, and U is the set of all students in the class, then the complement of A would be the set of all male students in the class. The complement of set A is written A' (read: "A-prime"). The Venn diagram of Figure 8.3 shows a set B. Its complement, B', is shown in color.

Example 5 Let $U = \{1, 2, 3, 4, 5, 6, 7\}$, $A = \{1, 3, 5, 7\}$, and $B = \{3, 4, 6\}$. Find each of the following sets.
(a) A'

Set A' contains the elements of U that are not in A.

$$A' = \{2, 4, 6\}$$

(b) $B' = \{1, 2, 5, 7\}$
(c) $\emptyset' = U$ and $U' = \emptyset$ **7**

7 Let $U =$ {a, b, c, d, e, f, g}, with $K =$ {c, d, f, g} and $R =$ {a, c, d, e, g}. Find
(a) K',
(b) R'.

Answer:
(a) {a, b, e}
(b) {b, f}

Given two sets A and B, the set of all elements belonging to *both* set A and set B is called the **intersection** of the two sets, written $A \cap B$. For example, the elements that belong to both $A = \{1, 2, 4, 5, 7\}$ and $B = \{2, 4, 5, 7, 9, 11\}$ are 2, 4, 5, and 7, so that

$$A \cap B = \{1, 2, 4, 5, 7\} \cap \{2, 4, 5, 7, 9, 11\} = \{2, 4, 5, 7\}.$$

The Venn diagram of Figure 8.4 shows two sets A and B; their intersection, $A \cap B$, is shown in color.

Example 6 **(a)** $\{9, 15, 25, 36\} \cap \{15, 20, 25, 30, 35\} = \{15, 25\}$

The elements 15 and 25 are the only ones belonging to both sets.

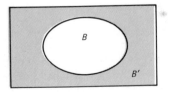

Figure 8.3

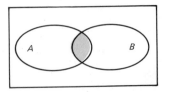

Figure 8.4

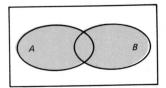

Figure 8.5

8 Find the following.
(a) $\{1, 2, 3, 4\} \cap$
    $\{3, 5, 7, 9\}$
(b) $\{a, b, c, d, e, g\} \cap$
    $\{a, c, e, g, h, j\}$

Answer:
(a) $\{3\}$
(b) $\{a, c, e, g\}$

(b) $\{-2, -3, -4, -5, -6\} \cap \{-4, -3, -2, -1, 0, 1, 2\} =$
$\{-4, -3, -2\}$ **8**

Two sets that have no elements in common are called **disjoint sets.** For example, there are no elements common to both $\{50, 51, 54\}$ and $\{52, 53, 55, 56\}$, so that these two sets are disjoint, and

$$\{50, 51, 54\} \cap \{52, 53, 55, 56\} = \emptyset.$$

The result of this example can be generalized: For any sets A and B,

if A and B are disjoint sets then $A \cap B = \emptyset$.

The set of all elements belonging to set A *or* set B is the **union** of the two sets, written $A \cup B$. For example,

$$\{1, 3, 5\} \cup \{3, 5, 7, 9\} = \{1, 3, 5, 7, 9\}.$$

The Venn diagram of Figure 8.5 shows two sets A and B, with their union $A \cup B$ shown in color.

Example 7 **(a)** Find the union of $\{1, 2, 5, 9, 14\}$ and $\{1, 3, 4, 8\}$.

Begin by listing the elements of the first set, $\{1, 2, 5, 9, 14\}$. Then include any elements from the second set that are not already listed. Doing this gives

$$\{1, 2, 5, 9, 14\} \cup \{1, 3, 4, 8\} = \{1, 2, 3, 4, 5, 8, 9, 14\}.$$

9 Find the following.
(a) $\{a, b, c\} \cup \{a, c, e\}$
(b) $\{a, c, d, e, f\} \cup$
    $\{a, b, c, d, g\}$

Answer:
(a) $\{a, b, c, e\}$
(b) $\{a, b, c, d, e, f, g\}$

(b) $\{1, 3, 5, 7\} \cup \{2, 4, 6\} = \{1, 2, 3, 4, 5, 6, 7\}$. **9**

Finding the complement of a set, the intersection of two sets, or the union of two sets are examples of *set operations*. These set operations are ways of combining sets to get other sets, just as operations on numbers, such as addition, subtraction, multiplication, and division, usually produce other numbers.

The set operations are summarized below.

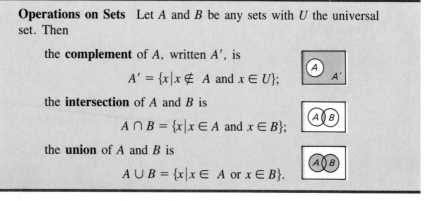

Operations on Sets Let A and B be any sets with U the universal set. Then

the **complement** of A, written A', is

$$A' = \{x \mid x \notin A \text{ and } x \in U\};$$

the **intersection** of A and B is

$$A \cap B = \{x \mid x \in A \text{ and } x \in B\};$$

the **union** of A and B is

$$A \cup B = \{x \mid x \in A \text{ or } x \in B\}.$$

Example 8 The table below gives the current dividend, price to earnings ratio (the quotient of the price per share and the annual earnings per share), and price change at the end of a day for six companies, as listed on the New York Stock Exchange.

Stock	Dividend	Price to earnings ratio	Price change
ATT	1.20	17	$-\frac{1}{2}$
GE	2.32	14	$+1\frac{3}{4}$
Hershey	1.5	13	$-1\frac{1}{8}$
IBM	4.4	16	$+3\frac{3}{8}$
Mobil	2.2	14	$-\frac{1}{2}$
PepsiCo	.64	17	$+\frac{1}{2}$

Let set A include all stocks with a dividend greater than $2: B$ all stocks with a price to earnings ratio of at least 16, and C all stocks with a positive price change. Find the following.

(a) A'

Set A' contains all the listed stocks outside set A, those with a dividend less than or equal to $2, so

$$A' = \{\text{ATT, Hershey, PepsiCo}\}.$$

(b) $A \cap B$

The intersection of A and B will contain those stocks that offer a dividend greater than $2 *and* have a price to earnings ratio of at least 16.

$$A \cap B = \{\text{IBM}\}$$

(c) $A \cup C$

The union of A and C contains all stocks with a dividend greater than $2 or a positive price change (or both).

$$A \cup C = \{\text{GE, IBM, Mobil, PepsiCo}\} \quad \boxed{10}$$

10 If $U =$
$\{-3, -2, -1, 0, 1, 2, 3\}$,
$A = \{-1, -2, -3\}$, and
$B = \{-2, 0, 2\}$, find
(a) $A \cup B$
(b) B'
(c) $A \cap B$.

Answer:
(a) $\{-3, -2, -1, 0, 2\}$
(b) $\{-3, -1, 1, 3\}$
(c) $\{-2\}$

8.1 Exercises

Write true *or* false *for each of the following.*

1. $3 \in \{2, 5, 7, 9, 10\}$

2. $6 \in \{-2, 6, 9, 5\}$

3. $1 \in \{3, 4, 5, 1, 11\}$

4. $12 \in \{19, 17, 14, 13, 12\}$

5. $9 \notin \{2, 1, 5, 8\}$

6. $3 \notin \{7, 6, 5, 4\}$

7. $\{2, 5, 8, 9\} = \{2, 5, 9, 8\}$

8. $\{3, 0, 9, 6, 2\} = \{2, 9, 0, 3, 6\}$

9. $\{5, 8, 9\} = \{5, 8, 9, 0\}$

10. $\{3, 7, 12, 14\} = \{3, 7, 12, 14, 0\}$

11. {all counting numbers less than 6}
$= \{1, 2, 3, 4, 5, 6\}$

12. {all whole numbers greater than 7 and less than 10}
$= \{8, 9\}$

13. {all whole numbers not greater than 4} $= \{0, 1, 2, 3\}$

14. {all counting numbers not greater than 3}
= {0, 1, 2}

15. $\{x \mid x \text{ is a whole number}, x \le 5\} = \{0, 1, 2, 3, 4, 5\}$

16. $\{x \mid x \text{ is an integer}, -3 \le x < 4\}$
= {−3, −2, −1, 0, 1, 2, 3, 4}

17. $\{x \mid x \text{ is an odd integer}, 6 \le x \le 18\}$
= {7, 9, 11, 15, 17}

18. $\{x \mid x \text{ is an even counting number}, x \le 9\}$
= {0, 2, 4, 6, 8}

19. {5, 7, 9, 19} ∩ {7, 9, 11, 15} = {7, 9}

20. {8, 11, 15} ∩ {8, 11, 19, 20} = {8, 11}

21. {2, 1, 7} ∪ {1, 5, 9} = {1}

22. {6, 12, 14, 16} ∪ {6, 14, 19} = {6,14}

23. {3, 2, 5, 9} ∩ {2, 7, 8, 10} = {2}

24. {8, 9, 6} ∪ {9, 8, 6} = {8,9}

25. {3, 5, 9, 10} ∩ ∅ = {3, 5, 9, 10}

26. {3, 5, 9, 10} ∪ ∅ = {3, 5, 9, 10}

27. {1, 2, 4} ∪ {1, 2, 4} = {1, 2, 4}

28. {1, 2, 4} ∩ {1, 2, 4} = ∅

29. ∅ ∪ ∅ = ∅

30. ∅ ∩ ∅ = ∅

Let A = {2, 4, 6, 8, 10, 12}, *B* = {2, 4, 8, 10}, *C* = {4, 10, 12}, *D* = {2, 10}, *and U* = {2, 4, 6, 8, 10, 12, 14}. *Write* true *or* false *for each of the following. (See Examples 2–4.)*

31. $A \subset U$ **32.** $C \subseteq U$

33. $D \subset B$ **34.** $D \subseteq A$

35. $A \subset B$ **36.** $B \subset C$

37. $\emptyset \subset A$ **38.** $\emptyset \subset \emptyset$

39. {4, 8, 10} $\subset B$ **40.** {0, 2} $\subset D$

41. $D \not\subset B$ **42.** $A \not\subset C$

43. There are exactly 32 subsets of *A*.

44. There are exactly 16 subsets of *B*.

45. There are exactly 6 subsets of *C*.

46. There are exactly 4 subsets of *D*.

Find the number of subsets for each of the following sets. (See Examples 3 and 4.)

47. {4, 5, 6}   **48.** {3, 7, 9, 10}

49. {5, 9, 10, 15, 17}   **50.** {6, 9, 1, 4, 3, 2}

51. ∅   **52.** {0}

53. $\{x \mid x \text{ is a counting number between 6 and 12}\}$

54. $\{x \mid x \text{ is a whole number between 8 and 12}\}$

Let U = {2, 3, 4, 5, 7, 9}, *X* = {2, 3, 4, 5}, *Y* = {3, 5, 7, 9}, *and Z* = {2, 4, 5, 7, 9}. *Find each of the following sets. (See Examples 5–7.)*

55. $X \cap Y$ **56.** $X \cup Y$

57. $Y \cup Z$ **58.** $Y \cap Z$

59. $X \cup U$ **60.** $Y \cap U$

61. X' **62.** Y'

63. $X' \cap Y'$ **64.** $X' \cap Z$

65. $Z' \cap \emptyset$ **66.** $Y' \cup \emptyset$

67. $X \cup (Y \cap Z)$ **68.** $Y \cap (X \cup Z)$

Let U = {all students in this school},
M = {all students taking this course},
N = {all students taking accounting},
P = {all students taking zoology}.

Describe each of the following sets in words.

69. M' **70.** $M \cup N$ **71.** $N \cap P$

72. $N' \cap P'$ **73.** $M \cup P$ **74.** $P' \cup M'$

Natural Science *Work the following problems. (See Example 8.)*

75. The lists below show some symptoms of an overactive thyroid and an underactive thyroid.

Underactive thyroid	Overactive thyroid
Sleepiness, s	Insomnia, i
Dry hands, d	Moist hands, m
Intolerance of cold, c	Intolerance of heat, h
Goiter, g	Goiter, g

(a) Find the smallest possible universal set *U* that includes all the symptoms listed.

Let *N* be the set of symptoms for an underactive thyroid, and let *O* be the set of symptoms for an overactive thyroid. Find each of the following sets.

(b) O' (c) N' (d) $N \cap O$

(e) $N \cup O$ (f) $N \cap O'$

76. The following list includes the producers of most (93%) of the hazardous wastes in the United States.

Industry	% (in 1977)
Inorganic Chemicals (1)	11
Organic Chemicals (2)	34
Electroplating (3)	12
Petroleum Refining (4)	5
Smelting and Refining (5)	26
Textiles Dyeing and Finishing (6)	5

Let x be the number associated with an industry in the table above. The universal set is $U = \{1, 2, 3, 4, 5, 6\}$. List the elements of the following subsets of U.
(a) $\{x|x$ is an industry that produces more than 15% of the wastes$\}$.
(b) $\{x|x$ is an industry that produces less than 5% of the wastes$\}$.
(c) Find all subsets of U containing industries that produce a total of at least 75% of the waste.

8.2 Venn Diagrams

Venn diagrams were used in the last section to help in understanding set union and intersection. The rectangular region in a Venn diagram represents the universal set, U. Including only a single set, A, inside the universal set, as in Figure 8.6, divides U into two regions. Region 1 represents those elements outside set A, while region 2 represents those elements belonging to set A. (The numbering of these regions is arbitrary.)

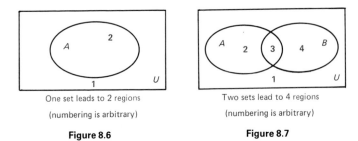

One set leads to 2 regions
(numbering is arbitrary)

Figure 8.6

Two sets lead to 4 regions
(numbering is arbitrary)

Figure 8.7

The Venn diagram of Figure 8.7 shows two sets inside U. These two sets divide the universal set into four regions. As labeled in Figure 8.7, region 1 includes those elements outside of both set A and set B. Region 2 includes those elements belonging to A and not to B. Region 3 includes those elements belonging to both A and B. Which elements belong to region 4? (Again, the labeling is arbitrary.)

Example 1 Draw Venn diagrams similar to Figure 8.7 and shade the regions representing the following sets.
(a) $A' \cap B$
Set A' contains all the elements outside set A. As labeled in Figure 8.7, A' is made up of regions 1 and 4. Set B is made up of the elements in regions 3 and 4. The intersection of sets A' and B, the set $A' \cap B$, is made up of the elements in the region common to regions 1 and 4 and regions 3 and 4. The result, region 4, is shaded in Figure 8.8.

Draw Venn diagrams for the following.
(a) $A \cup B'$
(b) $A' \cap B'$

Answer:

(a)

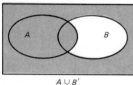

$A \cup B'$

(b)

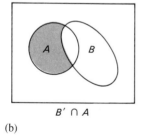

$A' \cap B'$

2 Draw a Venn diagram for the following.
(a) $B' \cap A$
(b) $(A \cup B)' \cap C$

Answer:

(a)

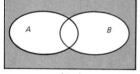

$B' \cap A$

(b)

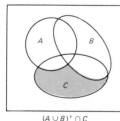

$(A \cup B)' \cap C$

(b) $A' \cup B'$

Again, set A' is represented by regions 1 and 4, while B' is made up of regions 1 and 2. Find $A' \cup B'$ by identifying the elements belonging to either regions 1 and 4 or to regions 1 and 2. The result, regions 1, 2, and 4, is shaded in Figure 8.9. **1**

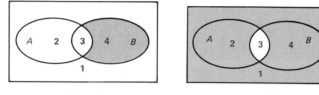

Figure 8.8   **Figure 8.9**

Venn diagrams can be drawn with three sets inside U. These three sets divide the universal set into eight regions which can be numbered (arbitrarily) as in Figure 8.10.

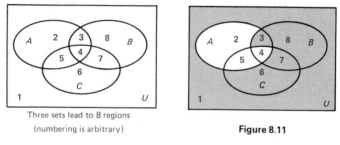

Three sets lead to 8 regions
(numbering is arbitrary)

Figure 8.10   **Figure 8.11**

Example 2 Shade $A' \cup (B \cap C')$ on a Venn diagram.

First find $B \cap C'$. Set B is made up of regions of 3, 4, 7, and 8, while C' is made up of regions 1, 2, 3, and 8. The overlap of these regions, the set $B \cap C'$, is made up of regions 3 and 8. Set A' is made up of regions 1, 6, 7, and 8. The union of regions 3 and 8 and regions 1, 6, 7, and 8 includes regions 1, 3, 6, 7, and 8, which are shaded in Figure 8.11. **2**

Venn diagrams can be used to solve problems that result from surveying groups of people. As an example, suppose a researcher collecting data on a group of 50 surfers found

32 were blonde,

28 had blue eyes,

15 were blue-eyed blondes.

The researcher wishes to use this data to answer the following questions.

(a) How many were not blonde?

(b) How many were neither blue-eyed nor blonde?

(c) How many were blonde but not blue-eyed?

Since some of the surfers are listed in both categories, some in only one category, and some in neither category, a Venn diagram like the one in Figure 8.12(a) can help to clarify the information. In Figure 8.12(b), place 15 in the region common to both blonde and blue eyes, since there are 15 surfers with both characteristics. Of the 32 blondes, 15 also have blue eyes, so place $32 - 15 = 17$ in the region for blonde but not blue-eyed. Since 15 of the 28 with blue eyes are also blonde, write $28 - 15 = 13$ in the region for blue-eyed but not blonde. Finally, of the 50 surfers, $50 - 17 - 15 - 13 = 5$ have neither blond hair nor blue eyes, so put 5 in the region outside both circles. Now the questions can be answered:

(a) $13 + 5 = 18$ were not blonde;

(b) 5 were neither blue-eyed nor blonde;

(c) 17 were blonde but not blue-eyed. ③

③ Place numbers in the regions on a Venn diagram if the data on the 50 surfers showed

29 blondes,
17 with blue eyes,
10 blue-eyed blondes.

Answer:

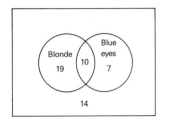

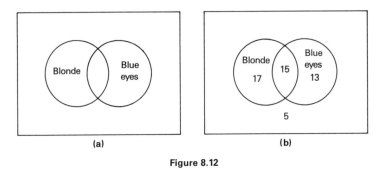

(a)   (b)

Figure 8.12

Example 3 A group of 60 freshman business students at a large university was surveyed, with the following results.

19 of the students read *Business Week;*

18 read *The Wall Street Journal;*

50 read *Fortune;*

13 read *Business Week* and *The Journal;*

11 read *The Journal* and *Fortune;*

13 read *Business Week* and *Fortune;*

9 read all three.

Use this data to answer the following questions.

(a) How many students read none of the publications?

(b) How many read only *Fortune?*

(c) How many read *Business Week* and *The Journal,* but not *Fortune?*

Many of the students are listed more than once in the data above. For example, some of the 50 students who read *Fortune* also read *Business Week*. The 9 students who read all three are counted in the 13 who read *Business Week* and *Fortune,* and so on.

Again use a Venn diagram to clarify the data. Since 9 students read all three publications, begin by placing 9 in the area in Figure 8.13 that belongs to all three regions. Of the 13 students who read *Business Week* and *Fortune,* 9 also read *The Journal.* Therefore, only $13 - 9 = 4$ read just *Business Week* and *Fortune.* Place the number 4 in the area of Figure 8.13 common to *Business Week* and *Fortune* readers. In the same way, place 4 in the region common only to *Business Week* and *The Journal,* and 2 in the region common only to *Fortune* and *The Journal.*

A total of 19 students read *Business Week.* However, $4 + 9 + 4 = 17$ readers have already been placed in the region representing *Business Week.* The balance of this region will contain only $19 - 17 = 2$ students. These 2 students read *Business Week* only—not *Fortune* and not *The Journal.* In the same way, 3 students read only *The Journal* and 35 read only *Fortune.*

A total of $2 + 4 + 3 + 4 + 9 + 2 + 35 = 59$ students are placed in the three regions of Figure 8.13. Since 60 students were surveyed, $60 - 59 = 1$ student reads none of the three publications and 1 is placed outside all three regions.

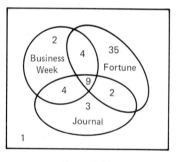

Figure 8.13

Figure 8.13 can now be used to answer the questions asked above.

(a) Only 1 student reads none of the three publications.

(b) From Figure 8.13, there are 35 students who read only Fortune.

(c) The overlap of the regions representing *Business Week* and *The Journal* shows that 4 students read *Business Week* and *The Journal* but not *Fortune.* ■4

4 In the example about the three publications, how many students read exactly
(a) 1,
(b) 2
of the publications?

Answer:
(a) 40
(b) 10

Example 4 Jeff Friedman is a section chief for an electric utility company. The employees in his section cut down tall trees, climb poles, and splice wire. Friedman reported the following information to the management of the utility.

Of the 100 employees in my section,

45 can cut tall trees;

50 can climb poles;

57 can splice wire;

28 can cut trees and climb poles;

20 can climb poles and splice wire;

25 can cut trees and splice wire;

11 can do all three;

9 can't do any of the three (management trainees).

The data supplied by Friedman lead to the numbers shown in Figure 8.14. Add all the numbers from the regions to get the total number of Friedman's employees:

$$9 + 3 + 14 + 23 + 11 + 9 + 17 + 13 = 99.$$

Friedman claimed to have 100 employees, but his data indicates only 99. The management decided that Friedman didn't qualify as a section chief, and reassigned him as a nightshift meter reader in Guam. (Moral: He should have taken this course.) **5**

As in the examples above, usually the best way to begin solving these problems is to start in the innermost region, the intersection of the three categories and then work outward.

5 In Example 4, suppose 46 employees can cut tall trees. Then how many
(a) can only cut tall trees?
(b) can cut trees or climb poles?
(c) can cut trees or climb poles or splice wire?

Answer:
(a) 4
(b) 68
(c) 91

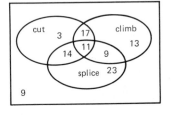

Figure 8.14

8.2 Exercises

Use a Venn diagram similar to Figure 8.7 to show each of the following sets. (See Example 1.)

1. $B \cap A'$

2. $A \cup B'$

3. $A' \cup B$

4. $A' \cap B'$

5. $B' \cup (A' \cap B')$

6. $(A \cap B) \cup B'$

7. U'

8. $\emptyset'$

Use a Venn diagram similar to Figure 8.10 to show each of the following sets. (See Example 2.)

9. $(A \cap B) \cap C$

10. $(A \cap C') \cup B$

11. $A \cap (B \cup C')$

12. $A' \cap (B \cap C)$

13. $(A' \cap B') \cap C$

14. $(A \cap B') \cup C$

15. $(A \cap B') \cap C$

16. $A' \cap (B' \cup C)$

Use Venn diagrams to answer the following questions. (See Examples 3 and 4.)

17. Jeff Friedman, of Example 4 in the text, was again reassigned, this time to the home economics department of the electric utility. He interviewed 140 people in a suburban shopping center to find out some of their cooking habits. He obtained the following results. Should he be reassigned yet one more time?

> 58 use microwave ovens;
>
> 63 use electric ranges;
>
> 58 use gas ranges;
>
> 19 use microwave ovens and electric ranges;
>
> 17 use microwave ovens and gas ranges;
>
> 4 use both gas and electric ranges;
>
> 1 uses all three;
>
> 2 cook only with solar energy.

18. Toward the middle of the harvesting season, peaches for canning come in three types, earlies, lates, and extra lates, depending on the expected date of ripening. During a certain week, the following data was recorded at a fruit delivery station.

> 34 trucks went out carrying early peaches;
>
> 61 had late peaches;
>
> 50 had extra lates;
>
> 25 had earlies and lates;
>
> 30 had lates and extra lates;
>
> 8 had earlies and extra lates;
>
> 6 had all three;
>
> 9 had only figs (no peaches at all).

(a) How many trucks had only late variety peaches?
(b) How many had only extra lates?
(c) How many had only one type of peach?
(d) How many trucks in all went out during the week?

19. Social Science A recent nationwide survey revealed the following data concerning unemployment in the United States. Out of the total population of people who are working or seeking work,

> 9% is unemployed;
>
> 11% is black;
>
> 30% is younger than 18;
>
> 1% is unemployed black youths;

> 5% of the unemployed are younger than 18;
>
> 7% are employed black adults;
>
> 23% of the nonblack youths are employed.

What percent are
(a) not black?
(b) employed adults?
(c) unemployed adults?
(d) unemployed black adults?
(e) unemployed non-black youths?

20. Country-western songs emphasize three basic themes: love, prison, and trucks. A survey of the local country-western radio station produced the following data.

> 12 songs were about a truck driver who was in love while in prison;
>
> 13 about a prisoner in love;
>
> 28 about a person in love;
>
> 18 about a truck driver in love;
>
> 3 about a truck driver in prison who was not in love;
>
> 2 about a prisoner who was not in love and did not drive a truck;
>
> 8 about a person out of jail who was not in love, and did not drive a truck;
>
> 16 about truck drivers who were not in prison.

(a) How many songs were surveyed?
Find the number of songs about
(b) truck drivers;
(c) prisoners;
(d) truck drivers in prison;
(e) people not in prison;
(f) people not in love.

21. Natural Science After a genetics experiment, the number of pea plants having certain characteristics was tallied, with the results as follows.

> 22 were tall;
>
> 25 had green peas;
>
> 39 had smooth peas;
>
> 9 were tall and had green peas;
>
> 17 were tall and had smooth peas;
>
> 20 had green peas and smooth peas;
>
> 6 had all three characteristics;
>
> 4 had none of the characteristics.

(a) Find the total number of plants counted.
(b) How many plants were tall and had peas which were neither smooth nor green?
(c) How many plants were not tall but had peas which were smooth and green?

22. Natural Science Human blood can contain either no antigens, the A antigen, the B antigen, or both the A and B antigens. A third antigen, called the Rh antigen, is important in human reproduction, and again may or may not be present in an individual. Blood is called type A-positive if the individual has the A and Rh, but not the B antigen. A person having only the A and B antigens is said to have type AB-negative blood. A person having only the Rh antigen has type O-positive blood. Other blood types are defined in a similar manner. Identify the blood type of the individual in regions (a)-(g) of the Venn diagram.

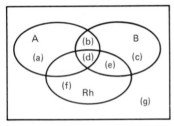

23. Natural Science (Use the diagram from Exercise 22.) In a certain hospital, the following data was recorded.

25 patients had the A antigen;

17 had the A and B antigens;

27 had the B antigen;

22 had the B and Rh antigens;

30 had the Rh antigen;

12 had none of the antigens;

16 had the A and Rh antigens;

15 had all three antigens.

How many patients
(a) were represented?
(b) had exactly one antigen?
(c) had exactly two antigens?
(d) had O-positive blood?
(e) had AB-positive blood?
(f) had B-negative blood?
(g) had O-negative blood?
(h) had A-positive blood?

24. A survey of 80 sophomores at a western college showed that

36 take English;

32 take history;

32 take political science;

16 take political science and history;

16 take history and English;

14 take political science and English;

6 take all three.

How many students
(a) take English and neither of the other two?
(b) take none of the three courses?
(c) take history, but neither of the other two?
(d) take political science and history, but not English?
(e) do not take political science?

25. The following table shows the number of people in a certain small town in Georgia who fit in the given categories.

Age	Drink vodka (V)	Drink bourbon (B)	Drink gin (G)	Totals
20–25 (Y)	40	15	15	70
26–35 (M)	30	30	20	80
Over 35 (O)	10	50	10	70
Totals	80	95	45	220

Using the letters given in the table, find the number of people in each of the following sets.
(a) $Y \cap V$ (b) $M \cap B$
(c) $M \cup (B \cap Y)$ (d) $Y' \cap (B \cup G)$
(e) $O' \cup G$ (f) $M' \cap (V' \cap G')$

26. The following table shows the results of a survey in a medium-sized town in Tennessee. The survey asked questions about the investment habits of local citizens.

Age	Stocks (S)	Bonds (B)	Savings accounts (A)	Totals
18–29 (Y)	6	2	15	23
30–49 (M)	14	5	14	33
50 or over (O)	32	20	12	64
Totals	52	27	41	120

Using the letters given in the table, find the number of people in each of the following sets.

(a) $Y \cap B$

(b) $M \cup A$

(c) $Y \cap (S \cup B)$

(d) $O' \cup (S \cup A)$

(e) $(M' \cup O') \cap B$

Let $n(A)$ *represent the number of elements in set A. (For example, if* $A = \{x, y, z\}$, *then* $n(A) = 3$.) *Use Venn diagrams to answer the following questions.*

27. If $n(A) = 5$, $n(B) = 8$, and $n(A \cap B) = 4$, what is $n(A \cup B)$?

28. If $n(A) = 12$, $n(B) = 27$, and $n(A \cup B) = 30$, what is $n(A \cap B)$?

29. Suppose $n(B) = 7$, $n(A \cap B) = 3$, and $n(A \cup B) = 20$. What is $n(A)$?

30. Suppose $n(A \cap B) = 5$, $n(A \cup B) = 35$, and $n(A) = 13$. What is $n(B)$?

8.3 Permutations and Combinations

After making do with your old automobile for several years, you finally decide to replace it with a new, small, super-economy model. You drive over to Ned's New Car Emporium to choose the car that's just right for you. Once there, you find that you can select from 5 models, each with 4 power options, a choice of 8 exterior color combinations and 3 interior colors.

How many different new cars are available to you? Problems of this sort are best solved by means of the counting principles discussed in this section. These counting methods are very useful in probability.

Let us begin with a simpler example. If there are three roads from town A to town B and two roads from town B to town C, in how many ways can someone travel from A to C by way of B? For each of the three roads from A there are two different routes leading from B to C, making $3 \cdot 2 = 6$ different trips, as shown in Figure 8.15.

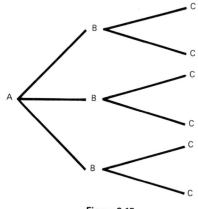

Figure 8.15

This example illustrates a general principle of counting called the *multiplication principle*.

Multiplication Principle Suppose n choices must be made, with

$$m_1 \text{ ways to make choice 1,}$$
$$m_2 \text{ ways to make choice 2,}$$

and so on, with

$$m_n \text{ ways to make choice } n.$$

Then there are

$$m_1 \cdot m_2 \cdots m_n$$

different ways to make the entire sequence of choices.

Using this multiplication principle, there are

$$5 \cdot 4 \cdot 8 \cdot 3 = 480$$

ways of selecting a new car at Ned's.

Example 1 A combination lock can be set to open to any three-letter sequence. How many such sequences are possible?

Since there are 26 letters in the alphabet, there are 26 choices for each of the three letters, and, by the multiplication principle, $26 \cdot 26 \cdot 26 = 17,576$ different sequences. ■

Example 2 A teacher has 5 different books to be arranged side by side. How many different arrangements are possible?

Five choices will be made, one for each space that will hold a book. Any of the five possible books could be chosen for the first space. There are four possible choices for the second space, since one book has already been placed in the first space, three possible choices for the third space, and so on. By the multiplication principle, the number of different possible arrangements (sequence of choices) is $5 \cdot 4 \cdot 3 \cdot 2 \cdot 1 = 120$. ■

1 (a) In how many ways can 6 business tycoons line up their golf carts at the country club?

(b) How many ways can 4 pupils be seated in a row with 4 seats?

Answer:

(a) $6 \cdot 5 \cdot 4 \cdot 3 \cdot 2 \cdot 1 = 720$

(b) $4 \cdot 3 \cdot 2 \cdot 1 = 24$

The use of the multiplication principle often leads to products such as $5 \cdot 4 \cdot 3 \cdot 2 \cdot 1$. For convenience, the symbol $n!$ (read "n factorial"), defined as follows, is used for such products.

---

n-factorial For any natural number n,

$$n! = n(n - 1)(n - 2) \ldots (3)(2)(1).$$

With this symbol, the product $5 \cdot 4 \cdot 3 \cdot 2 \cdot 1$ can be written as $5!$. Also, $3! = 3 \cdot 2 \cdot 1 = 6$. The definition of $n!$ could be used to show that $n[(n - 1)!] = n!$ for all natural numbers $n \geq 2$. It is helpful if this result also holds for $n = 1$. This

2 Evaluate:

(a) 4!

(b) 6!

(c) 7!

(d) 1!

Answer:

(a) 24

(b) 720

(c) 5040

(d) 1

3 In how many ways can 3 of 7 items be arranged?

Answer:

$7 \cdot 6 \cdot 5 = 210$

can only happen if 0! is defined to equal 1:

$$0! = 1. \quad \boxed{2}$$

Example 3 Suppose the teacher in Example 2 wishes to place only 3 of the 5 books on his desk. How many arrangements of 3 books are possible?

The teacher again has 5 ways to fill the first space, 4 ways to fill the second space, and 3 ways to fill the third. Since he wants to use only 3 books, there are only 3 spaces to be filled giving $5 \cdot 4 \cdot 3 = 60$ arrangements. ▪ 3

Permutations The answer 60 in Example 3 is called the number of permutations of 5 things taken 3 at a time. A **permutation** of n elements taken r at a time is any arrangement or ordering, *without repetition,* of the r elements. With permutations, *order* is important. Each change in the order of a particular set of r elements results in a new permutation. For example, ABC and BCA are two different permutations of the letters A, B, and C.

The number of permutations of n things taken r at a time (with $r \leq n$) is written $P(n, r)$. Based on the work in Example 3 above, $P(5, 3) = 5 \cdot 4 \cdot 3 = 60$. Factorial notation can be used to express this product as follows.

$$5 \cdot 4 \cdot 3 = 5 \cdot 4 \cdot 3 \cdot \frac{2 \cdot 1}{2 \cdot 1} = \frac{5 \cdot 4 \cdot 3 \cdot 2 \cdot 1}{2 \cdot 1} = \frac{5!}{2!} = \frac{5!}{(5-3)!}$$

This example illustrates the general rule stated below, which is a direct consequence of the multiplication principle.

Permutations If $P(n, r)$ (where $r \leq n$) is the number of permutations of n elements taken r at a time, then

$$P(n, r) = \frac{n!}{(n-r)!}.$$

---

4 Find the number of permutations of

(a) 5 things taken 2 at a time;

(b) 9 things taken 3 at a time.

Find each of the following.

(c) $P(3, 1)$

(d) $P(7, 3)$

(e) $P(12, 2)$

Answer:

(a) 20

(b) 504

(c) 3

(d) 210

(e) 132

Find $P(n, r)$, either with the rule above or with direct application of the multiplication principle, as in the following example.

Example 4 Find the number of permutations of 8 elements taken 3 at a time.

Since there are 3 choices to be made, the multiplication principle gives $P(8, 3) = 8 \cdot 7 \cdot 6 = 336$. Alternatively, by the formula for $P(n, r)$,

$$P(8, 3) = \frac{8!}{(8-3)!} = \frac{8!}{5!} = \frac{8 \cdot 7 \cdot 6 \cdot 5 \cdot 4 \cdot 3 \cdot 2 \cdot 1}{5 \cdot 4 \cdot 3 \cdot 2 \cdot 1} = 8 \cdot 7 \cdot 6 = 336. \quad ▪ \; 4$$

Example 5 Find each of the following.

(a) The number of permutations of the letters A, B, and C

Find

$$P(3, 3) = \frac{3!}{(3-3)!} = \frac{3!}{0!} = \frac{3!}{1} = 3! = 3 \cdot 2 \cdot 1 = 6,$$

by the formula for $P(n, r)$ with both n and r equal to 3.

The 6 permutations (or arrangements) are

$$ABC, \ ACB, \ BAC, \ BCA, \ CAB, \ CBA.$$

(b) The number of permutations possible using just two of the letters A, B, and C Find $P(3, 2)$:

$$P(3, 2) = \frac{3!}{(3-2)!} = \frac{3!}{1!} = 3! = 6,$$

the same answer as in part (a). This is because, in the case of $P(3, 3)$, after the first two choices are made, the third is determined. ▪ **5**

⑤ Find the number of permutations of the letters C, O, D, and E

(a) using all the letters;

(b) using 2 of the 4 letters.

Answer:

(a) 24

(b) 12

Combinations   In Example 3, we found that there are 60 ways that a teacher can arrange 3 of 5 different books on a desk. That is, there are 60 permutations of 5 things taken 3 at a time. Suppose now that the teacher does not wish to arrange the books on his desk, but rather wishes to choose, at random, any 3 of the 5 books to give to a book sale to raise money for his school. In how many ways can he do this?

At first glance, we might say 60 again, but this is incorrect. The number 60 counts all possible *arrangements* of 3 books chosen from 5. However, the following arrangements would all lead to the same set of three books being given to the book sale.

mystery-biography-textbook	biography-textbook-mystery
mystery-textbook-biography	textbook-biography-mystery
biography-mystery-textbook	textbook-mystery-biography

The list shows 6 different *arrangements* of 3 books, but only one subset of 3 books selected from the 5 books, for the book sale. This subset is called a **combination.** The number of combinations of 5 things taken 3 at a time is written $\binom{5}{3}$. Since they are subsets, combinations are *not ordered*.

To evaluate $\binom{5}{3}$, start with the $5 \cdot 4 \cdot 3$ *permutations* of 5 things taken 3 at a time. Since combinations are unordered, find the number of combinations by dividing the number of permutations by the number of ways each group of 3 can be ordered—that is, by $3!$.

⑥ Evaluate $\dfrac{P(n, r)}{r!}$ for the following values.

(a) $n = 6, r = 2$

(b) $n = 8, r = 4$

(c) $n = 4, r = 1$

Answer:

(a) 15

(b) 70

(c) 4

$$\binom{5}{3} = \frac{5 \cdot 4 \cdot 3}{3!} = \frac{5 \cdot 4 \cdot 3}{3 \cdot 2 \cdot 1} = 10$$

There are 10 ways that the teacher can choose 3 books at random for the book sale.

Generalizing this discussion gives the formula for the number of combinations of n elements taken r at a time, written $\binom{n}{r}$:

$$\binom{n}{r} = \frac{P(n, r)}{r!}. \quad \boxed{6}$$

A more useful version of this formula can be found using the result for $P(n, r)$ given above.

$$\binom{n}{r} = \frac{P(n, r)}{r!}$$

$$= \frac{n!}{(n - r)!} \cdot \frac{1}{r!}$$

$$= \frac{n!}{(n - r)!r!}$$

This last form is the most useful for calculation. The steps above lead to the following result.

> **Combinations** If $\binom{n}{r}$ denotes the number of combinations of n elements taken r at a time, where $r \le n$, then
>
> $$\binom{n}{r} = \frac{n!}{(n - r)!r!}.$$

Table 9 at the back of the book gives the values of $\binom{n}{r}$ for $n \le 20$. Many calculators have a key for $n!$ and some are programmed to give the values of $\binom{n}{r}$ and $P(n, r)$.

Example 6 How many committees of 3 people can be formed from a group of 8 people?

A committee is an unordered group, so find $\binom{8}{3}$. By the formula for combinations,

$$\binom{8}{3} = \frac{8!}{5!3!} = \frac{8 \cdot 7 \cdot 6 \cdot 5 \cdot 4 \cdot 3 \cdot 2 \cdot 1}{5 \cdot 4 \cdot 3 \cdot 2 \cdot 1 \cdot 3 \cdot 2 \cdot 1} = \frac{8 \cdot 7 \cdot 6}{3 \cdot 2 \cdot 1} = 56. \quad \blacksquare \; \boxed{7}$$

Example 7 From a group of 30 employees, 3 are to be selected to work on a special project.

(a) In how many different ways can the employees be selected?

Here the number of 3-element combinations that can be formed from a set of 30 elements is needed. (We want combinations and not permutations, since order within the group of 3 does not matter.)

$$\binom{30}{3} = \frac{30!}{27!3!} = \frac{30(29)(28)(27) \ldots (3)(2)(1)}{27(26)(25) \ldots (2)(1)(3)(2)(1)}$$

$$= \frac{30(29)(28)}{3(2)(1)}$$

$$= 4060$$

There are 4060 ways to select the project group.

(b) In how many ways can the group of 3 be selected if it has been decided that one particular person must work on the project?

7 Find the following.

(a) $\binom{6}{3}$

(b) $\binom{5}{1}$

(c) $\binom{7}{0}$

Answer:
(a) 20
(b) 5
(c) 1

Since one person has already been selected for the project, the problem is reduced to selecting two more from the remaining 29 employees.

$$\binom{29}{2} = \frac{29!}{27!2!} = 406$$

8 Five orchids from a collection of twenty are to be selected for a flower show.
(a) In how many ways can this be done?
(b) In how many different ways can the group of five be selected if two particular orchids must be included?

Answer:

(a) $\binom{20}{5} = 15{,}504$

(b) $\binom{18}{3} = 816$

8 In this case, the project group can be selected in 406 ways. **8**

The formulas for permutations and combinations given in this section will be very useful in solving probability problems in later sections. Any difficulty in using these formulas usually comes from being unable to differentiate between them. Both permutations and combinations give the number of ways to choose r objects from a set of n objects. The differences between permutations and combinations are outlined below.

Permutations Different orderings or arrangements of the r objects are different permutations.	**Combinations** Each choice or subset of r objects gives one combination. Order within the r objects does not matter.
$$P(n, r) = \frac{n!}{(n - r)!}$$	$$\binom{n}{r} = \frac{n!}{(n - r)!r!}$$

In the next examples, concentrate on recognizing which formulas should be applied.

Example 8 For each of the following problems, tell whether permutations or combinations should be used to solve the problem.
(a) How many four-digit code numbers are possible if no digits are repeated?

Since changing the order of the four digits results in a different code, use permutations.

(b) A sample of 3 is randomly selected from a batch of 15 items. How many different samples are possible?

The order in which the 3 items are selected is not important. The sample is unchanged if the items are rearranged, so combinations should be used.

(c) In a baseball conference with 8 teams, how many games must be played so that each team plays every other team exactly once?

Selection of two teams for a game is an *unordered* subset of 2 from the set of 8 teams. Use combinations again.

9 Solve the problems in Example 8.

Answer:
(a) 5040
(b) 455
(c) 28
(d) 360

(d) In how many ways can 4 patients be assigned to 6 hospital rooms?

The room assignments are an *ordered* selection of 4 rooms from the 6 rooms. Exchanging the rooms of any two patients within a selection of 4 rooms gives a different assignment, so permutations should be used. **9**

Example 9 A manager must select 4 employees for promotion: 12 employees are eligible.

(a) In how many ways can the four be chosen?

Since there is no reason to differentiate among the 4 who are selected, use combinations.

$$\binom{12}{4} = \frac{12!}{4!8!} = 495$$

(b) In how many ways can 4 employees be chosen (from 12) to be placed in 4 different jobs?

In this case, once a group of 4 is selected, they can be assigned in many different ways (or arrangements) to the 4 jobs. Therefore, this problem requires permutations.

$$P(12, 4) = \frac{12!}{8!} = 11,880 \quad \boxed{10}$$

10 A salesman has the names of 6 prospects.

(a) In how many ways can he arrange his schedule if he calls on all 6?

(b) In how many ways can he do it if he decides to call on only 4 of the 6?

Answer:

(a) 720

(b) 360

11 In how many ways can a full house of eights and aces (3 eights and 2 aces) be dealt?

Answer:

24

Example 10 In how many ways can a full house of aces and eights (3 aces and 2 eights) be dealt in five-card poker?

The arrangement of the three aces or the two eights does not matter, so use combinations and the multiplication principle. There are $\binom{4}{3}$ ways to get 3 aces from the 4 aces in the deck, and $\binom{4}{2}$ ways to get 2 eights. The number of ways to get 3 aces and 2 eights is

$$\binom{4}{3} \cdot \binom{4}{2} = 4 \cdot 6 = 24. \quad \boxed{11}$$

Example 11 Five cards are dealt from a standard 52-card deck (Figure 8.16).

(a) How many such hands have all face cards?

Figure 8.16

The face cards are the king, queen, and jack of each suit. Since there are four suits, there are 12 face cards. The arrangement of the five cards is not important, so use combinations and find

$$\binom{12}{5} = \frac{12!}{5!7!} = 792.$$

(b) How many such hands have all cards of the same suit?

The total number of ways that 5 cards of a particular suit of 13 cards can be dealt is $\binom{13}{5}$. The arrangement of the five cards is not important, so use combinations. Since there are four different suits, by the multiplication principle, there are

$$4 \cdot \binom{13}{5} = 4 \cdot 1287 = 5148$$

12 In how many ways can 5 red cards be dealt?

Answer:
65,780

ways to deal 5 cards of the same suit. **12**

It should be stressed that not all counting problems lend themselves to either permutations or combinations. Whenever a tree diagram or the multiplication principle can be used directly, as in the example at the beginning of this section, then use it.

8.3 Exercises

Evaluate the following factorials, permutations, and combinations.

1. $P(4, 2)$ **2.** $3!$ **3.** $\binom{8}{3}$ **4.** $7!$

5. $P(8, 1)$ **6.** $\binom{8}{1}$ **7.** $4!$ **8.** $P(4, 4)$

9. $\binom{12}{5}$ **10.** $\binom{10}{8}$ **11.** $P(13, 2)$ **12.** $P(12, 3)$

Use a computer to find values for Exercises 13-20.

13. $P(25, 12)$ **14.** $P(38, 17)$ **15.** $P(14, 5)$

16. $P(17, 12)$ **17.** $\binom{21}{10}$ **18.** $\binom{34}{25}$

19. $\binom{25}{16}$ **20.** $\binom{30}{15}$

Use the multiplication principle to solve the following problems. (See Examples 1–3.)

21. How many different types of homes are available if a builder offers a choice of 5 basic plans, 3 roof styles, and 2 exterior finishes?

22. An auto manufacturer produces 7 models, each available in 6 different colors, with 4 different upholstery fabrics, and 5 interior colors. How many varieties of the auto are available?

23. How many different 4-letter radio station call letters can be made
  (a) if the first letter must be K·or W and no letter may be repeated?
  (b) if repeats are allowed (but the first letter is K or W)?
  (c) How many of the 4-letter call letters (starting with K or W) with no repeats end in R?

24. A menu offers a choice of 3 salads, 8 main dishes, and 5 desserts. How many different meals are possible?

25. A couple has narrowed down the choice of a name for their new baby to 3 first names and 5 middle names. How many different first and middle name arrangements are possible?

26. A concert to raise money for an economics prize is to consist of 5 works: 2 overtures, 2 sonatas, and a piano concerto. In how many ways can the program be arranged?

27. How many different license numbers consisting of 3 letters followed by 3 digits are possible?

28. How many different license plate numbers can be formed using 3 letters followed by 3 digits if no repeats are allowed?

29. How many license plate numbers (see Exercise 28) are possible is there are no repeats and either numbers or letters can come first?

30. How many 7-digit telephone numbers are possible if the first digit cannot be zero and
   (a) only odd digits may be used;
   (b) the telephone number must be a multiple of 10 (that is, it must end in zero);
   (c) the telephone number must be a multiple of 100;
   (d) the first three digits are 481;
   (e) no repetitions are allowed?

Use permutations to solve each of the following problems. (See Examples 4 and 5.)

31. In an experiment on social interaction, 6 people will sit in 6 seats in a row. In how many ways can this be done?

32. In how many ways can 7 of 10 monkeys be arranged in a row for a genetics experiment?

33. A business school gives courses in typing, shorthand, transcription, business English, technical writing, and accounting. In how many ways can a student arrange a schedule if 3 courses are taken?

34. If your college offers 400 courses, 20 of which are in mathematics, and your counselor arranges your schedule of 4 courses by random selection, how many schedules are possible that do not include a math course?

35. In a club with 15 members, how many ways can a slate of 3 officers consisting of president, vice-president, and secretary/treasurer be chosen?

36. A baseball team has 20 players. How many 9-player batting orders are possible?

Use combinations to solve each of the following problems. (See Examples 6 and 7.)

37. How many different two-card hands can be dealt from an ordinary deck (52 cards)?

38. How many different thirteen-card bridge hands can be dealt from an ordinary deck?

39. Five cards are marked with the numbers 1, 2, 3, 4, and 5, then shuffled, and two cards are drawn. How many different two-card combinations are possible?

40. Five cards are drawn from an ordinary deck. In how many ways is it possible to draw
   (a) all queens;
   (b) all face cards (face cards are the Jack, Queen, and King);
   (c) no face card;
   (d) exactly 2 face cards;
   (e) 1 heart, 2 diamonds, and 2 clubs.

41. If a baseball coach has 5 good hitters and 4 poor hitters on the bench and chooses 3 players at random, in how many ways can he choose at least 2 good hitters?

42. An economics club has 30 members. If a committee of 4 is to be selected, in how many ways can it be done?

For each of the following problems, decide if it is a permutations or combinations problem, and then solve the problem. (See Examples 8–11.)

43. In how many ways can an employer select 2 new employees from a group of 4 applicants?

44. Hal's Hamburger Palace sells hamburgers with cheese, relish, lettuce, tomato, mustard, or catsup. How many different kinds of hamburgers can be made using any three of the extras?

45. In a club with 8 men and 11 women members, how many 5-member committees can be chosen that have
   (a) all men;   (b) all women;
   (c) 3 men and 2 women?

46. A contest requires matching 8 baby pictures of celebrities with their names. How many different entries are possible?

47. A group of 3 students is to be selected from a group of 12 students to take part in a class in cell biology.
   (a) In how many ways can this be done?
   (b) In how many ways can the group which will *not* take part be chosen?

48. The coach of the Morton Valley Softball Team has 6 good hitters and 8 poor hitters. He chooses three hitters at random.
   (a) In how many ways can he choose 2 good hitters and 1 poor hitter?
   (b) In how many ways can he choose all good hitters?

49. In a game of musical chairs, 12 children will sit in 11 chairs (one will be left out). How many seatings are possible?

50. Marbles are drawn without replacement from a bag containing 15 marbles.
(a) How many samples of 2 marbles can be drawn?
(b) How many samples of 4 marbles can be drawn?
(c) If the bag contains 3 yellow, 4 white, and 8 blue marbles, how many samples of 2 marbles can be drawn in which both marbles are blue?

51. A city council is composed of 5 liberals and 4 conservatives. A delegation of 3 is to be selected to attend a convention.
(a) How many delegations are possible?
(b) How many delegations could have all liberals?
(c) How many delegations could have 2 liberals and 1 conservative?
(d) If one member of the council serves as mayor, how many delegations which include the mayor are possible?

52. In an experiment on plant hardiness, a researcher gathers 6 wheat plants, 3 barley plants, and 2 rye plants. She wishes to select 4 plants at random.
(a) In how many ways can this be done?
(b) In how many ways can this be done if 2 wheat plants must be included?

53. A group of 7 workers decides to send a delegation of 2 to their supervisor to discuss their grievances.
(a) How many delegations are possible?
(b) If it is decided that a particular employee must be in the delegation, how many different delegations are possible?
(c) If there are 2 women and 5 men in the group, how many delegations would include at least 1 woman?

54. From a pool of 7 secretaries, 3 are selected to be assigned to 3 managers. In how many ways can this be done?

55. From 10 names on a ballot, 4 will be elected to a political party committee. In how many ways can the committee of 4 be formed if each person will have a different responsibility?

56. There are 5 rotten apples in a crate of 25 apples.
(a) How many samples of 3 apples can be drawn from the crate?
(b) How many samples of 3 could be drawn in which all 3 are rotten?

(c) How many samples of 3 could be drawn in which there are two good apples and one rotten one?

57. A bag contains 5 black, 1 red, and 3 yellow jelly beans; you take 3 at random. How many samples are possible in which the jelly beans are
(a) all black; (b) all red;
(c) all yellow; (d) 2 black, 1 red;
(e) 2 black, 1 yellow; (f) 2 yellow, 1 black;
(g) 2 red, 1 yellow.

58. In how many ways can 5 out of 9 plants be arranged in a row on a window sill?

*If the n objects in a permutations problem are not all distinguishable, that is, there are n_1 of type 1, n_2 of type 2, and so on for r different types, then the number of **distinguishable permutations** is*

$$\frac{n!}{n_1! n_2! \ldots n_r!}.$$

Example: In how many ways can you arrange the letters in the word Mississippi?

This word contains 1 m, 4 i's, 4 s's, and 2 p's. To use the formula, let $n = 11$, $n_1 = 1$, $n_2 = 4$, $n_3 = 4$, $n_4 = 2$ to get

$$\frac{11!}{1! 4! 4! 2!} = 34,650$$

arrangements. The letters in a word with 11 *different* letters can be arranged in $11! = 39,916,800$ ways. ∎

59. Find the number of distinguishable permutations of the letters in each of the following words.
(a) initial (b) little (c) decreed

60. A printer has 5 A's, 4 B's, 2 C's, and 2 D's. How many different "words" are possible which use all these letters? (A "word" does not have to have any meaning here.)

61. Mike has 4 blue, 3 green, and 2 red books to arrange on a shelf.
(a) In how many ways can this be done if they can be arranged in any order?
(b) In how many ways if books of the same color are identical and must be grouped together?
(c) In how many distinguishable ways if books of the same color are identical but need not be grouped together?

62. A child has a set of differently shaped plastic objects. There are 3 pyramids, 4 cubes, and 7 spheres.

(a) In how many ways can she arrange them in a row if they are all different colors?

(b) In how many ways if the same shapes must be grouped?

(c) In how many distinguishable ways can they be arranged in a row if objects of the same shape are also the same color, but need not be grouped?

Use a computer to solve the following exercises.

63. How many 5-card poker hands are possible with a regular deck of 52 cards?

64. How many 5-card poker hands with all cards from the same suit are possible?

65. How many 13-card bridge hands are possible with a regular deck of 52 cards?

66. How many 13-card bridge hands with 4 aces are possible?

67. Natural Science Eleven drugs have been found to be effective in the treatment of a disease. It is believed that the sequence in which the drugs are administered is important in the effectiveness of the treatment. In how many orders can 5 of the 11 drugs be administered?

68. Natural Science A biologist is attempting to classify 52,000 species of insects by assigning 3 initials to each species. Is it possible to classify all the species in this way? If not, how many initials should be used?

8.4 Basic Properties of Probability

If you go to a supermarket and buy five pounds of peaches at 54¢ per pound, you can easily find the *exact* price of your purchase: $2.70. On the other hand, the produce manager of the market is faced with the problem of ordering peaches. The manager may have a good estimate of the number of pounds of peaches that will be sold during the day, but there is no way to know *exactly*. The number of pounds that customers will purchase during a day is **random:** it cannot be predicted exactly.

Many problems that come up in applications of mathematics involve random phenomena for which exact prediction is impossible. The best that can be done is to construct a mathematical model that gives the *probability* of the possible outcomes.

In probability, each repetition of an experiment is called a **trial.** The possible results of each trial are called **outcomes.** If the outcomes occur randomly (that is, if they cannot be predicted), then the experiment is a **random experiment.** Throughout this text, the term *experiment* will mean *random experiment.* An example of a probability experiment is the tossing of a coin. On each trial of the experiment (each toss), there are two possible outcomes, heads and tails, abbreviated *h*, and *t*, respectively. If the two outcomes, *h* and *t*, are equally likely to occur, then the coin is not "loaded" to favor one side over the other. Such a coin is called **fair.** For a coin that is not loaded, this "equally likely" assumption is made for each trial.

Since *two* equally likely outcomes are possible, *h* and *t*, and just *one* of them is heads, theoretically a coin tossed many, many times should come up heads approximately 1/2 of the time. Also, the more times the coin is tossed, the closer the occurrence of heads should be to 1/2.

Suppose an experiment could be repeated again and again under unchanging conditions. Suppose the experiment is repeated *n* times, and that a certain outcome happens *m* times. The ratio *m/n* is called the **relative frequency** of the outcome after *n* trials.

If this ratio m/n approaches closer and closer to some fixed number p as n gets larger and larger, then p is the **probability** of the outcome. If p exists, then

$$p \approx \frac{m}{n}$$

as n gets larger and larger.

This approach to probability is consistent with most people's intuitive feeling about the meaning of probability—a way of measuring the likelihood of a certain outcome. For example, since 1/4 of the cards in an ordinary deck are diamonds, we assume that if we repeatedly drew a card from a well-shuffled deck, kept track of whether it was a diamond or not, and then replaced the card in the deck, after a large number of repetitions of the experiment about 1/4 of the cards drawn would be diamonds. As another example, since there are two sides to a fair coin and either side is equally likely to land up on a toss of a coin, the probability of heads on a single toss of a fair coin is 1/2, written

$\boxed{1}$ Find $P(t)$.

$$P(h) = \frac{1}{2}. \quad \boxed{1}$$

Answer:

1/2

Example 1 Suppose the spinner of Figure 8.17 is spun.

(a) Find the probability that it will point to 1.

Since there are three equally likely outcomes possible on a single spin, and one of them is 1, then $P(1) = 1/3$.

(b) Find the probability that it will point to either 1 or 2.

$\boxed{2}$ Find the probability that the spinner of Figure 9.16

(a) points to 2;

(b) does not point to 1.

Since 1 and 2 are *two* of *three* equally likely possible outcomes, then $P(1 \text{ or } 2) = 2/3$. $\quad \boxed{2}$

Answer:

(a) 1/3

(b) 2/3

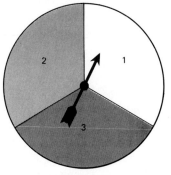

Figure 8.17

An ordinary die is a cube whose six faces show the numbers 1, 2, 3, 4, 5, and 6. If the die is not "loaded" to favor certain faces over others (a fair die), then any one of the six faces is equally likely to come up when the die is rolled.

Example 2 **(a)** If a single fair die is rolled, find the probability of rolling the number 4.

Since one out of six faces is a 4, $P(4) = 1/6$.

(b) Using the same die, find the probability of rolling an odd number.

Since three of the six faces show odd numbers, those with 1, 3, and 5,

$$P(\text{odd}) = \frac{3}{6} = \frac{1}{2}. \quad \boxed{3}$$

3 Suppose a single fair die is rolled. Find the probability of a
(a) 1;
(b) 2 or 3;
(c) 3, 4, 5, or 6.

Answer:
(a) 1/6
(b) 2/6 = 1/3
(c) 4/6 = 2/3

Sample Spaces and Events The set of all possible outcomes for an experiment is the **sample space** for the experiment. A sample space for the experiment of tossing a coin has two outcomes, heads and tails. If S represents this sample space, then

$$S = \{h, t\}.$$

Example 3 Give the sample space for each experiment.

(a) A spin of the spinner in Figure 8.16

The three outcomes are 1, 2, or 3, so the sample space is

$$\{1, 2, 3\}.$$

(b) For the purposes of a public opinion poll, people are classified as young, middle-aged, or older, and as male or female.

A sample space for this poll could be written as a set of ordered pairs:

$$\{(\text{young, male}), (\text{young, female}), (\text{middle-aged, male}),$$
$$(\text{middle-aged, female}), (\text{older, male}), (\text{older, female})\}.$$

(c) An experiment consists of studying the number of boys and girls in families with exactly three children. Let b represent *boy* and g represent *girl*.

A three-child family can have three boys, written *bbb*, three girls, *ggg*, or various combinations, such as *bgg*. A sample space with four outcomes (not equally likely) is

$$S_1 = \{3 \text{ boys}, 2 \text{ boys and } 1 \text{ girl}, 1 \text{ boy and } 2 \text{ girls}, 3 \text{ girls}\}.$$

Another sample space which considers the ordering of the births of the three children with, for example, *bgg* different from *gbg* or *ggb*, is

$$S_2 = \{bbb, bbg, bgb, gbb, bgg, gbg, ggb, ggg\}.$$

The second sample space, S_2, has equally likely outcomes; S_1 does not, since there is more than one way to get a family with 2 boys and 1 girl or a family of 2 girls and 1 boy, but only one way to get three boys or three girls. $\boxed{4}$

4 Write an equally likely sample space for the experiment of rolling a single fair die.

Answer:
{1, 2, 3, 4, 5, 6}

For the purpose of assigning probabilities, it is usually best to choose a sample space with equally likely outcomes.

An **event** is a subset of a sample space. If the sample space for tossing a coin is $S = \{h, t\}$, then one event is $E = \{h\}$, which represents the outcome "heads."

For the sample space of tossing a single fair die, {1, 2, 3, 4, 5, 6}, some possible events are:

the number showing is even   {2, 4, 6};

the number is 1   {1};

the number is less than 5   {1, 2, 3, 4};

the number is a multiple of 3   {3, 6}.

Example 4 For the sample space S_2 in Example 3(c), write the following events.
(a) Event H: the family has exactly two girls

Families can have exactly two girls with either *bgg, gbg,* or *ggb,* so that event H is

$$H = \{bgg,\ gbg,\ ggb\}.$$

(b) Event K: the three children are the same sex

Two outcomes satisfy this condition, all boys or all girls.

$$K = \{bbb,\ ggg\}.$$

(c) Event J: the family has three girls

Only *ggg* satisfies this condition, so

$$J = \{ggg\}. \ \blacksquare$$

In Example 4(c), event J had only one possible outcome, *ggg*. Such an event, with only one possible outcome, is a **simple event.** If event E equals the sample space S, then E is a **certain event.** If event $E = \emptyset$, then E is an **impossible event.** ⑤

We now need to assign to each event E in a sample space S a number called the *probability of the event,* which indicates the relative likelihood of the event.

For sample spaces with *equally likely* outcomes, the probability of an event can be found with the following *basic probability principle.*

> **Basic Probability Principle** Let a sample space S have n possible equally likely outcomes. Let event E contain m of these outcomes, all distinct. Then the probability that event E occurs, written $P(E)$, is
>
> $$P(E) = \frac{m}{n}.$$

⑤ Suppose a die is tossed. Write the following events.
(a) The number showing is less than 3.
(b) The number showing is 5.
(c) The number showing is 10.
(d) Which of the events in parts (a)–(c) is a simple event? A certain event? An impossible event?

Answer:
(a) {1, 2}
(b) {5}
(c) ∅
(d) Simple: (b); certain: none; impossible: (c)

Example 5 Suppose a single die is rolled. Use the sample space $S = \{1, 2, 3, 4, 5, 6\}$ and give the probability of each of the following events.
(a) E: the number rolled is even.

Here, $E = \{2, 4, 6\}$, a set with three elements. Since S contains 6 elements,

$$P(E) = \frac{3}{6} = \frac{1}{2}.$$

(b) F: the number rolled is greater than 4.

Since F contains two elements, 5 and 6,

$$P(F) = \frac{2}{6} = \frac{1}{3}.$$

(c) G: the number rolled is less than 10.

Event G is a certain event, with

$$G = \{1, 2, 3, 4, 5, 6\} \quad \text{and} \quad P(G) = \frac{6}{6} = 1.$$

(d) H: the number rolled is 8.

This event is impossible, so

$$P(H) = 0. \quad \boxed{6}$$

6 A fair die is rolled. Find the probability of rolling
(a) an odd number;
(b) 2, 4, 5, or 6;
(c) a number greater than 5;
(d) the number 7.

Answer:
(a) 1/2
(b) 2/3
(c) 1/6
(d) 0

Example 6 If a single playing card is drawn at random from an ordinary 52-card bridge deck, find the probability of each of the following events.

(a) Drawing an ace

There are four aces in the deck, so

$$P(\text{ace}) = \frac{4}{52} = \frac{1}{13}.$$

(b) Drawing a face card

Since there are 12 face cards,

$$P(\text{face card}) = \frac{12}{52} = \frac{3}{13}.$$

(c) Drawing a spade

The deck contains 13 spades, so

$$P(\text{spade}) = \frac{13}{52} = \frac{1}{4}.$$

7 A single playing card is drawn at random from an ordinary 52-card deck. Find the probability of drawing
(a) a queen;
(b) a diamond;
(c) a red card.

Answer:
(a) 1/13
(b) 1/4
(c) 1/2

(d) Drawing a spade or a heart

Besides the 13 spades, the deck contains 13 hearts giving

$$P(\text{spade or heart}) = \frac{26}{52} = \frac{1}{2}. \quad \boxed{7}$$

The definition of probability given above leads to the following *properties of probability*. (Recall that a simple event contains only one of the equally likely outcomes.)

Properties of Probability Let $S = \{s_1, s_2, s_3, \ldots, s_n\}$ be the sample space obtained from the union of the n distinct simple events $\{s_1\}, \{s_2\}, \{s_3\}, \ldots, \{s_n\}$ with associated probabilities $p_1, p_2, p_3, \ldots, p_n$. Then

1. $0 \leq p_1 \leq 1, 0 \leq p_2 \leq 1, \ldots, 0 \leq p_n \leq 1$;
   (All probabilities are between 0 and 1, inclusive.)
2. $p_1 + p_2 + p_3 + \cdots + p_n = 1$;
   (The sum of all probabilities for a sample space is 1.)
3. $P(S) = 1$;
4. $P(\emptyset) = 0$.

The Addition Principle Suppose event E is the union of several simple events, say

$$E = \{s_1, s_2, s_3\} = \{s_1\} \cup \{s_2\} \cup \{s_3\}.$$

$P(E)$, the probability of event E, is found by adding the probabilities for each of the simple events making up E. For $E = \{s_1, s_2, s_3\}$,

$$P(E) = P(\{s_1\}) + P(\{s_2\}) + P(\{s_3\}).$$

The generalization of this result is called the *addition principle:*

Addition Principle for Simple Events Suppose $E = \{s_1, s_2, s_3, \ldots, s_m\}$, where $\{s_1\}, \{s_2\}, \{s_3\}, \ldots, \{s_m\}$ are distinct simple events. Then

$$P(E) = P(\{s_1\}) + P(\{s_2\}) + P(\{s_3\}) + \cdots + P(\{s_m\}).$$

The addition rule *does not necessarily apply* to the addition of probabilities of events that are not simple. As an example, the sum of the probability of getting at least 4 on a single roll of a die, and the probability of getting an even number on a single roll, is *not* equal to the probability of getting at least 4 or an even number on a single roll. In other words, P(at least 4) $= P(4, 5,$ or $6) = 1/2$, P(even number) $= P(2, 4,$ or $6) = 1/2$, and P(at least 4 or even) $= P(2, 4, 5,$ or $6) = 2/3$, with P(at least 4) $+ P$(even number) $\neq P$(at least 4 or even).

Subjective Probability In many realistic problems, it is not possible to establish exact probabilities for events. Useful approximations, however, often can be found by using past experience as a guide to the future. The next example shows an approach to such **subjective probabilities.**

Example 7 The manager of a department store has decided to make a study on the size of purchases made by people coming into the store. To begin, he chooses a day that seems fairly typical and gathers the following data. (Purchases have been rounded to the nearest dollar, with sales tax ignored.)

Amount of purchase	Number of customers
$0	158
$1–$5	94
$6–$9	203
$10–$19	126
$20–$49	47
$50–$99	38
$100 and over	53

First, the manager adds the number of customers to find that 719 people came into the store that day. Of these 719 people, $126/719 \approx .175$ made a purchase of at least $10 but no more than $19. Also, $53/719 \approx .074$ of the customers spent $100 or more. For this day, the probability that a customer entering the store will spend from $10 to $19 is .175, and the probability that the customer will spend $100 or more is .074. Probabilities for the other purchase amounts can be assigned in the same way, giving the results below.

Amount of purchase	Probability
$0	.220
$1–$5	.131
$6–$9	.282
$10–$19	.175
$20–$49	.065
$50–$99	.053
$100 and over	.074
Total	1.000

From the table, .282 of the customers spend from $6 to $9, inclusive—over a quarter of the customers. Since this price range attracts so many customers, perhaps the store's advertising should emphasize items in, or near, this price range.

The manager should use this table of probabilities to help in predicting the results on other days only if the manager is reasonably sure that the day when the measurements were made is fairly typical of the other days the store is open. For example, on the last few days before Christmas, the probabilities might be quite different. **8**

A table of probabilities, as in Example 7, sets up a **probability distribution:** that is, for each possible outcome of an experiment, a number, called the probability of that outcome, is assigned. Probability distributions are discussed further in Section 10.4.

Another example of subjective probability occurs in weather forecasting. In one method, the forecaster compares atmospheric conditions with similar conditions in the past, and predicts the weather according to what happened then under such

8 A traffic engineer gathered the following data on the number of accidents at four busy intersections.

Intersection	Number of accidents
1	25
2	32
3	28
4	35

Use the data to determine the probabilities of accidents at these intersections.

Answer:

1	.21
2	.27
3	.23
4	.29

conditions. For example, if in the past, under certain kinds of atmospheric conditions, rain resulted 80 times out of 100, the forecaster will announce an 80% chance of rain.

One difficulty with subjective probability is that different people may assign different probabilities to the same event. Nevertheless, subjective probabilities can be assigned to many occurrences where the objective approach to probability cannot be used. In the next chapter the use of subjective probability is shown in more detail.

Odds Sometimes probability statements are given in terms of **odds,** a comparison of $P(E)$ with $P(E')$, where the event E' is the complement of event E.

Odds The **odds in favor** of an event E are defined as the ratio of $P(E)$ to $P(E')$, or

$$\frac{P(E)}{P(E')}, \qquad P(E') \neq 0.$$

Example 8 Suppose the weather forecaster says that the probability of rain tomorrow is 1/3. Find the odds in favor of rain tomorrow.

Let E be the event "rain tomorrow." Then E' is the event "no rain tomorrow." Since $P(E) = 1/3$, $P(E') = 2/3$. By definition,

$$\text{Odds in favor of rain} = \frac{\dfrac{1}{3}}{\dfrac{2}{3}} = \frac{1}{2}, \quad \text{written 1 to 2} \quad \text{or} \quad 1:2.$$

On the other hand, the odds that it will *not* rain are

$$\frac{\dfrac{2}{3}}{\dfrac{1}{3}} = \frac{2}{1}, \quad \text{written 2 to 1.} \quad \boxed{9}$$

9 Suppose $P(E) = 9/10$. Find the odds
(a) in favor of E;
(b) against E.

Answer:
(a) 9 to 1
(b) 1 to 9

If the odds in favor of an event are, say, 3 to 5, then the probability of the event is 3/8, while the probability of the complement of the event is 5/8. (Odds of 3 to 5 indicate 3 outcomes in favor of the event out of a total of 8 outcomes.) This example suggests the following generalization.

If the odds favoring event E are m to n, then

$$P(E) = \frac{m}{m + n} \quad \text{and} \quad P(E') = \frac{n}{m + n}.$$

---

10 If the odds in favor of event E are 1 to 5, find
(a) $P(E)$;
(b) $P(E')$.

Answer:
(a) 1/6
(b) 5/6

Example 9 The odds that a particular bid will be the low bid are 4 to 5. Find the probability that the bid will be the low bid.

Odds of 4 to 5 show 4 favorable chances out of $4 + 5 = 9$ chances altogether:

$$P(\text{bid will be low bid}) = \frac{4}{4 + 5} = \frac{4}{9}.$$

There is a 5/9 chance that the bid will *not* be the low bid. **10**

8.4 Exercises

Write a sample space with equally likely outcomes for each of the following experiments. (See Example 3.)

1. A two-headed coin is tossed once.

2. Two coins are tossed.

3. Three coins are tossed.

4. Slips of paper marked with the numbers 1, 2, 3, 4, and 5 are placed in a box. After being mixed, two slips are drawn.

5. An unprepared student takes a three-question true/false quiz in which he guesses the answer to all three questions.

6. A die is rolled and then a coin is tossed.

Write the following events in set notation and give the probability of each event. (See Examples 4–6.)

7. In the experiment of Exercise 2:
(a) both coins show the same face;
(b) at least one coin turns up heads.

8. In Exercise 1:
(a) the result of the toss is heads;
(b) the result of the toss is tails.

9. In Exercise 4:
(a) both slips are marked with even numbers;
(b) both slips are marked with odd numbers;
(c) both slips are marked with the same number;
(d) one slip is marked with an odd number, the other with an even number.

10. In Exercise 5:
(a) the student gets all three answers correct;
(b) he gets all three answers wrong;
(c) he gets exactly two answers correct;
(d) he gets at least one answer correct.

11. A single fair die is rolled. Find the probability (see Examples 2 and 5) of rolling
(a) the number 2;   (b) an odd number;
(c) a number less than 5;
(d) a number greater than 2.

12. A card is drawn from a well-shuffled deck of 52 cards. Find the probability (see Example 6) of drawing
(a) a 9;   (b) a black card;   (c) a black 9;
(d) a heart;   (e) the 9 of hearts;
(f) a face card.

13. A marble is drawn at random from a box containing 3 yellow, 4 white, and 8 blue marbles. Find the probability of drawing
(a) a yellow marble;   (b) a blue marble:
(c) a white marble.

An experiment is conducted for which the sample space is $S = \{s_1, s_2, s_3, s_4, s_5\}$. Which of the probability distributions in Exercise 14–19 is possible for this experiment? If a distribution is not possible, tell why.

14.

Outcomes	s_1	s_2	s_3	s_4	s_5
Probability	.09	.32	.21	.25	.13

15.

Outcomes	s_1	s_2	s_3	s_4	s_5
Probability	.92	.03	0	.02	.03

16.

Outcomes	s_1	s_2	s_3	s_4	s_5
Probability	$\frac{1}{3}$	$\frac{1}{4}$	$\frac{1}{6}$	$\frac{1}{8}$	$\frac{1}{10}$

17.

Outcomes	s_1	s_2	s_3	s_4	s_5
Probability	$\frac{1}{5}$	$\frac{1}{3}$	$\frac{1}{4}$	$\frac{1}{5}$	$\frac{1}{10}$

18.

Outcomes	s_1	s_2	s_3	s_4	s_5
Probability	.64	$-.08$	.30	.12	.02

19.

Outcomes	s_1	s_2	s_3	s_4	s_5
Probability	.05	.35	.5	.2	$-.3$

Which of the following are examples of subjective probability? (See Example 7.)

20. The probability of heads on five consecutive tosses of a coin

21. The probability that a freshman entering a four-year college will graduate with a degree

22. The probability that a person is allergic to penicillin

23. The probability of drawing an ace from a standard deck of 52 playing cards

24. The probability that an individual will get lung cancer from smoking cigarettes

25. A weather forecaster allowing a 70% chance of rain tomorrow

26. A gambler claiming that on a roll of a fair die, $P(\text{even}) = 1/2$

27. A surgeon giving a patient a 90% chance of full recovery

28. A bridge player having a 1/4 probability of being dealt a diamond

29. A forest ranger stating that the probability of a short fire season this year is only 3/10

The table below gives a certain golfer's probabilities of scoring at various ranges on a par-70 course.

Range	Probability
Below 60	.01
60–64	.08
65–69	.15
70–74	.28
75–79	.22
80–84	.08
85–89	.06
90–94	.04
95–99	.02
100 or more	.06

In a given round, find the probability that the golfer's score will be

30. 90 or higher;

31. below par of 70;

32. in the 70s;

33. in the 90s;

34. not in the 60s;

35. not in the 60s or 70s.

36. Find the odds in favor of the golfer shooting below par.

37. Find the odds against the golfer shooting in the 70s.

Francisco has set up the following probability distribution for the number of hours it will take him to finish his homework.

Hours	1	2	3	4	5	6
Probability	.05	.10	.20	.40	.10	.15

Find the probability that his homework will take

38. less than 3 hours;

39. 3 hours or less;

40. more than 2 hours;

41. at least 2 hours;

42. more than 1 hour and less than 5 hours;

43. 8 hours.

Work the following exercises on odds. (See Examples 8 and 9.)

44. Find the following for the experiment of Exercise 13.
   (a) What are the odds in favor of drawing a yellow marble?
   (b) What are the odds in favor of drawing a blue marble?
   (c) What are the odds in favor of drawing a white marble?

45. A single fair die is rolled. Find the odds in favor of rolling
   (a) the number 5;   **(b)** 3, 4, or 5;
   (c) 1, 2, 3, or 4;   **(d)** some number less than 2.

46. A baseball player with a batting average of .300 comes to bat. What are the odds in favor of his getting a hit?

47. In Exercise 4, what are the odds that sum of the numbers on the two slips of paper is 5?

48. If the odds that it will rain are 4 to 5, what is the probability of rain?

49. If the odds that a candidate will win an election are 3 to 2, what is the probability that the candidate will lose?

50. On page 134 of Roger Staubach's autobiography, *First Down, Lifetime to Go*, Staubach makes the following statement regarding his experience in Vietnam: "Odds against a direct hit are very low but when your life is in danger, you don't worry too much about the odds." Is this wording consistent with our definition of odds for and against? How could it have been said so as to be technically correct?

One way to solve a probability problem is to repeat the experiment (or a simulation of the experiment) many times, keeping track of the results. Then the probability can be estimated using the relative frequency m/n, where m favorable outcomes occur in n trials of an experiment. Use the computer to generate random numbers between 0 and 1. Assign the numbers x, $0 \leq x < .5$ to represent heads and the numbers $.5 \leq x < 1$ to represent tails. Then a series of, say, 5 of these random numbers represents the results of 5 coin tosses. For example, the 3 numbers .001234, .569241, .764219 represent heads, tails, tails.

Using random numbers to estimate probabilities is called the Monte Carlo *method of finding probabilities. Suppose a coin is tossed five times. Use the Monte Carlo method to approximate the probabilities in Exercises 51 and 52. Then calculate the theoretical probabilities using the methods of the text and compare the results.*

51. $P(4 \text{ heads})$

52. $P(2 \text{ heads}, 1 \text{ tail}, 2 \text{ heads})$ (in the order given)

Use the Monte Carlo method to approximate the following probabilities if four cards are drawn from a 52-card deck.

53. $P(\text{any two cards and then two kings})$

54. $P(\text{two kings})$

8.5 Probability of Alternative Events

Events are sets, so the union, intersection, and complements of events can be formed.

Example 1 Suppose a die is tossed; let E be the event "the number showing is more than 3," and let F be the event "the number showing is even." Then

$$E = \{4, 5, 6\} \quad \text{and} \quad F = \{2, 4, 6\}.$$

(a) $E \cap F$ is the event "the number showing on top is more than 3 *and* is even." Event $E \cap F$ contains the outcomes common to both E and F.

$$E \cap F = \{4, 6\}$$

(b) $E \cup F$ is the event "the number showing on top is more than 3 *or* is even." Event $E \cup F$ contains the outcomes of E or F or both.

$$E \cup F = \{2, 4, 5, 6\}$$

1 Give the following events for the experiment of Example 1 if $E = \{1, 3\}$ and $F = \{2, 3, 4, 5\}$.

(a) $E \cap F$
(b) $E \cup F$
(c) E'

Answer:
(a) $\{3\}$
(b) $\{1, 2, 3, 4, 5\}$
(c) $\{2, 4, 5, 6\}$

(c) Event E' is the event "the number showing on top is *not* more than 3"; E' contains the elements of the sample space that are not in E.

$$E' = \{1, 2, 3\}$$

The Venn diagrams of Figure 8.18 on the next page show the events $E \cap F$, $E \cup F$, and E'. **1**

Two events that cannot both occur at the same time, such as getting both a head and a tail on the same toss of a coin, are **mutually exclusive events.** Events A and B are mutually exclusive events if $A \cap B = \emptyset$.

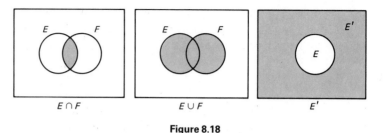

Figure 8.18

Example 2 Let $S = \{1, 2, 3, 4, 5, 6\}$, the sample space for tossing a die. Let $E = \{4, 5, 6\}$, and let $G = \{1, 2\}$. Then E and G are mutually exclusive events since they have no outcomes in common; $E \cap G = \emptyset$. See Figure 8.19. ▪

2 In Example 2 let $F = \{2, 4, 6\}$ and $K = \{1, 3, 5\}$. Are the following events mutually exclusive?
(a) F and K
(b) F and G

Answer:
(a) Yes
(b) No

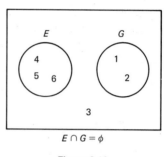

$E \cap G = \phi$

Figure 8.19

A summary of the set operations for events is given below.

Let E and F be events for a sample space S.

$E \cap F$ occurs when both E and F occur.
$E \cup F$ occurs when either E or F or both occur.
E' occurs when E does not.
E and F are mutually exclusive if $E \cap F = \emptyset$.

If E and F are mutually exclusive events, the probability of $E \cup F$ is given by the following addition principle, an extension of the addition principle for simple events given in the previous section.

Addition Principle For *mutually exclusive* events E and F,

$$P(E \cup F) = P(E) + P(F).$$

Example 3 Find the probability that a family with three children has at least two girls.

The sample space for this experiment is

$$\{ggg, \ ggb, \ gbg, \ bgg, \ gbb, \ bgb, \ bbg, \ bbb\}.$$

The event E, "the family has at least two girls," is the union of two mutually exclusive events, F: "the family has two girls" and G: "the family has three girls." Since $P(F) = 3/8$ and $P(G) = 1/8$, use the addition principle to find $P(E)$ as follows.

3 In Example 3, find the probability of no more than 2 girls.

Answer:
7/8

$$P(E) = P(F) + P(G) = \frac{3}{8} + \frac{1}{8} = \frac{4}{8} = \frac{1}{2} \quad \boxed{3}$$

For any event E from a sample space S,

$$E \cup E' = S \quad \text{and} \quad E \cap E' = \emptyset.$$

Since $E \cap E' = \emptyset$, events E and E' are mutually exclusive, so that

$$P(E \cup E') = P(E) + P(E').$$

However, since $E \cup E' = S$, the sample space, and $P(S) = 1$,

$$P(E \cup E') = P(E) + P(E') = 1,$$

which gives the following useful result.

Probability of Complements For any event E,

$$P(E) = 1 - P(E') \quad \text{and} \quad P(E') = 1 - P(E).$$

Example 4 In a particular experiment, $P(E) = 5/8$. Find $P(E')$

4 (a) Let $P(K) = 2/3$. Find $P(K')$.
(b) If $P(X') = 3/4$, find $P(X)$.

Answer:
(a) 1/3
(b) 1/4

$$P(E') = 1 - P(E) = 1 - \frac{5}{8} = \frac{3}{8}. \quad \boxed{4}$$

Example 5 Janet Passman has appointments for three job interviews. Based on her own assessment of her qualifications for these jobs and her knowledge of the job market, she assigns probabilities of being hired as follows: Company X, 2/5; Company Y, 3/10; and Company Z, 1/10. Assume that being hired by Company X, by Company Y, and by Company Z are mutually exclusive events.
(a) What is the probability that she will get none of these jobs?
There are four outcomes in the sample space: she is hired by Company X, she is hired by Company Y, she is hired by Company Z, or she is not hired at all. The probabilities of the four outcomes in the sample space must add up to 1. Therefore, the probability that she is not hired is

$$1 - \left(\frac{2}{5} + \frac{3}{10} + \frac{1}{10} \right) = \frac{2}{10} = \frac{1}{5}.$$

(b) Find the probability that she is hired by Company X or Company Y.
Since the two events are mutually exclusive, the probability of being hired by either Company X or Y is the sum

$$\frac{2}{5} + \frac{3}{10} = \frac{7}{10}.$$

(c) What is the probability that she is hired by one of the three companies?

The complement of this event is the event that she gets none of the jobs. From part (a), the probability of the complement is 1/5. Therefore, the desired probability is

$$1 - \frac{1}{5} = \frac{4}{5}. \quad \boxed{5} \; \boxed{6}$$

The addition rule given earlier in this section applies only to mutually exclusive events. For two events that are *not* mutually exclusive, use the following **extended addition principle.**

Extended Addition Principle For any events E and F, from a sample space S,

$$P(E \cup F) = P(E) + P(F) - P(E \cap F).$$

The reason the extended addition principle works can be seen by looking at a Venn diagram. In the diagram of Figure 8.20, the universal set is the sample space, and sets E and F correspond to events E and F. Set E includes regions 2 and 3; F includes regions 3 and 4; $E \cup F$ includes regions 2, 3, and 4; and $E \cap F$ includes just region 3. The probability of region 3 should not be counted twice, but, as region 3 is included in both E and F, it will be counted twice unless $P(E \cap F)$ is subtracted.

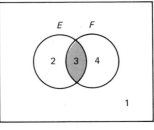

Figure 8.20

Venn diagrams can be useful in finding probabilities, as the next example shows.

Example 6 Susan is a college student who receives heavy sweaters from her aunt at the first sign of cold weather. The probability that a sweater is the wrong size is .47, the probability that it is a loud color is .59, and the probability that it is both the wrong size and a loud color is .31.

Let W represent the event "wrong size" and L represent "loud color". Place the given information on a Venn diagram, as in Figure 8.21, by starting with .31 in the intersection of L and W. Event W has probability .47. Since .31 has already been placed inside the intersection of W and L,

$$.47 - .31 = .16$$

---

5 In Example 5, find the probability that Janet Passman is
(a) not hired by X or Y;
(b) not hired by Z;
(c) not hired by X or Z.

Answer:
(a) 3/10
(b) 9/10
(c) 1/2

6 The probability that a salesperson will have weekly sales of certain amounts is as shown below.

Sales	Probability
$ 500	.25
1000	.40
1500	.20
2000	.10
2500	.05

Find the probability that the salesperson has sales
(a) less than $1500;
(b) more than $1000.

Answer:
(a) .65
(b) .35

goes inside region W, but outside the intersection of W and L. In the same way,

$$.59 - .31 = .28$$

goes inside the region for L, and outside the overlap.

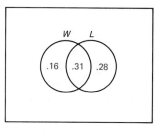

Figure 8.21

(a) Find the probability that the sweater is the correct size and not a loud color.
Using regions W and L, this event becomes $W' \cap L'$. From the Venn diagram of Figure 8.21, the labeled regions have probability

$$.16 + .31 + .28 = .75.$$

Since the entire region of the Venn diagram is assumed to have probability 1, the region outside W and L, or $W' \cap L'$, has probability

$$1 - .75 = .25.$$

The probability is .25 that the sweater is the correct size and not a loud color.
(b) Find the probability that the sweater is the correct size or is not loud.
The region $W' \cup L'$ has probability

$$.25 + .16 + .28 = .69. \quad \boxed{7}$$

Example 7 If a single card is drawn from an ordinary deck of cards, find the probability that it will be red or a face card.
Let R and F represent the events "red" and "face card" respectively. Then

$$P(R) = \frac{26}{52}, \qquad P(F) = \frac{12}{52}, \qquad \text{and} \qquad P(R \cap F) = \frac{6}{52}.$$

(There are six red face cards in the deck.) By the extended addition principle,

$$P(R \cup F) = P(R) + P(F) - P(R \cap F)$$
$$= \frac{26}{52} + \frac{12}{52} - \frac{6}{52}$$
$$= \frac{32}{52}$$
$$= \frac{8}{13}. \quad \boxed{8}$$

7 Suppose $P(A) = .2$, $P(B) = .5$, and $P(A \cap B) = .1$. Find the following.
(a) $P(A \cup B)$
(b) $P(A' \cup B)$

Answer:
(a) .6
(b) .9

8 A single card is drawn from an ordinary deck. Find the probability that it is black or a 9.

Answer:
7/13

Example 8 Suppose two fair dice are rolled. Find each of the following probabilities.

(a) The first die is 2 or the sum is 6 or 7.

The sample space for the throw of two dice is shown in Figure 8.22. The two events are labeled A and B. From the diagram,

$$P(A) = \frac{6}{36}, \qquad P(B) = \frac{11}{36}, \qquad \text{and} \qquad P(A \cap B) = \frac{2}{36}.$$

By the extended addition principle,

$$P(A \cup B) = P(A) + P(B) - P(A \cap B),$$

$$P(A \cup B) = \frac{6}{36} + \frac{11}{36} - \frac{2}{36} = \frac{15}{36} = \frac{5}{12}.$$

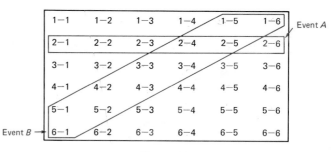

Figure 8.22

9 In the experiment of Example 8, find the following probabilities.
(a) The sum is 5 or the second die shows a 3.
(b) Both dice show the same number or the sum is at least 11.

Answer:
(a) 1/4
(b) 2/9

(b) The sum is 11 or the second die is 5.

$P(\text{sum is } 11) = 2/36$, $P(\text{second die is } 5) = 6/36$, and $P(\text{sum is } 11$ and second die is 5$) = 1/36$, so

$$P(\text{sum is } 11 \text{ or second die is } 5) = \frac{2}{36} + \frac{6}{36} - \frac{1}{36} = \frac{7}{36}. \quad \boxed{9}$$

Example 9 The personnel director at a manufacturing company has received 20 applications from people applying for a job as plant manager. Of these 20 people, 8 have MBA degrees, 9 have previous related experience, and 5 have both MBA degrees and experience. If the personnel director chooses a candidate at random, find the probability that the candidate has an MBA degree or previous related experience.

Use M for "has degree" and E for "has experience." As stated, $P(M) = 8/20$, $P(E) = 9/20$, and $P(M \cap E) = 5/20$, with

10 In Example 9, find the probability that an applicant has neither a degree nor experience.

Answer:
2/5

$$P(M \cup E) = \frac{8}{20} + \frac{9}{20} - \frac{5}{20} = \frac{12}{20} = \frac{3}{5}. \quad \boxed{10}$$

8.5 Exercises

Two dice are rolled. Find the probability of rolling each of the following sums. (See Example 8.)

1. 2   **2.** 4   **3.** 5   **4.** 6

5. 8   **6.** 9   **7.** 10   **8.** 13

9. 9 or more   **10.** Less than 7

11. Between 5 and 8   **12.** Not more than 5

13. Not less than 8   **14.** Between 3 and 7

Mrs. Elliott invites 10 relatives to a party: her mother, 2 uncles, 3 brothers, and 4 cousins. If the chances of any one guest arriving first are equally likely, find the probability that the first guest to arrive is

15. an uncle or brother;

16. a brother or cousin;

17. a brother or her mother;

18. an uncle or a cousin.

One card is drawn from an ordinary deck of cards. (See Example 7.) Find the probability of drawing

19. a 9 or a 10;

20. red or a 3;

21. a 9 or a black 10;

22. a heart or black;

23. less than a 4 (count aces as 1's);

24. a diamond or a 7;

25. a black card or an ace;

26. a heart or a card less than 5.

The numbers 1, 2, 3, 4, and 5 are written on slips of paper and two slips are drawn at random without replacement. Find each of the following probabilities.

27. The sum of the numbers drawn is 9

28. The numbers drawn are both even

29. The sum of the numbers drawn is at most 5

30. One of the numbers is even or greater than 3

31. The first number is 2 or the sum is 6

32. The sum is 5 or the second number is 2

A basket contains 6 red apples, and 4 yellow apples. A sample of 3 apples is drawn. Find the probability that the sample contains

33. all red apples;

34. all yellow apples;

35. 2 yellow and 1 red apple;

36. more red than yellow apples.

Use Venn diagrams to work Exercises 37–44. (See Example 6.)

37. Suppose $P(E) = .26$, $P(F) = .41$, and $P(E \cap F) = .17$. Find each of the following.
(a) $P(E \cup F)$ (b) $P(E' \cap F)$
(c) $P(E \cap F')$ (d) $P(E' \cup F')$

38. Let $P(Z) = .42$, $P(Y) = .38$, and $P(Z \cup Y) = .61$. Find each of the following.
(a) $P(Z' \cap Y')$ (b) $P(Z' \cup Y')$
(c) $P(Z' \cup Y)$ (d) $P(Z \cap Y')$

39. Social Science Fifty students in a Texas school were interviewed with the following results: 45 spoke Spanish, 10 spoke Vietnamese, and 8 spoke both languages. Find the probability that a randomly selected student from this group
(a) speaks both languages;
(b) speaks neither language;
(c) speaks only one of the two languages.

40. Social Science A study on body types gave the following results: 45% were short, 25% were short and overweight, and 24% were tall and not overweight. Find the probability that a person is
(a) overweight;
(b) short, but not overweight;
(c) tall and overweight.

41. Social Science The following data were gathered for 130 adult U.S. workers: 55 were women, 3 women earned more than $40,000, 62 men earned less than $40,000. Find the probability that an individual is
(a) a woman earning less than $40,000;
(b) a man earning more than $40,000;
(c) a man or is earning more than $40,000;
(d) a woman or is earning less than $40,000.

42. Social Science A survey of 100 people about their music expenditures gave the following information: 38 bought rock music, 20 were teenagers who bought rock music, 26 were teenagers. Find the probability that a person
(a) is a teenager who buys non-rock music;
(b) buys rock music or is a teenager;
(c) is not a teenager;
(d) is not a teenager, but buys rock music.

43. Suppose that 8% of a certain batch of calculators have a defective case, and that 11% have defective batteries. Also, 3% have both a defective case and defective batteries. A calculator is selected from the batch at random. Find the probability that the calculator has a good case and good batteries.

44. A student believes that her probability of passing accounting is .74, of passing mathematics is .39, and that her probability of passing both courses is .25. Find the probability that the student passes at least one course.

Natural Science *Color blindness is an inherited characteristic which is sex-linked, so that it is more common in males than in females. If M represents male and C represents red-green color blindness, using the relative frequencies of the incidence of males and red-green color blindness as probabilities, $P(C) = .049$, $P(M \cap C) = .042$, $P(M \cup C) = .534$. Find the following. (Hint: Use a Venn diagram with two circles labeled M and C.)*
45. $P(C')$ **46.** $P(M)$ **47.** $P(M')$
48. $P(M' \cap C')$ **49.** $P(C \cap M')$ **50.** $P(C \cup M')$

Work the following problems.

Management *The table below shows the probability that a customer of a department store will make a purchase in the indicated range.*

Cost	Probability
Below $2	.07
$2–$4.99	.18
$5–$9.99	.21
$10–$19.99	.16
$20–$39.99	.11
$40–$69.99	.09
$70–$99.99	.07
$100–$149.99	.08
$150 or over	.03

Find the probability that a customer makes a purchase which is
51. less than $5;   **52.** $10 to $69.99;
53. $20 or more;   **54.** more than $4.99;
55. less than $100;   **56.** $100 or more.

57. Social Science The table below shows the probability of a person accumulating credit card charges over a 12-month period.

Charges	Probability
Under $100	.31
$100–$499	.18
$500–$999	.18
$1000–$1999	.13
$2000–$2999	.08
$3000–$4999	.05
$5000–$9999	.06
$10000 or more	.01

Find the probability that a person's total charges during the period are
(a) $500 or more;
(b) less than $1000;
(c) $500 to $2999;
(d) $3000 or more.

58. Social Science The table below gives the probability that a person has life insurance in the indicated range.

Amount of insurance	Probability
None	.17
Less than $10,000	.20
$10,000–$24,999	.17
$25,000–$49,999	.14
$50,000–$99,999	.15
$100,000–$199,999	.12
$200,000 or more	.05

Find the probability that an individual has the following amounts of life insurance.
(a) Less than $10,000
(b) $10,000 to $99,999
(c) $50,000 or more
(d) Less than $50,000 or $100,000 or more

59. Natural Science The results of a study relating blood cholesterol level to coronary disease are given below.

Cholesterol level	Probability of coronary disease
Under 200	.10
200–219	.15
220–239	.20
240–259	.26
Over 259	.29

Find the probability of coronary disease if the cholesterol level is
(a) less than 240;
(b) 220 or more;
(c) from 200 to 239;
(d) from 220 to 259.

Natural Science *Gregor Mendel, an Austrian monk, was the first to use probability in the study of genetics. In an effort to understand the mechanism of character transmittal from one generation to the next in plants, he counted the number of occurrences of various characteristics. Mendel found that the flower color in certain pea plants obeyed this scheme:*

Pure red crossed with pure white produces red.

The red offspring received from its parents genes for both red (R) and white (W), but in this case red is dominant *and white* recessive, *so the offspring exhibits the color red. However, the offspring still carries both genes, and when two such offspring are crossed, several things can happen in the third generation as shown in the table below, which is called a* Punnet square.

	2nd parent R	2nd parent W
1st parent R	RR	RW
1st parent W	WR	WW

This table shows the possible combinations of genes. Use the fact that red is dominant over white to find

60. P(red); **61.** P(white).

Natural Science *Mendel found no dominance in snapdragons, with one red gene and one white gene producing pink-flowered offspring. These second generation pinks, however, still carry one red and one white gene, and when they are crossed, the next generation still yields the Punnet square above. Find*

62. P(red); **63.** P(pink); **64.** P(white).

(Mendel verified these probability ratios experimentally with large numbers of observations, and did the same for many character units other than flower color. The importance of his work, published in 1866, was not recognized until 1900.)

Natural Science *In most animals and plants, it is very unusual for the number of main parts of the organism (arms, legs, toes, flower petals, etc.) to vary from generation to generation. Some species, however, have* meristic variability, *in which the number of certain body parts varies from generation to generation. One researcher* studied the front feet of certain guinea pigs and produced the probabilities shown below.*

$$P\text{(only four toes, all perfect)} = .77$$
$$P\text{(one imperfect toe and four good ones)} = .13$$
$$P\text{(exactly five good toes)} = .10$$

Find the probability of having

65. no more than four good toes;

66. five toes, whether perfect or not.

Approximate the following probabilities, using the Monte Carlo method. (See Section 8.4, Exercises 51–54.)

67. A jeweler received 8 identical watches each in a box marked with the series number of the watch. An assistant, who does not know that the boxes are marked, is told to polish the watches and then put them back in the boxes. She puts them in the boxes at random. What is the probability that she gets at least one watch in the right box?

68. A check room attendant has 10 hats but has lost the numbers identifying them. If he gives them back randomly, what is the probability that at least 2 of the hats are given back correctly?

*From ''An Analysis of Variability in Guinea Pigs'' by J. R. Wright, from *Genetics,* Vol. 19, 1934, pp. 506–36. Reprinted by permission.

8.6 Conditional Probability

The training manager for a large stockbrokerage firm has noticed that some of the firm's brokers use the firm's research advice, while other brokers tend to go with their own feelings of which stocks will go up. To see if the research department is better than just the feelings of the brokers, the manager conducted a survey of 100 brokers, with results as shown in the following table.

	Picked stocks that went up	Didn't pick stocks that went up	Totals
Used research	30	15	45
Didn't use research	30	25	55
Totals	60	40	100

Letting A represent the event "picked stocks that went up," and letting B represent the event "used research," $P(A)$, $P(A')$, $P(B)$, and $P(B')$ can be found. ①

① Use the data in the table to find
(a) $P(A)$,
(b) $P(B)$,
(c) $P(A')$,
(d) $P(B')$.

Answer:
(a) .6
(b) .45
(c) .4
(d) .55

Suppose we want to find the probability that a broker using research will pick stocks that go up. From the table above, of the 45 brokers who use research, there are 30 who picked stocks that went up, so

$$P(\text{broker who uses research picks stocks that go up}) = \frac{30}{45} = .667.$$

This is a different number than the probability that a broker picks stocks that go up, .6, (from Problem 1 at the side) since we have additional information (the broker uses research) which reduced the sample space. In other words, we found the probability that a broker picks stocks that go up, A, given the additional information that the broker uses research, B. This is called the *conditional probability* of event A, given that event B has occurred, written $P(A|B)$. In the example above,

$$P(A|B) = \frac{30}{45},$$

which can be written as

$$P(A|B) = \frac{\dfrac{30}{100}}{\dfrac{45}{100}} = \frac{P(A \cap B)}{P(B)},$$

where $P(A \cap B)$ represents the probability that A and B will both occur. This result is used as the definition of conditional probability.

> The **conditional probability** of an event E given event F, written $P(E|F)$, is
>
> $$P(E|F) = \frac{P(E \cap F)}{P(F)}, \qquad P(F) \neq 0.$$

Example 1 Use the information in the chart at the beginning of this section, along with the definition above, to find the following probabilities.

(a) $P(B|A)$

By the definition of conditional probability,

$$P(B|A) = \frac{P(B \cap A)}{P(A)}.$$

In the example, $P(B \cap A) = 30/100$, and $P(A) = 60/100$.

$$P(B|A) = \frac{\dfrac{30}{100}}{\dfrac{60}{100}} = \frac{1}{2}$$

If a broker picked stocks that went up, then the probability is 1/2 that the broker used research.

(b) $P(A'|B)$

$$P(A'|B) = \frac{P(A' \cap B)}{P(B)} = \frac{\dfrac{15}{100}}{\dfrac{45}{100}} = \frac{1}{3}$$

(c) $P(B'|A')$

$$P(B'|A') = \frac{P(B' \cap A')}{P(A')} = \frac{\dfrac{25}{100}}{\dfrac{40}{100}} = \frac{5}{8} \quad \boxed{2}$$

Venn diagrams can be used to illustrate problems in conditional probability. A Venn diagram for Example 1, in which the probabilities are used to indicate the number in the set defined by each region, is shown in Figure 8.23. In the diagram, $P(B|A)$ is found by reducing the sample space to just set A. Then $P(B|A)$ is the ratio of the number in that part of set B which is also in A to the number in set A, or $.3/.6 = .5$.

---

2 The table shows the results of a survey of a buffalo herd.

	Males	Females	Totals
Adults	500	1300	1800
Calves	520	500	1020
Totals	1020	1800	2820

Let M represent "male" and A represent "adult." Find each of the following.
(a) $P(M|A)$
(b) $P(M'|A)$
(c) $P(A|M')$
(d) $P(A'|M)$

Answer:
(a) 5/18
(b) 13/18
(c) 13/18
(d) 26/51

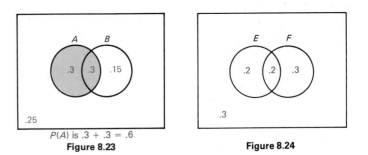

$P(A)$ is $.3 + .3 = .6$
Figure 8.23

Figure 8.24

Example 2   Given $P(E) = .4$, $P(F) = .5$, and $P(E \cup F) = .7$, find $P(E|F)$.

Find $P(E \cap F)$ first. Then use a Venn diagram to find $P(E|F)$. By the extended addition principle,

$$P(E \cup F) = P(E) + P(F) - P(E \cap F)$$

or

$$.7 = .4 + .5 - P(E \cap F)$$

$$P(E \cap F) = .2.$$

Now use the probabilities to indicate the number in each region of the Venn diagram in Figure 8.24. $P(E|F)$ is the ratio of the probability of that part of E which is in F to the probability of F or

3   Find $P(F|E)$ if $P(E) = .3$, $P(F) = .4$, and $P(E \cup F) = .6$.
Answer:
1/3

$$P(E|F) = \frac{P(E \cap F)}{P(F)} = \frac{.2}{.5} = \frac{2}{5}. \quad \boxed{3}$$

Product Rule   Starting with the definition of conditional probability given above,

$$P(E|F) = \frac{P(E \cap F)}{P(F)}, \qquad (P(F) \neq 0),$$

and multiplying both sides of the equation by $P(F)$ gives the following **product rule** of probability.

Product Rule of Probability   For any events E and F,

$$P(E \cap F) = P(F) \cdot P(E|F) = P(E) \cdot P(F|E).$$

The product rule gives a method for finding the probability that events E and F both occur as illustrated by the next few examples.

Example 3   In a class with 2/5 women and 3/5 men, 25% of the women are business majors. Find the probability that a student chosen at random from the class is a women business major.

Let B and W represent the events "business major" and "women", respectively. We want to find $P(B \cap W)$. By the product rule,

$$P(B \cap W) = P(W) \cdot P(B|W).$$

From the given information, $P(W) = 2/5 = .4$ and $P(B|W) = .25$. Then

$$P(B \cap W) = .4(.25) = .10. \quad \boxed{4}$$

The next examples show how a tree diagram is used to find conditional probabilities.

Example 4 A company needs to hire a new director of advertising. It has decided to try to hire away either person A or person B, assistant advertising directors for its major competitor. To decide between A and B, the company does research on the campaigns managed by A or B (none are managed by both), and finds that A is in charge of twice as many advertising campaigns as B. Also, A's campaigns have satisfactory results three out of four times, while B's campaigns have satisfactory results in only two out of five times. Suppose one of the competitor's advertising campaigns (managed by A or B) is selected randomly. Find the probabilities of the following events.

(a) A is in charge of an advertising campaign that produces satisfactory results.

First construct a tree diagram which shows the various possible outcomes for this experiment, as shown in Figure 8.25 on the next page. Since A does twice as many jobs as B, the probabilities of A and B having the job are 2/3 and 1/3 respectively, as shown on the first stage of the tree. The second stage shows four different conditional probabilities. The ratings for the advertising campaigns are S (satisfactory) and U (unsatisfactory). For example, along the branch from B to S,

$$P(S|B) = \frac{2}{5}$$

since B has satisfactory results in two out of five campaigns. At each point where the tree branches, the sum of the probabilities is 1. Each of the four composite branches in the tree is numbered, with its probability given on the right. These four branches represent mutually exclusive events. The event that A has a campaign with a satisfactory result (event A ∩ S) is associated with branch 1, so

$$P(A \cap S) = \frac{1}{2}.$$

(b) B runs the campaign and produces satisfactory results.

This event, B ∩ S, is shown on branch 3:

$$P(B \cap S) = \frac{2}{15}.$$

(c) The campaign is satisfactory.

The result S combines branches 1 and 3, so

$$P(S) = \frac{1}{2} + \frac{2}{15} = \frac{19}{30}.$$

4 In a litter of puppies, 3 were female and 4 were male. Half the males were black. Find the probability that a puppy chosen at random from the litter would be a black male.

Answer:

2/7

(d) The campaign is unsatisfactory.

Event U combines branches 2 and 4, so

$$P(U) = \frac{1}{6} + \frac{1}{5} = \frac{11}{30}.$$

Alternatively, $P(U) = 1 - P(S) = 1 - 19/30 = 11/30$.

(e) Either A runs the campaign or the results are satisfactory (or both).

Event A combines branches 1 and 2, while event S combines branches 1 and 3, so use branches 1, 2, and 3.

$$P(A \cup S) = \frac{1}{2} + \frac{1}{6} + \frac{2}{15} = \frac{4}{5} \quad \boxed{5}$$

5 Find each of the following probabilities for Example 4.

(a) A did the work, given that the results are satisfactory.

(b) $P(B|U)$

Answer:

(a) 15/19

(b) 6/11

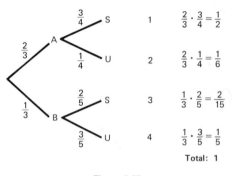

Figure 8.25

The next three examples could be worked with combinations, as explained earlier. In this section we show an alternate approach, using tree diagrams and conditional probability.

Example 5 From a box containing 1 red, 3 white, and 2 green marbles, two marbles are drawn one at a time without replacing the first before the second is drawn. Find the probability that one white and one green marble are drawn.

A tree diagram showing the various possible outcomes is given in Figure 8.26. In this diagram, *W* represents the event "drawing a white marble" and *G* represents "drawing a green marble." On the first draw, $P(W \text{ on the 1st}) = 3/6 = 1/2$ because three of the six marbles in the box are white. On the second draw, $P(G \text{ on the 2nd}|W \text{ on the 1st}) = 2/5$. One white marble has been removed, leaving 5, of which 2 are green.

We want to find the probability of drawing one white marble and one green marble. Two events satisfy this condition: drawing a white marble first and then a green one (branch 2 of the tree diagram), or drawing a green marble first and then a white one (branch 4). For branch 2,

6 Find the probability of drawing a green marble and then a white marble.

Answer:
1/5

$$P(W \text{ on 1st}) \cdot P(G \text{ on 2nd} | W \text{ on 1st}) = \frac{1}{2} \cdot \frac{2}{5} = \frac{1}{5}.$$ **6**

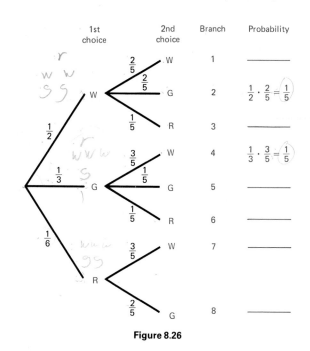

Figure 8.26

Since these two events are mutually exclusive, the final probability is the sum of the two probabilities.

$$P(\text{one } W, \text{one } G) = P(W \text{ on 1st}) \cdot P(G \text{ on 2nd} | W \text{ on 1st})$$

7 In Example 5 find the probability of drawing 1 white and 1 red marble.

Answer:
1/5

$$+ P(G \text{ on 1st}) \cdot P(W \text{ on 2nd} | G \text{ on 1st}) = \frac{2}{5} \quad \text{**7**}$$

Example 6 Two cards are drawn without replacement from an ordinary deck (52 cards). Find the probability that the first card is a heart and the second card is red.

Start with the tree diagram of Figure 8.27. On the first draw, since there are 13 hearts in the 52 cards, the probability of drawing a heart first is 13/52 = 1/4. On the second draw, since a heart has been drawn already, there are 25 red cards in the remaining 51 cards. Thus the probability of drawing a red card on the second draw, given that the first is a heart, is 25/51. By the product rule of probability

8 Find the probability of drawing a heart on the first draw and a black card on the second, if two cards are drawn without replacement.

Answer:
13/102 or .1275.

$$P(\text{heart on 1st and red on 2nd})$$
$$= P(\text{heart on 1st}) \cdot P(\text{red on 2nd} | \text{heart on 1st})$$
$$= \frac{1}{4} \cdot \frac{25}{51} = \frac{25}{204} = .1225. \quad \text{**8**}$$

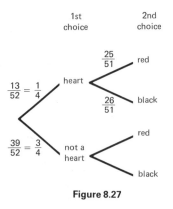

Figure 8.27

Independent Events Suppose a fair coin is tossed and shows heads. The probability of heads on the next toss is still 1/2; the fact that heads was the result on a given toss has no effect on the outcome of the next toss. Coin tosses are **independent events,** since the outcome of one toss does not help decide the outcome of the next toss. Rolls of a fair die are independent events; the fact that a 2 came up on one roll does not increase our knowledge of the outcome of the next roll. On the other hand, the events "today is cloudy" and "today is rainy" are **dependent events;** if it is cloudy, then there is an increased chance of rain.

If events E and F are independent, then the knowledge that E has occurred gives no (probability) information about the occurrence or nonoccurrence of event F. That is, $P(F)$ is exactly the same as $P(F|E)$, or

$$P(F|E) = P(F).$$

In fact, this is the formal definition of independent events:

E and F are **independent events** if

$$P(F|E) = P(F) \quad \text{or} \quad P(E|F) = P(E).$$

Using this definition, the product rule can be simplified for independent events.

Product Rule for Independent Events If E and F are independent events, then

$$P(E \cap F) = P(E) \cdot P(F).$$

Example 7 A calculator requires a key-stroke assembly and a logic circuit. Assume that 99% of the key-stroke assemblies are satisfactory and 97% of the logic circuits are satisfactory. Find the probability that a finished calculator will be satisfactory.

If the failure of a key-stroke assembly and the failure of a logic circuit are independent events, then

$$P(\text{satisfactory calculator})$$
$$= P(\text{satisfactory key-stroke assembly})$$
$$\cdot P(\text{satisfactory logic circuit})$$
$$= (.99)(.97) \approx .96.$$

9 Find the probability of getting 4 successive heads on 4 tosses of a fair coin.

Answer:
1/16

The probability of a defective calculator is $1 - .96 = .04$. **9**

Example 8 When black-coated mice are crossed with brown-coated mice, a pair of genes, one from each parent, determines the coat color of the offspring. Let *b* represent the gene for brown and *B* the gene for black. If a mouse carries either one *B* gene and one *b* gene (*Bb* or *bB*) or two *B* genes (*BB*), the coat will be black. If the mouse carries two *b* genes (*bb*), the coat will be brown. Find the probability that a mouse, born to a brown-coated female and a black-coated male who is known to carry the *Bb* combination, will be brown.

To be brown-coated, the offspring must receive one *b* gene from each parent. The brown-coated parent carries two *b* genes, so that the probability of getting one *b* gene from the mother is 1. The probability of getting one *b* gene from the black-coated father is 1/2. Therefore, since these are independent events, the probability of a brown-coated offspring from these parents is $1 \cdot 1/2 = 1/2$. **10**

10 Find the probability that the mice of Example 8 will have a black-coated offspring.

Answer:
1/2

It is common for students to confuse the ideas of *mutually exclusive* events and *independent* events. Events *E* and *F* are mutually exclusive if $E \cap F = \emptyset$. For example, if a family has exactly one child, the only possible outcomes are $B = \{\text{boy}\}$ and $G = \{\text{girl}\}$. These two events are mutually exclusive. However, the events are *not* independent, since $P(G|B) = 0$ (if a family with only one child has a boy, the probability it has a girl is then 0). Since $P(G|B) \neq P(G)$, the events are not independent.

Of all the families with exactly *two* children, the events $G_1 = \{\text{first child is a girl}\}$ and $G_2 = \{\text{second child is a girl}\}$ are independent, since $P(G_2|G_1)$ equals $P(G_2)$. However, G_1 and G_2 are not mutually exclusive, since $G_1 \cap G_2 = \{\text{both children are girls}\} \neq \emptyset$.

The only way to show that two events *E* and *F* are independent is to show that $P(F|E) = P(F)$ [or $P(E|F) = P(E)$].

8.6 Exercises

If a single fair die is rolled, find the probability of rolling

1. 2, given that the number rolled was odd;

2. 4, given that the number rolled was even;

3. an even number, given that the number rolled was 6.

If two fair dice are rolled (recall the 36-outcome sample space), find the probability of rolling.

4. a sum of 8, given the sum was greater than 7;

5. a sum of 6, given the roll was a "double" (two identical numbers);

6. a double, given that the sum was 9.

If two cards are drawn without replacement from an ordinary deck (see Example 6), find the probability that

7. the second is a heart, given that the first is a heart;

8. the second is black, given that the first is a spade;

9. the second is a face card, given that the first is a jack;

10. the second is an ace, given that the first is not an ace;

11. a jack and a 10 are drawn;

12. an ace and a 4 are drawn;

13. two black cards are drawn;

14. two hearts are drawn.

Social Science *Use a tree diagram in Exercises 15–33. (See Examples 4–6.)*
Suppose 20% of the population is 65 or over, 26% of those 65 or over have loans, and 53% of those under 65 have loans. Find the probability that a person

15. is 65 or over and has a loan;

16. has a loan.

In a certain area, 15% of the population are joggers and 40% of the joggers are women. If 55% of those who do not jog are women, find the probability that an individual from that community is

17. a woman jogger;   **18.** not a jogger.

A survey has shown that 52% of the women in a certain community work outside the home. Of these women, 64% are married, while 86% of the women who do not work outside the home are married. Find the probability that a woman in that community is

19. married;

20. a single working woman.

Two-thirds of the population is on a diet at least occasionally. Of this group, 4/5 drink diet soft drinks, while 1/2 of the rest of the population drink diet soft drinks. Find the probability that a person

21. drinks diet soft drinks;

22. diets but does not drink diet soft drinks.

A smooth-talking young man has a 1/3 probability of talking a policeman out of giving him a speeding ticket. The probability that he is stopped for speeding during a given weekend is 1/2. Find the probability that

23. he will receive no tickets on a given weekend;

24. he will receive no tickets on three consecutive weekends.

Slips of paper marked with the digits 1, 2, 3, 4, and 5 are placed in a box and mixed well. If two slips are drawn (without replacement), find the probability that

25. the first is even and the second is odd;

26. the first is a 3 and the second a number greater than 3;

27. both are even;

28. both are marked 3.

Two marbles are drawn without replacement from a jar with four black and three white marbles. Find the probability that

29. both are white;

30. both are black;

31. the second is white given that the first is black;

32. the first is black and the second is white;

33. one is black and the other is white.

The Midtown Bank has found that most customers at the tellers' windows either cash a check or make a deposit. The chart below indicates the transactions for one teller for one day.

	Cash check	No check	Totals
Make deposit	50	20	70
No deposit	30	10	40
Totals	80	30	110

Letting C represent "cashing a check" and D represent "making a deposit," express each of the following probabilities in words and find its value. (See Example 1.)

34. $P(C|D)$ **35.** $P(D'|C)$ **36.** $P(C'|D')$

37. $P(C'|D)$ **38.** $P[(C \cap D)']$

A pet shop has 10 puppies, 6 of them males. There are 3 beagles (1 male), 1 cocker spaniel (male), and 6 poodles. Construct a table similar to the one above and find the probability that one of these puppies, chosen at random, is

39. a beagle;

40. a beagle, given that it is a male;

41. a male, given that it is a beagle;

42. a cocker spaniel, given that it is a female;

43. a poodle, given that it is a male;

44. a female, given that it is a beagle.

A bicycle factory runs two assembly lines, A and B. If 95% of line A's products pass inspection, while only 90% of line B's products pass inspection, and 60% of the factory's bikes come off assembly line B (the rest off A), find the probability that one of the factory's bikes did not pass inspection and came off

45. assembly line A;   **46.** assembly line B.

47. Natural Science Both of a certain pea plant's parents had a gene for red and a gene for white flowers. (See Exercises 60–64 in Section 8.5.) If the offspring has red flowers, find the probability that it combined a gene for red and a gene for white (rather than two for red).

Assume that boy and girl babies are equally likely.

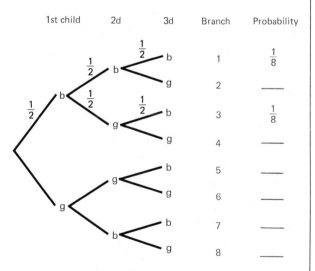

1st child	2d	3d	Branch	Probability

Fill in the remaining probabilities on the tree diagram above and use the information to find the probability that a family with three children has all girls, given that

48. the first is a girl;   **49.** the third is a girl;

50. the second is a girl;   **51.** at least two are girls;

52. at least one is a girl.

Natural Science *The following table shows frequencies for red-green color blindness, where M represents male and C represents color-blind.*

	M	M'	Totals
C	.042	.007	.049
C'	.485	.466	.951
Totals	.527	.473	1.000

Use this table to find the following probabilites.

53. $P(M)$ **54.** $P(C)$ **55.** $P(M \cap C)$

56. $P(M \cup C)$ **57.** $P(M|C)$ **58.** $P(M'|C)$

59. Are the events C and M described above dependent? [Recall that two events E and F are dependent if $P(E|F) \neq P(E)$.]

60. A scientist wishes to determine if there is any dependence between color blindness (C) and deafness (D). Given the probabilities listed in the table below, what should his findings be? (See Exercises 53–59.)

	D	D'	Totals
C	.0004	.0796	.0800
C'	.0046	.9154	.9200
Totals	.0050	.9950	1.0000

The Motor Vehicle Department has found that the probability of a person passing the test for a driver's license on the first try is .75. The probability that an individual who fails on the first test will pass on the second try is .80, and the probability that an individual who fails the first and second tests will pass the third time is .70. Find the probability that an individual

61. fails both the first and second tests;

62. will fail three times in a row;

63. will require at least two tries to pass the text.

According to a booklet put out by EastWest Airlines, 98% of all scheduled EastWest flights actually take place. (The other flights are cancelled due to weather, equipment problems, and so on.) Assume that the event that a given flight takes place is independent of the event that another flight takes place. (See Example 7.)

64. Elizabeth Thornton plans to visit her company's branch offices; her journey requires three separate flights on EastWest. What is the probability that all these flights will take place?

65. Based on the reasons we gave for a flight to be can-celled, how realistic is the assumption of independence that we made?

66. In one area, 4% of the population drives a luxury car. However, 17% of the CPAs drive a luxury car. Are the events "drive a luxury car" and "person is a CPA" independent?

67. Corporations such as banks, where a computer is essen-tial to day-to-day operations, often have a second, backup computer in case of failure by the main com-puter. Suppose that there is a .003 chance that the main

computer will fail in a given time period and a .005 chance that the backup computer will fail while the main computer is being repaired. Assume these failures represent independent events, and find the fraction of the time that the corporation can assume it will have computer service. How realistic is our assumption of independence?

68. A key component of a space rocket will fail with a probability of .03. How many such components must be used as backups to ensure the probability that at least one of the components will work is .999999?

Chapter 8   Review Exercises

Write true or false for each of the following.

1. $-6 \in \{8, 4, -3, -9, 6\}$ **2.** $8 \in \{3, 9, 7\}$

3. $2 \notin \{0, 1, 2, 3, 4\}$ **4.** $0 \in \{0, 1, 2, 3, 4\}$

5. $\{3, 4, 5\} \subset \{2, 3, 4, 5, 6\}$

6. $\{1, 2, 5, 8\} \subset \{1, 2, 5, 10, 11\}$

7. $\{3, 6, 9, 10\} \subset \{3, 9, 11, 13\}$

8. $\emptyset \subset \{1\}$

9. $\{2, 8\} \not\subseteq \{2, 4, 6, 8\}$

10. $0 \subset \emptyset$

List the elements in the following sets.

11. $\{x \mid x$ is a counting number more than 8 and less than 5$\}$

12. $\{x \mid x$ is an integer, $-3 \le x < 1\}$

13. $\{$all counting numbers less than five$\}$

14. $\{$all whole numbers not greater than 2$\}$

Let $U = \{a, b, c, d, e, f, g\}$, $K = \{c, d, f, g\}$, *and* $R = \{a, c, d, e, g\}$. *Find the following.*

15. The number of subsets of K

16. The number of subsets of R

17. K' **18.** R' **19.** $K \cap R$

20. $K \cup R$ **21.** $(K \cap R)'$ **22.** $(K \cup R)'$

23. $\emptyset'$ **24.** U'

Let $U = \{$all employees of the K.O. Brown Company$\}$
   $A = \{$employees in the accounting department$\}$
   $B = \{$employees in the sales department$\}$

$C = \{$employees with at least 10 years' service$\}$
$D = \{$employees with an MBA degree$\}$.

Describe the following sets in words.

25. $A \cap C$ **26.** $B \cap D$ **27.** $A \cup D$

28. $A' \cap D$ **29.** $B' \cap C'$ **30.** $(B \cup C)'$

Draw Venn diagrams for Exercises 31–34.

31. $B \cup A'$ **32.** $A' \cap B$

33. $A' \cap (B' \cap C)$ **34.** $(A \cup B)' \cap C$

The following data apply to Exercises 35–38. A telephone survey of television viewers revealed that

   20 *watch situation comedies*

   19 *watch game shows*

   27 *watch movies*

   5 *watch both situation comedies and game shows*

   8 *watch both game shows and movies*

   10 *watch both situation comedies and movies*

   3 *watch all three*

   6 *watch none of these.*

35. How many viewers were interviewed?

36. How many viewers watch comedies and movies but not game shows?

37. How many viewers watch only movies?

38. How many viewers watch comedies and game shows but not movies?

39. In how many ways can 6 taxicabs line up at the train station?

40. In how many ways can a sample of 3 oranges be taken from a bag of a dozen oranges?

41. In how many ways can a selection of 2 pictures from a group of 5 pictures be arranged in a row on a wall?

42. In how many ways can the 5 pictures of Exercise 41 be arranged if one must be first?

43. In a Chinese restaurant the menu lists eight items in column A and six items in column B. To order a dinner, the diner is told to select three from column A and two from column B. How many dinners are possible?

44. A spokesperson is to be selected from each of three departments in a small college. If there are seven people in the first department, five in the second department, and four in the third department, how many different groups of three representatives are possible?

Write sample spaces for the following.

45. A die is rolled.

46. A card is drawn from a deck containing only the thirteen spades.

47. The weight of a person is measured to the nearest half pound; the scale will not measure more than 300 pounds.

48. A coin is tossed four times.

An urn contains five balls labeled 3, 5, 7, 9, and 11, while a second urn contains four red and two green balls. An experiment consists of pulling one ball from each urn, in turn. Write each of the following.

49. The sample space

50. Event E, the first ball is greater than 5

51. Event F, the second ball is green

52. Are the outcomes in the sample space equally likely?

A company sells typewriters and copiers. Let E be the event "a customer buys a typewriter," and let F be the event "a customer buys a copier." Write each of the following using ∩, ∪, or ', as necessary.

53. A customer buys neither.

54. A customer buys at least one.

When a single card is drawn from an ordinary deck, find the probability that it will be

55. a heart;

56. a red queen;

57. a face card;

58. black or a face card;

59. red, given it is a queen;

60. a jack, given it is a face card;

61. a face card, given it is a king.

Find the odds in favor of a card drawn being

62. a club;

63. a black jack;

64. a red face card or a queen.

A sample shipment of five swimming pool filters is chosen at random. The probability of exactly 0, 1, 2, 3, 4, or 5 filters being defective is given in the following table.

Number defective	0	1	2	3	4	5
Probability	.31	.25	.18	.12	.08	.06

Find the probability that the following number of filters is defective.

65. No more than 3

66. At least 3

The square shows the four possible (equally likely) combinations when both parents are carriers of the sickle cell anemia trait. Each carrier parent has normal cells (N) and trait cells (T).

	2nd parent	
	N_2	T_2
1st parent N_1		N_1T_2
T_1		

67. Complete the table.

68. If the disease occurs only when two trait cells combine, find the probability that a child born to these parents will have sickle cell anemia.

69. The child will carry the trait but not have the disease if a normal cell combines with a trait cell. Find this probability.

70. Find the probability that the child is neither a carrier nor has the disease.

Find the probability for the following sums when two fair dice are rolled.

71. 8

72. 0

73. At least 10

74. No more than 5

75. Odd and greater than 8

76. 12, given it is greater than 10

77. 7, given that one die is 4

78. At least 9, given that one die is 5

Suppose $P(E) = .51$, $P(F) = .37$, and $P(E \cap F) = .22$. Find each of the following probabilities.

79. $P(E \cup F)$

80. $P(E \cap F')$

81. $P(E' \cup F)$

82. $P(E' \cap F')$

The table below shows the results of a survey of 1000 new or used car buyers of a certain model car.

	Satisfied	Not satisfied	Totals
New	300	100	400
Used	450	150	600
Totals	750	250	1000

Let S represent the event "satisfied," and N the event "bought a new car." Find each of the following.

83. $P(N \cap S)$

84. $P(N \cup S')$

85. $P(N|S)$

86. $P(N'|S)$

87. $P(S|N')$

88. $P(S'|N')$

Case 10

Drug Search Probabilities—Smith Kline and French Laboratories*

In searching for a new drug with commercial possibilities, drug company researchers use the ratio

$$N_S:N_A:N_D:1.$$

That is, if the company gives preliminary screening to N_S substances, it may find that N_A of them are worthy of further study, with N_D of these surviving into full scale development. Finally, 1 of the substances will result in a marketable drug. Typical numbers used by Smith Kline and French Laboratories in planning research budgets might be $2000:30:8:1$.

Exercises

1. Suppose a compound has been chosen for preliminary study. Use the ratio $2000:30:8:1$ to find the probability that the compound will survive and become a marketable drug.

2. Find the probability that the compound will not lead to a marketable drug.

3. Suppose a such compounds receive preliminary screening. Set up the probability that none of them produces a marketable drug. (Assume independence throughout these exercises.)

4. Use your results from Exercise 3 to set up the probability that at least one of the drugs will prove marketable.

5. Suppose now that N scientists are employed in the preliminary screening, and that each scientist can screen c compounds per year. Set up the probability that no marketable drugs will be discovered in a year.

6. Set up the probability that at least one marketable drug will be discovered.

For the following exercises, evaluate your answer in Exercise 6, which should have been $1 - (1999/2000)^{Nc}$, for the following values of N and c. Use a calculator with a y^x key, or a computer.

7. $N = 100$, $c = 6$ **8.** $N = 25$, $c = 10$

*From "Scientific Manpower Allocation to New Drug Screening Program" by E. B. Pyle III, B. Douglas, G. W. Ebright, W. J. Westlake, and A. D. Bender, from *Management Science*, Vol. 19, No. 12, August 1973. Copyright © 1973 The Institute of Management Sciences. Reprinted by permission.

Case 11

Making a First Down

A first down is desirable in football—it guarantees four more plays by the team making it, assuming no score or turnover occurs in the plays. After getting a first down, a team can get another by advancing the ball at least ten yards. During the four plays given by a first down, a team's position will be indicated by a phrase such as "third and 4," which means that the team has already had two of its four plays, and that 4 more yards are needed to get the 10 yards necessary for another first down. An article in a management journal* offers the following results for 189 games of a recent National Football League season. "Trials" represents the number of times a team tried to make a first down, given that it was currently playing either a third or a fourth down. Here *n* represents the number of yards still needed for a first down.

n	Trials	Successes	Probability of making first down with n yards to go
1	543	388	
2	327	186	
3	356	146	
4	302	97	
5	336	91	

Exercises

1. Use the basic probability definition to complete the table.

2. Why is the sum of the answers in Exercise 1 not equal to 1?

*From "Optimal Strategies of Fourth Down" by Virgil Carter and Robert Machols from *Management Science,* vol. 24, no. 16, December 1978. Copyright © 1978 The Institute of Management Sciences. Reprinted by permission.

9 Further Topics in Probability

This chapter introduces some important applications of probability which illustrate its usefulness in such diverse fields as management, medicine, insurance, and the military.

9.1 Bayes' Formula

Suppose the probability that a person gets lung cancer, given that the person smokes a pack or more of cigarettes daily, is known. For a research project, it might be necessary to know the probability that a person smokes a pack or more of cigarettes daily, given that the person has lung cancer. More generally, if $P(E|F)$ is known for two events E and F, can $P(F|E)$ be found? It turns out that it can, using the formula to be developed in this section. To find this formula, start with the product rule:

$$P(E \cap F) = P(E) \cdot P(F|E) \qquad \text{or} \qquad P(F \cap E) = P(F) \cdot P(E|F).$$

From the fact that $P(E \cap F) = P(F \cap E)$,

$$P(E) \cdot P(F|E) = P(F) \cdot P(E|F),$$

or

[1] Use formula (1) to find $P(F|E)$ if $P(F) = 1/4$, $P(E) = 1/3$, and $P(E|F) = 1/2$.

Answer:
3/8

$$P(F|E) = \frac{P(F) \cdot P(E|F)}{P(E)}. \quad \boxed{1} \tag{1}$$

Given the two events E and F, if E occurs, then either F also occurs or F' also occurs. The probabilities of $E \cap F$ and $E \cap F'$ can be expressed as follows.

$$P(E \cap F) = P(F) \cdot P(E|F)$$
$$P(E \cap F') = P(F') \cdot P(E|F')$$

Since $(E \cap F) \cup (E \cap F') = E$ (because F and F' together form the sample space),

$$P(E) = P(E \cap F) + P(E \cap F')$$

or

$$P(E) = P(F) \cdot P(E|F) + P(F') \cdot P(E|F').$$

From this, equation (1) produces the following result, a special case of Bayes' Formula, which is generalized later in this section.

Bayes' Formula (Special Case)

$$P(F|E) = \frac{P(F) \cdot P(E|F)}{P(F) \cdot P(E|F) + P(F') \cdot P(E|F')}. \qquad (2)$$

2 Use the special case of Bayes' formula to find $P(F|E)$ if $P(F) = .2$, $P(E|F) = .1$, and $P(E|F') = .3$.
[Hint: $P(F') = 1 - P(F)$.]

Answer:
$1/13 \approx .077$

2

Example 1 For a given length of time, the probability of worker error on a certain production line is .1, the probability that an accident will occur when there is a worker error is .3, and the probability that an accident will occur when there is no worker error is .2. Find the probability of a worker error if there is an accident.

Let A represent "accident" and E represent "worker error." From the information above,

$$P(E) = .1, \qquad P(A|E) = .3, \qquad \text{and} \qquad P(A|E') = .2.$$

These probabilities are shown in the tree diagram in Figure 9.1.

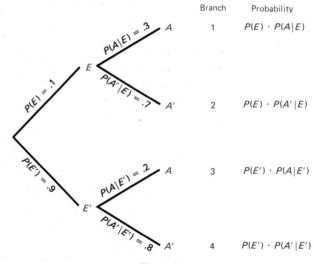

Figure 9.1

Find $P(E|A)$ using equation (2) above:

$$P(E|A) = \frac{P(E) \cdot P(A|E)}{P(E) \cdot P(A|E) + P(E') \cdot P(A|E')}$$

$$= \frac{(.1)(.3)}{(.1)(.3) + (.9)(.2)} = \frac{1}{7}. \quad \boxed{3}$$

3 In Example 1, find $P(E'|A)$.

Answer:
6/7

Equation (2) above can be generalized to more than two possibilities with the tree diagram of Figure 9.2. This diagram shows the paths that can produce an event E. Events $F_1, F_2, \ldots, F_n$ are mutually exclusive events whose union is the sample space, and E is an event that has occurred. See Figure 9.3.

Branch   Probability

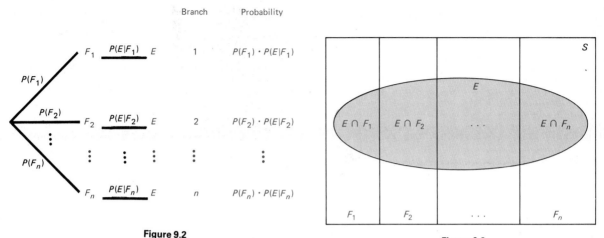

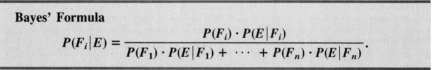

Figure 9.2

Figure 9.3

Find the probability $P(F_i|E)$, where $1 \leq i \leq n$, by dividing the probability for the branch containing $P(E|F_i)$ by the sum of the probabilities of all the branches producing event E. That is,

Bayes' Formula

$$P(F_i|E) = \frac{P(F_i) \cdot P(E|F_i)}{P(F_1) \cdot P(E|F_1) + \cdots + P(F_n) \cdot P(E|F_n)}.$$

This result is known as **Bayes' formula,** after the Reverend Thomas Bayes, whose paper on probability was published a little over two hundred years ago.

The statement of Bayes' formula can be daunting. Actually, it is easier to remember the formula by thinking of the tree diagram that produced it. Go through the following steps.

Using Bayes' Formula

Step 1 Start a tree diagram with branches representing events F_1, $F_2, \ldots, F_n$. Label each branch with its corresponding probability.

Step 2 From the end of each of these branches, draw a branch for event E. Label this branch with the probability of getting to it, or $P(E|F_i)$.

Step 3 There are now n different paths that result in event E. Next to each path, put its probability—the product of the probabilities that the first branch occurs, $P(F_i)$, and that the second branch occurs, $P(E|F_i)$: that is, $P(F_i) \cdot P(E|F_i)$.

Step 4 $P(F_i|E)$ is found by dividing the probability of the branch for F_i by the sum of the probabilities of all the branches producing event E.

Example 2 Based on past experience, a company knows that an experienced machine operator (one or more years of experience) will produce a defective item 1% of the time. People with some experience (up to one year) have a 2.5% defect rate, while new people have a 6% defect rate. At any one time, the company has 60% experienced employees, 30% with some experience, and 10% new employees. Suppose a particular item is defective; find the probability that it was produced by a new operator.

Let E represent the event "item is defective," with F_1 representing "item was made by an experienced operator," F_2 "item was made by a person with some experience," and F_3 "item was made by a new employee." Then

$$P(F_1) = .60 \qquad P(E|F_1) = .01$$
$$P(F_2) = .30 \qquad P(E|F_2) = .025$$
$$P(F_3) = .10 \qquad P(E|F_3) = .06.$$

We need to find $P(F_3|E)$, the probability that an item was produced by a new operator, given that it is defective. First, draw a tree diagram using the given information, as in Figure 9.4.

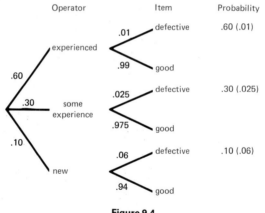

Figure 9.4

Find $P(F_3|E)$ using the bottom branch of the tree in Figure 9.4: divide the probability for this branch by the sum of the probabilities of all the branches leading to E (a defective item).

$$P(F_3|E) = \frac{.10(.06)}{.60(.01) + .30(.025) + .10(.06)} = \frac{.006}{.0195} = \frac{4}{13} \quad \boxed{4}$$

4 In Example 2 find
(a) $P(F_1|E)$;
(b) $P(F_2|E)$.

Answer:
(a) 4/13
(b) 5/13

After working Problem 4 at the side, check that $P(F_1|E) + P(F_2|E) + P(F_3|E) = 1$. (That is, the defective item was made by *someone*.)

Example 3 A manufacturer buys items from six different suppliers. The fraction of the total number of items obtained from each supplier, along with the probability that an item purchased from that supplier is defective, is shown in the following table.

Supplier	Fraction of total supplied	Probability of defective
1	.05	.04
2	.12	.02
3	.16	.07
4	.23	.01
5	.35	.03
6	.09	.05

Find the probability that a defective item came from supplier 5.

Let F_1 be the event that an item came from supplier 1, with F_2, F_3, F_4, F_5, and F_6 defined in a similar manner. Let E be the event that an item is defective. We want to find $P(F_5|E)$. Use the probabilities in the table above to prepare a tree diagram. By Bayes' formula,

$$P(F_5|E) =$$

$$\frac{(.35)(.03)}{(.05)(.04) + (.12)(.02) + (.16)(.07) + (.23)(.01) + (.35)(.03) + (.09)(.05)}$$

$$= \frac{.0105}{.0329} \approx .319.$$

There is about a 32% chance that a defective item came from supplier 5. ▪ ⑤

⑤ In Example 3 find the probability that the defective item came from
(a) supplier 3;
(b) supplier 6.

Answer:
(a) .340
(b) .137

9.1 Exercises

For two events, M and N, $P(M) = .4$, $P(N|M) = .3$, and $P(N|M') = .4$. Find each of the following. (See Example 1.)

1. $P(M|N)$ **2.** $P(M'|N)$

For mutually exclusive events R_1, R_2, R_3, $P(R_1) = .05$, $P(R_2) = .6$, and $P(R_3) = .35$. Also, $P(Q|R_1) = .40$, $P(Q|R_2) = .30$, and $P(Q|R_3) = .60$. Find each of the following. (See Example 2.)

3. $P(R_1|Q)$ **4.** $P(R_2|Q)$

5. $P(R_3|Q)$ **6.** $P(R_1'|Q)$

Suppose three jars have the following contents: 2 black balls and 1 white ball in the first; 1 black ball and 2 white balls in the second; 1 black ball and 1 white ball in the third. If the probability of selecting one of the three jars is 1/2, 1/3, and 1/6, respectively, find the probability that if a white ball is drawn, it came from the

7. second jar;   **8.** third jar.

Management Of all the people applying for a certain job, 70% are qualified, and 30% are not. The personnel manager claims that she approves qualified people 85% of the time; she approves an unqualified person 20% of the time. Find each of the following probabilities.

9. A person is qualified if he or she was approved by the manager.

10. A person is unqualified if he or she was approved by the manager.

Management A building contractor buys 70% of his cement from supplier A, and 30% from supplier B. A total of

90% of the bags from A *arrive undamaged, while 95% of the bags from B come undamaged. Give the probability that a damaged bag is from supplier*

11. A; **12.** B.

Management *The probability that a customer of a local department store will be a "slow pay" is .02. The probability that a "slow pay" will make a large down payment when buying a refrigerator is .14. The probability that a person who is not a "slow pay" will make a large down payment when buying a refrigerator is .50. Suppose a customer makes a large down payment on a refrigerator. Find the probability that the customer is*

13. a "slow pay"; **14.** not a "slow pay."

Management *Companies A, B, and C produce 15%, 40%, and 45% respectively of the major appliances sold in a certain area. In that area, 1% of the Company A appliances, 1 1/2% of the Company B appliances, and 2% of the Company C appliances need service within the first year. Suppose an appliance that needs service within the first year is chosen at random; find the probability that it was manufactured by Company*

15. A; **16.** B.

Management *On a given weekend in the fall, a tire company can buy television advertising time for a college football game, a baseball game, or a professional football game. If the company sponsors the college game, there is a 70% chance of a high rating, a 50% chance if they sponsor a baseball game, and a 60% chance if they sponsor a professional football game. The probability of the company sponsoring these various games is .5, .2, and .3 respectively. Suppose the company does get a high rating; find the probability that it sponsored*

17. a college game;

18. a professional football game.

Management *According to a business publication, there is a 50% chance of a booming economy next summer, a 20% chance of a mediocre economy, and a 30% chance of a recession. The probabilities that a particular investment strategy will produce a huge profit under each of these possibilities are .1, .6, and .3 respectively. Suppose it turns out that the strategy does produce huge profits; find the probability that the economy was*

19. booming; **20.** in recession.

Natural Science *In a test for toxemia, a disease that affects pregnant women, the woman lies on her left side and then rolls over on her back. The test is considered positive if there is a 20 mm rise in her blood pressure within one minute. The results have produced the following probabilities, where T represents having toxemia at some time during the pregnancy, and N represents a negative test.*

$$P(T'|N) = .90 \quad \text{and} \quad P(T|N') = .75$$

Assume that $P(N') = .11$, and find each of the following.

21. $P(N|T)$ **22.** $P(N'|T)$

23. Natural Science The probability that a person with certain symptoms has hepatitis is .8. The blood test used to confirm this diagnosis gives positive results for 90% of those who have the disease and 5% of those without the disease. What is the probability that an individual with the symptoms who reacts positively to the test has hepatitis?

24. Social Science In a certain county, the Democrats have 53% of the registered voters, 12% of whom are under 21. The Republicans have 47% of all registered voters, of whom 10% are under 21. If Kay is a registered voter who is under 21, what is the probability that she is a Democrat?

Social Science *The following table shows the fraction of the population in various income levels, as well as the probability that a person from that income level will take an airline flight within the next year.*

Income Level	Proportion of population	Probability of a flight during the next year
$0–$5999	12.8%	.04
$6000–$9999	14.6	.06
$10,000–$14,999	18.5	.07
$15,000–$19,999	17.8	.09
$20,000–$24,999	13.9	.12
$25,000 and over	22.4	.13

If a person is selected at random from an airline flight, find the probability that the person has the following income level. (See Example 3.)

25. $10,000–$14,999 **26.** $25,000 and over

Social Science *The following table shows the proportion of people over 18 who are in various age categories, along* with the probability that a person in a given age category will vote in a general election.

Age	Proportion of voting age population	Probability of a person of this age voting
18–21	11.0%	.48
22–24	7.6	.53
25–44	37.6	.68
45–64	28.3	.64
65 or over	15.5	.74

Suppose a voter is picked at random. Find the probability that the voter is in the following age categories.

27. 18–21

28. 65 or over

9.2 Applications of Counting

Permutations and combinations—ways of counting the number of outcomes for various kinds of experiments—were discussed in Chapter 8. Now these counting methods are used to help solve probability problems. The solution to many of these problems depends on the basic probability principle of Section 8.4:

> Let a sample space S have n possible equally likely outcomes, and let event E contain m of these outcomes, all distinct elements. The probability that event E occurs is
>
> $$P(E) = \frac{m}{n}.$$

Example 1 From a group of 22 employees, 4 are to be selected to present a list of grievances to management.

(a) In how many ways can this be done?

Four employees from a group of 22 can be selected in $\binom{22}{4}$ ways. From Section 8.3,

$$\binom{22}{4} = \frac{22!}{4!18!} = \frac{22(21)(20)(19)}{4(3)(2)(1)} = 7315.$$

There are 7315 ways to choose 4 people from 22.

(b) One of the 22 employees is Michael Branson; the group agrees that he *must* be one of the four people chosen. If the 4 people are chosen at random, find the probability that Branson will be among those selected.

The required probability will equal the quotient of the number of ways a group of 4 including Branson can be selected and the total number of ways a group of any 4 employees can be selected. Find the number of ways the group of 4 can include Branson by finding the number of ways that the additional three employees can be chosen. The 3 are chosen from 21 employees; this can be done in

$$\binom{21}{3} = \frac{21!}{3!18!} = \frac{21(20)(19)}{3(2)(1)} = 1330$$

ways. Using this result and the result from part (a), the probability that Branson will be one of the four employees chosen is thus

$$P(\text{Branson is chosen}) = \frac{1330}{7315} \approx .182. \quad \blacksquare$$

1. A jar contains one white and 4 red jelly beans. (a) What is the probability that out of two jelly beans selected at random from the jar, one will be white? (b) What is the probability of choosing 3 red jelly beans?

Answer:
(a) 2/5
(b) 2/5

Example 2 When shipping diesel engines abroad, it is common to pack 12 engines in one container which is then loaded on a rail car and sent to a port. Suppose that a company has received complaints from its customers that many of the engines arrive in nonworking condition. To help solve this problem, the company decides to make a spot check of containers after loading—the company will test three engines from a container at random; if any of the three is nonworking, the container will not be shipped until each engine in it is checked. Suppose a given container has two nonworking engines, and find the probability that the container will not be shipped.

The container will not be shipped if the sample of three engines contains one or two defective engines. Thus, letting $P(1 \text{ defective})$ represent the probability of exactly 1 defective engine in the sample,

$$P(\text{not shipping}) = P(1 \text{ defective}) + P(2 \text{ defectives}).$$

There are $\binom{12}{3}$ ways to choose the 3 engines for testing:

$$\binom{12}{3} = \frac{12!}{3!9!} = \frac{12(11)(10)}{3(2)(1)} = 220.$$

There are $\binom{2}{1}$ ways of choosing one defective engine from the two in the container, and for each of these ways, there are $\binom{10}{2}$ ways of choosing two good engines from among the ten in the container. By the multiplication principle, there are

$$\binom{2}{1}\binom{10}{2}$$

2. Calculate $\binom{2}{1}\binom{10}{2}$.

Answer:
90

ways of choosing a sample of 3 engines containing one defective. ②

Using the result from Problem 2 at the side,

$$P(1 \text{ defective}) = \frac{90}{220}.$$

There are $\binom{2}{2}$ ways of choosing two defective engines from the two defective engines in the container, and $\binom{10}{1}$ ways of choosing one good engine from among the ten good engines, giving

3. Calculate $\binom{2}{2}\binom{10}{1}$.

Answer:
10

$$\binom{2}{2}\binom{10}{1}$$

ways of choosing a sample of three engines containing two defectives. ③

Then, using the result from Problem 3 at the side,

$$P(2 \text{ defectives}) = \frac{10}{220}$$

and

$$P(\text{not shipping}) = P(1 \text{ defective}) + P(2 \text{ defectives})$$
$$= \frac{90}{220} + \frac{10}{220} = \frac{100}{220} \approx .455.$$

The probability is $1 - .455 = .545$ that the container *will* be shipped, even though it has two defective engines. The management must decide if this probability is acceptable; if not, it may be necessary to test more than three engines from a container.

Instead of finding the sum $P(1 \text{ defective}) + P(2 \text{ defectives})$, the result in Example 2 could be found by calculating $1 - P(\text{no defectives})$.

$$P(\text{not shipping}) = 1 - P(\text{no defectives in sample})$$

$$= 1 - \frac{\binom{2}{0}\binom{10}{3}}{\binom{12}{3}}$$

$$= 1 - \frac{1(120)}{220}$$

$$= 1 - \frac{120}{220} = \frac{100}{220} \approx .455 \quad \blacksquare$$

4 In Example 2, if a sample of two engines is tested, what is the probability that the container will not be shipped?

Answer:
.318

Example 3 In a common form of the card game *poker,* a hand of five cards is dealt to each player from a deck of 52 cards. There are a total of

$$\binom{52}{5} = \frac{52!}{5!47!} = 2{,}598{,}960$$

such hands possible. Find each of the following probabilities.
(a) A hand containing only hearts, called a *heart flush*
    There are 13 hearts in a deck; there are

$$\binom{13}{5} = \frac{13!}{5!8!} = \frac{13(12)(11)(10)(9)}{5(4)(3)(2)(1)} = 1287$$

different hands containing only hearts. The probability of a heart flush is

$$P(\text{heart flush}) = \frac{1287}{2{,}598{,}960} = \frac{33}{66{,}640} \approx .000495.$$

(b) A flush of any suit
There are four suits to a deck, so

$$P(\text{flush}) = 4 \cdot P(\text{heart flush}) = 4 \cdot \frac{33}{66,640} \approx .00198.$$

(c) A full house of aces and eights (three aces and two eights)
There are $\binom{4}{3}$ ways to choose 3 aces from among the 4 in the deck, and $\binom{4}{2}$ ways to choose 2 eights.

$$P(\text{3 aces, 2 eights}) = \frac{\binom{4}{3} \cdot \binom{4}{2}}{2,598,960} = \frac{1}{108,290} \approx .00000923$$

(d) Any full house (3 cards of one value, 2 of another)
There are 13 values in a deck, so there are 13 choices for the first value mentioned, leaving 12 choices for the second value (order *is* important here, since a full house of aces and eights, for example, is not the same as a full house of eights and aces). Since, from part (c), the probability of any particular full house is 1/108,290,

$$P(\text{full house}) = 13 \cdot 12 \cdot \left(\frac{1}{108,290} \right) = \frac{156}{108,290} \approx .00144. \quad \boxed{6}$$

5 Find the probability of being dealt a poker hand with four kings.

Answer:
.00001847

Example 4   A music teacher has three violin pupils, Fred, Carl, and Helen. For a recital, the teacher selects a first violinist and a second violinist. The third pupil will play with the others, but not solo. If the teacher selects randomly, what is the probability that Helen is first violinist, Carl is second violinist, and Fred does not solo?

Use permutations to find the number of arrangements in the sample space.

$$P(3, 3) = 3! = 6$$

The six arrangements are equally likely, since the teacher will select randomly. Thus, the required probability is 1/6. $\quad \boxed{6}$

6 An office manager must select 4 employees, A, B, C, and D, to work on a project. Each employee will be responsible for a specific task and one will be the coordinator. If the manager assigns the tasks randomly, what is the probability that B is the coordinator?

Answer:
1/4

Example 5   Suppose a group of n people is in a room. Find the probability that at least two of the people have the same birthday.
"Same birthday" refers to the month and the day, not necessarily the same year. Also, ignore leap years, and assume that each day in the year is equally likely as a birthday. First find the probability that *no two people* among five people have the same birthday. There are 365 different birthdays possible for the first of the five people, 364 for the second (so that the people have different birthdays), 363 for the third, and so on. The number of ways the five people can have different birthdays is thus the number of permutations of 365 things (days) taken 5 at a time or

$$P(365, 5) = 365 \cdot 364 \cdot 363 \cdot 362 \cdot 361.$$

The number of ways that the five people can have the same or different birthdays is

$$365 \cdot 365 \cdot 365 \cdot 365 \cdot 365 = (365)^5.$$

Finally, the *probability* that none of the five people have the same birthday is

$$\frac{P(365, 5)}{(365)^5} = \frac{365 \cdot 364 \cdot 363 \cdot 362 \cdot 361}{365 \cdot 365 \cdot 365 \cdot 365 \cdot 365} \approx .973.$$

The probability that at least two of the five people *do* have the same birthday is $1 - .973 = .027$.

This result can be extended for more than five people. In general, the probability that no two people among n people have the same birthday is

$$\frac{P(365, n)}{(365)^n}.$$

The probability that at least two of the n people *do* have the same birthday is

$$1 - \frac{P(365, n)}{(365)^n}. \quad \boxed{7}$$

[7] Evaluate

$$1 - \frac{P(365, n)}{(365)^n}$$

for

(a) $n = 3$;

(b) $n = 6$.

Answer:

(a) .008

(b) .040

The following table shows this probability for various values of n.

Number of people, n	Probability that two have the same birthday
5	.027
10	.117
15	.253
20	.411
22	.476
23	.507
25	.569
30	.706
35	.814
40	.891
50	.970
366	1

8 Set up (do not calculate) the probability that at least 2 of the 7 astronauts in the Mercury project have the same birthday.

Answer:

$1 - P(365, 7)/365^7$

The probability that 2 people among 23 have the same birthday is .507, a little more than half. Many people are surprised at this result—somehow it seems that a larger number of people should be required. **8**

9.2 Exercises

Management *A shipment of 9 typewriters contains 2 defectives. Find the probability that a sample of the following size, drawn from the 9, will not contain a defective. (See Example 2.)*

1. 1   **2.** 2   **3.** 3   **4.** 4

Management *Refer to Example 2. The management feels that the probability of .545 that a container will be shipped even though it contains two defectives is too high. They*

decide to increase the sample size chosen. Find the probability that a container will be shipped even though it contains two defectives if the sample size is increased to

5. 4;   **6.** 5.

A basket contains 6 red apples and 4 yellow apples. A sample of 3 apples is drawn. Find the probability that the sample contains each of the following. (See Example 1.)

7. All red apples

8. All yellow apples

9. 2 yellow and 1 red apple

10. More red than yellow apples

Two cards are drawn at random from an ordinary deck of 52 cards.

11. How many two card hands are possible?

Find the probability that the two card hand contains

12. two aces;

13. at least one ace;

14. all spades;

15. two cards of the same suit;

16. only face cards;

17. no face cards;

18. no card higher than 8 (count ace as 1).

Twenty-six slips of paper are each marked with a different letter of the alphabet, and placed in a basket. A slip is pulled out, its letter recorded (in the order in which the slip was drawn), and the slip replaced. This is done five times. Find the probabilities that the "word" formed

19. is "chuck";

20. starts with p;

21. has all different letters;

22. contains no x, y, or z.

Find the probability of the following hands at poker. Assume aces are either high or low. (See Example 3.)

23. Royal flush (five highest cards of a single suit)

24. Straight flush (five in a row in a single suit, but not a royal flush)

25. Four of a kind (four cards of the same value)

26. Straight (five cards in a row, not all of the same suit with ace either high or low)

A bridge hand is made up of 13 cards from a deck of 52. Set up the probability that a hand chosen at random

27. contains only hearts;

28. has four aces;

29. contains exactly three aces and exactly three kings;

30. has six of one suit, five of another, and two of another.

Work the following exercises. (See Examples 4 and 5.)

31. Set up the probability that at least two of the thirty-nine presidents of the United States have had the same birthday.

32. Estimate the probability that at least two of the one hundred U.S. senators have the same birthday.

33. Give the probability that two of the 435 members of the House of Representatives have the same birthday.

34. Show that the probability that in a group of n people *exactly one* pair have the same birthday is

$$\binom{n}{2} \cdot \frac{P(365, n - 1)}{(365)^n}.$$

35. To win a contest, a player must match four movie stars with their baby pictures. Suppose this is done at random. Find the probability of getting no matches correct; of getting exactly two correct.

36. A contractor has hired a decorator to send three different sofas out to the contractor's model homes each week for a year. The contractor does not want exactly the same three sofas sent out twice. Find the minimum number of sofas that the decorator will need.

37. An elevator has four passengers and stops at seven floors. It is equally likely that a person will get off at any one of the seven floors. Find the probability that no two passengers leave at the same floor.

The rest of these exercises involve the idea of a circular permutation: the number of ways of arranging distinct objects in a circle. The number of ways of arranging n distinct objects in a line is n!, but there are (n − 1)! ways for arranging the n items in a circle (see Exercise 44). Find the number of ways of arranging the following number of distinct items in a circle.

38. 4   **39.** 7   **40.** 10

41. Suppose that eight people sit at a circular table. Find the probability that two particular people are sitting next to each other.

42. A key ring contains seven keys: one black, one gold, and five silver. If the keys are arranged at random on the ring, find the probability that the black key is next to the gold key.

43. A circular table for a board of directors has ten seats for the ten attending members of the board. The chairman of the board always sits closest to the window. The vice president for sales, who is currently out of favor, will sit three positions to the chairman's left, since the chairman doesn't see so well out of his left eye. The chairman's daughter-in-law will sit opposite him. All other members take seats at random. Find the probability that a particular other member will sit next to the chairman.

44. Show that the number of ways of arranging n distinct items in a circle is $(n - 1)!$.

45. Rework Exercises 23–26 using the Monte Carlo method (see 8.4 Exercises 51–54) to approximate the

answers with $n = 25$. Since each hand has 5 cards, you will need $25 \cdot 5 = 125$ random numbers to "look at" 25 hands. Compare these experimental results with the theoretical results.

46. Rework Exercises 27–30 using the Monte Carlo method (see 8.4 Exercises 51–54) to approximate the answers with $n = 20$. Since each hand has 13 cards, you will need $20 \cdot 13 = 260$ random numbers to "look at" 20 hands.

9.3 Bernoulli Trials

Many probability problems are concerned with experiments in which an event is repeated many times. For example, we might want to find the probability of getting 7 heads in 8 tosses of a coin, or hitting a target 6 times out of 6, or finding 1 defective item in a sample of 15 items. Probability problems of this kind are called **repeated trials** problems, or **Bernoulli trials.** In each case, some outcome is designated a success, and any other outcome is considered a failure. Thus, if the probability of success in a single trial is p, the probability of failure will be $1 - p$. Repeated trials problems, or **binomial problems,** must satisfy the following conditions.

Bernoulli Trials

1. The same experiment is repeated several times.
2. There are only two possible outcomes, success or failure.
3. The probability of success remains the same for each trial.
4. The repeated trials are independent.

Consider the solution of a typical problem of this type: finding the probability of getting 5 ones on 5 rolls of a die. The probability of getting a one on 1 roll is 1/6, while the probability of any other result is 5/6.

$$P(5 \text{ ones on } 5 \text{ rolls}) = P(1) \cdot P(1) \cdot P(1) \cdot P(1) \cdot P(1) = \left(\frac{1}{6}\right)^5$$

$$\approx .00013$$

S	S	S	S	F
S	S	S	F	S
S	S	F	S	S
S	F	S	S	S
F	S	S	S	S

Now, let us find the probability of getting a one exactly 4 times in 5 rolls of the die. The desired outcome for this experiment can occur in more than one way as shown at the side, where S represents getting a success (a one) and F represents getting a failure (any other result). The probability of each of these five outcomes is

$$\left(\frac{1}{6}\right)^4\left(\frac{5}{6}\right).$$

Since the five outcomes represent mutually exclusive alternative events, add the five probabilities.

$$P(\text{4 ones in 5 rolls}) = 5\left(\frac{1}{6}\right)^4\left(\frac{5}{6}\right) = \frac{5^2}{6^5} \approx .0032$$

The probability of rolling a one exactly 3 times in 5 rolls of a die can be computed in the same way. The probability of 3 successes and 2 failures will be

$$\left(\frac{1}{6}\right)^3\left(\frac{5}{6}\right)^2.$$

Again the desired outcome can occur in more than one way. Let the set $\{1, 2, 3, 4, 5\}$ represent the first, second, third, fourth, and fifth tosses. The number of 3-element subsets of this set will correspond to the number of ways in which 3 successes and 2 failures can occur. Using combinations, there are $\binom{5}{3}$ such subsets. Since $\binom{5}{3} = 5!/(3!2!) = 10$,

$$P(\text{3 ones in 5 rolls}) = 10\left(\frac{1}{6}\right)^3\left(\frac{5}{6}\right)^2 = \frac{250}{6^5} \approx .032. \quad \boxed{1}$$

The results found above suggest the following generalization.

> **Probability in a Bernoulli Experiment**   If p is the probability of success in a single trial of a Bernoulli experiment, the probability of x successes and $n - x$ failures in n independent repeated trials of the experiment is
>
> $$\binom{n}{x}p^x(1 - p)^{n-x}.$$

Example 1   The advertising agency that handles the Diet Supercola account believes that 40% of all consumers prefer this product over its competitors. Suppose a random sample of six people is chosen. Assume that all responses are independent of each other. Find the probability of the following.

(a) Exactly 3 of the 6 people prefer Diet Supercola.

Use the result from the box above. In this example, $p = P(\text{success}) = P(\text{prefer Diet Supercola}) = .4$. The sample is made up of six people, so $n = 6$. To find the probability that exactly 3 people prefer this drink, let $x = 3$.

$$P(\text{exactly 3}) = \binom{6}{3}(.4)^3(1 - .4)^{6-3}$$

$$= 20(.4)^3(.6)^3 \qquad \text{Use Table 9}$$

$$= 20(.064)(.216)$$

$$= .27648$$

$\boxed{1}$ Find the probability of rolling a one

(a) exactly twice in 5 rolls of a die;

(b) exactly once in 5 rolls of a die.

Answer:

(a) .161

(b) .402

2 Eighty percent of all students at a certain school ski. If a sample of 5 students at this school is selected, and if their responses are independent, find the probability that exactly

(a) 1 of the 5 students skis;

(b) 4 of the 5 students ski.

Answer:

(a) $\binom{5}{1}(.8)^1(.2)^4 = .0064$

(b) $\binom{5}{4}(.8)^4(.2)^1 = .4096$

3 Find the probability of getting 2 fours in 8 tosses of a die.

Answer:

$\binom{8}{2}\left(\frac{1}{6}\right)^2\left(\frac{5}{6}\right)^6 \approx .2605$

4 Five percent of the clay pots fired in a certain way are defective. Find the probability of getting exactly 2 defective pots in a sample of 12.

Answer:

$\binom{12}{2}(.05)^2(.95)^{10} \approx .0988$

5 In Example 4, find the probability that out of 10 customers

(a) 6 or more buy the shoe;

(b) no more than 5 buy the shoe.

Answer:

(a) .0473490

(b) .9526510

(b) None of the 6 people prefers Diet Supercola.

Let $x = 0$.

$$P(\text{exactly } 0) = \binom{6}{0}(.4)^0(1 - .4)^6$$

$$= 1(1)(.6)^6$$

$$\approx .0467 \quad \boxed{2}$$

Example 2 Find the probability of getting exactly 7 heads in 8 tosses of a fair coin.

The probability of success, getting a head in a single toss, is 1/2, so the probability of failure, getting a tail, is 1/2.

$$P(\text{7 heads in 8 tosses}) = \binom{8}{7}\left(\frac{1}{2}\right)^7\left(\frac{1}{2}\right)^1 = 8\left(\frac{1}{2}\right)^8 = .03125 \quad \boxed{3}$$

Example 3 Assuming that selection of items for a sample can be treated as independent trials, find the probability of the occurrence of one defective item in a random sample of 15 items from a production line, if the probability that an item is defective is .01.

The probability of success (a defective item) is .01, while the probability of failure (an acceptable item) is .99. Thus,

$$P(\text{1 defective in 15 items}) = \binom{15}{1}(.01)^1(.99)^{14}$$

$$= 15(.01)(.99)^{14}$$

$$\approx .130. \quad \boxed{4}$$

Example 4 A new style of shoe is sweeping the country. In one area, 30% of all the shoes sold are of this new style. Assume that these sales are independent events. Find the following probabilities.

(a) Out of 10 customers in a shoe store, at least 8 buy the new shoe style.

For at least 8 people out of 10 to buy the shoe, it must be sold to 8, 9, or 10 people. This means

$$P(\text{at least 8}) = P(8) + P(9) + P(10)$$

$$= \binom{10}{8}(.3)^8(.7)^2 + \binom{10}{9}(.3)^9(.7)^1 + \binom{10}{10}(.3)^{10}(.7)^0$$

$$\approx .0014467 + .0001378 + .0000059$$

$$= .0015904.$$

(b) Out of 10 customers in a shoe store, no more than 7 buy the new shoe style.

"No more than 7" means 0, 1, 2, 3, 4, 5, 6, or 7 people buy the shoe. We could add $P(0)$, $P(1)$, and so on, but it is easier to use the formula $P(E) = 1 - P(E')$. The complement of "no more than 7" is "8 or more," so

$$P(\text{no more than 7}) = 1 - P(\text{8 or more})$$

$$\approx 1 - .0015904 \qquad \text{Answer from part (a)}$$

$$= .9984096. \quad \boxed{5}$$

9.3 Exercises

Suppose that a family has 5 children and that the probability of having a girl is 1/2. Find the probability that the family will have

1. exactly 2 girls;

2. exactly 3 girls;

3. no girls;

4. no boys;

5. at least 4 girls;

6. at least 3 boys;

7. no more than 3 boys;

8. no more than 4 girls.

A die is rolled 12 times. Find the probability of rolling

9. exactly 12 ones;

10. exactly 6 ones;

11. exactly 1 one;

12. exactly 2 ones;

13. no more than 3 ones;

14. no more than 1 one.

A coin is tossed 5 times. Find the probability of getting

15. all heads;

16. exactly 3 heads;

17. no more than 3 heads;

18. at least 3 heads.

Management *A factory tests a random sample of 20 transistors for defectives. The probability that a particular transistor will be defective has been established by past experience to be .05.*

19. What is the probability that there are no defectives in the sample?

20. What is the probability that the number of defectives in the sample is at most 2?

Management *A company gives prospective employees a 6-question multiple-choice test. Each question has 5 possible answers, so that there is a 1/5 or 20% chance of answering a question correctly just by guessing. Find the probability of answering*

21. exactly 2 questions correctly;

22. no questions correctly;

23. at least 4 correctly;

24. no more than 3 correctly.

25. Five out of the 50 clients of a certain stockbroker will lose their life savings as a result of his advice. Find the probability that in a sample of 3 clients, exactly 1 loses all his money. (Assume independence.)

Management *According to a recent article in a business publication, only 20% of the population of the United States*

has never had a Big Mac hamburger at McDonald's. Assume independence and find the probability that in a random sample of 10 people

26. exactly 2 never had a Big Mac;

27. exactly 5 never had a Big Mac;

28. 3 or fewer never had a Big Mac;

29. 4 or more *have* had a Big Mac.

Natural Science *A new drug cures 70% of the people taking it. Suppose 20 people take the drug; find the probability that*

30. exactly 18 are cured;

31. exactly 17 are cured;

32. at least 17 are cured;

33. at least 18 are cured.

In a 10-question multiple-choice biology test with 5 choices for each question, an unprepared student guesses on each item. Find the probability that he answers

34. exactly 6 questions correctly;

35. exactly 7 correctly;

36. at least 8 correctly;

37. fewer than 8 correctly.

Natural Science *Assume that the probability that a person will die within a month after a certain operation is 20%. Find the probability that in 3 such operations*

38. all 3 people survive;

39. exactly 1 person survives;

40. at least 2 people survive;

41. no more than 1 person survives.

Natural Science *Six mice from the same litter, all suffering from a vitamin A deficiency are fed a certain dose of carrots. If the probability of recovery under such treatment is .70, find the probability that*

42. none recover;

43. exactly 3 of the 6 recover;

44. all recover;

45. no more than 3 recover.

46. Natural Science In an experiment on the effects of a radiation dose on cells, a beam of radioactive particles is aimed at a group of 10 cells. Find the probability that 8 of the cells will be hit by the beam, if the probability that any single cell will be hit is .6. (Assume independence.)

47. Natural Science The probability of a mutation of a given gene under a dose of 1 roentgen of radiation is approximately 2.5×10^{-7}. What is the probability that in 10,000 genes, at least 1 mutation occurs?

48. Natural Science A new drug being tested causes a serious side effect in 5 out of 100 patients. What is the probability that no side effects occur in a sample of 10 patients taking the drug?

Calculate each of the following probabilities.

49. A flu vaccine has a probability of 80% of preventing a person who is inoculated from getting flu. A county health office inoculates 134 people. What is the probability that
(a) exactly 10 of them get the flu?
(b) no more than 10 get the flu?
(c) none of them get the flu?

50. The probability that a male will be color-blind is .042. What is the probability that in a group of 53 men.
(a) exactly 5 are color-blind?
(b) no more than 5 are color-blind?
(c) at least 1 is color-blind?

51. The probability that a certain machine turns out a defective item is .05. What is the probability that in a run of 75 items
(a) exactly 5 defectives are produced?
(b) no defectives are produced?
(c) at least 1 defective is produced?

52. A company is taking a survey to find out if people like their product. Their last survey indicated that 70% of the population like their product. Based on that, of a sample of 58 people, what is the probability that
(a) all 58 like the product?
(b) from 28 to 30 (inclusive) like the product?

9.4 Markov Chains

This section introduces a special type of repeated trials experiment, called a **Markov chain,** where the outcome of a trial depends only on the outcome of the previous trial. That is, given the present state of the system, future states are independent of past states. Such experiments are common enough in applications to make their study worthwhile.

Markov chains are named after the Russian mathematician A. A. Markov, 1856–1922, who started the theory of such processes. To see how Markov chains work, we look at an example.

Example 1 A small town has only two dry cleaners, Johnson and NorthClean. Johnson's manager desires to increase the firm's market share by an extensive advertising campaign. After the campaign, a market research firm finds that there is a probability of .8 that a Johnson customer will bring his next batch of dirty items to Johnson, and a .35 chance that a NorthClean customer will switch to Johnson for his next batch. Assume that the probability that a customer comes to a given cleaners depends only on where the last load of clothes was taken. If there is an .8 chance that a Johnson customer will return to Johnson, then there must be a $1 - .8 = .2$ chance that the customer will switch to NorthClean. In the same way, there is a

$1 - .35 = .65$ chance that a NorthClean customer will return to NorthClean. These probabilities give the following *transition matrix*.

Second load

Johnson   NorthClean

First load   Johnson $\begin{bmatrix} .8 & .2 \\ .35 & .65 \end{bmatrix}$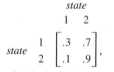
NorthClean

1 Given the transition matrix

state
1   2

state $\begin{array}{c} 1 \\ 2 \end{array} \begin{bmatrix} .3 & .7 \\ .1 & .9 \end{bmatrix}$,

(a) what is the probability of changing from state 1 to state 2?

(b) what does the number .1 represent?

Answer:

(a) .7

(b) The probability of changing from state 2 to state 1

A **transition matrix** has the following features.

1. It is square, since all possible states must be used both as rows and as columns;

2. all entries are between 0 and 1, inclusive, because all entries represent probabilities;

3. the sum of the entries in any row must be 1, since the numbers in the row give the probability of changing from the state at the left to one of the states indicated across the top.

Example 2 Suppose that when the new promotional campaign began, Johnson had 40% of the market and NorthClean had 60%. Use the tree diagram of Figure 9.5 to find how these proportions would change after another week of advertising.

Add the numbers indicated with arrows to find the proportion of people taking their cleaning to Johnson after one week.

$$.32 + .21 = .53$$

Similarly, the proportion taking their cleaning to NorthClean is

$$.08 + .39 = .47.$$

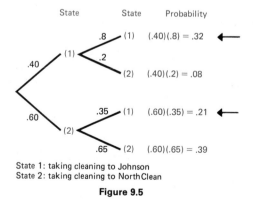

State 1: taking cleaning to Johnson
State 2: taking cleaning to NorthClean

Figure 9.5

The initial distribution of 40% and 60% becomes, after one week, 53% and

47%. These distributions can be written as the probability vectors

$$[.40 \quad .60] \quad \text{and} \quad [.53 \quad .47].$$

A **probability vector** is a matrix of only one row, having nonnegative entries with the sum of the entries equal to 1.

The results from the tree diagram above are exactly the same as the result of multiplying the initial probability vector by the transition matrix:

$$[.4 \quad .6] \begin{bmatrix} .8 & .2 \\ .35 & .65 \end{bmatrix} = [.53 \quad .47].$$

Find the market share after two weeks by multiplying the vector $[.53 \quad .47]$ and the transition matrix. ②

2 Find the product

$$[.53 \quad .47] \begin{bmatrix} .8 & .2 \\ .35 & .65 \end{bmatrix}.$$

Answer:
$[.59 \quad .41]$ (rounded)

The product from Problem 2 at the side (with the numbers rounded to the nearest hundredth) shows that after two weeks of advertising, Johnson's share of the market has increased to 59%. To get the share after three weeks, multiply the probability vector $[.59 \quad .41]$ and the transition matrix. Continuing this process gives each cleaner's share of the market after any number of weeks. ③

3 Find each cleaner's market share after 3 weeks.

Answer:
$[.62 \quad .38]$

The following table gives the market share for each cleaner at the end of week n for several values of n.

Week	Johnson	NorthClean
Start	.4	.6
1	.53	.47
2	.59	.41
3	.62	.38
4	.63	.37
5	.63	.37
12	.64	.36

The results seem to approach the probability vector $[.64 \quad .36]$. This vector is called the **equilibrium vector** or **fixed vector** for the given transition matrix. The equilibrium vector gives a long range prediction—the shares of the market will stabilize (under the same conditions) at 64% for Johnson and 36% for NorthClean. ■

Starting with some other initial probability vector and going through the steps above would give the same equilibrium vector. In fact, the long range trend is the same no matter what the initial vector is. The long range trend depends only on the transition matrix, not on the initial distribution.

One of the many applications of Markov chains is in finding these long-range predictions. It is not possible to make long-range predictions with all transition matrices, but there is a large set of transition matrices for which long-range predictions *are* possible. Such predictions are always possible with *regular transition matrices*. A transition matrix is **regular** if some power of the matrix contains all positive entries. A Markov chain is a **regular Markov chain** if its transition matrix is regular.

Example 3 Decide if the following transition matrices are regular.

(a) $A = \begin{bmatrix} .75 & .25 & 0 \\ 0 & .5 & .5 \\ .6 & .4 & 0 \end{bmatrix}$

Square the matrix A by multiplying it by itself.

$$A^2 = \begin{bmatrix} .5625 & .3125 & .125 \\ .3 & .45 & .25 \\ .45 & .35 & .2 \end{bmatrix}$$

All entries in A^2 are positive, so that matrix A is regular.

(b) $B = \begin{bmatrix} .5 & 0 & .5 \\ 0 & 1 & 0 \\ 0 & 0 & 1 \end{bmatrix}$

Find various powers of B.

$$B^2 = \begin{bmatrix} .25 & 0 & .75 \\ 0 & 1 & 0 \\ 0 & 0 & 1 \end{bmatrix} \qquad B^3 = \begin{bmatrix} .125 & 0 & .875 \\ 0 & 1 & 0 \\ 0 & 0 & 1 \end{bmatrix}$$

$$B^4 = \begin{bmatrix} .0625 & 0 & .9375 \\ 0 & 1 & 0 \\ 0 & 0 & 1 \end{bmatrix}$$

Further powers of B will still give the same zero entries, so that no power of matrix B contains all positive entries. For this reason, B is not regular. **4**

4 Decide if the following transition matrices are regular.

(a) $\begin{bmatrix} 0 & 1 \\ 1 & 0 \end{bmatrix}$

(b) $\begin{bmatrix} .3 & .7 \\ 1 & 0 \end{bmatrix}$

Answer:
(a) No
(b) Yes

The equilibrium vector can be found without doing all the work shown above. Let $V = [v_1 \quad v_2]$ represent the desired vector. Let P represent the transition matrix. By definition, V is the fixed probability vector if $VP = V$. In our example,

$$[v_1 \quad v_2] \begin{bmatrix} .8 & .2 \\ .35 & .65 \end{bmatrix} = [v_1 \quad v_2].$$

Since $[v_1 \quad v_2]$ is a probability vector, $v_1 + v_2 = 1$. Multiply on the left to get

$$[.8v_1 + .35v_2 \quad .2v_1 + .65v_2] = [v_1 \quad v_2].$$

Set corresponding entries from the two matrices equal to get

$$.8v_1 + .35v_2 = v_1 \qquad .2v_1 + .65v_2 = v_2.$$

Simplify each of these equations.

$$-.2v_1 + .35v_2 = 0 \qquad .2v_1 - .35v_2 = 0$$

These last two equations are really the same. (The equations in the system obtained from $VP = V$ are always dependent.) To find the values of v_1 and v_2, recall that

$V = [v_1 \quad v_2]$ is a probability vector, so that

$$v_1 + v_2 = 1.$$

Find v_1 and v_2 by solving the system

$$-.2v_1 + .35v_2 = 0$$
$$v_1 + v_2 = 1.$$

From the second equation, $v_1 = 1 - v_2$. Substitute for v_1 in the first equation:

$$-.2(1 - v_2) + .35v_2 = 0. \quad \boxed{5}$$

Since $v_2 = .364$ (from Problem 5 at the side) and $v_1 = 1 - v_2$, then $v_1 = 1 - .364 = .636$ and the equilibrium vector is $[.636 \quad .364]$.

Example 4 Find the equilibrium vector for the transition matrix

$$P = \begin{bmatrix} \frac{1}{3} & \frac{2}{3} \\ \frac{3}{4} & \frac{1}{4} \end{bmatrix}.$$

Look for a vector $[v_1 \quad v_2]$ such that

$$[v_1 \quad v_2] \begin{bmatrix} \frac{1}{3} & \frac{2}{3} \\ \frac{3}{4} & \frac{1}{4} \end{bmatrix} = [v_1 \quad v_2].$$

Multiply on the left, obtaining the two equations

$$\frac{1}{3}v_1 + \frac{3}{4}v_2 = v_1 \qquad \frac{2}{3}v_1 + \frac{1}{4}v_2 = v_2.$$

Simplify each of these equations.

$$-\frac{2}{3}v_1 + \frac{3}{4}v_2 = 0 \qquad \frac{2}{3}v_1 - \frac{3}{4}v_2 = 0$$

Since $v_1 + v_2 = 1$, then $v_1 = 1 - v_2$. Substitute this into either of the two given equations. Using the first equation gives

$$-\frac{2}{3}(1 - v_2) + \frac{3}{4}v_2 = 0$$

$$-\frac{2}{3} + \frac{2}{3}v_2 + \frac{3}{4}v_2 = 0$$

$$-\frac{2}{3} + \frac{17}{12}v_2 = 0$$

$$\frac{17}{12}v_2 = \frac{2}{3}$$

$$v_2 = \frac{8}{17},$$

and $v_1 = 1 - 8/17 = 9/17$. The equilibrium vector is $[9/17 \quad 8/17]$. $\quad \blacksquare$ $\boxed{6}$

$\boxed{5}$ Solve the equation for v_2. Round to the nearest thousandth.

Answer:

.364

$\boxed{6}$ Find the equilibrium vector for the transition matrix

$$P = \begin{bmatrix} .3 & .7 \\ .5 & .5 \end{bmatrix}.$$

Answer:

$[5/12 \quad 7/12]$

Example 5 The probability that a complex assembly line works correctly depends on whether or not the line worked correctly the last time it was used. The various probabilities are as given in the following transition matrix.

$$\begin{array}{cc} & \text{Works properly now} \quad \text{Does not} \\ \begin{array}{l} \text{Worked properly before} \\ \text{Did not} \end{array} & \begin{bmatrix} .9 & .1 \\ .7 & .3 \end{bmatrix} \end{array}$$

Find the long-range probability that the assembly line will work properly.

Begin by finding the equilibrium vector $[v_1 \quad v_2]$, where

$$[v_1 \quad v_2]\begin{bmatrix} .9 & .1 \\ .7 & .3 \end{bmatrix} = [v_1 \quad v_2].$$

Multiplying on the left and setting corresponding entries equal gives the equations

$$.9v_1 + .7v_2 = v_1 \quad \text{and} \quad .1v_1 + .3v_2 = v_2$$

or

$$-.1v_1 + .7v_2 = 0 \quad \text{and} \quad .1v_1 - .7v_2 = 0.$$

Substitute $v_1 = 1 - v_2$ in the first of these equations to get

$$-.1(1 - v_2) + .7v_2 = 0$$
$$-.1 + .1v_2 + .7v_2 = 0$$
$$-.1 + .8v_2 = 0$$
$$.8v_2 = .1$$
$$v_2 = \frac{1}{8},$$

and $v_1 = 1 - 1/8 = 7/8$. The equilibrium vector is $[7/8 \quad 1/8]$. In the long run, the company can expect the assembly line to run properly 7/8 of the time. **7**

7 In Example 3, suppose the company modifies the line so that the transition matrix becomes

$$\begin{bmatrix} .95 & .05 \\ .8 & .2 \end{bmatrix}.$$

Find the long range probability that the assembly line will work properly.

Answer:
16/17

The examples suggest the following generalization. If a regular Markov chain has transition matrix P and initial probability vector v, the products

$$vP$$
$$vPP = vP^2$$
$$vP^2P = vP^3$$
$$\vdots$$
$$vP^n$$

lead to the equilibrium vector V. The results of this section can be summarized as follows.

Suppose a regular Markov chain has a transition matrix P.

1. For any initial probability vector v, the products vP^n approach a unique vector V as n gets larger and larger. Vector V is called the **equilibrium** or **fixed vector.**

2. Vector V has the property that $VP = V$.

3. Find V by solving a system of equations obtained from the matrix equation $VP = V$ and from the fact that the sum of the entries of V is 1.

9.4 Exercises

Which of the following could be a probability vector?

1. $\begin{bmatrix} \frac{2}{3} & \frac{1}{3} \end{bmatrix}$

2. $\begin{bmatrix} \frac{1}{2} & 1 \end{bmatrix}$

3. $\begin{bmatrix} 0 & 1 \end{bmatrix}$

4. $\begin{bmatrix} .1 & .1 \end{bmatrix}$

5. $\begin{bmatrix} .4 & .2 & 0 \end{bmatrix}$

6. $\begin{bmatrix} \frac{1}{4} & \frac{1}{8} & \frac{5}{8} \end{bmatrix}$

Which of the following could be a transition matrix, by definition?

7. $\begin{bmatrix} .5 & 0 \\ 0 & .5 \end{bmatrix}$

8. $\begin{bmatrix} \frac{2}{3} & \frac{1}{3} \\ 1 & 0 \end{bmatrix}$

9. $\begin{bmatrix} \frac{1}{4} & \frac{3}{4} \\ \frac{1}{2} & \frac{1}{2} \end{bmatrix}$

10. $\begin{bmatrix} \frac{1}{4} & \frac{3}{4} & 0 \\ 2 & 0 & 1 \\ 1 & \frac{2}{3} & 3 \end{bmatrix}$

11. $\begin{bmatrix} \frac{1}{3} & \frac{1}{3} & \frac{1}{3} \\ 0 & 1 & 0 \\ \frac{1}{2} & 0 & \frac{1}{2} \end{bmatrix}$

12. $\begin{bmatrix} \frac{1}{3} & \frac{1}{2} & 1 \\ 0 & 1 & 0 \\ \frac{1}{2} & 0 & \frac{1}{2} \end{bmatrix}$

13. $\begin{bmatrix} \frac{1}{3} & \frac{1}{2} & 1 \\ \frac{1}{3} & 0 & 0 \\ \frac{1}{3} & \frac{1}{2} & 0 \end{bmatrix}$

14. $\begin{bmatrix} .9 & .1 & 0 \\ .1 & .6 & .3 \\ 0 & .3 & .7 \end{bmatrix}$

Which of the following transition matrices are regular? (See Example 3.)

15. $\begin{bmatrix} .2 & .8 \\ .9 & .1 \end{bmatrix}$

16. $\begin{bmatrix} .22 & .78 \\ .43 & .57 \end{bmatrix}$

17. $\begin{bmatrix} 1 & 0 \\ .6 & .4 \end{bmatrix}$

18. $\begin{bmatrix} .55 & .45 \\ 0 & 1 \end{bmatrix}$

19. $\begin{bmatrix} 0 & 1 & 0 \\ .4 & .2 & .4 \\ 1 & 0 & 0 \end{bmatrix}$

20. $\begin{bmatrix} .3 & .5 & .2 \\ 1 & 0 & 0 \\ .5 & .1 & .4 \end{bmatrix}$

Find the equilibrium vector for each of the following transition matrices. (See Examples 4 and 5.)

21. $\begin{bmatrix} \frac{1}{4} & \frac{3}{4} \\ \frac{1}{2} & \frac{1}{2} \end{bmatrix}$

22. $\begin{bmatrix} \frac{2}{3} & \frac{1}{3} \\ \frac{1}{8} & \frac{7}{8} \end{bmatrix}$

23. $\begin{bmatrix} .3 & .7 \\ .4 & .6 \end{bmatrix}$

24. $\begin{bmatrix} .8 & .2 \\ .1 & .9 \end{bmatrix}$

25. $\begin{bmatrix} .1 & .1 & .8 \\ .4 & .4 & .2 \\ .1 & .2 & .7 \end{bmatrix}$

26. $\begin{bmatrix} .5 & .2 & .3 \\ .1 & .4 & .5 \\ .2 & .2 & .6 \end{bmatrix}$

27. $\begin{bmatrix} .25 & .35 & .4 \\ .1 & .3 & .6 \\ .55 & .4 & .05 \end{bmatrix}$

28. $\begin{bmatrix} .16 & .28 & .56 \\ .43 & .12 & .45 \\ .86 & .05 & .09 \end{bmatrix}$

29. Management Years ago, about 10% of all cars sold were small, while 90% were large. This has changed drastically over the years; 20% of small car owners have switched to large cars, while 60% of large car owners have switched to small cars.

(a) Write a transition matrix for this information.

(b) Write a probability vector for the initial distribution of cars.

(c) What percent of the market will each class of automobile have in 4 years?

(d) What is the long-range prediction?

30. Management Each month, a sales manager classifies her salespeople as low, medium, or high producers. There is a .4 chance that a low producer one month will become a medium producer the following month, and a .1 chance that a low producer will become a high producer. A medium producer will become a low or high producer, respectively, with probabilities .25 and .3. A high producer will become a low or medium producer, respectively, with probabilities .05 and .4. Find the long-range trend for the proportion of low, medium, and high producers.

31. Management An insurance company classifies its drivers into three groups: G_0 (no accidents), G_1 (one accident), and G_2 (more than one accident). The probability that a driver in G_0 will stay in G_0 after one year is .85, that he will become a G_1 is .10, and that he will become a G_2 is .05. A driver in G_1 cannot move to G_0 (this insurance company has a long memory). There is an .80 probability that a G_1 driver will stay in G_1 and a .20 probability that he will become a G_2. A driver in G_2 must stay in G_2.
(a) Write a transition matrix using this information. Suppose that the company accepts 50,000 new policyholders, all of whom are in group G_0. Find the number in each group
(b) after 1 year;   **(c)** after 2 years;
(d) after 3 years;   **(e)** after 4 years.
(f) Find the equilibrium vector here. Interpret your result.

32. Management The difficulty with the mathematical model of Exercise 31 is that no "grace period" is provided; there should be a certain probability of moving from G_1 or G_2 back to G_0 (say, after four years with no accidents). A new system with this feature might produce the following transition matrix.

$$\begin{bmatrix} .85 & .10 & .05 \\ .15 & .75 & .10 \\ .10 & .30 & .60 \end{bmatrix}$$

Suppose that when this new policy is adopted, the company has 50,000 policyholders in group G_0. Find the number of these in each group
(a) after 1 year;   **(b)** after 2 years;
(c) after 3 years.
(d) Find the equilibrium vector here. Interpret your result.

33. Management Research done by the Gulf Oil Corporation* produced the following transition matrix for the probability that a person with one form of home heating would switch to another.

Will switch to

		Oil	Gas	Electric
	Oil	.825	.175	0
Now has	Gas	.060	.919	.021
	Electric	.049	0	.951

The current share of the market held by these three types of heat is given by [.26   .60   .14]. Find the share of the market held by each type of heat after
(a) 1 year;   **(b)** 2 years;   **(c)** 3 years.
(d) What is the long-range prediction?

34. Natural Science A certain genetic defect is carried only by males. Suppose the probability is .95 that a male offspring will have the defect if his father did, and .10 that a male offspring will have it if his father did not. Find the long-range prediction for the fraction of the males in the population who will have this defect.

35. Natural Science A large group of mice is kept in a cage having connected compartments A, B, and C. Mice in compartment A move to B with probability .3 and to C with probability .4. Mice in B move to A or C with probabilities of .15 and .55, respectively. Mice in C move to A and B with probabilities of .3 and .6, respectively. Find the long-range prediction for the fraction of mice in each of the compartments.

36. Social Science In a survey investigating change in housing patterns in one urban area, it was found that 75% of the population lived in single-family dwellings and 25% in multiple housing of some kind. Five years later, in a follow-up survey, of those who had been living in single-family dwellings, 90% still did so, but 10% had moved to multiple-family dwellings. Of those in multiple housing, 95% were still living in that type

---

*From "Forecasting Market Shares of Alternative Home-Heating Units by Markov Process Using Transition Probabilities Estimates from Aggregate Time Series Data" by Ali Ezzati, from *Management Science,* Vol. 21, No. 4, December 1974. Copyright © 1974 The Institute of Management Sciences. Reprinted by permission.

of housing, while 5% had moved to single-family dwellings. Assume that these trends continue.
(a) Write a transition matrix for this information.
(b) Write a probability vector for the initial distribution of housing.
(c) What percent of the population can be expected in each category five years later?
(d) What is the long-range prediction?

37. **Social Science** At the end of June, in a presidential election year, 40% of the electorate was registered as liberal, 45% as conservative, and 15% as independent. Over a one-month period, the liberals retained 80% of their constituency, while 15% switched to conservative and 5% to independent. The conservatives retained 70%, lost 20% to the liberals, and 10% to the independents. The independents retained 60% and lost 20% each to the conservatives and liberals. Assume that these trends continue.
(a) Write a 3 × 3 transition matrix showing the information of the problem.
(b) Write a probability vector for the initial distribution.
(c) Find the percent of each type of voter at the end of July.
(d) Find the percent of each type of voter at the end of August.
(e) Find the percent of each type of voter at the end of September.
(f) Find the long-range prediction.

38. **Social Science** In one state, a land-use survey showed that 35% of the land was used for agricultural purposes, while 10% was urban. Ten years later, of the agricultural land, 15% had become urban and 80% had remained agricultural. (The remainder lay idle.) Of the idle land, 20% had become urbanized and 10% had been converted for agricultural use. Of the urban land, 90% remained urban and 10% was idle. Assume that these trends continue.
(a) Write a transition matrix for this information.
(b) Write a probability vector for the initial distribution of land.
(c) What is the land-use pattern after twenty years?
(d) What is the long-range prediction for land use in this state?

39. **Social Science** At one liberal arts college, students are classified as humanities majors, science majors, or

undecideds. There is a 20% chance that a humanities major will change to a science major from one year to the next, and a 45% chance that a humanities major will change to undecided. A science major will change to humanities with probability .15, and to undecided with probability .35. An undecided will switch to humanities or science with probabilities of .5 and .3 respectively. Find the long-range prediction for the fraction of students in each of these three majors.

40. **Natural Science** The weather in a certain spot is classified as fair, cloudy without rain, or rainy. A fair day is followed by a fair day 60% of the time, and by a cloudy day 25% of the time. A cloudy day is followed by a cloudy day 35% of the time, and by a rainy day 25% of the time. A rainy day is followed by a cloudy day 40% of the time, and by another rainy day 25% of the time. Find the long range prediction for the proportion of fair, cloudy, and rainy days.

For each of the following transition matrices, find the first five powers of the matrix. Then find the probability that state 2 changes to state 4 after 5 repetitions of the experiment.

41. $\begin{bmatrix} .1 & .2 & .2 & .3 & .2 \\ .2 & .1 & .1 & .2 & .4 \\ .2 & .1 & .4 & .2 & .1 \\ .3 & .1 & .1 & .2 & .3 \\ .1 & .3 & .1 & .1 & .4 \end{bmatrix}$

42. $\begin{bmatrix} .3 & .2 & .3 & .1 & .1 \\ .4 & .2 & .1 & .2 & .1 \\ .1 & .3 & .2 & .2 & .2 \\ .2 & .1 & .3 & .2 & .2 \\ .1 & .1 & .4 & .2 & .2 \end{bmatrix}$

43. A company with a new training program classified each employee in one of the four states: s_1, never in the program; s_2, currently in the program; s_3, discharged; s_4, completed the program. The transition matrix for this company is given below.

$$\begin{array}{c} \\ s_1 \\ s_2 \\ s_3 \\ s_4 \end{array} \begin{array}{cccc} s_1 & s_2 & s_3 & s_4 \end{array} \\ \begin{bmatrix} .4 & .2 & .05 & .35 \\ 0 & .45 & .05 & .5 \\ 0 & 0 & 1.0 & 0 \\ 0 & 0 & 0 & 1.0 \end{bmatrix}$$

(a) What percent of employees who had never been in the program (state s_1) completed the program (state s_4) after the program had been offered five times?

(b) If the initial percent of employees in each state was [.5 .5 0 0], find the corresponding percents after the program had been offered four times.

Find the equilibrium vector for the transition matrices in Exercises 44 and 45 by taking powers of the matrix.

44. $\begin{bmatrix} .1 & .2 & .2 & .3 & .2 \\ .2 & .1 & .1 & .2 & .4 \\ .2 & .1 & .4 & .2 & .1 \\ .3 & .1 & .1 & .2 & .3 \\ .1 & .3 & .1 & .1 & .4 \end{bmatrix}$

45. $\begin{bmatrix} .3 & .2 & .3 & .1 & .1 \\ .4 & .2 & .1 & .2 & .1 \\ .1 & .3 & .2 & .2 & .2 \\ .2 & .1 & .3 & .2 & .2 \\ .1 & .1 & .4 & .2 & .2 \end{bmatrix}$

46. Find the long range prediction for the percent of employees in each state for the company training program from Exercise 43.

9.5 Expected Value

In a recent year, a citizen of the United States could expect to complete 12 years of school, to earn $20,091 per year, and to live in a household of 2.7 persons. What is meant here by "expect"? Many people have completed less than 12 years of school, while many others have completed more. Many people earn less than $20,091 per year; others earn more. The idea of a household of 2.7 members is a little hard to swallow. These numbers all refer to *averages*. When the term "expect" is used in this way, it refers to *mathematical expectation,* which is a kind of average.

Suppose we wish to find the expected repair time (to the nearest hour) required to repair a certain machine, given the following information.

Repair time (to nearest hour)	1	2	3	4
Probability that repair time is required	.22	.29	.32	.17

The table shows, for example, that the probability that repairs will take 3 hours is .32. It might be tempting to find the expected number of hours by averaging the numbers 1, 2, 3, and 4. This won't work, however, since the numbers do not occur equally often; for example, a 3-hour job is more likely than a 1-hour job. The differing probabilities of occurrence can be taken into account with a **weighted average,** found by multiplying each repair time in the table by its corresponding probability.

$$\text{Expected number of hours} = 1(.22) + 2(.29) + 3(.32) + 4(.17)$$
$$= .22 + .58 + .96 + .68$$
$$= 2.44$$

Based on the given data, the expected number of hours required to repair the machine is 2.44. ①

The idea of a weighted average is used to define *expected value*, or *mathematical expectation*, as follows.

① Find the weighted average.

(a)

x	2	3	4	5
P(x)	.3	.2	.4	.1

(b)

x	2	4	6	8
P(x)	.1	.4	.2	.3

Answer:
(a) 3.3
(b) 5.4

Expected Value Suppose an experiment has n outcomes, $x_1, x_2, x_3, \ldots, x_n$, with corresponding probabilities of $p_1, p_2, p_3, \ldots, p_n$. Then the expected value of the outcomes is

$$x_1 p_1 + x_2 p_2 + x_3 p_3 + \cdots + x_n p_n.$$

Example 1 The local church decides to raise money by raffling a microwave oven worth $400. A total of 2000 tickets are sold at $1 each. Find the expected value of winning for a person who buys one ticket in the raffle.

Here the possible outcomes are the possible amounts of net winnings, where net winnings = amount of winning − cost of ticket. The net winnings of the person winning the oven are $400 (amount of winning) − $1 (cost of ticket) = $399. The net winnings for each losing ticket are $0 − $1 = −$1.

The probability of winning is 1/2000, while the probability of losing is 1999/2000. This information is shown in the following chart.

Outcome (net winnings)	$399	−$1
Probability	$\dfrac{1}{2000}$	$\dfrac{1999}{2000}$

The expected net winnings for a person buying one ticket are

$$399\left(\frac{1}{2000}\right) + (-1)\left(\frac{1999}{2000}\right) = \frac{399}{2000} - \frac{1999}{2000} = -\frac{1600}{2000} = -.80.$$

On the average, a person buying one ticket in the raffle will lose $.80, or 80¢.

It is not possible to lose 80¢ in this raffle—you either lose $1, or you win a $400 prize. However, if you bought tickets in many such raffles over a long period of time, you would lose 80¢ per ticket, on the average. ②

② Suppose you buy one of 1000 tickets at 10¢ each in a lottery where the first prize is $50. What are your expected net winnings?

Answer:
−5¢

Example 2 What is the expected number of girls in a family with three children?

Some families with three children will have 0 girls, others will have 1 girl, and so on. We need to find the probabilities associated with 0, 1, 2, or 3 girls in a family of three children. These probabilties can be found by first writing the sample space S of all possible three-child families: $S = \{ggg, ggb, bgg, gbb, bgb, bbg, bbb, gbg\}$. This sample space leads to the probabilities shown in the following chart, assuming that the probability of a girl at each birth is 1/2.

Outcome (number of girls)	0	1	2	3
Probability	$\dfrac{1}{8}$	$\dfrac{3}{8}$	$\dfrac{3}{8}$	$\dfrac{1}{8}$

The expected number of girls can now be found by multiplying each outcome (number of girls) by its corresponding probability and finding the sum of these values.

$$\text{Expected number of girls} = 0 \cdot \frac{1}{8} + 1 \cdot \frac{3}{8} + 2 \cdot \frac{3}{8} + 3 \cdot \frac{1}{8}$$

$$= \frac{3}{8} + \frac{6}{8} + \frac{3}{8}$$

$$= \frac{12}{8} = \frac{3}{2} = 1\frac{1}{2}$$

On the average, a three-child family will have 1 1/2 girls. ▪

3 Find the expected outcome for throwing one die. Begin by completing this table.

Outcome	1	2	3	4	5	6
Probability	$\frac{1}{6}$					

Answer:

All probabilities are 1/6; expected outcome is 3 1/2.

Example 3 Each day Donna and Mary toss a coin to see who buys the coffee (40¢ a cup). One tosses and the other calls the outcome. If the person who calls the outcome is correct, the other buys the coffee; otherwise the caller pays. Find Donna's expected winnings.

Let's assume that an honest coin is used, that Mary tosses the coin, and that Donna calls the outcome. The possible results and corresponding probabilities are shown in the tree diagram of Figure 9.6. Donna wins a 40¢ cup of coffee whenever the results and calls match, and loses a 40¢ cup when there is no match. Her expected winnings are

$$E = (.40)\left(\frac{1}{4}\right) + (-.40)\left(\frac{1}{4}\right) + (-.40)\left(\frac{1}{4}\right) + (.40)\left(\frac{1}{4}\right) = 0.$$

On the average, over the long run, Donna neither wins nor loses. ▪

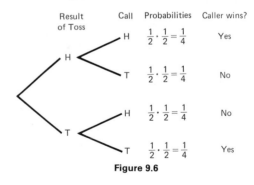

Figure 9.6

A game with an expected value of 0 (such as the one of Example 3) is called a **fair game.** Casinos do not offer fair games. If they did, they would win on the average $0, and have a hard time paying the help! Casino games have expected winnings for the house that vary from 1 1/2¢ per dollar to 60¢ per dollar. Exercises 14–17 below show how to find the expected value for certain games of chance.

The idea of expected value can be very useful in decision making, as shown by the next example.

4 A person can take one of two jobs. With job A, there is a 50% chance of making $60,000 per year after 5 years and a 50% change of making $30,000. With job B, there is a 30% chance of making $90,000 per year after 5 years and a 70% chance of making $20,000. Based strictly on expected value, which job should be taken?

Answer:
Job A has an expected salary of $45,000; job B, $41,000. Take job A.

Example 4 At age 50, you receive a letter from the Mutual of Mauritania Insurance Company. According to the letter, you must tell the company immediately which of the following two options you will choose: take $20,000 at age 60 (if you are alive, $0 otherwise) or $30,000 at age 70 (again, if you are alive, $0 otherwise). Based only on the idea of expected value, which should you choose?

Life insurance companies have constructed elaborate tables showing the probability of a person living a given number of years into the future. Suppose the most recent such table shows that the probability of living from age 50 to age 60 is .88, while the probability of living from age 50 to age 70 is .64. The expected values of the two options are given below.

First option: $(20,000)(.88) + (0)(.12) = 17,600$
Second option: $(30,000)(.64) + (0)(.36) = 19,200$

Based strictly on expected values, choose the second option.   **4**

9.5 Exercises

Find the expected value in Exercises 1–4. Identify any fair games. (See Examples 1–3.)

1.

Outcome	2	3	4	5
Probability	.1	.4	.3	.2

2.

Outcome	4	6	8	10
Probability	.4	.4	.05	.15

3.

Outcome	9	12	15	18	21
Probability	.14	.22	.36	.18	.10

4.

Outcome	30	32	36	38	44
Probability	.31	.30	.29	.06	.04

5. A raffle offers a first prize of $100, and two second prizes of $40 each. One ticket costs $1, and 500 tickets are sold. Find the expected winnings for a person who buys one ticket.

6. A raffle offers a first prize of $1000, two second prizes of $300 each, and twenty prizes of $10 each. If 10,000 tickets are sold at 50¢ each, find the expected winnings for a person buying one ticket.

Many of the following exercises use the idea of combinations discussed in the last chapter.

7. If 3 marbles are drawn from a bag containing 3 yellow and 4 white marbles, what is the expected number of yellow marbles in the sample?

8. If 5 apples in a barrel of 25 apples are known to be rotten, what is the expected number of rotten apples in a sample of 2 apples?

9. A delegation of 3 is selected from a city council made up of 5 liberals and 4 conservatives.
(a) What is the expected number of liberals on the committee?
(b) What is the expected number of conservatives?

10. From a group of 2 women and 5 men, a delegation of 2 is selected. Find the expected number of women in the delegation.

11. In a club with 20 senior and 10 junior members, what is the expected number of junior members on a 3-member committee?

12. If 2 cards are drawn at one time from a deck of 52 cards, what is the expected number of diamonds?

13. Suppose someone offers to pay you $5 if you draw 2 diamonds in the game of Exercise 12. He says that you should pay 50¢ for the chance to play. Is this a fair game?

Find the expected winnings for each of the following games of chance.

14. In one form of roulette, you bet $1 on "even." If one of the 18 even numbers comes up, you get your dollar back, plus another one. If one of the 20 noneven (18 odd, 0, and 00) numbers comes up, you lose.

15. In another form of roulette, there are only 19 noneven numbers (no 00).

16. Numbers is an illegal game in which you bet $1 on any three-digit number from 000 to 999. If your number comes up, you get $500.

17. In Keno, the house has a pot containing 80 balls, each marked with a different number from 1 to 80. You buy a ticket for $1 and mark one of the 80 numbers on it. The house then selects 20 numbers at random. If your number is among the 20, you get $3.20 (for a net winning of $2.20).

18. Use the assumptions of Example 3 to find Mary's expected winnings. If Mary tosses and Donna calls, is it still a fair game?

19. Suppose one day Mary brings a two-headed coin and uses it to toss for the coffee. Since Mary tosses, Donna calls.
(a) Is this still a fair game?
(b) What is Donna's expected gain if she calls heads?
(c) What is it if she calls tails?

In Exercises 20 and 21, assume that the probability of a girl at each birth is 1/2.

20. Find the expected number of girls in a family of four children.

21. Find the expected number of boys in a family of five children.

22. Jack must choose at age 40 to inherit $25,000 at age 50 (if he is still alive) or $30,000 at age 55 (if he is still alive). If the probabilities for a person of age 40 to live to be 50 and 55 are .90 and .85, respectively, which choice gives him the larger expected inheritance?

23. Management An insurance company has written 100 policies of $10,000, 500 of $5000, and 1000 policies of $1000 on people of age 20. If experience shows that the probability of dying during the twentieth year of life is .001, how much can the company expect to pay out during the year the policies were written?

24. Management A builder is considering a job which promises a profit of $30,000 with a probability of .7 or a loss (due to bad weather, strikes, and such) of $10,000 with a probability of .3. What is her expected profit?

25. Management Experience has shown that a ski lodge will be full (160 guests) during the Christmas holidays if there is a heavy snow pack in December, while a light snowfall in December means that it will have only 90 guests. What is the expected number of guests if the probability for a heavy snow in December is .40? (Assume that there must either be a light snowfall or a heavy snowfall.)

26. Management A magazine distributor offers a first prize of $100,000, two second prizes of $40,000 each, and two third prizes of $10,000 each. A total of 2,000,000 entries are received in the contest. Find the expected winnings if you submit one entry to the contest. If it would cost you 25¢ in time, paper, and stamps to enter, would it be worth it?

27. Management A local car dealer gets complaints about his cars, as shown in the following table.

Number of complaints per day	0	1	2	3	4	5	6
Probability	.01	.05	.15	.26	.33	.14	.06

Find the expected number of complaints per day.

28. Management Levi Strauss and Company* uses expected value to help its salespeople rate their accounts. For each account, a salesperson estimates potential additional volume and the probability of getting it. The product of these gives the expected value of the potential, which is added to the existing volume. The totals are then classified as A, B, or C as follows: below $40,000, class C; between $40,000 and $55,000, class B; above $55,000, class A. Complete the chart on the next page for one of its salespeople.

29. At the end of play in a major golf tournament, two players, an "old pro" and a "new kid," are tied. Sup-

---

*This example was supplied by James McDonald, Levi Strauss and Company, San Francisco.

Account number	Existing volume	Potential additional volume	Probability of additional volume	Expected value of potential	Existing Volume + expected volume of potential	Class
1	$15,000	$10,000	.25	$2,500	$17,500	C
2	40,000	0	—	—	40,000	C
3	20,000	10,000	.20			
4	50,000	10,000	.10			
5	5,000	50,000	.50			
6	0	100,000	.60			
7	30,000	20,000	.80			

pose first prize is $80,000 and second prize is $20,000. Find the expected winnings for the old pro if
(a) both players are of equal ability;
(b) the new kid will freeze up, giving the old pro a 3/4 chance of winning.

30. Natural Science In a certain animal species, the probability that a healthy adult female will have no offspring in a given year is .31, while the probability of 1, 2, 3, or 4 offspring are respectively .21, .19, .17, and .12. Find the expected number of offspring.

31. A recent McDonald's contest offered cash prizes and probabilities of winning on one visit, as shown in the following table. Suppose you spend $1 to buy a bus pass that lets you go to 25 different McDonald's and pick up entry forms. Find the expected value of winning a prize.

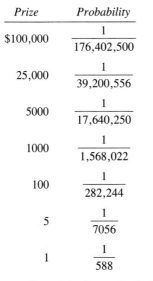

Prize	Probability
$100,000	$\dfrac{1}{176,402,500}$
25,000	$\dfrac{1}{39,200,556}$
5000	$\dfrac{1}{17,640,250}$
1000	$\dfrac{1}{1,568,022}$
100	$\dfrac{1}{282,244}$
5	$\dfrac{1}{7056}$
1	$\dfrac{1}{588}$

32. Natural Science One of the few methods that can be used in an attempt to cut the severity of a hurricane is to *seed* the storm. In this process, silver iodide crystals are dropped into the storm. Unfortunately, silver iodide crystals sometimes cause the storm to *increase* its speed. Wind speeds may also increase or decrease even with no seeding. The probabilities and amounts of property damage shown in the tree diagram below are from an article by R. A. Howard, J. E. Matheson, and D. W. North, "The Decision to Seed Hurricanes."*

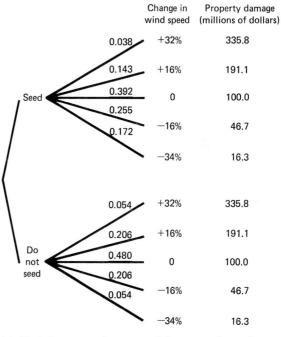

	Change in wind speed	Property damage (millions of dollars)
Seed		
0.038	+32%	335.8
0.143	+16%	191.1
0.392	0	100.0
0.255	−16%	46.7
0.172	−34%	16.3
Do not seed		
0.054	+32%	335.8
0.206	+16%	191.1
0.480	0	100.0
0.206	−16%	46.7
0.054	−34%	16.3

(a) Find the expected amount of damage under each option, "seed" and "do not seed."
(b) To minimize total expected damage, what option should be chosen?

*From "The Decision to Seed Hurricanes," Howard, R. A., *Science,* Vol. 176, pp. 1191–1202, Fig. 7, 16 June 1972. Copyright 1972 by the American Association for the Advancement of Science. Reprinted by permission.

9.6 Decision Making

John F. Kennedy once remarked he had assumed that as president it would be difficult to choose between distinct, opposite alternatives when a decision needed to be made. Actually, however, he found that such decisions were easy to make; the hard decisions came when he was faced with choices that were not as clear-cut. Most decisions fall in this last category—decisions which must be made under conditions of uncertainty. In the previous section we saw how to use expected values to help make a decision. These ideas are extended in this section, where we consider decision making in the face of uncertainty. Let us begin with an example.

Freezing temperatures are endangering the orange crop in central California. A farmer can protect his crop by burning smudge pots—the heat from the pots keeps the oranges from freezing. However, burning the pots is expensive; the cost is $2000. The farmer knows that if he burns smudge pots he will be able to sell his crop for a net profit (after smudge pot costs are deducted) of $5,000, provided that the freeze does develop and wipes out other orange crops in California. If he does nothing he will either lose $1000 in labor costs if it does freeze, or make a profit of $4800 if it does not freeze. (If it does not freeze, there will be a large supply of oranges, and thus his profit will be lower than if there was a small supply.)

What should the farmer do? He should begin by carefully defining the problem. First, he must decide on the **states of nature,** the possible alternatives over which he has no control. Here there are two: freezing temperatures, or no freezing temperatures. Next, the farmer should list the things he can control—his actions or **strategies.** He has two possible strategies: to use smudge pots or not. The consequences of each action under each state of nature, called **payoffs,** are summarized in a **payoff matrix,** as shown below. The payoffs in this case are the profits for each possible combination of events.

States of nature

| | Freeze | No freeze |

Strategies of farmer Use smudge pots $\begin{bmatrix} \$5000 & \$2800 \\ -\$1000 & \$4800 \end{bmatrix}$
Do not use pots

To get the $2800 entry in the payoff matrix, use the profit if there is no freeze, $4800, and subtract the $2000 cost of using the pots. [1]

Once the farmer makes the payoff matrix, what then? The farmer might be an optimist (some might call him a gambler); in this case he might assume that the best will happen and go for the biggest number on the matrix ($5000). For this profit, he must adopt the strategy "use smudge pots."

On the other hand, if the farmer is a pessimist, he would want to minimize the worst thing that could happen. If he uses smudge pots, the worst thing that could happen to him would be a profit of $2800, which will result if there is no freeze. If

[1] Explain how each of the following payoffs in the matrix were obtained.
(a) −$1000
(b) $5000

Answer:
(a) If it freezes and smudge pots are not used, the farmer's profit is −$1000, for labor costs.
(b) If it freezes and smudge pots are used, the farmer makes a profit of $5000.

he does not use smudge pots, he might face a loss of $1000. To minimize the worst, he once again should adopt the strategy "use smudge pots."

Suppose the farmer decides that he is neither an optimist nor a pessimist, but would like further information before choosing a strategy. For example, he might call the weather forecaster and ask for the probability of a freeze. Suppose the forecaster says that this probability is only .1. What should the farmer do? He should recall the discussion of expected value from the previous section and work out the expected profit for each of his two possible strategies. If the probability of a freeze is .1, then the probability that there is no freeze is .9. This information leads to the following expected values.

If smudge pots are used: $5000(.1) + 2800(.9) = 3020$

If no smudge pots are used: $-1000(.1) + 4800(.9) = 4220$

Here the maximum expected profit, $4220, is obtained if smudge pots are *not* used. ②

As the example shows, the farmer's beliefs about the probabilities of a freeze affect his choice of strategies.

Example 1 A small Christmas card manufacturer must decide in February about the type of cards she should emphasize in her fall line of cards. She has three possible strategies: emphasize modern cards, emphasize old-fashioned cards, or a mixture of the two. Her success is dependent on the state of the economy in December. If the economy is strong, she will do well with her modern cards, while in a weak economy people long for the old days and buy old-fashioned cards. In an in-between economy, her mixture of lines would do the best. She first prepares a payoff matrix for all three possibilities. The numbers in the matrix represent her profits in thousands of dollars.

States of nature

		Weak Economy	In-between	Strong Economy
Strategies	Modern	40	85	120
	Old-fashioned	106	46	83
	Mixture	72	90	68

(a) If the manufacturer is an optimist, she should aim for the biggest number on the matrix, 120 (representing $120,000 in profit). Her strategy in this case would be to produce modern cards.

(b) A pessimistic manufacturer wants to find the worst of all bad things that can happen. If she produces modern cards, the worst that can happen is a profit of $40,000. For old-fashioned cards, the worst is a profit of $46,000, while the worst that can happen from a mixture is a profit of $68,000. Her strategy here is to use a mixture.

② What should the farmer do if the probability of a freeze is .6? What is his expected profit?

Answer:
Use smudge pots, $4120

3 Suppose the manufacturer reads another article which gives the following predictions: 35% chance of a weak economy, 25% chance of an in-between economy, and a 40% chance of a strong economy. What is the best strategy now? What is the expected profit?

Answer:
Modern, $83,250

(c) Suppose the manufacturer reads an article in *The Wall Street Journal* that claims that leading experts believe there is a 50% chance of a weak economy at Christmas, a 20% chance of an in-between economy, and a 30% chance of a strong economy. The manufacturer can now use this information to find her expected profit for each possible strategy.

Modern:	$40(.50) + 85(.20) + (120)(.30) = 73$
Old-fashioned:	$106(.50) + 46(.20) + 83(.30) = 87.1$
Mixture:	$72(.50) + 90(.20) + 68(.30) = 74.4$

Here the best strategy is old-fashioned cards; the expected profit is 87.1, or $87,100. **3**

9.6 Exercises

1. An investor has $20,000 to invest in stocks. She has two possible strategies: buy conservative blue-chip stocks or buy highly speculative stocks. There are two states of nature: the market goes up or the market goes down. The following payoff matrix shows the net amounts she will have under the various circumstances.

$$
\begin{array}{c}
\text{Market up} \quad \text{Market down} \\
\begin{array}{l}
\text{Buy blue-chip} \\
\text{Buy speculative}
\end{array}
\left[
\begin{array}{cc}
\$25,000 & \$18,000 \\
\$30,000 & \$11,000
\end{array}
\right]
\end{array}
$$

What should the investor do if she is
(a) an optimist? **(b)** a pessimist?
(c) Suppose there is a .7 probability of the market going up. What is the best strategy? What is the expected profit?
(d) What is the best strategy if the probability of a market rise is .2?

2. Management A developer has $100,000 to invest in land. He has a choice of two parcels (at the same price), one on the highway and one on the coast. With both parcels, his ultimate profit depends on whether he faces light opposition from environmental groups or heavy opposition. He estimates that the payoff matrix is as follows (the numbers represent his profit).

$$
\begin{array}{c}
\textit{Opposition} \\
\text{Light} \qquad \text{Heavy} \\
\begin{array}{l}
\text{Highway} \\
\text{Coast}
\end{array}
\left[
\begin{array}{cc}
\$70,000 & \$30,000 \\
\$150,000 & -\$40,000
\end{array}
\right]
\end{array}
$$

What should the developer do if he is
(a) an optimist? **(b)** a pessimist?
(c) Suppose the probability of heavy opposition is .8. What is his best strategy? What is the expected profit?
(d) What is the best strategy if the probability of heavy opposition is only .4?

3. Hillsdale College has sold out all tickets for a jazz concert to be held in the stadium. If it rains, the show will have to be moved to the gym, which has a much smaller seating capacity. The dean must decide in advance whether to set up the seats and the stage in the gym or in the stadium, or both, just in case. The payoff matrix below shows the net profit in each case.

	States of nature	
	Rain	No rain
Set up in stadium	$-\$1550$	$\$1500$
Strategies Set up in gym	$\$1000$	$\$1000$
Set up both	$\$750$	$\$1400$

What strategy should the dean choose if she is
(a) an optimist? **(b)** a pessimist?
(c) If the weather forecaster predicts rain with a probability of .6, what strategy should she choose to maximize expected profit? What is the maximum expected profit?

4. Management An analyst must decide what fraction of the items produced by a certain machine are defective. He has already decided that there are three possibilities for the fraction of defective items: .01, .10, and .20. He may recommend two courses of action: repair the machine or make no repairs. The payoff matrix below represents the *costs* to the company in each case, in hundreds of dollars.

States of nature

		.01	.10	.20
Strategies	Repair	130	130	130
	No repair	25	200	500

What strategy should the analyst recommend if he is
(a) an optimist?   (b) a pessimist?
(c) Suppose the analyst is able to estimate probabilities for the three states of nature as follows.

Fraction of defectives	Probability
.01	.70
.10	.20
.20	.10

Which strategy should he recommend? Find the expected cost to the company if this strategy is chosen.

5. **Management** The research department of the Allied Manufacturing Company has developed a new process which it believes will result in an improved product. Management must decide whether or not to go ahead and market the new product. The new product may be better than the old or it may not be better. If the new product is better, and the company decides to market it, sales should increase by $50,000. If it is not better, and they replace the old product with the new product on the market, they will lose $25,000 to competitors. If they decide not to market the new product they will lose $40,000 if it is better, and research costs of $10,000 if it is not.
(a) Prepare a payoff matrix.
(b) If management believes the probability that the new product is better to be .4, find the expected profits under each strategy and determine the best action.

6. **Management** A businessman is planning to ship a used machine to his plant in Nigeria. He would like to use it there for the next four years. He must decide whether or not to overhaul the machine before sending it. The cost of overhaul is $2600. If the machine fails when in operation in Nigeria, it will cost him $6000 in

lost production and repairs. He estimates the probability that it will fail at .3 if he does not overhaul it, and .1 if he does overhaul it. Neglect the possiblity that the machine might fail more than once in the four years.
(a) Prepare a payoff matrix.
(b) What should the businessman do to minimize his expected costs?

7. **Management** A contractor prepares to bid on a job. If all goes well, his bid should be $30,000, which will cover his costs plus his usual profit margin of $4500. However, if a threatened labor strike actually occurs, his bid should be $40,000 to give him the same profit. If there is a strike and he bids $30,000, he will lose $5500. If his bid is too high, he may lose the job entirely, while if it is too low, he may lose money.
(a) Prepare a payoff matrix.
(b) If the contractor believes that the probability of a strike is .6, how much should he bid?

8. **Natural Science** A community is considering an anti-smoking campaign.* The city council will choose one of three possible strategies: a campaign for everyone over age 10 in the community, a campaign for youths only, or no campaign at all. The two states of nature are a true cause-effect relationship between smoking and cancer and no cause-effect relationship. The costs to the community (including loss of life and productivity) in each case are as shown below.

States of nature

Strategies	Cause-effect relationship	No cause-effect relationship
Campaign for all	$100,000	$800,000
Campaign for youth	$2,820,000	$20,000
No campaign	$3,100,100	$0

What action should the city council choose if it is
(a) optimistic?   (b) pessimistic?
(c) If the Director of Public Health estimates that the probability of a true cause-effect relationship is .8, which strategy should the city council choose?

*This problem is based on an article by B. G. Greenberg in the September 1969 issue of the *Journal of the American Statistical Association*.

Sometimes the numbers (or payoffs) in a payoff matrix do not represent money (profits or costs, for example), but utility. A **utility** is a number which measures the satisfaction (or lack of it) that results from a certain action. The numbers must be assigned by each individual, depending on how he or she feels about a situation. For example, one person might assign a utility of +20 for a week's vacation in San Francisco, with −6 being assigned if the vacation were moved to Sacramento. Work the following problems in the same way as those above.

9. **Political Science** A politician must plan her reelection strategy. She can emphasize jobs or she can emphasize the environment. The voters can be concerned about jobs or about the environment. A payoff matrix showing the utility of each possible outcome is shown below.

$$\begin{array}{c} & & \textit{Voters} \\ & & \text{Jobs} \quad \text{Environment} \\ \textit{Candidate} & \begin{array}{c} \text{Jobs} \\ \text{Environment} \end{array} & \begin{bmatrix} +25 & -10 \\ -15 & +30 \end{bmatrix} \end{array}$$

The political analysts feel that there is a .35 chance that

the voters will emphasize jobs. What strategy should the candidate adopt? What is its expected utility?

10. In an accounting class, the instructor permits the students to bring a calculator or a reference book (but not both) to an examination. The examination itself can emphasize either numerical problems or definitions. In trying to decide which aid to take to an examination, a student first decides on the utilities shown in the following payoff matrix.

$$\begin{array}{c} & & \textit{Exam emphasizes} \\ & & \text{Numbers} \quad \text{Definitions} \\ \textit{Student chooses} & \begin{array}{c} \text{Calculator} \\ \text{Book} \end{array} & \begin{bmatrix} +50 & 0 \\ +10 & +40 \end{bmatrix} \end{array}$$

(a) What strategy should the student choose if the probability that the examination will emphasize numbers is .6? What is the expected utility in this case?

(b) Suppose the probability that the examination emphasizes numbers is .4. What strategy should be chosen by the student?

Chapter 9   Review Exercises

1. For events E and F, $P(E) = .2$, $P(F|E) = .3$, and $P(F|E') = .2$. Find each of the following.
   (a) $P(E|F)$ (b) $P(E|F')$

Of the appliance repair shops listed in the phone book, 80% are competent and 20% are not. A competent shop can repair an appliance correctly 95% of the time; an incompetent shop can repair an appliance correctly 60% of the time. Suppose an appliance was repaired correctly. Find the probability that it was repaired by

2. a competent shop;   3. an incompetent shop.

Suppose an appliance was repaired incorrectly. Find the probability that it was repaired by

4. a competent shop;   5. an incompetent shop.

6. Box A contains 5 red balls and 1 black ball; box B contains 2 red and 3 black balls. A box is chosen at random, and a ball is selected from it. The probability of choosing box A is 3/8. If the selected ball is black, what is the probability that it came from box A?

7. Find the probability that the ball in Exercise 6 came from box B, given that it is red.

8. A manufacturer buys items from four different suppliers. The fraction of the total number of items that is obtained from each supplier, along with the probability that an item purchased from that supplier is defective, is shown in the following table.

Supplier	Fraction of total supplied	Probability of defective
1	.17	.04
2	.39	.02
3	.35	.07
4	.09	.03

(a) Find the probability that a defective item came from supplier 4.

(b) Find the probability that a defective item came from supplier 2.

A basket contains 4 black, 2 blue, and 5 green balls. A sample of 3 balls is drawn. Find the probability that the sample contains

9. all black balls;

10. all blue balls;

11. 2 black balls and 1 green ball;

12. 2 black balls;

13. 2 green and 1 blue ball;

14. 1 blue ball.

Suppose two cards are drawn without replacement from an ordinary deck of 52 cards. Find the probability that

15. both cards are red;

16. both cards are spades;

17. at least one card is a spade;

18. the cards are a king and an ace.

Suppose a family plans six children, and the probability that a particular child is a girl is 1/2. Find the probability that the family will have

19. exactly 3 girls;   20. all girls;

21. at least 4 girls;   22. no more than 2 boys.

A certain machine used to manufacture screws produces a defective rate of .01. A random sample of 20 screws is selected. Find the probability that the sample contains

23. exactly 4 defective screws;

24. exactly 3 defective screws;

25. no more than 4 defective screws.

26. Set up the probability that the sample has 12 or more defective screws. (Do not evaluate.)

Which of the following are transition matrices?

27. $\begin{bmatrix} .4 & .6 \\ 1 & 0 \end{bmatrix}$ 28. $\begin{bmatrix} -.2 & 1.2 \\ .8 & .2 \end{bmatrix}$

29. $\begin{bmatrix} .8 & .2 & 0 \\ 0 & 1 & 0 \\ .1 & .4 & .5 \end{bmatrix}$ 30. $\begin{bmatrix} .6 & .2 & .3 \\ .1 & .5 & .4 \\ .3 & .3 & .4 \end{bmatrix}$

31. Currently, 35% of all hot dogs sold in one area are made by Dogkins, while 65% are made by Long Dog. Suppose that Dogkins starts a heavy advertising campaign, with the campaign producing the following transition matrix.

After campaign

$$\text{Before campaign} \quad \begin{array}{c} \text{Dogkins} \\ \text{Long Dog} \end{array} \begin{array}{cc} \text{Dogkins} & \text{Long Dog} \\ \begin{bmatrix} .8 & .2 \\ .4 & .6 \end{bmatrix} \end{array}$$

(a) Find the share of the market for each company after the campaign.

(b) Find the share of the market for each company after three such campaigns.

(c) Predict the long-range market share for Dogkins.

32. A credit card company classifies its customers in three groups: nonusers in a given month, light users, and heavy users. The transition matrix for these states is

$$\begin{array}{c} \text{Nonuser} \\ \text{Light} \\ \text{Heavy} \end{array} \begin{array}{ccc} \text{Nonuser} & \text{Light} & \text{Heavy} \\ \begin{bmatrix} .8 & .15 & .05 \\ .25 & .55 & .2 \\ .04 & .21 & .75 \end{bmatrix} \end{array}$$

Suppose the initial distribution for the three states is [.4 .4 .2]. Find the distribution after

(a) 1 month;   (b) 2 months.

(c) What is the long-range prediction for the distribution of users?

33. A medical researcher is studying the risk of heart attack in men. She first divides men into three weight categories, thin, normal, and overweight. By studying the ancestors, children, and grandchildren of these men, the researcher comes up with the following transition matrix.

$$\begin{array}{c} \text{Thin} \\ \text{Normal} \\ \text{Overweight} \end{array} \begin{array}{ccc} \text{Thin} & \text{Normal} & \text{Overweight} \\ \begin{bmatrix} .3 & .5 & .2 \\ .2 & .6 & .2 \\ .1 & .5 & .4 \end{bmatrix} \end{array}$$

Suppose that the distribution of men by weight is initially given by [.2 .55 .25]. Find the distribution after

(a) 1 generation;   (b) 2 generations.

(c) Find the long-range prediction for the distribution of weights.

34. Patients in groups of five were given a new treatment for a fatal disease. The experiment was repeated ten times with the following results.

Number who survived	Probability
0	.1
1	.1
2	.2
3	.3
4	.3
5	0

Find the expected number who survive.

35. You pay $6 to play a game where you will roll a die, with payoffs as follows: $8 for a 6, $7 for a 5, $4 otherwise. What is your mathematical expectation? Is the game fair?

36. A lottery has a first prize of $5000, two second prizes of $1000 each, and two $100 third prizes. A total of ten thousand tickets is sold, at $1 each. Find the expected winnings of a person buying one ticket.

37. Find the expected number of girls in a family of five children.

38. A developer can buy a piece of property that will produce a profit of $16,000 with probability .7, or a loss of $9000 with probability .3. What is the expected profit?

39. Game boards for a recent United Airlines contest could be obtained by sending a self-addressed stamped envelope to a certain address. The prize was a ticket for any city to which United flies. Assume that the value of the ticket was $1000 (we might as well go first class), and that the probability that a particular game board would win was 1/4000. If the stamps to enter the contest cost 30¢ and the envelopes are 1¢ each, find the expected winnings for a person ordering one game board.

40. Three cards are drawn from a standard deck of 52 cards.
(a) What is the expected number of aces?
(b) What is the expected number of clubs?

41. Suppose someone offers to pay you $100 if you draw three cards from a standard deck of 52 cards and all the cards are clubs. What should you pay for the chance to win if it is a fair game?

42. In labor-management relations, both labor and management can adopt either a friendly or a hostile attitude. The results are shown in the following payoff matrix.

The numbers give the wage gains made by an average worker.

		Management Friendly	Hostile
Labor	Friendly	$600	$800
	Hostile	$400	$950

(a) Suppose the chief negotiator for labor is an optimist. What strategy should he choose?
(b) What strategy should he choose if he is a pessimist?
(c) The chief negotiator for labor feels that there is a 70% chance that the company will be hostile. What strategy should he adopt? What is the expected payoff?
(d) Just before negotiations begin, a new management is installed in the company. There is only a 40% chance that the new management will be hostile. What strategy should be adopted by labor?

43. A candidate for city council can come out in favor of a new factory, be opposed to it, or waffle on the issue. The change in votes for the candidate depends on what her opponent does, with payoffs as shown.

		Opponent Favors	Waffles	Opposes
Candidate	Favors	0	−1000	−4000
	Waffles	1000	0	−500
	Opposes	5000	2000	0

(a) What should the candidate do if she is an optimist?
(b) What should she do if she is a pessimist?
(c) Suppose the candidate's campaign manager feels there is a 40% chance that the opponent will favor the plant, and a 35% chance that he will waffle. What strategy should the candidate adopt? What is the expected change in the number of votes?
(d) The opponent conducts a new poll which shows strong opposition to the new factory. This changes the probability he will favor the factory to 0 and the probability he will waffle to .7. What strategy should our candidate adopt? What is the expected change in the number of votes now?

Case 12

Medical Diagnosis

When a patient is examined, information, typically incomplete, is obtained about his state of health. Probability theory provides a mathematical model appropriate for this situation, as well as a procedure for quantitatively interpreting such partial information to arrive at a reasonable diagnosis.*

To do this, list the states of health that can be distinguished in such a way that the patient can be in one and only one state at the time of the examination. Each state of health H is associated with a number $P(H)$ between 0 and 1 such that the sum of all these numbers is 1. This number $P(H)$ represents the probability, before examination, that a patient is in the state of health H, and $P(H)$ may be chosen subjectively from medical experience, using any information available prior to the examination. The probability may be most conveniently established from clinical records; that is, a mean probability is established for patients in general, although the number would vary from patient to patient. Of course, the more information that is brought to bear in establishing $P(H)$, the better the diagnosis.

For example, limiting the discussion to the condition of a patient's heart, suppose there are exactly 3 states of health, with probabilities as follows.

	State of health H	$P(H)$
H_1	patient has a normal heart	.8
H_2	patient has minor heart irregularities	.15
H_3	patient has a severe heart condition	.05

Having selected $P(H)$, the information of the examination is processed. First, the results of the examination must be classified. The examination itself consists of observing the state of a number of characteristics of the patient. Let us assume that the examination for a heart condition consists of a stethoscope examination and a cardiogram. The outcome of such an examination, C, might be one of the following:

C_1—stethoscope shows normal heart and cardiogram shows normal heart;

C_2—stethoscope shows normal heart and cardiogram shows minor irregularities;

and so on.

It remains to assess for each state of health H the conditional probability $P(C|H)$ of each examination outcome C using only the knowledge that a patient is in a given state of health. (This may be based on the medical knowledge and clinical experience of the doctor.) The conditional probabilities $P(C|H)$ will not vary from patient to patient, so that they may be built into a diagnostic system, although they should be reviewed periodically.

Suppose the result of the examination is C_1. Let us assume the following probabilities.

$$P(C_1|H_1) = .9$$
$$P(C_1|H_2) = .4$$
$$P(C_1|H_3) = .1$$

Now, for a given patient, the appropriate probability associated with each state of health H, after examination, is $P(H|C)$ where C is the outcome of the examination. This can be calculated by using Bayes' theorem. For example, to find $P(H_1|C_1)$—that is, the probability that the patient has a normal heart given that the examination showed a normal

---

*From "Probabilistic Medical Diagnosis," Roger Wright, *Some Mathematical Models in Biology,* Robert M. Thrall, ed., (The University of Michigan, 1967), by permission of Robert M. Thrall.

stethoscope examination and a normal cardiogram—we use Bayes' theorem as follows:

$P(H_1|C_1)$

$$= \frac{P(C_1|H_1)P(H_1)}{P(C_1|H_1)P(H_1) + P(C_1|H_2)P(H_2) + P(C_1|H_3)P(H_3)}$$

$$= \frac{(.9)(.8)}{(.9)(.8) + (.4)(.15) + (.1)(.05)} \approx .92.$$

Hence, the probability is about .92 that the patient has a normal heart on the basis of the examination results. This means that in 8 out of 100 patients, some abnormality will be present and not be detected by the stethoscope or the cardiogram.

Exercises

1. Find $P(H_2|C_1)$.

2. Assuming the following probabilities, find $P(H_1|C_2)$:
   $P(C_2|H_1) = .2$, $P(C_2|H_2) = .8$, $P(C_2|H_3) = .3$.

3. Assuming the probabilities of Exercise 2, find $P(H_3|C_2)$.

Case 13

Optimal Inventory for a Service Truck

For many different items it is difficult or impossible to take the item to a central repair facility when service for the item is required. Washing machines, large television sets, office copiers, and computers are only a few examples of such items. Service for items of this type is commonly performed by sending a repair person to the item, with the person driving to the item in a truck containing various parts that might be required in repairing the item. Ideally, the truck should contain all the parts that might be required in repairing the item. However, most parts would be needed only infrequently, so that inventory costs for the parts would be high.

An optimum policy for deciding on the parts to stock on a truck would require that the probability of not being able to repair an item without a trip back to the warehouse for needed parts be as low as possible, consistent with minimum inventory costs. An analysis similar to the one below was developed at the Xerox Corporation.*

To set up a mathematical model for deciding on the optimum truck stocking policy, let us assume that a broken machine might require one of 5 different parts (we could assume any number of different parts—we use 5 to simplify the notation). Suppose also that the probability that a particular machine requires part 1 is p_1, that it requires part 2 is p_2, and so on. Assume also that the failure of different part types are independent, and that at most one part of each type is used on a given job.

Suppose that, on the average, a repair person makes N service calls per time period. If the repair person is unable to make a repair because at least one of the parts is unavailable, there is a penalty cost, L, corresponding to wasted time for the repair person, an extra trip to the parts depot, customer unhappiness, and so on. For each of the parts carried on the truck, an average inventory cost is incurred. Let H_i be the average inventory cost for part i, where $1 \leq i \leq 5$.

*From "Optimal Inventories Based on Job Completion Rate for Repairs Requiring Multiple Items" by Stephen Smith, John Chambers, and Eli Shlifer from *Management Science,* vol. 26, no. 8, August 1980. Copyright © 1980 The Institute of Management Sciences. Reprinted by permission.

Let M_1 represent a policy of carrying only part 1 on the repair truck, M_{24} represent a policy of carrying only parts 2 and 4, with M_{12345} and M_0 representing policies of carrying all parts and no parts, respectively.

For policy M_{35}, carrying parts 3 and 5 only, the expected cost per time period per repair person, written $C(M_{35})$, is

$$C(M_{35}) = (H_3 + H_5) + NL[1 - (1 - p_1)(1 - p_2)(1 - p_4)].$$

(The expression in brackets represents the probability of needing at least one of the parts not carried, 1, 2, or 4 here.) As further examples, $C(M_{125})$ is

$$C(M_{125}) = (H_1 + H_2 + H_5) + NL[1 - (1 - p_3)(1 - p_4)],$$

while

$$C(M_{12345}) = (H_1 + H_2 + H_3 + H_4 + H_5) + NL[1 - 1]$$
$$= H_1 + H_2 + H_3 + H_4 + H_5,$$

and
$$C(M_0) = NL[1 - (1 - p_1)(1 - p_2)(1 - p_3)(1 - p_4)(1 - p_5)].$$

To find the best policy, evaluate $C(M_0)$, $C(M_1)$, . . ., $C(M_{12345})$ and choose the smallest result. (A general solution method is in the *Management Science* paper.)

Example: Suppose that, for a particular item, only 3 possible parts might need to be replaced. By studying past records of failures of the item, and finding necessary inventory costs, suppose that the following values have been found.

p_1	p_2	p_3	H_1	H_2	H_3
.09	.24	.17	$15	$40	$9

Suppose $N = 3$ and L is $54. Then, as an example,

$$C(M_1) = H_1 + NL[1 - (1 - p_2)(1 - p_3)]$$
$$= 15 + 3(54)[1 - (1 - .24)(1 - .17)]$$
$$= 15 + 3(54)[1 - (.76)(.83)]$$
$$= 15 + 59.81$$
$$= 74.81.$$

Thus, if policy M_1 is followed (carrying only part 1 on the truck), the expected cost per repair person per time period is $74.81. Also,

$$C(M_{23}) = H_2 + H_3 + NL[1 - (1 - p_1)]$$
$$= 40 + 9 + 3(54)[.09]$$
$$= 63.58,$$

so that M_{23} is a better policy than M_1. By finding the expected values for all other possible policies (see the exercises below), the optimum policy may be chosen. ■

Exercises

1. Refer to the example above and find each of the following.
   (a) $C(M_0)$ (b) $C(M_2)$ (c) $C(M_3)$
   (d) $C(M_{12})$ (e) $C(M_{13})$ (f) $C(M_{123})$.

2. Which policy leads to lowest expected cost?

3. In the example above, $p_1 + p_2 + p_3 = .09 + .24 + .17 = .50$. Why is it not necessary that the probabilities add to 1?

4. Suppose an item to be repaired might need one of n different parts. How many different policies would then need to be evaluated?

10 Statistics

Statistics deals with the collection and summarization of data, and methods of drawing conclusions about a population based on data from a sample of the population. Statistical models have become increasingly useful in a variety of fields—for example, manufacturing, government, agriculture, medicine, the social sciences, and in all types of research. This chapter gives a brief introduction to some of the key topics from statistical theory.

Often, a researcher wishes to learn something about a characteristic of a population, but because the population is very large or mobile, it is not possible to examine all of its elements. Instead, a limited sample drawn from the population is studied to determine the characteristics of the population. For these inferences to be correct, the sample chosen must be a **random sample.** Random samples are representative of the population because they are chosen so that every element of the population is equally likely to be selected. A hand dealt from a well-shuffled deck of cards is a random sample. In other situations, a random sample can be selected by assigning a number to each element in the population and then by using some chance method to select those numbers that will be included in the sample. This method, using birth dates, was used by the Selective Service for the draft. Random sampling typically relies on a chance device.

10.1 Frequency Distributions

Once a sample has been chosen and all data of interest are collected, the data must be organized so that conclusions may be more easily drawn. One method of organization is to group the data into intervals; equal intervals are usually chosen. For example, suppose the data are the scores of 30 students on a mathematics test, as shown below.

60	90	76	81	59	46
48	61	57	78	86	65
63	54	68	93	71	78
79	67	75	87	76	74
86	57	62	95	80	70

The highest score on the list is 95 and the lowest is 46; one convenient way to group this data is in intervals of size 10, starting with 40–49, and ending with 90–99. This gives an interval for each score in the data and results in six equal intervals of a convenient size. Too many intervals of smaller size would not simplify the data enough, while too few intervals of larger size would conceal information that the data might provide. A rule of thumb is to use from six to fifteen intervals.

In Table 1 below, the six intervals chosen above are listed in the first column, the number of grades falling into each interval are tallied in the second column, and the tallies are totaled and their frequency is then entered in the third column. Table 1 is called a **grouped frequency distribution.**

Table 1

Interval	Tally	Frequency
40–49	||	2
50–59	||||	4
60–69	ⅡⅠ ||	7
70–79	ⅡⅠ ||||	9
80–89	ⅡⅠ	5
90–99	|||	3
		30 Total

The frequency distribution in Table 1 shows information about the data that might not have been noticed before. For example, the interval with the largest number of scores is 70–79, and 16 scores (more than half) were between 59 and 80. Also, the number of scores in each interval increases rather evenly (up to 9) and then decreases at about the same pace. However, some information is lost; for example, we no longer know how many students had a score of 48.

Example 1 Suppose 30 business executives give their recommendations as to the number of college units in management that a business major should have. Use the following results to make a grouped frequency distribution for this data. Use intervals 0–4, 5–9, and so on.

3	25	22	16	0	9	14	8	34	21
15	12	9	3	8	15	20	12	28	19
17	16	23	19	12	14	29	13	24	18

First tally the number of college units falling into each interval. Then total the tallies in each interval, as in Table 2.

Table 2

Interval	Tally	Frequency
0–4	|||	3
5–9	||||	4
10–14	ⅡⅠ |	6
15–19	ⅡⅠ |||	8
20–24	ⅡⅠ	5
25–29	|||	3
30–34	|	1
		30 Total ∎

1 An accounting firm selected 24 personal tax returns prepared by a certain incompetent tax preparer. The number of errors per return were as follows.

8 12 0 6 10 8 0 14
8 12 14 16 4 14 7 11
9 12 17 5 11 21 22 19

Prepare a grouped frequency distribution for this data. Use intervals 0–4, 5–9, and so on.

Answer:

Interval	Frequency
0–4	3
5–9	7
10–14	9
15–19	3
20–24	2
Total	24

The information in a grouped frequency distribution can be displayed in a special kind of bar graph called a **histogram.** In a histogram, interval sizes correspond to widths of the bars, and since equal intervals are used, the bars in a histogram are all the same width. The heights of the bars are determined by the frequencies. A histogram for the data in Table 1 is shown in Figure 10.1.

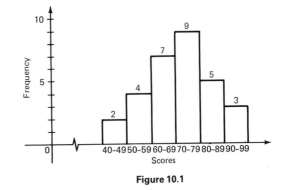

Figure 10.1

A **frequency polygon** is another form of graph that illustrates a grouped frequency distribution. The polygon is formed by joining consecutive midpoints of the tops of the histogram bars with straight line segments. The midpoints of the first and last bars are joined to endpoints on the horizontal axis where the next midpoint would appear. A frequency polygon for the data of Table 1 is illustrated in Figure 10.2.

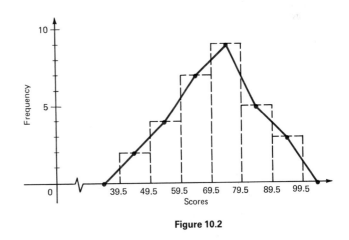

Figure 10.2

Table 3

Interval	Frequency
0–4	3
5–9	4
10–14	6
15–19	8
20–24	5
25–29	3
30–34	1

Example 2 The grouped frequency distribution of college units shown in Table 3 was found in Example 1. Draw a histogram and a frequency polygon for this distribution.

First draw a histogram, shown in black in Figure 10.3. To get a frequency polygon, connect consecutive midpoints of the tops of the bars. The frequency polygon is shown in color in Figure 10.3. **2**

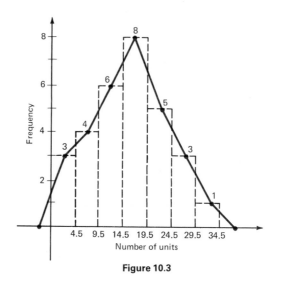

Figure 10.3

2 In problem 1 at the side, the following grouped frequency distribution was found.

Interval	Frequency
0–4	3
5–9	7
10–14	9
15–19	3
20–24	2

Make a histogram and a frequency polygon for this distribution.

Answer:

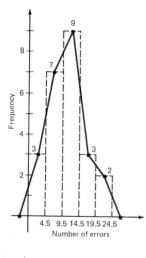

We can obtain further information from Table 1 by adding a new column, called *cumulative frequency,* as shown in Table 4. The entries in this new column give the cumulative number of scores up to and including those in a given interval. For example, Table 4 shows that 13 of the scores were below 70, 22 were below 80, and 30 (all scores) were below 100. Table 4 is called a **cumulative frequency distribution.**

Table 4

Interval	Frequency	Cumulative frequency
40–49	2	2
50–59	4	6
60–69	7	13
70–79	9	22
80–89	5	27
90–99	3	30

A graph which illustrates the cumulative frequencies, called a **cumulative frequency polygon,** is shown in Figure 10.4. Note that the points are not at the midpoints of the intervals, as in a frequency polygon, but at the beginning of each interval, to correspond with the information that the cumulative frequency provides.

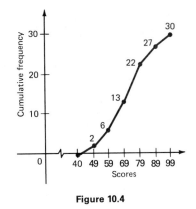

Figure 10.4

Example 3 Complete a cumulative frequency distribution for the data of Example 2. Draw a cumulative frequency polygon.

The first number in the cumulative frequency column is the same as the first frequency, the second number is the same as the sum of the first two frequencies, and so on.

3 Make a cumulative frequency distribution for the data of problem 2 at the side. Draw a cumulative frequency polygon.

Answer:

Interval	Fre-quency	Cumulative frequency
0–4	3	3
5–9	7	10
10–14	9	19
15–19	3	22
20–24	2	24

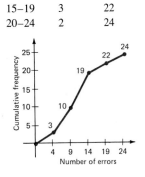

Table 5

Interval	Frequency	Cumulative frequency
0–4	3	3
5–9	4	$3 + 4 = 7$
10–14	6	$7 + 6 = 13$
15–19	8	$13 + 8 = 21$
20–24	5	$21 + 5 = 26$
25–29	3	$26 + 3 = 29$
30–34	1	$29 + 1 = 30$

The cumulative frequency polygon is shown in Figure 10.5. **3**

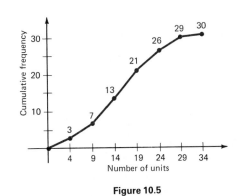

Figure 10.5

10.1 Exercises

1. **Social Science** The histogram below shows the percent of the U.S. population in each age group in 1980.* What percent of the population was in each of the following age groups?
   (a) 10–19
   (b) 60–69
   (c) What age group has the largest percent of the population?

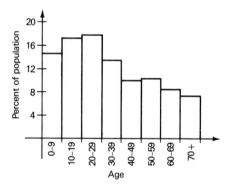

2. **Social Science** The histogram below shows the estimated percent of the U.S. population in each age group in the year 2000.* What percent of the population is estimated to be in each of the following age groups then?
   (a) 20–29
   (b) 70+
   (c) What age group has the largest percent of the population?
   (d) Compare the histogram in Exercise 1 with the histogram below. What seems to be true of the U.S. population?

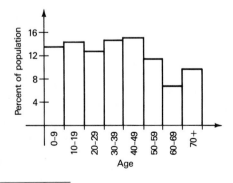

For exercises 3–6

(a) *group the data as indicated;*
(b) *prepare a frequency distribution with a column for intervals, frequencies, and cumulative frequencies;*
(c) *construct a histogram;*
(d) *construct a frequency polygon;*
(e) *construct a cumulative frequency polygon.*

(See Examples 1–3.)

3. The data below give the number of college units completed by 30 of the EZ Life Company's employees. Group the data into six intervals, starting with 0–24.

74	133	4	127	20	30
103	27	139	118	138	121
149	132	64	141	130	76
42	50	95	56	65	104
4	140	12	88	119	64

4. The scores of 80 students on a business law test were as follows. Group the data into seven intervals, starting with 30–39.

79	71	78	87	69	50	63	51
60	46	65	65	56	88	94	56
74	63	87	62	84	76	82	67
59	66	57	81	93	93	54	88
55	69	78	63	63	48	89	81
98	42	91	66	60	70	64	70
61	75	82	65	68	39	77	81
67	62	73	49	51	76	94	54
83	71	94	45	73	95	72	66
71	77	48	51	54	57	69	87

5. **Natural Science** The daily high temperatures in Tucson for the month of July for one year were as follows. Group the data using 70–74 as the first interval.

79	84	88	96	102	104	110	108	106	106
104	99	97	92	94	90	82	74	72	83
85	92	100	99	101	107	111	102	97	94	92

6. **Natural Science** The heights of 40 women (in centimeters) are as follows. Group the data using 140–149 as the first interval.

174 190 172 182 179 186 171 152 174 185
180 170 160 173 163 177 165 157 149 167
169 182 178 158 182 169 181 173 183 176
170 162 159 147 150 192 179 165 148 188

7. Social Science The frequency with which letters occur in a large sample of any written language does not vary much. Therefore, determining the frequency of each letter in a coded message is usually the first step in deciphering it. The percent frequencies of the letters in the English language are as follows.

Letter	%	Letter	%
E	13	L	3.5
T	9	C, U, M	3
A, O	8	F, P, Y	2
N	7	W, G, B	1.5
I, R	6.5	V	1
S, H	6	K, X, J	0.5
D	4	Q, Z	0.2

Use the introductory paragraph of this exercise as a sample of the English language. Find the percent frequency for each letter in the sample. Compare your results with the frequencies given above.

8. Social Science The following message is written in a code in which the frequency of the symbols is the main key to the solution.

)?$- -8$))y*$+8506$*$3 \times 6;4?$*7*$\&\times$*-6.48()985)?
$(8+2:;48)81\&?(;46$*3)y*$;48\&(+8$(*$509+\&8$()8=8
$(5$*$-8-5$(81?098;4\&+)\&15$*$50:$)6$)6$*$;?6;6\&$*$0?-7

(a) Find the frequency of each symbol.
(b) By comparing the high-frequency symbols with the high-frequency letters in English, and the low-frequency symbols with the low-frequency letters, try to decipher the message. (Hint: Look for repeated two-symbol combinations and double letters for added clues. Try to identify vowels first.)

9. Get five coins and toss them all at once. Keep track of the number of heads. Repeat this experiment 64 times. Make a histogram of your results.

10. Use the results of Exercise 9 to make a frequency polygon.

11. The frequency distribution below gives the theoretical results of tossing 5 coins 64 times. Use this distribution to make a frequency polygon. Compare it to the one you got in Exercise 10.

Number of heads	Frequency
0	2
1	10
2	20
3	20
4	10
5	2

12. Toss six coins a total of 64 times. Keep track of the number of heads. Make a histogram of your results.

13. Make a frequency polygon using the results of Exercise 12.

14. The frequency distribution below gives the theoretical results of tossing 6 coins 64 times. Use the distribution to make a frequency polygon. Compare it to the one you made in Exercise 13.

Number of heads	Frequency
0	1
1	6
2	15
3	20
4	15
5	6
6	1

10.2 Measures of Central Tendency

In summarizing data, it is often useful to have a single number, a typical or "average" number, which is representative of the entire collection of data. Such numbers are referred to as **measures of central tendency.** Everyone is familiar with some type of average. For example, we compare our heights and weights to those of the typical or "average" person on weight charts. Students are quite familiar with the "class average" and their own "average" at any time in a given course.

Mean These kinds of averages usually refer to the arithmetic average, or mean. The **arithmetic mean** (the **mean**) of a set of numbers is the sum of the numbers, divided by the total number of numbers. Write the sum of n numbers x_1, x_2, x_3, . . . , x_n, in a compact way, using **summation notation,** also called **sigma notation.** With the Greek letter Σ (sigma), the sum

$$x_1 + x_2 + x_3 + \cdots + x_n \text{ is written}$$

$$x_1 + x_2 + x_3 + \cdots + x_n = \sum_{i=1}^{n} x_i.$$

① Let $x_1 = 2$, $x_2 = 3$, $x_3 = 7$, and $x_4 = 9$. Find

(a) $\displaystyle\sum_{i=1}^{4} x_i$

(b) $\displaystyle\sum_{i=1}^{4} x_i^2$

Answer:

(a) 21

(b) 143

①

In statistics, $\sum_{i=1}^{n} x_i$ is often abbreviated as just $\Sigma(x)$. The symbol $\bar{x}$ (read x-bar) is used to represent the mean.

Mean The mean of the n numbers, x_1, x_2, x_3, . . . , x_n, is

$$\bar{x} = \frac{\sum (x)}{n}.$$

Example 1 Daily sales at Lupe's Laundry last week were \$86, \$103, \$118, \$117, \$126, \$158, and \$149. Find the mean, $\bar{x}$, of these sales.

Let $x_1 = 86$, $x_2 = 103$, and so on. ($n = 7$ since there are 7 numbers.) Then

$$\bar{x} = \text{mean} = \frac{\sum (x)}{n} = \frac{86 + 103 + 118 + 117 + 126 + 158 + 149}{7}$$

$$= \frac{857}{7}$$

$$\bar{x} \approx 122.43.$$

The mean sales at the laundry were \$122.43 per day, rounded to the nearest cent.

② Find the mean of the following list of numbers.

\$25.12	\$42.58
\$76.19	\$32
\$81.11	\$26.41
\$19.76	\$59.32
\$71.18	\$21.03

Answer:

\$45.47

②

The mean of data that have been arranged in a frequency distribution is found in a similar way. For example, suppose the following data are collected.

Value	Frequency
84	2
87	4
88	7
93	4
99	3
	20 Total

The value 84 appears twice, 87 four times, and so on. To find the mean, first add 84 two times, 87 four times, and so on; or get the same result faster by multiplying 84 by 2, 87 by 4, and so on, and then by adding the results. Dividing the sum by 20, the total of the frequencies, gives the mean.

$$\bar{x} = \frac{(84 \cdot 2) + (87 \cdot 4) + (88 \cdot 7) + (93 \cdot 4) + (99 \cdot 3)}{20}$$

$$= \frac{168 + 348 + 616 + 372 + 297}{20}$$

$$= \frac{1801}{20}$$

$$\bar{x} = 90.05$$

Example 2 Find the mean for the data shown in the following frequency distribution.

Value	Frequency	Value × Frequency
30	6	30 · 6 = 180
32	9	32 · 9 = 288
33	7	33 · 7 = 231
37	12	37 · 12 = 444
42	6	42 · 6 = 252
	40 Total	1395 Total

A new column, "Value × Frequency," has been added to the frequency distribution. Adding the products from this column gives a total of 1395. The total from the frequency column is 40. The mean is

$$\bar{x} = \frac{1395}{40} = 34.875. \quad \boxed{3}$$

The mean of grouped data is found in a similar way. For grouped data, intervals are used, rather than single values. To calculate the mean, it is assumed that all these values are located at the midpoint of the interval. The letter x is used to represent the midpoints and f represents the frequencies, as shown in the next example.

Example 3 Find the mean for the following grouped frequency distribution (Table 1 of the previous section).

Interval	Midpoint, x	Frequency, f	Product, xf
40–49	44.5	2	89
50–59	54.5	4	218
60–69	64.5	7	451.5
70–79	74.5	9	670.5
80–89	84.5	5	422.5
90–99	94.5	3	283.5
		30 Total	2135 Total

[3] Find $\bar{x}$ for the following frequency distribution.

Value	Frequency
7	2
9	3
11	6
13	4
15	1
17	4

Answer:
$\bar{x} = 12.1$

An additional column for the midpoint of each interval has been included. Get the numbers in this column by adding the endpoints of each interval and dividing by 2. For the interval 40–49, the midpoint is (40 + 49)/2 = 44.5. Get the numbers in the product column on the right by multiplying frequencies and corresponding midpoints. Finally, divide the total of the product column by the total of the frequency column to get

$$\bar{x} = \frac{2135}{30} = 71.2 \text{ (to the nearest tenth).} \quad \blacksquare$$

The formula for the mean of a grouped frequency distribution is given below.

> **Mean of a Grouped Distribution** The mean of a distribution where x represents the midpoints, f the frequencies, and $n = \Sigma(f)$, is
>
> $$\bar{x} = \frac{\Sigma\,(xf)}{n}.$$

4 Find the mean of the following grouped frequency distribution.

Interval	Frequency
0–4	6
5–9	4
10–14	7
15–19	3

Answer:

8.75

4

Median Asked by a reporter to give the average height of the players on his team, the Little League coach lined up his 15 players by increasing height. He picked out the player in the middle and pronounced this player to be of average height. This kind of average, called the **median,** is defined as the middle entry in a set of data arranged in either increasing or decreasing order. If there is an even number of entries, the median is defined to be the mean of the two center entries.

Odd number of entries	*Even number of entries*
8	2
7	3
Median = 4	4⎤ Median = $\frac{4 + 7}{2}$ = 5.5
3	7⎦
1	9̶
	12

The procedure for finding the median of a grouped frequency distribution is more complicated. We omit it here because it is more common to find the mean when working with grouped frequency distributions.

Example 4 Find the median for the following lists of numbers.
(a) 11, 12, 17, 20, 23, 28, 29

The median is the middle number, in this case, 20. (Note that the numbers are already arranged in numerical order.) In this list, three numbers are smaller than 20 and three are larger.

(b) 15, 13, 7, 11, 19, 30, 39, 5, 10

First arrange the numbers in numerical order, from smallest to largest.

$$5, \quad 7, \quad 10, \quad 11, \quad 13, \quad 15, \quad 19, \quad 30, \quad 39$$

The middle number can now be determined: the median is 13.

(c) 47, 59, 32, 81, 74, 153

Write the numbers in numerical order.

$$32, \quad 47, \quad 59, \quad 74, \quad 81, \quad 153$$

There are six numbers here; the median is the mean of the two middle numbers, or

5 Find the median for each of the following lists of numbers.

(a) 12, 15, 17, 19, 35, 42, 58

(b) 28, 68, 7, 15, 47, 59, 13, 74, 32, 25

Answer:

(a) 19

(b) 30

$$\text{median} = \frac{59 + 74}{2} = \frac{133}{2} = 66\frac{1}{2}. \quad \boxed{5}$$

In some situations, the median gives a truer representative or typical element of the data than the mean. For example, suppose in an office there are 10 salespersons, 4 secretaries, the sales manager, and Ms. Daly, who owns the business. Their annual salaries are as follows: secretaries, $15,000 each; salespersons, $25,000 each; manager, $35,000; and owner, $200,000. The mean salary is

$$\bar{x} = \frac{(15,000)4 + (25,000)10 + 35,000 + 200,000}{16} = \$34,062.50.$$

However, since 14 people earn less than $34,062.50 and only 2 earn more, this does not seem very representative. The median salary is found by ranking the salaries by size: $15,000, $15,000, $15,000, $15,000, $25,000, $25,000, . . . , $200,000. Since there are 16 salaries (an even number) in the list, the mean of the 8th and 9th entries will give the value of the median. The 8th and 9th entries are both $25,000, so the median is $25,000. In this example, the median gives a truer average than the mean.

Mode Sue's scores on ten class quizzes include one 7, two 8's, six 9's, and one 10. She claims that her average grade on quizzes is 9, because most of her scores are 9's. This kind of "average," found by selecting the most frequent entry, is called the **mode.** Similarly, in a grouped frequency distribution, the interval with the greatest frequency is the **modal class.**

Example 5 Find the mode for each list of numbers.

6 Find the mode for each of the following lists of numbers.

(a) 29, 35, 29, 18, 29, 56, 48

(b) 13, 17, 19, 20, 20, 13, 25, 27, 13, 20

(c) 512, 546, 318, 729, 854, 253

Answer:

(a) 29

(b) 13 and 20

(c) No mode

(a) 57, 38, 55, 55, 80, 87, 98

The number 55 occurs more often than any other, and is the mode. It is not necessary to place the numbers in numerical order when looking for the mode.

(b) 182, 185, 183, 185, 187, 187, 189

Both 185 and 187 occur twice. This list has *two* modes.

(c) 10,708 11,519, 10,972, 17,546, 13,905, 12,182

No number occurs more than once. This list has no mode. **6**

The mode has the advantages of being easily found and not being influenced by extremes—data which are very large or very small compared to the rest of the data. It is often used in samples where the data to be "averaged" are not numerical. A major disadvantage to the mode is that there may be more than one, in case of ties, or there may be no mode at all when all entries occur with the same frequency.

The mean is the most commonly used measure of central tendency. Its advantages are that it is easy to compute, it takes all the data into consideration, and it is reliable—that is, repeated samples are likely to give very similar means. A disadvantage of the mean is that it is influenced by extreme values, as illustrated in the salary example above.

The median can be easy to compute and is influenced very little by extremes. Like the mode, the median can be found in situations where the data are not numerical. For example, in a taste test, people are asked to rank five soft drinks from the one they like best to the one they like least. The combined rankings then produce an ordered sample from which the median can be identified. A disadvantage of the median is the need to rank the data in order; this can be difficult when the number of items is large.

10.2 Exercises

Find the mean for each list of numbers. Round to the nearest tenth. (See Example 1.)

1. 8, 10, 16, 21, 25

2. 44, 41, 25, 36, 67, 51

3. 130, 141, 149, 152, 158, 163, 139, 170

4. 42, 48, 54, 62, 69, 75, 90, 94

5. 21,900, 22,850, 24,930, 29,710, 28,340, 40,000

6. 38,500, 39,720, 42,183, 21,982, 43,250

7. 9.4, 11.3, 10.5, 7.4, 9.1, 8.4, 9.7, 5.2, 1.1, 4.7

8. 30.1, 42.8, 91.6, 51.2, 88.3, 21.9, 43.7, 51.2

9. .06, .04, .05, .08, .03, .14, .18, .29, .07, .01

10. .31, .09, .08, .22, .46, .51, .48, .42, .53, .42

Find the mean for each of the following. Round to the nearest tenth. (See Example 2.)

11.
Value	Frequency
3	4
5	2
9	1
12	3

12.
Value	Frequency
9	3
12	5
15	1
18	1

13.
Value	Frequency
12	4
13	2
15	5
19	3
22	1
23	5

14.
Value	Frequency
25	1
26	2
29	5
30	4
32	3
33	5

15.
Value	Frequency
104	6
112	14
115	21
119	13
123	22
127	6
132	9

16.
Value	Frequency
246	2
291	4
295	3
304	8
307	9
319	2

Find the median for each of the following lists of numbers. Don't forget to first place the numbers in numerical order if necessary. (See Example 4.)

17. 12, 18, 32, 51, 58, 92, 106

18. 596, 604, 612, 683, 719

19. 100, 114, 125, 135, 150, 172

20. 298, 346, 412, 501, 515, 521, 528, 621

21. 32, 58, 97, 21, 49, 38, 72, 46, 53

22. 1072, 1068, 1093, 1042, 1056, 1005, 1009

23. 576, 578, 542, 551, 565, 525, 590, 559

24. 7, 15, 28, 3, 14, 18, 46, 59, 1, 2, 9, 21

25. 28.4, 9.1, 3.4, 27.6, 59.8, 32.1, 47.6, 29.8

26. .6, .4, .9, 1.2, .3, 4.1, 2.2, .4, .7, .1

Find the mode or modes for each of the following lists of numbers. (See Example 5.)

27. 4, 9, 8, 6, 9, 2, 1, 3

28. 21, 32, 46, 32, 49, 32, 49

29. 80, 72, 64, 64, 72, 53, 64

30. 97, 95, 94, 95, 94, 97, 97

31. 74, 68, 68, 68, 75, 75, 74, 74, 70

32. 158, 162, 165, 162, 165, 157, 163

33. 5, 9, 17, 3, 2, 8, 19, 1, 4, 20

34. 12, 15, 17, 18, 21, 29, 32, 74, 80

35. 6.1, 6.8, 6.3, 6.3, 6.9, 6.7, 6.4, 6.1, 6.0

36. 12.75, 18.32, 19.41, 12.75, 18.30, 19.45, 18.33

For grouped data, we give the modal class *(interval containing the most data values). Give the mean and modal class for each of the following collections of grouped data. (See Example 3.)*

37.

College units	Frequency	College units	Frequency
0–24	4	75–99	3
25–49	3	100–124	5
50–74	6	125–149	9

38. (You will need to use Exercise 5, Section 10.1, to obtain the frequencies for this distribution before finding the mean and modal class.)

Temperature	Frequency	Temperature	Frequency
70–74		95–99	
75–79		100–104	
80–84		105–109	
85–89		110–114	
90–94			

39. Use the distribution of Exercise 4, Section 10.1.

40. Use the distribution of Exercise 6, Section 10.1.

Social Science *The table below (from an article in* Business Week, *March 3, 1986, page 80)* gives the expected increase of workers (in thousands) from 1984 to 1995 in the jobs listed. The weekly earnings in a recent year for each job are also given.*

Find the mean for each of the following.

41. The increase in number of jobs (in thousands)

42. The weekly earnings

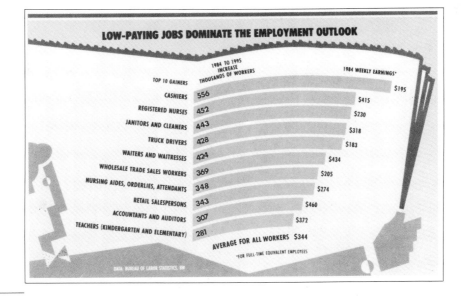

LOW-PAYING JOBS DOMINATE THE EMPLOYMENT OUTLOOK

TOP 10 GAINERS	1984 TO 1995 INCREASE THOUSANDS OF WORKERS	1984 WEEKLY EARNINGS*
CASHIERS	556	$195
REGISTERED NURSES	452	$415
JANITORS AND CLEANERS	443	$230
TRUCK DRIVERS	428	$318
WAITERS AND WAITRESSES	424	$183
WHOLESALE TRADE SALES WORKERS	369	$434
NURSING AIDES, ORDERLIES, ATTENDANTS	348	$205
RETAIL SALESPERSONS	343	$274
ACCOUNTANTS AND AUDITORS	307	$460
TEACHERS (KINDERGARTEN AND ELEMENTARY)	281	$372
AVERAGE FOR ALL WORKERS		$344

*FOR FULL-TIME EQUIVALENT EMPLOYEES

DATA: BUREAU OF LABOR STATISTICS, BW

The table below gives the average monthly temperatures, in degrees Fahrenheit, for a certain area.

Month	Maximum	Minimum
January	39	16
February	39	18
March	44	21
April	50	26
May	60	32
June	69	37
July	79	43
August	78	42
September	70	37
October	51	31
November	47	24
December	40	20

Find the mean for each of the following.

43. The maximum temperature

44. The minimum temperature

*U.S. wheat prices and production figures for a recent 10-year period are given below.**

Year	Price ($ per bushel)	Production (millions of bushels)
1976	2.70	2200
1977	2.30	2000
1978	2.95	1750
1979	3.80	2200
1980	3.90	2400
1981	3.60	2800
1982	3.55	2800
1983	3.50	2450
1984	3.35	2600
1985	3.20	2750

Find the mean for each of the following.

45. Price per bushel of wheat

46. Wheat production

Find the mean, median, and mode for each set of ungrouped data.

47. The weight gains of 10 experimental rats fed on a special diet were −1, 0, −3, 7, 1, 1, 5, 4, 1, 4.

48. A sample of 7 measurements of the thickness of a copper wire were .010, .010, .009, .008, .007, .009, .008.

49. The times in minutes that 12 patients spent in a doctor's office were 20, 15, 18, 22, 10, 12, 16, 17, 19, 21, 23, 13.

50. The scores on a 10-point botany quiz were 6, 6, 8, 10, 9, 7, 6, 5, 6, 8, 3.

10.3 Measures of Variation

The mean gives a measure of central tendency of a list of numbers, but tells nothing about the *spread* of the numbers in the list. For example, look at the following three samples.

I	3	5	6	3	3
II	4	4	4	4	4
III	10	1	0	0	9

*Source: U.S. Department of Agriculture.

Each of these three samples has a mean of 4, and yet they are quite different; the amount of dispersion or variation within the samples is different. Therefore, in addition to a measure of central tendency, another kind of measure is needed that describes how much the numbers vary.

The largest number in Sample I is 6, while the smallest is 3, a difference of 3. In Sample II this difference is 0; in Sample III it is 10. The difference between the largest and smallest number in a sample is called the **range,** one example of a measure of variation. The range of Sample I is 3, of Sample II is 0, and of Sample III is 10. The range has the advantage of being very easy to compute and gives a rough estimate of the variation among the data in the sample. However, it depends only on the two extremes and tells nothing about how the other data are distributed between the extremes.

Example 1 Find the range for each list of numbers.
(a) 12, 27, 6, 19, 38, 9, 42, 15
The highest number here is 42; the lowest is 6. The range is the difference of these numbers, or

$$42 - 6 = 36.$$

(b) 74, 112, 59, 88, 200, 73, 92, 175

$$\text{Range} = 200 - 59 = 141 \quad \blacksquare$$

1 Find the range for the numbers

159, 283, 490, 390, 375, 297.

Answer:
331

The most useful measure of variation is the *standard deviation*. Before defining it, however, we must find the **deviations from the mean,** the differences found by subtracting the mean from each number in a distribution.

Example 2 Find the deviations from the mean for the numbers

$$32, \quad 41, \quad 47, \quad 53, \quad 57.$$

Adding these numbers and dividing by 5 gives a mean of 46. To find the deviations from the mean, subtract 46 from each number in the list. For example, the first deviation from the mean is $32 - 46 = -14$; the last is $57 - 46 = 11$.

2 Find the deviations from the mean for each set of numbers.
(a) 19, 25, 36, 41, 52, 61
(b) 6, 9, 5, 11, 3, 2

Answer:
(a) Mean is 39; deviations are
−20, −14, −3, 2, 13, 22
(b) Mean is 6; deviations from the mean are 0, 3, −1, 5, −3, −4

Number	Deviations from mean
32	−14
41	−5
47	1
53	7
57	11

To check your work, find the sum of these deviations. It should always equal 0. (The answer is always 0 because the positive and negative numbers cancel each other.) **2**

To find a measure of variation, we might be tempted to use the mean of the deviations. However, as mentioned above, this number is always 0, no matter how widely the data are dispersed. This problem is avoided by first squaring each deviation.

Number	Deviation from mean	Square of deviation
32	-14	196
41	-5	25
47	1	1
53	7	49
57	11	121

Now we are ready to define the **standard deviation:** it is the square root of the mean of the squares of the deviations. However, in calculating the standard deviation, since the sample usually has less variation than the entire population, a correction is made by dividing the sum of the squares of the deviations by $n - 1$ instead of n. In the distribution of Example 2, where $n = 5$, the corrected mean of the squares of the deviations is

$$\frac{196 + 25 + 1 + 49 + 121}{5 - 1} = \frac{392}{4} = 98.$$

The standard deviation is the square root of this number, or $\sqrt{98}$. The letter s is used to represent standard deviation. For the numbers 32, 41, 47, 53, 57,

$$s = \sqrt{98} = 9.9 \qquad \text{(to the nearest tenth).}$$

The square root of 98 can be found in a table or by using a calculator.

Given a list of n numbers, the standard deviation of these numbers is found by using the following formula.*

> **Standard Deviation** The standard deviation of the n numbers, $x_1, x_2, x_3, \ldots, x_n$, with mean $\bar{x}$, is
>
> $$s = \sqrt{\frac{\sum (x - \bar{x})^2}{n - 1}}.$$

Example 3 Find the standard deviation of the numbers

$$7, \quad 9, \quad 18, \quad 22, \quad 27, \quad 29, \quad 32, \quad 40.$$

The mean of the numbers is

$$\frac{7 + 9 + 18 + 22 + 27 + 29 + 32 + 40}{8} = 23.$$

---

*Some calculators which compute variance and standard deviation use n, and others use $n - 1$. Be sure to check how your calculator works before using it for the exercises.

Arrange the work in columns as follows.

Number	Deviation from mean	Square of deviation
7	-16	256
9	-14	196
18	-5	25
22	-1	1
27	4	16
29	6	36
32	9	81
40	17	289
		900 Total

Divide the total of the last column by $n - 1$, where n is the number of squares in the list. In this case $n = 8$, so the quotient is

$$\frac{900}{7} = 128.6,$$

(rounded) and the standard deviation is

$$\sqrt{128.6} = 11.3. \quad \boxed{3}$$

The square of the standard deviation, s^2, is called the **variance.** In Example 3 above, the variance is $(11.3)^2$ or 128.6.

For data in a grouped frequency distribution, a different formula for the standard deviation is used.

> **Standard Deviation for a Grouped Distribution** The standard deviation for a distribution with mean $\bar{x}$, where x is an interval midpoint with frequency f, and $n = \Sigma f$, is
>
> $$s = \sqrt{\frac{\Sigma (x - \bar{x})^2 f}{n - 1}}.$$

The formula indicates that the product $(x - \bar{x})^2 f$ is to be found for each interval. Then, these products are summed, and the sum is divided by one less than the total frequency, that is by $n - 1$. The square root of this result is s, the standard deviation.

Example 4 Find s for the grouped data of Example 3, Section 10.2.

Begin by including columns for $x - \bar{x}$, $(x - \bar{x})^2$, and $(x - \bar{x})^2 \cdot f$. Recall from Section 10.2 that $\bar{x} = 71.2$.

3 Find the standard deviation of each set of numbers. The deviations from the mean were found in problem 2 at the side.

(a) 19, 25, 36, 41, 52, 61

(b) 6, 9, 5, 11, 3, 2

Answer:

(a) 15.9

(b) 3.5

Interval	f	x	$x - \bar{x}$	$(x - \bar{x})^2$	$(x - \bar{x})^2 \cdot f$
40–49	2	44.5	−26.7	712.89	1425.78
50–59	4	54.5	−16.7	278.89	1115.56
60–69	7	64.5	−6.7	44.89	314.23
70–79	9	74.5	3.3	10.89	98.01
80–89	5	84.5	13.3	176.89	884.45
90–99	3	94.5	23.3	542.89	1628.67
	30				5466.70 Totals

4 Find the standard deviation for the grouped data that follows. (Hint: $\bar{x} = 28.5$)

Value	Frequency
20–24	3
25–29	2
30–34	4
35–39	1

Answer:
$\sqrt{28.06} = 5.3$

Use the formula above to find s.

$$s = \sqrt{\frac{\sum (x - \bar{x})^2 \cdot f}{n - 1}} = \sqrt{\frac{5466.70}{29}} = \sqrt{188.5} = 13.7 \quad \blacksquare 4$$

10.3 Exercises

Find the range and standard deviation for each of the following sets of numbers. (See Examples 1 and 3.)

1. 6, 8, 9, 10, 12

2. 12, 15, 19, 23, 26

3. 7, 6, 12, 14, 18, 15

4. 4, 3, 8, 9, 7, 10, 1

5. 42, 38, 29, 74, 82, 71, 35

6. 122, 132, 141, 158, 162, 169, 180

7. 241, 248, 251, 257, 252, 287

8. 51, 58, 62, 64, 67, 71, 74, 78, 82, 93

9. 3, 7, 4, 12, 15, 18, 19, 27, 24, 11

10. 15, 42, 53, 7, 9, 12, 28, 47, 63, 14

11. 21, 28, 32, 42, 51

12. 76, 78, 92, 104, 111

Find the standard deviation for the following grouped data.

13. (From Exercise 3, Section 10.1)

College units	Frequency
0–24	4
25–49	3
50–74	6
75–99	3
100–124	5
125–149	9

14. (From Exercise 4, Section 10.1)

Scores	Frequency
30–39	1
40–49	6
50–59	13
60–69	22
70–79	17
80–89	13
90–99	8

15. The data of Exercise 5, Section 10.1

16. The data of Exercise 6, Section 10.1

An application of standard deviation is given by **Chebyshev's theorem.** (*P. L. Chebyshev was a Russian mathematician who lived from 1821 to 1894.) His theorem applies to* any *distribution of numbers. It states:*

For any distribution of numbers, at least $1 - 1/k^2$ of the numbers lie within k standard deviations of the mean.

Example: For any distribution, at least

$$1 - \frac{1}{3^2} = 1 - \frac{1}{9} = \frac{8}{9}$$

of the numbers lie within 3 standard deviations of the mean. ■

Find the fraction of all the numbers of a data set lying within the following numbers of standard deviations from the mean.

17. 2   **18.** 4   **19.** 5

In a certain distribution of numbers, the mean is 50 with a standard deviation of 6. Use Chebyshev's theorem to tell what percent of the numbers are

20. between 38 and 62;   **21.** between 32 and 68;

22. between 26 and 74;   **23.** between 20 and 80;

24. less than 38 or more than 62;

25. less than 32 or more than 68;

26. less than 26 or more than 74.

27. **Management** The Forever Power Company conducted tests on the life of its batteries and those of a competitor (Brand X) with the following results for samples of 10 batteries of each brand.

	Hours of use
Forever Power	20 22 22 25 26 27 27 28 30 35
Brand X	15 18 19 23 25 25 28 30 34 38

Compute the mean and standard deviation for each sample. Compare the means and standard deviations of the two brands and then answer the questions below.
(a) Which batteries have a more uniform life in hours?
(b) Which batteries have the highest average life in hours?

28. **Management** The weekly wages of the seven employees of Harold's Hardware Store are $150, $160, $190, $210, $250, and $1000.
(a) Find the mean and standard deviation of this distribution.
(b) How many of the seven employees earn within one standard deviation of the mean? How many earn within two standard deviations of the mean?

Management *The Quaker Oats Company conducted a survey to determine if a proposed premium, to be included in their cereal, was appealing enough to generate new sales.* Four cities were used as test markets, where the cereal was distributed with the premium, and four cities as control markets, where the cereal was distributed without the premium. The eight cities were chosen on the basis of*

---

*This example was supplied by Jeffery S. Berman, Senior Analyst, Marketing Information, Quaker Oats Company.

their similarity in terms of population, per capita income, and total cereal purchase volume. The results were as follows.

		Percent change in average market shares per month
	1	+18
Test cities	2	+15
	3	+7
	4	+10
	1	+1
Control cities	2	−8
	3	−5
	4	0

29. Find the mean of the change in market share for the four test cities.

30. Find the mean of the change in market share for the four control cities.

31. Find the standard deviation of the change in market share for the test cities.

32. Find the standard deviation of the change in market share for the control cities.

33. Find the difference between the mean of Exercise 29 and the mean of Exercise 30. This represents the estimate of the percent change in sales due to the premium.

34. The two standard deviations from Exercise 31 and Exercise 32 were used to calculate an "error" of ±7.95 for the estimate in Exercise 33. With this amount of error what is the smallest and largest estimate of the increase in sales? [Hint: Use the answer to Exercise 33.]

On the basis of the results of Exercise 34, the company decided to mass produce the premium and distribute it nationally.

Use a computer to solve the problems in Exercises 35–38.

35. Twenty-five laboratory rats, used in an experiment to test the food value of a new product, made the following weight gains in grams.

5.25	5.03	4.90	4.97	5.03
5.12	5.08	5.15	5.20	4.95
4.90	5.00	5.13	5.18	5.18
5.22	5.04	5.09	5.10	5.11
5.23	5.22	5.19	4.99	4.93

Find the mean gain and the standard deviation of the gains.

36. An assembly-line machine turns out washers with the following thicknesses in mm.

1.20	1.01	1.25	2.20	2.58	2.19
1.29	1.15	2.05	1.46	1.90	2.03
2.13	1.86	1.65	2.27	1.64	2.19
2.25	2.08	1.96	1.83	1.17	2.24

Find the mean and standard deviation of these thicknesses.

37. The prices of pork bellies futures on the Chicago Mercantile Exchange over a period of several weeks were as follows.

48.25	48.50	47.75	48.45	46.85
47.10	46.50	46.90	46.60	47.00
46.35	46.65	46.85	47.20	46.60

47.00	45.00	45.15	44.65	45.15
46.25	45.90	46.10	45.82	45.70
47.05	46.95	46.90	47.15	47.10

Find the mean and standard deviation.

38. A medical laboratory tested 21 samples of human blood for acidity on the pH scale with the following results.

7.1	7.5	7.3	7.4	7.6	7.2	7.3
7.4	7.5	7.3	7.2	7.4	7.3	7.5
7.5	7.4	7.4	7.1	7.3	7.4	7.4

Find the mean and standard deviation.

10.4 Normal Distribution

A **probability distribution** assigns to each outcome of an experiment the probability of the occurrence of that outcome. The probabilities are determined theoretically or, in an experimental situation, by using frequencies. Probability distributions were first mentioned in Section 8.4, and, as shown there, are usually presented as tables.

Example 1   A bank wishes to improve its services to the public. The manager decides to begin by finding the number of minutes the tellers spend on each transaction. The transaction times then may take any of the values 0, 1, 2, 3, ... Suppose the manager found the times for 75 transactions, with results as shown in Table 1 below. Give a probability distribution for these results.

Table 1

Minutes	Frequency
1	3
2	5
3	9
4	12
5	15
6	11
7	10
8	6
9	3
10	1
	75 Total

Table 2

Minutes	Probability
1	.04
2	.07
3	.12
4	.16
5	.20
6	.15
7	.13
8	.08
9	.04
10	.01
	1.00 Total

To get a probability distribution, use the relative frequencies as probabilities. A total of 75 transaction times were found, so each frequency should be divided by

75 to get the probabilities shown in Table 2. Some of the results are rounded to the nearest hundredth. The total of the probabilities should be 1, since all possible outcomes are included. **1**

Figure 10.6(a) shows a histogram and frequency polygon for the information in Table 2. The heights of the bars are the probabilities. The transaction times in Example 1 were given to the nearest minute. Theoretically at least, they could have been timed to the nearest tenth of a minute, or hundredth of a minute, or even more accurately. In each case, a histogram and frequency polygon could be drawn. If the times are measured with smaller and smaller units, there are more bars in the histogram, and the frequency polygon begins to look more and more like the curve in Figure 10.6(b) instead of a polygon. Actually it is possible for the transaction times to take on any real number value greater than 0. A distribution in which the outcomes can take any real number value within some interval is a **continuous distribution.** The graph of a continuous distribution is a curve.

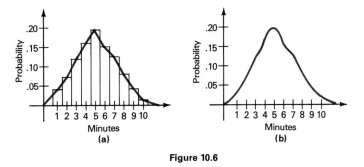

Figure 10.6

The distribution of heights (in inches) of college freshmen women is another example of a continuous distribution, since these heights include infinitely many possible measurements, such as 53, 58.5, 66.3, 72.666, . . . , and so on. Figure 10.7 shows the continuous distribution of heights of college freshmen women. Here the most frequent heights occur near the center of the interval shown.

Another continuous curve, which approximates the distribution of yearly incomes in the United States, is given in Figure 10.8. The graph shows that the most frequent incomes are grouped near the low end of the interval. This kind of distribution, where the peak is not at the center, is called **skewed.**

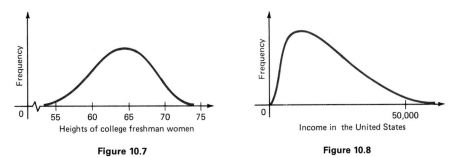

Figure 10.7   **Figure 10.8**

Many different experiments produce probability distributions which come from a very important class of continuous probability distributions called *normal distributions*. Each normal probability distribution has associated with it a bell-shaped curve. The normal distributions are important because many natural and social phenomena have approximately normal distributions. The distribution shown in Figure 10.7 is approximately a normal distribution, while the one of Figure 10.8 is not. The distribution of the lengths of the leaves of a certain tree would approximate a normal distribution, as should the distribution of the actual weights of cereal boxes that have an average weight of 16 ounces.

There are actually many normal distributions, some tall and thin, others short and wide, as shown in Figure 10.9, but the area beneath the curve is always equal to 1. Each normal distribution is determined by its mean and standard deviation. A small standard deviation leads to a tall, narrow curve like the one in the center of Figure 10.9. A large standard deviation produces a flat wide curve like the one on the right in Figure 10.9.

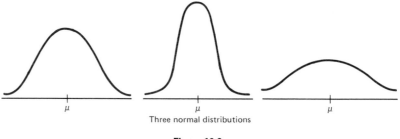

Three normal distributions

Figure 10.9

To distinguish work with probability distributions from the frequency distributions studied earlier, the Greek letters μ (mu) for the mean and σ (sigma) for the standard deviation are used.

Work with normal distributions is simplified by using the **standard normal distribution** which has $\mu = 0$ and $\sigma = 1$. Since the total area under a normal curve is equal to 1, parts of the area can be used to determine probabilities. In Figure 10.10, the shaded area under the curve from a to b represents the probability that an observed data value will be between a and b. The areas under portions of the standard normal curve have been calculated, and can be found by consulting Table 10 at the back of the book.

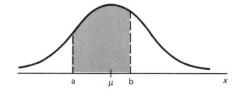

Figure 10.10

Example 2 Use Table 10 to find the fraction of all the scores in a standard normal distribution that lie between the mean and the following standard deviations from the mean.

(a) 1 standard deviation above the mean

In Table 10, z is used to represent the number of standard deviations from the mean, written as a decimal to the nearest hundredth. Here $z = 1.00$. Find 1.00 in the z column; you should find .3413 next to it. This entry means that .3413 or 34.13% of all values lie between the mean and one standard deviation above the mean. In other words, the shaded area of Figure 10.11 represents 34.13% of the total area under the normal curve.

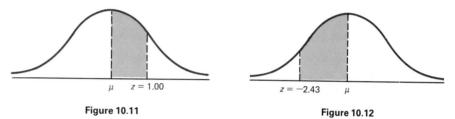

Figure 10.11   Figure 10.12

2 Find the percent of area between the mean and
(a) 1.51 standard deviations above the mean;
(b) 2.04 standard deviations below the mean.

In (c) and (d) find the percent of area in the shaded region.
(c)

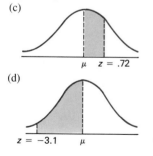

(d)

Answer:
(a) 43.45%
(b) 47.93%
(c) 26.42%
(d) 49.9%

(b) 2.43 standard deviations below the mean

Here, $z = -2.43$. The negative sign indicates that we want z standard deviations *below* the mean. Since we are interested only in area, look for 2.43 in the z column of Table 10. This will lead to the entry .4925. A total of .4925 or 49.25% of all values lie between the mean and 2.43 standard deviations below the mean. This region is shaded in Figure 10.12. **2**

Example 3 Find the following areas under the standard normal curve.

(a) The area between 1.41 standard deviations *below* the mean and 2.25 standard deviations *above* the mean

First, draw a sketch showing the desired area, as in Figure 10.13 on the next page. From Table 10, the area between the mean and 1.41 standard deviations below the mean is .4207. Also, the area from the mean to 2.25 standard deviations above the mean is .4878. As the sketch shows, the total desired area can be found by *adding* these numbers.

$$\begin{array}{r} .4207 \\ +.4878 \\ \hline .9085 \end{array}$$

The shaded area in Figure 10.13 represents 90.85% of the total area under the normal curve.

(b) The area between .58 standard deviations above the mean and 1.94 standard deviations above the mean

Figure 10.14 shows the desired area. The area between the mean and .58 standard deviations above the mean is .2190. The area between the mean and 1.94

standard deviations above the mean is .4738. As the sketch shows, the desired area is found by *subtracting* the two areas.

$$
\begin{array}{r}
.4738 \\
-.2190 \\
\hline
.2548
\end{array}
$$

The shaded area of Figure 10.14 represents 25.48% of the total area under the normal curve.

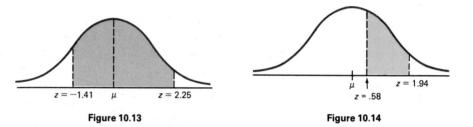

z = −1.41 μ z = 2.25	μ ↑ z = 1.94 z = .58
Figure 10.13	**Figure 10.14**

3 Find the following normal curve areas as percents of the total area.
(a) Between .31 standard deviations below the mean and 1.01 standard deviations above the mean

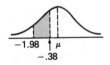

−.31 μ   1.01

(b) Between .38 and 1.98 standard deviations below the mean

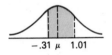

−1.98 ↑ μ
     −.38

(c) To the right of 1.49 standard deviations above the mean

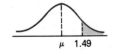

μ   1.49

Answer:
(a) 46.55%
(b) 32.82%
(c) 6.81%

(c) The area to the right of 2.09 standard deviations above the mean
    The total area under a normal curve is 1. Thus, the total area to the right of the mean is 1/2, or .5000. From Table 10, the area from the mean to 2.09 standard deviations above the mean is .4817. The area to the right of 2.09 standard deviations is found by subtracting .4817 from .5000.

$$
\begin{array}{r}
.5000 \\
-.4817 \\
\hline
.0183
\end{array}
$$

A total of 1.83% of the total area is to the right of 2.09 standard deviations above the mean. Figure 10.15 shows the desired area. **3**

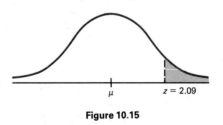

μ   z = 2.09

Figure 10.15

    The standard normal distribution is used to find probabilities for other normal distributions by converting given *x*-values, called **z-scores,** to comparable values in the standard normal distribution.

Area Under a Normal Curve If a normal distribution has mean μ and standard deviation σ, the percent of the area under the associated normal curve that is to the left of the value x is exactly the same as the percent of the area to the left of

$$z = \frac{x - \mu}{\sigma}$$

under the standard normal curve.

Example 4 Suppose it has been determined that the sales force for Dixie Office Supplies drives an average of $\mu = 1200$ miles per month in a company car, with a standard deviation of $\sigma = 150$ miles. Assume that the number of miles driven by a salesperson is closely approximated by a normal distribution.

(a) Find the probability that a driver travels between 1200 and 1600 miles per month.

First find the number of standard deviations above the mean that corresponds to 1600 miles by finding the z-score for 1600.

For $x = 1600$,

$$z = \frac{x - \mu}{\sigma}$$

$$= \frac{1600 - 1200}{150} \qquad \text{Let } x = 1600, \ \mu = 1200, \ \sigma = 150$$

$$= \frac{400}{150}$$

$$z = 2.67$$

Since $\mu = 1200$, the z-score for $x = 1200$ is 0. See Figure 10.16. From Table 10, .4962, or 49.62% of all drivers travel between 1200 and 1600 miles per month, so the required probability is .4962.

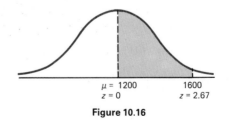

$\mu = 1200$
$z = 0$

1600
$z = 2.67$

Figure 10.16

(b) Find the probability that a driver travels between 1000 and 1500 miles per month.

As shown in Figure 10.17, z-scores for both $x = 1000$ and $x = 1500$ are needed.

4 The heights of female sophomore college students at one school have $\mu = 172$ centimeters, with $\sigma = 10$ centimeters. Find the probability that the height of such a student is

(a) between 172 cm and 185 cm;

(b) between 160 cm and 180 cm;

(c) less than 165 cm.

Answer:

(a) .4032

(b) .6730

(c) .2420

For $x = 1000$,

$$z = \frac{1000 - 1200}{150}$$

$$= \frac{-200}{150}$$

$$= -1.33$$

For $x = 1500$,

$$z = \frac{1500 - 1200}{150}$$

$$= \frac{300}{150}$$

$$= 2.00$$

From the table, $z = 1.33$ leads to an area of .4082, while $z = 2.00$ corresponds to .4773. A total of $.4082 + .4773 = .8855$, or 88.55%, of all drivers travel between 1000 and 1500 miles per month. From this, the probability that a driver travels between 1000 and 1500 miles per month is .8855. **4**

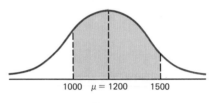

Figure 10.17

10.4 Exercises

Find the percent of area under a normal curve between the mean and the following number of standard deviations from the mean. (See Example 3.)

1. 2.50   **2.** 1.68   **3.** .45   **4.** .81

5. -1.71 **6.** -2.04 **7.** 3.11   **8.** 2.80

Find the percent of the total area under the normal curve between the following z-scores.

9. $z = 1.41$ and $z = 2.83$

10. $z = .64$ and $z = 2.11$

11. $z = -2.48$ and $z = -.05$

12. $z = -1.74$ and $z = -1.02$

13. $z = -3.11$ and $z = 1.44$

14. $z = -2.94$ and $z = -.43$

15. $z = -.42$ and $z = .42$

16. $z = -1.98$ and $z = 1.98$

Find a z-score satisfying each of the following conditions. (Hint: Use Table 10 backwards.)

17. 5% of the total area is to the right of z.

18. 1% of the total area is to the left of z.

19. 15% of the total area is to the left of z.

20. 25% of the total area is to the right of z.

Management *A certain type of light bulb has an average life of 500 hours, with a standard deviation of 100 hours. The length of life of the bulb can be closely approximated by a normal curve. An amusement park buys and installs 10,000 such bulbs. Find the total number that can be expected to last*

21. at least 500 hours;

22. less than 500 hours;

23. between 500 and 650 hours;

24. between 300 and 500 hours;

25. between 650 and 780 hours;

26. between 290 and 540 hours;

27. less than 740 hours;

28. more than 300 hours;

29. more than 790 hours;

30. less than 410 hours.

Management *A box of oatmeal must contain* 16 *ounces. The machine that fills the oatmeal boxes is set so that, on the average, a box contains* 16.5 *ounces. The boxes filled by the machine have weights that can be closely approximated by a normal curve. What percent of the boxes filled by the machine are underweight if the standard deviation is*

31. .5 ounce;   **32.** .3 ounce;

33. .2 ounce;   **34.** .1 ounce?

The chickens at Colonel Thompson's Ranch have a mean weight of 1850 *grams with a standard deviation of* 150 *grams. The weights of the chickens are closely approximated by a normal curve. Find the percent of all chickens weighing*

35. more than 1700 grams;

36. less than 1800 grams;

37. between 1750 grams and 1900 grams;

38. between 1600 grams and 2000 grams;

39. less than 1550 grams;

40. more than 2100 grams.

Natural Science *In nutrition, the recommended daily allowance of vitamins is a number set by the government as a guide to an individual's daily vitamin intake. Actually, vitamin needs vary drastically from person to person, but the needs are very closely approximated by a normal curve. To calculate the recommended daily allowance, the government first finds the average need for vitamins among people in the population, and the standard deviation. The recommended daily allowance is then defined as the mean plus* 2.5 *times the standard deviation.*

41. What percentage of the population will receive adequate amounts of vitamins under this plan?

Find the recommended daily allowance for the following vitamins.

42. Mean = 1800 units, standard deviation = 140 units

43. Mean = 159 units, standard deviation = 12 units

44. Mean = 1200 units, standard deviation = 92 units

Assume the following distributions are all normal, and use the areas under the normal curve given in Table 10 to answer the questions.

45. Management A machine produces bolts with an aver-

age diameter of .25 inches and a standard deviation of .02 inches. What is the probability that a bolt will be produced with a diameter greater than .3 inches?

46. Management The mean monthly income of the trainees of an engineering firm is $1200 with a standard deviation of $200. Find the probability that an individual trainee earns less than $1000 per month.

47. Management A machine which fills quart milk cartons is set up to average 32.2 ounces per carton, with a standard deviation of 1.2 ounces. What is the probability that a filled carton will contain less than 32 ounces of milk?

48. The average contribution to the campaign of Polly Potter, a candidate for city council, was $50 with a standard deviation of $15. How many of the 200 people who contributed to Potter's campaign gave between $30 and $100?

49. Management At the Discount Market, the average weekly grocery bill is $52.25 with a standard deviation of $15.50. What are the largest and smallest amounts spent by the middle 50% of this market's customers?

50. Natural Science The mean clotting time of blood is 7.45 seconds with a standard deviation of 3.6 seconds. What is the probability that an individual's blood clotting time will be less than 7 seconds or greater than 8 seconds?

51. The average size of the fish in Lake Amotan is 12.3 inches with a standard deviation of 4.1 inches. Find the probability of catching a fish there which is longer than 18 inches.

52. To be graded extra large, an egg must weigh at least 2.2 ounces. If the average weight for an egg is 1.5 ounces with a standard deviation of .4 ounces, how many of five dozen eggs would you expect to grade extra large?

A teacher gives a test to a large group of students. The results are closely approximated by a normal curve. The mean is 74, *with a standard deviation of* 6. *The teacher wishes to give A's to the top* 8% *of the students and F's to the bottom* 8%. *A grade of B is given to the next* 15%, *with D's given similarly. All other students get C's. Find the bottom cutoff (rounded to the nearest whole number) for the following grades. (Hint: Use Table 10 backwards.)*

53. A   **54.** B   **55.** C   **56.** D

Use a computer to find the following probabilities by finding the comparable area under a standard normal curve. (*See Examples 2 and 3.*)

57. *z* is between 1.372 and 2.548

58. *z* is between -2.751 and 1.693

59. *z* is greater than -2.476

60. *z* is less than 1.692

61. *z* is less than $-.4753$

62. *z* is greater than .2509

Use a computer to find the following probabilities for a distribution with a mean of 35.693 and a standard deviation of 7.104. (*See Example 4.*)

63. *x* is between 12.275 and 28.432

64. *x* is greater than 38.913

65. *x* is less than 17.462

66. *x* is between 17.462 and 53.106

10.5 The Binomial Distribution

In many practical situations, experiments have only two possible outcomes: *success* or *failure*. Examples of such experiments, called *binomial trials,* or *Bernoulli trials,* include tossing a coin (perhaps *h* would be called a success, with *t* a failure); rolling a die with the two outcomes being, for instance, 5, a success, and a number other than 5, a failure; or choosing a radio from a large batch and deciding if the radio is defective or not. (Bernoulli trials were first discussed in Section 9.3.)

A *binomial distribution* is a probability distribution that satisfies the following properties. The experiment is a series of independent trials with only two outcomes possible, success and failure. The probability of each outcome must be constant from trial to trial.

As an example, suppose a die is tossed 5 times. Identify a result of 1 or 2 as a success, with any other result a failure. Since each trial (each toss) can result in a success or a failure, the result of the 5 tosses can be any number of successes from 0 through 5. These six possible outcomes are not equally likely. The various probabilities can be found with the following result from Section 9.3:

$$P(x) = \binom{n}{x} p^x (1-p)^{n-x},$$

where *n* is the number of trials, *x* is the number of successes, *p* is the probability of success in a single trial, and $P(x)$ is the probability that *x* of the *n* trials result in success. In this example, $n = 5$ and $p = 1/3$. The distribution for this experiment is given in the table at the side.

The mean μ of a probability distribution is defined to be the expected value of *x*. Recall from Chapter 9 that expected value is found by adding the products of *x* and $P(x)$. For the distribution above,

x	$P(x)$
0	$\binom{5}{0}\left(\frac{1}{3}\right)^0\left(\frac{2}{3}\right)^5 = \dfrac{32}{243}$
1	$\binom{5}{1}\left(\frac{1}{3}\right)^1\left(\frac{2}{3}\right)^4 = \dfrac{80}{243}$
2	$\binom{5}{2}\left(\frac{1}{3}\right)^2\left(\frac{2}{3}\right)^3 = \dfrac{80}{243}$
3	$\binom{5}{3}\left(\frac{1}{3}\right)^3\left(\frac{2}{3}\right)^2 = \dfrac{40}{243}$
4	$\binom{5}{4}\left(\frac{1}{3}\right)^4\left(\frac{2}{3}\right)^1 = \dfrac{10}{243}$
5	$\binom{5}{5}\left(\frac{1}{3}\right)^5\left(\frac{2}{3}\right)^0 = \dfrac{1}{243}$

$$\mu = 0\left(\frac{32}{243}\right) + 1\left(\frac{80}{243}\right) + 2\left(\frac{80}{243}\right) + 3\left(\frac{40}{243}\right) + 4\left(\frac{10}{243}\right) + 5\left(\frac{1}{243}\right)$$

$$= \frac{405}{243} = 1\frac{2}{3}.$$

For a binomial distribution, which is a special kind of probability distribution, this method for finding the mean reduces to the formula, $\mu = np$, where n is the number of trials and p is the probability of success for a single trial. Using this simplified formula, the computation of the mean in the example above is

$$\mu = np = 5\left(\frac{1}{3}\right) = \frac{5}{3},$$

which agrees with the result obtained using expected value.

As mentioned earlier, the symbol σ is used to represent the standard deviation of a probability distribution; the variance, then, is σ^2. Like the mean, the variance, σ^2, of a theoretical probability distribution is an expected value—the expected value of the squared deviations from the mean, $(x - \mu)^2$. To find the variance for the example given above, first find the quantities $(x - \mu)^2$. (Recall, the mean μ is 5/3.) The results are given in the table at the side.

Now find σ^2 by adding the products $[(x - \mu)^2][P(x)]$.

x	$P(x)$	$x - \mu$	$(x - \mu)^2$
0	$\frac{32}{243}$	$-\frac{5}{3}$	$\frac{25}{9}$
1	$\frac{80}{243}$	$-\frac{2}{3}$	$\frac{4}{9}$
2	$\frac{80}{243}$	$\frac{1}{3}$	$\frac{1}{9}$
3	$\frac{40}{243}$	$\frac{4}{3}$	$\frac{16}{9}$
4	$\frac{10}{243}$	$\frac{7}{3}$	$\frac{49}{9}$
5	$\frac{1}{243}$	$\frac{10}{3}$	$\frac{100}{9}$

$$\sigma^2 = \frac{25}{9}\left(\frac{32}{243}\right) + \frac{4}{9}\left(\frac{80}{243}\right) + \frac{1}{9}\left(\frac{80}{243}\right)$$

$$+ \frac{16}{9}\left(\frac{40}{243}\right) + \frac{49}{9}\left(\frac{10}{243}\right) + \frac{100}{9}\left(\frac{1}{243}\right)$$

$$= \frac{10}{9}$$

Find the standard deviation σ by evaluating $\sqrt{10/9}$ to get $\sqrt{10}/3$, or approximately 1.05.

The formula for the variance, like that for the mean, is greatly simplified for a binomial distribution. The complicated procedure above for finding σ^2 and σ reduces to the formulas

$$\sigma^2 = np(1 - p), \quad \text{and} \quad \sigma = \sqrt{np(1 - p)}.$$

Substituting the appropriate values for n and p from the example into this new formula gives

$$\sigma^2 = 5\left(\frac{1}{3}\right)\left(\frac{2}{3}\right) = \frac{10}{9},$$

which agrees with the value found above. ①
A summary of these results is given on the following page.

① Find μ and σ for a binomial distribution having $n = 120$ and $p = 1/6$.

Answer:
$\mu = 20$; $\sigma = 4.08$

> **Binomial Distribution** Suppose an experiment is a series of n independent repeated trials, where the probability of a success in a single trial is always p. Let x be the number of successes in the n trials. Then the probability that exactly x successes will occur in n trials is given by
>
> $$\binom{n}{x} p^x (1-p)^{n-x}.$$
>
> The mean μ and variance σ^2 of a binomial distribution are respectively
>
> $$\mu = np \quad \text{and} \quad \sigma^2 = np(1-p).$$
>
> The standard deviation is
>
> $$\sigma = \sqrt{np(1-p)}.$$

Example 1 The probability that a plate picked at random from the assembly line in a china factory will be defective is .01. A sample of three is to be selected. Write the distribution for the number of defective plates in the sample, and give its mean and standard deviation.

Since three plates will be selected, the possible number of defective plates ranges from 0 to 3. Here, n (the number of trials) is 3; p (the probability of selecting a defective on a single trial) is .01. The distribution and the probability of each outcome are shown below.

x	$P(x)$
0	$\binom{3}{0}(.01)^0(.99)^3 = .970$
1	$\binom{3}{1}(.01)(.99)^2 = .029$
2	$\binom{3}{2}(.01)^2(.99) = .0003$
3	$\binom{3}{3}(.01)^3(.99)^0 = .000001$

The mean of the distribution is

$$\mu = np = 3(.01) = .03.$$

The standard deviation is

$$\sigma = \sqrt{np(1-p)} = \sqrt{3(.01)(.99)} = \sqrt{.0297} = .17. \quad \boxed{2}$$

Binomial distributions are very useful, but the probability calculations become difficult if n is large. However, the standard normal distribution of the previous section (which shall be referred to as the normal distribution from now on) can be used to get a good approximation of the binomial distributions.

2 The probability that a can of beer from a certain plant is defective is .005. A sample of 4 cans is selected at random. Write a distribution for the number of defective cans in the sample, and give its mean and standard deviation.

Answer:

x	$P(x)$
0	.961
1	.020
2	.015
3	.00002
4	.00000000

$\mu = .02, \sigma = .14$

As an example, to show how the normal distribution is used, consider the distribution for the expected number of heads if one coin is tossed 15 times. This is a binomial distribution with $p = 1/2$ and $n = 15$. The distribution is shown at the side. The mean of the distribution is

$$\mu = np = 15\left(\frac{1}{2}\right) = 7.5.$$

The standard deviation is

$$\sigma = \sqrt{np(1 - p)}$$
$$= \sqrt{15\left(\frac{1}{2}\right)\left(1 - \frac{1}{2}\right)}$$
$$= \sqrt{15\left(\frac{1}{2}\right)\left(\frac{1}{2}\right)}$$
$$= \sqrt{3.75} \approx 1.94.$$

In Figure 10.18 the normal curve with $\mu = 7.5$ and $\sigma = 1.94$ has been superimposed over the histogram of the distribution.

x	$P(x)$
0	.00003
1	.00046
2	.00320
3	.01389
4	.04166
5	.09164
6	.15274
7	.19638
8	.19638
9	.15274
10	.09164
11	.04166
12	.01389
13	.00320
14	.00046
15	.00003

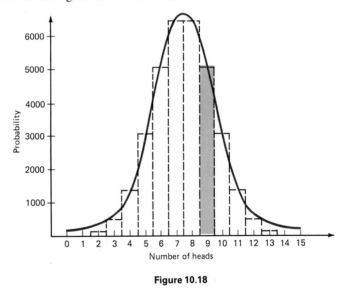

Figure 10.18

As shown in the distribution, the probability of getting exactly 9 heads in the 15 tosses is approximately .153. This answer is about the same fraction that would be found by dividing the area of the bar in color in Figure 10.18 by the total area of all 16 bars in the graph. (Some of the bars at the extreme left and right ends of the graph are too short to show up.)

As the graph suggests, the area in color is approximately equal to the area under the normal curve from $x = 8.5$ to $x = 9.5$. The normal curve is higher than the top of the bar in the left half but lower in the right half.

To find the area under the normal curve from $x = 8.5$ to $x = 9.5$, first find z-scores, as in the previous section. Use the mean and the standard deviation for the distribution, which we have already calculated, to get z-scores for $x = 8.5$ and $x = 9.5$.

For $x = 8.5$	For $x = 9.5$
$z = \dfrac{8.5 - 7.5}{1.94}$	$z = \dfrac{9.5 - 7.5}{1.94}$
$= \dfrac{1.00}{1.94}$	$= \dfrac{2.00}{1.94}$
$z \approx .52$	$z \approx 1.03$

From Table 10, $z = .52$ gives an area of .1985, while $z = 1.03$ gives .3485. Find the required result by subtracting these two numbers.

$$.3485 - .1985 = .1500$$

This answer, .1500, is close to the exact answer, .153, found above. ③

Example 2 About 6% of the bolts produced by a certain machine are defective.
(a) Find the probability that in a sample of 100 bolts, 3 or fewer are defective.
      This problem satisfies the conditions of the definition of a binomial distribution, so the normal curve approximation can be used. First find the mean and the standard deviation using $n = 100$ and $p = 6\% = .06$.

$$\mu = 100(.06) \qquad \sigma = \sqrt{100(.06)(1 - .06)}$$
$$\mu = 6 \qquad\qquad = \sqrt{100(.06)(.94)}$$
$$\qquad\qquad\qquad\qquad = \sqrt{5.64}$$
$$\qquad\qquad\qquad\qquad \sigma \approx 2.37$$

As the graph of Figure 10.19 shows, the area to the left of $x = 3.5$ (since we want 3 or fewer defective bolts) must be found. For $x = 3.5$,

$$z = \frac{3.5 - 6}{2.37} = \frac{-2.5}{2.37} \approx -1.05.$$

From the table, $z = -1.05$ corresponds to an area of .3531. Find the area to the left of z by subtracting .3531 from .5000 to get .1469. The probability of 3 or fewer defective bolts in a set of 100 bolts is .1469, or 14.69%.

<table>
<tr><td>

③ Use the normal distribution to find the probability of getting exactly the following number of heads in 15 tosses of a coin.

(a) 7
(b) 10

Answer:
(a) .1985
(b) .0909

</td></tr>
</table>

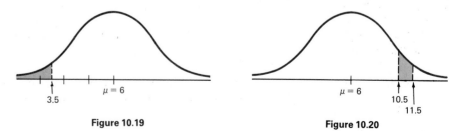

Figure 10.19   **Figure 10.20**

(b) Find the probability of exactly 11 defective bolts in a sample of 100 bolts. As Figure 10.20 shows, the area between $x = 10.5$ and $x = 11.5$ must be found.

4 About 9% of the transistors produced by a certain factory are defective. Find the probability that in a sample of 200, the following numbers of transistors will be defective. (Hint: $\sigma = 4.05$)
(a) Exactly 11
(b) 16 or fewer
(c) More than 14

Answer:
(a) .0226
(b) .3557
(c) .8051

If $x = 10.5$, then $z = \dfrac{10.5 - 6}{2.37} = 1.90.$

If $x = 11.5$, then $z = \dfrac{11.5 - 6}{2.37} = 2.32.$

Look in Table 10; $z = 1.90$ gives an area of .4713, while $z = 2.32$ yields .4898. The final answer is the difference of these numbers, or

$$.4898 - .4713 = .0185.$$

There is about a 1.85% chance of having exactly 11 defective bolts. **4**

The normal curve approximation to a binomial distribution is usually quite accurate, especially for practical problems. As a rule of thumb, the normal curve approximation can be used as long as both np and $n(1 - p)$ are at least 5.

10.5 Exercises

In Exercises 1–6, several binomial experiments are described. For each one, answer the questions presented by (a) giving the distribution; (b) finding the mean; (c) finding the standard deviation. (See Example 1.)

1. A die is rolled six times and the number of 1's that come up is tallied. Write the distribution of 1's that can be expected to occur.

2. A 6-item multiple choice test has four possible answers for each item. If a student selects all his answers randomly, how many can he expect to answer correctly? With what standard deviation?

3. Management To maintain quality control on the production line, the Bright Lite Company randomly selects three light bulbs each day for testing. Experience has shown a defective rate of .02. Write the distribution for the number of defectives in the daily samples.

4. In a taste test, each member of a panel of four is given two glasses of Super Cola, one made using the old formula and one with the new formula, and asked to identify the new formula. Assuming the panelists operate independently, write the distribution of the number of successful identifications, if each judge actually guesses.

5. The probability that a radish seed will germinate is .7. Joe's mother gives him four seeds to plant. Write the distribution for the number of seeds which he can expect to germinate.

6. Natural Science Five patients in Ward 8 of Memorial Hospital have a disease with a known mortality rate of .1. Write the distribution of the number who can be expected to survive.

Natural Science *Work the following exercises involving binomial experiments.*

7. The probability that an infant will die in the first year of life is about .025. In a group of 500 babies, what are the mean and standard deviation of the number of babies who can be expected to die in their first year of life?

8. The probability that a particular kind of mouse will have a brown coat is 1/4. In a litter of 8, assuming independence, how many could be expected to have a brown coat? With what standard deviation?

9. A certain drug is effective 80% of the time. Give the mean and standard deviation of the number of patients using the drug who recover, out of a group of 64 patients.

10. The probability that a newborn infant will be a girl is .49. If 50 infants are born on Susan B. Anthony's birthday, how many can be expected to be girls? With what standard deviation?

For the remaining exercises, use the normal curve approximation to a binomial distribution. (See Example 2.)

Suppose 16 *coins are tossed. Find the probability of getting exactly*

11. 8 heads;

12. 7 heads;

13. 10 tails;

14. 12 tails.

Suppose 1000 *coins are tossed. Find the probability of getting each of the following. (Hint:* $\sqrt{250} = 15.8$)

15. Exactly 500 heads

16. Exactly 510 heads

17. 480 heads or more

18. Fewer than 470 tails

19. Fewer than 518 heads

20. More than 550 tails

A die is tossed 120 *times. Find the probability of getting each of the following. (Hint:* $\sigma = 4.08$)

21. Exactly 20 fives

22. Exactly 24 sixes

23. Exactly 17 threes

24. Exactly 22 twos

25. More than 18 threes

26. Fewer than 22 sixes

Management *Two percent of the hamburgers sold at Tom's Burger Queen are defective. Tom sold 10,000 burgers last week. Find the probability that among these burgers*

27. fewer than 170 were defective;

28. more than 222 were defective.

Natural Science *A new drug cures 80% of the patients to whom it is administered. It is given to 25 patients. Find the probability that among these patients*

29. exactly 20 are cured;

30. exactly 23 are cured;

31. all are cured;

32. no one is cured;

33. 12 or fewer are cured;

34. between 17 and 23 are cured.

35. Rework Exercises 49–52 of Section 9.3 using the normal curve to approximate the binomial probabilities. Compare your answers for the binomial distribution with the results found in Section 9.3.

36. A coin is tossed 100 times. Find the probability of
   (a) exactly 50 heads;   (b) at least 55 heads;
   (c) no more than 40 heads.

Chapter 10   Review Exercises

In Exercises 1 and 2, (a) write a frequency distribution; (b) draw a histogram; (c) draw a frequency polygon.

1. The following numbers give the sales in dollars for the lunch hour at a local hamburger store for the last twenty Fridays. (Use intervals 450–474, 475–499, and so on.)

   480 451 501 478 512 473 509 515 458 566
   516 535 492 558 488 547 461 475 492 471

2. The number of units carried in one semester by the students in a business mathematics class was as follows. (Use intervals of 9–10, 11–12, 13–14, 15–16.)

   10   9 16 12 13 15 13 16 15 11 13
   12 12 15 12 14 10 12 14 15 15 13

Find the mean for each of the following.

3. 41, 60, 67, 68, 72, 74, 78, 83, 90, 97

4. 105, 108, 110, 115, 106, 110, 104, 113, 117

5.
Interval	Frequency
10–19	6
20–29	12
30–39	14
40–49	10
50–59	8

6.
Interval	Frequency
40–44	2
45–49	5
50–54	7
55–59	10
60–64	4
65–69	1

Find the median and the mode (or modes) for each of the following.

7. 32, 35, 36, 44, 46, 46, 59

8. 38, 36, 42, 44, 38, 36, 48, 35

Find the modal class for the distributions of

9. Exercise 5 above;

10. Exercise 6 above.

Find the range and standard deviation for each of the following distributions.

11. 14, 17, 18, 19, 32

12. 26, 43, 51, 29, 37, 56, 29, 82, 74, 93

Find the standard deviation for the following.

13. Exercise 5 above

14. Exercise 6 above

15. The annual returns of two stocks for three years are given below.

	1986	1987	1988
Stock I	11%	−1%	14%
Stock II	9%	5%	10%

(a) Find the mean and standard deviation for each stock over the three-year period.
(b) If you are looking for security with an 8% return, which of these two stocks would you choose?

16. The weight gains of 2 groups of 10 rats fed on two different experimental diets were as follows.

Weight gains

Diet A	1	0	3	7	1	1	5	4	1	4
Diet B	2	1	1	2	3	2	1	0	1	0

Compute the mean and standard deviation for each group and compare them to answer the questions below.
(a) Which diet produced the greatest mean gain?
(b) Which diet produced the most consistent gain?

17. A probability distribution has an expected value of 28 and a standard deviation of 4. Use Chebyshev's Theorem (Section 10.3, Exercises 17–26) to decide what percent of the distribution is
(a) between 20 and 36;
(b) less than 23.2 or greater than 32.8.

18. (a) Find the percent of the area under a normal curve within 2.5 standard deviations of the mean.
(b) Compare your answer to part (a) with the result using Chebyshev's Theorem.

Find the following areas under the standard normal curve.

19. Between $z = 0$ and $z = 1.27$

20. Between $z = -1.88$ and $z = 2.10$

21. To the left of $z = -.41$

22. To the right of $z = 2.50$

On standard IQ tests, the mean is 100, with a standard deviation of 15. The results are very close to fitting a normal curve. Suppose an IQ test is given to a very large group of people. Find the percentage of those people whose IQ score is

23. more than 130;   **24.** less than 85;

25. between 85 and 115.

26. A machine which fills quart milk cartons is set to fill them with 32.1 oz. of milk. If the actual contents of the cartons vary normally, with a standard deviation of .1 oz., what percent of the cartons contain less than a quart (32 oz.)?

27. The probability that a can of peaches from a certain cannery is defective is .005. A sample of 4 cans is selected at random. Write a distribution for the number of defective cans in the sample, and give its mean and standard deviation.

28. Use the normal curve to approximate the probability of getting the following number of heads in 15 tosses of a coin.
(a) Exactly 7   (b) Between 7 and 10
(c) At least 10

29. The probability that a small business will go bankrupt in its first year is .21. For 50 such small businesses, use the normal curve to approximate the following probabilities.
(a) Exactly 8 go bankrupt
(b) No more than 2 go bankrupt

The average resident of a certain Eastern suburb spends 42 minutes per day commuting, with a standard deviation of 12 minutes. Find the percent of all residents of this suburb who commute

30. at least 50 minutes per day;

31. no more than 35 minutes per day;

32. between 32 and 40 minutes per day;

33. between 38 and 60 minutes per day.

About 6% of the frankfurters produced by a certain machine are overstuffed, and thus defective. Find the probability that, in a sample of 500 frankfurters,

34. 25 or fewer are overstuffed (Hint: $\sigma = 5.3$);

35. exactly 30 are overstuffed.

36. Find a z-score such that 8% of the area under the curve is to the right of z.

An area which is infested with fruit flies is to be sprayed with a chemical which is known to be 98% effective for each application. Assume a sample of 100 flies is checked. Set up the calculation; don't solve.

37. Find the probability that 95% of the flies are killed in one application.

38. Find the probability that at least 95% of the flies are killed in one application.

39. Find the probability that at least 90% of the flies are killed in one application.

40. Find the probability that all the flies are killed in one application.

Case 14

Inventory Control

A department store must control its inventory carefully.* It should not reorder too often, because it then builds up a large warehouse full of merchandise, which is expensive to hold. On the other hand, it must reorder sufficiently often to be sure of having sufficient stock to meet customer demand. The company desires a simple chart that can be used by its employees to determine the best possible time to reorder merchandise. The merchandise level on hand will be checked periodically. At the end of each period, if the level on hand is less than some predetermined level given in the chart, which considers sales rate and waiting time for orders, the item will be reordered. The example uses the following variables.

F = frequency of stock review (in weeks)

r = acceptable risk of being out of stock (in percent)

P = level of inventory at which reordering should occur

L = waiting time for order to arrive (in weeks)

S = sales rate (in units per week)

M = minimum level of merchandise to guarantee that the probability of being out of stock is no higher than r

z = z-score (from Table 10 at back of book) corresponding to r

Goods should not be reordered until inventory on hand has declined to a level less than the rate of sales per week, S, times the sum of the number of weeks until the next stock review, F, and the expected waiting time in weeks, L, plus a minimum level of merchandise, M, necessary to guarantee that the probability of being out of stock is no higher than r. That is, $P = S(F + L) + M$.

To find M, which depends on S, F, and L, assume that both sales and waiting time are normally distributed. With this assumption, using a formula from more advanced statistics courses gives $M = z\sqrt{2S(F + L)}$, where z is the z-score (from Table 10) corresponding to r, and $\sqrt{2S(F + L)}$

is the standard deviation of normally distributed deviations in sales rate and waiting time.

Combining these two formulas,

$$P = S(F + L) + z\sqrt{2S(F + L)}.$$

Suppose the firm wishes to be 95% sure of having goods to sell, so that $r = 5\%$. From Table 10, $z = 1.64$. Hence, for $r = 5\%$,

$$P = S(F + L) + 1.64\sqrt{2S(F + L)}.$$

Based on this formula, the following chart was prepared.

Reorder Levels (P)

Reorder merchandise when inventory on hand
falls below the levels given in the chart.

$r = 5\%$, $F = 1$ week

Rate of sales (S) (units/week)	Waiting time in weeks for order (L)					
	1	2	3	4	5	6
9	28	39	50	61	71	81
10	30	43	55	66	78	89
11	33	46	59	72	85	97
12	35	50	64	78	92	105
13	38	54	69	84	99	113
14	40	57	83	89	105	121
15	43	61	78	95	112	129

$r = 5\%$, $F = 4$ weeks

Rate of sales (S) (units/week)	Waiting time in weeks for order (L)					
	1	2	3	4	5	6
1	10	12	13	15	16	17
2	17	20	23	25	28	30
3	24	28	32	35	39	43
4	30	35	40	45	50	55

To use the chart, for example, if an item is reviewed every 4 weeks ($F = 4$), the waiting time for a reorder is 3 weeks ($L = 3$), and the rate of sales is 2 per week ($S = 2$), then the item should be reordered when the number on hand falls below 23.

---

*Example supplied by Leonard W. Cooper, Operations Research Project Director, Federated Department Stores.

Exercises

1. Suppose an item is reviewed weekly, and the waiting time for a reorder is 4 weeks. If the average sales per week of the item is 12 units, and the current inventory level is 85 units, should it be reordered? What if the inventory level is 50 units?

2. Suppose an item is reviewed every four weeks. If orders require a 5-week waiting time, and if sales average 3 units per week, should the item be reordered if current inventory is 50 units? What if current inventory is 30?

11 Differential Calculus

The graph of Figure 11.1 shows the profit earned by a firm from the production of x hundred items. The graph shows that profit increases quickly to the point where x = 3. At this point, the profit is $4000. For the firm to produce more than x = 3 (or 300) items, a second factory must be reopened, with the additional costs for the opening causing profits to fall. The graph shows that only after 600 items are produced does profit again reach $4000.

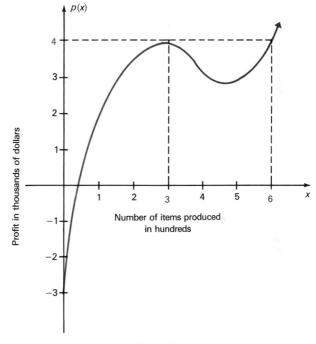

Figure 11.1

Optimum profit depends on sales—in the example above, if the sales manager says that only 500 items can be sold, then the optimum profit is found when x = 3 and the second factory should not be opened. On the other hand, if a sudden demand means that well over 600 items can be sold, then the second factory should be opened.

This chapter will show how one branch of calculus, called **differential calculus,** is used to find the optimum values of functions giving profits, costs, nutritional values, and so on. In addition, it will show how differential calculus is used to describe *rates of change* in management, social science, and biology.

483

Differential calculus depends on the *derivative,* which is introduced in Section 4 of this chapter. The definition of derivative requires the idea of a limit, which is discussed in the next section.

11.1 Limits

The idea of a limit arises when we consider a function $y = f(x)$ defined on some interval and ask "What happens to the values of $f(x)$ as x gets closer and closer to some fixed number?" For example, Figure 11.2 shows the graph of a function f. The arrowheads suggest that as x gets closer and closer to the number 5 (on the x-axis), the values of $f(x)$ get closer and closer to the number 4 (on the y-axis). That is, the limit of $f(x)$ as x approaches 5 is the number 4, written

$$\lim_{x \to 5} f(x) = 4.$$

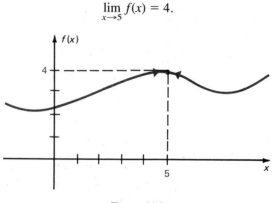

Figure 11.2

Example 1 **(a)** The graph in Figure 11.3 is of a function g. The open circle at $(3, -1)$ means that the function is not defined at $x = 3$. Since $g(3)$ is not defined, it is meaningless to ask about the value of g *at* 3. However, we can still ask how the values of g behave when x is *close to* 3. The graph suggests that as x gets closer and

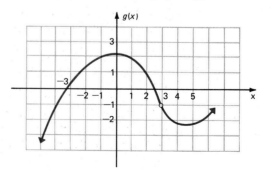

Figure 11.3

1 Find $\lim\limits_{x\to 2} f(x)$ for the following.

(a)

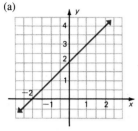

(b)

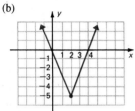

Answer:

(a) 4

(b) −5

2 Use the graph of Figure 11.4 to find the following limits.

(a) $\lim\limits_{x\to 7} h(x)$

(b) $\lim\limits_{x\to -1} h(x)$

(c) $\lim\limits_{x\to 9} h(x)$

Answer:

(a) 5

(b) −4

(c) 5

3 Graph $f(x) = 1/x$ and find the following limits.

(a) $\lim\limits_{x\to 1} f(x)$

(b) $\lim\limits_{x\to -3} f(x)$

(c) $\lim\limits_{x\to 2} f(x)$

Answer:

(a) 1

(b) −1/3

(c) 1/2

closer to 3 (but doesn't *equal* 3), the values of $g(x)$ get closer and closer to −1. As above, this is written

$$\lim_{x\to 3} g(x) = -1.$$

For values of x near 3, the function g never takes on the value −1. However, as x gets as close as desired to 3, the values of g get as close as desired to −1.

This example suggests that $x \to a$ means that x approaches as close as desired to a, without necessarily reaching a.

(b) The graph in Figure 11.3 also shows that as x gets closer and closer to −3, the values of $g(x)$ get closer and closer to 0, or

$$\lim_{x\to -3} g(x) = 0.$$

Here $g(-3)$ is defined and equals 0. **1**

Example 2 **(a)** The graph of the function h in Figure 11.4 shows that as x approaches 4 from the left, the values of $h(x)$ approach 3. However, as x gets closer and closer to 4 from the right, the values of $h(x)$ approach 5. Since the values of $h(x)$ approach two different numbers depending on whether x approaches 4 from the right or left,

$$\lim_{x\to 4} h(x) \text{ does not exist.}$$

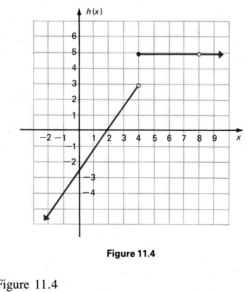

Figure 11.4

(b) Also from Figure 11.4

$$\lim_{x\to 6} h(x) = 5 \quad \text{and} \quad \lim_{x\to 8} h(x) = 5. \quad \text{**2** **3**}$$

The examples above suggest the following intuitive definition of limit.

> **Limit of a Function** If the values $f(x)$ of a function f can be made as close as desired to a single number L for all values of x sufficiently close to the number a, with $x \neq a$, then L is the limit of $f(x)$ as x approaches a, written
>
> $$\lim_{x \to a} f(x) = L.$$

This definition is *intuitive* because we have not defined "as close as desired." A more formal definition of limit would be needed for proving the rules for limits given later in the section.

Some Observations on Limits

1. If a limit exists, it is unique. The values of a function cannot approach two different limits at the same time.
2. A limit need not always exist. (See Figure 11.5 at $x = -4$ and at $x = 3$.)
3. A limit may exist at $x = a$, even if the function is not defined there. (See Figure 11.5 at $x = 4$.)

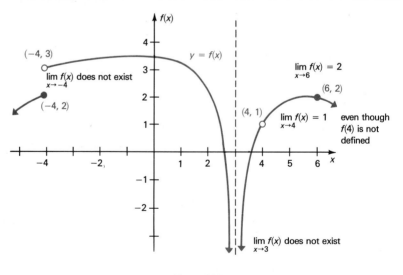

Figure 11.5

In the examples above, limits were found by examining a graph. Other methods for finding limits include, for example, looking at several values of the function as x approaches a given number.

Example 3 Find $\lim\limits_{x \to 1} \dfrac{x^2 - 4}{x - 2}$.

Find this limit by evaluating $f(x)$ for several values of x close to 1 on either side of 1. The results for several different values of x are shown in the following table.

Closer and closer to 1

x	0.8	0.9	0.99	0.9999	1.0000001	1.0001	1.001	1.01	1.05	1.1
$f(x)$	2.8	2.9	2.99	2.9999	3.0000001	3.0001	3.001	3.01	3.05	3.1

Closer and closer to 3

4 Find $\lim\limits_{x \to 4} \dfrac{x^2 - 4}{x - 2}$ by completing the following table. A calculator would be helpful.

x	$f(x)$
3.8	
3.9	
3.99	
3.999	
4.001	
4.1	
4.2	

Answer:

5.8; 5.9; 5.99; 5.999; 6.001; 6.1; 6.2; limit is 6

The table suggests that the values of $f(x)$ get closer and closer to 3 as the values of x get closer and closer to 1, or

$$\lim_{x \to 1} \frac{x^2 - 4}{x - 2} = 3. \quad \boxed{4}$$

Example 4 Find $\lim\limits_{x \to 2} \dfrac{x^2 - 4}{x - 2}$.

The value $x = 2$ is not in the domain of the function, since 2 makes the denominator equal 0. However, though $f(2)$ does not exist, f may still have a limit as x approaches 2. (Recall Example 1 (a) above; the limit existed at $x = 3$ even though the function was not defined at $x = 3$.) Make a table of values, choosing values of x close to 2.

Closer and closer to 2

x	1.8	1.9	1.99	1.9999	2.0000001	2.00001	2.001	2.05	2.1
$f(x)$	3.8	3.9	3.99	3.9999	4.0000001	4.00001	4.001	4.05	4.1

Closer and closer to 4

5 Complete the following table of values to find $\lim\limits_{x \to 3} \dfrac{x^2 - 9}{x - 3}$.

x	$f(x)$
2.9	
2.99	
2.999	
3.001	
3.01	
3.1	

Answer:

5.9; 5.99; 5.999; 6.001; 6.01; 6.1; limit is 6

The table suggests that the values of $f(x)$ get closer and closer to 4 as x gets closer and closer to 2, and so

$$\lim_{x \to 2} \frac{x^2 - 4}{x - 2} = 4. \quad \boxed{5}$$

In the last two examples, you may have noticed that, for $x \ne 2$,

$$\frac{x^2 - 4}{x - 2} = \frac{(x + 2)(x - 2)}{x - 2} = x + 2.$$

The values in the tables could have been found more easily by evaluating $x + 2$ instead of $(x^2 - 4)/(x - 2)$; the limits could also have been found in this way. For example,

$$\lim_{x \to 2} \frac{x^2 - 4}{x - 2} = \lim_{x \to 2} (x + 2).$$

As x gets closer and closer to 2, the expression $x + 2$ will get closer and closer to $2 + 2 = 4$, with

$$\lim_{x \to 2} \frac{x^2 - 4}{x - 2} = \lim_{x \to 2} (x + 2) = 4.$$

A graph of $f(x) = (x^2 - 4)/(x - 2)$ is shown in Figure 11.6.

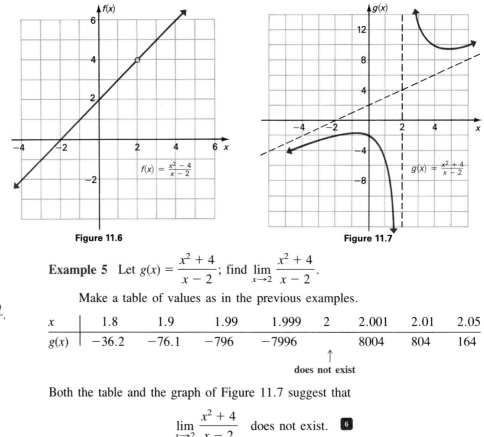

Figure 11.6 **Figure 11.7**

Example 5 Let $g(x) = \dfrac{x^2 + 4}{x - 2}$; find $\lim\limits_{x \to 2} \dfrac{x^2 + 4}{x - 2}$.

Make a table of values as in the previous examples.

x	1.8	1.9	1.99	1.999	2	2.001	2.01	2.05
$g(x)$	-36.2	-76.1	-796	-7996		8004	804	164

↑
does not exist

Both the table and the graph of Figure 11.7 suggest that

$$\lim_{x \to 2} \frac{x^2 + 4}{x - 2} \quad \text{does not exist.} \quad \boxed{6}$$

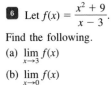

6 Let $f(x) = \dfrac{x^2 + 9}{x - 3}$.

Find the following.

(a) $\lim\limits_{x \to 3} f(x)$

(b) $\lim\limits_{x \to 0} f(x)$

Answer:

(a) Does not exist

(b) -3

The rules for limits given in the following box are often helpful in evaluating limits.

Rules for Limits For any real number a, assume that $\lim\limits_{x \to a} f(x)$ and $\lim\limits_{x \to a} g(x)$ both exist; then

(a) if k is a constant real number, $\lim\limits_{x \to a} k = k$;

(b) $\lim\limits_{x \to a} [f(x) + g(x)] = \lim\limits_{x \to a} f(x) + \lim\limits_{x \to a} g(x)$

(the limit of a sum equals the sum of the limits);

(c) $\lim\limits_{x \to a} [f(x) \cdot g(x)] = [\lim\limits_{x \to a} f(x)] \cdot [\lim\limits_{x \to a} g(x)]$

(the limit of a product equals the product of the limits);

(d) $\lim\limits_{x \to a} \dfrac{f(x)}{g(x)} = \dfrac{\lim\limits_{x \to a} f(x)}{\lim\limits_{x \to a} g(x)}$ if $\lim\limits_{x \to a} g(x) \neq 0$

(the limit of a quotient equals the quotient of the limits);

(e) $\lim\limits_{x \to a} [f(x)]^n = [f(a)]^n$ for any real number n for which $[f(a)]^n$ exists;

(f) $\lim\limits_{x \to a} f(x) = \lim\limits_{x \to a} g(x)$ if $f(x) = g(x)$ for $x \neq a$.

---

7 Use the limit rules to find the following.

(a)
$\lim\limits_{x \to 4} (3x - 9)$

(b)
$\lim\limits_{x \to -1} (2x^2 - 4x + 1)$

(c)
$\lim\limits_{x \to 0} (9x^3 - 8x^2 + 6x + 4)$

Answer:

(a) 3

(b) 7

(c) 4

8 Find the following.

(a) $\lim\limits_{x \to 2} \sqrt{3x + 3}$

(b) $\lim\limits_{x \to -4} \sqrt{8 - 7x}$

(c) $\lim\limits_{x \to 2} \sqrt{x^2 + 3x + 4}$

Answer:

(a) 3

(b) 6

(c) $\sqrt{14}$

Example 6 Find $\lim\limits_{x \to 2} (4x^2 - 2x + 3)$.

By rule (b) above,

$$\lim_{x \to 2} (4x^2 - 2x + 3) = \lim_{x \to 2} 4x^2 + \lim_{x \to 2} (-2x) + \lim_{x \to 2} 3.$$

Using rules (a), (c), and (e) gives

$$\lim_{x \to 2} 4x^2 + \lim_{x \to 2} (-2x) + \lim_{x \to 2} 3$$

$$= (\lim_{x \to 2} 4)(\lim_{x \to 2} x^2) + (\lim_{x \to 2} -2)(\lim_{x \to 2} x) + \lim_{x \to 2} 3$$

$$= (4)(2^2) + (-2)(2) + 3$$

$$= 15. \quad \boxed{7}$$

Example 7 Find $\lim\limits_{x \to 9} \sqrt{2x - 2}$.

Use rule (e), with $n = 1/2$, to get

$$\lim_{x \to 9} \sqrt{2x - 2} = \sqrt{2 \cdot 9 - 2} = \sqrt{18 - 2} = \sqrt{16} = 4. \quad \boxed{8}$$

Sometimes the rules of limits alone are not enough to find a limit. As the next example shows, the function may need to be expressed in a different form first.

Example 8 Find $\lim\limits_{x \to 4} \dfrac{\sqrt{x} - 2}{x - 4}$.

As x approaches 4, both the numerator and the denominator approach 0, giving the meaningless expression 0/0. Rationalize the *numerator* by multiplying numerator and denominator by $\sqrt{x} + 2$, to get

$$\frac{\sqrt{x} - 2}{x - 4} \cdot \frac{\sqrt{x} + 2}{\sqrt{x} + 2} = \frac{\sqrt{x} \cdot \sqrt{x} - 2\sqrt{x} + 2\sqrt{x} - 4}{(x - 4)(\sqrt{x} + 2)}$$

$$= \frac{x - 4}{(x - 4)(\sqrt{x} + 2)} \qquad \sqrt{x} \cdot \sqrt{x} = x$$

$$= \frac{1}{\sqrt{x} + 2}.$$

Now use the rules for limits.

$$\lim_{x \to 4} \frac{\sqrt{x} - 2}{x - 4} = \lim_{x \to 4} \frac{1}{\sqrt{x} + 2} = \frac{1}{\sqrt{4} + 2} = \frac{1}{2 + 2} = \frac{1}{4}$$

9 Find the following.

(a) $\lim\limits_{x \to 1} \dfrac{\sqrt{x} - 1}{x - 1}$

(b) $\lim\limits_{x \to 9} \dfrac{\sqrt{x} - 3}{x - 9}$

Answer:

(a) 1/2

(b) 1/6

Example 9 The graph of Figure 11.8 shows the profit from producing x units of a certain product in a given day. If one shift of workers can produce no more than 25 items, a second shift must be called to work when more than 25 items are needed. This second shift increases the fixed costs and causes a drop in profits (indicated by a break in the graph at $x = 25$) unless many more than 25 items are needed. (The graph shows that on 30 items, for example, the profit is about $340, and that production must increase to 35 items before profit returns to $400 and the second shift produces an increase in profits.) Profit when production is near 25 items

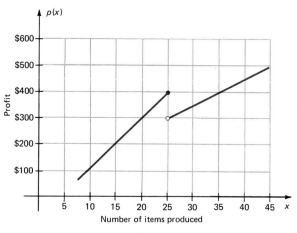

Figure 11.8

depends on whether or not the second shift has been called in. As the graph suggests,

10 Use the graph of Figure 11.8 to find the following.

(a) $\lim\limits_{x \to 20} p(x)$

(b) $\lim\limits_{x \to 30} p(x)$

Answer:

(a) $300

(b) $340

$$\lim_{x \to 25} p(x) \quad \text{does not exist.} \quad \boxed{10}$$

Suppose a small pond normally contains 12 units of dissolved oxygen in a fixed volume of water. Suppose also that at time $t = 0$ a quantity of organic waste is introduced into the pond, with the oxygen concentration t weeks later given by

$$f(t) = \frac{12t^2 - 15t + 12}{t^2 + 1}.$$

For example, after 2 weeks, the pond contains

$$f(2) = \frac{12 \cdot 2^2 - 15 \cdot 2 + 12}{2^2 + 1} = \frac{30}{5} = 6$$

units of oxygen; and after 4 weeks,

$$f(4) = \frac{12 \cdot 4^2 - 15 \cdot 4 + 12}{4^2 + 1} \approx 8.5$$

units. Choosing several values of t and finding the corresponding values of $f(t)$ leads to the graph in Figure 11.9.

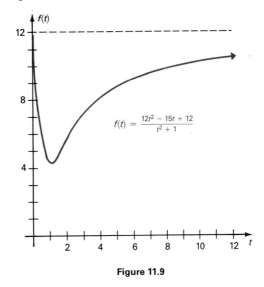

Figure 11.9

The graph suggests that as time goes on, the oxygen level returns closer and closer to the original 12 units. Because of this, the limit of $f(t)$ as t increases without bound is 12, written

$$\lim_{t \to \infty} f(t) = 12.$$

The symbol ∞ is read "infinity," while $t \to \infty$ is read "t increases without bound." (Also, $-\infty$ is read "negative infinity," with $t \to -\infty$ read "t decreases without bound.")

The graphs of $f(x) = 1/x$ (in black), and $g(x) = 1/x^2$ (in color), shown in Figure 11.10 lead to examples of such *limits at infinity*.

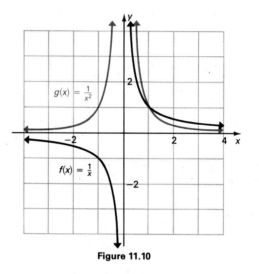

Figure 11.10

The graphs indicate that both $\lim\limits_{x \to \infty} (1/x) = 0$ and $\lim\limits_{x \to \infty} (1/x^2) = 0$, suggesting the following result.

> For any positive real number n and any real number k,
>
> $$\lim_{x \to \infty} \frac{k}{x^n} = 0 \qquad \text{and} \qquad \lim_{x \to -\infty} \frac{k}{x^n} = 0.$$

Example 10 Find the following limits.

(a) $\lim\limits_{x \to \infty} \dfrac{8x + 6}{3x - 1}$

By the rule given above, $\lim\limits_{x \to \infty} k/x^n = 0$. To use this rule with this quotient, divide each term in the numerator and denominator by x. Then use the limit theorems.

$$\lim_{x \to \infty} \frac{8x + 6}{3x - 1} = \lim_{x \to \infty} \frac{\dfrac{8x}{x} + \dfrac{6}{x}}{\dfrac{3x}{x} - \dfrac{1}{x}} = \lim_{x \to \infty} \frac{8 + \dfrac{6}{x}}{3 - \dfrac{1}{x}} = \frac{8 + 0}{3 - 0} = \frac{8}{3}$$

(b) $\displaystyle\lim_{x\to\infty} \frac{4x^2 - 6x + 3}{2x^2 - x + 4}$

Divide numerator and denominator by x^2, the highest power of x.

$$\lim_{x\to\infty} \frac{4x^2 - 6x + 3}{2x^2 - x + 4} = \lim_{x\to\infty} \frac{4 - \dfrac{6}{x} + \dfrac{3}{x^2}}{2 - \dfrac{1}{x} + \dfrac{4}{x^2}} = \frac{4 - 0 + 0}{2 - 0 + 0} = \frac{4}{2} = 2$$

11 Find the following limits.

(a) $\displaystyle\lim_{x\to\infty} \frac{4x^2 + 6x}{9x^2 - 3}$

(b) $\displaystyle\lim_{x\to\infty} \frac{3x^2 + 4x + 5}{8x^3 + 2x + 1}$

(c) $\displaystyle\lim_{x\to\infty} \frac{4x^2 + 9x + 1}{2x + 3}$

Answer:

(a) 4/9

(b) 0

(c) Does not exist

(c) $\displaystyle\lim_{x\to\infty} \frac{3x + 2}{4x^2 - 1} = \lim_{x\to\infty} \frac{\dfrac{3}{x} + \dfrac{2}{x^2}}{4 - \dfrac{1}{x^2}} = \frac{0 + 0}{4 - 0} = \frac{0}{4} = 0$

Here the highest power of x is x^2, which is divided into each term in the numerator and denominator.

(d) $\displaystyle\lim_{x\to\infty} \frac{9x^2 - 1}{3x + 5} = \lim_{x\to\infty} \frac{9 - \dfrac{1}{x^2}}{\dfrac{3}{x} + \dfrac{5}{x^2}} = \frac{9 - 0}{0 + 0} = \frac{9}{0}$

The result 9/0 is undefined; this means that this limit does not exist.

11.1 Exercises

In each of the following, use the graph to determine the value of the indicated limit. (See Examples 1 and 2.)

1. $\displaystyle\lim_{x\to 3} f(x)$

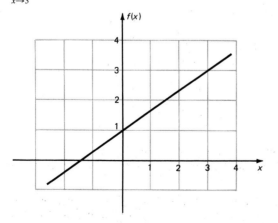

2. $\displaystyle\lim_{x\to 2} F(x)$

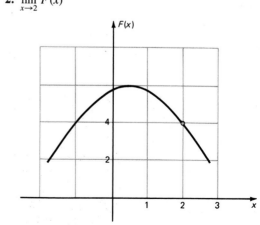

3. $\displaystyle\lim_{x \to -2} f(x)$

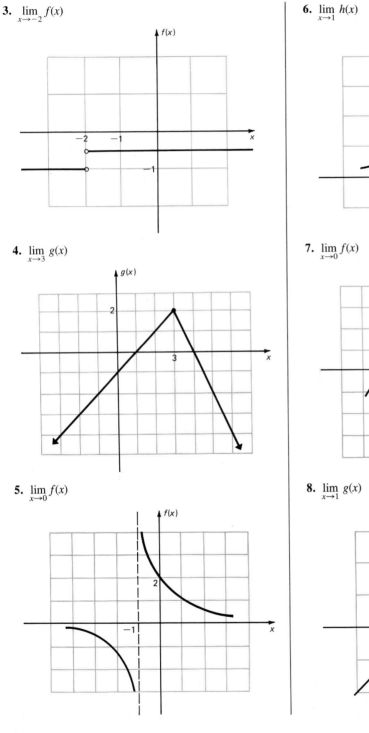

4. $\displaystyle\lim_{x \to 3} g(x)$

5. $\displaystyle\lim_{x \to 0} f(x)$

6. $\displaystyle\lim_{x \to 1} h(x)$

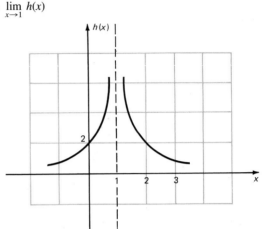

7. $\displaystyle\lim_{x \to 0} f(x)$

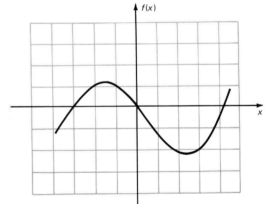

8. $\displaystyle\lim_{x \to 1} g(x)$

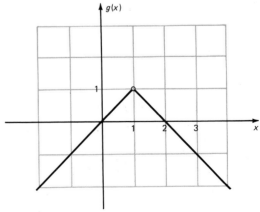

9. $\lim\limits_{x \to 3} F(x)$

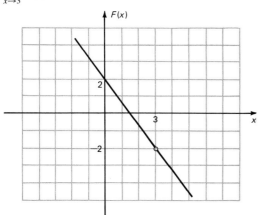

10. $\lim\limits_{x \to 0} f(x)$

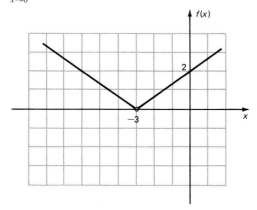

11. $\lim\limits_{x \to \infty} f(x)$

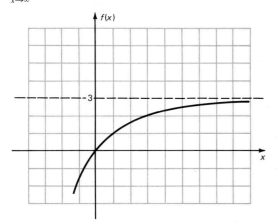

12. $\lim\limits_{x \to \infty} g(x)$

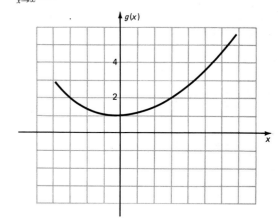

In each of the following, complete the table and use the results to find $\lim\limits_{x \to a} f(x)$. You will need a calculator with a $\sqrt{x}$ key for Exercises 19 and 20. (See Examples 3–5.)

13. $h(x) = x^2 + 2x + 1$, find $\lim\limits_{x \to 2} h(x)$.

x	1.9	1.99	1.999	2.001	2.01	2.1
$h(x)$			8.994001	9.006001		

14. $f(x) = 2x^2 - 4x + 3$, find $\lim\limits_{x \to 1} f(x)$.

x	.9	.99	.999	1.001	1.01	1.1
$f(x)$			1.000002	1.000002		

15. $g(x) = 2/x$, find $\lim\limits_{x \to 0} g(x)$.

x	$-.1$	$-.01$	$-.001$	.001	.01	.1
$g(x)$						

16. $h(x) = -5/x$, find $\lim\limits_{x \to 0} h(x)$.

x	$-.1$	$-.01$	$-.001$	.001	.01	.1
$h(x)$						

17. $k(x) = \dfrac{x^3 - 2x - 4}{x - 2}$, find $\lim\limits_{x \to 2} k(x)$.

x	1.9	1.99	1.999	2.001	2.01	2.1
$k(x)$						

18. $f(x) = \dfrac{2x^3 + 3x^2 - 4x - 5}{x + 1}$, find $\lim\limits_{x \to -1} f(x)$.

x	-1.1	-1.01	-1.001	$-.999$	$-.99$	$-.9$
$f(x)$						

19. $h(x) = \dfrac{\sqrt{x} + 2}{x - 1}$, find $\lim\limits_{x \to 1} h(x)$.

x	.9	.99	.999	1.001	1.01	1.1
$h(x)$						

20. $f(x) = \dfrac{\sqrt{x} - 3}{x - 3}$, find $\lim\limits_{x \to 3} f(x)$.

x	2.999	2.99	2.9	3.001	3.01	3.1
$f(x)$						

Use algebra and the rules for limits as needed to find the following limits that exist. (See Examples 6–8.)

21. $\lim\limits_{x \to 2} (2x^3 + 5x^2 + 2x + 1)$

22. $\lim\limits_{x \to -1} (4x^3 - x^2 + 3x - 1)$

23. $\lim\limits_{x \to 3} \dfrac{5x - 6}{2x + 1}$

24. $\lim\limits_{x \to -2} \dfrac{2x + 1}{3x - 4}$

25. $\lim\limits_{x \to 1} \dfrac{2x^2 - 6x + 3}{3x^2 - 4x + 2}$

26. $\lim\limits_{x \to 2} \dfrac{-4x^2 + 6x - 8}{3x^2 + 7x - 2}$

27. $\lim\limits_{x \to 3} \dfrac{x^2 - 9}{x - 3}$

28. $\lim\limits_{x \to -2} \dfrac{x^2 - 4}{x + 2}$

29. $\lim\limits_{x \to -2} \dfrac{x^2 - x - 6}{x + 2}$

30. $\lim\limits_{x \to 5} \dfrac{x^2 - 3x - 10}{x - 5}$

31. $\lim\limits_{x \to 3} \sqrt{x^2 - 4}$

32. $\lim\limits_{x \to 3} \sqrt{x^2 - 5}$

33. $\lim\limits_{x \to 4} \dfrac{-6}{(x - 4)^2}$

34. $\lim\limits_{x \to -2} \dfrac{3x}{(x + 2)^3}$

35. $\lim\limits_{x \to 0} \dfrac{x^3 - 4x^2 + 8x}{2x}$

36. $\lim\limits_{x \to 0} \dfrac{-x^5 - 9x^3 + 8x^2}{5x}$

37. $\lim\limits_{x \to 0} \dfrac{[1/(x + 3)] - 1/3}{x}$

38. $\lim\limits_{x \to 0} \dfrac{[-1/(x + 2)] + 1/2}{x}$

39. $\lim\limits_{x \to 25} \dfrac{\sqrt{x} - 5}{x - 25}$

40. $\lim\limits_{x \to 36} \dfrac{\sqrt{x} - 6}{x - 36}$

41. $\lim\limits_{x \to 5} \dfrac{\sqrt{x} - \sqrt{5}}{x - 5}$

42. $\lim\limits_{x \to 8} \dfrac{\sqrt{x} - \sqrt{8}}{x - 8}$

43. $\lim\limits_{h \to 0} \dfrac{(x + h)^2 - x^2}{h}$

44. $\lim\limits_{h \to 0} \dfrac{(x + h)^3 - x^3}{h}$

45. Management The graph below shows the profit from the daily production of x thousand kilograms of an industrial chemical. Use the graph to find the following limits. (See Example 9.)

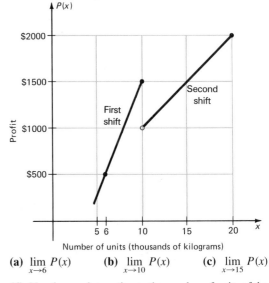

Number of units (thousands of kilograms)

(a) $\lim\limits_{x \to 6} P(x)$ **(b)** $\lim\limits_{x \to 10} P(x)$ **(c)** $\lim\limits_{x \to 15} P(x)$

(d) Use the graph to estimate the number of units of the chemical that must be produced before the second shift is profitable.

46. Management According to some economists, when the price of an essential commodity (such as gasoline) rises rapidly, consumption drops slowly at first. However, if the price rise continues, a "tipping" point may be reached, at which consumption takes a sudden, substantial drop. Suppose the graph on the right shows the consumption of gasoline, $G(t)$, in millions of gallons, in a certain area. We assume that the price is rising rapidly. Here t is time in months after the price began rising. Use the graph to find the following.

(a) $\lim\limits_{t \to 12} G(t)$ (b) $\lim\limits_{t \to 16} G(t)$

(c) The tipping point (in months)

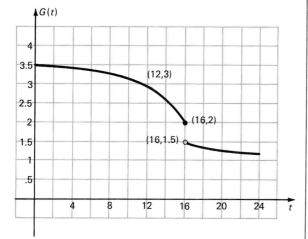

Find each of the following limits that exist. (*See Example 10.*)

47. $\lim\limits_{x \to \infty} \dfrac{3x}{5x - 1}$

48. $\lim\limits_{x \to \infty} \dfrac{5x}{3x - 1}$

49. $\lim\limits_{x \to \infty} \dfrac{2x + 3}{4x - 7}$

50. $\lim\limits_{x \to \infty} \dfrac{8x + 2}{2x - 5}$

51. $\lim\limits_{x \to \infty} \dfrac{x^2 + 2x}{2x^2 - 2x + 1}$

52. $\lim\limits_{x \to \infty} \dfrac{x^2 + 2x - 5}{3x^2 + 2}$

53. $\lim\limits_{x \to \infty} \dfrac{3x^3 + 2x - 1}{2x^4 - 3x^3 - 2}$

54. $\lim\limits_{x \to \infty} \dfrac{2x^2 + 11x - 10}{5x^3 + 3x^2 + 2x}$

55. $\lim\limits_{x \to \infty} \dfrac{2x^4 + 3x + 4}{x^2 - 4x + 1}$

56. $\lim\limits_{x \to \infty} \dfrac{2x^2 - 1}{3x^4 + 2}$

57. Political Science Members of a legislature must often vote repeatedly on the same issue. As time goes on, the members may change their vote. A formula* for the chance that a legislator will vote *yes* on the nth roll call vote on the same issue is

$$p_n = \frac{1}{2} + \left(p_0 - \frac{1}{2}\right)(1 - 2p)^n,$$

where n is the number of roll calls taken, p_n is the chance that the member will vote *yes* on the nth roll call vote, p is the chance that the legislator will change his or her position on successive roll calls, and p_0 is the chance that the member favors the issue at the beginning (p_n, p, and p_0 are always between 0 and 1, inclusive). Suppose that $p_0 = .7$ and $p = .2$. Find p_n for
(a) $n = 2$;   (b) $n = 4$;   (c) $n = 8$.
(d) Find $\lim\limits_{n \to \infty} p_n$.

Use a table of values to find the following limits.

58. $\lim\limits_{x \to 1.3} \dfrac{x^3 - 2x^2 + 5x - 8}{2x^4 - 3x^2}$

59. $\lim\limits_{x \to 5.2} \dfrac{x^4 - 6x^3 - 10x + 5}{4x^3 + 6x^2 - 12x - 6}$

60. $\lim\limits_{x \to 0} \dfrac{x^3 - 4x^2 + 2x}{x^4 + 5x^3 + 4x}$

61. $\lim\limits_{x \to 1} \dfrac{3.1x^3 + 5.2\sqrt{x}}{-2x^3 + 12.3x - 2\sqrt{x}}$

11.2 Continuity

Intuitively speaking, a function is **continuous** at a point if the graph of the function can be drawn in the vicinity of the point without lifting the pencil from the paper. If a function is not continuous at a point, it is **discontinuous** at the point. For example, the function shown in Figure 11.11 (a) is discontinuous at $x = 2$ because of the

*See John W. Bishir and Donald W. Drewes, *Mathematics in the Behavioral and Social Sciences* (New York: Harcourt Brace Jovanovich, 1970), p. 538.

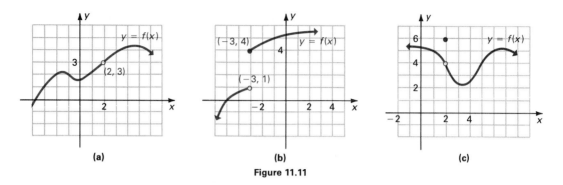

Figure 11.11

① Find any points of discontinuity for the following functions.

(a)

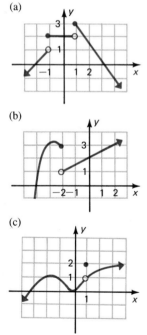

(b)

(c)

Answer:

(a) −1, 1

(b) −2

(c) 1

"hole" in the graph at (2, 3). The function of Figure 11.11 (b) is discontinuous at $x = -3$ because of the "jump" in the graph at $x = -3$. Finally, the function in Figure 11.11 (c) is discontinuous when $x = 2$. Even though $f(2)$ is defined (here $f(2) = 6$), the value of $f(2)$ is not "close" to the values of $f(x)$ for values of x "close" to 2. That is,

$$\lim_{x \to 2} f(x) \neq f(2). \quad ①$$

Now that we have seen some different ways for a function to have a point of discontinuity, look again at Figure 11.11(b). The function graphed in Figure 11.11(b), which is *not* continuous when $x = -3$, *is* continuous when $x = -5$. This difference comes from the fact that $f(-5)$ is defined, so that a point on the graph corresponds to $x = -5$. Second, $\lim_{x \to -5} f(x)$ exists. As x approaches -5 from either side of -5, the values of $f(x)$ approach one particular number, $f(-5)$, which is 0. These examples suggest the following definition.

Definition of Continuity at a Point   A function f is **continuous** at $x = c$ if

(a) $f(c)$ is defined;

(b) $\lim_{x \to c} f(x)$ exists;

(c) $\lim_{x \to c} f(x) = f(c)$.

If f is not continuous at $x = c$, it is **discontinuous** there.

Example 1   Tell why the following functions are discontinuous at the indicated points.

(a) $f(x)$ at $x = 3$ in Figure 11.12

The open circle on the graph of Figure 11.12 at the point where $x = 3$ means that $f(3)$ does not exist. Because of this, part (a) of the definition fails.

(b) $h(x)$ at $x = 0$ in Figure 11.13

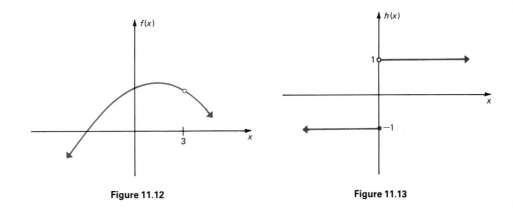

Figure 11.12   Figure 11.13

The graph of Figure 11.13 shows that $h(0) = -1$. Also, as x approaches 0 from the left, $h(x)$ is -1. However, as x approaches 0 from the right, $h(x)$ is 1. As mentioned in Section 11.1, for a limit to exist at a particular value of x, the values of $h(x)$ must approach a single number. Since no single number is approached by the values of $h(x)$ as x approaches 0, $\lim\limits_{x \to 0} h(x)$ does not exist, and part (b) of the definition fails.

(c) $g(x)$ at $x = 4$ in Figure 11.14

2 Tell why the following functions are discontinuous at the indicated points.

(a)

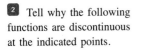
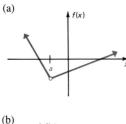

(b)

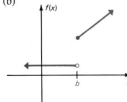

Answer:
(a) $f(a)$ does not exist
(b) $\lim\limits_{x \to b} f(x)$ does not exist

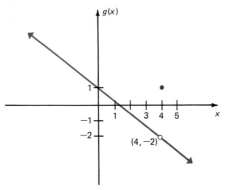

Figure 11.14

In Figure 11.14 the heavy dot above 4 shows that $g(4)$ is defined. In fact, $g(4) = 1$. However, the graph also shows that

$$\lim_{x \to 4} g(x) = -2,$$

so $\lim\limits_{x \to 4} g(x) \neq g(4)$, and part (c) of the definition fails. **2**

When discussing the continuity of a function, it is often helpful to use **interval notation,** described as follows.

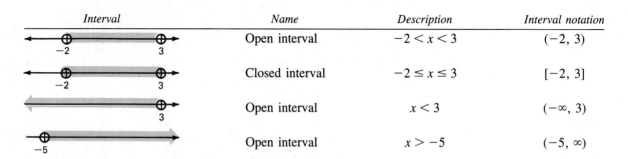

Interval	Name	Description	Interval notation
	Open interval	$-2 < x < 3$	$(-2, 3)$
	Closed interval	$-2 \le x \le 3$	$[-2, 3]$
	Open interval	$x < 3$	$(-\infty, 3)$
	Open interval	$x > -5$	$(-5, \infty)$

③ Write each of the following using interval notation.
(a)
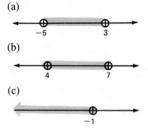
(b)

(c)

Answer: (a) $(-5, 3)$
(b) $[4, 7]$ (c) $(-\infty, -1]$

④ Are the functions with graphs as shown continuous on the indicated intervals?
(a) $(-4, -2)$; $(-3, 0)$
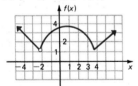

(b) $(-1, 1)$; $(0, 2)$

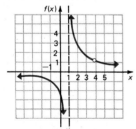

Answer:
(a) Yes; no (b) Yes; no

The symbol ∞ does not represent a number; ∞ is used for convenience in interval notation to indicate that the interval extends without bound in the positive direction. Also, $-\infty$ indicates no bound in the negative direction. ③

Continuity at a point was defined above; *continuity on an open interval* is discussed next.

> If a function is continuous at each point of an open interval, it is said to be **continuous on the open interval.**

Example 2 Is the function of Figure 11.15 continuous on the following open intervals?
(a) $(-2, -1)$

The function is discontinuous only at $x = -2, 0,$ and 2. Thus, it is continuous at every point of the open interval $(-2, -1)$, and is continuous on the open interval.
(b) $(1, 3)$

Since this interval includes the point of discontinuity $x = 2$, the function is not continuous on the open interval $(1, 3)$. ④

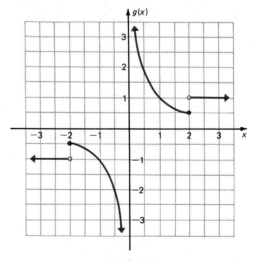

Figure 11.15

Example 3 A trailer rental firm charges a flat $4 to rent a hitch. The trailer itself is rented for $11 per day or fraction of a day. Let $C(x)$ represent the cost of renting a hitch and trailer for x days.

(a) Graph C.

The charge for one day is $4 for the hitch and $11 for the trailer, or $15. In fact, if $0 < x \le 1$, then $C(x) = 15$. To rent the trailer for more than one day, but not more than two days, the charge is $4 + 2 \cdot 11 = 26$ dollars. For any value of x satisfying $1 < x \le 2$, $C(x) = 26$. Also, if $2 < x \le 3$, then $C(x) = 37$. These results lead to the graph of Figure 11.16.

(b) Find any points of discontinuity for C.

As the graph suggests, C is discontinuous at 1, 2, 3, 4, and all other positive integers. **5**

5 Suppose the cost is $2.25 to mail a package weighing up to one pound plus $.50 for each additional pound or fraction of a pound. Let $P(x)$ represent the cost of mailing a package weighing x pounds. Find any points of discontinuity for P.

Answer:

$x = 2.25, 2.75, 3.25, 3.75, \ldots$

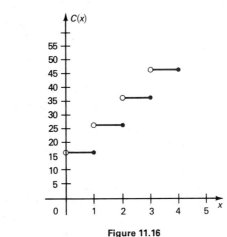

Figure 11.16

11.2 Exercises

Find all points of discontinuity for the following.

1.

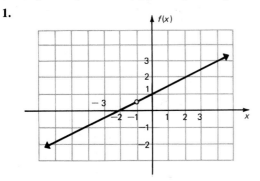

2.

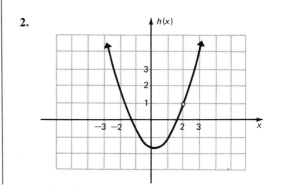

3.

4.

5.

6.

7.

8.

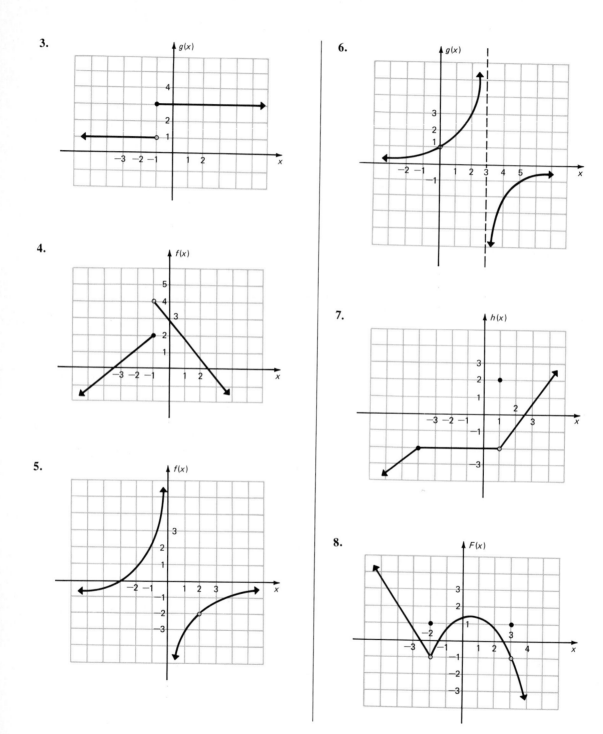

Are the functions defined as follows continuous at the given values of x? (See Examples 1 and 2.)

9. $f(x) = \dfrac{2}{x-3}$; $x = 0, x = 3$

10. $f(x) = \dfrac{6}{x}$; $x = 0, x = -1$

11. $g(x) = \dfrac{1}{x(x-2)}$; $x = 0, x = 2, x = 4$

12. $h(x) = \dfrac{-2}{3x(x+5)}$; $x = 0, x = 3, x = -5$

13. $h(x) = \dfrac{1+x}{(x-3)(x+1)}$; $x = 0, x = 3, x = -1$

14. $g(x) = \dfrac{-2x}{(2x+1)(3x+6)}$; $x = 0, x = -\dfrac{1}{2}$, $x = -2$

15. $k(x) = \dfrac{5+x}{2+x}$; $x = 0, x = -2, x = -5$

16. $f(x) = \dfrac{4-x}{x-9}$; $x = 0, x = 4, x = 9$

17. $g(x) = \dfrac{x^2-4}{x-2}$; $x = 0, x = 2, x = -2$

18. $h(x) = \dfrac{x^2-25}{x+5}$; $x = 0, x = 5, x = -5$

19. $p(x) = x^2 - 4x + 11$; $x = 0, x = 2, x = -1$

20. $q(x) = -3x^3 + 2x^2 - 4x + 1$; $x = -2, x = 3, x = 1$

21. $p(x) = \dfrac{|x+2|}{x+2}$; $x = -2, x = 0, x = 2$

22. $r(x) = \dfrac{|5-x|}{x-5}$; $x = -5, x = 0, x = 5$

23. Management A company charges $1.20 per pound for a certain fertilizer on all orders not over 100 pounds, and $1 per pound for orders over 100 pounds. Let $F(x)$ represent the cost for buying x pounds of the fertilizer. (See Example 3.)
(a) Find $F(80)$. (b) Find $F(150)$.
(c) Graph $y = F(x)$.
(d) Where is F discontinuous?

24. Management The cost to transport a mobile home depends on the distance, x, in miles that the home is moved. Let $C(x)$ represent the cost to move a mobile home x miles. One firm charges as follows.

Cost per mile	Distance in miles
$2	if $0 < x \le 150$
1.50	if $150 < x \le 400$
1.25	if $400 < x$

(a) Find $C(130)$. (b) Find $C(210)$.
(c) Find $C(350)$. (d) Find $C(500)$.
(e) Graph $y = C(x)$.
(f) For what positive values of x is C discontinuous?

25. Management Recently, a car rental firm charged $30 a day to rent a car for a rental period of 1 through 5 days. Days 6 and 7 were then "free," with days 8 through 12 again $30 each. Let $A(t)$ represent the average cost per day to rent the car for t days, where $0 < t \le 12$. (The average cost for t days is the total cost for t days, divided by the number of days.) Find each of the following.
(a) $A(4)$ (b) $A(5)$ (c) $A(6)$ (d) $A(7)$
(e) $A(8)$ (f) $\lim\limits_{t \to 5} A(t)$ (g) $\lim\limits_{t \to 6} A(t)$

26. Social Science With certain skills (such as music) learning is rapid at first and then levels off. Sudden insights may cause learning to speed up sharply. A typical graph of such learning is shown in the figure. Where is the function discontinuous?

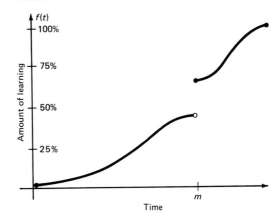

27. Management The graph below shows the interest rate for business loans for the business days of a recent month. Where is the graph discontinuous?

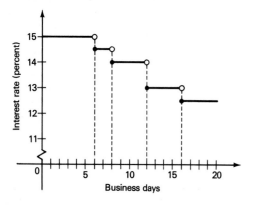

28. Natural Science Suppose a gram of ice is at a temperature of $-100°C$. The graph below shows the temperature of the ice as an increasing number of calories of heat are applied. It takes 80 calories to melt one gram of ice at $0°C$ into water, and 539 calories to boil one gram of water at $100°C$ into steam. Where is this graph discontinuous?

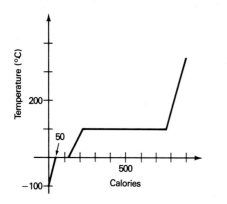

Write each of the following in interval notation.

29.

30.

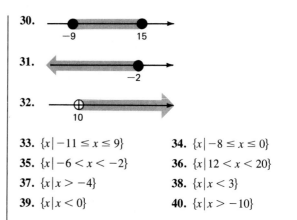

31.

32.

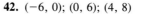

33. $\{x \mid -11 \leq x \leq 9\}$ **34.** $\{x \mid -8 \leq x \leq 0\}$

35. $\{x \mid -6 < x < -2\}$ **36.** $\{x \mid 12 < x < 20\}$

37. $\{x \mid x > -4\}$ **38.** $\{x \mid x < 3\}$

39. $\{x \mid x < 0\}$ **40.** $\{x \mid x > -10\}$

Are the functions graphed as shown continuous on the indicated intervals? (See Example 2.)

41. $(-3, 0)$; $(0, 3)$; $(0, 4)$

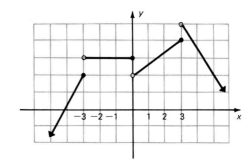

42. $(-6, 0)$; $(0, 6)$; $(4, 8)$

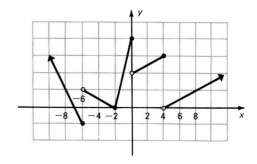

43. $(-12, 6)$; $(0, 6)$; $(6, 12)$

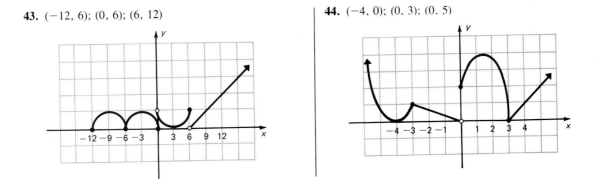

44. $(-4, 0)$; $(0, 3)$; $(0, 5)$

11.3   Rates of Change

One of the main applications of calculus is describing how one variable changes in relation to another. For example, a person in business wants to know how profit changes with respect to advertising, while a person in medicine wants to know how a patient's reaction to a drug changes with respect to the dose.

The graph in Figure 11.17 shows the profit $P(x)$ for a certain product when x hundred dollars is spent on advertising. The graph suggests that as advertising expenditures increase, profit also increases, but the rate of increase slows, as shown by the "flattening" of the profit curve. For example, when spending on advertising increases from \$0 to \$100, profit increases by \$1000 (from \$1000 to \$2000), but a further expenditure of \$200 on advertising is necessary for the next \$1000 increase in profit.

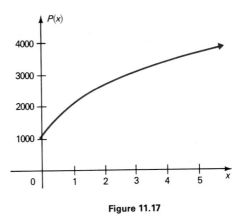

Figure 11.17

The **average rate of change** of profit with respect to advertising on some interval is defined as the change in profit divided by the change in advertising costs on that interval. For example, on the interval [1, 3],

$$\text{Average rate of change} = \frac{\text{Change in profit}}{\text{Change in advertising costs}}$$

$$= \frac{3000 - 2000}{3 - 1} = \frac{1000}{2} = 500.$$

On the interval [1, 3], an increase of \$100 in advertising produces an average increase of \$500 in profit.

Example 1 The graph of Figure 11.18 shows the profit, $P(x)$, in hundreds of thousands of dollars, from a new video game x months after its introduction on the market. The graph shows that profit increases until the game has been on the market for 25 months, and then decreases as the popularity of the game goes down. Here the average rate of change of profit with respect to *time* on some interval is defined as the change in profit divided by the change in time on the interval.

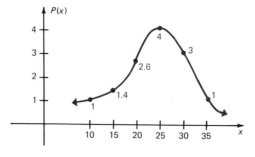

Figure 11.18

(a) On the interval from 15 to 25 months the average rate of change of the profits from the game is given by

$$\frac{4 - 1.4}{25 - 15} = \frac{2.6}{10} = .26.$$

On the average, from 15 months to 25 months, each month shows an increase in profit of .26 hundred thousand dollars, or \$26,000.

(b) From 25 months to 30 months, the average rate of change is

$$\frac{3 - 4}{30 - 25} = \frac{-1}{5} = -.20,$$

with each month in this interval showing an average *decline* in profits of .20 hundred thousand dollars, or \$20,000. ①

(c) From 10 months to 35 months, the average rate of change of profit is

$$\frac{1 - 1}{35 - 10} = \frac{0}{25} = 0, \quad \text{or} \quad \$0.$$ ∎

① Use Figure 11.18 and find the average rate of change of the profit if time changes from
(a) $x = 15$ to $x = 20$;
(b) $x = 20$ to $x = 25$;
(c) $x = 25$ to $x = 35$;
(d) $x = 30$ to $x = 35$.

Answer:
(a) .24 hundred thousand, or \$24,000
(b) .28 hundred thousand, or \$28,000
(c) −.3 hundred thousand, or −\$30,000
(d) −.4 hundred thousand, or −\$40,000

> **Average Rate of Change** The average rate of change for a function $f(x)$ as x changes from a to b, where $a < b$, is
>
> $$\frac{f(b) - f(a)}{b - a}.$$

As Example 1(c) suggests, finding the average rate of change of a function over a large interval can lead to answers that aren't very helpful. The results are often more useful if the average rate of change is found over a fairly narrow interval, with the usefulness generally increasing as the width of the interval decreases. In fact, the most useful result comes from taking the limit of the average rate of change as the width of the interval approaches 0.

As an example, suppose that for a certain firm, the profit $P(x)$, in thousands of dollars, is related to the volume of production by the mathematical model

$$P(x) = 16x - x^2,$$

where x represents the number of units produced. Assume that the company can produce any nonnegative number of units. (A *unit* here can represent any nonnegative number of a product; x might represent thousands or tens of thousands of an item.) A graph of the profit function is shown in Figure 11.19. The graph shows that an increase in production from 1 unit to 2 units will increase profit more than an increase in production from 6 units to 7 units. An increase from 7 units to 8 units produces very little increase in profit, while an increase in production from 8 units to 9 units actually produces a decline in total profit.

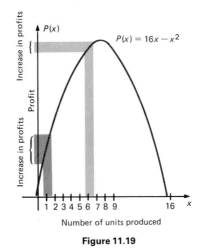

Figure 11.19

Using the profit function $P(x) = 16x - x^2$, the profit from producing 1 unit is

$$P(1) = 16(1) - 1^2 = 15,$$

or $15,000. The profit from producing 2 units is

$$P(2) = 16(2) - 2^2 = 28,$$

or $28,000. This 1-unit increase of production (from 1 unit to 2 units) increases profit by $28 - 15 = 13$ or $13,000. ②

②
(a) Find the profit from producing 4 units.
(b) Find the increase in profit if production increases from 1 unit to 4 units.

Answer:
(a) 48 or $48,000
(b) 33 or $33,000

As Problem 2 at the side shows, if production increases from 1 unit to 4 units, profit increases by $33,000. The average rate of increase in profit for this 3-unit increase in production is found by dividing the change in profits by the change in production:

$$\text{Average rate of increase} = \frac{P(4) - P(1)}{4 - 1} = \frac{48 - 15}{3} = \frac{33}{3} = 11$$

or $11,000 per unit.

We have seen that an increase in production from 1 unit to 4 units leads to an average rate of increase in profit of $11,000 per unit, while an increase from 1 unit to 2 units leads to an average rate of increase in profit of $13,000 per unit. In general, if production increases from 1 unit to $1 + h$ units, where h is a small positive number, what happens to the average rate of increase in profit? To find out, make a table which shows the average rate of increase of profit for some selected values of h.

Average Rate of Change in Profit
(in Thousands of Dollars) for a Production Level of 1

h	$-.1$	$-.001$	$.001$	$.1$
$1 + h$	.9	.999	1.001	1.1
$P(1 + h)$	13.59	14.985999	15.013999	16.39
$P(1)$	15	15	15	15
Change in profit $= P(1 + h) - P(1)$	-1.41	$-.014001$	$.013999$	1.39
Average rate of change in profit $=$				
$\dfrac{P(1 + h) - P(1)}{h}$	14.1	14.001	13.999	13.9

Average rate of change approaches 14

The numbers in the bottom row of the table are found by dividing the increase in profit (row 5) by the change in production (row 1)—that is, by evaluating the quotient

$$\frac{P(1 + h) - P(1)}{(1 + h) - 1} = \frac{P(1 + h) - P(1)}{h}$$

for the various values of h, with $h \neq 0$.

The table suggests that as h approaches 0 (or as production gets closer and closer to 1), the average rate of increase in profit approaches the limit 14. This limit,

$$\lim_{h \to 0} \frac{P(1 + h) - P(1)}{h},$$

is called the **instantaneous rate of change,** or just the **rate of change,** of the profit at a production level of $x = 1$.

The rate of change in profit varies as the level of production changes. However, at the exact instant when production is 1 unit, the rate of change in profit is $14,000 per unit. ③

③ Make a table similar to the one in the text and find the instantaneous rate of change of profit at $x = 4$.

Answer:

8

11.3 Exercises

1. **Natural Science** The graph shows the population in millions of bacteria t minutes after a bactericide is introduced into the culture. Find the average rate of change of population with respect to time for the following time intervals.

   (a) 1 to 2   (b) 1 to 3   (c) 2 to 3
   (d) 2 to 5   (e) 3 to 4   (f) 4 to 5

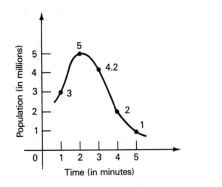

Time (in minutes)

2. **Management** The graph shows the total sales in thousands of dollars from the distribution of x thousand catalogs. Find the average rate of change of sales with respect to the number of catalogs distributed for the following changes in x.

   (a) 10 to 20
   (b) 10 to 40
   (c) 20 to 30
   (d) 30 to 40

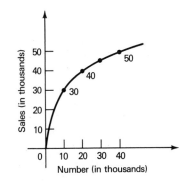

Number (in thousands)

3. **Management** The graph shows annual sales (in units) of a typical product. Sales increase slowly at first to some peak, hold steady for a while, and then decline as the product goes out of style. Find the average annual rate of change in sales for the following changes in years.

   (a) 1 to 3   (b) 2 to 4   (c) 3 to 6
   (d) 5 to 7   (e) 7 to 9   (f) 8 to 11
   (g) 9 to 10   (h) 10 to 12

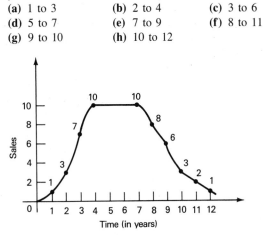

Time (in years)

4. Management The graph below, from the November 25, 1985 issue of *Fortune*, page 164,* shows the stock prices since 1982 for the Hawaiian firm Alexander & Baldwin (compared to the Standard and Poor's 500 index). Estimate the average rate of change for this firm's stock prices over the following intervals.
(a) End of 1982 to end of 1983
(b) End of 1983 to October 1985
(c) End of March 1985 to end of June 1985
(d) Beginning of 1984 to end of June 1984

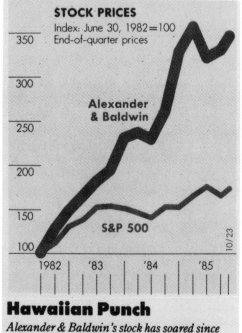

STOCK PRICES
Index: June 30, 1982=100
End-of-quarter prices

Hawaiian Punch

Alexander & Baldwin's stock has soared since investors began to focus on its real estate.

5. Natural Science The graph shows the temperature in degrees Celsius as a function of the altitude h in feet when an inversion layer is over southern California. (An inversion layer is formed when air at a higher altitude, say 3000 feet, is warmer than air at sea level, even though air normally is cooler with increasing altitude.) Find the average rate of change in temperature as

the altitude changes from
(a) 1000 to 3000 feet; **(b)** 1000 to 5000 feet;
(c) 3000 to 9000 feet; **(d)** 1000 to 9000 feet.

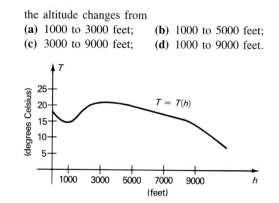

6. A car is moving along a straight test track. The position in feet of the car, $s(t)$, at various times t is measured, with the following results.

t (seconds)	0	2	6	10	12	18	20
$s(t)$ (feet)	0	10	14	18	30	36	40

Find the average velocity (velocity equals distance divided by time) as the time in seconds changes from
(a) 0 to 6; **(b)** 2 to 10; **(c)** 2 to 12;
(d) 6 to 10; **(e)** 6 to 18; **(f)** 12 to 20.

7. The second graph on the next page* shows the number of airline accidents (excluding the Soviet Union) over a period of several years. Estimate the average rate of change of the number of accidents from
(a) 1950 to 1972;
(b) 1972 to 1980;
(c) 1980 to 1985;
(d) 1984 to 1985.

8. The fourth graph on the next page* shows the fatality rate, in deaths per 100 million passenger-kilometers flown. Estimate the average rate of change of the fatality rate from
(a) 1950 to 1985;
(b) 1955 to 1960;
(c) 1970 to 1985.

---

*Graph, "Hawaiian Punch," from *Fortune*, November 25, 1985. Copyright © 1985 by Time Inc. All rights reserved.

*Two graphs, "Accidents" and "Fatality rate" from "Why Safety Records Look Good Though Deaths Go Up" from *Fortune*, October 14, 1985, p. 22. Source: International Civil Aviation Organization. Reprinted by permission.

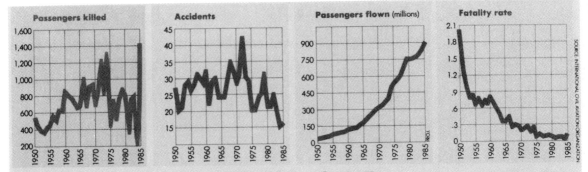

Why Safety Records Look Good Though Deaths Go Up

Deaths among airline passengers outside the Soviet Union (for which statistics aren't available) broke the previous annual record by September this year. Airlines estimate that they will carry 900 million passengers, a new high, this year. Accidents are fewer than in any year except 1984. But planes are big and the toll per accident is up. So is the fatality rate, figured in deaths per 100 million passenger-kilometers flown.

Find the average rate of change for the functions defined as follows over the given closed intervals.

9. $y = x^2 + 2x$ between $x = 0$ and $x = 3$

10. $y = -4x^2 - 6$ between $x = 2$ and $x = 5$

11. $y = 2x^3 - 4x^2 + 6x$ between $x = -1$ and $x = 1$

12. $y = -3x^3 + 2x^2 - 4x + 1$ between $x = 0$ and $x = 1$

13. $y = \sqrt{x}$ between $x = 1$ and $x = 4$

14. $y = \sqrt{3x - 2}$ between $x = 1$ and $x = 2$

15. $y = \dfrac{1}{x - 1}$ between $x = -2$ and $x = 0$

16. $y = \dfrac{-5}{2x - 3}$ between $x = 2$ and $x = 4$

17. Management Suppose that the total cost in dollars to produce x items is given by

$$C(x) = 2x^2 - 5x + 6.$$

Find the average rate of change of cost as x increases from

(a) 2 to 4; **(b)** 2 to 3.

(c) Use $C(x) = 2x^2 - 5x + 6$ to complete the table below.

h	1	.1	.01	.001	.0001
$2 + h$	3	2.1	2.01	2.001	2.0001
$C(2 + h)$	9	4.32	4.0302	4.003002	4.00030002
$C(2 + h) - C(2)$	5	.32	.0302	.003002	.00030002
$\dfrac{C(2 + h) - C(2)}{h}$	5	3.2			

(d) Use the bottom row of the chart to find

$$\lim_{h \to 0} \frac{C(2 + h) - C(2)}{h}.$$

18. Use the rules for limits given in Section 11.1 to find the limit of Exercise 17(d),

$$\lim_{h \to 0} \frac{C(2 + h) - C(2)}{h}.$$

From Exercise 17, also find

$$\lim_{h \to 0} \frac{C(6 + h) - C(6)}{h} \quad \text{and} \quad \lim_{h \to 0} \frac{C(10 + h) - C(10)}{h}.$$

19. In Exercise 17, what is the instantaneous rate of change of cost with respect to the number of items produced when $x = 2$? (This number, called the **marginal cost** at $x = 2$, will be discussed in more detail later; it is the approximate cost of producing the third item.)

20. Redo the chart of Exercise 17. This time, change the second line to $4 + h$, the third line to $C(4 + h)$, and so on. Find

$$\lim_{h \to 0} \frac{C(4 + h) - C(4)}{h}.$$

11.4 The Derivative

Tangent Lines   One of the key problems of calculus is to find a line **tangent** to the graph of a function. (This tangent line turns out to be very useful in applications.) The tangent line to a curve is defined in this section. In geometry, a tangent line to a circle is a line that touches the circle at only one point. See Figure 11.20. Tangent lines to more complicated curves are not as easy to define. In Figure 11.21 we would probably agree that the lines at P_1, P_3, and P_4 are tangent lines to the curve, while the lines at P_2 and P_5 are not.

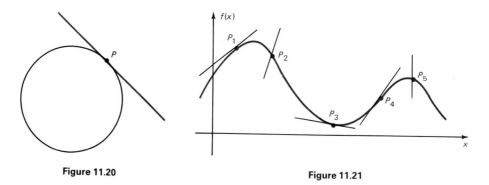

Figure 11.20   **Figure 11.21**

The definition of a tangent line is obtained from the definition of the slope of a line. Begin with a fixed point R, with coordinates $(x_0, f(x_0))$ on the graph of a function $y = f(x)$, as in Figure 11.22. Choose a different point S on the graph and draw the line through R and S: this line is called a **secant line.** If S has coordinates $(x_0 + h, f(x_0 + h))$, then by the definition of slope, the slope of the secant line RS is

$$\text{Slope of secant} = \frac{f(x_0 + h) - f(x_0)}{x_0 + h - x_0} = \frac{f(x_0 + h) - f(x_0)}{h}.$$

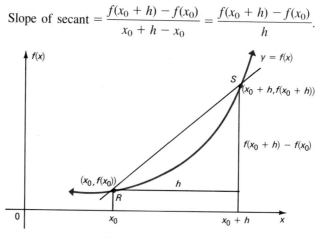

Figure 11.22

As h approaches 0, point S will slide along the curve, getting closer and closer to the fixed point R. See Figure 11.23 which shows successive positions S_1, S_2, S_3, S_4 of the point S. If the slopes of these secant lines approach a limit as h approaches 0, then this limit is defined to be the slope of the tangent line at point R.

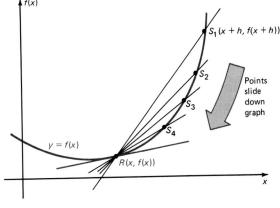

Figure 11.23

Tangent Line The tangent line to the graph of $y = f(x)$ at the point $(x_0, f(x_0))$ is the line through this point having slope

$$\lim_{h \to 0} \frac{f(x_0 + h) - f(x_0)}{h},$$

provided this limit exists. If this limit does not exist, then there is no tangent at the point.

The slope of this line at a point is also called the **slope of the curve** at the point.

Example 1 Find the slope of the tangent line to the graph of $y = x^2 + 2$ when $x = -1$. Find the equation of the tangent line.
    Use the definition above, with $f(x) = x^2 + 2$ and $x_0 = -1$. The slope of the tangent line is

$$\text{Slope of tangent} = \lim_{h \to 0} \frac{f(x_0 + h) - f(x_0)}{h}$$

$$= \lim_{h \to 0} \frac{[(-1 + h)^2 + 2] - [(-1)^2 + 2]}{h}$$

$$= \lim_{h \to 0} \frac{[1 - 2h + h^2 + 2] - [1 + 2]}{h}$$

$$= \lim_{h \to 0} \frac{-2h + h^2}{h} = \lim_{h \to 0} (-2 + h) = -2.$$

The slope of the tangent line at $(-1, f(-1)) = (-1, 3)$ is -2. The equation of the tangent line can be found with the point-slope form of the equation of a line from Chapter 2.

$$y - y_1 = m(x - x_1)$$
$$y - 3 = -2[x - (-1)]$$
$$y - 3 = -2(x + 1)$$
$$y - 3 = -2x - 2$$
$$y = -2x + 1$$

Figure 11.24 shows a graph of $f(x) = x^2 + 2$, along with a graph of the tangent line at $x = -1$. **1**

1 Rework Example 1 for $x = 2$.

Answer:
Slope is 4;
equation is $y = 4x - 2$.

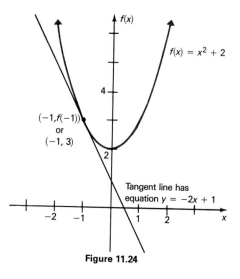

Figure 11.24

The Derivative So far, the limit

$$\lim_{h \to 0} \frac{f(x_0 + h) - f(x_0)}{h}$$

has been used in two different ways—as the instantaneous rate of change and as the slope of a tangent line. This particular limit is so important that it is given a special name, the *derivative*. The limit above gives the definition of the derivative at the point $(x_0, f(x_0))$. This definition is generalized to apply to *any* point $(x, f(x))$.

Definition of the Derivative The derivative of the function f, at x, written $f'(x)$, is defined as

$$f'(x) = \lim_{h \to 0} \frac{f(x + h) - f(x)}{h},$$

provided this limit exists.

An alternate notation, y', is sometimes used for the derivative of the function $y = f(x)$. A new function $y' = f'(x)$ can be obtained from a function $y = f(x)$. This new function has as domain all the points at which the limit above exists, and its value at x is $f'(x)$. This new function $y' = f'(x)$ is called the **derivative** of $y = f(x)$. If x is a value in the domain of f, and if $f'(x)$ exists, then f is **differentiable** at x.

The next few examples show how to use the definition to find the derivative of a function by means of a four-step procedure summarized after Example 4.

Example 2 Let $f(x) = x^2$.
(a) Find the derivative.

By definition, for all values of x where the following limit exists, the derivative is

$$f'(x) = \lim_{h \to 0} \frac{f(x + h) - f(x)}{h}.$$

Use the following sequence of steps to evaluate this limit.

Step 1 Find $f(x + h)$.
Replace x with $x + h$ in the equation for $f(x)$. Simplify the result.

$$f(x) = x^2$$
$$f(x + h) = (x + h)^2$$
$$= x^2 + 2xh + h^2$$

Step 2 Find $f(x + h) - f(x)$.
Since $f(x) = x^2$,

$$f(x + h) - f(x) = x^2 + 2xh + h^2 - x^2$$
$$= 2xh + h^2.$$

Step 3 Form and simplify the quotient $\dfrac{f(x + h) - f(x)}{h}$.

$$\frac{f(x + h) - f(x)}{h} = \frac{2xh + h^2}{h} = \frac{h(2x + h)}{h} = 2x + h$$

Step 4 Find the limit of the result in step 3 as h approaches 0.

$$f'(x) = \lim_{h \to 0} \frac{f(x + h) - f(x)}{h} = \lim_{h \to 0} (2x + h) = 2x + 0 = 2x$$

From the last step, the derivative of $f(x) = x^2$ is $f'(x) = 2x$.

(b) Calculate and interpret $f'(3)$.

$$f'(3) = 2(3) = 6$$

The number 6 is the slope of the tangent line to the graph of $f(x) = x^2$ at the point where $x = 3$, that is, at $(3, f(3)) = (3, 9)$. See Figure 11.25 on the following page. **2**

2 Let $f(x) = -2x^2 + 7$.
Find the following.
(a) $f(x + h)$
(b) $f(x + h) - f(x)$
(c) $\dfrac{f(x + h) - f(x)}{h}$
(d) $f'(x)$
(e) $f'(4)$
(f) $f'(0)$

Answer:
(a) $-2x^2 - 4xh - 2h^2 + 7$
(b) $-4xh - 2h^2$
(c) $-4x - 2h$
(d) $-4x$
(e) -16
(f) 0

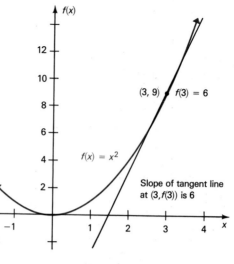

Figure 11.25

In Example 2(b), do not confuse $f(3)$ and $f'(3)$. The value $f(3)$ is found by substituting 3 for x in $f(x)$; doing so gives $f(3) = 3^2 = 9$. On the other hand, $f'(3)$ is the slope of the tangent line to the curve at $x = 3$; as Example 2(b) shows, $f'(3) = 2 \cdot 3 = 6$.

Example 3 Let $f(x) = 2x^3 + 4x$. Find $f'(x)$. Then find $f'(2)$ and $f'(-3)$. Go through the four steps used above.

Step 1 Find $f(x + h)$ by replacing x with $x + h$.
$$f(x + h) = 2(x + h)^3 + 4(x + h)$$
$$= 2(x^3 + 3x^2h + 3xh^2 + h^3) + 4(x + h)$$
$$= 2x^3 + 6x^2h + 6xh^2 + 2h^3 + 4x + 4h.$$

Step 2 $f(x + h) - f(x) = 2x^3 + 6x^2h + 6xh^2 + 2h^3 + 4x + 4h - 2x^3 - 4x$
$$= 6x^2h + 6xh^2 + 2h^3 + 4h.$$

Step 3 $\dfrac{f(x + h) - f(x)}{h} = \dfrac{6x^2h + 6xh^2 + 2h^3 + 4h}{h}$
$$= \dfrac{h(6x^2 + 6xh + 2h^2 + 4)}{h}$$
$$= 6x^2 + 6xh + 2h^2 + 4.$$

Step 4 Now use the rules for limits to get

$$f'(x) = \lim_{h \to 0} \frac{f(x + h) - f(x)}{h} = \lim_{h \to 0} (6x^2 + 6xh + 2h^2 + 4)$$
$$= 6x^2 + 6x(0) + 2(0)^2 + 4 = 6x^2 + 4.$$

Use this result to find $f'(2)$ and $f'(-3)$.

$$f'(2) = 6 \cdot 2^2 + 4 = 28$$
$$f'(-3) = 6 \cdot (-3)^2 + 4 = 58 \quad \boxed{3}$$

3 Let $f(x) = -3x^3 + 2x$.
Find the following.
(a) $f(x + h)$
(b) $f(x + h) - f(x)$
(c) $\dfrac{f(x + h) - f(x)}{h}$
(d) $f'(x)$
(e) $f'(-1)$

Answer:
(a) $-3x^3 - 9x^2h - 9xh^2$
$\qquad - 3h^3 + 2x + 2h$
(b) $-9x^2h - 9xh^2$
$\qquad - 3h^3 + 2h$
(c) $-9x^2 - 9xh$
$\qquad - 3h^2 + 2$
(d) $-9x^2 + 2$
(e) -7

Example 4 Let $f(x) = 4/x$. Find $f'(x)$.

Step 1 $f(x + h) = \dfrac{4}{x + h}$

Step 2 $f(x + h) - f(x) = \dfrac{4}{x + h} - \dfrac{4}{x}$

$= \dfrac{4x - 4(x + h)}{x(x + h)}$ Find a common denominator

$= \dfrac{4x - 4x - 4h}{x(x + h)}$ Simplify the numerator

$= \dfrac{-4h}{x(x + h)}$

Step 3 $\dfrac{f(x + h) - f(x)}{h} = \dfrac{\dfrac{-4h}{x(x + h)}}{h}$

$= \dfrac{-4h}{x(x + h)} \cdot \dfrac{1}{h}$ Invert and multiply

$= \dfrac{-4}{x(x + h)}$

Step 4 $f'(x) = \lim\limits_{h \to 0} \dfrac{f(x + h) - f(x)}{h} = \lim\limits_{h \to 0} \dfrac{-4}{x(x + h)}$

$= \dfrac{-4}{x(x + 0)}$

$= \dfrac{-4}{x(x)}$

$= \dfrac{-4}{x^2}$ **4**

4 Let $f(x) = -5/x$. Find the following.
(a) $f(x + h)$
(b) $f(x + h) - f(x)$
(c) $\dfrac{f(x + h) - f(x)}{h}$
(d) $f'(x)$
(e) $f'(-1)$

Answer:

(a) $\dfrac{-5}{x + h}$

(b) $\dfrac{5h}{x(x + h)}$

(c) $\dfrac{5}{x(x + h)}$

(d) $\dfrac{5}{x^2}$

(e) 5

The four steps used when finding the derivative of $f(x)$, denoted $f'(x)$, are summarized below.

Finding $f'(x)$ from the Definition of the Derivative

Step 1 Find $f(x + h)$.

Step 2 Find $f(x + h) - f(x)$.

Step 3 Divide by h to get $\dfrac{f(x + h) - f(x)}{h}$.

Step 4 Let $h \to 0$; $f'(x) = \lim\limits_{h \to 0} \dfrac{f(x + h) - f(x)}{h}$ if this limit exists.

The next example shows how to find a point on a graph where the slope of the tangent line equals a given number.

Example 5 Find all points on the graph of $f(x) = x^2 + 6x + 5$ where the slope of the tangent line is 0.

Go through the four steps given above to find that

$$f'(x) = 2x + 6.$$

Find all values of x where the slope of the tangent is 0 by finding values of x that make $f'(x) = 0$.

$$f'(x) = 0$$
$$2x + 6 = 0$$
$$2x = -6$$
$$x = -3$$

When $x = -3$, the slope of the tangent is 0. Since horizontal lines have a slope of 0, the tangent to the graph of $f(x) = x^2 + 6x + 5$ at $x = -3$ is horizontal, as shown in Figure 11.26. When $x = -3$, then $f(-3) = -4$, making the slope of the tangent line 0 at the point $(-3, -4)$. The equation of this tangent line is $y = -4$. **5**

5 Find all points on the graph of $f(x) = x^2 - x$ where the slope of the tangent line is 0.

Answer:
$(1/2, -1/4)$

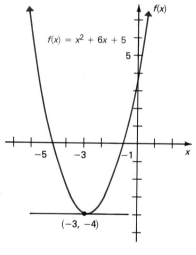

$f(x) = x^2 + 6x + 5$

$(-3, -4)$

Figure 11.26

The derivative was defined with the limit

$$\lim_{h \to 0} \frac{f(x + h) - f(x)}{h}.$$

This is exactly the same limit used in the previous section to define the instantaneous rate of change of one variable with respect to another.

> **Instantaneous Rate of Change** For a function $y = f(x)$, the instantaneous rate of change of y with respect to x is given by the derivative $f'(x)$.

From now on, we will use "rate of change" to mean "instantaneous rate of change."

Example 6 The cost in dollars to manufacture x electronic calculators is given by $C(x) = x^3 - 4x^2 + 3x$. Find the rate of change of cost with respect to the number manufactured when 5 electronic calculators are made.

The rate of change of cost is given by the derivative of the cost function. Going through the four steps for finding $f'(x)$ leads to

$$C'(x) = 3x^2 - 8x + 3.$$

When $x = 5$,

$$C'(5) = 3(5)^2 - 8(5) + 3 = 38.$$

At the point where exactly 5 electronic calculators are made, the rate of change of cost is $38. **6**

6 The revenue from the sale of x can openers is given by $R(x) = 2x^3 - 4x^2 + x$. Find the rate of change of revenue when
(a) $x = 2$;
(b) $x = 8$.
(Hint: $R'(x) = 6x^2 - 8x + 1$.)

Answer:
(a) 9
(b) 321

Existence of the Derivative The definition of the derivative included the phrase "provided this limit exists." If the limit used to define the derivative does not exist for some value of x, then the derivative does not exist at that x-value, although it may exist at other points. For example, a derivative cannot exist at a point where the function itself is not defined. Since the definition of the derivative involves $f(x)$, if there is no function value for a particular value of x, there can be no tangent line for that value. This is the case in Example 4 where $f(x) = 4/x$. Since division by 0 is not defined, this function is not defined at $x = 0$, and so there is no tangent line when $x = 0$.

Derivatives also do not exist at "corners" or "sharp points" on a graph. For example, Figure 11.27 shows the graph of $f(x) = |x|$ (recall: $|x|$ represents the absolute value of x). By the definition of derivative, the derivative at x is given by

$$\lim_{h \to 0} \frac{f(x + h) - f(x)}{h},$$

provided this limit exists. To find the derivative at 0 for $f(x) = |x|$, replace x with 0 and $f(x)$ with $|0|$ to get

$$\lim_{h \to 0} \frac{|0 + h| - |0|}{h} = \lim_{h \to 0} \frac{|h|}{h}.$$

If $h > 0$, then $|h| = h$, and

$$\frac{|h|}{h} = \frac{h}{h} = 1,$$

while if $h < 0$, then $|h| = -h$, and

$$\frac{|h|}{h} = \frac{-h}{h} = -1.$$

For positive values of h, the quotient $|h|/h$ is 1, while it is -1 for negative values of h. At $x = 0$, there are *two* tangent lines—one with slope of 1 and one with slope of -1. For this reason,

$$\lim_{h \to 0} \frac{|h|}{h} \quad \text{does not exist,}$$

and there is no derivative at 0. However, the derivative of $f(x) = |x|$ exists for all other values of x.

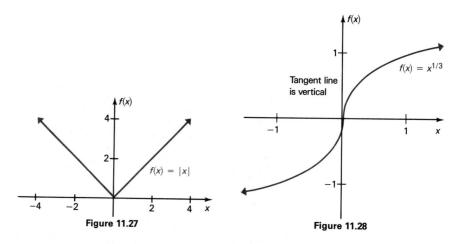

Figure 11.27

Figure 11.28

A graph of $f(x) = x^{1/3}$ is shown in Figure 11.28. As the graph suggests, the tangent line is vertical when $x = 0$. Since a vertical line has an undefined slope, the derivative of $f(x) = x^{1/3}$ cannot exist when $x = 0$, although it exists for all other values of x.

Figure 11.29 summarizes the various ways that a derivative can fail to exist.

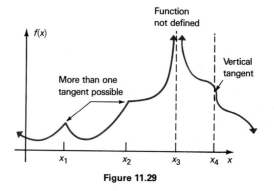

Figure 11.29

11.4 Exercises

Find the slope of the tangent line to each of the following curves when x has the given value. (Hint for Exercise 5: In Step 3, multiply numerator and denominator by $\sqrt{16 + h} + \sqrt{16}$.) (See Example 1.)

1. $f(x) = -4x^2 + 11x; \quad x = -2$

2. $f(x) = 6x^2 - 4x; \quad x = -1$

3. $f(x) = -\dfrac{2}{x}; \quad x = 4$ **4.** $f(x) = \dfrac{6}{x}; \quad x = -1$

5. $f(x) = \sqrt{x}; \quad x = 16$ **6.** $f(x) = -3\sqrt{x}; \quad x = 1$

Find the equation of the tangent line to each of the following curves when x has the given value. (See Example 1.)

7. $f(x) = x^2 + 2x; \quad x = 3$

8. $f(x) = 6 - x^2; \quad x = -1$

9. $f(x) = \dfrac{5}{x}; \quad x = 2$

10. $f(x) = -\dfrac{3}{(x + 1)}; \quad x = 1$

11. $f(x) = 4\sqrt{x}; \quad x = 9$

12. $f(x) = \sqrt{x}; \quad x = 25$

Estimate the slope of the tangent line at the given point (x, y) in each of the following graphs.

13.

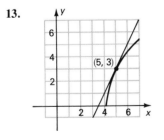

14.

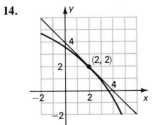

15.

16.

17.

18.

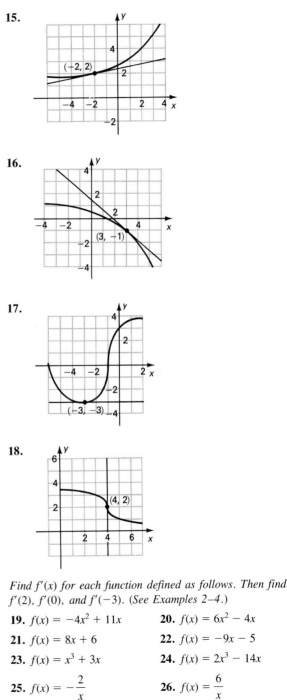

Find $f'(x)$ for each function defined as follows. Then find $f'(2)$, $f'(0)$, and $f'(-3)$. (See Examples 2–4.)

19. $f(x) = -4x^2 + 11x$ **20.** $f(x) = 6x^2 - 4x$

21. $f(x) = 8x + 6$ **22.** $f(x) = -9x - 5$

23. $f(x) = x^3 + 3x$ **24.** $f(x) = 2x^3 - 14x$

25. $f(x) = -\dfrac{2}{x}$ **26.** $f(x) = \dfrac{6}{x}$

27. $f(x) = \dfrac{4}{x - 1}$

28. $f(x) = \dfrac{3}{x + 2}$

29. $f(x) = \sqrt{x}$

30. $f(x) = -3\sqrt{x}$

Find all points where the functions graphed as shown do not have derivatives.

31.

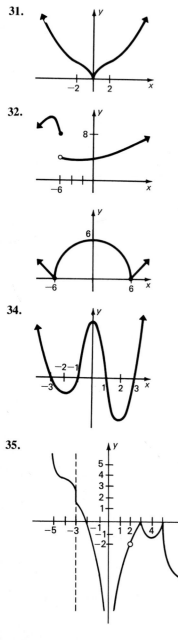

32.

34.

35.

36.

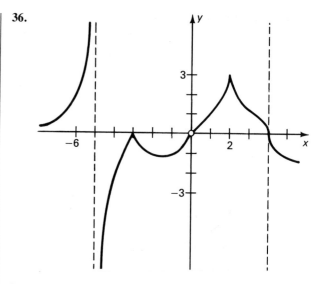

Find all points (if there are any) where the functions defined as follows have tangent lines with a slope of 0. (In Exercises 41–44 it is necessary to factor.) (See Example 5.)

37. $f(x) = 9x^2$

38. $f(x) = -2x^2$

39. $f(x) = 8x^2 + 32x$

40. $f(x) = 2x^2 + 6x$

41. $f(x) = 2x^3 - 3x^2 - 12x$

42. $f(x) = 2x^3 - 6x^2 + 6x$

43. $f(x) = 4x^3 + 24x^2$

44. $f(x) = 4x^3 - 18x^2$

45. $f(x) = \dfrac{1}{x}$

46. $f(x) = \dfrac{-2}{x - 1}$

Work the following exercises.

47. Management Suppose the demand for a certain item is given by

$$D(x) = -2x^2 + 4x + 6,$$

where x represents the price of the item. Find the rate of change of demand with respect to price for the following values of x. (See Example 6.)

(a) $x = 3$ **(b)** $x = 6$

48. Management Suppose the profit from the expenditure of x thousand dollars on advertising is given by

$$P(x) = 1000 + 32x - 2x^2.$$

Find the rate of change of profit with respect to the expenditure on advertising for the following amounts. In each case, decide if the firm should increase the expenditure.

(a) $x = 8$ **(b)** $x = 6$ **(c)** $x = 12$ **(d)** $x = 20$

49. Natural Science A biologist estimates that when a bactericide is introduced into a culture of bacteria, the number of bacteria present is given by

$$B(t) = 1000 + 50t - 5t^2,$$

where $B(t)$ is the number of bacteria, in millions, present at time t, measured in hours. Find the rate of change of the number of bacteria with respect to time for the following values of t.

(a) $t = 2$ **(b)** $t = 3$ **(c)** $t = 4$
(d) $t = 5$ **(e)** $t = 6$ **(f)** $t = 8$

50. Natural Science When does the population of bacteria in Exercise 49 start to decline?

51. Natural Science (Continuation of Exercise 5 from the previous section.) The graph below shows the temperature in degrees Celsius as a function of the altitude h in feet when an inversion layer is over southern California. Estimate the derivatives of $T(h)$ at the marked points.

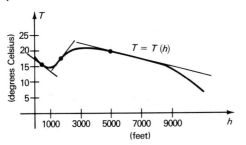

52. Natural Science In a research study, the population of a certain shellfish in an area at time t was closely approximated by the graph below. Estimate the derivative at the marked points.

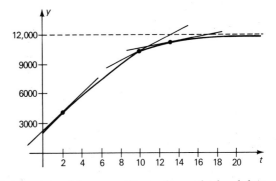

Find the slope of the tangent line to the graph of each function defined as follows.

53. $f(x) = \dfrac{x^2 + 4.9}{3x^2 - 1.7x}$ at $x = 4.9$

54. $f(x) = \dfrac{3x - 7.2}{x^2 + 5x + 2.3}$ at $x = -6.2$

55. $f(x) = \sqrt{x^3 + 4.8}$ at $x = 8.17$

56. $f(x) = \sqrt{6x^2 + 3.5}$ at $x = 3.49$

57. $f(x) = \dfrac{2.8}{\sqrt{x} + 1.2}$ at $x = 2.65$

58. $f(x) = \dfrac{-3.7}{4.1 + 2\sqrt{x}}$ at $x = 1.23$

11.5 Techniques for Finding Derivatives

The process of finding the derivative of a function is called **differentiation.** In the previous section the symbols y' and $f'(x)$ were used to represent the derivative of the function defined by $y = f(x)$. Sometimes it is important to show that the derivative is taken with respect to a particular variable; for example, if y is a function of x, use the notation

$$\frac{dy}{dx}$$

for the derivative of y with respect to x.

Another notation is used to write a derivative without functional symbols such as f or f'. With this notation, the derivative of $y = 2x^3 + 4x$, for example, which we found in the last section to be $y' = 6x^2 + 4$, would be written

$$\frac{d}{dx}[2x^3 + 4x] = 6x^2 + 4,$$

or

$$D_x[2x^3 + 4x] = 6x^2 + 4.$$

Either $\dfrac{d}{dx}[f(x)]$ or $D_x[f(x)]$ represents the derivative of the function f with respect to x. ①

① Use the results of Exercises 19–30 in the previous section to find each of the following.

(a) $\dfrac{d}{dx}[x^3 + 3x]$

(b) $\dfrac{d}{dx}\left[\dfrac{-2}{x}\right]$

(c) $D_x\left[\dfrac{4}{x-1}\right]$

(d) $D_x[-3\sqrt{x}]$

Answer:

(a) $3x^2 + 3$

(b) $\dfrac{2}{x^2}$

(c) $\dfrac{-4}{(x-1)^2}$

(d) $\dfrac{-3}{2\sqrt{x}}$

Notations for the Derivative The derivative of $y = f(x)$ may be written in any of the following ways:

$$f'(x), \qquad y', \qquad \frac{dy}{dx}, \qquad \frac{d}{dx}[f(x)], \qquad \text{or} \qquad D_x[f(x)].$$

Variables other than x often are used as the independent variable. For example, if $y = f(t)$ gives population growth as a function of time, then the derivative of y with respect to t could be written

$$f'(t), \qquad \frac{dy}{dt}, \qquad \frac{d}{dt}[f(t)], \qquad \text{or} \qquad D_t[f(t)].$$

In this section the definition of the derivative,

$$f'(x) = \lim_{h \to 0} \frac{f(x+h) - f(x)}{h},$$

is used to develop some rules for finding derivatives more easily than by the four-step process of the previous section.

The first rule tells how to find the derivative of a constant function $f(x) = k$, where k is a constant real number. Since $f(x + h)$ is also k, by definition $f'(x)$ is

$$f'(x) = \lim_{h \to 0} \frac{f(x+h) - f(x)}{h}$$

$$= \lim_{h \to 0} \frac{k - k}{h} = \lim_{h \to 0} \frac{0}{h}$$

$$= \lim_{h \to 0} 0 = 0,$$

proving the following rule.

Constant Rule If $f(x) = k$ where k is any real number, then

$$f'(x) = 0.$$

(The derivative of a constant is 0.)

(For reference, a summary of all the formulas for derivatives is given at the end of Section 11.8.)

Figure 11.30 illustrates this constant rule; it shows a graph of the horizontal line $y = k$. At any point P on this line, the tangent line at P is the line itself. Since a horizontal line has a slope of 0, the slope of the tangent line is 0. This agrees with the result above: the derivative of a constant is 0.

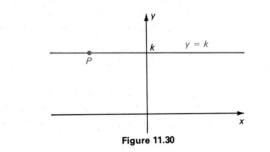

Figure 11.30

2 Find the derivatives of the following.
(a) $y = -4$
(b) $f(x) = \pi$
(c) $y = 0$

Answer:
(a) 0
(b) 0
(c) 0

Example 1
(a) If $f(x) = 9$, then $f'(x) = 0$.
(b) If $y = \pi$, then $y' = 0$.
(c) If $y = 2^3$, then $dy/dx = 0$. **2**

Functions of the form $y = x^n$, where n is a real number, occur often in applications. To find a rule for the derivative of $y = x^n$, first use the definition to get the derivative for various values of n. These results, shown in the table at the side, suggest the following rule.

Function	Derivative
$y = x^2$	$y' = 2x^1$
$y = x^3$	$y' = 3x^2$
$y = x^4$	$y' = 4x^3$
$y = x^5$	$y' = 5x^4$
$y = x^{-1}$	$y' = -1 \cdot x^{-2}$
	$= -1/x^2$
$y = x^{1/2}$	$y' = \frac{1}{2}x^{-1/2}$
	$= 1/(2x^{1/2})$

Power Rule If $f(x) = x^n$ for a real number n, then

$$f'(x) = n \cdot x^{n-1}.$$

(The derivative of $f(x) = x^n$ is found by decreasing the exponent by 1 and multiplying the result by the original exponent n.)

For a proof of the power rule, see Exercise 73 below.

Example 2 Find derivatives of the following functions.
(a) $y = x^6$
Multiply x^{6-1} by 6.

$$y' = 6 \cdot x^{6-1} = 6x^5$$

(b) If $y = x^1$, then $y' = 1 \cdot x^{1-1} = 1 \cdot x^0 = 1 \cdot 1 = 1$. (Recall that $x^0 = 1$ if $x \neq 0$.)

(c) If $y = t^{-3}$, then $y' = -3 \cdot t^{-3-1} = -3 \cdot t^{-4} = -\dfrac{3}{t^4}$.

(d) $D_t[t^{4/3}] = \dfrac{4}{3}t^{4/3-1} = \dfrac{4}{3}t^{1/3}$.

(e) $y = \sqrt{x}$

Replace $\sqrt{x}$ by the equivalent expression $x^{1/2}$. Then

$$y' = \frac{1}{2} \cdot x^{(1/2)-1} = \frac{1}{2}x^{-1/2} = \frac{1}{2x^{1/2}} = \frac{1}{2\sqrt{x}}.$$

The next rule tells how to find the derivative of the product of a constant and a function.

Constant Times a Function Let k be a real number. Then the derivative of $y = k \cdot f(x)$ is

$$y' = k \cdot f'(x).$$

(The derivative of a constant times a function is the constant times the derivative of the function.)

Example 3 Find the derivatives of the following functions.

(a) $y = 8x^4$

Since the derivative of $f(x) = x^4$ is $f'(x) = 4x^3$,

$$y' = 8(4x^3) = 32x^3.$$

(b) $y = -\dfrac{3}{4}t^{12}$

$$\frac{dy}{dt} = -\frac{3}{4}(12t^{11}) = -9t^{11}$$

(c) If $y = 15x$, then $y' = 15(1) = 15$.

(d) $D_x[10x^{3/2}] = 10\left(\dfrac{3}{2}x^{1/2}\right) = 15x^{1/2}$.

(e) If $y = 1000t^{.4}$, then $y' = 1000(.4t^{-.6}) = 400t^{-.6}$.

(f) $y = \dfrac{6}{x}$

Replace $\dfrac{6}{x}$ with $6 \cdot \dfrac{1}{x}$, or $6x^{-1}$. Then

$$y' = 6(-1x^{-2}) = -6x^{-2} = -\frac{6}{x^2}.$$

---

3 Find the derivatives of the following.

(a) $y = x^4$
(b) $y = x^{17}$
(c) $y = x^{-2}$
(d) $y = t^{-5}$
(e) $y = t^{3/2}$

Answer:

(a) $y' = 4x^3$
(b) $y' = 17x^{16}$
(c) $y' = -2/x^3$
(d) $y' = -5/t^6$
(e) $y' = (3/2)t^{1/2}$

---

4 Find the derivatives of the following.

(a) $y = 12x^3$
(b) $f(x) = 30t^7$
(c) $y = -35t$
(d) $y = 10x^{.2}$
(e) $y = 5\sqrt{x}$
(f) $y = -10/t$

Answer:

(a) $36x^2$
(b) $210t^6$
(c) -35
(d) $2x^{-.8}$
(e) $(5/2)x^{-1/2}$ or $5/(2\sqrt{x})$
(f) $10t^{-2}$ or $10/t^2$

The final rule in this section is for the derivative of a function that is a sum or difference of functions.

Sum or Difference Rule If $y = f(x) + g(x)$, and both $f'(x)$ and $g'(x)$ exist, then

$$y' = f'(x) + g'(x);$$

if $y = f(x) - g(x)$, then

$$y' = f'(x) - g'(x).$$

(The derivative of a sum or difference of two functions is the sum or difference of the derivatives of the functions.)

For a proof of this rule, see Exercise 74 below. This rule can be generalized to sums and differences with more than two terms.

Example 4 Find the derivatives of the following functions.
(a) $y = 6x^3 + 15x^2$

Let $f(x) = 6x^3$ and $g(x) = 15x^2$; then $y = f(x) + g(x)$. Since $f'(x) = 18x^2$ and $g'(x) = 30x$,

$$\frac{dy}{dx} = 18x^2 + 30x.$$

(b) $p(t) = 8t^4 - 6\sqrt{t} + \dfrac{5}{t}$

Rewrite $p(t)$ as $p(t) = 8t^4 - 6t^{1/2} + 5t^{-1}$; then $p'(t) = 32t^3 - 3t^{-1/2} - 5t^{-2}$,

which may be written as $p'(t) = 32t^3 - \dfrac{3}{\sqrt{t}} - \dfrac{5}{t^2}$. ⑤

As shown in the previous section, the derivative of a function gives the rate of change of the function.

Example 5 A tumor has the approximate shape of a cone. See Figure 11.31. The radius of the tumor is fixed by the bone structure at 2 centimeters, but the tumor is growing along the height of the cone. The formula for the volume of a cone is $V = \frac{1}{3}\pi r^2 h$, where r is the radius of the base and h is the height of the cone. Find the rate of change in the volume of the tumor with respect to the height.

To emphasize that the rate of change of the volume is found with respect to the height, use the symbol dV/dh for the derivative. For this tumor, r is fixed at 2 cm. By substituting 2 for r,

$$V = \frac{1}{3}\pi r^2 h \qquad \text{becomes} \qquad V = \frac{1}{3}\pi \cdot 2^2 \cdot h \qquad \text{or} \qquad V = \frac{4}{3}\pi h.$$

⑤ Find the derivatives of the following.
(a) $y = -4x^5 - 8x + 6$
(b) $y = 32x^5 - 100x^2 + 12x$
(c) $y = 8t^{3/2} + 2t^{1/2}$
(d) $f(t) = -\sqrt{t} + 6/t$

Answer:
(a) $y' = -20x^4 - 8$
(b) $y' = 160x^4 - 200x + 12$
(c) $y' = 12t^{1/2} + t^{-1/2}$
 or $12t^{1/2} + 1/t^{1/2}$
(d) $f'(t) = -1/(2\sqrt{t}) - 6/t^2$

Since $4\pi/3$ is a constant,

$$\frac{dV}{dh} = \frac{4\pi}{3} \approx 4.2 \text{ cu cm per cm.}$$

For each additional centimeter that the tumor grows in height, its volume will increase approximately 4.2 cubic centimeters. ▪

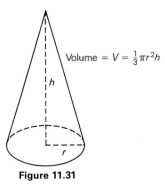

Volume $= V = \frac{1}{3}\pi r^2 h$

Figure 11.31

In business and economics the rates of change of such variables as cost, revenue, and profits are important considerations. Economists use the word **marginal** to refer to rates of change: for example, *marginal cost* refers to the rate of change of cost. Since the derivative of a function gives the rate of change of the function, a marginal cost (or revenue, or profit) function is found by taking the derivative of the cost (or revenue, or profit) function. Roughly speaking, the marginal cost at some level of production x is the cost to produce the $(x + 1)$st item. (Similar statements could be made for revenue or profit.)

Example 6   Suppose that the total cost in hundreds of dollars to produce x thousand barrels of beer is given by

$$C(x) = 4x^2 + 100x + 500.$$

Find the marginal cost for the following values of x.

(a) $x = 5$

Find the marginal cost by first finding the derivative of the total cost function:

$$C'(x) = 8x + 100.$$

When $x = 5$,

$$C'(5) = 8(5) + 100 = 140.$$

After 5 thousand barrels of beer have been produced, the cost to produce 1 thousand more barrels will be approximately 140 hundred dollars, or $14,000. The *actual* cost to produce one thousand more barrels is $C(6) - C(5)$:

$$C(6) - C(5) = (4 \cdot 6^2 + 100 \cdot 6 + 500) - (4 \cdot 5^2 + 100 \cdot 5 + 500)$$
$$= 1244 - 1100$$
$$= 144 \text{ hundred dollars or } \$14,400.$$

6 A balloon is spherical. The formula for the volume of a sphere is $V = (4/3)\pi r^3$, where r is the radius of the sphere. Find the following.
(a) dV/dr
(b) The rate of change of the volume when $r = 3$ inches

Answer:
(a) $4\pi r^2$
(b) 36π cubic inches per inch

7 The cost to produce x units of wheat is given by $C(x) = 5000 + 20x + 10\sqrt{x}$. Find the marginal cost when

(a) $x = 9$;
(b) $x = 16$;
(c) $x = 25$.

Answer:
(a) 65/3
(b) 85/4
(c) 21

8 The revenue from the sale of x units of ice cream is given by $R(x)$, where $R(x) = 800x + 10x^2$. Find the marginal revenue when

(a) $x = 10$;
(b) $x = 25$;
(c) $x = 32$.

Answer:
(a) 1000
(b) 1300
(c) 1440

In this example, the actual marginal cost and the approximation are fairly close.

(b) $x = 30$

After 30 thousand barrels have been produced, the cost to produce 1 thousand more barrels will be approximately

$$C'(30) = 8(30) + 100 = 340,$$

or $34,000.

Management must be careful to keep track of marginal costs. If the marginal cost of producing an extra unit exceeds the revenue received from its sale, the company will lose money on that unit. **7** **8**

The definitions of marginal cost, revenue, and profit are summarized below.

Marginal Cost, Revenue, and Profit

If $C(x)$ is the total cost to produce x units, then $C'(x)$ is the **marginal cost,** the approximate cost of the $(x + 1)$st unit.

If $R(x)$ is the total revenue from selling x units, then $R'(x)$ is the **marginal revenue,** the approximate revenue from the $(x + 1)$st unit.

If $P(x) = R(x) - C(x)$ is the total profit from selling x units, then $P'(x) = R'(x) - C'(x)$ is the **marginal profit,** the approximate profit on the $(x + 1)$st unit.

11.5 Exercises

Find the derivatives of the functions defined as follows. (See Examples 1–4.)

1. $f(x) = 9x^2 - 8x + 4$

2. $f(x) = 10x^2 + 4x - 9$

3. $y = 10x^3 - 9x^2 + 6x$

4. $y = 3x^3 - x^2 - 12x$

5. $y = x^4 - 5x^3 + 9x^2 + 5$

6. $y = 3x^4 + 11x^3 + 2x^2 - 4x$

7. $f(x) = 6x^{1.5} - 4x^{.5}$

8. $f(x) = -2x^{2.5} + 8x^{.5}$

9. $y = -15x^{3.2} + 2x^{1.9}$

10. $y = 18x^{1.6} - 4x^{3.1}$

11. $y = 24t^{3/2} + 4t^{1/2}$

12. $y = -24t^{5/2} - 6t^{1/2}$

13. $y = 8\sqrt{x} + 6x^{3/4}$

14. $y = -100\sqrt{x} - 11x^{2/3}$

15. $g(x) = 6x^{-5} - x^{-1}$

16. $y = -2x^{-4} + x^{-1}$

17. $y = -4x^{-2} + 3x^{-3}$

18. $g(x) = 8x^{-4} - 9x^{-2}$

19. $y = 10x^{-2} + 3x^{-4} - 6x$

20. $y = x^{-5} - x^{-2} + 5x^{-1}$

21. $f(t) = \dfrac{6}{t} - \dfrac{8}{t^2}$

22. $f(t) = \dfrac{4}{t} + \dfrac{2}{t^3}$

23. $y = \dfrac{9 - 8x + 2x^3}{x^4}$

24. $y = \dfrac{3 + x - 7x^4}{x^6}$

25. $f(x) = 12x^{-1/2} - 3x^{1/2}$

26. $r(x) = -30x^{-1/2} + 5x^{1/2}$

27. $p(x) = -10x^{-1/2} + 8x^{-3/2}$

28. $h(x) = x^{-1/2} - 14x^{-3/2}$

29. $y = \dfrac{6}{\sqrt[4]{x}}$

30. $y = \dfrac{-2}{\sqrt[3]{x}}$

31. $y = \dfrac{-5t}{\sqrt[3]{t^2}}$

32. $g(t) = \dfrac{9t}{\sqrt{t^3}}$

Find each of the following.

33. $\dfrac{dy}{dx}$ if $y = 8x^{-5} - 9x^{-4}$

34. $\dfrac{dy}{dx}$ if $y = -3x^{-2} - 4x^{-5}$

35. $D_x\left[9x^{-1/2} + \dfrac{2}{x^{3/2}}\right]$

36. $D_x\left[\dfrac{8}{\sqrt[4]{x}} - \dfrac{3}{\sqrt[3]{x^3}}\right]$

37. $f'(-2)$ if $f(x) = 6x^2 - 4x$

38. $f'(3)$ if $f(x) = 9x^3 - 8x^2$

39. $f'(4)$ if $f(t) = 2\sqrt{t} - \dfrac{3}{\sqrt{t}}$

40. $f'(8)$ if $f(t) = -5\sqrt[3]{t} + \dfrac{6}{\sqrt[3]{t}}$

For the functions defined as follows, find any points where the tangent line is horizontal. (Recall: find the x-values of points where the derivative is 0.)

41. $y = 6x^2 - 20x + 5$

42. $y = 9x^2 - 18x - 4$

43. $y = x^3 - \dfrac{5}{2}x^2 + 2x - 1$

44. $y = \dfrac{4}{3}x^3 - 13x^2 + 12x - 6$

45. $y = \dfrac{2}{3}x^3 - \dfrac{3}{2}x^2 - 5x + 8$

46. $y = x^3 - 2x^2 - 15x + 6$

47. $y = x^4 - 16x$

48. $y = x^4 - 256x$

49. $y = \dfrac{1}{x}$

50. $y = \dfrac{-4}{x^2}$

Find the slope of the tangent line to the graph of each function defined as follows at the given value of x. Find the equation of the tangent line in Exercises 51 and 52.

51. $y = x^4 - 5x^3 + 2;\quad x = 2$

52. $y = -2x^5 - 7x^3 + 8x^2;\quad x = 1$

53. $y = 3x^{3/2} - 2x^{1/2};\quad x = 9$

54. $y = -2x^{1/2} + x^{3/2};\quad x = 4$

55. $y = 5x^{-1} - 2x^{-2};\quad x = 3$

56. $y = -x^{-3} + x^{-2};\quad x = 1$

Work the following exercises. (See Examples 5 and 6.)

57. Management If the price of a product is given by

$$P(x) = \dfrac{-1000}{x} + 1000,$$

where x represents the demand for the product, find the rate of change of price when the demand is 10.

58. Management Often sales of a new product grow rapidly at first and then level off with time. This is the case with the sales represented by the function

$$S(t) = 100 - 100t^{-1},$$

where t represents time in years. Find the rate of change of sales for the following values of t.
(a) $t = 1$ **(b)** $t = 10$

59. Natural Science Suppose $P(t) = 100/t$ represents the percent of acid in a chemical solution after t days of exposure to an ultraviolet light source. Find the percent of acid in the solution after the following number of days.
(a) 1 day **(b)** 100 days
(c) Find and interpret $P'(100)$.

60. Natural Science   Insulin affects the glucose, or blood sugar, level of some diabetics according to the function

$$G(x) = -.2x^2 + 450,$$

where $G(x)$ is the blood sugar level one hour after x units of insulin are injected. (This mathematical model is only approximate and valid only for values of x less than about 40.)
(a) Find $G(0)$.   (b) Find $G(25)$.

61. Use the function G of Exercise 60 and find dG/dx for the following values of x. Interpret your answers.
(a) $x = 10$ (b) $x = 25$

62. Natural Science   A short length of blood vessel has a cylindrical shape. The volume of a cylinder is given by $V = \pi r^2 h$. Suppose an experimental device is set up to measure the volume of blood in a blood vessel of fixed length 80 mm as the radius changes.
(a) Find dV/dr.
Suppose a drug is administered which causes the blood vessel to expand. Evaluate dV/dr for the following values of r.
(b) 4 mm   (c) 6 mm   (d) 8 mm

63. Natural Science   In an experiment testing methods of sexually attracting male insects to sterile females, equal numbers of males and females of a certain species are permitted to intermingle. Assume that

$$M(t) = 4t^{3/2} + 2t^{1/2}$$

approximates the number of matings observed among the insects in an hour, where t is the temperature in degrees Celsius. (This formula is only valid for certain temperature ranges.) Find each of the following.
(a) $M(16)$ (b) $M(25)$
(c) The rate of change of M when $t = 16$

64. Management   The profit in dollars from the sale of x expensive cassette recorders is

$$P(x) = x^3 - 5x^2 + 7x + 10.$$

Find the marginal profit for the following values of x.
(a) $x = 4$ (b) $x = 8$ (c) $x = 10$ (d) $x = 12$

65. Management   The total cost to produce x handcrafted weathervanes is

$$C(x) = 100 + 8x - x^2 + 4x^3.$$

Find the marginal cost for the following values of x.
(a) $x = 0$ (b) $x = 4$ (c) $x = 6$ (d) $x = 8$

66. Management   An analyst has found that a company's costs and revenues for one product are given by

$$C(x) = 2x \quad \text{and} \quad R(x) = 6x - \frac{x^2}{1000},$$

respectively, where x is the number of items produced.
(a) Find the marginal cost function.
(b) Find the marginal revenue function.
(c) Using the fact that profit is the difference between revenue and costs, find the marginal profit function.
(d) What value of x makes marginal profit equal 0?
(e) Find the profit when the marginal profit is 0.

(As we shall see in the next chapter, this process is used to find *maximum* profit.)

Physical Science   *It can be shown that the instantaneous velocity of a particle moving in a straight line is given by*

$$\lim_{h \to 0} \frac{s(t + h) - s(t)}{h},$$

where $s(t)$ gives the position of the particle at time t. Since this limit is the derivative of $s(t)$, the velocity of a particle is given by $s'(t)$. If $v(t)$ represents velocity at time t, then $v(t) = s'(t)$. For each of the following position functions, find (a) $v(t)$; (b) the velocity when $t = 0, t = 5,$ and $t = 10$.

67. $s(t) = 6t + 5$

68. $s(t) = 9 - 2t$

69. $s(t) = 11t^2 + 4t + 2$

70. $s(t) = 25t^2 - 9t + 8$

71. $s(t) = 4t^3 + 8t^2$

72. $s(t) = -2t^3 + 4t^2 - 1$

73. Perform each step and give reasons for your results in the following proof that the derivative of $y = x^n$ is $y' = n \cdot x^{n-1}$. (We prove this result only for positive integer values of n, but it is valid for all values of n.)
(a) Recall the binomial theorem from algebra:

$$(p + q)^n = p^n + n \cdot p^{n-1}q$$
$$+ \frac{n(n - 1)}{2} p^{n-2}q^2 + \cdots + q^n.$$

Let $y = (x + h)^n$, and evaluate $(x + h)^n$.
(b) Find the quotient $\dfrac{(x + h)^n - x^n}{h}$.
(c) Use the definition of derivative to find y'.

74. Perform each step and give reasons for your result in the proof that the derivative of $y = f(x) + g(x)$ is $y' = f'(x) + g'(x)$.

(a) Let $s(x) = f(x) + g(x)$. Show that

$$s'(x) = \lim_{h \to 0} \frac{[f(x + h) + g(x + h)] - [f(x) + g(x)]}{h}.$$

(b) Show that $s'(x)$ equals

$$\lim_{h \to 0} \left[\frac{f(x + h) - f(x)}{h} + \frac{g(x + h) - g(x)}{h} \right].$$

(c) Finally, show that $s'(x) = f'(x) + g'(x)$.

11.6 Derivatives of Products and Quotients

The derivative of a sum of two functions is the sum of the derivatives of the functions. What about products? Is the derivative of a product equal to the product of the derivatives? Let's try an example.

$$\text{Let} \quad g(x) = 2x + 3 \quad \text{and} \quad k(x) = 3x^2,$$
$$\text{so that} \quad g'(x) = 2 \quad \text{and} \quad k'(x) = 6x.$$

Let $f(x)$ be the product of g and k; that is, $f(x) = (2x + 3)(3x^2) = 6x^3 + 9x^2$. By the rules of the preceding section, $f'(x) = 18x^2 + 18x$. On the other hand, $g'(x) = 2$ and $k'(x) = 6x$, with the product $g'(x) \cdot k'(x) = 2(6x) = 12x \neq f'(x)$. In this example, the derivative of a product is *not* equal to the product of the derivatives, nor is this usually the case.

The rule for finding derivatives of products is given below.

Product Rule If $f(x) = g(x) \cdot k(x)$, and if both $g'(x)$ and $k'(x)$ exist, then

$$f'(x) = g(x) \cdot k'(x) + k(x) \cdot g'(x).$$

(The derivative of a product of two functions is the first function times the derivative of the second, plus the second function times the derivative of the first.)

To sketch the proof of the product rule, let

$$f(x) = g(x) \cdot k(x).$$

Then $f(x + h) = g(x + h) \cdot k(x + h)$, and, by definition, $f'(x)$ is

$$f'(x) = \lim_{h \to 0} \frac{f(x + h) - f(x)}{h}$$

$$= \lim_{h \to 0} \frac{g(x + h) \cdot k(x + h) - g(x) \cdot k(x)}{h},$$

provided this limit exists.

Now an algebraic procedure is used: add and subtract $g(x + h) \cdot k(x)$ in the numerator, giving

$$f'(x) = \lim_{h \to 0} \frac{g(x + h) \cdot k(x + h) - g(x + h) \cdot k(x) + g(x + h) \cdot k(x) - g(x) \cdot k(x)}{h}$$

$$= \lim_{h \to 0} \frac{g(x + h)[k(x + h) - k(x)] + k(x)[g(x + h) - g(x)]}{h}$$

$$= \lim_{h \to 0} g(x + h) \left[\frac{k(x + h) - k(x)}{h} \right] + \lim_{h \to 0} k(x) \left[\frac{g(x + h) - g(x)}{h} \right]. \qquad (*)$$

If g' and k' both exist, then

$$\lim_{h \to 0} \frac{g(x + h) - g(x)}{h} = g'(x) \qquad \text{and} \qquad \lim_{h \to 0} \frac{k(x + h) - k(x)}{h} = k'(x).$$

The fact that g' exists can be used to prove that

$$\lim_{h \to 0} g(x + h) = g(x), \qquad \text{and} \qquad \lim_{h \to 0} k(x) = k(x).$$

Substituting these results into equation (*) gives

$$f'(x) = g(x) \cdot k'(x) + k(x) \cdot g'(x),$$

the desired result.

Example 1 Let $f(x) = (2x + 3)(3x^2)$. Use the product rule to find $f'(x)$.

Here f is given as the product of $g(x) = 2x + 3$ and $k(x) = 3x^2$. Use the product rule and the fact that $g'(x) = 2$ and $k'(x) = 6x$.

$$f'(x) = g(x) \cdot k'(x) + k(x) \cdot g'(x)$$
$$= (2x + 3)(6x) + (3x^2)(2)$$
$$= 12x^2 + 18x + 6x^2 \qquad \text{Use algebra to simplify}$$
$$f'(x) = 18x^2 + 18x$$

This result is the same as that found at the beginning of the section. ▮**1**

Example 2 Find the derivative of $y = (\sqrt{x} + 3)(x^2 - 5x)$.

Let $g(x) = \sqrt{x} + 3 = x^{1/2} + 3$, and $k(x) = x^2 - 5x$. Then, by the product rule,

$$y' = g(x) \cdot k'(x) + k(x) \cdot g'(x)$$
$$= (x^{1/2} + 3)(2x - 5) + (x^2 - 5x)\left(\frac{1}{2} x^{-1/2} \right).$$

Simplify by multiplying and combining terms.

$$y' = (2x)(x^{1/2}) + 6x - 5x^{1/2} - 15 + (x^2)\left(\frac{1}{2} x^{-1/2} \right) - (5x)\left(\frac{1}{2} x^{-1/2} \right)$$

$$= 2x^{3/2} + 6x - 5x^{1/2} - 15 + \frac{1}{2} x^{3/2} - \frac{5}{2} x^{1/2}$$

$$= \frac{5}{2} x^{3/2} + 6x - \frac{15}{2} x^{1/2} - 15. \qquad ▮\textbf{2}$$

▮**1** Use the product rule to find the derivatives of the following.
(a) $f(x) = (5x^2 + 6)(3x)$
(b) $g(x) = (8x)(4x^2 + 5x)$

Answer:
(a) $45x^2 + 18$
(b) $96x^2 + 80x$

▮**2** Find the derivatives of the following.
(a) $f(x) = (x^2 - 3)(\sqrt{x} + 5)$
(b) $g(x) = (\sqrt{x} + 4)(5x^2 + x)$

Answer:
(a) $\frac{5}{2} x^{3/2} + 10x - \frac{3}{2} x^{-1/2}$

(b) $\frac{25}{2} x^{3/2} + 40x +$

$\frac{3}{2} x^{1/2} + 4$

The problems above could have been worked by multiplying out the original functions. The product rule would then not be needed. However, in the next section we shall see products of functions where this cannot be done—where the product rule is essential.

What about *quotients* of functions? Find the derivative of the quotient of two functions with the next result.

Quotient Rule If $f(x) = \dfrac{g(x)}{k(x)}$ and both $g'(x)$ and $k'(x)$ exist, and if $k(x) \neq 0$, then

$$f'(x) = \frac{k(x) \cdot g'(x) - g(x) \cdot k'(x)}{[k(x)]^2}.$$

(The derivative of a quotient is the denominator times the derivative of the numerator, minus the numerator times the derivative of the denominator, all over the denominator squared.)

The proof of the quotient rule is similar to that of the product rule.

Example 3 Let $f(x) = \dfrac{2x - 1}{4x + 3}$. Find $f'(x)$.

Let $g(x) = 2x - 1$, with $g'(x) = 2$. Also, let $k(x) = 4x + 3$, and $k'(x) = 4$. Then, by the quotient rule,

$$f'(x) = \frac{k(x) \cdot g'(x) - g(x) \cdot k'(x)}{[k(x)]^2}$$

$$= \frac{(4x + 3)(2) - (2x - 1)(4)}{(4x + 3)^2}$$

$$= \frac{8x + 6 - 8x + 4}{(4x + 3)^2}$$

$$f'(x) = \frac{10}{(4x + 3)^2}. \quad \boxed{3}$$

3 Find the derivatives of the following.

(a) $f(x) = \dfrac{3x + 7}{5x + 8}$

(b) $g(x) = \dfrac{2x + 11}{5x - 1}$

Answer:

(a) $\dfrac{-11}{(5x + 8)^2}$

(b) $\dfrac{-57}{(5x - 1)^2}$

Example 4 Find $D_x\left(\dfrac{x^{-1} - 2}{x^{-2} + 4}\right)$.

Use the quotient rule.

$$D_x\left(\frac{x^{-1} - 2}{x^{-2} + 4}\right) = \frac{(x^{-2} + 4)(-x^{-2}) - (x^{-1} - 2)(-2x^{-3})}{(x^{-2} + 4)^2}$$

$$= \frac{-x^{-4} - 4x^{-2} + 2x^{-4} - 4x^{-3}}{(x^{-2} + 4)^2}$$

$$= \frac{x^{-4} - 4x^{-3} - 4x^{-2}}{(x^{-2} + 4)^2}$$

In the next chapter derivatives are used to solve practical problems. That work is easier if the derivatives are first simplified by writing with only positive exponents. This derivative, for example, can be simplified if the negative exponents are removed by multiplying numerator and denominator by x^4, and using $(x^{-2} + 4)^2 = x^{-4} + 8x^{-2} + 16$.

$$\frac{x^{-4} - 4x^{-3} - 4x^{-2}}{(x^{-2} + 4)^2} = \frac{1 - 4x - 4x^2}{x^4(x^{-4} + 8x^{-2} + 16)} = \frac{1 - 4x - 4x^2}{1 + 8x^2 + 16x^4}$$ **4**

4 Find each derivative. Write answers with positive exponents.

(a) $D_x\left[\dfrac{x^{-2} - 1}{x^{-1} + 2}\right]$

(b) $D_x\left[\dfrac{2 + x^{-1}}{x^{-3} + 1}\right]$

Answer:

(a) $\dfrac{-1 - 4x - x^2}{x^2 + 4x^3 + 4x^4}$

(b) $\dfrac{2x + 6x^2 - x^4}{1 + 2x^3 + x^6}$

Suppose $y = C(x)$ gives the total cost to manufacture x items. Then the *average cost per item* is found by dividing the total cost by the number of items. The rate of change of average cost is called the **marginal average cost,** and is the derivative of the average cost.

> **Average Cost** If the total cost to manufacture x items is given by $C(x)$, then the **average cost per item** is
>
> $$\text{Average cost} = \frac{C(x)}{x}.$$
>
> The **marginal average cost** is the derivative of the average cost function.

A company naturally would be interested in making this average cost as small as possible. We will see in the next chapter that this can be done by taking the derivative of $C(x)/x$. This derivative often can be found by using the quotient rule, as the next example shows.

Example 5 The total cost in thousands of dollars to manufacture x electrical generators is given by $C(x)$, where

$$C(x) = -x^3 + 15x^2 + 1000.$$

(a) Find the average cost per generator.

The average cost is given by the total cost divided by the number of items, or

$$\frac{C(x)}{x} = \frac{-x^3 + 15x^2 + 1000}{x}.$$

(b) Find the marginal average cost.

The marginal average cost is the derivative of the average cost function. Find this derivative with the quotient rule.

$$\frac{d}{dx}\left[\frac{C(x)}{x}\right] = \frac{x(-3x^2 + 30x) - (-x^3 + 15x^2 + 1000)(1)}{x^2}$$

$$= \frac{-3x^3 + 30x^2 + x^3 - 15x^2 - 1000}{x^2}$$

$$= \frac{-2x^3 + 15x^2 - 1000}{x^2}$$ **5**

5 The total revenue in thousands of dollars from the sale of x dozen CB radios is given by

$$R(x) = 32x^2 + 7x + 80.$$

(a) Find the average revenue.

(b) Find the marginal average revenue.

Answer:

(a) $\dfrac{32x^2 + 7x + 80}{x}$

(b) $\dfrac{32x^2 - 80}{x^2}$

11.6 Exercises

Use the product rule to find the derivative of the functions defined as follows. (See Examples 1 and 2.) (In Exercises 13–16, use the fact that $p^2 = p \cdot p$.)

1. $y = 3x(2x - 5)$

2. $y = -4x(8x + 1)$

3. $y = (2x - 5)(x + 4)$

4. $y = (3x + 7)(x - 1)$

5. $y = (8x - 2)(3x + 9)$

6. $y = (4x + 1)(7x + 12)$

7. $y = (3x^2 + 2)(2x - 1)$

8. $y = (5x^2 - 1)(4x + 3)$

9. $y = (x^2 + x)(3x - 5)$

10. $y = (2x^2 - 6x)(x + 2)$

11. $y = (9x^2 + 7x)(x^2 - 1)$

12. $y = (2x^2 - 4x)(5x^2 + 4)$

13. $y = (2x - 5)^2$

14. $y = (7x - 6)^2$

15. $y = (x^2 - 1)^2$

16. $y = (3x^2 + 2)^2$

17. $y = (x + 1)(\sqrt{x} + 2)$

18. $y = (2x - 3)(\sqrt{x} - 1)$

19. $y = (5\sqrt{x} - 1)(2\sqrt{x} + 1)$

20. $y = (-3\sqrt{x} + 6)(4\sqrt{x} - 2)$

Use the quotient rule to find the derivatives of the functions defined as follows. (See Examples 3 and 4.)

21. $y = \dfrac{x + 1}{2x - 1}$

22. $y = \dfrac{3x - 5}{x - 4}$

23. $y = \dfrac{7x + 1}{3x + 8}$

24. $y = \dfrac{6x - 11}{8x + 1}$

25. $y = \dfrac{2}{3x - 5}$

26. $y = \dfrac{-4}{2x - 11}$

27. $y = \dfrac{5 - 3x}{4 + x}$

28. $y = \dfrac{9 - 7x}{1 - x}$

29. $y = \dfrac{x^2 + x}{x - 1}$

30. $y = \dfrac{x^2 - 4x}{x + 3}$

31. $y = \dfrac{x - 2}{x^2 + 1}$

32. $y = \dfrac{4x + 11}{x^2 - 3}$

33. $y = \dfrac{3x^2 + x}{2x^2 - 1}$

34. $y = \dfrac{-x^2 + 6x}{4x^2 + 1}$

35. $y = \dfrac{x^2 - 4x + 2}{x + 3}$

36. $y = \dfrac{x^2 + 7x - 2}{x - 2}$

37. $y = \dfrac{2x^{-1} + 3}{x^{-2} - 4}$

38. $y = \dfrac{5 - 3x^{-2}}{2 + x^{-2}}$

39. $y = \dfrac{\sqrt{x}}{x - 1}$

40. $y = \dfrac{\sqrt{x}}{2x + 3}$

41. $y = \dfrac{5x + 6}{\sqrt{x}}$

42. $y = \dfrac{9x - 8}{\sqrt{x}}$

Work the following exercises.

43. Management The total cost to produce x units of perfume is given by

$$C(x) = 9x^2 - 4x + 8.$$

Find the average cost per unit to produce (see Example 5)

(a) 10 units;   (Find $C(10)/10$.)

(b) 20 units;

(c) x units.

(d) Find the derivative of the average cost function.

44. Management The total profit from selling x units of mathematics textbooks is given by

$$P(x) = 10x^2 - 5x - 18.$$

Find the average profit from selling

(a) 8 units;   **(b)** 15 units;   **(c)** x units.

(d) Find the marginal average profit.

45. Management Suppose you are the manager of a trucking firm, and one of your drivers reports that, according to her calculations, her truck burns fuel at the rate of

$$G(x) = \frac{1}{200}\left(\frac{800}{x} + x\right)$$

gallons per mile when traveling at x miles per hour on a smooth, dry road. (If a driver did report this to you, it would probably be a good idea to make the driver your mathematical modeler.)

(a) If the driver tells you that she wants to travel 20 miles per hour, what should you tell her? (Hint:

Take the derivative of G and evaluate it for $x = 20$. Then interpret your results.)

(b) If the driver wants to go 40 miles per hour, what should you say? (Hint: Find $G'(40)$. In Example 3 of Section 12.3 we will find the speed that produces the lowest possible cost.)

46. Natural Science Assume that the total number of bacteria in millions present at a certain time t is given by

$$N(t) = (t - 10)^2(2t) + 50.$$

(a) Find $N'(t)$.

At what rate is the population of bacteria changing
(b) when $t = 8$?   **(c)** when $t = 11$?
(d) The answer in part (b) is negative, while the answer in part (c) is positive. What does this mean in terms of the population of bacteria?

47. Natural Science When a certain drug is introduced into a muscle, the muscle responds by contracting. The amount of contraction, s, in millimeters, is related to the concentration of the drug, x, in milliliters, by

$$s(x) = \frac{x}{m + nx},$$

where m and n are constants.
(a) Find $s'(x)$.
(b) Evaluate $s'(x)$ when $x = 50$, $m = 10$, and $n = 3$.

48. Social Science According to the psychologist L. L. Thurstone, the number of facts of a certain type that are remembered after t hours is given by

$$f(t) = \frac{kt}{at - m},$$

where k, m, and a are constants. Find $f'(t)$ if $a = 99$, $k = 90$, $m = 90$, and
(a) $t = 1$;   **(b)** $t = 10$.

11.7 The Chain Rule

Suppose an oil well off the Gulf Coast is leaking, with the leak spreading oil in a circle over the surface. At any time t, in minutes, after the beginning of the leak, the radius of the circular oil slick is $r(t) = 4t$ feet. Since $A(r) = \pi r^2$ gives the area of a circle of radius r, express the area as a function of time by substituting $4t$ for r in $A(r) = \pi r^2$ to get

$$A(r) = \pi r^2$$
$$A(r(t)) = \pi(4t)^2 = 16\pi t^2.$$

The function $A(r(t))$ is a **composite function,** one in which another function is substituted to give a function of a function.

1 Let $f(x) = \sqrt{x + 4}$ and $g(x) = x^2 + 5x + 1$. Find $f[g(x)]$ and $g[f(x)]$.

Answer:

$f[g(x)] = \sqrt{x^2 + 5x + 5}$,
$g[f(x)] = (\sqrt{x + 4})^2$
$\qquad + 5\sqrt{x + 4} + 1$
$\qquad = x + 5 + 5\sqrt{x + 4}$

Example 1   Let $f(x) = x^{1/2}$ and $g(x) = 5x + 9$. Find the composite functions $f[g(x)]$ and $g[f(x)]$.

To find $f[g(x)]$, replace each x in $f(x)$ with $g(x)$.

$$f[g(x)] = f[5x + 9] = (5x + 9)^{1/2}$$

Also,

$$g[f(x)] = g[x^{1/2}] = 5x^{1/2} + 9.$$

In this example, $f[g(x)] \neq g[f(x)]$. **1**

In the example at the beginning of this section, the radius of the circular oil slick was given by

$$r(t) = 4t, \quad \text{with} \quad \frac{dr}{dt} = 4.$$

The area of the oil slick was given by

$$A(r) = \pi r^2, \quad \text{with} \quad \frac{dA}{dr} = 2\pi r.$$

As these derivatives show, the radius is increasing 4 times as fast as the time t, and the area is increasing $2\pi r$ times as fast as the radius r. It seems reasonable, then, that the area is increasing $2\pi r \cdot 4 = 8\pi r$ times as fast as time. That is,

$$\frac{dA}{dt} = \frac{dA}{dr} \cdot \frac{dr}{dt} = 2\pi r \cdot 4 = 8\pi r.$$

Since $r = 4t$,

$$\frac{dA}{dt} = 8\pi(4t) = 32\pi t.$$

To check this, use the fact that

$$A = 16\pi t^2, \quad \text{with} \quad \frac{dA}{dt} = 32\pi t,$$

which is the same result found above. The product used above,

$$\frac{dA}{dt} = \frac{dA}{dr} \cdot \frac{dr}{dt},$$

is an example of the **chain rule**, used to find the derivative of a *composite function*.

Chain Rule If y is a function of u, say $y = f(u)$, and if u is a function of x, say $u = g(x)$, then $y = f(u) = f[g(x)]$, and

$$\frac{dy}{dx} = \frac{dy}{du} \cdot \frac{du}{dx}.$$

One way to remember the chain rule is to pretend that dy/du and du/dx are fractions, with du "canceling out." The proof of the chain rule requires advanced concepts, and is not given here.

Example 2 Find dy/dx if $y = (3x^2 - 5x)^{1/2}$.
　　Let $y = u^{1/2}$, and $u = 3x^2 - 5x$. Then

$$\frac{dy}{dx} = \frac{dy}{du} \cdot \frac{du}{dx}$$

$$= \frac{1}{2}u^{-1/2} \cdot (6x - 5).$$

Replacing u with $3x^2 - 5x$ gives

$$\frac{dy}{dx} = \frac{1}{2}(3x^2 - 5x)^{-1/2}(6x - 5) = \frac{6x - 5}{2(3x^2 - 5x)^{1/2}}.$$ **2**

2 Use the chain rule to find dy/dx if $y = 10(2x^2 + 1)^4$.

Answer:

$160x(2x^2 + 1)^3$

An alternate statement of the chain rule, in terms of composite functions, is given below.

> **The Chain Rule (alternate form)** Let $y = f[g(x)]$. Then, if all indicated derivatives exist,
>
> $$y' = f'[g(x)] \cdot g'(x).$$
>
> (To find the derivative of $f[g(x)]$, find the derivative of $f(x)$, replace each x with $g(x)$, and then multiply the result by the derivative of $g(x)$.)

Example 3 Use the chain rule to find the derivative of $y = \sqrt{15x^2 + 1}$.

Since $y = (15x^2 + 1)^{1/2}$, let $f(x) = x^{1/2}$ and $g(x) = 15x^2 + 1$. Then $y = f[g(x)]$, and

$$y' = f'[g(x)] \cdot g'(x).$$

Here $f'(x) = \frac{1}{2}x^{-1/2}$, with $f'[g(x)] = \frac{1}{2}[g(x)]^{-1/2} = \frac{1}{2}(15x^2 + 1)^{-1/2}$, and

$$y' = \frac{1}{2}[g(x)]^{-1/2} \cdot g'(x) = \frac{1}{2}(15x^2 + 1)^{-1/2} \cdot (30x)$$

$$= \frac{15x}{(15x^2 + 1)^{1/2}}.$$ **3**

3 Use the alternate form of the chain rule to find the derivative of $y = \sqrt{14x + 1}$.

Answer:

$\dfrac{7}{\sqrt{14x + 1}}$

While the chain rule is essential for finding the derivatives of some of the functions discussed later, the derivatives of the algebraic functions we have seen so far can be found by the following *generalized power rule,* a special case of the chain rule for functions defined as $y = u^n$, where u is a function of x.

> **Generalized Power Rule** Let u be a function of x, and let $y = u^n$, for any real number n. Then
>
> $$y' = n \cdot u^{n-1} \cdot u'.$$
>
> (The derivative of $y = u^n$ is found by decreasing the exponent on u by 1 and multiplying the result by the exponent n and by the derivative of u with respect to x.)

Example 4

(a) Use the generalized power rule to find the derivative of $y = (3 + 5x)^2$.

Let $u = 3 + 5x$, and $n = 2$. Then $u' = 5$. By the generalized power rule,

$$y' = \frac{dy}{dx} = n \cdot u^{n-1} \cdot u'$$

$$
\begin{array}{cccc}
n & u & n-1 & u' \\
\downarrow & \downarrow & \downarrow & \downarrow
\end{array}
$$
$$= 2 \cdot (3 + 5x)^{2-1} \cdot 5$$
$$= 10(3 + 5x)$$
$$= 30 + 50x.$$

(b) Find y' if $y = (3 + 5x)^{-3/4}$.

Use the generalized power rule with $n = -\dfrac{3}{4}$, $u = 3 + 5x$, $u' = 5$.

$$y' = -\frac{3}{4}(3 + 5x)^{(-3/4)-1}(5)$$

$$= -\frac{15}{4}(3 + 5x)^{-7/4} \quad \boxed{4}$$

This result could not have been found by any of the rules given earlier.

Example 5 Find the derivative of the following.

(a) $y = 2(7x^2 + 5)^4$

Let $u = 7x^2 + 5$. Then $u' = 14x$, and

$$
\begin{array}{cccc}
n & u & n-1 & u' \\
\downarrow & \downarrow & \downarrow & \downarrow
\end{array}
$$
$$y' = 2 \cdot 4 (7x^2 + 5)^{4-1} \cdot 14x$$
$$= 112x(7x^2 + 5)^3.$$

(b) $y = \sqrt{9x + 2}$

Write $y = \sqrt{9x + 2}$ as $y = (9x + 2)^{1/2}$. Then

$$y' = \frac{1}{2}(9x + 2)^{-1/2}(9) = \frac{9}{2}(9x + 2)^{-1/2}.$$

The derivative also can be written as

$$y' = \frac{9}{2(9x + 2)^{1/2}} \quad \text{or} \quad y' = \frac{9}{2\sqrt{9x + 2}}. \quad \boxed{5}$$

Example 6 Suppose the revenue, $R(x)$, produced by selling x units of a biological nutrient is given by

$$R(x) = 4\sqrt{3x + 1} + \frac{4}{x}.$$

Side column:

$\boxed{4}$ Find dy/dx for the following.

(a) $y = (2x + 5)^6$

(b) $y = (4x^2 - 7)^3$

(c) $f(x) = \sqrt{3x^2 - x}$

(d) $g(x) = (2 - x^4)^{-3}$

Answer:

(a) $12(2x + 5)^5$

(b) $24x(4x^2 - 7)^2$

(c) $\dfrac{6x - 1}{2\sqrt{3x^2 - x}}$

(d) $\dfrac{12x^3}{(2 - x^4)^4}$

$\boxed{5}$ Find dy/dx for the following.

(a) $y = 12(x^2 + 6)^5$

(b) $y = 8(4x^2 + 2)^{3/2}$

Answer:

(a) $120x(x^2 + 6)^4$

(b) $96x(4x^2 + 2)^{1/2}$

Assume that $x > 0$. Find the marginal revenue when $x = 5$.

As mentioned earlier, the marginal revenue is given by the derivative of the revenue function. By the generalized power rule,

$$R'(x) = \frac{6}{\sqrt{3x + 1}} - \frac{4}{x^2}.$$

When $x = 5$,

$$R'(5) = \frac{6}{\sqrt{3(5) + 1}} - \frac{4}{5^2}$$

$$= \frac{6}{\sqrt{16}} - \frac{4}{25} = \frac{6}{4} - \frac{4}{25} = 1.34.$$

After 5 units have been sold, the sale of one more unit will increase revenue by about \$1.34. **6**

6 Use the function of Example 6 and find the marginal revenue when
(a) $x = 8$;
(b) $x = 16$;
(c) $x = 20$.

Answer:
(a) 1.14
(b) .84
(c) .76

Sometimes both the generalized power rule and the product or quotient rule must be used, as the next example shows.

Example 7 Find the derivative of $y = 4x(3x + 5)^5$.

Write $4x(3x + 5)^5$ as the product

$$4x \cdot (3x + 5)^5.$$

Find the derivative of $(3x + 5)^5$ by letting $u = 3x + 5$. Now use the product rule and the generalized power rule.

7 Find the derivatives of the following.
(a) $y = 6x(x + 2)^2$
(b) $y = -9x(2x^2 + 1)^3$

Answer:
(a) $6(x + 2)(3x + 2)$
(b) $-9(2x^2 + 1)^2(14x^2 + 1)$

$$\overset{\text{derivative of } (3x+5)^5}{} \qquad \overset{\text{derivative of } 4x}{}$$

$$y' = 4x[\overbrace{5(3x + 5)^4 \cdot 3}] + (3x + 5)^5\overbrace{(4)}$$
$$= 60x(3x + 5)^4 + 4(3x + 5)^5$$
$$= 4(3x + 5)^4[15x + (3x + 5)^1] \qquad \text{Factor out the greatest}$$
$$= 4(3x + 5)^4(18x + 5). \quad \boxed{7} \qquad\qquad \text{common factor}$$

Example 8 Find the derivative of $y = \dfrac{(3x + 2)^7}{x - 1}$.

Use the quotient rule and the generalized power rule.

8 Find the derivatives of the following.
(a) $y = \dfrac{(2x + 1)^3}{3x}$

(b) $y = \dfrac{(x - 6)^5}{3x - 5}$

Answer:
(a) $\dfrac{(2x + 1)^2(4x - 1)}{3x^2}$

(b) $\dfrac{(x - 6)^4(12x - 7)}{(3x - 5)^2}$

$$\frac{dy}{dx} = \frac{(x - 1)[7(3x + 2)^6 \cdot 3] - (3x + 2)^7(1)}{(x - 1)^2}$$

$$= \frac{21(x - 1)(3x + 2)^6 - (3x + 2)^7}{(x - 1)^2}$$

$$= \frac{(3x + 2)^6[21(x - 1) - (3x + 2)]}{(x - 1)^2}$$

$$= \frac{(3x + 2)^6[21x - 21 - 3x - 2]}{(x - 1)^2}$$

$$\frac{dy}{dx} = \frac{(3x + 2)^6(18x - 23)}{(x - 1)^2} \quad \boxed{8}$$

11.7 Exercises

Find $f[g(x)]$ and $g[f(x)]$. (See Example 1.)

1. $f(x) = x^{1/3}$; $g(x) = 2x + 5$

2. $f(x) = x^{-3/4}$; $g(x) = 8x - 3$

3. $f(x) = \dfrac{2}{\sqrt{x}}$; $g(x) = 4 - 3x$

4. $f(x) = \dfrac{5}{\sqrt[4]{x}}$; $g(x) = 9 + 5x$

5. $f(x) = 5x^2 - x$; $g(x) = 5x + 3$

6. $f(x) = 2x + 3x^2$; $g(x) = 9x - 1$

7. $f(x) = \sqrt{x + 5}$; $g(x) = x^2 + 7$

8. $f(x) = 8 - x^2$; $g(x) = \sqrt{3 - x}$

Write each of the functions, defined as follows, as a composite of two functions. (There may be more than one way to do this.)

9. $y = (3x - 7)^{1/3}$

10. $y = (5 - x)^{2/5}$

11. $y = \sqrt{9 - 4x}$

12. $y = -\sqrt{13 + 7x}$

13. $y = \dfrac{\sqrt{x} + 3}{\sqrt{x} - 3}$

14. $y = \dfrac{2}{\sqrt{x} + 5}$.

15. $y = (x^{1/2} - 3)^2 + (x^{1/2} - 3) + 5$

16. $y = (x^2 + 5x)^{1/3} - 2(x^2 + 5x)^{2/3} + 7$

Find the derivatives of the functions defined as follows. (See Examples 2–5.)

17. $y = (2x + 9)^2$

18. $y = (8x - 3)^2$

19. $y = 6(5x - 1)^3$

20. $y = -8(3x + 2)^3$

21. $y = -2(12x^2 + 4)^3$

22. $y = 5(3x^2 - 5)^3$

23. $y = 9(x^2 + 5x)^4$

24. $y = -3(x^2 - 5x)^4$

25. $y = 12(2x + 5)^{3/2}$

26. $y = 45(3x - 8)^{3/2}$

27. $y = -7(4x^2 + 9x)^{3/2}$

28. $y = 11(5x^2 + 6x)^{3/2}$

29. $y = 8\sqrt{4x + 7}$

30. $y = -3\sqrt{7x - 1}$

31. $y = -2\sqrt{x^2 + 4x}$

32. $y = 4\sqrt{2x^2 + 3}$

Use the product or quotient rule to find the derivatives of the functions defined as follows. (See Examples 7–8.)

33. $y = 4x(2x + 3)^2$

34. $y = -6x(5x - 1)^2$

35. $y = (x + 2)(x - 1)^2$

36. $y = (3x + 1)^2(x + 4)$

37. $y = 5(x + 3)^2(x - 1)^2$

38. $y = -9(x + 4)^2(2x - 3)^2$

39. $y = (x + 1)^2\sqrt{x}$

40. $y = (3x + 5)^2\sqrt{x}$

41. $y = \dfrac{1}{(x - 4)^2}$

42. $y = \dfrac{-5}{(2x + 1)^2}$

43. $y = \dfrac{(x + 3)^2}{x - 1}$

44. $y = \dfrac{(x - 6)^2}{x + 4}$

45. $y = \dfrac{x^2 + 4x}{(x + 2)^2}$

46. $y = \dfrac{3x^2 - x}{(x - 1)^2}$

47. $y = (x^{1/2} + 1)(x^{1/2} - 1)^{1/2}$

48. $y = (3 - x^{2/3})(x^{2/3} + 2)^{1/2}$

Work the following exercises.

49. Management Assume that the total revenue from the sale of x television sets is given by

$$R(x) = 1000\left(1 - \frac{x}{500}\right)^2.$$

Find the marginal revenue for the following values of x. (See Example 6.)
(a) $x = 100$ **(b)** $x = 150$ **(c)** $x = 200$ **(d)** $x = 400$
(e) Find the average revenue from the sale of x sets.
(f) Find the marginal average revenue.

50. Natural Science The total number of bacteria in millions present in a culture is given by

$$N(t) = 2t(5t + 9)^{1/2} + 12,$$

where t represents time in hours after the beginning of an experiment. Find the rate of change of the population of bacteria with respect to time when
(a) $t = 0$;   **(b)** $t = 7/5$;   **(c)** $t = 8$;   **(d)** $t = 11$.

51. Natural Science To test an individual's use of calcium, a small amount of radioactive calcium is injected into the person's bloodstream with measurements of the calcium remaining in the bloodstream made each day for several days. Suppose the amount of the calcium remaining in the bloodstream in milligrams per cubic centimeter t days after the initial injection is approximated by

$$C(t) = \frac{1}{2}(2t + 1)^{-1/2}.$$

Find the rate of change of C with respect to time when
(a) $t = 0$; **(b)** $t = 4$; **(c)** $t = 6$; **(d)** $t = 7.5$.

52. Natural Science The strength of a person's reaction to a certain drug is given by

$$R(Q) = Q\left(C - \frac{Q}{3}\right)^{1/2},$$

where Q represents the quantity of the drug given to the patient and C is a constant.
(a) The derivative $R'(Q)$ is called the *sensitivity* to the drug. Find $R'(Q)$.
(b) Find $R'(Q)$ if $Q = 87$ and $C = 59$.

11.8 Derivatives of Exponential and Logarithmic Functions

Exponential and logarithmic functions were examined in detail in Chapter 4 of this book. Much of the work with these functions involved the number e. (Recall: to seven decimal places, $e = 2.7182818$.) The number e can be obtained as a limit:

$$e = \lim_{m \to \infty} \left(1 + \frac{1}{m}\right)^m.$$

(See the discussion of continuous compounding of interest in Chapter 4 to see how this limit is obtained.) Our work with logarithms mainly involved *natural logarithms:* those to base e. For review, graphs of $y = e^x$ and $y = \ln x$ are shown in Figure 11.32.

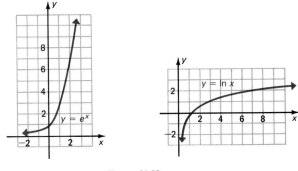

Figure 11.32

In this section formulas will be developed for the derivatives of $y = e^x$ and $y = \ln x$. Find the derivative of $y = \ln x$, assuming $x > 0$, by going back to the definition of the derivative. If $y = f(x)$, then

$$y' = f'(x) = \lim_{h \to 0} \frac{f(x + h) - f(x)}{h},$$

provided this limit exists. Here, $f(x) = \ln x$, with

$$y' = f'(x) = \lim_{h \to 0} \frac{\ln (x + h) - \ln x}{h}.$$

This limit can be found using properties of logarithms.

$$y' = \lim_{h \to 0} \frac{\ln (x + h) - \ln x}{h} = \lim_{h \to 0} \frac{\ln \left(\dfrac{x + h}{x} \right)}{h}$$

$$= \lim_{h \to 0} \left[\frac{1}{h} \ln \left(\frac{x + h}{x} \right) \right].$$

Since $\log_a x^r = r \log_a x$, the last result can be written as

$$y' = \lim_{h \to 0} \ln \left(\frac{x + h}{x} \right)^{1/h}$$

$$= \lim_{h \to 0} \ln \left(1 + \frac{h}{x} \right)^{1/h}.$$

Now make a substitution: let $m = x/h$, so that $h/x = 1/m$. As $h \to 0$, with x fixed, $m \to \infty$ (since $x > 0$), so

$$\lim_{h \to 0} \ln \left(1 + \frac{h}{x} \right)^{1/h} = \lim_{m \to \infty} \ln \left(1 + \frac{1}{m} \right)^{m/x}$$

$$= \lim_{m \to \infty} \ln \left[\left(1 + \frac{1}{m} \right)^m \right]^{1/x}.$$

By the property $\log_a x^r = r \log_a x$, this becomes

$$= \lim_{m \to \infty} \left[\frac{1}{x} \cdot \ln \left(1 + \frac{1}{m} \right)^m \right].$$

Since x is fixed here (a constant), by the rule (c) for limits this last limit becomes

$$\lim_{m \to \infty} \left[\frac{1}{x} \cdot \ln \left(1 + \frac{1}{m} \right)^m \right] = \left(\lim_{m \to \infty} \frac{1}{x} \right) \lim_{m \to \infty} \left[\ln \left(1 + \frac{1}{m} \right)^m \right]$$

$$= \frac{1}{x} \cdot \lim_{m \to \infty} \left[\ln \left(1 + \frac{1}{m} \right)^m \right]$$

$$= \frac{1}{x} \cdot \ln \left[\lim_{m \to \infty} \left(1 + \frac{1}{m} \right)^m \right] \qquad \text{Continuity of } \ln x$$

$$= \frac{1}{x} \ln e \qquad \text{Box at beginning of section}$$

$$= \frac{1}{x} \cdot 1 = \frac{1}{x}. \qquad \ln e = 1$$

By this result, if $y = \ln x$, for $x > 0$, then $y' = 1/x$.

The chain rule gives a more general statement. Recall:

If $y = f(g(x))$, then $y' = f'(g(x)) \cdot g'(x)$.   Chain rule

Let $y = \ln |g(x)|$. Then

$$y' = D_x \ln |g(x)| \cdot g'(x) = \frac{1}{g(x)} \cdot g'(x) = \frac{g'(x)}{g(x)}.$$

Using this result, if $y = \ln (-x)$, for $x < 0$, the derivative is

$$y' = \frac{1}{-x}(-1) = \frac{1}{x},$$

the same as the derivative of $y = \ln x$. For this reason, absolute value bars are used in the summary below.

If $y = \ln |x|$, then $y' = \dfrac{1}{x}$.

If $y = \ln |g(x)|$, then $y' = \dfrac{g'(x)}{g(x)}$.

Example 1 Find the derivative of the functions defined as follows.

(a) $y = \ln |5x|$

Let $g(x) = 5x$, so that $g'(x) = 5$. From the box above,

$$y' = \frac{g'(x)}{g(x)} = \frac{5}{5x} = \frac{1}{x}.$$

(b) $y = \ln |3x^2 - 4x|$

$$y' = \frac{6x - 4}{3x^2 - 4x}$$

(c) $y = 3x \ln x^2$

Since $3x \ln x^2$ is the product of $3x$ and $\ln x^2$, use the product rule.

$$y' = (3x)\left[\frac{d}{dx} \ln x^2\right] + (\ln x^2)\left[\frac{d}{dx} 3x\right]$$

$$= 3x\left(\frac{2x}{x^2}\right) + (\ln x^2)(3)$$

$$= 6 + 3 \ln x^2$$

$$= 6 + \ln (x^2)^3 \qquad \text{Property of logarithms}$$

$$y' = 6 + \ln x^6$$

Alternatively, write the answer as $y' = 6 + 6 \ln x$. **1**

Find the derivative of $y = e^x$ by starting with $y = e^x$ and taking natural logarithms on each side.

$$y = e^x$$

$$\ln y = \ln e^x$$

$$= x \cdot \ln e \qquad \text{Property of logarithms}$$

$$\ln y = x \qquad\qquad \ln e = 1$$

1 Find y' for the following.

(a) $y = \ln |7 + x|$

(b) $y = \ln |4x^2|$

(c) $y = \ln |8x^3 - 3x|$

(d) $y = x^2 \ln |x|$

Answer:

(a) $y' = \dfrac{1}{7 + x}$

(b) $y' = \dfrac{2}{x}$

(c) $y' = \dfrac{24x^2 - 3}{8x^3 - 3x}$

(d) $y' = x(1 + 2 \ln |x|)$

Now differentiate implicitly: take the derivative of each side. Use the chain rule on the left.

$$D_x \ln y = D_x x$$

$$\frac{1}{y} \cdot \frac{dy}{dx} = 1$$

or

$$\frac{dy}{dx} = y.$$

Since $y = e^x$,

$$\frac{dy}{dx} = e^x, \qquad \text{or} \qquad y' = e^x.$$

This result shows one of the main reasons for the widespread use of e as a base— $y = e^x$ is its own derivative. Similarly, by the chain rule, if $y = e^{g(x)}$, then

$$y' = [D_x e^{g(x)}] \cdot g'(x) = e^{g(x)} \cdot g'(x).$$

Summaries of these results follow.

If $y = e^x$, then $y' = e^x$.

If $y = e^{g(x)}$, then $y' = g'(x) \cdot e^{g(x)}$.

Example 2 Find derivatives of the functions defined as follows.

(a) $y = 4e^{5x}$

Let $g(x) = 5x$, with $g'(x) = 5$. Then

$$y' = 4 \cdot 5e^{5x} = 20e^{5x}.$$

(b) $y = 3e^{-4x}$

$$y' = 3(-4e^{-4x}) = -12e^{-4x}$$

(c) $y = 10e^{3x^2}$

$$y' = 6x(10e^{3x^2}) = 60xe^{3x^2} \quad \boxed{2}$$

Example 3 Let $y = e^x \cdot \ln |x|$. Find y'.

Use the product rule.

$$y' = e^x \cdot \frac{1}{x} + \ln |x| \cdot e^x$$

$$y' = e^x \left(\frac{1}{x} + \ln |x| \right) \quad \boxed{3}$$

Example 4 Let $y = \dfrac{100{,}000}{1 + 100e^{-.3x}}$. Find y'.

◼ 2 Find each derivative.
(a) $y = 3e^{12x}$
(b) $y = -6e^{(-10x+1)}$
(c) $y = e^{-x^2}$

Answer:
(a) $y' = 36e^{12x}$
(b) $y' = 60e^{(-10x+1)}$
(c) $y' = -2xe^{-x^2}$

◼ 3 Find each derivative.
(a) $y = 3x^2 e^x$
(b) $y = -9xe^{x^2}$

Answer:
(a) $y' = 3x^2 e^x + 6xe^x$
(b) $y' = -18x^2 e^{x^2} - 9e^{x^2}$

Use the quotient rule.

$$y' = \frac{(1 + 100e^{-.3x})(0) - 100{,}000(-30e^{-.3x})}{(1 + 100e^{-.3x})^2}$$

$$= \frac{3{,}000{,}000e^{-.3x}}{(1 + 100e^{-.3x})^2} \quad \boxed{4}$$

4 Find each derivative.

(a) $y = \dfrac{e^x}{1 + x}$

(b) $y = \dfrac{\ln |x|}{2 - 3x}$

(c) $y = \dfrac{10{,}000}{1 + 2e^x}$

Answer:

(a)

$y' = \dfrac{xe^x}{(1 + x)^2}$

(b)

$y' = \dfrac{2 - 3x + 3x \cdot \ln |x|}{x(2 - 3x)^2}$

(c)

$y' = \dfrac{-20{,}000e^x}{(1 + 2e^x)^2}$

Example 5 Often a population, or the sales of a certain product, will start growing slowly, then grow more rapidly, and then gradually level off. Such growth can often be approximated by a mathematical model of the form

$$f(x) = \frac{b}{1 + ae^{kx}}$$

for appropriate constants a, b, and k. For example, suppose that the sales of a new product can be approximated for its first few years on the market by

$$S(x) = \frac{100{,}000}{1 + 100e^{-.3x}},$$

where x is time in years since the introduction of the product. Find the rate of change of the sales when $x = 4$.

The derivative was given in the previous example. Using this derivative,

$$S'(4) = \frac{3{,}000{,}000e^{-.3(4)}}{[1 + 100e^{-.3(4)}]^2}$$

$$= \frac{3{,}000{,}000e^{-1.2}}{(1 + 100e^{-1.2})^2}.$$

Using a calculator, or Table 3 in the back of the book, $e^{-1.2} \approx .301$, with

$$S'(4) \approx \frac{3{,}000{,}000(0.301)}{[1 + 100(.301)]^2} = \frac{903{,}000}{(1 + 30.1)^2} = \frac{903{,}000}{967.21} \approx 934.$$

The rate of change of sales at time $x = 4$ is an increase of about 934 units per year. **5**

5 Suppose a deer population is given by

$$f(x) = \frac{10{,}000}{1 + 2e^x},$$

where x is time in years. (See problem 4 at the side.) Find the rate of change of the population when

(a) $x = 0$;

(b) $x = 5$.

(c) Is the population increasing or decreasing?

Answer:

(a) About -2200

(b) About -33

(c) Decreasing

For reference, the definition of the derivative and the rules for derivatives developed in this chapter are listed below.

Definition of the Derivative The derivative of the function f, at x, written $f'(x)$, is defined as

$$f'(x) = \lim_{h \to 0} \frac{f(x + h) - f(x)}{h}$$

provided this limit exists. Alternative notations for the derivative include dy/dx, $D_x y$, $D_x[f(x)]$, and y'.

Rules for Derivatives (Assume all indicated derivatives exist.)

Constant Function If $f(x) = k$, where k is any real number, then

$$f'(x) = 0.$$

Power Rule If $f(x) = x^n$, for any real number n, then

$$f'(x) = n \cdot x^{n-1}.$$

Constant Times a Function Let k be a real number. Then the derivative of $y = k \cdot f(x)$ is

$$y' = k \cdot f'(x).$$

Sum or Difference Rule If $y = f(x) \pm g(x)$, then

$$y' = f'(x) \pm g'(x).$$

Product Rule If $f(x) = g(x) \cdot k(x)$, then

$$f'(x) = g(x) \cdot k'(x) + k(x) \cdot g'(x).$$

Quotient Rule If $f(x) = \dfrac{g(x)}{k(x)}$, and $k(x) \neq 0$, then

$$f'(x) = \frac{k(x) \cdot g'(x) - g(x) \cdot k'(x)}{[k(x)]^2}.$$

Chain Rule If y is a function of u, say $y = f(u)$, and if u is a function of x, say $u = g(x)$, then $y = f(u) = f[g(x)]$, and

$$\frac{dy}{dx} = \frac{dy}{du} \cdot \frac{du}{dx}.$$

Chain Rule (alternate form) Let $y = f[g(x)]$. Then

$$y' = f'[g(x)] \cdot g'(x).$$

Generalized Power Rule Let u be a function of x, and let $y = u^n$ for any real number n. Then

$$y' = n \cdot u^{n-1} \cdot u'.$$

Natural Logarithmic Function If $y = \ln |g(x)|$, then

$$y' = \frac{g'(x)}{g(x)}.$$

Exponential Function If $y = e^{g(x)}$, then

$$y' = g'(x) \cdot e^{g(x)}.$$

11.8 Exercises

Find derivatives of the functions defined as follows. (See Examples 1–4.)

1. $y = e^{4x}$

2. $y = e^{-2x}$

3. $y = -6e^{-2x}$

4. $y = 8e^{4x}$

5. $y = -8e^{2x}$

6. $y = .2e^{5x}$

7. $y = -16e^{x+1}$

8. $y = -4e^{-.1x}$

9. $y = e^{x^2}$

10. $y = e^{-x^2}$

11. $y = 3e^{2x^2}$

12. $y = -5e^{4x^3}$

13. $y = 4e^{2x^2-4}$

14. $y = -3e^{3x^2+5}$

15. $y = xe^x$

16. $y = x^2e^{-2x}$

17. $y = (x - 3)^2e^{2x}$

18. $y = (3x^2 - 4x)e^{-3x}$

19. $y = \ln |3 - x|$

20. $y = \ln |1 + x^2|$

21. $y = \ln |2x^2 - 7x|$

22. $y = \ln |-8x^2 + 6x|$

23. $y = \ln \sqrt{x + 5}$

24. $y = \ln \sqrt{2x + 1}$

25. $y = \ln |(2x + 7)(3x - 2)|$

26. $y = \ln |(x^2 - 1)(3x + 8)|$

27. $y = \ln \left| \dfrac{5x - 1}{2x + 4} \right|$

28. $y = \ln \left| \dfrac{6 - x}{3x + 5} \right|$

29. $y = \ln (x^4 + 5x^2)^{3/2}$

30. $y = \ln |(5x^3 - 2x)^{3/2}|$

31. $y = -3x \ln |x + 2|$

32. $y = (3x + 1) \ln |x - 1|$

33. $y = x^2 \ln |x|$

34. $y = x \ln |2 - x^2|$

35. $y = \dfrac{x^2}{e^x}$

36. $y = \dfrac{e^x}{2x + 1}$

37. $y = (2x^3 - 1) \ln |x|$

38. $y = \dfrac{\ln |x|}{x^3}$

39. $y = \dfrac{\ln |x|}{4x + 7}$

40. $y = \dfrac{-2 \ln |x|}{3x - 1}$

41. $y = \dfrac{\ln |x|}{e^x}, \ x > 0$

42. $y = \dfrac{x^2}{e^x}$

43. $y = \dfrac{3x^2}{\ln |x|}$

44. $y = \dfrac{x^3 - 1}{2 \ln |x|}$

45. $y = (\ln |x + 1|)^4$

46. $y = \sqrt{\ln |x - 3|}$

47. $y = \dfrac{e^x}{\ln |x|}$

48. $y = \dfrac{e^x - 1}{\ln |x|}$

49. $y = \dfrac{e^x + e^{-x}}{x}$

50. $y = \dfrac{e^x - e^{-x}}{x}$

51. $y = e^{x^2} \ln |x|$

52. $y = e^{2x-1} \ln |2x - 1|$

53. $y = \dfrac{5000}{1 + 10e^{.4x}}$

54. $y = \dfrac{600}{1 - 50e^{.2x}}$

55. $y = \dfrac{10,000}{9 + 4e^{-.2x}}$

56. $y = \dfrac{500}{12 + 5e^{-.5x}}$

57. $y = \ln |\ln |x||$

58. $y = (\ln 4)(\ln |3x|)$

Work the following exercises. (See Example 5.)

59. Management Often, sales of a new product grow rapidly and then level off with time. Such a situation can be represented by an equation of the form

$$S(t) = 100 - 90e^{-.3t},$$

where t represents time in years and $S(t)$ represents sales. Find the rate of change of sales when
(a) $t = 1$; **(b)** $t = 10$.

60. Management Suppose $P(x) = e^{-.02x}$ represents the proportion of cars manufactured by a given company that are still free of defects after x months of use. Find the proportion of cars free of defects after
(a) 1 month; **(b)** 10 months; **(c)** 100 months.
(d) Calculate and interpret $P'(100)$.

61. Natural Science The concentration of pollutants, in grams per liter, in the east fork of the Big Weasel River is approximated by

$$P(x) = .04e^{-4x},$$

where x is the number of miles downstream from a paper mill that the measurement is taken. Find
(a) $P(.5)$; **(b)** $P(1)$; **(c)** $P(2)$.

Find the rate of change of the concentration with respect to distance at
(d) $x = .5$; **(e)** $x = 1$; **(f)** $x = 2$.

62. Management Based on data in a car magazine, we constructed the mathematical model

$$y = 100e^{-.03045t}$$

for the percent of cars of a certain type still on the road after t years. Find the percent of cars on the road after the following number of years.
(a) 0 **(b)** 2 **(c)** 4 **(d)** 6

Find y' for the following values of t.

(e) 0   (f) 2   (g) 4   (h) 6

63. Recall from Chapter 5 the formula for interest compounded continuously:

$$A = Pe^{ni},$$

where A is the final amount on deposit when P dollars are deposited at a rate of interest i compounded continuously for n years. Assume $P = 100$, and $n = 5$. Find A' for the following values of i.

(a) .04   (b) .08   (c) .12   (d) .16

64. Use the formula $A = Pe^{ni}$ from Exercise 63; let $P = 100$ and $i = .12$. Find A' for the following values of n.

(a) 2   (b) 5   (c) 10   (d) 20

65. Natural Science Suppose that the population of a certain collection of rare Brazilian ants is given by

$$P(t) = 1000e^{.2t},$$

where t represents the time in days. Find the rate of change of the population when $t = 2$; when $t = 8$.

66. Natural Science Assume that the amount of a radioactive substance present at time t is given by

$$A(t) = 500e^{-.25t}$$

grams. Find the rate of change of the quantity present when

(a) $t = 0$;   (b) $t = 4$;   (c) $t = 6$;   (d) $t = 10$.

67. Natural Science Consider an experiment in which equal numbers of male and female insects of a certain species are permitted to intermingle. Assume that

$$M(t) = (e^{.1t} + 1) \ln \sqrt{t}$$

represents the number of matings observed among the insects in an hour, where t is the temperature in degrees Celsius. (Note: The formula is an approximation at best and holds only for specific temperature intervals.) Find

(a) $M(15)$;   (b) $M(25)$.

(c) Find the rate of change of $M(t)$ when $t = 15$.

68. Social Science According to work by the psychologist C. L. Hull, the strength of a habit is a function of the number of times the habit is repeated. If N is the number of repetitions and $H(N)$ is the strength of the habit, then

$$H(N) = 1000(1 - e^{-kN}),$$

where k is a constant. Find $H'(N)$ if $k = .1$ and

(a) $N = 10$;   (b) $N = 100$;   (c) $N = 1000$.

(d) Show that $H'(N)$ is always positive. What does this mean?

Graph the functions defined as follows by plotting points in the given intervals.

69. $y = x + \ln x$, $(0, 9]$

70. $y = \ln x - \dfrac{1}{x}$, $(0, 9]$

71. $y = e^x \ln x$, $(0, 5]$

72. $y = \dfrac{5 \ln 10x}{e^{.5x}}$, $(0, 10]$

Chapter 11   Review Exercises

Decide if the limits in Exercises 1–18 exist. If a limit exists, find its value.

1. $\lim\limits_{x \to -3} f(x)$

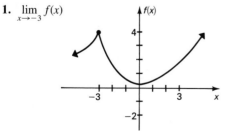

2. $\lim\limits_{x \to -1} g(x)$

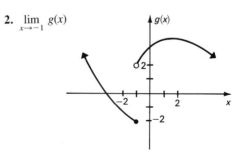

3. $\lim\limits_{x \to \infty} f(x)$

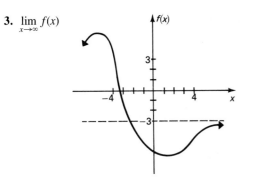

4. $\lim\limits_{x \to 0} \dfrac{x^2 - 5}{2x}$

5. $\lim\limits_{x \to -1} (2x^2 + 3x + 5)$

6. $\lim\limits_{x \to 2} (-x^2 + 4x + 1)$

7. $\lim\limits_{x \to 6} \dfrac{2x + 5}{x - 3}$

8. $\lim\limits_{x \to 3} \dfrac{2x + 5}{x - 3}$

9. $\lim\limits_{x \to 4} \dfrac{x^2 - 16}{x - 4}$

10. $\lim\limits_{x \to 2} \dfrac{x^2 + 3x - 10}{x - 2}$

11. $\lim\limits_{x \to -4} \dfrac{2x^2 + 3x - 20}{x + 4}$

12. $\lim\limits_{x \to 3} \dfrac{3x^2 - 2x - 21}{x - 3}$

13. $\lim\limits_{x \to 9} \dfrac{\sqrt{x} - 3}{x - 9}$

14. $\lim\limits_{x \to 16} \dfrac{\sqrt{x} - 4}{x - 16}$

15. $\lim\limits_{x \to \infty} \dfrac{x^2 + 5}{5x^2 - 1}$

16. $\lim\limits_{x \to \infty} \dfrac{x^2 + 6x + 8}{x^3 + 2x + 1}$

17. $\lim\limits_{x \to \infty} \left(\dfrac{3}{4} + \dfrac{2}{x} - \dfrac{5}{x^2} \right)$

18. $\lim\limits_{x \to \infty} \left(\dfrac{9}{x^4} + \dfrac{1}{x^2} - 3 \right)$

Find all points of discontinuity for the following.

19.

20.

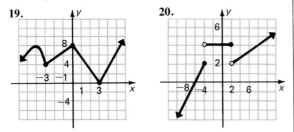

Are the functions defined as follows continuous at the given points?

21. $f(x) = \dfrac{6x + 1}{2x - 3}$: $\quad x = \dfrac{3}{2}$; $x = -\dfrac{1}{6}$; $x = 0$

22. $f(x) = \dfrac{2}{x(x + 4)}$: $\quad x = 2$; $x = 0$; $x = -4$

23. $f(x) = \dfrac{-5}{3x(2x - 1)}$: $\quad x = -5$; $x = 0$; $x = -\dfrac{1}{3}$; $x = \dfrac{1}{2}$

24. $f(x) = \dfrac{2 - 3x}{(1 + x)(2 - x)}$: $\quad x = \dfrac{2}{3}$; $x = -1$; $x = 2$; $x = 0$

25. $f(x) = \dfrac{x - 6}{x + 5}$: $\quad x = 6$; $x = -5$; $x = 0$

26. $f(x) = \dfrac{x^2 - 9}{x + 3}$: $\quad x = 3$; $x = -3$; $x = 0$

27. $f(x) = x^2 + 3x - 4$: $\quad x = 1$; $x = -4$; $x = 0$

28. $f(x) = 2x^2 - 5x - 3$: $\quad x = -\dfrac{1}{2}$; $x = 3$; $x = 0$

29. A company charges $1.50 per pound when a certain chemical is bought in lots of 125 pounds or less, with a price per pound of $1.35 if more than 125 pounds are purchased. Let $C(x)$ represent the cost to purchase x pounds. Find
(a) $C(100)$; **(b)** $C(125)$; **(c)** $C(140)$.
(d) Graph $y = C(x)$.
(e) Where is C discontinuous?

30. Use the information of Exercise 29 and find the average cost per pound if the following number of pounds are bought.
(a) 100 **(b)** 125 **(c)** 140 **(d)** 200

Write each of the following in interval notation.

31.

32.

33. $\{x | -4 \le x \le 2\}$

34. $\{x | x > 3\}$

35. $\{x | x \le -1\}$

36. $\{x | -6 \le x < 2\}$

Use the graph to find the average rate of change of f on the following intervals.

37. $x = 0$ to $x = 4$

38. $x = 2$ to $x = 8$

39. $x = 2$ to $x = 4$

40. $x = 0$ to $x = 6$

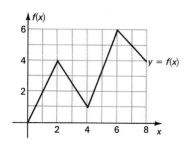

Find the average rate of change for the functions defined as follows.

41. $y = 6x^2 + 2$, from $x = 1$ to $x = 4$

42. $y = -2x^3 - x^2 + 5$, from $x = -2$ to $x = 6$

43. $y = \dfrac{-6}{3x - 5}$, from $x = 4$ to $x = 9$

44. $y = \dfrac{x + 4}{x - 1}$, from 2 to 5

Use the definition of the derivative to find the derivative of the functions defined as follows.

45. $y = 4x + 3$ **46.** $y = 5x^2 + 6x$

47. $y = -x^3 + 7x$ **48.** $y = 11x^2 - x^3$

Find the slope of the tangent line to the given curve at the given value of x. Find the equation of each tangent line.

49. $y = x^2 - 6x$; at $x = 2$

50. $y = 8 - x^2$; at $x = 1$

51. $y = \dfrac{3}{x - 1}$; at $x = -1$

52. $y = \dfrac{-2}{x + 5}$; at $x = -2$

53. $y = (3x^2 - 5x)(2x)$; at $x = -1$

54. $y = \dfrac{3}{x^2 - 1}$; at $x = 2$

55. $y = \sqrt{6x - 2}$; at $x = 3$

56. $y = -\sqrt{8x + 1}$; at $x = 3$

Find all values of x (if any) where the graphs of the following functions have tangent lines with slope 0.

57. $f(x) = -5x^2$ **58.** $f(x) = 4x^2 - 6x$

59. $f(x) = 3x - 9x^2$ **60.** $f(x) = \dfrac{x^3}{3} + \dfrac{x^2}{2} - 2x$

61. $f(x) = -\dfrac{6}{x}$ **62.** $f(x) = x^3 - 4x^2$

Work the following exercises.

63. Suppose that the profit in cents from selling x pounds of potatoes is given by

$$P(x) = 15x + 25x^2.$$

Find the marginal profit when

(a) $x = 6$;   (b) $x = 20$;   (c) $x = 30$.

64. In Exercise 63, find the average profit function and the marginal average profit function.

65. The sales of a company are related to its expenditure on research by

$$S(x) = 1000 + 50\sqrt{x} + 10x,$$

where $S(x)$ gives sales in millions when x thousand dollars are spent on research. Find the rate of change of sales, dS/dx, when

(a) $x = 9$;   (b) $x = 16$;   (c) $x = 25$.

66. Suppose that the profit in hundreds of dollars from selling x units of a product is given by $P(x)$, where

$$P(x) = \dfrac{x^2}{x - 1}.$$

Find the marginal profit when

(a) $x = 4$;   (b) $x = 12$;   (c) $x = 20$.

67. A company finds its sales are related to the amount spent on training programs by

$$T(x) = \dfrac{1000 + 50x}{x + 1},$$

where $T(x)$ is sales in hundreds of thousands of dollars when x thousand dollars are spent on training. Find the rate of change of sales, $T'(x)$, when

(a) $x = 9$;   (b) $x = 19$.

68. Waverly Products has determined that its profits are related to advertising expenditures by the function

$$P(x) = 5000 + 16x - 3x^2,$$

where $P(x)$ is the profit when x hundred dollars are spent on advertising.

(a) Find the marginal profit function.

(b) At an expenditure of \$1000 ($x = 10$), is the profit increasing or decreasing?

Find the derivative of each of the following.

69. $y = 5x^2 - 7x - 9$

70. $y = x^3 - 4x^2$

71. $y = 6x^{7/3}$

72. $y = -3x^{-2}$

73. $f(x) = x^{-3} + \sqrt{x}$

74. $f(x) = 6x^{-1} - 2\sqrt{x}$

75. $y = (3t^2 + 7)(t^3 - t)$

76. $y = (-5t + 4)(t^3 - 2t^2)$

77. $y = 4\sqrt{x}(2x - 3)$

78. $y = -3\sqrt{x}(8 - 5x)$

79. $g(t) = -3t^{1/3}(5t + 7)$

80. $p(t) = 8t^{3/4}(7t - 2)$

81. $y = 12x^{-3/4}(3x + 5)$

82. $y = 15x^{-3/5}(x + 6)$

83. $k(x) = \dfrac{3x}{x + 5}$

84. $r(x) = \dfrac{-8}{2x + 1}$

85. $y = \dfrac{\sqrt{x} - 1}{x + 2}$

86. $y = \dfrac{\sqrt{x} + 6}{x - 3}$

87. $y = \dfrac{x^2 - x + 1}{x - 1}$

88. $y = \dfrac{2x^3 - 5x^2}{x + 2}$

89. $f(x) = (3x - 2)^4$

90. $k(x) = (5x - 1)^6$

91. $y = \sqrt{2t - 5}$

92. $y = -3\sqrt{8t - 1}$

93. $y = 3x(2x + 1)^3$

94. $y = 4x^2(3x - 2)^5$

95. $r(t) = \dfrac{5t^2 - 7t}{(3t + 1)^3}$

96. $s(t) = \dfrac{t^3 - 2t}{(4t - 3)^4}$

97. $y = \dfrac{x^2 + 3x - 10}{x - 2}$

98. $y = \dfrac{x^2 - x - 6}{x - 3}$

99. $y = -6e^{2x}$

100. $y = 8e^{.5x}$

101. $y = e^{-2x^3}$

102. $y = -4e^{x^2}$

103. $y = 5x \cdot e^{2x}$

104. $y = -7x^2 \cdot e^{-3x}$

105. $y = \ln |2 + x^2|$

106. $y = \ln |5x + 3|$

107. $y = \dfrac{\ln |3x|}{x - 3}$

108. $y = \dfrac{\ln |2x - 1|}{x + 3}$

Find each of the following.

109. $D_x\left[\dfrac{\sqrt{x} + 1}{\sqrt{x} - 1}\right]$

110. $D_x\left[\dfrac{2x + \sqrt{x}}{1 - x}\right]$

111. $\dfrac{dy}{dt}$ if $y = \sqrt{t^{1/2} + t}$

112. $\dfrac{dy}{dx}$ if $y = \dfrac{\sqrt{x} - 1}{x}$

113. $f'(1)$ if $f(x) = \dfrac{\sqrt{8 + x}}{x + 1}$

114. $f'(-2)$ if $f(t) = \dfrac{2 - 3t}{\sqrt{2 + t}}$

12

Applications of the Derivative

The graph of Figure 12.1 shows the population of deer on the Kaibab Plateau on the North Rim of the Grand Canyon between 1905 and 1930. In the early years of the 20th century, hunters were very effective in reducing the population of predators. With few predators, the deer population increased rapidly and this increase in population caused a depletion of food resources. Deer population peaked around 1925, and then rapidly dropped.

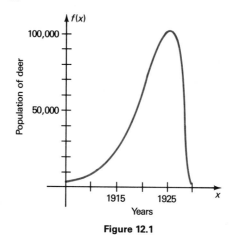

Figure 12.1

Given a graph such as the one in Figure 12.1, maximum and minimum values can usually be read directly. (Also, we saw earlier how to find maximum and minimum values for quadratic functions by finding the vertex of a parabola.) For more complicated functions with graphs that may be difficult to draw, derivatives can be used to find maximum and minimum values. The procedure for doing this is discussed in the first part of this chapter.

12.1 Maxima and Minima

Business people are interested in finding a production level that will produce a minimum cost of manufacturing an item, while biologists must predict the maximum size of a population of organisms. The need to find a maximum or a minimum value of a function comes up in a great many different situations. Calculus can be a very useful tool for finding these maxima or minima.

Let c be a number in the domain of a function f.

1. $f(c)$ is a **relative maximum** for f if there exists an open interval (a, b) containing c such that

$$f(x) \leq f(c)$$

for all x in (a, b);

2. $f(c)$ is a **relative minimum** for f if there exists an open interval (a, b) containing c such that

$$f(x) \geq f(c)$$

for all x in (a, b).

 Identify the x-values of all points where these graphs have relative maxima or relative minima.

(a)

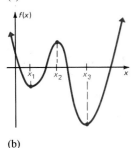

(b)

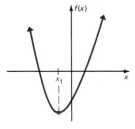

(c)

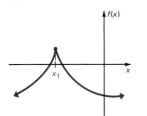

Answer:
(a) Relative maximum at x_2, relative minima at x_1 and x_3
(b) No relative maximum, relative minimum at x_1
(c) Relative maximum at x_1, no relative minimum

Think of a relative maximum as a peak (not necessarily the highest peak) and a relative minimum as a valley (not necessarily the lowest valley) on the graph of the function. A function has a **relative extremum** at c if it has either a relative maximum or a relative minimum there.

Example 1 Identify the x-values of all points where the graph of Figure 12.2 has relative maxima or relative minima.

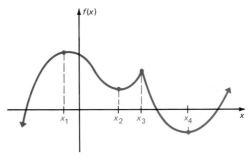

Figure 12.2

The graph has relative maxima at x_1 and x_3 and relative minima at x_2 and x_4.

How can the equation for a function be used to find its extreme values? Figure 12.3 on the next page suggests a method. The function graphed in Figure 12.3 has relative maxima when $x = x_2$ or $x = x_4$, and relative minima when $x = x_1$ or $x = x_3$. The tangents at the points having x-values, x_1, x_2, and x_3, respectively, are drawn in Figure 12.3. All three of these tangents are horizontal and have a slope of 0. Since the derivative gives the slope of a tangent line, one way to find relative extrema for a function is to find all points where the derivative (the slope of the tangent) is 0. These points *might* lead to relative extrema.

However, a function can have relative extrema even when the derivative does not exist. For example, the function of Figure 12.3 also has a relative maximum when $x = x_4$, even though no derivative exists at the point where $x = x_4$ (since no tangent line can be drawn).

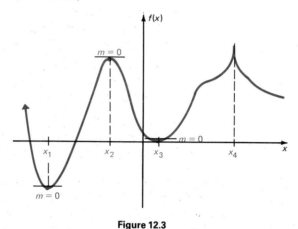

Figure 12.3

In summary, when looking for relative extrema, check all values that make the derivative equal 0 and all values where the function exists but the derivative does not exist. Any of these values might lead to relative extrema. (A rough sketch of the graph of the function is often enough to tell if an extremum has been found.) The points where a derivative equals 0 or does not exist (but the function is defined) are called **critical points** of the function. The following result summarizes these facts.

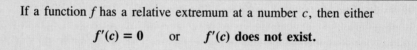

If a function f has a relative extremum at a number c, then either

$$f'(c) = 0 \quad \text{or} \quad f'(c) \text{ does not exist.}$$

Be very careful not to get this result backward. It does *not* say that a function has relative extrema at all critical points of the function. For example, Figure 12.4

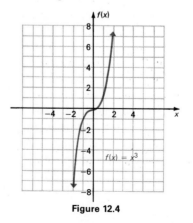

Figure 12.4

2 The functions defined as follows have critical points as indicated. Graph each and decide if the critical points lead to a relative maximum, a relative minimum, or neither.

(a)
$f(x) = 6x^2 + 24x + 20$;
$(-2, -4)$

(b)
$f(x) = (1/3)x^3 - x^2 - 15x + 6$;
$(-3, 33)$, $(5, -157/3)$

Answer:

(a) Relative minimum at $x = -2$, relative minimum is $f(-2) = -4$

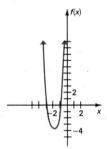

(b) Relative maximum at $x = -3$, relative maximum is $f(-3) = 33$; relative minimum at $x = 5$, minimum is $f(5) = -157/3$

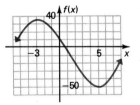

shows the graph of $f(x) = x^3$. The derivative, $f'(x) = 3x^2$, is 0 when $x = 0$, so that the function has a critical point when $x = 0$. However, as suggested by the graph of Figure 12.4, $f(x) = x^3$ has neither a relative maximum nor a relative minimum at $x = 0$ (or anywhere else, for that matter). **2**

First Derivative Test   Suppose all critical points have been found for some function f. How can we then tell if these critical points lead to relative maxima, relative minima, or neither a relative maximum nor a relative minimum? One way is suggested by the graph in Figure 12.5.

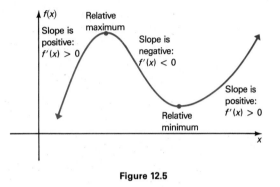

Figure 12.5

As the graph in Figure 12.5 suggests, the slope of a tangent line just to the left of a relative minimum is negative, while the slope is positive just to the right. The opposite is true for relative maxima. We have seen that it is possible to discover if the slope of a tangent line is positive or negative by finding whether the derivative of the function is positive or negative. These results are summarized as the *first derivative test* for locating maxima and minima.

First Derivative Test   Let c be a critical point for a function f. Suppose that f is differentiable on (a, b), and that c is the only critical point of f in (a, b).

1. $f(c)$ is a **relative maximum** of f if the derivative $f'(x)$ is positive in an open interval extending left from c and negative in an open interval extending right from c.

2. $f(c)$ is a **relative minimum** of f if the derivative $f'(x)$ is negative in an interval extending left from c and positive in an interval extending right from c.

This test is called the first derivative test to set it apart from the second derivative test, discussed later.

The sketches in the following table show how the first derivative test works. Assume the same conditions on a, b, and c as given above.

$f(x)$ has:	Sign of f' in (a, c)	Sign of f' in (c, b)	Sketch
Relative maximum	+	−	
Relative minimum	−	+	
No relative extrema	+	+	
No relative extrema	−	−	

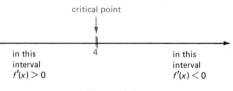

Example 2 Find all relative maxima and minima for the functions defined as follows.

(a) $f(x) = -x^2 + 8x + 3$

Here $f'(x) = -2x + 8$. This derivative equals 0 when $x = 4$. Use the first derivative test to decide if $x = 4$ leads to a relative maximum or minimum. Look at two intervals, one to the left of 4, and one to the right of 4. (See Figure 12.6.)

Figure 12.6

3 Find all relative extrema
for the functions defined as
follows.
(a) $f(x) = 2x^2 - 8x + 1$
(b) $f(x) = 3 + 4x - 3x^2$

Answer:
(a) Relative minimum
at $x = 2$, minimum is
$f(2) = -7$
(b) Relative maximum at
$x = 2/3$, maximum is
$f(2/3) = 13/3$

Decide on the sign of the derivative to the left of 4 by choosing any number in this interval, such as 3. Since $f'(3) = -2 \cdot 3 + 8 = 2$, which is positive, $f'(x) > 0$ for *all* values of x in this interval. Choose a similar test point to the right of 4 to see that $f'(x) < 0$ in that interval. Since $f'(x) > 0$ to the left of 4, and $f'(x) < 0$ to the right of 4, the function f has a relative *maximum* at 4 by part one of the first derivative test. This maximum is $f(4) = 19$. 3

(b) $f(x) = 2x^3 - 3x^2 - 72x + 15$

The derivative is $f'(x) = 6x^2 - 6x - 72$. Set this derivative equal to 0.

4 Find all relative maxima or relative minima for the following functions.
(a) $f(x) = 5x^2 - 20x + 3$
(b) $f(x) = x^3 + 4x^2 - 3x + 5$

Answer:

(a) Relative minimum at 2, minimum is $f(2) = -17$
(b) Relative maximum at −3, maximum is $f(-3) = 23$; relative minimum at 1/3, minimum is $f(1/3) = 121/27$

$$6x^2 - 6x - 72 = 0$$
$$6(x^2 - x - 12) = 0$$
$$6(x - 4)(x + 3) = 0$$
$$x - 4 = 0 \quad \text{or} \quad x + 3 = 0$$
$$x = 4 \qquad\qquad x = -3$$

Use the first derivative test by considering the three intervals shown in Figure 12.7. Choose the values −4, 0, and 5 as test points to find the signs of the values of the derivative in each interval, as shown in the figure. Since $f'(x) > 0$ to the left of −3, and $f'(x) < 0$ just to the right, f has a relative maximum when $x = -3$. This relative maximum is $f(-3) = 150$. Also, $f'(x) < 0$ just to the left of 4, and $f'(x) > 0$ to the right, so that f has a relative minimum when $x = 4$. The value of this relative minimum is $f(4) = -193$. Use this information on extrema and plot points as needed to get the graph of f shown in Figure 12.8. **4**

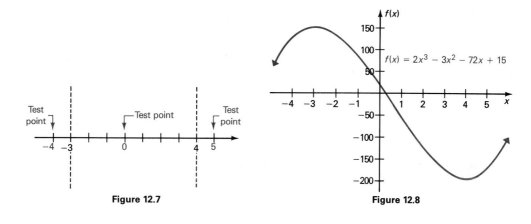

Figure 12.7

$f(x) = 2x^3 - 3x^2 - 72x + 15$

Figure 12.8

(c) $f(x) = 6x^{2/3} - 4x$
   Find $f'(x)$:

$$f'(x) = 4x^{-1/3} - 4 = \frac{4}{x^{1/3}} - 4.$$

The derivative fails to exist when $x = 0$, but the function itself is defined when $x = 0$, making 0 a critical value for f. Find other critical values by setting $f'(x) = 0$.

$$f'(x) = \frac{4}{x^{1/3}} - 4 = 0$$

$$\frac{4}{x^{1/3}} = 4$$

$$4 = 4x^{1/3}$$

$$1 = x^{1/3}$$

$$1 = x$$

The critical values of the function, 0 and 1, are used to locate intervals on a number line, as in Figure 12.9.

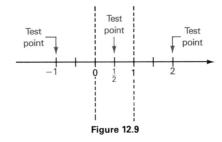

Figure 12.9

Use the test points indicated in Figure 12.9 to get $f'(-1) = -8$, $f'(1/2) \approx 1.04$, and $f'(2) \approx -.83$. Then by the first derivative test, f has a relative maximum at $x = 1$; the value of this relative maximum is $f(1) = 2$. Also, f has a relative minimum at $x = 0$; this relative minimum is $f(0) = 0$.

Using the extrema found above, and plotting a few other points as needed, gives the graph of f shown in Figure 12.10. ■

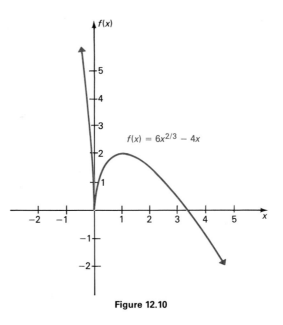

$$f(x) = 6x^{2/3} - 4x$$

Figure 12.10

The next example shows how to find relative maxima or minima for an exponential function.

Example 3 Find all relative maxima or relative minima for $f(x) = (2x + 1)e^{-x}$. First take the derivative, using the product rule.

$$f'(x) = (2x + 1)(-e^{-x}) + e^{-x}(2)$$
$$= -2x \cdot e^{-x} - e^{-x} + 2e^{-x}$$
$$= -2x \cdot e^{-x} + e^{-x}$$
$$= e^{-x}(-2x + 1)$$

Set the derivative equal to 0.

$$e^{-x}(-2x + 1) = 0$$

Since e^{-x} is never 0, this derivative can equal 0 only when $-2x + 1 = 0$, or when $x = 1/2$.

To decide if there is a maximum or a minimum at $x = 1/2$, use the first derivative test. Since $f'(0) > 0$ and $f'(1) < 0$, there is a relative maximum of $2/e^{1/2} \approx 1.2$ at $x = 1/2$. **5**

We can now summarize the steps in finding relative maxima or relative minima.

To find relative extrema for a function f,

Step 1   find the derivative, $f'(x)$;

Step 2   find all critical points c (those points where $f'(c) = 0$ or $f'(c)$ does not exist but $f(c)$ does exist);

Step 3   use the first derivative test to identify relative maxima or minima.

Absolute Maxima and Minima   As the examples above illustrate, a function may well have several relative maxima or relative minima. However, a function never has more than one absolute maximum or absolute minimum.

Let f be a function defined on some interval. Let c be a number in the interval.

1. $f(c)$ is the **absolute maximum** of f on the interval if

$$f(x) \leq f(c)$$

for every x in the interval.

2. $f(c)$ is the **absolute minimum** of f on the interval if

$$f(x) \geq f(c)$$

for every x in the interval.

The function graphed in Figure 12.11 on the next page has an absolute maximum at x_1 and an absolute minimum at x_2. The function graphed in Figure 12.12 has neither an absolute maximum nor an absolute minimum. **6**

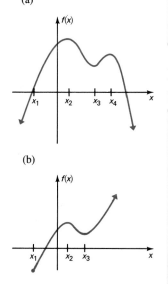

5 Find any relative maxima or relative minima for the following.
(a) $f(x) = 3e^x$
(b) $f(x) = x^2 e^{5x}$

Answer:
(a) None
(b) Relative minimum of 0 at $x = 0$; relative maximum of about .02 at $x = -2/5$

6 Identify the location of any absolute extrema for these graphs.
(a)

(b)

Answer:
(a) Absolute maximum at x_2, no absolute minimum
(b) Absolute minimum at x_1, no absolute maximum

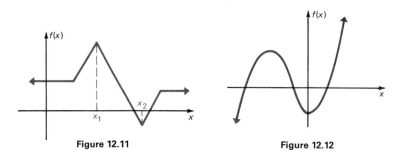

Figure 12.11   Figure 12.12

One of the main reasons for the importance of absolute extrema is given by the following *extreme value theorem* (which is proved in more advanced courses).

Extreme Value Theorem   A function f continuous on a closed interval $[a, b]$ will have both an absolute maximum and an absolute minimum on the interval.

This extreme value theorem guarantees the existence of absolute extrema. Use the following steps to find these extrema.

To find absolute extrema for a function f continuous on an interval $[a, b]$:

Step 1   Find f' and solve $f'(x) = 0$ to find all critical points c_1, c_2, . . . in (a, b). Also, identify the critical points where f' does not exist;

Step 2   Evaluate $f(c_1)$, $f(c_2)$, . . . for all critical points in (a, b);

Step 3   Evaluate $f(a)$ and $f(b)$ for the endpoints a and b of the interval $[a, b]$;

Step 4   The largest value found in Steps 2 or 3 is the absolute maximum for f on $[a, b]$; the smallest value found is the absolute minimum.

Example 4   Find the absolute maximum and the absolute minimum for $f(x) = x^2 - 4x + 7$ defined on the closed interval $[-3, 4]$.

Look first for critical points of the function in the interval $(-3, 4)$. Here $f'(x) = 2x - 4$, and $2x - 4 = 0$ when $x = 2$. This x-value is in the interval $(-3, 4)$. By the first derivative test, $x = 2$ leads to a relative minimum; the relative minimum is $f(2) = 3$. Now check the endpoints of the interval on which the function is defined.

Endpoint	*Value of function*
-3	$f(-3) = 28$
4	$f(4) = 7$

Since 3, the relative minimum, is the smallest of the three numbers 3, 7, and 28, the value 3 is the absolute minimum. The largest of the three numbers 3, 7, and 28 is

28, which occurs at $x = -3$; 28 is the absolute maximum, and it occurs at the endpoint of the interval.

On the given domain the function never has a value smaller than 3 or larger than 28. A graph of $f(x) = x^2 - 4x + 7$ defined on the closed interval $[-3, 4]$ is shown in Figure 12.13. **7**

7 Find the absolute maximum and absolute minimum values of
(a) $f(x) = -x^2 + 4x - 8$ on $[-4, 4]$;
(b) $f(x) = 2x^2 - 6x + 6$ on $[0, 5]$.

Answer:
(a) Absolute maximum of -4 at $x = 2$, absolute minimum of -40 at $x = -4$
(b) Absolute maximum of 26 at $x = 5$, absolute minimum of $3/2$ at $x = 3/2$

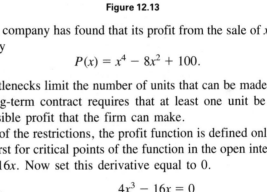

Figure 12.13

Example 5 A company has found that its profit from the sale of x units of an auto part is given by

$$P(x) = x^4 - 8x^2 + 100.$$

Production bottlenecks limit the number of units that can be made to no more than 5, while a long-term contract requires that at least one unit be made. Find the maximum possible profit that the firm can make.

Because of the restrictions, the profit function is defined only for the domain $[1, 5]$. Look first for critical points of the function in the open interval $(1, 5)$. Here $f'(x) = 4x^3 - 16x$. Now set this derivative equal to 0.

8 For a certain firm, the cost to produce x units of an item is given by
$C(x) = x^3 - 15x^2 + 48x + 50$.
Because of production problems, the function is defined only for the domain $[1, 6]$. Find the maximum and minimum costs that the firm will face.

Answer:
Maximum of 94 when $x = 2$, minimum of 14 when $x = 6$

$$4x^3 - 16x = 0$$
$$4x(x^2 - 4) = 0$$
$$4x(x + 2)(x - 2) = 0$$

$4x = 0$ or	$x + 2 = 0$ or	$x - 2 = 0$
$x = 0$ or	$x = -2$ or	$x = 2$

Since $x = 0$ and $x = -2$ are not in the interval $(1, 5)$, disregard them. Now evaluate the function at the remaining critical point 2, and at the endpoints of the domain, 1 and 5.

x-value	Value of function
1	93
2	84
5	525 ← **absolute maximum**

Maximum profit of \$525 occurs when 5 units are made. **8**

12.1 Exercises

Find the location and value of all relative maxima and relative minima for the functions graphed as follows. (See Example 1.)

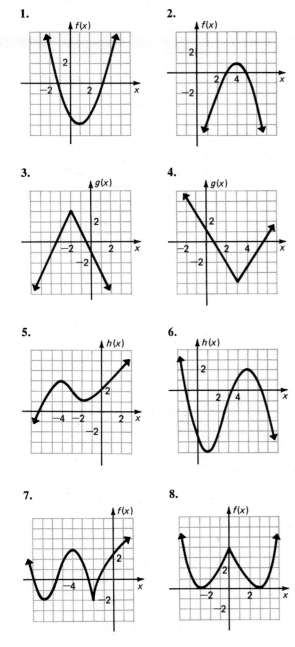

1.

2.

3.

4.

5.

6.

7.

8.

Find the x-values of all points where the functions defined as follows have any relative extrema. Find the value of any relative extrema. (See Examples 2 and 3.)

9. $f(x) = x^2 + 12x - 8$

10. $f(x) = x^2 - 4x + 6$

11. $f(x) = 8 - 6x - x^2$

12. $f(x) = 3 - 4x - 2x^2$

13. $f(x) = x^3 + 6x^2 + 9x - 8$

14. $f(x) = x^3 + 3x^2 - 24x + 2$

15. $f(x) = -\dfrac{4}{3}x^3 - \dfrac{21}{2}x^2 - 5x + 8$

16. $f(x) = -\dfrac{2}{3}x^3 - \dfrac{1}{2}x^2 + 3x - 4$

17. $f(x) = 2x^3 - 21x^2 + 60x + 5$

18. $f(x) = 2x^3 + 15x^2 + 36x - 4$

19. $f(x) = x^4 - 18x^2 - 4$

20. $f(x) = x^4 - 8x^2 + 9$

21. $f(x) = -(8 - 5x)^{2/3}$

22. $f(x) = (2 - 9x)^{2/3}$

23. $f(x) = 2x + 3x^{2/3}$

24. $f(x) = 3x^{5/3} - 15x^{2/3}$

25. $f(x) = x - \dfrac{1}{x}$

26. $f(x) = x^2 + \dfrac{1}{x}$

27. $f(x) = \dfrac{x^2}{x^2 + 1}$

28. $f(x) = \dfrac{x^2}{x - 3}$

29. $f(x) = \dfrac{x^2 - 2x + 1}{x - 3}$

30. $f(x) = \dfrac{x^2 - 6x + 9}{x + 2}$

31. $f(x) = -xe^x$

32. $f(x) = xe^{-x}$

33. $f(x) = x^2 e^{-x}$

34. $f(x) = (x + 4)e^{-2x}$

35. $f(x) = e^x + e^{-x}$

36. $f(x) = -x^2 e^x$

37. $f(x) = x \cdot \ln |x|$

38. $f(x) = x - \ln |x|$

Find the location of all absolute maxima and absolute minima (if any) for the functions graphed as follows.

39.

40.

41.

42.

43.

44.

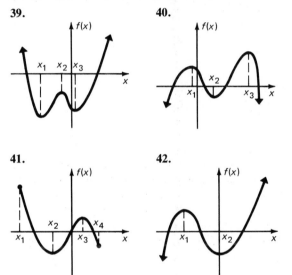

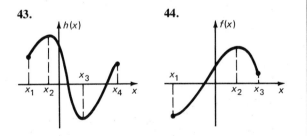

Find the location of all absolute extrema for the functions having domains as specified in Exercises 45–66. A calculator will be helpful for many of these problems. (See Example 4.)

45. $f(x) = x^2 + 6x + 2$; $[-4, 0]$

46. $f(x) = x^2 - 4x + 1$; $[-6, 5]$

47. $f(x) = 5 - 8x - 4x^2$; $[-5, 1]$

48. $f(x) = 9 - 6x - 3x^2$; $[-4, 3]$

49. $f(x) = x^3 - 3x^2 - 24x + 5$; $[-3, 6]$

50. $f(x) = x^3 - 6x^2 + 9x - 8$; $[0, 5]$

51. $f(x) = \frac{1}{3}x^3 - \frac{1}{2}x^2 - 6x + 3$; $[-4, 4]$

52. $f(x) = \frac{1}{3}x^3 + \frac{3}{2}x^2 - 4x + 1$; $[-5, 2]$

53. $f(x) = x^4 - 32x^2 - 7$; $[-5, 6]$

54. $f(x) = x^4 - 18x^2 + 1$; $[-4, 4]$

55. $f(x) = \frac{1}{1 + x}$; $[0, 2]$

56. $f(x) = \frac{-2}{x + 3}$; $[1, 4]$

57. $f(x) = \frac{8 + x}{8 - x}$; $[4, 6]$

58. $f(x) = \frac{1 - x}{3 + x}$; $[0, 3]$

59. $f(x) = \frac{x}{x^2 + 2}$; $[0, 4]$

60. $f(x) = \frac{x - 1}{x^2 + 1}$; $[1, 5]$

61. $f(x) = (x^2 + 18)^{2/3}$; $[-3, 3]$

62. $f(x) = (x^2 + 4)^{1/3}$; $[-2, 2]$

63. $f(x) = (x + 1)(x + 2)^2$; $[-4, 0]$

64. $f(x) = (x - 3)(x - 1)^3$; $[-2, 3]$

65. $f(x) = \frac{1}{\sqrt{x^2 + 1}}$; $[-1, 1]$

66. $f(x) = \frac{3}{\sqrt{x^2 + 4}}$; $[-2, 2]$

Work the following exercises. (See Example 5.)

67. Social Science A professor has found that the number of biology students attending class is approximated by

$$S(x) = -x^2 + 20x + 80,$$

where x is the number of daily hours that the student union is open. Find the number of hours that the union should be open so that the number of students attending class is a maximum. Find the maximum number of such students.

68. The number of people visiting Timberline Ski Lodge on Washington's Birthday is approximated by

$$W(x) = -x^2 + 60x + 180,$$

where x is the total snowfall in inches for the previous week. Find the snowfall that will produce the maximum number of visitors. Find the maximum number of visitors.

69. Management The total profit in thousands of dollars from the sale of x hundred thousand automobile tires is approximated by

$$P(x) = -x^3 + 9x^2 + 120x - 400.$$

(Assume $x \geq 5$.) Find the number of hundred thousands of tires that must be sold to maximize profit. Find the maximum profit.

70. Management A company has found through experience that increasing its advertising also increases its sales, up to a point. The company believes that the mathematical model connecting profit in dollars, $P(x)$, and expenditures on advertising in dollars, x, is

$$P(x) = 80 + 108x - x^3.$$

(a) Find the expenditure on advertising that leads to maximum profit.
(b) Find the maximum profit.

71. From information given in a recent business publication we constructed the mathematical model

$$M(x) = -\frac{1}{45}x^2 + 2x - 20, \qquad 30 \leq x \leq 65$$

to represent the miles per gallon used by a certain car at a speed of x miles per hour. Find the absolute maximum miles per gallon and the absolute minimum.

72. For a certain compact car,

$$M(x) = -.018x^2 + 1.24x + 6.2, \qquad 30 \leq x \leq 60$$

represents the miles per gallon obtained at a speed of x miles per hour. Find the speed that produces the absolute maximum miles per gallon and the absolute minimum.

73. Natural Science The number of salmon swimming upstream to spawn is approximated by

$$S(x) = -x^3 + 3x^2 + 360x + 5000,$$

where x represents the temperature of the water in degrees Celsius. (This function is valid only if $6 \leq x \leq 20$.) Find the water temperature that produces the maximum number of salmon swimming upstream.

74. Natural Science The microbe concentration, $B(x)$, in appropriate units, of Lake Tom depends approximately on the oxygen concentration, x, again in appropriate units, according to the function

$$B(x) = x^3 - 7x^2 - 160x + 1800.$$

(a) Find the oxygen concentration that will lead to the minimum microbe concentration.
(b) What is the minimum concentration?

75. Management Assume that total revenue received from the sale of x items is given by

$$R(x) = 30 \cdot \ln(2x + 1),$$

while the total cost to produce x items is $C(x) = x/2$. Find the number of items that should be manufactured so that profit, $R(x) - C(x)$, is a maximum.

76. Suppose dots and dashes are transmitted over a telegraph line so that dots occur a fraction p of the time (where $0 \leq p \leq 1$) and dashes occur a fraction $1 - p$ of the time. The *information content* of the telegraph line is given by $I(p)$, where

$$I(p) = -p \cdot \ln p - (1 - p) \cdot \ln(1 - p).$$

(a) Show that $I'(p) = -\ln p + \ln(1 - p)$.
(b) Let $I'(p) = 0$ and find the value of p that maximizes $I(p)$. In the long run, what fraction of the dots and dashes should be dots?

In many practical applications, only approximate locations of maxima or minima are needed (or, in fact, can even be found). To find the approximate maximum and minimum of each function on the given domain, use a computer to evaluate the function at intervals of .1. Then estimate the absolute maximum and absolute minimum values of the function.

77. $f(x) = x^3 - 9x^2 + 18x - 7$, $[0, 1.5]$

78. $f(x) = x^3 - 3x^2 - 12x + 4$, $[-1.5, 0]$ and $[3, 4]$

79. $f(x) = x^3 + 6x^2 - 6x + 3$, $[-5, -4]$ and $[0, 1]$

80. $f(x) = x^3 - 3x^2 - 12x + 7$, $[-2, -1]$ and $[2, 3]$

12.2 The Second Derivative Test

The derivative of the derivative of a function is called the **second derivative** of the function. The derivative of the second derivative gives the **third derivative.** This process could be continued getting *fourth derivatives, fifth derivatives,* and so on. For example, if $f(x) = x^4 + 2x^3 + 3x^2 - 5x + 7$, then

$$f'(x) = 4x^3 + 6x^2 + 6x - 5, \qquad \text{(the first derivative of } f\text{)}$$
$$f''(x) = 12x^2 + 12x + 6, \qquad \text{(the second derivative of } f\text{)}$$
$$f'''(x) = 24x + 12, \qquad \text{(the third derivative of } f\text{)}$$
$$f^{(4)}(x) = 24, \qquad \text{(the fourth derivative of } f\text{)}$$
$$f^{(5)}(x) = 0. \qquad \text{(the fifth derivative of } f\text{)}$$

The second derivative of $y = f(x)$ can be written with any of the following notations:

$$f''(x), \qquad y'', \qquad \frac{d^2y}{dx^2}, \qquad \text{or} \qquad D_x^2[f(x)].$$

The third derivative can be written in a similar way. For $n \geq 4$, the nth derivative is written $f^{(n)}(x)$.

Example 1 Find the second derivative of the functions defined as follows.

(a) $y = 8x^3 - 9x^2 + 6x + 4$

Here $y' = 24x^2 - 18x + 6$. The second derivative is the derivative of y', or

$$y'' = 48x - 18.$$

(b) $y = \dfrac{4x + 2}{3x - 1}$

Use the quotient rule to find y'.

$$y' = \frac{(3x - 1)(4) - (4x + 2)(3)}{(3x - 1)^2} = \frac{12x - 4 - 12x - 6}{(3x - 1)^2} = \frac{-10}{(3x - 1)^2}$$

Use the quotient rule again to find y''.

$$y'' = \frac{(3x - 1)^2(0) - (-10)(2)(3x - 1)(3)}{[(3x - 1)^2]^2}$$

$$= \frac{60(3x - 1)}{(3x - 1)^4}$$

$$y'' = \frac{60}{(3x - 1)^3}$$

1 Find the second derivatives of the following.

(a) $y = -9x^3 + 8x^2 + 11x - 6$

(b) $y = -2x^4 + 6x^2$

(c) $y = \dfrac{x + 2}{5x - 1}$

(d) $y = e^x + \ln x$

Answer:

(a) $y'' = -54x + 16$

(b) $y'' = -24x^2 + 12$

(c) $y'' = \dfrac{110}{(5x - 1)^3}$

(d) $y'' = e^x - \dfrac{1}{x^2}$

2 Let $f(x) = 4x^3 - 12x^2 + x - 1$. Find

(a) $f''(0)$;

(b) $f''(4)$;

(c) $f''(-2)$.

Answer:

(a) -24

(b) 72

(c) -72

(c) $y = xe^x$

Using the product rule gives

$$y' = x \cdot e^x + e^x \cdot 1 = xe^x + e^x.$$

Differentiate this result to get y''.

$$y'' = (xe^x + e^x) + e^x$$
$$= xe^x + 2e^x$$
$$y'' = (x + 2)e^x \quad \boxed{1}$$

Example 2 Let $f(x) = x^3 + 6x^2 - 9x + 8$. Find the following.

(a) $f''(0)$

Here $f'(x) = 3x^2 + 12x - 9$, so that $f''(x) = 6x + 12$. Then

$$f''(0) = 6(0) + 12 = 12.$$

(b) $f''(-3) = 6(-3) + 12 = -6$ \quad \boxed{2}

The derivative gives the slope of a tangent line, while the derivative of the derivative (the second derivative) gives the rate of change of the slope. As we have seen, a point just to the left of a relative minimum yields a tangent line with a negative slope. Thus, moving from left to right along the graph in a neighborhood of a relative minimum, the slope of the tangent line is increasing, from negative to zero to positive. See Figure 12.14. Hence, the rate of change of the slope (the derivative of the slope function) should be positive. An argument based on this discussion leads to the *second derivative test* for maxima and minima.

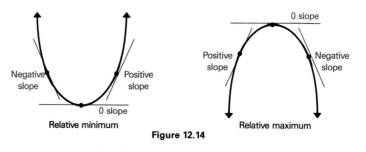

Figure 12.14

Second Derivative Test Let c be a critical point for the function f. Suppose f' exists for every point in an open interval containing c.

1. If $f''(c) > 0$, then $f(c)$ is a **relative minimum.**
2. If $f''(c) < 0$, then $f(c)$ is a **relative maximum.**
3. If $f''(c) = 0$, the test gives **no information.**

Example 3 Find all relative maxima and relative minima for

$$f(x) = 4x^3 + 7x^2 - 10x + 8.$$

First, find the critical points. Here $f'(x) = 12x^2 + 14x - 10$. Set this derivative equal to 0.

$$12x^2 + 14x - 10 = 0$$
$$2(6x^2 + 7x - 5) = 0$$
$$2(3x + 5)(2x - 1) = 0$$

$$3x + 5 = 0 \qquad \text{or} \qquad 2x - 1 = 0$$
$$3x = -5 \qquad\qquad\qquad 2x = 1$$
$$x = -\frac{5}{3} \qquad\qquad\qquad x = \frac{1}{2}$$

Now use the second derivative test. The second derivative is $f''(x) = 24x + 14$. Since the critical points are $-5/3$ and $1/2$,

$$f''\left(-\frac{5}{3}\right) = 24\left(-\frac{5}{3}\right) + 14 = -40 + 14 = -26 < 0,$$

so that $-5/3$ leads to a relative maximum of $f(-5/3) = 691/27 \approx 25.593$. Also,

$$f''\left(\frac{1}{2}\right) = 24\left(\frac{1}{2}\right) + 14 = 12 + 14 = 26 > 0,$$

so $1/2$ gives a relative minimum of $f(1/2) = 21/4 = 5.25$. ∎

A further application of second derivatives is given in the next example.

Example 4 Suppose the revenue from the sale of x units of a certain product is

$$R(x) = 6x^3 + 8x^2 + 9.$$

Then $R'(x) = 18x^2 + 16x$ is called the *velocity of the revenue,* and $R''(x) = 36x + 16$ is called the *acceleration of the revenue.* ▪4▪

The second derivative test works only for those critical points c that make $f'(c) = 0$. This test does not work for those critical points c for which $f'(c)$ does not exist (since $f''(c)$ would not exist either). Also, the second derivative test does not work for critical points c that make $f''(c) = 0$. In both of these cases, use the first derivative test.

▪3▪ Find all relative maxima and relative minima for the following functions. Use the second derivative test.
(a) $f(x) = 6x^2 + 12x + 1$
(b) $f(x) = x^3 - 3x^2 - 9x + 8$

Answer:
(a) Relative minimum of -5 at $x = -1$
(b) Relative maximum of 13 at $x = -1$, relative minimum of -19 at $x = 3$

▪4▪ In Example 4, find
(a) $R'(3)$;
(b) $R'(10)$;
(c) $R''(3)$;
(d) $R''(10)$.

Answer:
(a) 210
(b) 1960
(c) 124
(d) 376

12.2 Exercises

For each of the following, find f''. Then find $f''(0)$, $f''(2)$, and $f''(-3)$. (See Examples 1 and 2.)

1. $f(x) = 3x^3 - 4x + 5$
2. $f(x) = x^3 + 4x^2 + 2$
3. $f(x) = 3x^4 - 5x^3 + 2x^2$
4. $f(x) = -x^4 + 2x^3 - x^2$
5. $f(x) = (x + 4)^3$
6. $f(x) = (x - 2)^3$
7. $f(x) = \dfrac{2x + 1}{x - 2}$
8. $f(x) = \dfrac{x + 1}{x - 1}$

9. $f(x) = \dfrac{x^2}{1 + x}$ **10.** $f(x) = \dfrac{-x}{1 - x^2}$

11. $f(x) = \sqrt{x + 4}$ **12.** $f(x) = \sqrt{2x + 9}$

13. $f(x) = 5x^{3/5}$ **14.** $f(x) = -2x^{2/3}$

15. $f(x) = 2e^x$ **16.** $f(x) = \ln(2x - 3)$

For each of the following, find $f'''(x)$, the third derivative of f, and $f^{(4)}(x)$, the fourth derivative.

17. $f(x) = -x^4 + 2x^2 + 8$

18. $f(x) = 2x^4 - 3x^3 + x^2$

19. $f(x) = 4x^5 + 6x^4 - x^2 + 2$

20. $f(x) = 3x^5 - x^4 + 2x^3 - 7x$

21. $f(x) = \dfrac{x - 1}{x + 2}$ **22.** $f(x) = \dfrac{x + 1}{x}$

23. $f(x) = 5e^{2x}$ **24.** $f(x) = 2 + e^{-x}$

25. $f(x) = \ln|x|$ **26.** $f(x) = \dfrac{1}{x}$

27. $f(x) = x \ln|x|$ **28.** $f(x) = \dfrac{\ln|x|}{x}$

Find any critical points for the functions defined as follows, and use the second derivative test to decide if the critical points lead to relative maxima or relative minima. (See Example 3.) If $f''(a) = 0$ for a critical point $x = a$, then the second derivative test gives no information. In this case, use the first derivative test instead.

29. $f(x) = x^2 + 6x - 3$

30. $f(x) = -x^2 - 4x + 5$

31. $f(x) = 12 - 8x + 4x^2$

32. $f(x) = 6 + 4x + x^2$

33. $f(x) = x^3 + 2x^2 - 5$

34. $f(x) = 8x^3 - 9x^2 + 3$

35. $f(x) = -4x^3 - 15x^2 + 18x + 5$

36. $f(x) = -2x^3 - 3x^2 - 72x + 1$

37. $f(x) = \dfrac{2}{3}x^3 + \dfrac{1}{2}x^2 - x - \dfrac{1}{4}$

38. $f(x) = \dfrac{2}{3}x^3 + \dfrac{3}{2}x^2 - 2x + 1$

39. $f(x) = -2x^3 + 1$

40. $f(x) = (x - 1)^4$

41. $f(x) = x^4 - 8x^2$

42. $f(x) = x^4 - 32x^2 + 7$

43. $f(x) = x + \dfrac{3}{x}$ **44.** $f(x) = x - \dfrac{1}{x}$

45. $f(x) = \dfrac{x^2 + 9}{2x}$ **46.** $f(x) = \dfrac{x^2 + 16}{2x}$

47. $f(x) = \dfrac{2 - x}{2 + x}$ **48.** $f(x) = \dfrac{x + 2}{x - 1}$

Work the following exercises.

49. Management   The profit from the sale of x units of a product is

$$p(x) = 1000 - 12x + 10x^2 - x^3.$$

Find the number of units that leads to a maximum profit. What is the maximum profit?

50. Management   Because of raw material shortages, it is increasingly expensive to produce fine cigars. In fact, the profit in thousands of dollars from producing x hundred thousand cigars is approximated by

$$P(x) = x^3 - 23x^2 + 120x + 60.$$

Find the levels of production that will lead to the maximum profit and to the minimum profit.

51. Natural Science   The number of mosquitoes, $M(x)$, in millions, in a certain area depends on the April rainfall, x, measured in inches, approximately as follows:

$$M(x) = 50 - 32x + 14x^2 - x^3.$$

Find the amount of rainfall that will produce the maximum number of mosquitoes.

52. Natural Science   An autocatalytic chemical reaction is one in which the product being formed causes the rate of formation to increase. The rate of a certain autocatalytic reaction is given by

$$V(x) = 12x(100 - x),$$

where x is the quantity of the product present and 100 represents the quantity of chemical present initially. For what value of x is the rate of the reaction a maximum?

53. Natural Science The percent of concentration of a certain drug in the bloodstream x hours after the drug is administered is given by

$$K(x) = \frac{3x}{x^2 + 4}.$$

For example, after one hour the concentration is given by $K(1) = 3(1)/(1^2 + 4) = (3/5)\% = .6\% = .006$.

(a) Find the time at which concentration is a maximum.

(b) Find the maximum concentration.

54. Natural Science The percent of concentration of another drug in the bloodstream x hours after the drug is administered is given by

$$K(x) = \frac{4x}{3x^2 + 27}.$$

(a) Find the time at which the concentration is a maximum.

(b) Find the maximum concentration.

55. Natural Science Assume that the number of bacteria, in millions, present in a certain culture at time t is given by

$$R(t) = t^2(t - 18) + 96t + 1000.$$

(a) At what time before 8 hours will the population be maximized?

(b) Find the maximum population.

56. Management The revenue of a small firm, in thousands of dollars, is given by

$$R(x) = 2x^3 - 8x^2 + 4x,$$

where x is the number of units of product that the firm sells. Find the following. (See Example 4.)

(a) Velocity of revenue when $x = 8$

(b) When $x = 12$

(c) Acceleration of revenue when $x = 4$

(d) When $x = 8$

12.3 Applications of Maxima and Minima

Finding extrema for realistic problems requires an accurate mathematical model of the problem. For example, suppose the number of units that can be produced on a production line can be closely approximated by the mathematical model

$$T(x) = 8x^{1/2} + 2x^{3/2} + 50,$$

where x is the number of employees on the line. Once this mathematical model has been established, it can be used to produce information about the production line. For example, the derivative $T'(x)$ could be used to estimate the change in production resulting from the addition of an extra worker to the line.

In writing the mathematical model itself, keep track of any restrictions on the values of the variables. For example, since x represents the number of employees on a production line, x must certainly be restricted to the positive integers, or perhaps to a few common fractional values (we can conceive of halftime employees, but probably not 1/32-time employees).

On the other hand, to apply the tools of calculus to obtain an extremum for some function, the function must be defined and be meaningful at every real-number point in some interval. Because of this, the answer obtained by using a mathematical model of a practical problem might be a number that is not feasible in the setting of the problem.

Usually, the requirement that a function be continuous is of theoretical interest only. In most cases, calculus gives results that *are* acceptable in a given situation. And if the methods of calculus are used on a function f and lead to the conclusion that $80\sqrt{2}$ units should be produced in order to get the lowest possible cost, it is usually only necessary to calculate $f(80\sqrt{2})$ and then compare this result to various values of $f(x)$, where x is an acceptable number close to $80\sqrt{2}$. The lowest of these values of $f(x)$ then gives minimum cost. In most cases, the result obtained will be very close to the theoretical minimum.

This section gives several examples showing applications of calculus to extrema problems. Solve these examples with the following steps.

Solving Applied Problems

Step 1 Read the problem carefully. Make sure you understand what is given and what is unknown.

Step 2 If possible, sketch a diagram. Label the various parts.

Step 3 Decide on the variable whose values must be maximized or minimized. Express that variable as a function of *one* other variable.

Step 4 Find the critical points for the function of Step 3. Test each of these for maxima or minima.

Step 5 Check for extrema at any endpoints of the domain of the function of Step 3.

Example 1 For a semiconductor manufacturer to sell x new style memory chips, the price per chip must be

$$p(x) = 4 - \frac{x}{25}, \qquad x \text{ in } [0, 100].$$

(a) Find an expression for the total revenue $T(x)$ from the sale of x memory chips.

The total revenue is given by the product of the number of chips sold, x, and the price per chip, $4 - x/25$, so

$$T(x) = x\left(4 - \frac{x}{25}\right) = 4x - \frac{x^2}{25}.$$

(b) Find the value of x that leads to maximum revenue.

Here
$$T'(x) = 4 - \frac{2x}{25}.$$

Set this derivative equal to 0: $4 - \dfrac{2x}{25} = 0$

$$-\frac{2x}{25} = -4$$
$$2x = 100$$
$$x = 50.$$

1 A total of x earthmovers will be sold if the price, in thousands of dollars, is given by
$$p(x) = 32 - \frac{x}{8}.$$
Find
(a) an expression for the total revenue $R(x)$;
(b) the value of x that leads to maximum revenue;
(c) the maximum revenue.

Answer:
(a) $R(x) = x(32 - x/8)$
  $= 32x - x^2/8$
(b) $x = 128$
(c) $2048 thousand, or $2,048,000

2 An investor has built a series of self-storage units near a group of apartment houses. She must now decide on the monthly rental. From past experience, she feels that 200 units will be rented for $15 per month, with 5 additional rentals for each $.25 reduction in the rental price. Let x be the number of $.25 reductions in the price and find
(a) an expression for the number of units rented;
(b) an expression for the price per unit;
(c) an expression for the total revenue;
(d) the value of x leading to maximum revenue;
(e) the maximum revenue.

Answer:
(a) $200 + 5x$ (b) $15 - .25x$
(c) $(200 + 5x)(15 - .25x) =$
$3000 + 25x - 1.25x^2$
(d) $x = 10$ (e) $3125

Use the second derivative test to decide if $x = 50$ leads to maximum revenue or minimum revenue. Since $T'(x) = 4 - 2x/25$, the second derivative is $T''(x) = -2/25$, which is negative for all x, so $x = 50$ gives maximum revenue, as desired.
(c) Find the maximum revenue.

$$T(50) = 4(50) - \frac{50^2}{25} = 200 - 100 = 100 \text{ dollars.} \quad \blacksquare$$

Example 2 When Yuppies Galore, Inc. charges $600 for a seminar on management techniques, it attracts 1000 people. For each $20-decrease in the charge, an additional 100 people will attend the seminar.
(a) Find an expression for the total revenue if there are x of the $20-decreases in the price.

The price charged will be

$$\text{Price per person} = 600 - 20x,$$

and the number of people in the seminar will be

$$\text{Number of people} = 1000 + 100x.$$

The total revenue, $R(x)$, is given by the product of the price and the number of people attending, or

$$R(x) = (600 - 20x)(1000 + 100x)$$
$$= 600,000 + 40,000x - 2000x^2.$$

(b) Find the value of x that maximizes revenue.
Set the derivative, $R'(x) = 40,000 - 4000x$, equal to 0.

$$40,000 - 4000x = 0$$
$$-4000x = -40,000$$
$$x = 10$$

Since $R''(x) = -4000$, $x = 10$ leads to maximum revenue.
(c) Find the maximum revenue.
Use $R(x)$ from part (a); let $x = 10$.

$$R(10) = 600,000 + 40,000(10) - 2000(10)^2 = 800,000$$

dollars. For this level of revenue, $1000 + 100(10) = 2000$ people must attend the seminar; each person will pay $600 - 20(10) = $400. \quad **2**

Example 3 A truck burns fuel at the rate of

$$G(x) = \frac{1}{200}\left(\frac{800 + x^2}{x}\right)$$

gallons per mile when traveling x miles per hour on a straight, level road. If fuel costs $2 per gallon, find the speed that will produce the minimum total cost for a

1000-mile trip. Find the minimum total cost. (This is an extension of Exercise 45 in Section 11.6.)

The total cost of the trip, in dollars, is the product of the number of gallons per mile, the number of miles, and the cost per gallon. If $C(x)$ represents this cost, then

$$C(x) = \left[\frac{1}{200} \left(\frac{800 + x^2}{x} \right) \right] (1000)(2)$$

$$= \frac{2000}{200} \left(\frac{800 + x^2}{x} \right)$$

$$C(x) = \frac{8000 + 10x^2}{x}.$$

Find the value of x that will minimize cost by first finding the derivative:

$$C'(x) = \frac{10x^2 - 8000}{x^2}.$$

Set this derivative equal to 0.

$$\frac{10x^2 - 8000}{x^2} = 0$$

$$10x^2 - 8000 = 0$$

$$10x^2 = 8000$$

$$x^2 = 800$$

Take the square root on both sides to get

$$x \approx \pm 28.3 \text{ mph.}$$

Reject -28.3 as a speed, leaving $x = 28.3$ as the only critical value. Find $C''(x)$ to see if 28.3 leads to a minimum cost.

$$C''(x) = \frac{16,000}{x^3}$$

Since $C''(28.3) > 0$, the critical value 28.3 leads to a minimum. (It is not necessary to actually calculate $C''(28.3)$; just check that $C''(28.3) > 0$.)

The minimum total cost is

$$C(28.3) = \frac{8000 + 10(28.3)^2}{28.3} = 565.69 \text{ dollars.} \quad \blacksquare 3$$

3 A diesel generator burns fuel at the rate of

$$G(x) = \frac{1}{48} \left(\frac{300}{x} + 2x \right)$$

gallons per hour when producing x thousand kilowatt hours of electricity. Suppose that fuel costs $2.25 a gallon and find the value of x that leads to minimum total cost if the generator is operated for 32 hours. Find the minimum cost.

Answer:

$x = \sqrt{150} \approx 12.2$,

minimum cost is $73.50

Example 4 An open box is to be made by cutting squares from each corner of a 12 inch by 12 inch piece of metal and then folding up the sides. What size square should be cut from each corner in order to produce the box of maximum volume?

Let x represent the length of a side of the square that is cut from each corner, as shown in Figure 12.15. The width of the box that is to be built will be $12 - 2x$, while the length is also $12 - 2x$. The volume of the box is given by the product of the length, width, and height. In this example, the volume, $V(x)$, depends on x:

$$V(x) = x(12 - 2x)(12 - 2x)$$
$$= 144x - 48x^2 + 4x^3.$$

Find the value of x that will maximize volume.

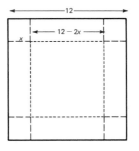

Figure 12.15

Start with $V'(x) = 144 - 96x + 12x^2$. Set this derivative equal to 0.

$$12x^2 - 96x + 144 = 0$$
$$12(x^2 - 8x + 12) = 0$$
$$12(x - 2)(x - 6) = 0$$

$$x - 2 = 0 \quad \text{or} \quad x - 6 = 0$$
$$x = 2 \quad \text{or} \quad x = 6$$

Since $V''(x) = -96 + 24x$, then $V''(2) = -96 + 24(2) = -48$, which is negative. Thus, $x = 2$ leads to maximum volume; the maximum volume is

$$V(2) = 144(2) - 48(2)^2 + 4(2)^3 = 128$$

cubic inches. The domain of $V(x)$ is all numbers in the interval $[0, 6]$. Both the endpoints of this domain, $x = 0$ and $x = 6$, lead to *minimum* volumes of 0. ◼ **4**

4 An open box is to be made by cutting squares from each corner of a 20 cm by 32 cm piece of metal and folding up the sides. Let x represent the length of the side of the square to be cut out. Find
(a) an expression for the volume of the box, $V(x)$;
(b) $V'(x)$;
(c) the value of x that leads to maximum volume (Hint: The solutions of the equation $V'(x) = 0$ are 4 and 40/3);
(d) the maximum volume.

Answer:
(a) $V(x) = 640x - 104x^2 + 4x^3$
(b) $V'(x) = 640 - 208x + 12x^2$
(c) $x = 4$
(d) $V(4) = 1152$ cubic centimeters

Economic Lot Size Suppose that a company manufactures a certain constant number of units of a product per year. Assume that the product can be manufactured in a number of batches of equal size during the year. The company could manufacture the item only once per year to minimize setup costs, but they would then have much higher warehouse costs. On the other hand, they might make many small batches, but this would increase setup costs. Calculus can be used to find the number of batches per year that should be manufactured in order to minimize total cost.

Figure 12.16 on the next page shows several of the possibilities for a product having an annual demand of 12,000 units. The top graph shows the results if only one batch of the product is made annually: an average of 6000 items will be held in a warehouse. If four batches (of 3000 each) are made at equal time intervals during a year, the average number of units in the warehouse falls to only 1500. If twelve batches are made, an average of 500 items will be in the warehouse.

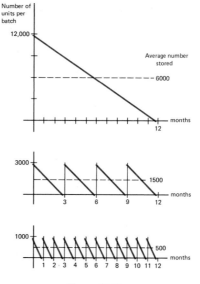

Figure 12.16

The solution is found using the following variables:

x = number of batches to be manufactured annually

k = cost of storing one unit of the product for one year

f = fixed setup cost to manufacture the product

g = variable cost of manufacturing a single unit of the product

M = total number of units produced annually.

The company has two types of costs associated with the production of its product: a cost associated with manufacturing the item, and a cost associated with storing the finished product.

First consider manufacturing costs. During a year the company will produce x batches of the product, or M/x units of the product per batch. Each batch has a fixed cost f and a variable cost g per unit, so that the manufacturing cost per batch is

$$f + g\left(\frac{M}{x}\right).$$

Since there are x batches per year, the total annual manufacturing cost is

$$\left[f + g\left(\frac{M}{x}\right)\right]x. \qquad (1)$$

Now find the storage cost. Since each batch consists of M/x units, and since demand is assumed to be constant, it is common to assume an average inventory of

$$\frac{1}{2}\left(\frac{M}{x}\right) = \frac{M}{2x}$$

units per year. The cost to store one unit of the product for a year is k, making a total storage cost of

$$k\left(\frac{M}{2x}\right) = \frac{kM}{2x}. \qquad (2)$$

The total production cost is the sum of the manufacturing and storage costs, or the sum of expressions (1) and (2). If $T(x)$ is the total cost of producing x batches,

$$T(x) = \left[f + g\left(\frac{M}{x}\right)\right]x + \frac{kM}{2x}.$$

Now find the value of x that will minimize $T(x)$. (Remember that f, g, k, and M are constants.) Find $T'(x)$:

$$T'(x) = f - \frac{kM}{2}x^{-2}.$$

Set this derivative equal to 0.

$$f - \frac{kM}{2}x^{-2} = 0$$

$$f = \frac{kM}{2x^2}$$

$$2fx^2 = kM$$

$$x^2 = \frac{kM}{2f}$$

$$x = \sqrt{\frac{kM}{2f}} \qquad (3)$$

The second derivative test can be used to show that $\sqrt{kM/(2f)}$ is the annual number of batches that gives minimum total production cost.

Example 5 A paint company has a steady annual demand for 24,500 cans of automobile primer. The cost accountant for the company says that it costs $2 to store one can of paint for one year and $500 to set up the plant for the production of the primer. Find the number of batches of primer that should be produced for the minimum total production cost.

Use equation (3) above.

$$x = \sqrt{\frac{kM}{2f}}$$

$$x = \sqrt{\frac{2(24,500)}{2(500)}} \qquad \text{Let } k = 2, \, M = 24,500, \, f = 500$$

$$x = \sqrt{49} = 7$$

Seven batches of primer per year will lead to minimum production costs. ▪5 ▪6

5 A manufacturer of business forms has an annual demand for 30,720 units of form letters to people delinquent in their payments of installment debt. It costs $5 per year to store one unit of the letters and $1200 to set up the machines to produce them. Find the number of batches that should be made annually to minimize total cost.

Answer:
8 batches

6 An office uses 576 cases of copy-machine paper during the year. It costs $3 per year to store one case. Each reorder costs $24. Find the number of orders that should be placed annually.

Answer:
6

12.3 Exercises

Exercises 1–8 involve maximizing and minimizing products and sums of numbers. Work all these problems using the steps shown in Exercise 1. Find only nonnegative solutions.

1. We want to find two numbers x and y such that $x + y = 100$ and the product $P = xy$ is as large as possible.
   (a) We know $x + y = 100$. Solve this equation for y.
   (b) Substitute this result for y into $P = xy$.
   (c) Find P'. Solve the equation $P' = 0$.
   (d) What are the two numbers?
   (e) What is the maximum value of P?

2. Find two numbers whose sum is 250 and whose product is as large as possible. What is the maximum product?

3. Find two numbers whose sum is 200 such that the sum of the squares of the two numbers is minimized.

4. Find two numbers whose sum is 30 such that the sum of the squares of the numbers is minimized.

5. Find numbers x and y, with x, $y \geq 0$, such that $x + y = 150$ and x^2y is maximized.

6. Find numbers x and y such that $x + y = 45$ and xy^2 is maximized.

7. Find numbers x and y, with x, $y \geq 0$, such that $x - y = 10$ and xy is minimized.

8. Find numbers x and y such that $x - y = 3$ and xy is minimized.

9. **Management** If the price charged for a candy bar is $p(x)$ cents, then x thousand candy bars will be sold in a certain city, where

$$p(x) = 100 - \frac{x}{10}.$$

   (a) Find an expression for the total revenue from the sale of x thousand candy bars. (Hint: Find the product of $p(x)$, x, and 1000.) (See Example 1.)
   (b) Find the value of x that leads to maximum revenue.
   (c) Find the maximum revenue.

10. **Management** The sale of cassette tapes of "lesser" performers is very sensitive to price. If a tape manufacturer chargers $p(x)$ dollars per tape, where

$$p(x) = 6 - \frac{x}{8},$$

   then x thousand tapes will be sold.

   (a) Find an expression for the total revenue from the sale of x thousand tapes. (Hint: Find the product of $p(x)$, x, and 1000.)
   (b) Find the value of x that leads to maximum revenue.
   (c) Find the maximum revenue.

11. A truck burns fuel at the rate of $G(x)$ gallons per mile, where

$$G(x) = \frac{1}{32}\left(\frac{64}{x} + \frac{x}{50}\right),$$

   while traveling x miles per hour.
   (a) If fuel costs $1.60 per gallon, find the speed that will produce minimum total cost for a 400-mile trip. (See Example 3.)
   (b) Find the minimum total cost.

12. A rock and roll band travels from engagement to engagement in a large bus. This bus burns fuel at the rate of $G(x)$ gallons per mile, where

$$G(x) = \frac{1}{50}\left(\frac{200}{x} + \frac{x}{15}\right),$$

   while traveling x miles per hour.
   (a) If fuel costs $2 per gallon, find the speed that will produce minimum total cost for a 250-mile trip.
   (b) Find the minimum total cost.

13. A farmer has 1200 meters of fencing. He wants to enclose a rectangular field bordering a river, with no fencing needed along the river. (See the sketch.) Let x represent the width of the field.

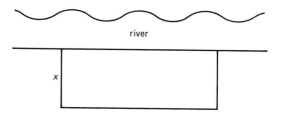

   (a) Write an expression for the length of the field.
   (b) Find the area of the field (area = length × width).
   (c) Find the value of x leading to the maximum area.
   (d) Find the maximum area.

14. Find the dimensions of the rectangular field of maximum area that can be made from 200 meters of fencing material. (This fence has four sides.)

15. Natural Science An ecologist is conducting a research project on breeding pheasants in captivity. She first must construct suitable pens. She wants a rectangular area with two additional fences down the center, as shown in the sketch. Find the maximum area she can enclose with 3600 meters of fencing.

16. A rectangular field is to be enclosed with a fence. One side of the field is against an existing fence, so that no fence is needed on that side. If material for the fence costs $2 per foot for the two ends and $4 per foot for the side parallel to the existing fence, find the dimensions of the field of largest area that can be enclosed for $1000.

17. A rectangular field is to be enclosed on all four sides with a fence. Fencing material costs $3 per foot for two opposite sides, and $6 per foot for the other two sides. Find the maximum area that can be enclosed for $2400.

18. Management The manager of an 80-unit apartment complex is trying to decide on the rent to charge. It is known from experience that at a rent of $200, all the units will be full. However, on the average, one additional unit will remain vacant for each $10 increase in rent.
(a) Let x represent the number of $10 increases. Find the amount of rent per apartment. (See Example 2.)
(b) Find the number of apartments rented.
(c) Find the total revenue from all rented apartments.
(d) What value of x leads to maximum revenue?
(e) What is the maximum revenue?

19. Management The manager of a peach orchard is trying to decide when to have the peaches picked. If they are picked now, the average yield per tree will be 100 pounds, which can be sold for 40¢ per pound. Past experience shows that the yield per tree will increase about 5 pounds per week, while the price will decrease about 2¢ per pound per week.
(a) Let x represent the number of weeks that the manager should wait. Find the income per pound.
(b) Find the number of pounds per tree.

(c) Find the total revenue from a tree.
(d) When should the peaches be picked in order to produce maximum revenue?
(e) What is the maximum revenue?

20. A local group of scouts has been collecting old aluminum beer cans for recycling. The group has already collected 12,000 pounds of cans, for which they could currently receive $4 per hundred pounds. The group can continue to collect cans at the rate of 400 pounds per day. However, a glut in the old-beer-can market has caused the recycling company to announce that it will lower its price, starting immediately, by $.10 per hundred pounds per day. The scouts can make only one trip to the recycling center. Find the best time for that single trip. What total income will be received?

21. Management In planning a small restaurant, it is estimated that a profit of $5 per seat will be made if the number of seats is between 60 and 80, inclusive. On the other hand, the profit on each seat will decrease by 5¢ for each seat above 80.
(a) Find the number of seats that will produce the maximum profit.
(b) What is the maximum profit?

22. A local club is arranging a charter flight to Hawaii. The cost of the trip is $425 each for 75 passengers, with a refund of $5 per passenger for each passenger in excess of 75.
(a) Find the number of passengers that will maximize the revenue received from the flight.
(b) Find the maximum revenue.

23. A television manufacturing firm needs to design an open-topped box with a square base. The box must hold 32 cubic inches. Find the dimensions of the box that can be built with the minimum amount of materials.

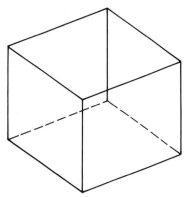

24. A closed box with a square base is to have a volume of 16,000 cubic centimeters. The material for the top and bottom of the box costs 3¢ per square centimeter, while the material for the sides costs 1.5¢ per square centimeter. Find the dimensions of the box that will lead to minimum total cost. What is the minimum total cost?

25. A company wishes to manufacture a box with a volume of 36 cubic feet that is open on top and that is twice as long as it is wide. Find the dimensions of the box produced from the minimum amount of material.

26. **Management** A mathematics book is to contain 36 square inches of printed matter per page, with margins of 1 inch along the sides, and 1 1/2 inches along the top and bottom. Find the dimensions of the page that will lead to the minimum amount of paper being used for a page.

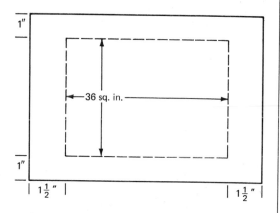

27. **Management** In Example 3 we found the speed in miles per hour that minimized cost when we considered only the cost of the fuel. Rework the problem taking into account the driver's salary of $8 per hour. (Hint: If the trip is 1000 miles at x miles per hour, the driver will be paid for $1000/x$ hours.)

28. **Management** Decide what you would do if your assistant brought you the following contract to sign:
    Your firm offers to deliver 300 tables to a dealer, at $90 per table, and to reduce the price per table on the entire order by 25¢ for each additional table over 300.
Find the dollar total involved in the largest possible transaction between you and the dealer; find the smallest possible dollar amount.

29. **Management** A company wishes to run a utility cable from point A on the shore (see the figure) to an installa-

tion at point B on the island. The island is 6 miles from the shore. It costs $400 per mile to run the cable on land and $500 per mile underwater. Assume that the cable starts at A and runs along the shoreline, then angles and runs underwater to the island. Find the point at which the line should begin to angle in order to yield the minimum total cost. (Hint: The length of the line underwater is $\sqrt{x^2 + 36}$.)

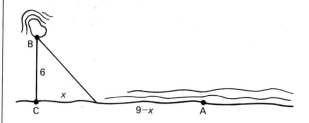

30. A hunter is at a point on a river bank. He wants to get to his cabin, located three miles north and eight miles west. (See the figure.) He can travel 5 miles per hour on the river but only 2 miles per hour on this very rocky land. How far up river should he go in order to reach the cabin in minimum time? (Hint: Distance = rate × time.)

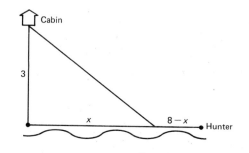

Management *The following exercises refer to economic lot size. (See Example 5.)*

31. Find the approximate number of batches that should be produced annually if 100,000 units are to be manufactured. It costs $1 to store a unit for one year and it costs $500 to set up the factory to produce each batch.

32. How many units per batch will be manufactured in Exercise 31?

33. A market has a steady annual demand for 16,800 cases of sugar. It costs $3 to store one case for one year. The market pays $7 for each order that is placed. Find the number of orders for sugar that should be placed each year.

34. Find the number of cases per order in Exercise 33.

35. The publisher of a best-selling book has an annual demand for 100,000 copies. It costs 50¢ to store one copy for one year. Setup of the press for a new printing costs $1000. Find the number of batches that should be printed annually.

36. A restaurant has an annual demand for 810 bottles of a California wine. It costs $1 to store one bottle for one year, and it costs $5 to place a reorder. Find the number of orders that should be placed annually.

12.4 Curve Sketching

Increasing and Decreasing Functions This section shows derivatives used as an aid in sketching the graph of a function. We begin by defining *increasing* and *decreasing* functions. Intuitively, a function is *increasing* if its graph goes *up* from left to right. A function is *decreasing* if its graph goes *down* from left to right. Graphs of increasing functions are shown in Figure 12.17(a)–(c), with graphs of decreasing functions in Figure 12.17(d)–(f). ①

① Write *increasing* or *decreasing* for the functions graphed as follows.

(a)

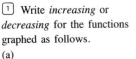

(b)

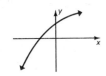

Answer:

(a) Increasing

(b) Decreasing

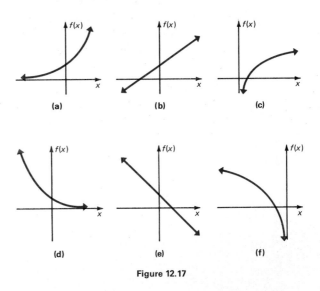

Figure 12.17

Increasing and decreasing functions are defined as follows.

> Let f be a function defined on some interval. Then for any two numbers x_1 and x_2 in the interval,
>
> 1. f is **increasing** on the interval if
> $$f(x_1) < f(x_2) \quad \text{whenever} \quad x_1 < x_2;$$
>
> 2. f is **decreasing** on the interval if
> $$f(x_1) > f(x_2) \quad \text{whenever} \quad x_1 < x_2.$$

Example 1 For which intervals is the function graphed in Figure 12.18 increasing? For which intervals is it decreasing?

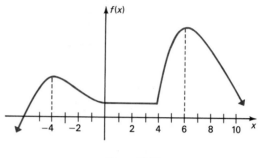

Figure 12.18

Moving from left to right, the function is increasing up to -4, then decreasing from -4 to 0, constant (neither increasing nor decreasing) from 0 to 4, increasing from 4 to 6, and finally, decreasing from 6 on. In interval notation, the function is increasing on $(-\infty, -4)$ and $(4, 6)$, decreasing on $(-4, 0)$ and $(6, \infty)$, and constant on $(0, 4)$. **2**

The derivative of a function can be used to tell whether a function is increasing or decreasing. We know that the derivative gives the slope of the tangent line to a curve at any point on the curve. If a function is increasing, the tangent line at any point on the graph will have positive slope, while if the function is decreasing, the slopes of the tangents will be negative. See Figure 12.19.

2 For what values of x is the function graphed as follows increasing? decreasing?

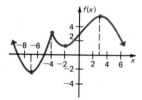

Answer:
Increasing on $(-7, -4)$ and $(-2, 3)$; decreasing on $(-\infty, -7)$, $(-4, -2)$, and $(3, \infty)$

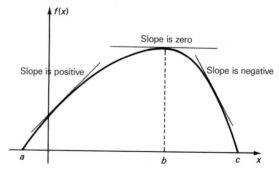

Figure 12.19

Suppose a function f has a derivative at each point in an open interval.

1. f is **increasing** on the interval if $f'(x) > 0$ for each x in the interval.
2. f is **decreasing** on the interval if $f'(x) < 0$ for each x in the interval.
3. f is **constant** on the interval if $f'(x) = 0$ for each x in the interval.

The derivative can change signs from positive to negative at points where $f'(x) = 0$, and also at points where $f'(x)$ does not exist. This suggests that the test for increasing and decreasing functions be applied as follows.

Step 1 Locate on a number line those values of x that make $f'(x) = 0$ and where $f'(x)$ does not exist. These points determine several open intervals.

Step 2 Choose a value of x in each of the intervals determined in Step 1. Use these values to decide if $f'(x) > 0$ or $f'(x) < 0$ in that interval.

Step 3 Use the test above to decide if f is increasing or decreasing on the interval.

Example 2 Find all intervals where functions defined as follows are increasing or decreasing. Locate all points where the tangent line to the graph is horizontal. Graph each function.

(a) $f(x) = 2x^2 + 6x - 5$

Follow the steps listed above. The derivative $f'(x) = 4x + 6$ is 0 when

$$4x + 6 = 0$$

$$x = -\frac{3}{2}.$$

There are no values of x where $f'(x)$ fails to exist. Locate $x = -3/2$ on a number line, as in Figure 12.20. This point determines two intervals, $(-\infty, -3/2)$ and $(-3/2, \infty)$.

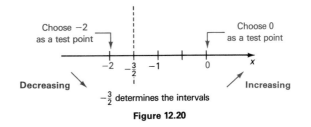

Figure 12.20

Now, choose any value of x in the interval $(-3/2, \infty)$. Choosing $x = 0$ gives

$$f'(0) = 4 \cdot 0 + 6 = 6,$$

which is positive. Since one value of x in this interval makes $f'(x) > 0$, all values will do so, with f therefore increasing on $(-3/2, \infty)$. Use -2 as a test point in the interval $(-\infty, -3/2)$:

$$f'(-2) = 4(-2) + 6 = -2,$$

which is negative. This means that f is decreasing on the interval $(-\infty, -3/2)$. The arrows in each interval in Figure 12.20 indicate where f is increasing or decreasing.

Since a horizontal line has a slope of 0, horizontal tangents can be found by solving the equation $f'(x) = 0$. As we saw above, $f'(x) = 0$ when $x = -3/2$.

Our only method of graphing most functions so far has been by plotting several points that lie on the graph. Now we have an additional tool: the test for determining where a function is increasing or decreasing. (Other tools are discussed later.) Using the intervals where $f(x) = 2x^2 + 6x - 5$ is increasing or decreasing, along with the fact (from Chapter 1) that the graph is a parabola, leads to the graph of Figure 12.21.

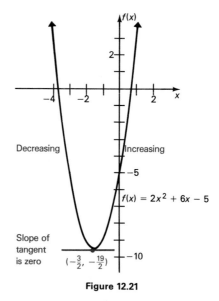

Figure 12.21

(b) $f(x) = x^3 + 3x^2 - 9x + 4$.

Here $f'(x) = 3x^2 + 6x - 9$. Set this derivative equal to 0, and solve the resulting equation by factoring.

$$3x^2 + 6x - 9 = 0$$
$$3(x^2 + 2x - 3) = 0$$
$$3(x + 3)(x - 1) = 0$$
$$x = -3 \quad \text{or} \quad x = 1$$

The tangent line is horizontal at $x = -3$ or $x = 1$. There are no values of x where $f'(x)$ fails to exist. Determine where the function is increasing or decreasing by locating -3 and 1 on a number line, as in Figure 12.22. (Be sure to place the values on the number line in numerical order.)

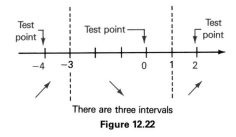

Figure 12.22

Choosing the test point -4 from the left interval gives

$$f'(-4) = 3(-4)^2 + 6(-4) - 9 = 15,$$

which is positive, so f is increasing on $(-\infty, -3)$. Selecting 0 from the middle interval gives $f'(0) = -9$, so f is decreasing on $(-3, 1)$. Finally, choosing 2 in the right-hand region gives $f'(2) = 15$, with f increasing on $(1, \infty)$. The intervals where f is increasing or decreasing are shown with arrows in Figure 12.22.

The information found above, along with point plotting as necessary, was used to draw the graph in Figure 12.23. ▪

3 Find all intervals where the following functions are increasing or decreasing.
(a) $f(x) = 2x^2 + 5x - 4$
(b) $f(x) = 6 + 3x - x^2$
(c) $f(x) = 4x^3 + 3x^2 - 18x + 1$

Answer:
(a) Increasing on $(-5/4, \infty)$; decreasing on $(-\infty, -5/4)$
(b) Increasing on $(-\infty, 3/2)$; decreasing on $(3/2, \infty)$
(c) Increasing on $(-\infty, -3/2)$ and $(1, \infty)$; decreasing on $(-3/2, 1)$

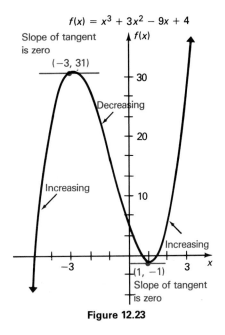

Figure 12.23

Concave Upward or Downward Intuitively, a graph is *concave upward* on an interval if it "holds water"; the graph is *concave downward* if it "spills water." (See Figure 12.24.)

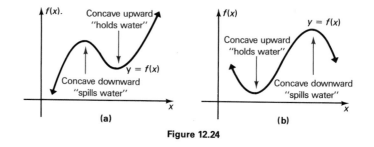

Figure 12.24

More precisely, a function is **concave upward** on an interval (a, b) if the graph of the function lies above its tangent line at each point of (a, b); a function is **concave downward** on (a, b) if the graph of the function lies below its tangent line at each point of (a, b). A point where a graph changes concavity is called a **point of inflection.** See Figure 12.25.

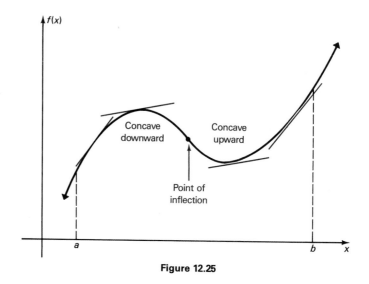

Figure 12.25

A function can be either increasing or decreasing, and either concave upward or concave downward. Examples of various combinations are shown in Figure 12.26.

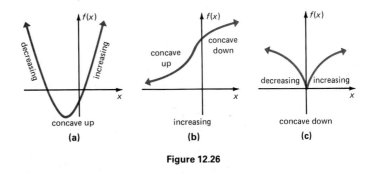

Figure 12.26

Figure 12.27 shows the graph of two functions that are concave upward on an interval (a, b). Several tangent lines are also shown. In Figure 12.27(a), the slopes of these tangent lines (moving from left to right) are first negative, then 0, and then positive. In Figure 12.27(b), the slopes are all positive, but getting larger. In both cases, the slopes are *increasing*. The slope at a point on a curve is given by the first derivative. The slope is increasing if the derivative of the slope function is positive—that is, if the derivative of the first derivative is positive. Since the derivative of the first derivative is the second derivative, a function has a graph that is concave upward on an interval if its second derivative is positive at each point of the interval.

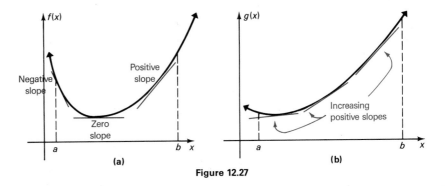

Figure 12.27

A similar result holds for functions that are concave downward, giving the following test.

> Let f be a function with derivatives f' and f'' existing at all points in an interval (a, b).
>
> 1. The function f is **concave upward** on (a, b) if $f''(x) > 0$ for all x in (a, b).
> 2. The function f is **concave downward** on (a, b) if $f''(x) < 0$ for all x in (a, b).

Example 3 Find the intervals where $f(x) = x^3 - 3x^2 + 5x - 4$ is concave upward or downward.

The first derivative is $f'(x) = 3x^2 - 6x + 5$, and the second derivative is $f''(x) = 6x - 6$. The function f is concave upward whenever $f''(x) > 0$, or

$$6x - 6 > 0$$
$$6x > 6$$
$$x > 1.$$

Also, f is concave downward if $f''(x) < 0$, or $x < 1$. Using interval notation, f is concave upward on $(1, \infty)$ and concave downward on $(-\infty, 1)$. A graph of f is shown in Figure 12.28. ▇

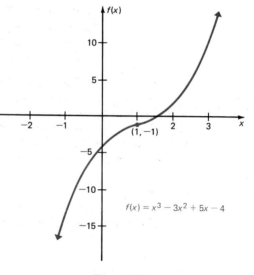

$$f(x) = x^3 - 3x^2 + 5x - 4$$

Figure 12.28

4 Find the intervals where the following are concave upward or concave downward. Identify any inflection points.

(a) $f(x) = 6x^3 - 24x^2 + 9x - 3$

(b) $f(x) = 2x^2 - 4x + 8$

Answer:

(a) Concave downward on $(-\infty, 4/3)$, concave upward on $(4/3, \infty)$; point of inflection is $(4/3, -175/9)$

(b) $f''(x) = 4$, which is always positive; function is always concave upward, no inflection point

In Example 3, the graph changed from being concave downard to being concave upward at the point where $x = 1$. As mentioned above, such a point where concavity changes is called a point of inflection. In Example 3, the point of inflection is $(1, f(1))$, or $(1, -1)$, and occurs at the value of x where $f''(x) = 0$. A point of inflection may also occur where $f''(x)$ does not exist.

Be careful: the fact that $f''(x_0) = 0$ does not mean that a point of inflection has been located. For example, if $f(x) = (x - 1)^4$, then $f''(x) = 12(x - 1)^2$, which is 0 when $x = 1$. However, the graph of $f(x) = (x - 1)^4$ is always concave upward and has no points of inflection. **4**

Example 4 The graph of Figure 12.29 shows the population of catfish in a commercial catfish farm as a function of time. As the graph shows, the population increases rapidly up to a point, and then increases at a slower rate. The horizontal dashed line shows that the population will approach some upper limit determined by the capacity of the farm. The point at which the rate of population growth starts to slow is the point of inflection for the graph.

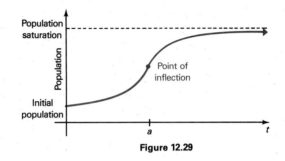

Figure 12.29

For the maximum yield of catfish, it turns out that harvest should take place at the point of fastest possible growth of the population—here, this is at the point of inflection. 5

5 Find the intervals where the function of Figure 12.29 is concave upward and concave downward.

Answer:
Concave upward on (0, *a*) and concave downward on (*a*, ∞)

Curve Sketching All the methods of this section, together with previous methods such as point plotting, finding asymptotes, and finding relative maxima and minima, assist in sketching the graphs of many different functions. This process, called **curve sketching,** uses these steps.

To sketch the graph of a function *f*:

Step 1 Find f'. Solve the equation $f'(x) = 0$. Also determine where f' does not exist (but f does). This locates any critical points. Find any relative extrema and determine where f is increasing or decreasing.

Step 2 Find f''. Solve the equation $f''(x) = 0$ and determine where f'' does not exist. This locates any points of inflection. Determine where f is concave upward or concave downward.

Step 3 Plot a few carefully chosen points as needed.

Example 5 Sketch the graph of $f(x) = x^3 - 2x^2 - 4x + 3$.

Use the derivative to find all intervals for which the function is increasing or decreasing.

$$f'(x) = 3x^2 - 4x - 4$$
$$= (3x + 2)(x - 2)$$

Verify that $f'(x) = 0$ when $x = -2/3$ or when $x = 2$. Use the second derivative test to check for maxima or minima. Here

$$f''(x) = 6x - 4,$$

which is positive when $x = 2$ and negative when $x = -2/3$, showing that $f(2)$ is a relative minimum and $f(-2/3)$ is a relative maximum.

The first derivative is positive on the intervals $(-\infty, -2/3)$ and $(2, \infty)$, so the function is increasing on these intervals. The first derivative is negative on the interval $(-2/3, 2)$, which shows that the function is decreasing on this interval.

The second derivative, $f''(x) = 6x - 4$, is 0 when $x = 2/3$. Choosing appropriate values to the right and left of 2/3 shows that the second derivative is negative for all values of x in the interval $(-\infty, 2/3)$, and positive for all x in $(2/3, \infty)$, with the graph concave downward on $(-\infty, 2/3)$, and concave upward on $(2/3, \infty)$. A summary of this information is given in the following chart. Using these facts, and plotting points as needed, sketch the graph as shown in Figure 12.30.

  Sketch the graphs.
(a) $f(x) = x^3 - 3x^2$
(b) $f(x) = x + 1/x$

Answer:
(a)

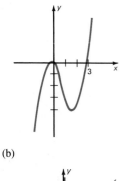

(b)

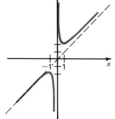

Interval	f'	f''	f
$(-\infty, -2/3)$	$+$		increasing
$(-2/3, 2)$	$-$		decreasing
$(2, \infty)$	$+$		increasing
$(-\infty, 2/3)$		$-$	concave downward
$(2/3, \infty)$		$+$	concave upward

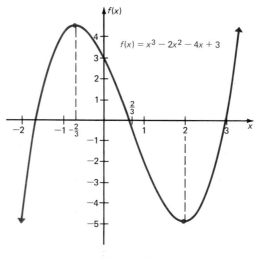

Figure 12.30

12.4 Exercises

Find the largest open intervals where the functions graphed as follows are increasing or decreasing. (See Example 1.) (Compare these exercises with Exercises 1–8 in Section 12.1.)

1.

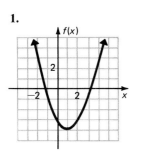

2.

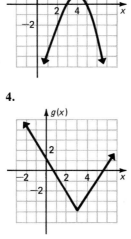

3.

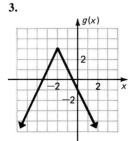

4.

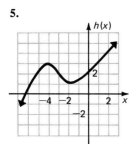

5.

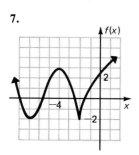

6.

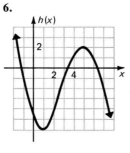

7.

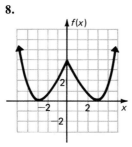

8.

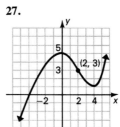

Find the largest open intervals where the functions defined as follows are increasing or decreasing. Graph each function. (See Example 2.)

9. $f(x) = x^2 + 12x - 6$

10. $f(x) = x^2 - 9x + 4$

11. $y = 5 + 9x - 3x^2$

12. $y = 3 + 4x - 2x^2$

13. $f(x) = 2x^3 - 3x^2 - 72x - 4$

14. $f(x) = 2x^3 - 3x^2 - 12x + 2$

15. $f(x) = 4x^3 - 15x^2 - 72x + 5$

16. $f(x) = 4x^3 - 9x^2 - 30x + 6$

17. $y = -3x + 6$

18. $y = 6x - 9$

19. $f(x) = \dfrac{x + 2}{x + 1}$

20. $f(x) = \dfrac{x + 3}{x - 4}$

21. $y = |x + 4|$

22. $y = -|x - 3|$

23. $f(x) = -\sqrt{x - 1}$

24. $f(x) = \sqrt{5 - x}$

25. $y = \sqrt{x^2 + 1}$

26. $y = x\sqrt{9 - x^2}$

Find the largest open intervals where the functions graphed or defined as follows are concave upward or concave downward. Find the location of any points of inflection. (See Example 3.)

27.

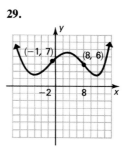

28.

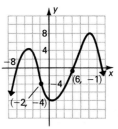

29.

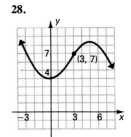

30.

31.

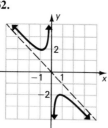

32.

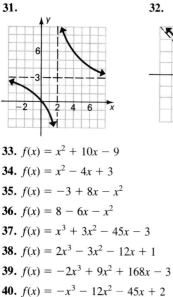

33. $f(x) = x^2 + 10x - 9$

34. $f(x) = x^2 - 4x + 3$

35. $f(x) = -3 + 8x - x^2$

36. $f(x) = 8 - 6x - x^2$

37. $f(x) = x^3 + 3x^2 - 45x - 3$

38. $f(x) = 2x^3 - 3x^2 - 12x + 1$

39. $f(x) = -2x^3 + 9x^2 + 168x - 3$

40. $f(x) = -x^3 - 12x^2 - 45x + 2$

41. $f(x) = \dfrac{3}{x - 5}$

42. $f(x) = \dfrac{-2}{x + 1}$

43. $f(x) = x(x + 5)^2$

44. $f(x) = -x(x - 3)^2$

Sketch the graphs of the following. Identify any points of inflection. (See Example 5.)

45. $f(x) = -x^2 - 10x - 25$

46. $f(x) = x^2 - 12x + 36$

47. $f(x) = 3x^3 - 3x^2 + 1$

48. $f(x) = 2x^3 - 4x^2 + 2$

49. $f(x) = -2x^3 - 9x^2 + 108x - 10$

50. $f(x) = -2x^3 - 9x^2 + 60x - 8$

51. $f(x) = 2x^3 + \dfrac{7}{2}x^2 - 5x + 3$

52. $f(x) = x^3 - \dfrac{15}{2}x^2 - 18x - 1$

53. $f(x) = (x + 3)^4$

54. $f(x) = x^3$

55. $f(x) = x^4 - 18x^2 + 5$

56. $f(x) = x^4 - 8x^2$

57. $f(x) = x + \dfrac{1}{x}$

58. $f(x) = 2x + \dfrac{8}{x}$

59. $f(x) = \dfrac{x^2 + 25}{x}$

60. $f(x) = \dfrac{x^2 + 4}{x}$

61. $f(x) = \dfrac{x - 1}{x + 1}$

62. $f(x) = \dfrac{x}{1 + x}$

Work the following exercises. (See Example 4.)

63. Management The accompanying figure shows the *product life cycle* graph, with typical products marked on it.*

(a) Where would you place home videotape recorders on this graph?

(b) Where would you place light bulbs?

(c) Which products are closest to the point of inflection on the left of the graph? What does the point of inflection mean here?

(d) Which products are closest to the point of inflection on the right of the graph? What does the point of inflection mean here?

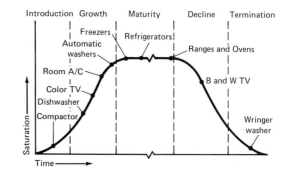

---

*From ''The Product Life Cycle: A Key to Strategic Marketing Planning'' in *MSU Business Topics* (Winter 1973), p. 30. Reprinted by permission of the publisher, Graduate School of Business Administration, Michigan State University.

64. Natural Science When a hardy new species is introduced into an area, the population often increases as shown.

Explain the significance of the following points on the graph.

(a) f_0 **(b)** $(a, f(a))$ **(c)** f_M

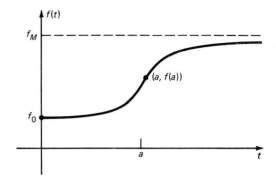

12.5 Multivariate Functions

If a company produces x items at a cost of $10 per item, then $C(x)$, the total cost of producing the items, is given by

$$C(x) = 10x.$$

The cost is a function of one independent variable. If the company produces two products, with x of one product at a cost of $10 each, and y of another product at a cost of $15 each, then the total cost to the firm is a function of *two* independent variables, x and y. By generalizing $f(x)$ notation, the total cost can be written as $C(x, y)$ where

$$C(x, y) = 10x + 15y.$$

When $x = 5$ and $y = 12$ the total cost is written $C(5, 12)$, with

$$C(5, 12) = 10 \cdot 5 + 15 \cdot 12 = 230.$$

This section and the next discuss such functions of more than one independent variable. It turns out that many of the ideas developed for functions of one variable also apply to functions of more than one variable. In particular, the fundamental idea of derivative generalizes in a very natural way to functions of more than one variable.

> $z = f(x, y)$ is a **function of two independent variables** if a unique value of z is obtained from each different ordered pair of real numbers (x, y). The variables x and y are **independent variables;** z is the **dependent variable.** The set of all ordered pairs of real numbers (x, y) such that $f(x, y)$ exists is the **domain** of f; the set of all values of $f(x, y)$ is the **range.**

Unless otherwise stated, assume here that the domain of a function of two independent variables is the largest set of ordered pairs of real numbers (x, y) for which $f(x, y)$ exists.

Functions of three independent variables, written $w = f(x, y, z)$, and of four independent variables, $v = f(w, x, y, z)$, and so on, can be defined in a similar way. Functions of more than one independent variable are called **multivariate functions.**

A function of one independent variable, $y = f(x)$, may be expressed as a set of ordered pairs (x, y). Similarly, a function of two independent variables, $z = f(x, y)$, is expressed as a set of **ordered triples,** (x, y, z). This idea can be extended to other multivariate functions with more independent variables.

Example 1 Let $z = f(x, y) = 8x^2 - 12xy + 6y^2 - 11$. Find the following.
(a) A typical ordered triple (x, y, z) that satisfies the given function
Choose any values for x and y, and then calculate the value of z. Suppose we let $x = 0$ and $y = 2$. Then

$$z = f(0, 2) = 8(0)^2 - 12(0)(2) + 6(2)^2 - 11 = 13.$$

The ordered triple is $(0, 2, 13)$.
(b) $f(-2, 1)$
Let $x = -2$ and $y = 1$.

$$f(-2, 1) = 8(-2)^2 - 12(-2)(1) + 6(1)^2 - 11 = 51 \quad \blacksquare$$

1 Let $g(x, y) = -4x/\sqrt{x^2 + y^2}$.
Find the following, where possible.
(a) $g(3, 0)$
(b) $g(-4, -3)$
(c) $g(0, 0)$

Answer:
(a) -4
(b) $16/5$
(c) Function not defined at $(0, 0)$—can't have a 0 denominator

Graphing Functions of Two Independent Variables Functions of one independent variable are graphed by using an x-axis and a y-axis to locate points in a plane. The plane determined by the x- and y-axes is called the *xy-plane*. A third axis is needed to graph functions of two independent variables—the z-axis, which goes through the origin in the xy-plane, and is perpendicular to both the x-axis and the y-axis. Figure 12.31 shows one possible way to draw the three axes. In Figure 12.31, the yz-plane is in the plane of the page, with the x-axis perpendicular to the plane of the page.

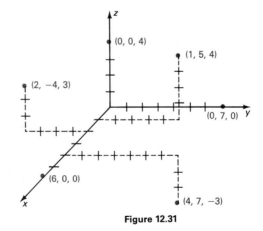

Figure 12.31

② Locate $P(3, 2, 4)$ on the coordinate system below.

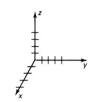

Answer:

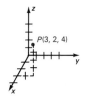

Just as ordered pairs were graphed earlier, we can now graph ordered triples of the form (x, y, z). For example, to locate the point corresponding to the ordered triple $(2, -4, 3)$, start at the origin and go 2 units along the positive x-axis. Then go 4 units in a negative direction (to the left), parallel to the y-axis. Finally, go up 3 units, parallel to the z-axis. The point representing $(2, -4, 3)$ and several other sample points are shown in Figure 12.31. The region of three-dimensional space where all coordinates are positive is called the **first octant**. ②

In Chapter 2 we found that the graph of $ax + by = c$ (where a and b are not both 0) is a straight line. This result generalizes to three dimensions.

> **Plane** The graph of
>
> $$ax + by + cz = d$$
>
> is a plane when a, b, c, and d are real numbers, with a, b, and c not all zero.

Example 2 Graph $2x + y + z = 6$.

By the result in the box, the graph of this equation is a plane. Earlier, we graphed straight lines by finding x- and y- intercepts. A similar idea helps graph a plane. To find the x-intercept, the point where the graph crosses the x-axis, let $y = 0$ and $z = 0$.

$$2x + 0 + 0 = 6$$
$$x = 3$$

The point $(3, 0, 0)$ is on the graph. Letting $x = 0$ and $z = 0$ gives the point $(0, 6, 0)$, while $x = 0$ and $y = 0$ lead to $(0, 0, 6)$. The plane through these three points includes the triangular surface shown in Figure 12.32. This region is the first-octant portion of the plane that is the graph of $2x + y + z = 6$. ∎

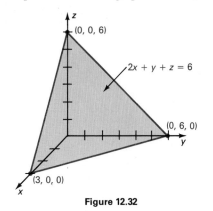

Figure 12.32

Example 3 Graph $x + z = 6$.

To find the x-intercept, let $z = 0$, giving $(6, 0, 0)$. If $x = 0$, we get the point $(0, 0, 6)$. Because there is no y in the equation $x + z = 6$, there can be no y-intercept. A plane that has no y-intercept is parallel to the y-axis. The first-octant portion of the graph of $x + z = 6$ is shown in Figure 12.33. ■

3 Describe each graph and give any intercepts.

(a) $2x + 3y - z = 4$

(b) $x + y = 3$

(c) $z = 2$

Answer:

(a) A plane; $(2, 0, 0)$, $(0, 4/3, 0)$, $(0, 0, -4)$

(b) A plane parallel to the z-axis; $(3, 0, 0)$, $(0, 3, 0)$

(c) A plane parallel to the x-axis and the y-axis; $(0, 0, 2)$

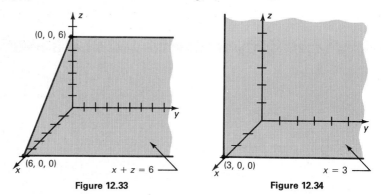

Figure 12.33   **Figure 12.34**

Example 4 Graph $x = 3$.

This graph, which goes through $(3, 0, 0)$, can have no y-intercept and no z-intercept. It is therefore a plane parallel to the y-axis and the z-axis and, therefore, to the yz plane. The first-octant portion of the graph is shown in Figure 12.34. **3**

Example 5 Graph $z = x^2 + y^2$.

If $x = 0$, the equation becomes $z = y^2$, which is the equation of the parabola in the yz-plane shown in color in Figure 12.35. Letting $y = 0$ gives $z = x^2$, which also has a parabola as its graph, the parabola in the xz-plane in Figure 12.35. Choosing other values for x and y leads to other parabolas on the surface of the three-dimensional figure. It can be shown that the graph of $z = x^2 + y^2$ is the bowl-shaped figure, called a **paraboloid,** that is shown in Figure 12.35. ■

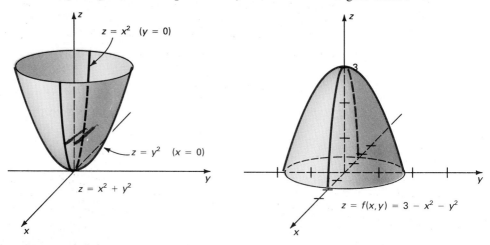

Figure 12.35   **Figure 12.36**

The graph of a function $z = f(x, y)$ is called a **surface.** It may be difficult to draw the surface resulting from a function of two variables. One useful way of picturing these graphs is by finding various **traces**—the curves that result when a surface is cut by a plane. For example, the two parabolas shown in Figure 12.35 are the yz-trace and xz-trace of the graph of $z = x^2 + y^2$.

Notice that the paraboloid in Figure 12.35 appears to have a relative minimum of 0 at the origin. Figure 12.36 shows a paraboloid with a relative maximum of 3 at $(x, y) = (0, 0)$. Maxima and minima of surfaces are discussed further in Section 12.7.

12.5 Exercises

1. Let $f(x, y) = 4x + 5y + 3$. Find the following. (See Example 1.)
   (a) $f(2, -1)$ (b) $f(-4, 1)$
   (c) $f(-2, -3)$ (d) $f(0, 8)$

2. Let $g(x, y) = -x^2 - 4xy + y^3$. Find the following.
   (a) $g(-2, 4)$ (b) $g(-1, -2)$
   (c) $g(-2, 3)$ (d) $g(5, 1)$

3. Let $h(x, y) = \sqrt{x^2 + 2y^2}$. Find the following.
   (a) $h(5, 3)$ (b) $h(2, 4)$
   (c) $h(-1, -3)$ (d) $h(-3, -1)$

4. Let $f(x, y) = \dfrac{\sqrt{9x + 5y}}{\log x}$. Find the following.

   (a) $f(10, 2)$ (b) $f(100, 1)$

   (c) $f(1000, 0)$ (d) $f\left(\dfrac{1}{10}, 5\right)$

Work the following exercises.

5. **Natural Science** The number of cows that can graze on a certain farm without causing overgrazing is approximated by

   $$C(x, y) = 9x + 5y - 4,$$

   where x is the number of acres of tall grass and y is the number of acres of short grass. Find the following.

   (a) $C(50, 0)$ (b) $C(30, 4)$
   (c) How many cows may graze if there are 80 acres of tall grass and 20 acres of short grass?
   (d) If the farm has 10 acres of short grass and 5 acres of tall grass, how many cows may graze?

6. **Management** The labor charge for assembling a precision camera is given by

   $$L(x, y) = 12x + 6y + 2xy + 40,$$

where x is the number of work hours required by a skilled worker and y is the number of hours required by a semiskilled worker. Find the following.

(a) $L(3, 5)$ (b) $L(5, 2)$
(c) If a skilled worker requires 7 hours and a semiskilled worker needs 9 hours, find the total labor charge.
(d) Find the total labor charge if a skilled worker needs 12 hours and a semiskilled worker requires 4 hours.

Use a calculator with a y^x key to work the following problems. *

7. **Natural Science** The oxygen consumption of a well-insulated mammal which is not sweating is approximated by

   $$m = \frac{2.5(T - F)}{w^{.67}},$$

   where T is the internal body temperature of the animal (in °C), F is the temperature of the outside of the animal's fur (in °C), and w is the animal's weight in kilograms. Find m for the following data.
   (a) $T = 38°$, $F = 6°$, $w = 32$ kg
   (b) $T = 40°$, $F = 20°$, $w = 43$ kg

8. **Natural Science** The surface area of a human (in square meters) is approximated by

   $$A = .202 \, W^{.425} H^{.725}$$

---

*From *Mathematics in Biology* by Duane J. Clow and N. Scott Urquhart. Copyright © 1974 by W. W. Norton & Company, Inc. Used by permission.

where W is the weight of the person in kilograms and H is the height in meters. Find A for the following data.

(a) $W = 72$, $H = 1.78$ **(b)** $W = 65$, $H = 1.40$
(c) $W = 70$, $H = 1.60$ **(d)** Find your own surface area.

Graph the first-octant portion of each of the following planes. (See Examples 2–4.)

9. $x + y + z = 6$
10. $x + y + z = 12$
11. $2x + 3y + 4z = 12$
12. $4x + 2y + 3z = 24$
13. $3x + 2y + z = 18$
14. $x + 3y + 2z = 9$
15. $x + y = 4$
16. $y + z = 5$
17. $x = 2$
18. $z = 3$

12.6 Partial Derivatives

The derivative of a function $y = f(x)$ gives the rate of change of y with respect to x. Suppose now we have a function of two independent variables, say $z = f(x, y)$, and want to know how a change in x or y will affect z. Find out by taking the derivative of $f(x, y)$ with respect to one variable, while holding the others constant.

As an example of this process, suppose that a small firm makes only two products, radios and cassette recorders, with the profit of the firm given by

$$P(x, y) = 40x^2 - 10xy + 5y^2 - 80,$$

where x is the number of units of radios sold and y is the number of units of cassette recorders sold. Lately, sales of radios have been steady at 10 units; only the sales of recorders vary. The management would like to know the marginal profit when $y = 12$.

Recall that marginal profit is given by the derivative of the profit function. Here, x is fixed at 10. Using this information, begin by finding a new function, $f(y) = P(10, y)$. Let $x = 10$ and get

$$f(y) = P(10, y) = 40(10)^2 - 10(10)y + 5y^2 - 80$$
$$= 3920 - 100y + 5y^2.$$

The function $f(y)$ shows the profit from the sale of y recorders, assuming that x is fixed at 10. To find the marginal profit when $y = 12$, first find the derivative df/dy.

$$\frac{df}{dy} = -100 + 10y$$

Letting $y = 12$ gives $-100 + 10(12) = -100 + 120 = 20$, the marginal profit when $y = 12$.

The derivative of the function $f(y)$ was taken with respect to y only; we assumed that x was fixed. To generalize this example, let $z = f(x, y)$. An intuitive definition of the partial derivatives of f with respect to x and y follows.

Partial Derivatives (Informal Definition)

The **partial derivative of f with respect to x** is the derivative of f obtained by treating x as a variable and y as a constant.

The **partial derivative of f with respect to y** is the derivative of f obtained by treating y as a variable and x as a constant.

The symbols $f_x(x, y)$ (no prime used), $\partial z/\partial x$, and $\partial f/\partial x$ are used to represent the partial derivative of $z = f(x, y)$ with respect to x, with similar symbols used for the partial derivative with respect to y. The symbol $f_x(x, y)$ is often abbreviated as just f_x, with $f_y(x, y)$ abbreviated f_y.

Generalizing from the definition of derivative given earlier, partial derivatives of a function $z = f(x, y)$ are defined as follows.

Partial Derivatives (Formal Definition)   Let $z = f(x, y)$ be a function of two variables. Let all indicated limits exist. Then the **partial derivative of f with respect to x** is

$$f_x = \frac{\partial f}{\partial x} = \lim_{h \to 0} \frac{f(x + h, y) - f(x, y)}{h};$$

the **partial derivative of f with respect to y** is

$$f_y = \frac{\partial f}{\partial y} = \lim_{h \to 0} \frac{f(x, y + h) - f(x, y)}{h}.$$

If the indicated limits do not exist, then the partial derivatives do not exist.

Similar definitions could be given for functions of more than two independent variables.

Example 1   Let $f(x, y) = 4x^2 - 9xy + 6y^3$. Find f_x and f_y.

To find f_x, treat y as a constant and x as a variable. This gives

$$f_x = 8x - 9y.$$

Now treat y as a variable and x as a constant, to find f_y.

$$f_y = -9x + 18y^2 \quad \boxed{1}$$

The next example shows how the chain rule can be used to find partial derivatives.

Example 2   Let $f(x, y) = \ln |x^2 + y|$. Find f_x and f_y.

Recall the formula for the derivative of a natural logarithm function. If $g(x) = \ln |x|$, then $g'(x) = 1/x$. Using this formula and the chain rule,

$$f_x = \frac{2x}{x^2 + y} \quad \text{and} \quad f_y = \frac{1}{x^2 + y}. \quad \boxed{2}$$

The notation

$$f_x(a, b) \quad \text{or} \quad \frac{\partial f}{\partial x}(a, b)$$

1 Find f_x and f_y.
(a) $f(x, y) = -x^2y + 3xy + 2xy^2$
(b) $f(x, y) = x^3 + 2x^2y + xy$

Answer:
(a) $f_x = -2xy + 3y + 2y^2$; $f_y = -x^2 + 3x + 4xy$
(b) $f_x = 3x^2 + 4xy + y$; $f_y = 2x^2 + x$

2 Find f_x and f_y.
(a) $f(x, y) = \ln |2x + 3y|$
(b) $f(x, y) = e^{xy}$

Answer:
(a) $f_x = \dfrac{2}{2x + 3y}$; $f_y = \dfrac{3}{2x + 3y}$
(b) $f_x = ye^{xy}$; $f_y = xe^{xy}$

represents the value of a partial derivative when $x = a$ and $y = b$, as shown in the next example.

Example 3 Let $f(x, y) = 2x^2 + 3xy^3 + 2y + 5$. Find the following.
(a) $f_x(-1, 2)$

First, find f_x by holding y constant.

$$f_x = 4x + 3y^3$$

Now let $x = -1$ and $y = 2$.

$$f_x(-1, 2) = 4(-1) + 3(2)^3 = -4 + 24 = 20$$

(b) $\dfrac{\partial f}{\partial y}(-4, -3)$

Since $\partial f/\partial y = 9xy^2 + 2$,

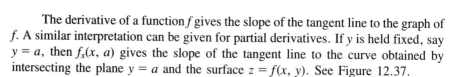

$$\frac{\partial f}{\partial y}(-4, -3) = 9(-4)(-3)^2 + 2 = 9(-36) + 2 = -322. \quad \blacksquare 3$$

3 Let $f(x, y) = x^2 + xy^2 + 5y - 10$. Find the following.
(a) $f_x(2, 1)$
(b) $\dfrac{\partial f}{\partial y}(-1, 0)$

Answer:
(a) 5
(b) 5

The derivative of a function f gives the slope of the tangent line to the graph of f. A similar interpretation can be given for partial derivatives. If y is held fixed, say $y = a$, then $f_x(x, a)$ gives the slope of the tangent line to the curve obtained by intersecting the plane $y = a$ and the surface $z = f(x, y)$. See Figure 12.37.

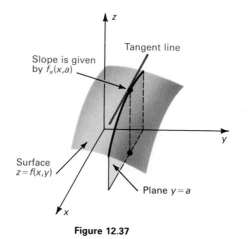

Figure 12.37

The derivative of $y = f(x)$ gives the rate of change of y with respect to x. In the same way, if $z = f(x, y)$, then f_x gives the rate of change of z with respect to x, provided that y is held constant.

Example 4 Suppose that the temperature of the water at the point on a river where a nuclear power plant discharges its hot waste water is approximated by

$$T(x, y) = 2x + 5y + xy - 40.$$

Here, x represents the temperature of the river water in degrees Celsius before it reaches the power plant and y is the number of megawatts (in hundreds) of electricity being produced by the plant.

(a) Find the temperature of the discharged water if the water reaching the plant has a temperature of 8°C and if 300 megawatts of electricity are being produced.

Here $x = 8$ and $y = 3$ (since 300 megawatts of electricity are being produced), with

$$T(8, 3) = 2(8) + 5(3) + 8(3) - 40 = 15.$$

The water at the outlet of the plant is at a temperature of 15°C. ④

(b) Find and interpret $T_x(9, 5)$.

First, find the partial derivative T_x.

$$T_x = 2 + y$$

This partial derivative gives the rate of change of T with respect to x. Replacing x with 9 and y with 5 gives

$$T_x(9, 5) = 2 + 5 = 7.$$

Just as marginal cost is the approximate cost of one more item, this result, 7, is the approximate change in temperature of the output water if input water temperature changes by 1 degree, from $x = 9$ to $x = 9 + 1 = 10$, while y remains constant.

(c) Find and interpret $T_y(9, 5)$.

The partial derivative T_y is

$$T_y = 5 + x.$$

This partial derivative gives the rate of change of T with respect to y, with

$$T_y(9, 5) = 5 + 9 = 14.$$

This result, 14, is the approximate change in temperature resulting from a 1-unit increase in production of electricity from $y = 5$ to $y = 5 + 1 = 6$ (600 megawatts), while x remains constant. ⑤

Earlier we found second derivatives, which are derivatives of derivatives. A similar method can be used to find **second-order partial derivatives,** which are defined as follows.

Second-Order Partial Derivatives For a function $z = f(x, y)$, if all indicated partial derivatives exist, then

$$\frac{\partial}{\partial x}\left(\frac{\partial z}{\partial x}\right) = \frac{\partial^2 z}{\partial x^2} = f_{xx} \qquad \frac{\partial}{\partial y}\left(\frac{\partial z}{\partial y}\right) = \frac{\partial^2 z}{\partial y^2} = f_{yy}$$

$$\frac{\partial}{\partial y}\left(\frac{\partial z}{\partial x}\right) = \frac{\partial^2 z}{\partial y \partial x} = f_{xy} \qquad \frac{\partial}{\partial x}\left(\frac{\partial z}{\partial y}\right) = \frac{\partial^2 z}{\partial x \partial y} = f_{yx}.$$

---

④ Find the temperature of the water at the outlet if the water has an initial temperature of 10°C and if 400 megawatts of electricity are being produced.

Answer:
40°C

⑤ Use the function of Example 4 to find and interpret the following.
(a) $T_x(5, 4)$
(b) $T_y(8, 3)$

Answer:
(a) $T_x(5, 4) = 6$; the approximate increase in temperature if the input temperature increases from 5 to 6 degrees
(b) $T_y(8, 3) = 13$; the approximate increase in temperature if the production of electricity increases from 300 to 400 megawatts

Often f_{xx} is used as an abbreviation for $f_{xx}(x, y)$, with f_{yy}, f_{xy}, and f_{yx} used in a similar way.

The notation

$$\frac{\partial^2 z}{\partial y \partial x},$$

for a mixed partial derivative can be misleading. As shown by the definition,

$$\frac{\partial^2 z}{\partial y \partial x} = \frac{\partial}{\partial y}\left(\frac{\partial z}{\partial x}\right),$$

which means f_x or $\partial z/\partial x$, the partial derivative of z with respect to x, is found first. Then this result, $\partial z/\partial x$, is differentiated with respect to y to get f_{xy} or $\partial^2 z/\partial y \partial x$. Notice that although

$$\frac{\partial^2 z}{\partial y \partial x} = f_{xy},$$

the order of x and y in the two expressions is reversed.

Example 5 Find all second-order partial derivatives for

$$f(x, y) = -4x^3 - 3x^2 y^3 + 2y^2.$$

First find f_x and f_y.

$$f_x = -12x^2 - 6xy^3 \quad \text{and} \quad f_y = -9x^2 y^2 + 4y$$

To find f_{xx}, take the partial derivative of f_x with respect to x.

$$f_{xx} = -24x - 6y^3$$

Take the partial derivative of f_y with respect to y; this gives f_{yy}.

$$f_{yy} = -18x^2 y + 4$$

Find f_{xy} by starting with f_x, then taking the partial derivative of f_x with respect to y.

$$f_{xy} = -18xy^2$$

Finally, find f_{yx} by starting with f_y; take its partial derivative with respect to x.

$$f_{yx} = -18xy^2 \quad \boxed{6}$$

Example 6 Let $f(x, y) = 2e^x - 8x^3 y^2$. Find all second-order partial derivatives. Here $f_x = 2e^x - 24x^2 y^2$ and $f_y = -16x^3 y$. [Recall: if $g(x) = e^x$, then $g'(x) = e^x$.] Now find the second partial derivatives.

$$f_{xx} = 2e^x - 48xy^2 \qquad f_{xy} = -48x^2 y$$
$$f_{yy} = -16x^3 \qquad f_{yx} = -48x^2 y \quad \boxed{7}$$

6 Let $f(x, y) = 4x^2 y^2 - 9xy + 8x^2 - 3y^4$. Find all second partial derivatives.

Answer:
$f_{xx} = 8y^2 + 16$;
$f_{yy} = 8x^2 - 36y^2$;
$f_{xy} = 16xy - 9$;
$f_{yx} = 16xy - 9$

7 Let $f(x, y) = 4e^{x+y} + 2x^3 y$. Find all second partial derivatives.

Answer:
$f_{xx} = 4e^{x+y} + 12xy$;
$f_{yy} = 4e^{x+y}$;
$f_{xy} = 4e^{x+y} + 6x^2$;
$f_{yx} = 4e^{x+y} + 6x^2$

In both our examples of second-order partial derivatives, $f_{xy} = f_{yx}$. It can be proved that this happens for many functions, including most functions found in applications, and all the functions in this book.

Partial derivatives of multivariate functions with more than two independent variables are found in a way similar to that for functions with two independent variables. For example, to find f_x for $w = f(x, y, z)$, treat y and z as constants and differentiate with respect to x.

Example 7 Let $f(x, y, z) = xy^2z + 2x^2y - 4xz^2$. Find f_x, f_y, f_z, f_{xy}, and f_{yz}.

$$f_x = y^2z + 4xy - 4z^2$$
$$f_y = 2xyz + 2x^2$$
$$f_z = xy^2 - 8xz$$

To find f_{xy}, differentiate f_x with respect to y.

$$f_{xy} = 2yz + 4x$$

In the same way, differentiate f_y with respect to z to get

$$f_{yz} = 2xy. \quad \boxed{8}$$

8 Let $f(x, y, z) = xyz + x^2yz + xy^2z^3$. Find f_x, f_y, f_z, and f_{xz}.

Answer:
$f_x = yz + 2xyz + y^2z^3$;
$f_y = xz + x^2z + 2xyz^3$;
$f_z = xy + x^2y + 3xy^2z^2$;
$f_{xz} = y + 2xy + 3y^2z^2$

12.6 Exercises

Find each of the following partial derivatives. (See Examples 1–3.)

1. Let $z = f(x, y) = 12x^2 - 8xy + 3y^2$. Find each of the following.

(a) $\dfrac{\partial z}{\partial x}$

(b) $\dfrac{\partial z}{\partial y}$

(c) $\left(\dfrac{\partial f}{\partial x}\right)(2, 3)$

(d) $f_y(1, -2)$

2. Let $z = g(x, y) = 5x + 9x^2y + y^2$. Find each of the following.

(a) $\dfrac{\partial g}{\partial x}$

(b) $\dfrac{\partial g}{\partial y}$

(c) $\left(\dfrac{\partial z}{\partial y}\right)(-3, 0)$

(d) $g_x(2, 1)$

Find f_x and f_y for the functions defined as follows. Then find $f_x(2, -1)$ and $f_y(-4, 3)$. Leave the answers in terms of e in Exercises 7–10 and 15–16. (See Examples 1–3.)

3. $f(x, y) = -2xy + 6y^3 + 2$

4. $f(x, y) = 4x^2y - 9y^2$

5. $f(x, y) = 3x^3y^2$

6. $f(x, y) = -2x^2y^4$

7. $f(x, y) = e^{x+y}$

8. $f(x, y) = 3e^{2x+y}$

9. $f(x, y) = -5e^{3x-4y}$

10. $f(x, y) = 8e^{7x-y}$

11. $f(x, y) = \dfrac{x^2 + y^3}{x^3 - y^2}$

12. $f(x, y) = \dfrac{3x^2y^3}{x^2 + y^2}$

13. $f(x, y) = \ln |1 + 3x^2y^3|$

14. $f(x, y) = \ln |2x^5 - xy^4|$

15. $f(x, y) = xe^{x^2y}$

16. $f(x, y) = y^2e^{(x+3y)}$

Find all second-order partial derivatives. (See Examples 5 and 6.)

17. $f(x, y) = 6x^3y - 9y^2 + 2x$

18. $g(x, y) = 5xy^4 + 8x^3 - 3y$

19. $R(x, y) = 4x^2 - 5xy^3 + 12y^2x^2$

20. $h(x, y) = 30y + 5x^2y + 12xy^2$

21. $r(x, y) = \dfrac{4x}{x + y}$

22. $k(x, y) = \dfrac{-5y}{x + 2y}$

23. $z = 4xe^y$

24. $z = -3ye^x$

25. $r = \ln |x + y|$

26. $k = \ln |5x - 7y|$

27. $z = x \ln |xy|$

28. $z = (y + 1) \ln |x^3y|$

Find f_x, f_y, f_z, and f_{yz} for the functions defined as follows. (See Example 7.)

29. $f(x, y, z) = x^2 + yz + z^4$

30. $f(x, y, z) = 3x^5 - x^2 + y^5$

31. $f(x, y, z) = \dfrac{6x - 5y}{4z + 5}$

32. $f(x, y, z) = \dfrac{2x^2 + xy}{yz - 2}$

33. $f(x, y, z) = \ln |x^2 - 5xz^2 + y^4|$

34. $f(x, y, z) = \ln |8xy + 5yz - x^3|$

Work the following exercises. (See Example 4.)

35. **Management** Suppose that the manufacturing cost of a precision electronic calculator is approximated by

$$M(x, y) = 40x^2 + 30y^2 - 10xy + 30,$$

where x is the cost of electronic chips and y is the cost of labor. Find the following.

(a) $M_y(4, 2)$ (b) $M_x(3, 6)$

(c) $\left(\dfrac{\partial M}{\partial x}\right)(2, 5)$ (d) $\left(\dfrac{\partial M}{\partial y}\right)(6, 7)$

36. **Management** The revenue from the sale of x units of a tranquilizer and y units of an antibiotic is given by

$$R(x, y) = 5x^2 + 9y^2 - 4xy.$$

(a) Suppose $x = 9$ and $y = 5$. What is the approximate effect on revenue if x is increased to 10, while y is fixed?

(b) Suppose $x = 9$ and $y = 5$. What is the approximate effect on revenue if y is increased to 6, while x is fixed?

*A **production function** $z = f(x, y)$ is a function which gives the quantity of an item produced, z, as a function of two other variables, x and y.*

37. **Management** The production function z for the United States was once estimated as

$$z = x^{.7}y^{.3},$$

where x stands for the amount of labor and y stands for the amount of capital. (This is an example of a **Cobb-Douglas production function.**) Find the marginal productivity of labor (find $\partial z/\partial x$) and of capital.

38. **Management** A similar production function for Canada is

$$z = x^{.4}y^{.6},$$

with x, y, and z as in Exercise 37. Find the marginal productivity of labor and of capital.

39. **Natural Science** The total number of matings per day between individuals of a certain species of grasshoppers is approximated by

$$M(x, y) = 2xy + 10xy^2 + 30y^2 + 20,$$

where x represents the temperature in °C and y represents the number of days since the last rain. Find the following.

(a) $\left(\dfrac{\partial M}{\partial x}\right)(20, 4)$ (b) $\left(\dfrac{\partial M}{\partial y}\right)(24, 10)$

(c) $M_x(17, 3)$ (d) $M_y(21, 8)$

40. **Natural Science** The oxygen consumption of a well-insulated mammal which is not sweating is approximated by

$$m = m(T, F, w) = \dfrac{2.5(T - F)}{w^{.67}} = 2.5(T - F)w^{-.67},$$

where T is the internal body temperature of the animal (in °C), F is the temperature of the outside of the animal's fur (in °C), and w is the animal's weight in kilograms.

(a) Find m_T.

(b) Suppose $T = 38°$, $F = 12°$, and $w = 30$ kg. Find $m_T(38, 12, 30)$.

(c) Find m_w.

(d) Suppose $T = 40°$, $F = 20°$, and $w = 40$ kg. Find $m_w(40, 20, 40)$.

(e) Find m_F.

(f) Suppose $T = 36°$, $F = 14°$, and $w = 25$ kg. Find $m_F(36, 14, 25)$.

41. Natural Science The surface area of a human, in square meters, is approximated by

$$A(W, H) = .202W^{.425}H^{.725},$$

where W is the weight of the person in kilograms and H is the height in meters.

(a) Find $\partial A/\partial W$.

(b) Suppose $W = 72$ and $H = 1.8$. Find
$$\left(\frac{\partial A}{\partial W}\right)(72, 1.8).$$

(c) Find $\partial A/\partial H$.

(d) Suppose $W = 70$ and $H = 1.6$. Find
$$\left(\frac{\partial A}{\partial H}\right)(70, 1.6).$$

42. Natural Science In one method of computing the quantity of blood pumped through the lungs in one minute, a researcher first finds each of the following (in milliliters).

$b =$ quantity of oxygen used by body in one minute

$a =$ quantity of oxygen per liter of blood that has just gone through the lungs

$v =$ quantity of oxygen per liter of blood that is about to enter the lungs

In one minute,

Amount of oxygen used

= amount of oxygen per liter

$\quad\times$ number of liters of blood pumped.

If C is the number of liters pumped through the blood in one minute, then

$$b = (a - v) \cdot C \quad \text{or} \quad C = \frac{b}{a - v}.$$

(a) Find C if $a = 160$, $b = 200$, and $v = 125$.

(b) Find C if $a = 180$, $b = 260$, and $v = 142$.

Find the following partial derivatives.

(c) $\partial C/\partial b$ \quad\quad\quad\quad **(d)** $\partial C/\partial v$

12.7 Maxima and Minima of Multivariate Functions

One of the most important applications of calculus is in finding maxima and minima for functions. Earlier, this idea was studied extensively for functions of a single independent variable, and now we find extrema for functions of two independent variables. In particular, we discuss an extension of the second derivative test which is used to identify maxima or minima. We begin with the definition of relative maxima and minima.

Relative Maxima and Minima Let (a, b) be the center of a circular region contained in the xy-plane. Then, for a function $z = f(x, y)$ defined for every (x, y) in the region,

1. $f(a, b)$ is a **relative maximum** if

$$f(a, b) \geq f(x, y)$$

for all points (x, y) in the circular region;

2. $f(a, b)$ is a **relative minimum** if

$$f(a, b) \leq f(x, y)$$

for all points (x, y) in the circular region.

As before, the word *extremum* is used for either a relative maximum or a relative minimum. Examples of a relative maximum and a relative minimum are given in Figures 12.38 and 12.39.

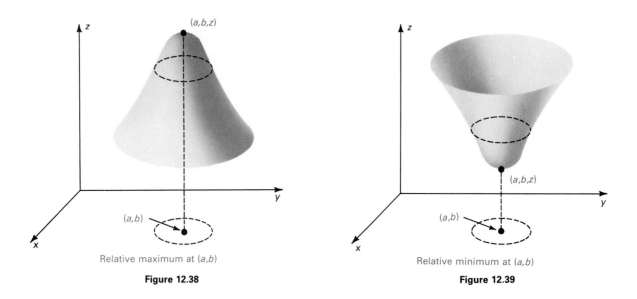

Relative maximum at (a,b)

Figure 12.38

Relative minimum at (a,b)

Figure 12.39

With functions of a single variable, we made a distinction between relative extrema and absolute extrema. It turns out that the methods for finding absolute extrema are quite involved for functions of two independent variables, so we shall discuss only relative extrema. However, in most practical applications the relative extrema coincide with the absolute extrema.

As suggested by Figure 12.40, at a relative maximum, the tangent line parallel to the x-axis has a slope of 0, as does the tangent line parallel to the y-axis. (Notice the similarity to functions of one variable.) That is, if the function $z = f(x, y)$ has a relative extremum at (a, b) then $f_x(a, b) = 0$ and $f_y(a, b) = 0$, as stated in the following theorem.

Location of Extrema Let a function $z = f(x, y)$ have a relative maximum or relative minimum at the point (a, b). Let $f_x(a, b)$ and $f_y(a, b)$ both exist. Then

$$f_x(a, b) = 0 \quad \text{and} \quad f_y(a, b) = 0.$$

Just as with functions of one variable, the fact that the slopes of the tangent lines are 0 is no guarantee that a relative extremum has been located. For example, Figure 12.41 shows the graph of $z = f(x, y) = x^2 - y^2$. Both $f_x(0, 0) = 0$ and $f_y(0, 0) = 0$, and yet $(0, 0)$ leads to neither a relative maximum nor a relative minimum for the function. The point $(0, 0, 0)$ on the graph of this function is called

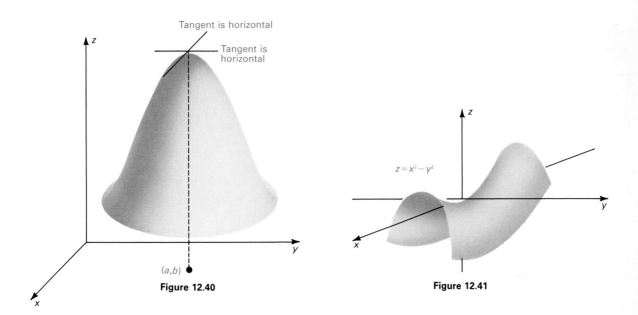

Figure 12.40

Figure 12.41

a **saddle point;** it is a minimum when approached from one direction but a maximum when approached from another direction. A saddle point is neither a maximum nor a minimum.

The theorem on location of extrema suggests a useful strategy for finding extrema. First, locate all points (a, b) where $f_x(a, b) = 0$ and $f_y(a, b) = 0$. Then test each of these points separately, using the test given after the next example. For a function $f(x, y)$, the points (a, b) such that $f_x(a, b) = 0$ and $f_y(a, b) = 0$ are called **critical points.**

Example 1 Find all critical points for

$$f(x, y) = 6x^2 + 6y^2 + 6xy + 36x - 5.$$

Critical points are all the points (a, b) such that $f_x(a, b) = 0$ and $f_y(a, b) = 0$. Here

$$f_x = 12x + 6y + 36 \quad \text{and} \quad f_y = 12y + 6x.$$

Set each of these two partial derivatives equal to 0.

$$12x + 6y + 36 = 0 \quad \text{and} \quad 12y + 6x = 0$$

These two equations make up a system of linear equations. To solve this system, rewrite $12y + 6x = 0$ as follows.

$$12y + 6x = 0$$
$$6x = -12y$$
$$x = -2y$$

Now substitute $-2y$ for x in the other equation.

$$12x + 6y + 36 = 0$$
$$12(-2y) + 6y + 36 = 0$$
$$-24y + 6y + 36 = 0$$
$$-18y + 36 = 0$$
$$-18y = -36$$
$$y = 2$$

The equation $x = -2y$ leads to $x = -2(2) = -4$. The solution of the system of equations is $(-4, 2)$. Since this is the only solution of the system, $(-4, 2)$ is the only critical point for the given function. By the theorem above, if the function of Example 1 has a relative extremum, it will occur at $(-4, 2)$. **1**

1 Find all critical points for the following.
(a) $f(x, y) = 4x^2 + 3xy + 2y^2 + 7x - 6y - 6$
(b) $f(x, y) = 4x^2 + 6xy + y^2 + 34x + 8y + 5$

Answer:
(a) $(-2, 3)$
(b) $(1, -7)$

The results of the next theorem can be used to decide whether $(-4, 2)$ in Example 1 leads to a relative maximum, a relative minimum, or neither.

Test for Relative Extrema For a function $z = f(x, y)$, let f_{xx}, f_{yy}, and f_{xy} all exist. Let (a, b) be a point for which

$$f_x(a, b) = 0 \quad \text{and} \quad f_y(a, b) = 0.$$

Define the number M by

$$M = f_{xx}(a, b) \cdot f_{yy}(a, b) - [f_{xy}(a, b)]^2.$$

Then

1. $f(a, b)$ is a **relative maximum** if $M > 0$ and $f_{xx}(a, b) < 0$.
2. $f(a, b)$ is a **relative minimum** if $M > 0$ and $f_{xx}(a, b) > 0$.
3. $f(a, b)$ is a **saddle point** (neither a maximum nor a minimum) if $M < 0$.
4. If $M = 0$, the test gives **no information.**

Example 2 The previous example showed that the only critical point for the function

$$f(x, y) = 6x^2 + 6y^2 + 6xy + 36x - 5$$

is $(-4, 2)$. Does $(-4, 2)$ lead to a relative maximum, a relative minimum, or neither?

Find out by using the test above. We already know that

$$f_x(-4, 2) = 0 \quad \text{and} \quad f_y(-4, 2) = 0.$$

Now find the various second partial derivatives used in finding M. From $f_x = 12x + 6y + 36$ and $f_y = 12y + 6x$,

$$f_{xx} = 12, \quad f_{yy} = 12, \quad \text{and} \quad f_{xy} = 6.$$

(If these second order partial derivatives had not all been constants, we would have had to evaluate them at the point $(-4, 2)$.) Now

$$M = f_{xx}(-4, 2) \cdot f_{yy}(-4, 2) - [f_{xy}(-4, 2)]^2 = 12 \cdot 12 - 6^2 = 108.$$

Since $M > 0$ and $f_{xx}(-4, 2) = 12 > 0$, part 2 of the theorem applies, showing that $f(x, y) = 6x^2 + 6y^2 + 6xy + 36x - 5$ has a relative minimum at $(-4, 2)$. This relative minimum is $f(-4, 2) = -77$. ◼

2 Find any relative maxima or minima for the functions defined in problem 1 at the side.

Answer:
(a) Relative minimum at $(-2, 3)$
(b) Neither a relative minimum nor a relative maximum at $(1, -7)$

Example 3 Find all points where the function

$$f(x, y) = 9xy - x^3 - y^3 - 6$$

has any relative maxima or relative minima.

First find any critical points. Here

$$f_x = 9y - 3x^2 \qquad \text{and} \qquad f_y = 9x - 3y^2.$$

Place each of these partial derivatives equal to 0.

$$
\begin{array}{ll}
f_x = 0 & f_y = 0 \\
9y - 3x^2 = 0 & 9x - 3y^2 = 0 \\
9y = 3x^2 & 9x = 3y^2 \\
3y = x^2 & 3x = y^2
\end{array}
$$

The final equation on the left, $3y = x^2$, can be rewritten as $y = x^2/3$. Substitute this into the final equation on the right.

$$3x = y^2 = \left(\frac{x^2}{3}\right)^2$$

$$3x = \frac{x^4}{9}$$

Rewrite this equation as

$$
\begin{array}{ll}
27x = x^4 & \\
x^4 - 27x = 0 & \\
x(x^3 - 27) = 0 & \text{Factor} \\
x = 0 \quad \text{or} \quad x^3 - 27 = 0 & \\
x^3 = 27 & \\
x = 3. &
\end{array}
$$

Use these values of x, along with the equation $3x = y^2$, to find y.

$$
\begin{array}{ll}
\text{If } x = 0, & \text{If } x = 3, \\
3x = y^2 & 3x = y^2 \\
3(0) = y^2 & 3(3) = y^2 \\
0 = y^2 & 9 = y^2 \\
0 = y & 3 = y \quad \text{or} \quad -3 = y.
\end{array}
$$

The points $(0, 0)$, $(3, 3)$ and $(3, -3)$ appear to be critical points; however, $(3, -3)$ does not make $f_x = 0$. The only possible relative extrema for $f(x, y) = 9xy - x^3 - y^3 - 6$ occur at the critical points $(0, 0)$ or $(3, 3)$. To identify any extrema, use the test. Here

$$f_{xx} = -6x, \qquad f_{yy} = -6y, \qquad \text{and} \qquad f_{xy} = 9.$$

Test each of the possible critical points.

For $(0, 0)$:

$f_{xx}(0, 0) = -6(0) = 0$

$f_{yy}(0, 0) = -6(0) = 0$

$f_{xy}(0, 0) = 9$

$M = 0 \cdot 0 - 9^2 = -81.$

Since $M < 0$, there is a saddle point at $(0, 0)$.

For $(3, 3)$:

$f_{xx}(3, 3) = -6(3) = -18$

$f_{yy}(3, 3) = -6(3) = -18$

$f_{xy}(3, 3) = 9$

$M = -18(-18) - 9^2 = 243.$

Here $M > 0$ and $f_{xx}(3, 3) = -18 < 0$; there is a relative maximum at $(3, 3)$. ■3

3 Find any relative extrema for $f(x, y) = \dfrac{2\sqrt{2}}{3}x^3 - xy + \dfrac{1}{3}y^3 - 10$.

Answer:

Relative minimum at $\left(\dfrac{1}{2}, \dfrac{\sqrt{2}}{2}\right)$

Example 4 A company is developing a new soft drink. The cost in dollars to produce a batch of the drink is approximated by

$$C(x, y) = 2200 + 27x^3 - 72xy + 8y^2,$$

where x is the number of kilograms of sugar per batch and y is the number of grams of flavoring per batch.

(a) Find the amounts of sugar and flavoring that result in minimum cost for a batch. Start with the following partial derivatives.

$$C_x = 81x^2 - 72y \qquad \text{and} \qquad C_y = -72x + 16y$$

Set each of these equal to 0 and solve for y.

$$81x^2 - 72y = 0 \qquad\qquad -72x + 16y = 0$$
$$-72y = -81x^2 \qquad\qquad 16y = 72x$$
$$y = \frac{9}{8}x^2 \qquad\qquad\qquad y = \frac{9}{2}x$$

Since $(9/8)x^2$ and $(9/2)x$ both are equal to y, they are equal to each other. Set $(9/8)x^2$ and $(9/2)x$ equal and solve the resulting equation for x.

$$\frac{9}{8}x^2 = \frac{9}{2}x$$
$$9x^2 = 36x$$
$$9x^2 - 36x = 0$$
$$9x(x - 4) = 0$$
$$9x = 0 \qquad \text{or} \qquad x - 4 = 0$$

The equation $9x = 0$ leads to $x = 0$ which is not a useful answer for our problem. Substitute $x = 4$ into $y = (9/2)x$ to find y.

$$y = \frac{9}{2}x = \frac{9}{2}(4) = 18$$

Now check to see if the critical point $(4, 18)$ leads to a relative minimum. For $(4, 18)$,

$$C_{xx} = 162x = 162(4) = 648, \qquad C_{yy} = 16, \qquad \text{and} \qquad C_{xy} = -72.$$

Also, $$M = (648)(16) - (-72)^2 = 5184.$$

Since $M > 0$ and $C_{xx}(4, 18) > 0$, the cost at $(4, 18)$ is a minimum.
(b) What is the minimum cost?

To find the minimum cost, go back to the cost function and evaluate $C(4, 18)$.

$$C(x, y) = 2200 + 27x^3 - 72xy + 8y^2$$
$$C(4, 18) = 2200 + 27(4)^3 - 72(4)(18) + 8(18)^2 = 1336$$

The minimum cost for a batch is \$1336.00. ■

12.7 Exercises

Find all points where the functions defined as follows have any relative extrema. Identify any saddle points. (See Examples 1–3.)

1. $f(x, y) = xy + x - y$

2. $f(x, y) = 4xy + 8x - 9y$

3. $f(x, y) = x^2 - 2xy + 2y^2 + x - 5$

4. $f(x, y) = x^2 + xy + y^2 - 6x - 3$

5. $f(x, y) = x^2 - xy + y^2 + 2x + 2y + 6$

6. $f(x, y) = x^2 + xy + y^2 + 3x - 3y$

7. $f(x, y) = x^2 + 3xy + 3y^2 - 6x + 3y$

8. $f(x, y) = 5xy - 7x^2 - y^2 + 3x - 6y - 4$

9. $f(x, y) = 4xy - 10x^2 - 4y^2 + 8x + 8y + 9$

10. $f(x, y) = x^2 + xy + 3x + 2y - 6$

11. $f(x, y) = x^2 + xy - 2x - 2y + 2$

12. $f(x, y) = x^2 + xy + y^2 - 3x - 5$

13. $f(x, y) = x^2 - y^2 - 2x + 4y - 7$

14. $f(x, y) = 4x + 2y - x^2 + xy - y^2 + 3$

15. $f(x, y) = 2x^3 + 3y^2 - 12xy + 4$

16. $f(x, y) = 5x^3 + 2y^2 - 60xy - 3$

17. $f(x, y) = x^2 + 4y^3 - 6xy - 1$

18. $f(x, y) = 3x^2 + 7y^3 - 42xy + 5$

19. $f(x, y) = e^{xy}$

20. $f(x, y) = x^2 + e^y$

21. $f(x, y) = 3xy - x^3 - y^3 + \dfrac{1}{8}$

22. $f(x, y) = \dfrac{3}{2}x - \dfrac{1}{2}x^3 - xy^2 + \dfrac{1}{16}$

23. $f(x, y) = x^4 - 2x^2 + y^2 + \dfrac{17}{16}$

24. $f(x, y) = 2y^3 - 3x^4 - 6x^2y + \dfrac{1}{16}$

25. $f(x, y) = x^4 - y^4 - 2x^2 + 2y^2 + \dfrac{1}{16}$

26. $f(x, y) = -x^4 + 4xy - 2y^2 + \dfrac{1}{16}$

Work the following exercises. (See Example 4.)

27. **Management** Suppose that the profit of a certain firm is approximated by

$$P(x, y) = 1000 + 24x - x^2 + 80y - y^2,$$

where x is the cost of a unit of labor and y is the cost of a unit of goods. Find values of x and y that maximize profit. Find the maximum profit.

28. The labor cost for manufacturing a precision camera can be approximated by

$$L(x, y) = \frac{3}{2}x^2 + y^2 - 2x - 2y - 2xy + 68,$$

where x is the number of hours required by a skilled craftsperson and y is the number of hours required by a semiskilled person. Find values of x and y that minimize the labor charge. Find the minimum labor charge.

29. The number of roosters that can be fed from x pounds of Super-Hen chicken feed and y pounds of Super-Rooster feed is given by

$$R(x, y) = 800 - 2x^3 + 12xy - y^2.$$

Find the number of pounds of each kind of feed that produces the maximum number of roosters.

30. The total profit from one acre of a certain crop depends on the amount spent on fertilizer, x, and hybrid seed, y, according to the model

$$P(x, y) = -x^2 + 3xy + 160x - 5y^2 + 200y + 2{,}600{,}000.$$

Find values of x and y that lead to maximum profit. Find the maximum profit.

31. The total cost to produce x units of electrical tape and y units of packing tape is given by

$$C(x, y) = 2x^2 + 3y^2 - 2xy + 2x - 126y + 3800.$$

Find the number of units of each kind of tape that should be produced so that the total cost is a minimum. Find the minimum total cost.

32. The total revenue from the sale of x spas and y solar heaters is approximated by

$$R(x, y) = 12 + 74x + 85y - 3x^2 - 5y^2 - 5xy.$$

Find the number of each that should be sold to produce maximum revenue. Find the maximum revenue.

Chapter 12   Review Exercises

Find the locations and the values of all relative maxima and relative minima for each of the following.

1. $f(x) = -x^2 + 4x - 8$

2. $f(x) = x^2 - 6x + 4$

3. $f(x) = 2x^2 - 8x + 1$

4. $f(x) = -3x^2 + 2x - 5$

5. $f(x) = 2x^3 + 3x^2 - 36x + 20$

6. $f(x) = 2x^3 + 3x^2 - 12x + 5$

7. $f(x) = -2x^3 - \frac{1}{2}x^2 + x - 3$

8. $f(x) = -\frac{4}{3}x^3 + x^2 + 30x - 7$

9. $f(x) = x^4 - \frac{4}{3}x^3 - 4x^2 + 1$

10. $f(x) = x^4 + \frac{8}{3}x^3 - 6x^2 + 1$

11. $f(x) = \dfrac{x - 1}{2x + 1}$

12. $f(x) = \dfrac{2x - 5}{x + 3}$

13. $f(x) = x \cdot e^x$

14. $f(x) = 3x \cdot e^{-x}$

15. $f(x) = \dfrac{e^x}{x - 1}$

16. $f(x) = (5x + 3)e^{-2x}$

Find the locations of all absolute maxima and absolute minima for the following functions on the given intervals.

17. $f(x) = -x^2 + 5x + 1$; $[1, 4]$

18. $f(x) = 4x^2 - 8x - 3$; $[-1, 2]$

19. $f(x) = x^3 + 2x^2 - 15x + 3$; $[-4, 2]$

20. $f(x) = -2x^3 - x^2 + 4x - 1$; $[-3, 1]$

Find the second derivatives of the following; then find $f''(1)$ and $f''(-3)$.

21. $f(x) = 3x^4 - 5x^2 - 11x$

22. $f(x) = \dfrac{5x - 1}{2x + 3}$

23. $f(x) = \dfrac{4 - 3x}{x + 1}$

24. $f(x) = -3e^{4x}$

25. $f(x) = e^{-x^2}$

26. $f(x) = \ln|6x + 1|$

Find the largest open intervals where the functions defined as follows are increasing or decreasing.

27. $f(x) = x^2 - 5x + 3$

28. $f(x) = -2x^2 - 3x + 4$

29. $f(x) = -x^3 - 5x^2 + 8x - 6$

30. $f(x) = 4x^3 + 3x^2 - 18x + 1$

31. $f(x) = \dfrac{6}{x - 4}$

32. $f(x) = \dfrac{5}{2x + 1}$

Find the largest open intervals where the following are concave upward or concave downward. Find the location of any points of inflection.

33. $f(x) = -4x^3 - x^2 + 4x + 5$

34. $f(x) = x^3 + \dfrac{5}{2}x^2 - 2x - 3$

35. $f(x) = x^4 + 2x^2$ **36.** $f(x) = 6x^3 - x^4$

37. $f(x) = \dfrac{x^2 - 4}{x}$ **38.** $f(x) = x + \dfrac{8}{x}$

39. $f(x) = \dfrac{2x}{3 - x}$ **40.** $f(x) = \dfrac{-4x}{1 + 2x}$

Work the following exercises.

41. Find two numbers whose sum is 25 and whose product is a maximum.

42. Find x and y such that $x = 2 + y$, and xy^2 is minimized.

43. Suppose the profit from a product is $P(x) = 16 + 4x - x^2$, where x is the price in dollars.
    (a) At what price will the maximum profit occur?
    (b) What is the maximum profit?

44. The total profit in dollars from the sale of x hundred boxes of candy is given by
$$P(x) = -x^3 + 10x^2 - 12x + 106.$$
    (a) Find the number of boxes of candy that should be sold in order to produce maximum profit.
    (b) Find the maximum profit.

45. The city park department is planning an enclosed play area in a new park. One side of the area will be against an existing building, with no fence needed there. Find the dimensions of the rectangular field of maximum area that can be enclosed with 900 meters of fence.

46. A company plans to package its product in a cylinder which is open at one end. The cylinder is to have a volume of 27π cubic inches. What radius should the circular bottom of the cylinder have to minimize the cost of the material? (Hint: The volume of a circular cylinder is $\pi r^2 h$ where r is the radius of the circular base and h is the height; the surface area is $2\pi rh + 2\pi r^2$.)

Find each of the following, given $f(x, y) = 2x^2 - x^3y + 46$.

47. $f(2, 3)$ **48.** $f(-1, 1)$

For each of the following, find f_x, f_y, f_{xx}, f_{yy}, and f_{xy}.

49. $f(x, y) = 8x^2 + 2xy + 3y - 17$

50. $f(x, y) = x^2y - 3x + 2xy + 5y^2 + 27$

51. $f(x, y) = x^2y^3 + 8xy$

52. $f(x, y) = \dfrac{y}{1 + x}$

Find all points where the functions defined as follows have maxima, minima, or saddle points.

53. $f(x, y) = x^2 + 5xy - 10x + 3y^2 - 12y$

54. $z = x^3 - 8y^2 + 6xy + 4$

55. $f(x, y) = \dfrac{1}{2}x^2 + \dfrac{1}{2}y^2 + 2xy - 5x - 7y + 10$

56. $f(x, y) = 6x^2 + 6y^2 + 6xy + 36x - 5$

57. $f(x, y) = x^2 + 7xy + 2y^2 - 22x - 36y + 40$

58. $z = 3x^3 - 12xy + 4y^2 + 18$

Work the following exercises.

59. The total cost to manufacture x units of doggy dishes and y units of kitty plates is given by
$$C(x, y) = x^2 + 5y^2 + 4xy - 70x - 164y + 1800.$$
    (a) Find values of x and y that lead to a minimum cost.
    (b) What is the minimum cost?

60. Find the dimensions of the rectangular field of maximum area that can be enclosed with 400 meters of fence.

Case 15

The Effect of Oil Price on the Optimal Speed of Ships*

With the major increase in fuel prices during the past decade, choosing an optimum speed for a large ocean-going ship has become a major concern of ship operators. Sailing slowly uses less fuel, but causes increased costs for salaries and interest on the value of the cargo.

The speed of sailing has a great effect on the amount of fuel used, the major expense in operating a ship. Past empirical studies have shown that the amount of fuel used by a ship is approximately proportional to the cube of the speed. This means that a 20% change in cruising speed produces about a 50% change in fuel consumption. As an example, a medium sized ship that burns 40 tons of fuel per day at a certain speed could save 50% of this amount, or 20 tons, with a 20% reduction in speed. At a cost of $250 per ton, this 20% speed reduction would save $5000 per day in fuel costs.

To find the optimal sailing speed for one leg of a journey (a one-way trip with a given cargo between two points), let

R = income from the leg
P = profit from the leg
D_s = number of days at sea in the leg
D_p = number of days in port in the leg
L = distance of the leg in nautical miles
V = actual cruising speed
V_0 = normal cruising speed
C = daily cost of the vessel
(excluding fuel for main engines)
F = actual daily fuel consumption in tons per day
F_0 = daily fuel consumption at normal cruising speed
F_c = cost of fuel in dollars per ton
V_m = minimum cruising speed.

The duration of the leg, D, is

$$D = D_s + D_p, \tag{1}$$

where

$$D_s = \frac{L}{24\,V}. \tag{2}$$

The profit from the leg is

$$P = R - (DC + FF_cD_s). \tag{3}$$

As we mentioned above, the amount of fuel used is proportional to the cube of the speed, or

$$F = \left(\frac{V}{V_0}\right)^3 \cdot F_0. \tag{4}$$

Our goal is to maximize the daily profit, Z, given by

$$Z = \frac{P}{D}.$$

Substituting equations (1), (2), and (4) into the equation for profit, (3), and dividing by D as given by equations (1) and (2) yields

$$Z = \frac{R - D_pC - \dfrac{LC}{24\,V} - \left(\dfrac{V}{V_0}\right)^3 F_0F_c\left(\dfrac{L}{24\,V}\right)}{D_p + \dfrac{L}{24\,V}} \tag{5}$$

It is not possible for a ship to sail at just any speed between 0 and its normal cruising speed, V_0. The design of the ship usually imposes some minimum cruising speed, V_m. This restriction forces V to be in the interval $[V_m, V_0]$.

To find the optimum value of V, set dZ/dV, from equation (5), equal to 0. This gives the cubic equation

$$V^3 + V^2\frac{L}{16D_p} - \frac{RV_0^3}{2F_0F_cD_p} = 0. \tag{6}$$

Any real solutions of this equation are potential critical values. As mentioned above, these critical values must lie in the interval $[V_m, V_0]$ to be useful. To maximize Z, therefore, we must use our work with absolute extrema and check all critical values in $[V_m, V_0]$, as well as both endpoints, V_m and V_0.

---

*Based on "The Effect of Oil Price on the Optimal Speed of Ships" by David Ronen from *Journal of the Operations Research Society*, Vol. 33, 1982. Copyright © 1982 Operational Research Society Ltd. Reprinted by permission.

Exercises

1. Find the optimum sailing speed for a voyage of 1536 miles, with 8 days in port, a normal cruising speed of 20 knots, a minimum speed of 10 knots, income from the leg of $86,400, fuel costs of $250 per ton, 50 tons of fuel consumed daily at normal cruising speed, and a daily cost for the vessel of $1000. (Hint: Try $V = 12$.)

2. Why does the variable C not appear in equation (6) above?

Case 16

A Total Cost Model for a Training Program*

In this application, we set up a mathematical model for determining the total costs in setting up a training program. Then we use calculus to find the time between training programs that produces the minimum total cost. The model assumes that the demand for trainees is constant and that the fixed cost of training a batch of trainees is known. Also, it is assumed that people who are trained, but for whom no job is readily available, will be paid a fixed amount per month while waiting for a job to open up.

The model uses the following variables.

D = demand for trainees per month

N = number of trainees per batch

C_1 = fixed cost of training a batch of trainees

C_2 = variable cost of training per trainee per month

C_3 = salary paid monthly to a trainee who has not yet been given a job after training

m = time interval in months between successive batches of trainees

t = length of training program in months

$Z(m)$ = total monthly cost of program

The total cost of training a batch of trainees is given by $C_1 + NtC_2$. However, $N = mD$, so that the total cost per batch is $C_1 + mDtC_2$.

After training, personnel are given jobs at the rate of D per month. Thus, $N - D$ of the trainees will not get a job the first month, $N - 2D$ will not get a job the second month, and so on. The $N - D$ trainees who do not get a job the first month produce total costs of $(N - D)C_3$, those not getting jobs during the second month produce costs of

$(N - 2D)C_3$, and so on. Since $N = mD$, the costs during the first month can be written as

$$(N - D)C_3 = (mD - D)C_3 = (m - 1)DC_3,$$

while the costs during the second month are $(m - 2)DC_3$, and so on. The total cost for keeping the trainees without a job is thus

$$(m - 1)DC_3 + (m - 2)DC_3 + (m - 3)DC_3 + \cdots + 2DC_3 + DC_3,$$

which can be factored to give

$$DC_3[(m - 1) + (m - 2) + (m - 3) + \cdots + 2 + 1].$$

The expression in brackets is the sum of the terms of an arithmetic sequence. Using formulas for arithmetic sequences, the expression in brackets can be shown to equal $m(m - 1)/2$, so that we have

$$DC_3\left[\frac{m(m - 1)}{2}\right] \tag{1}$$

as the total cost for keeping jobless trainees.

The total cost per batch is the sum of the training cost per batch, $C_1 + mDtC_2$, and the cost of keeping trainees without a proper job, given by (1). Since we assume that a batch of trainees is trained every m months, the total cost per month, $Z(m)$, is given by

$$Z(m) = \frac{C_1 + mDtC_2}{m} + \frac{DC_3\left[\dfrac{m(m - 1)}{2}\right]}{m}$$

$$= \frac{C_1}{m} + DtC_2 + DC_3\left(\frac{m - 1}{2}\right).$$

*Based on "A Total Cost Model for a Training Program" by P. L. Goyal and S. K. Goyal, Department of Mathematics and Computer Science, The Polytechnic of Wales, Treforest, Pontypridd. Used with permission.

Exercises

1. Find $Z'(m)$.

2. Solve the equation $Z'(m) = 0$.
   As a practical matter, it is usually required that m be a whole number. If m does not come out to be a whole number in Exercise 2, then m^+ and m^-, the two whole numbers closest to m, must be chosen. Calculate both $Z(m^+)$ and $Z(m^-)$; the smaller of the two provides the optimum value of Z.

3. Suppose a company finds that its demand for trainees is 3 per month, that a training program requires 12 months, that the fixed cost of training a batch of trainees is $15,000, that the variable cost per trainee per month is $100, and that trainees are paid $900 per month after training but before going to work. Use your result from Exercise 2 and find m.

4. Since m is not a whole number, find m^+ and m^-.

5. Calculate $Z(m^+)$ and $Z(m^-)$.

6. What is the optimum time interval between successive batches of trainees? How many trainees should be in a batch?

13 Integral Calculus

Calculus is divided into two broad areas: *differential calculus,* discussed in the last two chapters, and *integral calculus,* considered in this chapter. Like the derivative of a function, the definite integral of a function has many diverse applications. Geometrically, the derivative is related to the slope of the tangent line to a curve, while the definite integral is related to the area under a curve. Differential and integral calculus are related by the fundamental theorem of calculus, presented later in this chapter.

13.1 Antiderivatives

In Chapter 11 we found the derivative of a large variety of functions. Now we look at the reverse process. If some function f, such as $f(x) = 2x$, is the derivative of another function F, how can $F(x)$ be found? This reverse process is called finding *antiderivatives.*

If $F'(x) = f(x)$, then $F(x)$ is an **antiderivative** of $f(x)$.

Example 1 **(a)** If $F(x) = 10x$, then $F'(x) = 10$, so $F(x) = 10x$ is an antiderivative of $f(x) = 10$.
(b) For $F(x) = x^2$, $F'(x) = 2x$, making $F(x) = x^2$ an antiderivative of $f(x) = 2x$. ■

Example 2 Find an antiderivative of $f(x) = 5x^4$.
   To find a function $F(x)$ whose derivative is $5x^4$, work backwards. Recall that the derivative of x^n is nx^{n-1}. If

$$nx^{n-1} \quad \text{is} \quad 5x^4,$$

then $n - 1 = 4$ and $n = 5$, so x^5 is an antiderivative of $5x^4$.

   The function $F(x) = x^2$ is not the only function whose derivative is $f(x) = 2x$; for example, both $G(x) = x^2 + 2$ and $H(x) = x^2 - 7$ have $f(x) = 2x$ as a derivative. In fact, for any real number C, the function $f(x) = x^2 + C$ has $f(x) = 2x$ as its derivative. This means that there is a *family* or *class* of functions each of which is an antiderivative of $f(x)$. However, as stated below, if two functions $F(x)$ and $G(x)$ are each antiderivatives of $f(x)$, then $F(x)$ and $G(x)$ can differ only by a constant.

If $F(x)$ and $G(x)$ are both antiderivatives of $f(x)$, then there is a constant C such that

$$F(x) - G(x) = C.$$

(Two antiderivatives of a function can differ only by a constant.)

---

1 Find an antiderivative for each of the following.
(a) $3x^2$
(b) 5
(c) $8x^7$
(d) $\dfrac{1}{2}x^3$

Answer:
Only one possible antiderivative is given for each.
(a) x^3
(b) $5x$
(c) x^8
(d) $\dfrac{1}{8}x^4$

618

For example,

$$F(x) = x^2, \qquad G(x) = x^2 + 2, \qquad \text{and} \qquad H(x) = x^2 - 4$$

are all antiderivatives of $f(x) = 2x$, and any two of them differ only by a constant. The derivative of a function gives the slope of the tangent line at any x-value. The fact that these three functions have the same derivative, $f(x) = 2x$, means that their slopes at any particular value of x are the same, as shown in Figure 13.1.

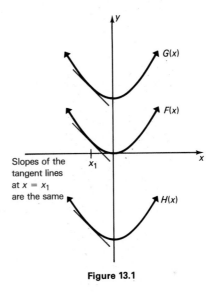

Figure 13.1

The family of all antiderivatives of the function f is indicated by

$$\int f(x) \, dx.$$

The symbol $\int$ is the **integral sign,** $f(x)$ is the **integrand,** and $\int f(x) \, dx$ is called an **indefinite integral,** the most general antiderivative of f. The process of finding $\int f(x) \, dx$ is called **integration.** The dx in the indefinite integral indicates that $\int f(x) \, dx$ is the "integral of $f(x)$ *with respect to x*" just as the symbol dy/dx denotes the "derivative of y with respect to x."

Indefinite Integral If $F'(x) = f(x)$, then

$$\int f(x) \, dx = F(x) + C,$$

for any real number C.

For example, using this notation,

$$\int (2x) \, dx = x^2 + C.$$

Working backwards, as above, to find an antiderivative is not very satisfactory. We need some rules for finding antiderivatives. Since finding an indefinite integral is the inverse of finding a derivative, each formula for derivatives leads to a rule for indefinite integrals.

As mentioned above, the derivative of x^n is found by multiplying x by n and reducing the exponent on x by 1. To find an indefinite integral—that is, to undo what was done—*increase* the exponent by 1 and *divide* by the new exponent, $n + 1$.

Power Rule for Indefinite Integrals   For any real number $n \neq -1$,

$$\int x^n \, dx = \frac{1}{n + 1} x^{n+1} + C.$$

This result can be verified by differentiating the expression on the right above:

$$\frac{d}{dx}\left(\frac{1}{n + 1} x^{n+1} + C\right) = \frac{n + 1}{n + 1} x^{(n+1)-1} + 0 = x^n.$$

(If $n = -1$, the expression in the denominator is 0, and the above rule cannot be used. We will see later how to find an antiderivative in this case.)

Example 3   Find each indefinite integral.
(a) $\int x^3 \, dx$

Use the power rule with $n = 3$.

$$\int x^3 \, dx = \frac{1}{3 + 1} x^{3+1} + C = \frac{1}{4} x^4 + C$$

(b) $\int \dfrac{1}{t^2} \, dt$

First, write $1/t^2$ as t^{-2}. Then

$$\int \frac{1}{t^2} \, dt = \int t^{-2} \, dt = \frac{t^{-1}}{-1} + C = \frac{-1}{t} + C.$$

(c) $\int \sqrt{u} \, du$

Since $\sqrt{u} = u^{1/2}$,

$$\int \sqrt{u} \, du = \int u^{1/2} \, du = \frac{1}{1/2 + 1} u^{3/2} + C = \frac{2}{3} u^{3/2} + C.$$

check this by differentiating $(2/3)u^{3/2} + C$; the derivative is $u^{1/2}$, the original function.

(d) $\int dx$

Writing dx as $1 \cdot dx$, and using the fact that $x^0 = 1$ for any nonzero number x,

$$\int dx = \int 1 \, dx = \int x^0 \, dx = \frac{1}{1}x^1 + C = x + C. \quad \boxed{2}$$

2 Find each of the following.
(a) $\int x^5 \, dx$
(b) $\int \sqrt[3]{x} \, dx$
(c) $\int 5 \, dx$

Answer:

(a) $\frac{1}{6}x^6 + C$

(b) $\frac{3}{4}x^{4/3} + C$

(c) $5x + C$

As shown in Chapter 11, the derivative of the product of a constant and a function is the product of the constant and the derivative of the function. A similar rule applies to indefinite integrals. Also, since derivatives of sums or differences are found term by term, indefinite integrals can also be found term by term. The next box states these rules for indefinite integrals.

Properties of Indefinite Integrals: Constant Multiple; Sum or Difference If all indicated integrals exist,

$$\int k \cdot f(x) \, dx = k \int f(x) \, dx, \text{ for any real number } k;$$

$$\int [f(x) \pm g(x)] \, dx = \int f(x) \, dx \pm \int g(x) \, dx.$$

Example 4 Find each of the following.
(a) $\int 2x^3 \, dx$

By the first result in the box above,

$$\int 2x^3 \, dx = 2 \int x^3 \, dx = 2\left(\frac{1}{4}x^4\right) + C = \frac{1}{2}x^4 + C.$$

Since C represents any real number, it is not necessary to multiply it by 2 in the next-to-last step.

(b) $\int \frac{12}{z^5} \, dz$

Use negative exponents.

$$\int \frac{12}{z^5} \, dz = \int 12z^{-5} \, dz$$

$$= 12 \int z^{-5} \, dz$$

$$= 12\left(\frac{z^{-4}}{-4}\right) + C$$

$$= -3z^{-4} + C$$

$$= \frac{-3}{z^4} + C.$$

(c) $\int(3z^2 - 4z + 5)\,dz$

By extending the sum or difference property given above to more than two terms,

$$\int (3z^2 - 4z + 5)\,dz = 3\int z^2\,dz - 4\int z\,dz + 5\int dz$$

$$= 3\left(\frac{1}{3}z^3\right) - 4\left(\frac{1}{2}z^2\right) + 5z + C$$

$$= z^3 - 2z^2 + 5z + C.$$

Only one constant C is needed in the answer: the three constants from integrating term by term are combined. **3**

Integration can always be checked by taking the derivative of the result. For instance, in Example 4(c) check that $z^3 - 2z^2 + 5z + C$ is the required indefinite integral by taking its derivative:

$$\frac{d}{dz}(z^3 - 2z^2 + 5z + C) = 3z^2 - 4z + 5.$$

The result is the original function to be integrated, so the work checks.

As shown in Chapter 11, the derivative of $f(x) = e^x$ is $f'(x) = e^x$. Also, the derivative of $f(x) = e^{kx}$ is $f'(x) = k \cdot e^{kx}$. These results lead to the following formulas for indefinite integrals of exponential functions.

Indefinite Integrals of Exponential Functions If k is a real number, $k \neq 0$, then

$$\int e^x\,dx = e^x + C;$$

$$\int e^{kx}\,dx = \frac{1}{k} \cdot e^{kx} + C.$$

---

3 Find each of the following.

(a) $\int(-6x^4)\,dx$

(b) $\int 9x^{2/3}\,dx$

(c) $\int \dfrac{8}{x^3}\,dx$

(d) $\int(5x^4 - 3x^2 + 6)\,dx$

(e) $\int(-2x^3 + 6x^2 - 3)\,dx$

(f) $\int\left(3\sqrt{x} + \dfrac{2}{x^2}\right)dx$

Answer:

(a) $-\dfrac{6}{5}x^5 + C$

(b) $\dfrac{27}{5}x^{5/3} + C$

(c) $-4x^{-2} + C$ or $-\dfrac{4}{x^2} + C$

(d) $x^5 - x^3 + 6x + C$

(e) $-\dfrac{1}{2}x^4 + 2x^3 - 3x + C$

(f) $2x^{3/2} - \dfrac{2}{x} + C$

4 Find each of the following.

(a) $\int(-4e^x)\,dx$

(b) $\int e^{3x}\,dx$

(c) $\int(e^{2x} - 2e^x)\,dx$

(d) $\int(-11e^{-x})\,dx$

Answer:

(a) $-4e^x + C$

(b) $\dfrac{1}{3}e^{3x} + C$

(c) $\dfrac{1}{2}e^{2x} - 2e^x + C$

(d) $11e^{-x} + C$

Example 5 Find each of the following.

(a) $\displaystyle\int 9e^x\,dx = 9\int e^x\,dx = 9e^x + C$

(b) $\displaystyle\int e^{9t}\,dt = \frac{1}{9}e^{9t} + C$

(c) $\displaystyle\int 3e^{(5/4)u}\,du = 3\left(\frac{1}{5/4}e^{(5/4)u}\right) + C$

$$= 3\left(\frac{4}{5}\right)e^{(5/4)u} + C$$

$$= \frac{12}{5}e^{(5/4)u} + C \quad \boxed{4}$$

The restriction $n \neq -1$ was necessary in the formula for $\int x^n \, dx$ since $n = -1$ made the denominator of $1/(n + 1)$ equal to 0. Find $\int x^n \, dx$ when $n = -1$, that is, find $\int x^{-1} \, dx$, by recalling the differentiation formula for the logarithmic function: the derivative of $f(x) = \ln |x|$, where $x \neq 0$, is $f'(x) = 1/x = x^{-1}$. This formula for the derivative of $f(x) = \ln |x|$ gives a formula for $\int x^{-1} \, dx$.

Indefinite Integral of x^{-1}

$$\int x^{-1} \, dx = \int \frac{1}{x} \, dx = \ln |x| + C, \qquad \text{where } x \neq 0.$$

The domain of the logarithm function is the set of positive real numbers. However, $y = x^{-1} = 1/x$ has as domain the set of all nonzero real numbers, so absolute value bars must be used in the antiderivative.

Example 6 Find each of the following.

(a) $\displaystyle\int \frac{4}{x} \, dx = 4 \int \frac{1}{x} \, dx = 4 \cdot \ln |x| + C$

(b) $\displaystyle\int \left(-\frac{5}{x} + e^{-2x}\right) dx = -5 \cdot \ln |x| - \frac{1}{2} e^{-2x} + C$ **5**

5 Find each of the following.
(a) $\int (-9/x) \, dx$
(b) $\int 12x^{-1} \, dx$
(c) $\int (8e^{4x} - 3x^{-1}) \, dx$

Answer:
(a) $-9 \cdot \ln |x| + C$
(b) $12 \cdot \ln |x| + C$
(c) $2e^{4x} - 3 \cdot \ln |x| + C$

Integrals have many applications. Often it is possible to find a function that expresses the rate of change of a quantity, rather than the quantity itself. Since the rate of change of a function is given by its derivative, if the derivative of a function is known, then the function itself can be found by integrating. For example, a marginal cost function is the derivative of a cost function, so a cost function must be an antiderivative of a marginal cost function. The next example illustrates this.

Example 7 Suppose a company has found that the marginal cost at a level of production of x thousand books is given by

$$C'(x) = \frac{50}{\sqrt{x}},$$

and the fixed cost (the cost to produce 0 books) is \$25,000. Find the cost function $C(x)$.

Write $50/\sqrt{x}$ as $50/x^{1/2}$ or $50x^{-1/2}$. With this,

$$\int \frac{50}{\sqrt{x}} \, dx = \int 50x^{-1/2} \, dx = 50(2x^{1/2}) + k = 100x^{1/2} + k.$$

6 The marginal cost at a level of production of x items is
$$C'(x) = 2x^3 + 6x - 5.$$
The fixed cost is \$800. Find the cost function $C(x)$.

Answer:
$C(x) = \frac{1}{2}x^4 + 3x^2 - 5x$
$\quad + 800$

Use k instead of C to avoid confusion with the C of the cost function. Find the value of k using the fact that $C(x)$ is 25,000 when $x = 0$.

$$C(x) = 100x^{1/2} + k$$
$$25,000 = 100 \cdot 0 + k$$
$$k = 25,000$$

With this result, the cost function is $C(x) = 100x^{1/2} + 25,000$. **6**

13.1 Exercises

Find each of the following. (See Examples 3–6.)

1. $\displaystyle\int 4x\, dx$

2. $\displaystyle\int 8x\, dx$

3. $\displaystyle\int 5t^2\, dt$

4. $\displaystyle\int 6x^3\, dx$

5. $\displaystyle\int 6\, dk$

6. $\displaystyle\int 2\, dy$

7. $\displaystyle\int (2z + 3)\, dz$

8. $\displaystyle\int (3x - 5)\, dx$

9. $\displaystyle\int (x^2 + 6x)\, dx$

10. $\displaystyle\int (t^2 - 2t)\, dt$

11. $\displaystyle\int (t^2 - 4t + 5)\, dt$

12. $\displaystyle\int (5x^2 - 6x + 3)\, dx$

13. $\displaystyle\int (4z^3 + 3z^2 + 2z - 6)\, dz$

14. $\displaystyle\int (12y^3 + 6y^2 - 8y + 5)\, dy$

15. $\displaystyle\int 5\sqrt{z}\, dz$

16. $\displaystyle\int t^{1/4}\, dt$

17. $\displaystyle\int (u^{1/2} + u^{3/2})\, du$

18. $\displaystyle\int (4\sqrt{v} - 3v^{3/2})\, dv$

19. $\displaystyle\int (15x\sqrt{x} + 2\sqrt{x})\, dx$

20. $\displaystyle\int (x^{1/2} - x^{-1/2})\, dx$

21. $\displaystyle\int (10u^{3/2} - 14u^{5/2})\, du$

22. $\displaystyle\int (56t^{5/2} + 18t^{7/2})\, dt$

23. $\displaystyle\int \left(\frac{1}{z^2}\right) dz$

24. $\displaystyle\int \left(\frac{4}{x^3}\right) dx$

25. $\displaystyle\int \left(\frac{1}{y^3} - \frac{1}{\sqrt{y}}\right) dy$

26. $\displaystyle\int \left(\sqrt{u} + \frac{1}{u^2}\right) du$

27. $\displaystyle\int (-9t^{-2} - 2t^{-1})\, dt$

28. $\displaystyle\int (8x^{-3} + 4x^{-1})\, dx$

29. $\displaystyle\int e^{2t}\, dt$

30. $\displaystyle\int e^{-3y}\, dy$

31. $\displaystyle\int 3e^{-.2x}$

32. $\displaystyle\int -4e^{.2v}\, dv$

33. $\displaystyle\int \left(\frac{3}{x} + 4e^{-.5x}\right) dx$

34. $\displaystyle\int \left(\frac{9}{x} - 3e^{-.4x}\right) dx$

35. $\displaystyle\int \frac{1 + 2t^3}{t}\, dt$

36. $\displaystyle\int \frac{2y^{1/2} - 3y^2}{y}\, dy$

37. $\displaystyle\int (e^{2u} + 4u)\, du$

38. $\displaystyle\int (v^2 - e^{3v})\, dv$

39. $\displaystyle\int (x + 1)^2\, dx$

40. $\displaystyle\int (2y - 1)^2\, dy$

41. $\displaystyle\int \frac{\sqrt{x} + 1}{\sqrt[3]{x}}\, dx$

42. $\displaystyle\int \frac{1 - 2\sqrt[3]{z}}{\sqrt[3]{z}}\, dz$

Find the cost function for each of the following marginal cost functions. (See Example 7.)

43. $C'(x) = 4x - 5$,   fixed cost is \$8

44. $C'(x) = 2x + 3x^2$,   fixed cost is \$15

45. $C'(x) = 0.2x^2 + 5x$,   fixed cost is \$10

46. $C'(x) = 0.8x^2 - x$,   fixed cost is \$5

47. $C'(x) = x^{1/2}$,   16 units cost \$40

48. $C'(x) = x^{2/3} + 2$,   8 units cost \$58

49. $C'(x) = x^2 - 2x + 3$,   3 units cost \$15

50. $C'(x) = x + \frac{1}{x^2}$,   2 units cost \$5.50

51. $C'(x) = \frac{1}{x} + 2x$,   7 units cost \$58.40

52. $C'(x) = 5x - \frac{1}{x}$,   10 units cost \$94.20

Work the following exercises.

53. **Management** The marginal profit of a small fast food stand is
$$P'(x) = -2x + 20,$$
where x is the sales volume in thousands of hamburgers. The "profit" is $-\$50$ when no hamburgers are sold. Find the profit function.

54. **Management** Suppose the marginal profit from the sale of x hundred items is
$$P'(x) = 4 - 6x + 3x^2,$$

and the profit on 0 items is $-\$40$. Find the profit function.

55. **Management** A mine begins producing at time $t = 0$. After t years, the mine is producing at the rate of

$$P'(t) = 10t - \frac{15}{\sqrt{t}}$$

tons per year. Find the total output of the mine after t years. (Assume no output when $t = 0$.)

56. The slope of the tangent line to a curve is given by

$$f'(x) = 6x^2 - 4x + 3.$$

If the point $(0, 1)$ is on the curve, find the equation of the curve.

57. Find the equation of the curve whose tangent line has a slope of

$$f'(x) = x^{2/3},$$

if the point $(1, 3/5)$ is on the curve.

58. **Biology** According to Fick's law, the diffusion of a solute across a cell membrane is given by

$$c'(t) = \frac{kA}{V}(C - c(t)), \qquad \textbf{(1)}$$

where A is the area of the cell membrane, V is the volume of the cell, $c(t)$ is the concentration inside the cell at time t, C is the concentration outside the cell, and k is a constant. If c_0 represents the concentration of the solute inside the cell when $t = 0$, then it can be shown that

$$c(t) = (c_0 - C)e^{-kAt/V} + C. \qquad \textbf{(2)}$$

(a) Use this last result to find $c'(t)$.
(b) Substitute back into equation (1) to show that (2) is indeed the correct antiderivative of (1).

13.2 Area and the Definite Integral

This section shows how to find the area of a region bounded on one side by a curve. Finding the area under a curve is very important in a great many applications. For example, the graph and caption in Figure 13.2 is from a test report in *Modern Photography** magazine, June 1982, page 111, on the Minolta X-700 camera. The area under the curve gives the total noise made by the camera.

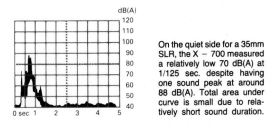

On the quiet side for a 35mm SLR, the X – 700 measured a relatively low 70 dB(A) at 1/125 sec. despite having one sound peak at around 88 dB(A). Total area under curve is small due to relatively short sound duration.

Figure 13.2

---

*From "Sounding out the Minolta X-700" in *Modern Photography*, June 1982. Copyright © 1981 by ABC Leisure Magazines, Inc. Reprinted by permission.

Figure 13.3 shows the region bounded by the lines $x = 0$, the x-axis, and the graph of

$$f(x) = \sqrt{4 - x^2}.$$

A very rough approximation of the area of this region can be found by using two rectangles as in Figure 13.4. The height of the rectangle on the left is $f(0) = 2$ and the height of the rectangle on the right is $f(1) = \sqrt{3}$. The width of each rectangle is 1, making the total area of the two rectangles

$$1 \cdot f(0) + 1 \cdot f(1) = 2 + \sqrt{3} \approx 3.7321 \text{ square units.}$$

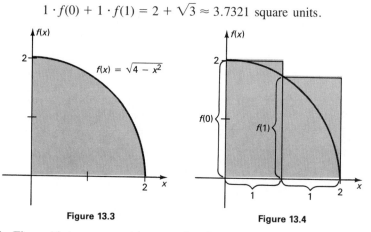

Figure 13.3 **Figure 13.4**

① Calculate the sum
$$\frac{1}{4} \cdot f(0) + \frac{1}{4} \cdot f\left(\frac{1}{4}\right) + \cdots$$
$$+ \frac{1}{4} \cdot f\left(\frac{7}{4}\right), \text{ using this}$$
information.

x	$f(x)$
0	2
$\frac{1}{4}$	1.98431
$\frac{1}{2}$	1.93649
$\frac{3}{4}$	1.85405
1	1.73205
$\frac{5}{4}$	1.56125
$\frac{3}{2}$	1.32288
$\frac{7}{4}$	.96825

Answer:
3.33982 square units

As Figure 13.4 suggests, this approximation is greater than the actual area. To improve the accuracy of the approximation, we could divide the interval from $x = 0$ to $x = 2$ into four equal parts, each of width $1/2$, as shown in Figure 13.5. As before, the height of each rectangle is given by the value of f at the left-hand side of the rectangle, and its area is the width, $1/2$, multiplied by the height. The total area of the four rectangles is

$$\frac{1}{2} \cdot f(0) + \frac{1}{2} \cdot f\left(\frac{1}{2}\right) + \frac{1}{2} \cdot f(1) + \frac{1}{2} \cdot f\left(1\frac{1}{2}\right)$$

$$= \frac{1}{2}(2) + \frac{1}{2}\left(\frac{\sqrt{15}}{2}\right) + \frac{1}{2}(\sqrt{3}) + \frac{1}{2}\left(\frac{\sqrt{7}}{2}\right)$$

$$= 1 + \frac{\sqrt{15}}{4} + \frac{\sqrt{3}}{2} + \frac{\sqrt{7}}{4} \approx 3.4957 \text{ square units.}$$

This approximation looks better, but it is still greater than the actual area desired. To improve the approximation, divide the interval from $x = 0$ to $x = 2$ into 8 parts with equal widths of $1/4$. (See Figure 13.6.) The total area of all these rectangles is

$$\frac{1}{4} \cdot f(0) + \frac{1}{4} \cdot f\left(\frac{1}{4}\right) + \frac{1}{4} \cdot f\left(\frac{1}{2}\right) + \frac{1}{4} \cdot f\left(\frac{3}{4}\right) + \frac{1}{4} \cdot f(1) + \frac{1}{4} \cdot f\left(\frac{5}{4}\right).$$

$$+ \frac{1}{4} \cdot f\left(\frac{3}{2}\right) + \frac{1}{4} \cdot f\left(\frac{7}{4}\right). \quad ①$$

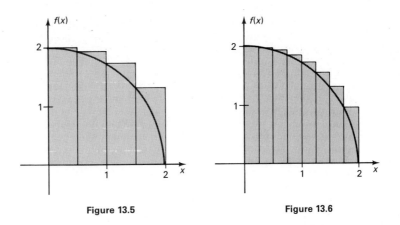

Figure 13.5   Figure 13.6

This process of approximating the area under a curve by using more and more rectangles to get a better and better approximation can be generalized. To do this, divide the interval from $x = 0$ to $x = 2$ into n equal parts. Each of these n intervals has width

$$\frac{2 - 0}{n} = \frac{2}{n},$$

so each rectangle has width $2/n$ and height determined by the function-value at the left side of the rectangle. A computer was used to find approximations to the area for several values of n in the table at the side.

The areas in the last column in the table are approximations of the area under the curve, above the x-axis, and between the lines $x = 0$ and $x = 2$. As n becomes larger and larger, the approximation is better and better, getting closer to the actual area. In this example, the exact area can be found by a formula from plane geometry. Write the given function as

n	Area
2	3.7321
4	3.4957
8	3.3398
10	3.3045
20	3.2285
50	3.1783
100	3.1512
500	3.1455

$$y = \sqrt{4 - x^2},$$

then square both sides to get

$$y^2 = 4 - x^2$$
$$x^2 + y^2 = 4,$$

the equation of a circle centered at the origin with radius 2. The region in Figure 13.3 is the quarter of this circle that lies in the first quadrant. The actual area of this region is one-quarter of the area of the entire circle, or

$$\frac{1}{4}\pi(2)^2 = \pi \approx 3.1416.$$

(An approximation for π was originally found by a process similar to this.)

It turns out that as the number of rectangles increases without bound, that is, as $n \to \infty$, for our region,

$$\lim_{n \to \infty} (\text{sum of areas of } n \text{ rectangles}) = \pi.$$

The area of the region under the graph of $f(x) = \sqrt{4 - x^2}$ and above the x-axis, between the lines $x = 0$ and $x = 2$, is defined to be this limit. This limit is the **definite integral** of $f(x) = \sqrt{4 - x^2}$ from $x = 0$ to $x = 2$, written

$$\int_0^2 \sqrt{4 - x^2}\, dx = \pi.$$

The 2 above the integral sign is the **upper limit** of integration, and 0 is the **lower limit** of integration. This use of the word "limit" has nothing to do with the limit of functions discussed earlier; it refers to the limits, or boundaries, of the region. In the next section we shall see how antiderivatives are used in finding the definite integral and thus the area under a curve.

At this point, some new notation and terminology is needed to write sums concisely. To indicate addition, or summation, the Greek letter sigma, Σ, is used as shown in the next example.

Example 1 Find the following sums.

(a) $\displaystyle\sum_{i=1}^{5} i$

Replace i in turn with the integers 1 through 5 and add the resulting terms.

$$\sum_{i=1}^{5} i = 1 + 2 + 3 + 4 + 5 = 15$$

(b) $\displaystyle\sum_{i=1}^{4} a_i = a_1 + a_2 + a_3 + a_4$

(c) $\displaystyle\sum_{i=1}^{3} (6x_i - 2)$, if $x_1 = 2$, $x_2 = 4$, $x_3 = 6$

Letting $i = 1$, 2, and 3 respectively gives

$$\sum_{i=1}^{3} (6x_i - 2) = (6x_1 - 2) + (6x_2 - 2) + (6x_3 - 2).$$

Now substitute the given values for x_1, x_2, and x_3.

$$\sum_{i=1}^{3} (6x_i - 2) = (6 \cdot 2 - 2) + (6 \cdot 4 - 2) + (6 \cdot 6 - 2)$$
$$= 10 + 22 + 34$$
$$= 66$$

(d) $\sum_{i=1}^{4} f(x_i)\, \Delta x$ if $f(x) = x^2$, $x_1 = 0$, $x_2 = 2$, $x_3 = 4$, $x_4 = 6$, and $\Delta x = 2$.

$$\sum_{i=1}^{4} f(x_i)\, \Delta x = f(x_1)\, \Delta x + f(x_2)\, \Delta x + f(x_3)\, \Delta x + f(x_4)\, \Delta x$$
$$= x_1^2\, \Delta x + x_2^2\, \Delta x + x_3^2\, \Delta x + x_4^2\, \Delta x$$
$$= 0^2(2) + 2^2(2) + 4^2(2) + 6^2(2)$$
$$= 0 + 8 + 32 + 72 = 112 \quad \boxed{2}$$

2 Find each sum.

(a) $\sum_{i=0}^{6} i$

(b) $\sum_{i=2}^{7} (3i + 1)$

(c) $\sum_{i=1}^{4} (5x_i - 1)$, if $x_1 = 9$,

$x_2 = 8$, $x_3 = 2$, $x_4 = 7$

(d) $\sum_{i=3}^{6} f(x_i)\, \Delta x$, if $f(x) =$

$3x^2 - 1$, $x_3 = 2$, $x_4 = 5$,

$x_5 = 1$, $x_6 = 5$, and $\Delta x = \dfrac{1}{5}$

Answer:
(a) 21
(b) 87
(c) 126
(d) 32.2

Now we can generalize the idea of area under the curve. Figure 13.7 shows the area bounded by the curve $y = f(x)$, the x-axis, and the vertical lines $x = a$ and $x = b$. To approximate this area, we might divide the region under the curve first into ten rectangles [Figure 13.7(a)] and then into 20 rectangles [Figure 13.7(b)]. The sums of the areas of the rectangles give approximations to the area under the curve.

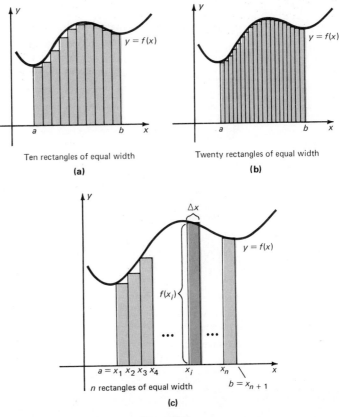

Ten rectangles of equal width
(a)

Twenty rectangles of equal width
(b)

n rectangles of equal width
(c)

Figure 13.7

To get a number that can be defined as the *exact* area, begin by dividing the interval from a to b into n pieces of equal width, using each of these n pieces as the base of a rectangle. [See Figure 13.7(c).] The endpoints of the n intervals are labeled $x_1, x_2, x_3, \ldots, x_{n+1}$, where $a = x_1$ and $b = x_{n+1}$. In the graph of Figure 13.7(c), the symbol Δx is used to represent the width of each of the intervals. For rectangles of equal width, Δx is always $\dfrac{b - a}{n}$. The darker rectangle is an arbitrary rectangle called the ith rectangle. Its area is the product of its length and width. Since the width of the ith rectangle is Δx and the length of the ith rectangle is given by the height $f(x_i)$,

$$\text{Area of } i\text{th rectangle} = f(x_i) \cdot \Delta x.$$

The total area under the curve is approximated by the sum of the areas of all n of the rectangles. Using the Σ symbol, the approximation to the total area becomes

$$\text{Area of all } n \text{ rectangles} = \sum_{i=1}^{n} f(x_i) \cdot \Delta x.$$

The exact area is defined to be the limit of this sum (if the limit exists) as the number of rectangles increases without bound.

Area Under a Curve If $f(x) \geq 0$ for all x in the interval $[a, b]$, then the area between $f(x)$, the x-axis, $x = a$, and $x = b$ is given by

$$\lim_{n \to \infty} \sum_{i=1}^{n} f(x_i) \, \Delta x = \int_a^b f(x) \, dx,$$

provided the limit exists, where $\Delta x = \dfrac{b - a}{n}$ and x_i is any value of x in the ith subinterval.

As indicated in this definition, any number in the ith interval can be used to find the height of the ith rectangle. Also, a more general definition is possible in which the rectangles do not necessarily all have the same width.

Example 2 Approximate $\int_0^4 2x \, dx$ by finding the area of the region under the graph of $f(x) = 2x$, above the x-axis, and between $x = 0$ and $x = 4$, using four rectangles of equal width whose heights are the values of the function at the midpoint of each rectangle.

As shown in Figure 13.8, the heights of the four rectangles, given by $f(x_i)$ for $i = 1, 2, 3, 4$, are as follows.

i	x_i	$f(x_i)$
1	$x_1 = .5$	$f(.5) = 1.0$
2	$x_2 = 1.5$	$f(1.5) = 3.0$
3	$x_3 = 2.5$	$f(2.5) = 5.0$
4	$x_4 = 3.5$	$f(3.5) = 7.0$

The width of each rectangle is $\Delta x = \dfrac{4 - 0}{4} = 1$. The sum of the areas of the four rectangles is

$$\sum_{i=1}^{4} f(x_i)\,\Delta x = f(x_1)\,\Delta x + f(x_2)\,\Delta x + f(x_3)\,\Delta x + f(x_4)\,\Delta x$$
$$= f(.5)\,\Delta x + f(1.5)\,\Delta x + f(2.5)\,\Delta x + f(3.5)\,\Delta x$$
$$= (1)(1) + (3)(1) + (5)(1) + (7)(1)$$
$$= 16. \quad \boxed{3}$$

3 Divide the region of Figure 13.8 into eight rectangles of equal width whose heights are the values of the function at the midpoint of each rectangle.

(a) Complete this table.

i	x_i	$f(x)$
1	.25	
2	.75	
3	1.25	
4	1.75	
5		
6		
7		
8		

(b) Use the results from the table to approximate $\int_0^4 2x\,dx$.

Answer:

(a)

i	x_i	$f(x)$
1	.25	.50
2	.75	1.50
3	1.25	2.50
4	1.75	3.50
5	2.25	4.50
6	2.75	5.50
7	3.25	6.50
8	3.75	7.50

(b) 16

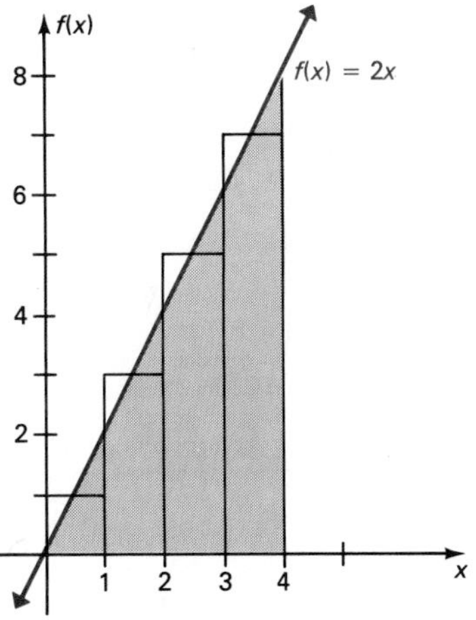

Figure 13.8

Using the formula for the area of a triangle, $A = (1/2)bh$, with b, the length of the base, equal to 4 and h, the height, equal to 8, gives

$$A = \frac{1}{2}bh = \frac{1}{2}(4)(8) = 16,$$

the exact value of the area. Our approximations equal the exact area in this case since the use of the midpoints of each subinterval distributed the error evenly above and below the graph. ■

Suppose the function $f(x) = x^2 + 20$ gives the marginal cost of some item at a particular x-value. Then $f(2) = 24$ gives the rate of change of cost at $x = 2$. That is, a unit change in x (at this point) will produce a change of approximately 24 units in the cost function. Also, $f(3) = 29$ means that a unit of change in x (when $x = 3$) will produce a change of approximately 29 units in the cost function.

To find the **total change** in the cost function as x changes from 2 to 3, we could divide the interval from 2 to 3 into n equal parts, using each part as the base of a rectangle as we did above. The area of each rectangle would approximate the change in cost at the x-value which is the left endpoint of the base of the rectangle. Then the sum of the areas of these rectangles would approximate the net total change in cost from $x = 2$ to $x = 3$. The limit of this sum as $n \to \infty$ would give the exact total change.

This result produces another application of the area under a curve: the area under the graph of the marginal cost function $f(x)$, above the x-axis, between $x = a$ and $x = b$ gives the net total change in the cost as x goes from a to b. Generalizing, the total change in a quantity can be approximated from the function which gives the rate of change of the quantity, using the same methods used to find the approximate area under a curve.

Example 3   Figure 13.9 shows the rate of change of the annual maintenance charges for a certain machine. Approximate the total maintenance charges over the 10-year life of the machine by using rectangles, dividing the interval from 0 to 10 into ten equal subdivisions. Each rectangle has width 1; if we use the left endpoint of each rectangle to determine the height of the rectangle, the approximation becomes

$$1 \cdot 0 + 1 \cdot 500 + 1 \cdot 750 + 1 \cdot 1800 + 1 \cdot 1800 + 1 \cdot 3000 + 1 \cdot 3000$$
$$+ 1 \cdot 3400 + 1 \cdot 4200 + 1 \cdot 5100 = 23{,}550.$$

About \$23,550 will be spent on maintenance over the 10-year life of the machine.   4

4 Use Figure 13.9 to estimate the maintenance change during
(a) the first 6 years of the machine's life;
(b) the first eight years.

Answer:
(a) \$7850
(b) \$14,250

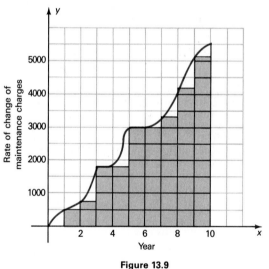

Figure 13.9

13.2   Exercises

Evaluate the following sums. (See Example 1.)

1. $\displaystyle\sum_{i=1}^{3} 3i$

2. $\displaystyle\sum_{i=1}^{6} (-5i)$

3. $\displaystyle\sum_{i=1}^{5} (2i + 7)$

4. $\displaystyle\sum_{i=1}^{10} (5i - 8)$

5. Let $x_1 = -5$,
$x_2 = 8$,
$x_3 = 7$,
$x_4 = 10$.
Find $\displaystyle\sum_{i=1}^{4} x_i$.

6. Let $x_1 = 10$,
$x_2 = 15$,
$x_3 = -8$,
$x_4 = -12$,
$x_5 = 0$.
Find $\displaystyle\sum_{i=1}^{5} x_i$.

7. Let $f(x) = x - 3$,
$x_1 = 4$,
$x_2 = 6$,
$x_3 = 7$.

Find $\sum_{i=1}^{3} f(x_i)$.

8. Let $f(x) = x^2 + 1$,
$x_1 = -2$,
$x_2 = 0$,
$x_3 = 2$,
$x_4 = 4$.

Find $\sum_{i=1}^{4} f(x_i)$.

9. Let $f(x) = 2x + 1$,
$x_1 = 0$,
$x_2 = 2$,
$x_3 = 4$,
$x_4 = 6$,
$\Delta x = 2$.

Find $\sum_{i=1}^{4} f(x_i) \Delta x$.

10. Let $f(x) = 1/x$,
$x_1 = 1/2$,
$x_2 = 1$,
$x_3 = 3/2$,
$x_4 = 2$,
$\Delta x = 1/2$.

Find $\sum_{i=1}^{4} f(x_i) \Delta x$.

Approximate the area under each given curve and above the x-axis on the given interval by using two rectangles. Let the height of the rectangle be given by the value of the function at the left side of the rectangle. Then repeat the process and approximate the area with four rectangles. (See Example 2.)

11. $f(x) = 3x + 2$, $[1, 5]$

12. $f(x) = -2x + 1$, $[-4, 0]$

13. $f(x) = x + 5$, $[2, 4]$

14. $f(x) = 3 + x$, $[1, 3]$

15. $f(x) = x^2$, $[1, 5]$

16. $f(x) = x^2$, $[0, 4]$

17. $f(x) = x^2 + 2$, $[-2, 2]$

18. $f(x) = -x^2 + 4$, $[-2, 2]$

19. $f(x) = e^x - 1$, $[0, 4]$

20. $f(x) = e^x + 1$, $[-2, 2]$

21. $f(x) = \dfrac{1}{x}$, $[1, 5]$

22. $f(x) = \dfrac{2}{x}$, $[1, 9]$

Work the following exercises. (See Example 2.)

23. Consider the region below $f(x) = x/2$, above the x-axis, between $x = 0$ and $x = 4$. Let x_i be the left endpoint of the ith subinterval.
(a) Approximate the area of the region using four rectangles.

(b) Approximate the area of the region using eight rectangles.
(c) Find $\int_0^4 f(x)\, dx$ by using the formula for the area of a triangle.

24. Find $\int_0^5 (5 - x)\, dx$ by using the formula for the area of a triangle.

Estimate the area under the curve by summing the area of rectangles. Let the function value at the left side of each rectangle give the height of the rectangle.

25. The graph below shows the rate of sales of new cars in a recent year. Estimate the total sales during that year. Use rectangles with a width of 1. (See Example 3.)

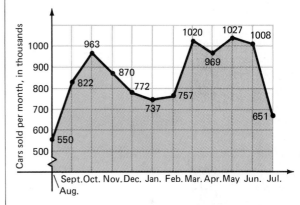

26. The graph below shows the rate of use of electrical energy (in kilowatt hours) in a certain city on a very hot day. Estimate the total usage of electricity on that day. Let the width of each rectangle be 2 hours.

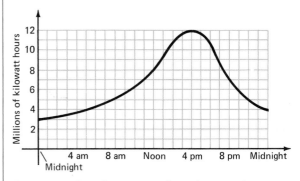

27. The graph on the next page shows the approximate concentration of alcohol in a person's bloodstream t hours after drinking 2 ounces of alcohol. Estimate the total amount of alcohol in the bloodstream by estimating the area under the curve. Use rectangles of width 1.

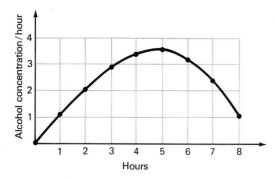

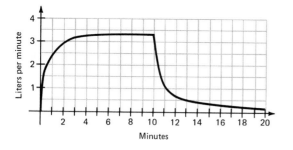

28. The graph below shows the rate of inhalation of oxygen by a person riding a bicycle very rapidly for 10 minutes. Estimate the total volume of oxygen inhaled in the first 20 minutes after the beginning of the ride. Use rectangles of width 1.

The next two graphs are from Road and Track *magazine.* The curve shows the velocity at time t, in seconds, when the car accelerates from a dead stop. To find the total distance traveled by the car in reaching 100 miles per hour, we must estimate the definite integral*

$$\int_0^T v(t)\, dt,$$

where T represents the number of seconds it takes for the car to reach 100 mph.

   Use the graphs to estimate this distance by adding the areas of rectangles with widths of 5 seconds. To adjust your answer to miles per hour, divide by 3600 (the number of seconds in an hour). You then have the number of miles that the car traveled in reaching 100 mph. Finally, multiply by 5280 feet per mile to convert the answers to feet.

---

*From *Road & Track,* April and May, 1978. Reprinted with permission of Road & Track.

29. Estimate the distance traveled by the Porsche 928, using the graph below.

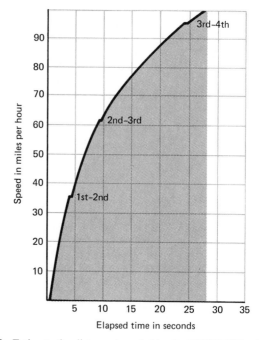

30. Estimate the distance traveled by the BMW 733i using the graph below.

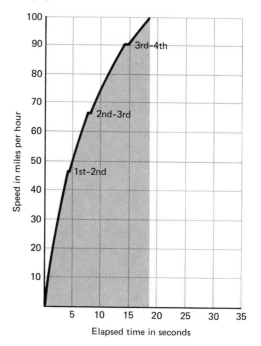

When doing research on the wolf-moose relationships in Michigan's Isle Royale National Park, the biologist Rolf Peterson needed a snow-depth index to help correlate populations with snow levels. The graph below is given a snow depth index of 1.0. The vertical scale gives snow depth in inches, and the horizontal scale gives months.†

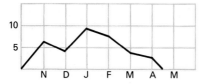

Find the snow depth index for the following years by forming the ratio of the areas under the given curves to the area under the curve above.

31.

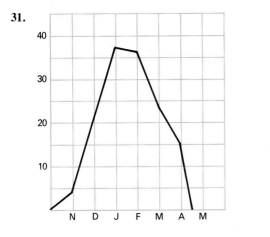

32.

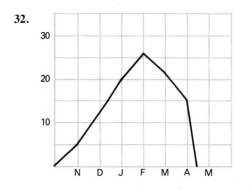

Use the method from the text to approximate the area between the x-axis and the graph of each function on the given interval.

33. $f(x) = x \ln x$; [1, 5]

34. $f(x) = x^2 e^{-x}$; [−1, 3]

35. $f(x) = \dfrac{\ln x}{x}$; [1, 5]

36. $f(x) = \dfrac{e^x - e^{-x}}{2}$; [0, 4]

13.3 The Fundamental Theorem of Calculus

The work from the last two sections can now be put together. We have seen that, if $f(x) \geq 0$,

$$\int_a^b f(x)\, dx$$

gives the area between the graph of $f(x)$ and the x-axis, from $x = a$ to $x = b$. We

† From *Wolf Ecology and Prey Relationships on Isle Royale* by Rolf Olin Peterson. Copyright © 1977 by the U.S. Government Printing Office, p. 187.

will see that this definite integral can be found using the antiderivatives discussed earlier. The connection between antiderivatives and definite integrals is given by the fundamental theorem of calculus.

Fundamental Theorem of Calculus Let f be continuous on the interval $[a, b]$, and let F be *any* antiderivative of f. Then

$$\int_a^b f(x)\ dx = F(b) - F(a) = F(x)\Big]_a^b.$$

The symbol $F(x)\big]_a^b$ is used to represent $F(b) - F(a)$. It is important to note that the fundamental theorem does not require $f(x) \geq 0$. The condition $f(x) \geq 0$ is necessary only when using the fundamental theorem to find area. ①

Example 1 Find $\int_0^3 (x^2 - 4)\ dx$.

First find an antiderivative of $x^2 - 4$.

$$F(x) = \int (x^2 - 4)\ dx$$

$$= \frac{x^3}{3} - 4x.$$

We let $C = 0$ since the fundamental theorem says that *any* antiderivative can be used. Then, by the fundamental theorem,

$$\int_0^3 (x^2 - 4)\ dx = \left(\frac{x^3}{3} - 4x\right)\Big]_0^3$$

$$= \left[\frac{3^3}{3} - 4(3)\right] - \left[\frac{0^3}{3} - 4(0)\right]$$

$$= -3.\ ②$$

Figure 13.10 shows the graph of $f(x) = x^2 - 4$ from Example 1. Here the value of the definite integral, -3, *does not* represent the area between the graph of $f(x)$ and the x-axis from $x = 0$ to $x = 3$. This is because $f(x) < 0$ in part of $[0, 3]$. The definite integral is a real number and represents area only when $f(x) \geq 0$ for all x in $[a, b]$.

Example 2 Find $\int_1^2 4x^3\ dx$.

Choosing x^4 as an antiderivative gives

$$\int_1^2 4x^3\ dx = x^4\Big]_1^2 = 2^4 - 1^4 = 16 - 1 = 15.$$

---

① Let $C(x) = x^3 + 4x^2 - x + 3$. Find the following.

(a) $C(x)\big]_1^5$

(b) $C(x)\big]_3^4$

(c) $C(x)\big]_{10}^{20}$

Answer:
(a) 216
(b) 64
(c) 8190

② Find each definite integral.

(a) $\int_6^9 8x\ dx$

(b) $\int_3^7 8x^3\ dx$

Answer:
(a) 180
(b) 4640

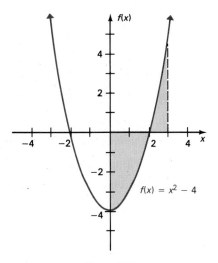

Figure 13.10

There is an infinite number of antiderivatives of $4x^3$, but all of these have the form $x^4 + C$, for a real number C. It doesn't matter which value of C is used:

$$\int_1^2 4x^3 \, dx = (x^4 + C)\Big]_1^2 = (2^4 + C) - (1^4 + C)$$
$$= 16 + C - (1 + C)$$
$$= 15.$$

The value of $\int_1^2 4x^3 \, dx$ is 15, no matter which value of C is used. For this reason, when finding a definite integral, choose the antiderivative with $C = 0$. **3**

Example 2 illustrates the difference between the definite integral and the indefinite integral. A definite integral is a real number; an indefinite integral is a family of functions—all the functions which are antiderivatives of a function f.

To see why the fundamental theorem of calculus is true for $f(x) \geq 0$, look at Figure 13.11 on the next page. Define the function $A(x)$ as the area between the x-axis and the graph of $y = f(x)$ from a to x. We first show that A is an antiderivative of f, that is, $A' = f$.

To do this, let h be a small positive number. Then $A(x + h) - A(x)$ is the shaded area of Figure 13.11. This area can be approximated with a rectangle having width h and height $f(x)$. The area of the rectangle is $h \cdot f(x)$, and

$$A(x + h) - A(x) \approx h \cdot f(x).$$

Dividing both sides by h gives

$$\frac{A(x + h) - A(x)}{h} \approx f(x).$$

3 Find each of the following.

(a) $\int_4^6 5z \, dz$

(b) $\int_2^5 8t^3 \, dt$

(c) $\int_1^9 \sqrt{z} \, dz$

Answer:

(a) 50

(b) 1218

(c) 52/3

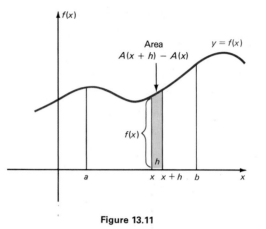

Figure 13.11

This approximation improves as h gets smaller and smaller. Take the limit on the left as h approaches 0, getting

$$\lim_{h \to 0} \frac{A(x+h) - A(x)}{h} = f(x).$$

This limit is simply $A'(x)$, so

$$A'(x) = f(x).$$

This result means that A is an antiderivative of f, as we set out to show.

Since $A(x)$ is the area under the curve $y = f(x)$ from a to x, $A(a) = 0$. The expression $A(b)$ is the area from a to b, the desired area. Since A is an antiderivative of f,

$$\int_a^b f(x)\, dx = A(x) \Big]_a^b = A(b) - A(a) = A(b) - 0 = A(b).$$

This suggests the proof of the fundamental theorem—for $f(x) \geq 0$ the area under the curve $y = f(x)$ from a to b is given by $\int_a^b f(x)\, dx$.

The fundamental theorem of calculus certainly deserves its name, which sets it apart as the most important theorem of calculus. It is the key connection between differential calculus and integral calculus, which originally were developed separately without knowledge of this connection between them.

The variable used in the integrand does not matter; that is,

$$\int_a^b f(x)\, dx = \int_a^b f(t)\, dt = \int_a^b f(u)\, du.$$

Each of these definite integrals represent the number $F(b) - F(a)$, where F is the antiderivative of f.

Key properties of definite integrals are listed below. Some of them are just restatements of properties from Section 13.1.

Properties of Definite Integrals For any real numbers a and b for which the definite integrals exist,

1. $\displaystyle\int_a^a f(x)\ dx = 0;$

2. $\displaystyle\int_a^b k \cdot f(x)\ dx = k \cdot \int_a^b f(x)\ dx,$ for any real constant k (constant multiple of a function);

3. $\displaystyle\int_a^b [f(x) \pm g(x)]\ dx = \int_a^b f(x)\ dx \pm \int_a^b g(x)\ dx$ (sum or difference of functions);

4. $\displaystyle\int_a^b f(x)\ dx = \int_a^c f(x)\ dx + \int_c^b f(x)\ dx,$ for any real number c.

If $f(x) \ge 0$, since the distance from a to a is 0, the first property says that the "area" under the graph of f bounded by $x = a$ and $x = a$ is 0. Also, since $\int_a^c f(x)\ dx$ represents the darker region in Figure 13.12, and $\int_c^b f(x)\ dx$ represents the lighter region,

$$\int_a^b f(x)\ dx = \int_a^c f(x)\ dx + \int_c^b f(x)\ dx,$$

as stated in the fourth property. While the figure shows $a < c < b$, the property is true for any value of c where both $f(x)$ and $F(x)$ are defined. Of course, the properties apply even if $f(x) < 0$.

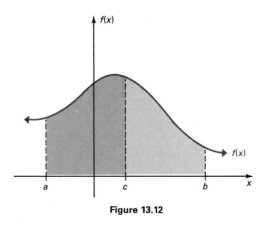

Figure 13.12

An algebraic proof is given here for the third property; proofs of the other properties are omitted. If $F(x)$ and $G(x)$ are antiderivatives of $f(x)$ and $g(x)$ respectively,

$$\int_a^b [f(x) + g(x)] \, dx = [F(x) + G(x)] \Big]_a^b$$

$$= [F(b) + G(b)] - [F(a) + G(a)]$$

$$= [F(b) - F(a)] + [G(b) - G(a)]$$

$$= \int_a^b f(x) \, dx + \int_a^b g(x) \, dx.$$

Example 3   Evaluate $\int_2^5 (6x^2 - 3x + 5) \, dx$.

Use the properties above, and the fundamental theorem, along with the power rule from Section 13.1.

$$\int_2^5 (6x^2 - 3x + 5) \, dx = 6 \int_2^5 x^2 \, dx - 3 \int_2^5 x \, dx + 5 \int_2^5 dx$$

$$= 2x^3 \Big]_2^5 - \frac{3}{2}x^2 \Big]_2^5 + 5x \Big]_2^5$$

$$= 2(5^3 - 2^3) - \frac{3}{2}(5^2 - 2^2) + 5(5 - 2)$$

$$= 2(125 - 8) - \frac{3}{2}(25 - 4) + 5(3)$$

$$= 234 - \frac{63}{2} + 15 = \frac{435}{2} \quad \boxed{4}$$

Example 4   Find $\int_1^2 \dfrac{dy}{y}$.

Using a result from Section 13.1,

$$\int_1^2 \frac{dy}{y} = \ln |y| \Big]_1^2 = \ln |2| - \ln |1|$$

$$= \ln 2 - \ln 1 \approx .6931 - 0 = .6931. \quad \boxed{5}$$

In the previous section we saw that, if $f(x) \geq 0$ in $[a, b]$, the definite integral $\int_a^b f(x) \, dx$ gives the area below the graph of the function $y = f(x)$, above the x-axis, and between the lines $x = a$ and $x = b$.

To see how to work around the requirement that $f(x) \geq 0$, look at Figure 13.13 which shows the graph of $f(x) = x^2 - 4$. As Figure 13.13 shows, the area bounded by the graph of f, the x-axis, and the vertical lines $x = 0$ and $x = 2$ lies below the x-axis. Using the fundamental theorem to find this area gives

$$\int_0^2 (x^2 - 4) \, dx = \left(\frac{x^3}{3} - 4x \right) \Big]_0^2 = \left(\frac{8}{3} - 8 \right) - (0 - 0) = \frac{-16}{3}.$$

The result is a negative number because $f(x)$ is negative for values of x in the interval $[0, 2]$. Since Δx is always positive, $f(x) < 0$ makes the product $f(x) \cdot \Delta x$ negative, so $\int_0^2 f(x) \, dx$ is negative. Since area is nonnegative, the required area is

4   Evaluate each definite integral.

(a) $\int_1^3 (x + 3x^2) \, dx$

(b) $\int_2^4 (6k^2 - 2k + 1) \, dk$

(c) $\int_3^5 (3t^2 - 4t + 2) \, dt$

Answer:

(a) 30

(b) 102

(c) 70

5   Evaluate the following.

(a) $\int_0^4 e^x \, dx$

(b) $\int_3^5 \dfrac{dx}{x}$

(c) $\int_1^6 e^x \, dx$

(d) $\int_2^8 \dfrac{4}{x} \, dx$

Answer:

(a) 53.59815

(b) .51083

(c) 400.71051

(d) 5.54518

given by $|-16/3|$ or $16/3$. Using a definite integral, the area could be written as

$$\left| \int_0^2 (x^2 - 4) \, dx \right|.$$

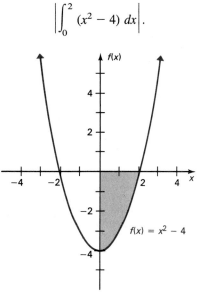

Figure 13.13

6 Find each area.

(a)

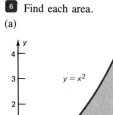

(b)

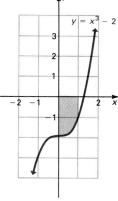

Answer:

(a) 8/3

(b) 7/4

Example 5 Find the area of the region between the x-axis and the graph of $f(x) = x^2 - 3x$ from $x = 1$ to $x = 3$.

The region is shown in Figure 13.14. Since the region lies below the x-axis, the area is given by

$$\left| \int_1^3 (x^2 - 3x) \, dx \right|.$$

By the fundamental theorem,

$$\int_1^3 (x^2 - 3x) \, dx = \left(\frac{x^3}{3} - \frac{3x^2}{2} \right) \Big]_1^3 = \left(\frac{27}{3} - \frac{27}{2} \right) - \left(\frac{1}{3} - \frac{3}{2} \right) = -\frac{10}{3}.$$

The required area is $|-10/3| = 10/3$ square units. **6**

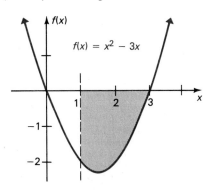

Figure 13.14

Example 6 Find the area between the x-axis and the graph of $f(x) = x^2 - 4$ from $x = 0$ to $x = 4$.

Figure 13.15 shows the required region. Part of the region is below the x-axis. The definite integral over that interval will have a negative value. To find the area, integrate the negative and positive portions separately and take the absolute value of the first result before combining the two results to get the total area. Start by finding the point where the graph crosses the x-axis. This is done by solving the equation

$$x^2 - 4 = 0.$$

The solutions of this equation are 2 and -2. The only solution in the interval $[0, 4]$ is 2. The total area of the region in Figure 13.15 is given by

$$\left| \int_0^2 (x^2 - 4)\,dx \right| + \int_2^4 (x^2 - 4)\,dx = \left| \left(\frac{1}{3}x^3 - 4x \right) \right]_0^2 \right| + \left(\frac{1}{3}x^3 - 4x \right) \right]_2^4$$

$$= \left| \frac{8}{3} - 8 \right| + \left(\frac{64}{3} - 16 \right) - \left(\frac{8}{3} - 8 \right)$$

$$= 16 \text{ square units.} \quad \boxed{7}$$

7 Find the area between the graph of $f(x) = x^2 - 4$ from
(a) $x = 1$ to $x = 5$ and
(b) $x = -2$ to $x = 3$.

Answer:
(a) 86/3
(b) 13

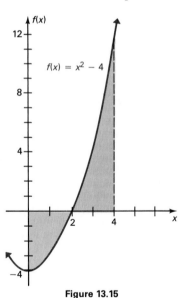

$f(x) = x^2 - 4$

Figure 13.15

In Example 6, incorrectly using one integral over the entire interval to find the area would give

$$\int_0^4 (x^2 - 4)\,dx = \left(\frac{x^3}{3} - 4x \right) \right]_0^4 = \left(\frac{64}{3} - 16 \right) - 0 = \frac{16}{3},$$

which is not the correct area. This definite integral represents no area, but is just a real number.

> **Finding Area** In summary, find the area bounded by $y = f(x)$, the vertical lines $x = a$ and $x = b$, and the x-axis with the following steps.
>
> *Step 1* Sketch a graph.
>
> *Step 2* Find any x-intercepts in $[a, b]$. These divide the total region into subregions.
>
> *Step 3* The definite integral will be *positive* for subregions above the x-axis and *negative* for subregions below the x-axis. Use separate integrals to find the areas of the subregions.
>
> *Step 4* The total area is the sum of the absolute values of the definite integrals for all of the subregions.

In the last section, we saw that the area under a rate of change function $f'(x)$ from $x = a$ to $x = b$ gives the total value of $f(x)$ on $[a, b]$. Now we can use the definite integral to solve these problems.

Example 7 The yearly rate of consumption of natural gas in trillions of cubic feet for a certain city is

$$C'(t) = t + e^{.01t},$$

when $t = 0$ corresponds to 1980. At this consumption rate, what is the total amount the city will use in the 10-year period of the eighties?

To find the consumption over the ten-year period from 1980 through 1989, use the definite integral.

$$\int_0^{10} (t + e^{.01t})\, dt = \left(\frac{t^2}{2} + \frac{e^{.01t}}{.01} \right)\Big]_0^{10}$$
$$= (50 + 100e^{.1}) - (0 + 100)$$
$$\approx -50 + 100(1.10517) \approx 60.5$$

Therefore, a total of about 60.5 trillion cubic feet of natural gas will be used during the eighties if the consumption rate remains the same. **8**

8 In Example 7, suppose a conservation campaign and higher prices cause the rate of consumption to be given by $C'(t) = \frac{1}{2}t + e^{.005t}$. Find the total amount of gas used in the eighties.

Answer:
About 35.25 trillion cubic feet

13.3 Exercises

Evaluate each of the following definite integrals. (See Examples 1–4.)

1. $\int_{-2}^{4} (-dp)$

2. $\int_{-4}^{1} 6x\, dx$

3. $\int_{-1}^{2} (3t - 1)\, dt$

4. $\int_{-2}^{2} (4z + 3)\, dz$

5. $\int_0^2 (5x^2 - 4x + 2)\, dx$

6. $\int_{-2}^{3} (-x^2 - 3x + 5)\, dx$

7. $\int_4^9 3\sqrt{u}\, du$

8. $\int_2^8 \sqrt{2r}\, dr$

9. $\int_0^1 2(t^{1/2} - t)\, dt$

10. $\int_0^4 -(3x^{3/2} + x^{1/2})\, dx$

11. $\int_{1}^{4} (5y\sqrt{y} + 3\sqrt{y})\, dy$

12. $\int_{4}^{9} (4\sqrt{r} - 3r\sqrt{r})\, dr$

13. $\int_{1}^{3} \frac{2}{x^2}\, dx$

14. $\int_{1}^{4} \frac{-3}{u^2}\, du$

15. $\int_{1}^{5} (5n^{-1} + n^{-3})\, dn$

16. $\int_{2}^{3} (3x^{-1} - x^{-4})\, dx$

17. $\int_{2}^{3} \left(2e^{-.1A} + \frac{3}{A}\right) dA$

18. $\int_{1}^{2} \left(\frac{-1}{B} + 3e^{.2B}\right) dB$

19. $\int_{1}^{2} \left(e^{5u} - \frac{1}{u^2}\right) du$

20. $\int_{.5}^{1} (p^3 - e^{4p})\, dp$

21. $\int_{-1}^{0} (2y - 3)^2\, dy$

22. $\int_{0}^{3} (4m + 2)^2\, dm$

23. $\int_{1}^{64} \frac{\sqrt{z} - 2}{\sqrt[3]{z}}\, dz$

24. $\int_{1}^{8} \frac{3 - y^{1/3}}{y^{2/3}}\, dy$

Use the definite integral to find the area between the x-axis and f(x) over the indicated interval. Check first to see if the graph crosses the x-axis in the given interval. (See Examples 5 and 6.)

25. $f(x) = 2x + 3;\quad [8, 10]$

26. $f(x) = 4x - 7;\quad [5, 10]$

27. $f(x) = 2 - 2x^2;\quad [0, 5]$

28. $f(x) = 9 - x^2;\quad [0, 6]$

29. $f(x) = x^2 + 4x - 5;\quad [-1, 3]$

30. $f(x) = x^2 - 6x + 5;\quad [-1, 4]$

31. $f(x) = x^3;\quad [-1, 3]$

32. $f(x) = x^3 - 2x;\quad [-2, 4]$

33. $f(x) = e^x - 1;\quad [-1, 2]$

34. $f(x) = 1 - e^{-x};\quad [-1, 2]$

35. $f(x) = \frac{1}{x};\quad [1, e]$

36. $f(x) = \frac{1}{x};\quad [e, e^2]$

Find the area of each shaded region.

37.

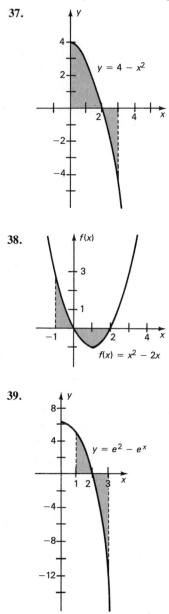

$y = 4 - x^2$

38.

$f(x)$

$f(x) = x^2 - 2x$

39.

$y = e^2 - e^x$

40.

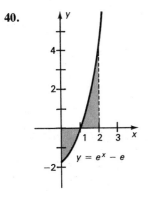

$y = e^x - e$

Work the following exercises. (See Example 7.)

41. Natural Science A chemical plant is polluting the air at a rate of

$$p(t) = 15t^{3/2} + 10$$

(in appropriate units), where t is time in months. How much pollution is produced by the plant in 25 months?

42. Management An oil tanker is leaking oil at the rate of $20t + 50$ barrels per hour, where t is time in hours after the tanker hits a hidden rock (when $t = 0$).
(a) Find the total number of barrels that the ship will leak on the first day.
(b) Find the number of barrels that the ship will leak on the second day.

43. Management De Win Enterprises has found that its expenditure rate per day (in hundreds of dollars) on a certain type of job is given by

$$E(x) = 4x + 2,$$

where x is the number of days since the start of the job.
(a) Find the total expenditure if the job takes 10 days.
(b) How much will be spent on the job from the 10th to the 20th day?
(c) If the company wants to spend no more than $5000 on the job, in how many days must they complete it?

44. Management De Win Enterprises (see Exercise 43) also knows that the rate of income per day (in hundreds of dollars) for the same job is

$$I(x) = 100 - x,$$

where x is the number of days since the job was started.
(a) Find the total income for the first 10 days.

(b) Find the income from the 10th to the 20th day.
(c) How many days must the job last for the total income to be at least $5000?

45. Natural Science The rate at which a substance grows is given by

$$R(x) = 200e^x,$$

where x is the time in days. What is the total accumulated growth after 2.5 days?

46. Natural Science A drug has a reaction rate in appropriate units of

$$R(t) = \frac{4}{t^2} + 1,$$

where t is measured in hours after the drug is administered. Find the total reaction to the drug
(a) from $t = 1/4$ to $t = 1$;
(b) from $t = 12$ to $t = 24$.

47. Management Most of the numbers in this exercise come from an article in *The Wall Street Journal*. Suppose that the rate of consumption of a natural resource is $c(t)$, where

$$c(t) = ke^{rt}.$$

Here t is time in years, r is constant, and k is the consumption in the year when $t = 0$. In 1984, Texaco sold 1.2 billion barrels of oil. Assume that $r = .04$.
(a) Write $c(t)$ for Texaco, letting $t = 0$ represent 1984.
(b) Set up a definite integral for the amount of oil that Texaco will sell in the next ten years.
(c) Evaluate the definite integral of part (b).
(d) Texaco has about 20 billion barrels of oil in reserve. To find the number of years that this amount will last, solve the equation

$$\int_0^T 1.2e^{.04t}\, dt = 20.$$

(e) Rework part (d) assuming that $r = .02$.

48. Management A mine begins producing at time $t = 0$. After t years, the mine is producing at the rate of

$$P'(t) = 10t - \frac{15}{\sqrt{t}}$$

tons per year. Write an expression for the total output of the mine from year 1 to year T.

49. Management The rate of consumption of one natural resource in one country is

$$C'(t) = 72e^{.014t},$$

where $t = 0$ corresponds to 1985. How much of the resource will be used, altogether, from 1985 to year T?

50. Management In Exercise 47, the rate of consumption of oil (in billions of barrels) by Texaco was given as

$$1.2e^{.04t},$$

where $t = 0$ corresponds to 1984. Find the total amount of oil used by Texaco from 1984 to year T. At this rate, how much will be used in 5 years?

13.4 Applications of Integrals

Given a function which represents the total maintenance charge for a machine from the time it is installed, the rate of maintenance at any time t is given by the derivative of the total maintenance charge. Also, the total maintenance function can be found by finding the antiderivative of the rate of maintenance function. As we have seen in this chapter, if we graph the function giving the rate of maintenance, then the total maintenance is given by the area under the curve. Example 1 shows how this works.

Example 1 Suppose a leasing company wants to decide on the yearly lease fee for a certain new typewriter. The company expects to lease the typewriter for 5 years, and it expects the rate of maintenance, $M(t)$ (in dollars), which it must supply, to be approximated by

$$M(t) = 10 + 2t + t^2,$$

1 Find the total maintenance charge for a lease of
(a) 1 year;
(b) 2 years.

Answer:
(a) $11.33
(b) $26.67

where t is the number of years the typewriter has been used. How much should the company charge for maintenance on the typewriter for the 5-year period?

The definite integral can be used to find the total maintenance charge the company can expect over the life of the typewriter. Figure 13.16 shows the graph of $M(t)$. The total maintenance charge for the 5-year period will be given by the shaded area of the figure, which can be found as follows.

$$\int_0^5 (10 + 2t + t^2)\, dt = \left(10t + t^2 + \frac{t^3}{3}\right)\Big]_0^5$$

$$= 50 + 25 + \frac{125}{3} - 0$$

$$\approx 116.67$$

2 Suppose the rate of change of cost in dollars to produce x items is given by

$$C(x) = x^2 - 8x + 15.$$

Find the total cost to produce the first 10 items.

Answer:
$83.33

The company can expect the total maintenance charge for 5 years to be about $117. Hence, the company should add about

$$\frac{\$117}{5} = \$23.40$$

to its annual lease price. **1** **2**

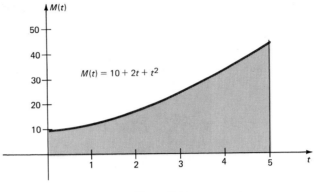

Figure 13.16

Example 2 Elizabeth Wilson, who runs a factory that makes signs, has been shown a new machine that staples the signs to the handles. She estimates that the rate of savings, $S(x)$, from the machine will be approximated by

$$S(x) = 3 + 2x,$$

where x represents the number of years the stapler has been in use. If the machine costs $70, would it pay for itself in 5 years?

We need to find the area under the rate of savings curve shown in Figure 13.17, between the lines $x = 0$, $x = 5$, and the x-axis. Using a definite integral gives

$$\int_0^5 (3 + 2x)\, dx = 3x + x^2 \Big]_0^5 = 40.$$

The total savings in five years is $40, so the machine will not pay for itself in this time period. **3**

3 Find the amount of savings over the first 6 years for the machine in Example 1.

Answer:

$54

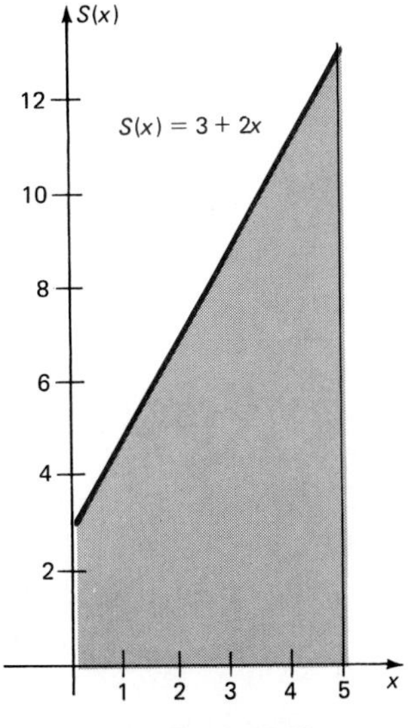

S(x)

$S(x) = 3 + 2x$

Figure 13.17

Since the machine costs a total of $70, it will pay for itself when the area under the savings curve of Figure 13.18 equals 70, or at a time t such that

$$\int_0^t (3 + 2x)\, dx = 70.$$

Evaluating the definite integral gives

$$3x + x^2 \bigg]_0^t = (3t + t^2) - (3 \cdot 0 + 0^2) = 3t + t^2.$$

Since the total savings must equal 70,

$$3t + t^2 = 70.$$

Solve this quadratic equation to verify that the machine will pay for itself in seven years.   ④

④ Find the number of years in which the machine in Example 2 will pay for itself if the rate of savings is $S(x) = 5 + 4x$.

Answer:

4.8 years

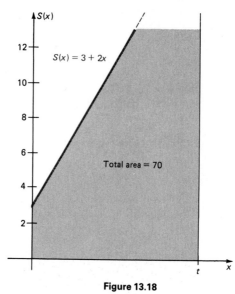

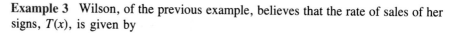

Figure 13.18

Example 3   Wilson, of the previous example, believes that the rate of sales of her signs, $T(x)$, is given by

$$T(x) = 15 + 10x,$$

where x is the number of years she has been in business. When can she expect to sell her 1000th sign?

Let t be the time at which Wilson will sell her 1000th sign. The area under the sales curve between $x = 0$ and $x = t$ gives the total sales during that period. Since the total sales must be 1000,

$$\int_0^t (15 + 10x)\, dx = 1000.$$

Since

$$\int_0^t (15 + 10x)\, dx = 15x + 5x^2 \Big]_0^t = 15t + 5t^2,$$

solve the equation

$$15t + 5t^2 = 1000$$
$$5t^2 + 15t - 1000 = 0.$$

Divide both sides of this equation by 5 to get

$$t^2 + 3t - 200 = 0,$$

from which (using the quadratic formula)

$$t = 12.72 \text{ years.}$$

Wilson can expect to sell her 1000th sign about twelve years and nine months after she goes into business. ▪5 ▪6

In Section 13.2 we saw that a definite integral can be used to find the area *under* a curve. This idea can be extended to find the area *between* two curves, as shown in the next example.

Example 4 A company is considering a new manufacturing process in one of its plants. The new process provides substantial initial savings, with the savings declining with time x according to the rate-of-savings function

$$S(x) = 100 - x^2,$$

where $S(x)$ is in thousands of dollars. At the same time, the cost of operating the new process increases with time x, according to the rate-of-cost function (in thousands of dollars)

$$C(x) = x^2 + \frac{14}{3}x.$$

(a) For how many years will the company realize savings?

Figure 13.19 on the next page shows the graphs of the rate-of-savings and the rate-of-cost functions. The rate-of-cost (marginal cost) is increasing, while the rate-of-savings (marginal savings) is decreasing. The company should use this new process until the difference between these quantities is zero—that is, until the time at which these graphs intersect. The graphs intersect when

$$C(x) = S(x),$$

or

$$100 - x^2 = x^2 + \frac{14}{3}x.$$

▪5 In Example 3 when will Wilson sell

(a) the 500th sign?
(b) the 1200th sign?

Answer:

(a) 8.61 years
(b) 14.06 years

▪6 Suppose the rate of sales in Example 3 is

$$T(x) = 40x - 350.$$

When will Wilson sell her 1000th sign?

Answer:

In 20 years

Solve this equation as follows.

$$0 = 2x^2 + \frac{14}{3}x - 100$$

$$= 3x^2 + 7x - 150 \qquad \text{Multiply by 3/2}$$

$$= (x - 6)(3x + 25) \qquad \text{Factor}$$

Set each factor equal to 0 and solve both equations to get

$$x = 6 \qquad \text{or} \qquad x = -\frac{25}{3}.$$

Only 6 is a meaningful solution here. The company should use the new process for 6 years.

(b) What will be the net total savings during this period?

Since the total savings over the six-year period is given by the area under the rate-of-savings curve and the total additional cost by the area under the rate-of-cost curve, the net total savings over the 6-year period is given by the area between the rate-of-cost and the rate-of-savings curves and the lines $x = 0$ and $x = 6$. This area can be evaluated with a definite integral as follows:

$$\text{Total savings} = \int_0^6 \left[(100 - x^2) - \left(x^2 + \frac{14}{3}x \right) \right] dx$$

$$= \int_0^6 \left(100 - \frac{14}{3}x - 2x^2 \right) dx$$

$$= 100x - \frac{7}{3}x^2 - \frac{2}{3}x^3 \Big]_0^6$$

$$= 100(6) - \frac{7}{3}(36) - \frac{2}{3}(216) = 372.$$

The company will save a total of $372,000 over the 6-year period. **7** **8**

7 In Example 4, find the total savings if pollution control regulations permit the new process for only 4 years.

Answer:

$320,000

8 If a company's marginal savings and marginal cost functions are

$$S(x) = 150 - 2x^2$$

and

$$C(x) = x^2 + 15x,$$

when $S(x)$ and $C(x)$ give amounts in thousands of dollars, find the following.
(a) Number of years until savings equal costs
(b) Total savings

Answer:
(a) 5
(b) $437,500

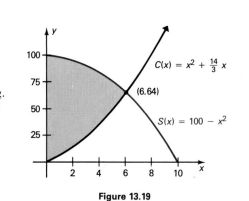

Figure 13.19

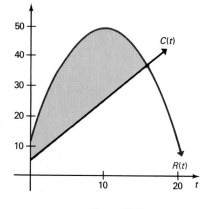

Figure 13.20

Example 5 A farmer has been using a new fertilizer that gives him a better yield, but because it exhausts the soil of other nutrients he must use other fertilizers in greater and greater amounts, so that his costs increase each year. The new fertilizer produces a rate of increase in revenue (in hundreds of dollars) given by

$$R(t) = -.4t^2 + 8t + 10,$$

where t is measured in years. The rate of increase in yearly costs (also in hundreds of dollars) due to use of the fertilizer is given by

$$C(t) = 2t + 5.$$

How long can the farmer profitably use the fertilizer? What will be his net increase in revenue over this period?

The farmer should use the new fertilizer until the marginal costs equal the marginal revenue. Find this point by solving the equation $R(t) = C(t)$ as follows.

$$-.4t^2 + 8t + 10 = 2t + 5$$
$$-4t^2 + 80t + 100 = 20t + 50$$
$$-4t^2 + 60t + 50 = 0$$

Use the quadratic formula to get

$$t = 15.8.$$

The new fertilizer will be profitable for about 15.8 years.

To find the total amount of additional revenue over the 15.8-year period, find the area between the graphs of the rate of revenue and the rate of cost functions, as shown in Figure 13.20.

$$\text{Total savings} = \int_0^{15.8} [R(t) - C(t)] \, dt$$

$$= \int_0^{15.8} [(-.4t^2 + 8t + 10) - (2t + 5)] \, dt$$

$$= \int_0^{15.8} (-.4t^2 + 6t + 5) \, dt$$

$$= \left(\frac{-.4t^3}{3} + \frac{6t^2}{2} + 5t \right) \Bigg]_0^{15.8}$$

$$= 302.01$$

The total savings will amount to about $30,000 over the 15.8-year period.

It is not realistic to say that the farmer will need to use the new process for 15.8 years—he will probably have to use it for 15 years or for 16 years. In this case, when the mathematical model produces results that are not in the domain of the function, it will be necessary to find the total savings after 15 years and after 16 years and then select the best result. ■

Example 6 Suppose the price, in cents, for a certain product is

$$p(x) = 900 - 20x - x^2,$$

when the demand for the product is x units. Also, suppose the function

$$p(x) = x^2 + 10x$$

gives the price, in cents, when the supply is x units. The graphs of both functions are shown in Figure 13.21 along with the equilibrium point at which supply and demand are equal. To find the equilibrium supply or demand x, solve the equation

$$900 - 20x - x^2 = x^2 + 10x$$
$$0 = 2x^2 + 30x - 900$$
$$0 = x^2 + 15x - 450.$$

The only positive solution of the equation is $x = 15$.

At the equilibrium point where supply and demand are both 15 units, the price is

$$p(15) = 900 - 20(15) - 15^2 = 375,$$

or $3.75.

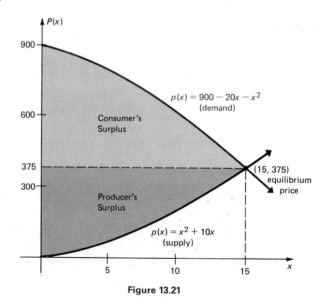

Figure 13.21

As the demand graph shows, there are consumers who are willing to pay more than the equilibrium price, so they benefit from the equilibrium price. Some benefit a lot, some less, and so on. Their total benefit, called the **consumer's surplus,** is represented by the area shown in Figure 13.21. The consumer's surplus is

$$\int_0^{15} (900 - 20x - x^2)\, dx - \int_0^{15} 375\, dx.$$

Evaluating these definite integrals gives

$$\left(900x - 10x^2 - \frac{1}{3}x^3\right)\Big]_0^{15} - 375x\Big]_0^{15}$$

$$= \left[900(15) - 10(15)^2 - \frac{1}{3}(15)^3 - 0\right] - [(375)(15) - 0]$$

$$= 4500.$$

Here the consumer's surplus is 4500 cents, or $45.

9 Given the demand function $p(x) = 12 - .07x$ and the supply function $p(x) = .05x$, where $p(x)$ is the price in dollars, find
(a) the equilibrium point;
(b) the consumer's surplus;
(c) the producer's surplus.

Answer:
(a) $x = 100$
(b) $350
(c) $250

On the other hand, some suppliers would have offered the product at a price below the equilibrium price, so they too gain from the equilibrium price. The total of the supplier's gains, called the **producer's surplus,** is also represented by an area shown in Figure 13.21. The producer's surplus is given by

$$\int_0^{15} 375 \, dx - \int_0^{15} (x^2 + 10x) \, dx = 375x\Big]_0^{15} - \left(\frac{1}{3}x^3 + 5x^2\right)\Big]_0^{15}$$

$$= 5625 - \left[\frac{1}{3}(15)^3 + 5(15)^2\right]$$

$$= 3375.$$

The producer's surplus is $33.75.   **9**

13.4 Exercises

Management *Work the following exercises. (See Examples 1–5.)*

1. A car-leasing firm must decide how much to charge for maintenance on the cars it leases. After careful study, the firm decides that the rate of maintenance, $M(x)$, on a new car will approximate

$$M(x) = 60(1 + x^2),$$

where x is the number of years the car has been in use. What total maintenance charge can the company expect for a two-year lease? What amount should be added to the monthly lease payments to pay for maintenance?

2. Using the function of Exercise 1, find the maintenance charge the company can expect during the third year. Find the total charge during the first three years. What monthly charge should be added to cover a three-year lease?

3. A company is considering a new manufacturing process. It knows that the rate of savings from the process, $S(t)$, will be about

$$S(t) = 1000(t + 2),$$

where t is the number of years the process has been in use. Find the total savings during the first year. Find the total savings during the first 6 years.

4. Assume that the new process in Exercise 3 costs $16,000. About when will it pay for itself?

5. A company is introducing a new product. Production is expected to grow slowly because of difficulties in the start-up process. It is expected that the rate of production, $P(x)$, will be approximated by

$$P(x) = 1000e^{.2x},$$

where x is the number of years since the introduction of the product. Will the company be able to supply 20,000 units during the first 4 years?

6. About when will the company of Exercise 5 be able to supply its 15,000th unit?

7. Suppose a company wants to introduce a new machine which will yield a rate of savings given by

$$S(x) = 294 - x^2,$$

where x is the number of years of operation of the machine, but will also entail costs of

$$C(x) = x^2 + \frac{1}{2}x.$$

(a) For how many years will it be profitable to use this new machine?

(b) What are the total net savings during the first year of use of the machine?

(c) What are the total net savings over the entire period of profitable life of the machine?

8. A farmer has been using a new fertilizer that gives him a better yield. Because it exhausts the soil of other nutrients, he must use other fertilizers in greater and greater amounts, so that his costs also increase each year. The new fertilizer produces an increase in rate of revenue (in hundreds of dollars) of

$$K(t) = -.4t^2 + 8t + 10,$$

where t is measured in years. The yearly rate of costs due to use of the fertilizer increase according to the function

$$C(t) = 2t + 5.$$

(a) How long can the farmer profitably use the fertilizer?

(b) What will his net savings be over this period?

9. De Win Enterprises had an expenditure rate (in hundreds of dollars) of $E(x) = 4x + 2$, and an income rate (in hundreds of dollars) of $I(x) = 100 - x$ on a particular job, where x was the number of days from the start of the job. Their profit on that job will equal total income less total expenditure. Profit will be maximized if the job ends at the optimum time, which is the point where the two curves meet. Find the following.

(a) The optimum number of days for the job to last.

(b) The total income for the optimum number of days.

(c) The total expenditure for the optimum number of days.

(d) The maximum profit for the job.

10. A factory has installed a new process that will produce an increased rate of revenue (in thousands of dollars) of

$$R(t) = -t^2 + 15.5t + 16,$$

where t is measured in years. The new process produces additional costs (in thousands of dollars) at the rate of

$$C(t) = .25t + 4.$$

(a) When will it no longer be profitable to use this new process?

(b) Find the total net savings.

Management *Work the following supply and demand exercises. (See Example 6.)*

11. Suppose the supply function for a certain commodity is given by

$$p(x) = 100 + 3x + x^2.$$

Suppose that supply and demand are in equilibrium when $x = 3$. Find the producer's surplus.

12. Find the consumer's surplus if the demand function for an item is given by

$$p(x) = 50 - x^2,$$

assuming supply and demand are in equilibrium at $x = 5$.

13. Find the consumer's surplus if the demand price for an item is given by

$$p(x) = -x^2 - 8x + 50,$$

if supply and demand are in equilibrium at $x = 3$.

14. Find the producer's surplus if the supply function for an item is

$$p(x) = \frac{1}{2}x^2 + 10x.$$

Assume supply and demand are in equilibrium when $x = 5$.

15. Find the producer's surplus if the supply function of some item is given by

$$p(x) = x^2 + 2x + 50.$$

Assume supply and demand are in equilibrium at $x = 20$.

16. Find the consumer's surplus if the demand for an item is given by

$$p(x) = -(x + 4)^2 + 66,$$

if supply and demand are in equilibrium at $x = 3$.

17. Find the consumer's surplus and the producer's surplus for an item having supply function

$$p(x) = x^2 + \frac{11}{4}x$$

and demand function

$$p(x) = 150 - x^2.$$

18. Suppose the supply function of a certain item is given by

$$p(x) = \frac{7}{5}x,$$

and the demand function is given by

$$p(x) = -\frac{3}{5}x + 10.$$

(a) Graph the supply and demand curves.
(b) Find the point at which supply and demand are in equilibrium.
(c) Find the consumer's surplus.
(d) Find the producer's surplus.

Work the following exercises.

19. Management In a recent inflationary period, costs of a certain industrial process, in millions of dollars, were increasing according to the function

$$i(t) = .45t^{3/2},$$

where $i(t)$ is the rate of increase in costs at time t measured in years. Find the total increase in costs during the first 4 years.

20. Management A new smog control device will reduce the output of oxides of sulfur from automobile exhausts. It is estimated that the rate of savings to the community from the use of this device will be approximated by

$$S(x) = -x^2 + 4x + 8,$$

where $S(x)$ is the rate of savings in millions of dollars after x years of use of the device. This new device cuts down on the production of oxides of sulfur, but causes an increase in the production of oxides of nitrogen. The additional costs to the community caused by this increase are approximated by

$$C(x) = \frac{3}{25}x^2,$$

where $C(x)$ is the additional cost in millions after x years.

(a) For how many years will it pay to use the new device?
(b) What will be the total savings over this period of time?

21. Management If a large truck has been driven x thousand miles, the rate of repair costs in dollars per mile is given by $R(x)$, where

$$R(x) = .05x^{3/2}.$$

Find the total repair costs if the truck is driven
(a) 100,000 miles;
(b) 400,000 miles.

22. Natural Science Pollution from a factory is entering a lake. The rate of concentration, $P(t)$, of the pollutant at time t, is given by

$$P(t) = 140t^{5/2},$$

where t is the number of years since the factory started spilling pollutant into the lake. Ecologists estimate that the lake can accept a total level of pollution of 4850 units before all the fish in the lake die. Can the factory operate for 4 years without killing all the fish in the lake?

23. Natural Science A program is started to clean up the lake in Exercise 22. The cleanup efforts result in changing the rate of concentration $P(t)$ of the pollutant at time t to

$$P(t) = 4000e^{-t/10},$$

where t is the number of years from the time the program is started. Suppose there are 2000 units of pollution in the lake when the cleanup program begins. Can the factory operate an additional 4 years without killing all the fish in the lake?

24. Natural Science From 1905 to 1920, most of the predators of the Kaibab Plateau of Arizona were killed by hunters. This allowed the deer population there to grow rapidly until they had depleted their food sources, which caused a rapid decline in population. The rate of change of this deer population during that time span is approximated by the function

$$D(t) = \frac{25}{2}t^3 - \frac{5}{8}t^4,$$

where t is the time in years $(0 \le t \le 25)$.
(a) Find the function for the deer population if there were 4000 deer in 1905 $(t = 0)$.
(b) What was the population in 1920?
(c) When was the population at a maximum?
(d) What was the maximum population?

25. Management After a new firm starts in business, it finds that its rate of profits (in hundreds of dollars) after t years of operation, is given by

$$P(t) = 6t^2 + 4t + 5.$$

Find the total profits in the first three years.

26. Management Find the profit in Exercise 25 in the fourth year of operation.

27. An oil tanker is leaking oil at the rate of $20t + 50$ barrels per hour, where t is time in hours after the tanker hits a hidden rock. Find the total number of barrels that the ship will leak on the first day.

28. Find the number of barrels that the ship of Exercise 27 will leak on the second day.

29. Management A worker new to a job will improve his efficiency with time so that it takes him fewer hours to produce an item with each day on the job up to a certain point. Suppose the rate of change of the number of hours it takes a worker in a certain factory to produce the xth item is given by

$$H(x) = 20 - 2x.$$

(a) What is the total number of hours required to produce the first 5 items?

(b) What is the total number of hours required to produce the first 10 items?

30. Natural Science After long study, tree scientists conclude that a eucalyptus tree will grow at the rate of $.2 + 4t^{-4}$ feet per year, when t is time in years. Find the number of feet that the tree will grow in the second year.

31. Natural Science Find the number of feet the tree in Exercise 30 will grow in the third year.

32. Natural Science For a certain drug, the rate of reaction in appropriate units is given by

$$R(t) = \frac{5}{t} + \frac{2}{t^2},$$

where t is measured in hours after the drug is administered. Find the total reaction to the drug

(a) from $t = 1$ to $t = 12$;

(b) from $t = 12$ to $t = 24$.

33. Social Science Suppose that all the people in a country are ranked according to their incomes, starting at the bottom. Let x represent the fraction of the community making the lowest income ($0 \leq x \leq 1$); $x = .4$, therefore, represents the lower 40% of all income producers. Let $I(x)$ represent the proportion of the total income earned by the lowest x of all people. Thus, $I(.4)$ represents the fraction of total income earned by the lowest 40% of the population. Suppose

$$I(x) = .9x^2 + .1x.$$

Find and interpret the following.

(a) $I(.1)$ **(b)** $I(.4)$ **(c)** $I(.6)$ **(d)** $I(.9)$

34. Social Science In Exercise 33, if income were distributed uniformly, we would have $I(x) = x$. The area between the curves $I(x) = x$ and the particular function $I(x)$ for a given country is called the *coefficient of inequality* for that country.

(a) Graph $I(x) = x$ and $I(x) = .9x^2 + .1x$ for $0 \leq x \leq 1$ on the same axes.

(b) Find the area between the curves.

13.5   Other Methods of Integration

Substitution In Section 13.1 we saw how to integrate a few simple functions. More complicated functions can sometimes be integrated by *substitution*. The technique depends on the idea of a differential. If $u = f(x)$, the *differential* of u, written du, is defined as

$$du = f'(x)\ dx.$$

For example, if $u = 6x^4$, then $du = 24x^3\ dx$. ①

Recall the chain rule for derivatives as used in the following example:

$$\frac{d}{dx}(x^2 - 1)^5 = 5(x^2 - 1)^4(2x) = 10x(x^2 - 1)^4.$$

① Find du for the following.

(a) $u = 9x$

(b) $u = 5x^3 + 2x^2$

(c) $u = e^{-2x}$

Answer:

(a) $du = 9dx$

(b) $du = (15x^2 + 4x)\ dx$

(c) $du = -2e^{-2x}\ dx$

As in this example, the result of using the chain rule is often a product of two functions. Because of this, functions formed by the product of two functions can sometimes be integrated by using the chain rule in reverse. In the example above, working backwards from the derivative gives

$$\int 10x(x^2 - 1)^4 \, dx = (x^2 - 1)^5 + C.$$

To find an antiderivative involving products, it often helps to make a substitution: let $u = x^2 - 1$, then $du = 2x \, dx$. Now substitute u for $x^2 - 1$ and du for $2x \, dx$ in the indefinite integral above.

$$\int 10x(x^2 - 1)^4 \, dx = \int 5 \cdot 2x(x^2 - 1)^4 \, dx$$

$$= 5 \int (x^2 - 1)^4 (2x \, dx)$$

$$= 5 \int u^4 \, du$$

This last integral can now be found by the power rule.

$$5 \int u^4 \, du = 5 \cdot \frac{1}{5} u^5 + C = u^5 + C$$

Finally, substitute $x^2 - 1$ for u.

$$\int 10x(x^2 - 1)^4 \, dx = (x^2 - 1)^5 + C$$

This method of integration is called **integration by substitution.** As shown above, it is simply the chain rule for derivatives in reverse. The results can always be verified by differentiation.

Example 1   Find $\int 6x(3x^2 + 4)^4 \, dx$.

A certain amount of trial and error may be needed to decide on the expression to set equal to u. The integrand must be written as two factors, one of which is the derivative of the other. In this example, if $u = 3x^2 + 4$, then $du = 6x \, dx$. Now substitute.

$$\int 6x(3x^2 + 4)^4 \, dx = \int (3x^2 + 4)^4 (6x \, dx) = \int u^4 \, du$$

Find this last indefinite integral.

$$\int u^4 \, du = \frac{u^5}{5} + C$$

Now replace u with $3x^2 + 4$.

$$\int 6x(3x^2 + 4)^4 \, dx = \frac{u^5}{5} + C = \frac{(3x^2 + 4)^5}{5} + C$$

To verify this result, find the derivative:

$$\frac{d}{dx}\left[\frac{(3x^2+4)^5}{5}+C\right]=\frac{5}{5}(3x^2+4)^4(6x)+0=(3x^2+4)^4(6x),$$

which is the original function. **2**

2 Find the following.
(a) $\int 8x(4x^2-1)^5\,dx$
(b) $\int 18x^2(6x^3-5)^{3/2}\,dx$

Answer:

(a) $\dfrac{(4x^2-1)^6}{6}+C$

(b) $\dfrac{2(6x^3-5)^{5/2}}{5}+C$

Example 2 Find $\int x^2\sqrt{x^3+1}\,dx$.

An expression raised to a power is usually a good choice for u, so because of the square root or 1/2 power, let $u=x^3+1$; then $du=3x^2\,dx$. The integrand does not contain the constant 3, which is needed for du. To take care of this, multiply by 3/3, placing 3 inside the integral sign and 1/3 outside.

$$\int x^2\sqrt{x^3+1}\,dx=\frac{1}{3}\int 3x^2\sqrt{x^3+1}\,dx=\frac{1}{3}\int\sqrt{x^3+1}(3x^2\,dx)$$

Now substitute u for x^3+1 and du for $3x^2\,dx$, and integrate.

$$\frac{1}{3}\int\sqrt{x^3+1}\,3x^2\,dx=\frac{1}{3}\int\sqrt{u}\,du=\frac{1}{3}\int u^{1/2}\,du$$

$$=\frac{1}{3}\cdot\frac{u^{3/2}}{3/2}+C=\frac{2}{9}u^{3/2}+C$$

Since $u=x^3+1$,

$$\int x^2\sqrt{x^3+1}\,dx=\frac{2}{9}(x^3+1)^{3/2}+C.\quad \textbf{3}$$

3 Find the following.
(a) $\int x(5x^2+6)^4\,dx$
(b) $\int x^2(x^3-2)^5\,dx$
(c) $\int x\sqrt{x^2+16}\,dx$

Answer:

(a) $\dfrac{1}{50}(5x^2+6)^5+C$

(b) $\dfrac{1}{18}(x^3-2)^6+C$

(c) $\dfrac{1}{3}(x^2+16)^{3/2}+C$

The substitution method given in the examples above will not always work. For example, we might try to find

$$\int x^3\sqrt{x^3+1}\,dx$$

by substituting $u=x^3+1$, so that $du=3x^2\,dx$. However, there is no constant which can be inserted inside the integral sign to give $3x^2$. This integral, and a great many others, cannot be evaluated by substitution. **4**

4 Find the following.
(a) $\int x(6x+2)^{1/2}\,dx$
(b) $\int 5x^2(8x^4+2)^3\,dx$

Answer:
Neither can be found by the substitution method.

Integration by substitution may also be needed to evaluate definite integrals. First find the necessary antiderivatives and then evaluate them at the upper and lower limits of integration.

Example 3 Evaluate $\displaystyle\int_1^4\frac{2x+5}{(x^2+5x)^2}\,dx$.

To evaluate this definite integral, first find the indefinite integral

$$\int\frac{2x+5}{(x^2+5x)^2}\,dx.$$

Let $u = x^2 + 5x$, so that $du = (2x + 5) \, dx$. This gives

$$\int \frac{2x + 5}{(x^2 + 5x)^2} \, dx = \int \frac{du}{u^2} = \int u^{-2} \, du$$

$$= \frac{u^{-1}}{-1} + C = \frac{-1}{u} + C.$$

Substituting $x^2 + 5x$ for u gives

$$\int \frac{2x + 5}{(x^2 + 5x)^2} \, dx = \frac{-1}{x^2 + 5x} + C.$$

This antiderivative can now be used to evaluate the given definite integral. Using the fundamental theorem of calculus,

$$\int_1^4 \frac{2x + 5}{(x^2 + 5x)^2} \, dx = \frac{-1}{x^2 + 5x} \Big]_1^4$$

$$= \frac{-1}{4^2 + 5 \cdot 4} - \frac{-1}{1^2 + 5 \cdot 1}$$

$$= \frac{-1}{36} + \frac{1}{6} = \frac{5}{36}.$$

Remember, it is not necessary to include the constant C when evaluating a definite integral. **5**

Recall the formula for $\dfrac{d}{dx}(e^u)$, where $u = f(x)$:

$$\frac{d}{dx}(e^u) = e^u \frac{d}{dx} u.$$

For example, if $u = x^2$ then $\dfrac{d}{dx} u = \dfrac{d}{dx}(x^2) = 2x$, and

$$\frac{d}{dx}(e^{x^2}) = e^{x^2} \cdot 2x.$$

Working backwards, if $u = x^2$, then $du = 2x \, dx$, so

$$\int e^{x^2} \cdot 2x \, dx = \int e^u \, du = e^u + C = e^{x^2} + C.$$

Summarizing, we have the following rule for the indefinite integral of e^u, where $u = f(x)$.

Indefinite Integral of e^u If $u = f(x)$, then

$$\int e^u \, du = e^u + C.$$

---

5 Evaluate each definite integral.

(a) $\int_0^3 x\sqrt{x^2 + 1} \, dx$

(b) $\int_0^2 z(z^2 + 1)^2 \, dz$

(c) $\int_0^2 (r + 2)\sqrt{r^2 + 4r} \, dr$

Answer:

(a) $\dfrac{1}{3}(10^{3/2} - 1)$

(b) $\dfrac{62}{3}$

(c) $\dfrac{12^{3/2}}{3}$

Example 4   Find $\int x^2 \cdot e^{x^3}\, dx$.

Let $u = x^3$, the exponent on e. Then $du = 3x^2\, dx$. Multiplying by 3/3 gives

$$\int x^2 \cdot e^{x^3}\, dx = \frac{1}{3} \int e^{x^3}(3x^2\, dx)$$

$$= \frac{1}{3} \int e^u\, du$$

$$= \frac{1}{3} e^u + C$$

$$= \frac{1}{3} e^{x^3} + C. \quad \boxed{6}$$

6 Find the following.
(a) $\int 8xe^{3x^2}\, dx$
(b) $\int x^2 e^{x^3}\, dx$

Answer:

(a) $\frac{4}{3} e^{3x^2} + C$

(b) $\frac{1}{3} e^{x^3} + C$

Recall that the antiderivative of $f(x) = 1/x$ is $\ln |x|$. The next example uses $\int x^{-1}\, dx = \ln |x| + C$, and the method of substitution.

Example 5   Find $\int \dfrac{(2x - 3)\, dx}{x^2 - 3x}$.

Let $u = x^2 - 3x$, so that $du = (2x - 3)\, dx$. Then

$$\int \frac{(2x - 3)\, dx}{x^2 - 3x} = \int \frac{du}{u} = \ln |u| + C = \ln |x^2 - 3x| + C. \quad \blacksquare$$

Generalizing from Example 5 gives the rule for finding the indefinite integral of u^{-1}, where $u = f(x)$.

Indefinite Integral of u^{-1}   If $u = f(x)$, then

$$\int u^{-1}\, du = \int \frac{du}{u} = \ln |u| + C.$$

7 Find the following.
(a) $\int \dfrac{4\, dx}{x - 3}$

(b) $\int \dfrac{(2x - 9)\, dx}{x^2 - 9x}$

(c) $\int \dfrac{(3x^2 + 8)\, dx}{x^3 + 8x + 5}$

Answer:
(a) $4 \ln |x - 3| + C$
(b) $\ln |x^2 - 9x| + C$
(c) $\ln |x^3 + 8x + 5| + C$

The substitution method is useful if the integral can be written in one of the following forms, where $u(x)$ is some function of x.

Substitution Method

Form of the Integral	*Form of the Antiderivative*
1. $\int [u(x)]^n \cdot u'(x)\, dx,\ n \neq -1$	$\dfrac{[u(x)]^{n+1}}{n + 1} + C$
2. $\int e^{u(x)} \cdot u'(x)\, dx$	$e^{u(x)} + C$
3. $\int \dfrac{u'(x)\, dx}{u(x)}$	$\ln |u(x)| + C$

Tables of Integrals As we noted above, there are many useful functions whose antiderivatives cannot be found by the methods we have discussed. However, many useful integrals can be found from a *table of integrals* such as Table 11 at the back of the book. The next examples show how to use this table.

8 Find the following.

(a) $\displaystyle\int \frac{4}{\sqrt{x^2 + 100}}\, dx$

(b) $\displaystyle\int \frac{-9}{\sqrt{x^2 - 4}}\, dx$

(c) $\displaystyle\int -6 \ln |x|\, dx$

Answer:

(a)

$4 \ln \left| \dfrac{x + \sqrt{x^2 + 100}}{10} \right| + C$

(b)

$-9 \ln \left| \dfrac{x + \sqrt{x^2 - 4}}{2} \right| + C$

(c)

$-6x(\ln |x| - 1) + C$

9 Find the following.

(a) $\displaystyle\int \frac{1}{x^2 - 4}\, dx$

(b) $\displaystyle\int \frac{-6}{x\sqrt{25 - x^2}}\, dx$

(c) $\displaystyle\int \frac{5}{x\sqrt{36 + x^2}}\, dx$

Answer:

(a)

$\dfrac{1}{4} \ln \left(\dfrac{x - 2}{x + 2} \right) + C$

$(x^2 > 4)$

(b)

$\dfrac{6}{5} \ln \left(\dfrac{5 + \sqrt{25 - x^2}}{x} \right) + C$

$(0 < x < 5)$

(c)

$-\dfrac{5}{6} \ln \left| \dfrac{6 + \sqrt{36 + x^2}}{x} \right|$

$+ C$

Example 6 Find $\displaystyle\int \frac{1}{\sqrt{x^2 + 16}}\, dx$.

By inspecting the table, we see that if $a = 4$, this antiderivative is the same as entry 5 of the table. Entry 5 of the table is

$$\int \frac{1}{\sqrt{x^2 + a^2}}\, dx = \ln \left| \frac{x + \sqrt{x^2 + a^2}}{a} \right| + C.$$

Substituting 4 for a in this entry, we get

$$\int \frac{1}{\sqrt{x^2 + 16}}\, dx = \ln \left| \frac{x + \sqrt{x^2 + 16}}{4} \right| + C.$$

This result could be verified by taking the derivative of the right-hand side of this last equation. **8**

Example 7 Find $\displaystyle\int \frac{8}{16 - x^2}\, dx$.

Convert this antiderivative into the one given in entry 7 of the table by writing the 8 in front of the integral sign (permissible only with constants) and by letting $a = 4$. Doing this gives

$$8 \int \frac{1}{16 - x^2}\, dx = 8 \left[\frac{1}{2 \cdot 4} \ln \left(\frac{4 + x}{4 - x} \right) \right] + C$$

$$= \ln \left(\frac{4 + x}{4 - x} \right) + C.$$

In entry 7 of the table, the condition $x^2 < a^2$ is given. Here $a = 4$, so the result given above is valid only for $x^2 < 16$, with the final answer written as

$$\int \frac{8}{16 - x^2}\, dx = \ln \left(\frac{4 + x}{4 - x} \right) + C, \qquad \text{for } x^2 < 16.$$

Because of the condition $x^2 < 16$, the expression in parentheses is always positive, so that absolute value bars are not needed **9**

Example 8 Find $\int \sqrt{9x^2 + 1}\, dx$.

This antiderivative seems most similar to entry 15 of the table. However, entry 15 requires that the coefficient of the x^2 term be 1. We can satisfy that requirement here by factoring out the 9.

10 Find the following.

(a) $\displaystyle\int \frac{3}{16x^2 - 1}\, dx$

(b) $\displaystyle\int \frac{-1}{100x^2 - 1}\, dx$

Answer:

(a)

$$\frac{3}{8}\ln\left|\frac{x - \frac{1}{4}}{x + \frac{1}{4}}\right| + C$$

$$\left(x^2 > \frac{1}{16}\right)$$

(b)

$$-\frac{1}{20}\ln\left|\frac{x - \frac{1}{10}}{x + \frac{1}{10}}\right| + C$$

$$\left(x^2 > \frac{1}{100}\right)$$

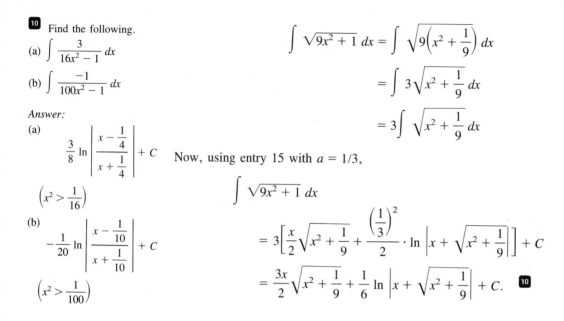

$$\int \sqrt{9x^2 + 1}\, dx = \int \sqrt{9\left(x^2 + \frac{1}{9}\right)}\, dx$$

$$= \int 3\sqrt{x^2 + \frac{1}{9}}\, dx$$

$$= 3\int \sqrt{x^2 + \frac{1}{9}}\, dx$$

Now, using entry 15 with $a = 1/3$,

$$\int \sqrt{9x^2 + 1}\, dx$$

$$= 3\left[\frac{x}{2}\sqrt{x^2 + \frac{1}{9}} + \frac{\left(\frac{1}{3}\right)^2}{2}\cdot\ln\left|x + \sqrt{x^2 + \frac{1}{9}}\right|\right] + C$$

$$= \frac{3x}{2}\sqrt{x^2 + \frac{1}{9}} + \frac{1}{6}\ln\left|x + \sqrt{x^2 + \frac{1}{9}}\right| + C. \quad \boxed{10}$$

13.5 Exercises

Use substitution to find the following indefinite integrals.
(See Examples 1, 2, 4 and 5.)

1. $\displaystyle\int 4(2x + 3)^4\, dx$

2. $\displaystyle\int (-4t + 1)^3\, dt$

3. $\displaystyle\int \frac{4}{(y - 2)^3}\, dy$

4. $\displaystyle\int \frac{-3}{(x + 1)^4}\, dx$

5. $\displaystyle\int \frac{2\, dm}{(2m + 1)^3}$

6. $\displaystyle\int \frac{3\, du}{\sqrt{3u - 5}}$

7. $\displaystyle\int \frac{2x + 2}{(x^2 + 2x - 4)^4}\, dx$

8. $\displaystyle\int \frac{6x^2\, dx}{(2x^3 + 7)^{3/2}}$

9. $\displaystyle\int z\sqrt{z^2 - 5}\, dz$

10. $\displaystyle\int r\sqrt{r^2 + 2}\, dr$

11. $\displaystyle\int (-4e^{2p})\, dp$

12. $\displaystyle\int 5e^{-.3g}\, dg$

13. $\displaystyle\int 3x^2 e^{2x^3}\, dx$

14. $\displaystyle\int re^{-r^2}\, dr$

15. $\displaystyle\int (1 - t)e^{2t - t^2}\, dt$

16. $\displaystyle\int (x^2 - 1)e^{x^3 - 3x}\, dx$

17. $\displaystyle\int \frac{e^{1/z}}{z^2}\, dz$

18. $\displaystyle\int \frac{e^{\sqrt{y}}}{2\sqrt{y}}\, dy$

19. $\displaystyle\int \frac{-8}{1 + 3x}\, dx$

20. $\displaystyle\int \frac{9}{2 + 5t}\, dt$

21. $\displaystyle\int \frac{dt}{2t + 1}$

22. $\displaystyle\int \frac{dw}{5w - 2}$

23. $\displaystyle\int \frac{v\, dv}{(3v^2 + 2)^4}$

24. $\displaystyle\int \frac{x\, dx}{(2x^2 - 5)^3}$

25. $\displaystyle\int \frac{x - 1}{(2x^2 - 4x)^2}\, dx$

26. $\displaystyle\int \frac{2x + 1}{(x^2 + x)^3}\, dx$

27. $\displaystyle\int \left(\frac{1}{r} + r\right)\left(1 - \frac{1}{r^2}\right)\, dr$

28. $\displaystyle\int \left(\frac{2}{A} - A\right)\left(\frac{-2}{A^2} - 1\right)\, dA$

29. $\displaystyle\int \frac{x^2 + 1}{(x^3 + 3x)^{2/3}}\, dx$

30. $\displaystyle\int \frac{B^3 - 1}{(2B^4 - 8B)^{3/2}}\, dB$

31. $\displaystyle\int p(p + 1)^5\, dp$

32. $\displaystyle\int x^3(1 + x^2)^{1/4}\, dx$

33. $\displaystyle\int t\sqrt{5t - 1}\, dt$

34. $\displaystyle\int 4r\sqrt{8 - r}\, dr$

35. $\int \dfrac{u}{\sqrt{u-1}}\, du$

36. $\int \dfrac{2x}{(x+5)^6}\, dx$

Use substitution to find the following definite integrals. (See Example 3.)

37. $\displaystyle\int_0^1 2x(x^2+1)^3\, dx$

38. $\displaystyle\int_0^1 y^2(y^3-4)^3\, dy$

39. $\displaystyle\int_0^4 (\sqrt{x^2+12x})(x+6)\, dx$

40. $\displaystyle\int_0^8 (\sqrt{x^2-6x})(x-3)\, dx$

41. $\displaystyle\int_{-1}^{e-2} \dfrac{t}{t^2+2}\, dt$

42. $\displaystyle\int_{-2}^0 \dfrac{-4x}{x^2+3}\, dx$

43. $\displaystyle\int_0^1 ze^{2z^2}\, dz$

44. $\displaystyle\int_1^2 x^2 e^{-x^3}\, dx$

Use the table of integrals to find each antiderivative. (See Examples 6–8.)

45. $\int \ln|4x|\, dx$

46. $\int \ln\left|\dfrac{3}{5}x\right|\, dx$

47. $\int \dfrac{-4}{\sqrt{x^2+36}}\, dx$

48. $\int \dfrac{9}{\sqrt{x^2+9}}\, dx$

49. $\int \dfrac{6}{x^2-9}\, dx$

50. $\int \dfrac{-12}{x^2-16}\, dx$

51. $\int \dfrac{-4}{x\sqrt{9-x^2}}\, dx$

52. $\int \dfrac{3}{x\sqrt{121-x^2}}\, dx$

53. $\int \dfrac{-2x}{3x+1}\, dx$

54. $\int \dfrac{6x}{4x-5}\, dx$

55. $\int \dfrac{2}{3x(3x-5)}\, dx$

56. $\int \dfrac{-4}{3x(2x+7)}\, dx$

57. $\int \dfrac{4}{4x^2-1}\, dx$

58. $\int \dfrac{-6}{9x^2-1}\, dx$

59. $\int \dfrac{3}{x\sqrt{1-9x^2}}\, dx$

60. $\int \dfrac{-2}{x\sqrt{1-16x^2}}\, dx$

61. $\int x^4 \ln|x|\, dx$

62. $\int 4x^2 \ln|x|\, dx$

63. $\int \dfrac{\ln|x|}{x^2}\, dx$

64. $\int \dfrac{-2\ln|x|}{x^3}\, dx$

65. $\int xe^{-2x}\, dx$

66. $\int xe^{3x}\, dx$

Work the following exercises.

67. Management Suppose the marginal revenue in hundred thousands of dollars from the sale of x jet planes is
$$R(x) = 2x(x^2+50)^3.$$
Find the total revenue from the sale of 3 planes.

68. Management The rate of expenditure for maintenance for a particular machine is given by
$$M(x) = \sqrt{x^2+12x}(2x+12),$$
where x is time measured in years. Find the total maintenance charge for the first 5 years of the machine's life.

69. Management The rate of growth of the profit (in millions of dollars) from a new technology is approximated by
$$p(x) = xe^{-x^2},$$
where x represents time measured in years. Find the total profit from the first 3 years that the new technology is in operation.

70. Natural Science If x milligrams of a certain drug are administered to a person, the rate of change in the person's temperature, in degrees Celsius, with respect to the dosage (the person's *sensitivity* to the drug) is given by
$$D(x) = \dfrac{2}{x+9}.$$
Find the total change in body temperature if 3 milligrams of the drug are administered.

13.6 Differential Equations and Applications

The rate of change with respect to time of many natural growth or decay processes can be expressed as a function of time by writing

$$\frac{dy}{dt} = kt \quad \text{or} \quad \frac{dy}{dt} = ky,$$

where t represents time, $y = f(t)$, and k is a constant. Any equation such as these, containing one or more of the derivatives of a function, is called a **differential equation.**

In most differential equations, it is convenient to represent the derivative as dy/dx, rather than y' or $f'(x)$. For a second derivative, the notation d^2y/dx^2 is used. Some other examples of differential equations include

$$\frac{dy}{dx} = x^{1/2}, \quad \frac{dy}{dx} = 2y^2, \quad \text{and} \quad \frac{d^2y}{dx^2} = 3x^2 + 4.$$

To solve a differential equation such as

$$\frac{dy}{dt} = kt, \tag{1}$$

we need an antiderivative of each side of the equation. Since the left side of the equation is the derivative of y with respect to t, the antiderivative on the left is just $y + C_1$. On the right side,

$$\int kt \, dt = \frac{1}{2}kt^2 + C_2,$$

since t is the variable and k is a constant. The solution of equation (1) is thus

$$y + C_1 = \frac{1}{2}kt^2 + C_2$$

or

$$y = \frac{1}{2}kt^2 + C_2 - C_1.$$

Replacing the constant $C_2 - C_1$ with the single constant C gives

$$y = \frac{1}{2}kt^2 + C. \tag{2}$$

Equation (2) is called the *general solution* of differential equation (1). (From now on we will add just one constant with the understanding that it represents the difference between the two constants obtained in the integrations.) ⬚

1 Find the general solution of
(a) $dy/dx = 4x$;
(b) $dy/dx = -x^3$;
(c) $dy/dx = 2x^2 - 5x$.

Answer:
(a) $y = 2x^2 + C$

(b) $y = -\frac{1}{4}x^4 + C$

(c) $y = \frac{2}{3}x^3 - \frac{5}{2}x^2 + C$

Example 1 The time-dating of dairy products depends on the solution of a differential equation. The rate of growth of bacteria in such products increases with time. If y is the number of bacteria (in thousands) present at a time t (in days), then the rate of growth of bacteria can be expressed as dy/dt and we have

$$\frac{dy}{dt} = kt,$$

where k is an appropriate constant. Suppose, for some product, $k = 10$, so that

$$\frac{dy}{dt} = 10t. \tag{3}$$

Further, suppose at $t = 0$, $y = 50$ (in thousands). Find an equation for y in terms of t.

   We need to solve the differential equation $dy/dt = 10t$. Find an antiderivative for each side and add just one constant. This gives

$$y = \int 10t \, dt,$$

$$y = \frac{10t^2}{2} + C,$$

or

$$y = 5t^2 + C,$$

the general solution of equation (3). To find a **particular solution,** where the value of C is known, use the fact that $y = 50$ when $t = 0$. This is called an **initial condition** or **boundary condition.** By substitution, we have

$$50 = 5(0)^2 + C$$
$$C = 50.$$

Thus, the particular solution is

$$y = 5t^2 + 50. \quad \boxed{2}$$

2 Suppose the maximum allowable value for y in Example 1 is 550. How should the product be dated?

Answer:
10 days from the date when $t = 0$

Example 2 Find the particular solution of

$$\frac{dy}{dx} = 2x + 5,$$

given that $y = 2$ when $x = -1$.

   Take antiderivatives on both sides of the equation to get

$$y = \frac{2x^2}{2} + 5x + C$$

$$y = x^2 + 5x + C.$$

Substituting for x and y gives

$$2 = (-1)^2 + 5(-1) + C$$
$$C = 6.$$

The particular solution is $y = x^2 + 5x + 6$. **3**

3 Find a particular solution to
(a) $dy/dx = 3x^2$, if $y = 4$ when $x = 1$;
(b) $dy/dx = 4x^3 - 3x^2 + 2x$ given that $y = 2$ when $x = 1$.

Answer:
(a) $y = x^3 + 3$
(b) $y = x^4 - x^3$
$\quad + x^2 + 1$

Example 3 Find the particular solution to

$$\frac{d^2y}{dx^2} = 2x + 4, \tag{4}$$

given that $y = 2$ when $x = 0$ and $y = -4$ when $x = 3$.
Integrating both sides with respect to x gives

$$\int \frac{d^2y}{dx^2}\, dx = \int (2x + 4)\, dx.$$

On the left,

$$\int \frac{d^2y}{dx^2}\, dx = \frac{dy}{dx} + C,$$

so that equation (4) leads to

$$\frac{dy}{dx} = x^2 + 4x + C_1$$

for some constant C_1. Integrate again to get

$$y = \frac{x^3}{3} + 2x^2 + C_1x + C_2 \tag{5}$$

for some constant C_2. Equation (5) is the general solution of equation (4). To find values of C_1 and C_2, use the given pairs of values. If $x = 0$, then $y = 2$, so that

$$2 = \frac{0^3}{3} + 2 \cdot 0^2 + C_1 \cdot 0 + C_2$$
$$C_2 = 2.$$

Using $C_2 = 2$, $x = 3$, and $y = -4$, we have

$$-4 = \frac{3^3}{3} + 2 \cdot 3^2 + C_1 \cdot 3 + 2$$
$$-4 = 29 + 3C_1$$
$$C_1 = \frac{-33}{3} = -11.$$

4 Find and check the particular solution to

$$\frac{d^2y}{dx^2} = x^2 - 4x,$$

given that $y = 5$ when $x = 0$ and $y = 1$ when $x = -1$.

Answer:

$$y = \frac{x^4}{12} - \frac{2x^3}{3} + \frac{57x}{12} + 5$$

The particular solution is thus

$$y = \frac{x^3}{3} + 2x^2 - 11x + 2.$$

To check this solution, differentiate twice. **4**

Differential equations have many practical applications, especially in economics and biology. For example, **marginal productivity** is the rate at which production changes (increases or decreases) for a unit change in investment. Thus, marginal productivity can be expressed as the first derivative of the function which gives production in terms of investment.

Example 4 Suppose the marginal productivity of a manufacturing process is given by

$$P'(x) = 3x^2 - 10, \tag{6}$$

where x is the amount of the investment in hundreds of thousands of dollars. If the process produces 100 units per month with its present investment of \$300,000 (that is, $x = 3$), by how much would production increase if the investment is increased to \$500,000?

To obtain an equation for production we can take antiderivatives on both sides of equation (6) to get

$$P(x) = x^3 - 10x + C.$$

To find C, use the given initial values: $P(x) = 100$ when $x = 3$.

$$100 = 3^3 - 10(3) + C$$
$$C = 103$$

Production is thus given by

$$P(x) = x^3 - 10x + 103,$$

and if investment is increased to \$500,000, production becomes

$$P(5) = 5^3 - 10(5) + 103 = 178.$$

An increase to \$500,000 in investment will increase production from 100 units to 178 units. **5**

5 In Example 4, if marginal productivity is changed to

$P'(x) = 3x^2 + 2x,$

with the same initial conditions,
(a) find an equation for production;
(b) find the increase in production if investment increases to \$500,000.

Answer:
(a) $P(x) = x^3 + x^2 + 64$
(b) Production goes from 100 units to 214 units, an increase of 114 units.

13.6 Exercises

Find general solutions for the following differential equations.

1. $\dfrac{dy}{dx} = 8x - 7$

2. $\dfrac{dy}{dx} = -x + 2$

3. $\dfrac{dy}{dx} = -2x + 3x^2$

4. $\dfrac{dy}{dx} = 6x^2 - 4x$

5. $\dfrac{dy}{dx} = -4 + 3x^3$

6. $\dfrac{dy}{dx} = 4x^3 + 3$

7. $\dfrac{dy}{dx} = e^{4x}$

8. $\dfrac{dy}{dx} = e^{3x}$

9. $\dfrac{dy}{dx} = 2e^{-6x}$

10. $\dfrac{dy}{dx} = 3e^{-2x}$

11. $\dfrac{dy}{dx} = x\sqrt{1 - x^2}$

12. $\dfrac{dy}{dx} = x^2(1 + 2x^3)^{1/2}$

13. $\dfrac{dy}{dx} = x^2 e^{-x^3}$

14. $\dfrac{dy}{dx} = x \cdot e^{-5x^2}$

15. $4x - 4 + \dfrac{dy}{dx} = 6x^2$

16. $-8x^3 - 3x^2 + \dfrac{1}{2} \cdot \dfrac{dy}{dx} = 3$

17. $\dfrac{d^2y}{dx^2} = -8x$

18. $\dfrac{d^2y}{dx^2} = 4x$

19. $\dfrac{d^2y}{dx^2} = 5 - 4x$

20. $\dfrac{d^2y}{dx^2} = 2 + 8x$

21. $\dfrac{d^2y}{dx^2} = 4x^2 - 2x$

22. $x^2 - \dfrac{d^2y}{dx^2} = 3x$

23. $5x - 3 + \dfrac{d^2y}{dx^2} = 0$

24. $\dfrac{d^2y}{dx^2} = 6 - x$

25. $\dfrac{d^2y}{dx^2} = e^x$

26. $\dfrac{d^2y}{dx^2} = 6e^{-x}$

Find particular solutions for the following differential equations. (See Examples 2 and 3.)

27. $\dfrac{dy}{dx} + 2x = 3x^2;\quad y = 2$ when $x = 0$

28. $\dfrac{dy}{dx} = 4x + 3;\quad y = -4$ when $x = 0$

29. $\dfrac{dy}{dx} = 5x + 2;\quad y = -3$ when $x = 0$

30. $\dfrac{dy}{dx} = 6 - 5x;\quad y = 6$ when $x = 0$

31. $\dfrac{dy}{dx} = 3x^2 - 4x + 2;\quad y = 3$ when $x = -1$

32. $\dfrac{dy}{dx} = 4x^3 - 3x^2 + x;\quad y = 0$ when $x = 1$

33. $\dfrac{d^2y}{dx^2} = 2x + 1;\quad y = 2$ when $x = 0;\quad y = 3$
when $x = -2$

34. $\dfrac{d^2y}{dx^2} = -3x + 2;\quad y = -3$ when $x = 0;$
$y = -19$ when $x = 3$

35. $\dfrac{d^2y}{dx^2} = e^x + 1;\quad y = 2$ when $x = 0;\quad y = \dfrac{3}{2}$
when $x = 1$

36. $\dfrac{d^2y}{dx^2} = -e^x - 2;\quad y = -2$ when $x = 0;\quad y = \dfrac{3}{2}$
when $x = 1$

37. $3x^2 - \dfrac{d^2y}{dx^2} = 2;\quad y = 4$ when $x = 0;\quad y = 6$
when $x = 2$

38. $\dfrac{d^2y}{dx^2} + 4x^2 = 1;\quad y = \dfrac{1}{6}$ when $x = 1;\quad y = -1$
when $x = 0$

Work the following exercises. (See Examples 1 and 6.)

39. Management In the time-dating of dairy products example discussed in the text, suppose the number of bacteria present at time $t = 0$ was 250.
(a) Find a particular solution of the differential equation.
(b) Find the number of days the product can be sold if the maximum value of y is 970.

40. Social Science Suppose the rate at which a rumor spreads—that is, the number of people who have heard the rumor over a period of time—increases with the number of days. If y is the number of people who have heard the rumor, then

$$\dfrac{dy}{dt} = kt,$$

where t is the time in days.
(a) If y is 0 when $t = 0$, and y is 100 when $t = 2$, find k.
(b) Using the value of k from part (a), find y when $t = 3;\ 5;\ 10$.

41. Management The rate of change of demand for a certain product is given by

$$\dfrac{dy}{dx} = -4x + 40,$$

where x represents the price of the item. Find the demand at the following price levels if $y = 6$ when $x = 0$.
(a) $x = 5$ **(b)** $x = 8$ **(c)** $x = 10$ **(d)** $x = 20$

42. Management The marginal profit of a certain company is given by

$$\frac{dy}{dx} = 32 - 4x,$$

where x represents the amount in hundreds of dollars spent on advertising. Find the profit for each of the following advertising expenditures if the profit is $1000 when nothing is spent for advertising.
(a) $x = 3$ **(b)** $x = 5$ **(c)** $x = 7$ **(d)** $x = 10$

43. Natural Science The rate at which the number of bacteria (in thousands) in a culture is changing after the introduction of a bactericide is given by

$$\frac{dy}{dx} = 50 - 10x,$$

where y is the number of bacteria (in thousands) present at time x. Find the number of bacteria present at each of the following times if there were 1000 thousand bacteria present at time $x = 0$.
(a) $x = 2$ **(b)** $x = 5$ **(c)** $x = 10$ **(d)** $x = 15$

44. Management Suppose the marginal cost of producing x copies of a book is given by

$$\frac{dy}{dx} = \frac{50}{\sqrt{x}}.$$

If $y = 25,000$ when $x = 0$, find the cost of producing the following number of books.
(a) 100 books **(b)** 400 books **(c)** 625 books

45. Natural Science Suppose a radioactive sample decays according to the relationship

$$\frac{dy}{dx} = -.08y,$$

where x is measured in years. Find the amount left after 5 years if $y = 20$ grams when $x = 0$.

46. Management A company has found that the rate at which a person new to the assembly line produces items is

$$\frac{dy}{dx} = 7.5e^{-.3x},$$

where x is the number of days the person has worked on the line. How many items can a new worker be expected to produce on the 8th day if he produces none when $x = 0$?

47. Social Science Assume the rate of change of the population of a certain city is given by

$$\frac{dy}{dt} = 6000e^{.06t},$$

where y is the population at time t, measured in years. If the population was 100,000 in 1970 (assume $t = 0$ in 1970), predict the population in 1990.

48. Management A company has found that the rate of change of sales during an advertising campaign is

$$\frac{dy}{dx} = 30e^{.03x},$$

where y is the sales in thousands at time x in months. If the sales were $200,000 when $x = 0$, find the sales at the end of 6 months.

Chapter 13   Review Exercises

Find each indefinite integral.

1. $\int (2x^3 - x)\, dx$

2. $\int (x^2 + 5x)\, dx$

3. $\int (6x^2 - 2x + 5)\, dx$

4. $\int (-x^3 + 2x^2 - 7x)\, dx$

5. $\int (x^{3/2} - x + x^{2/5})\, dx$

6. $\int (5x^{1/4} - 2x^{1/2} + x^2)\, dx$

7. $\int (\sqrt{x})^3 \, dx$

8. $\int (\sqrt{x})^{-3} \, dx$

9. $\int (1/x^3) \, dx$

10. $\int (-4/x^5) \, dx$

11. $\int \left(\dfrac{-2}{\sqrt{x}} - \dfrac{3}{x} \right) dx$

12. $\int \left(\dfrac{5}{\sqrt{x}} + \dfrac{6}{x} \right) dx$

13. $\int e^{-5x} \, dx$

14. $\int 2e^{4x} \, dx$

15. $\int (-5/x) \, dx$

16. $\int 3/(2x) \, dx$

Find the cost functions for the following marginal cost functions.

17. $C'(x) = 6x + 2$; fixed cost is $75

18. $C'(x) = .6x^2$; fixed cost is $90

19. $C'(x) = x^{3/4} + 2x$; 16 units cost $375

20. $C'(x) = 3x + 2/x^2$; 100 units cost $9500

In the study of bioavailability *in pharmacy, a drug is given to a patient. The level of concentration of the drug is then measured periodically, producing* blood level curves, *such as the ones in the figure. The areas under the curves give the total amount of the drug available to the patient.**

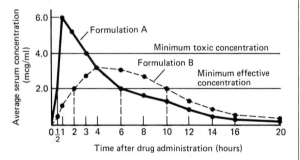

Find the total area under the curve for

21. Formulation A;

22. Formulation B.

Evaluate the following definite integrals.

23. $\int_1^2 (3x^2 + 5) \, dx$

24. $\int_1^6 (2x^2 + x) \, dx$

---

*These graphs are taken from *Basics of Bioavailability*, by D. J. Chodos and A. R. DeSantos. The UpJohn Company. Copyright © 1978.

25. $\int_1^5 (3x^{-2} + x^{-3}) \, dx$

26. $\int_2^3 (5x^{-2} + x^{-4}) \, dx$

27. $\int_1^3 2x^{-1} \, dx$

28. $\int_1^6 8x^{-1} \, dx$

29. $\int_0^4 2e^x \, dx$

30. $\int_1^6 \dfrac{5}{2} e^{4x} \, dx$

Work the following exercises.

31. The rate of change of sales of a new brand of tomato soup, in thousands, is given by

$$S(x) = \sqrt{x} + 2,$$

where x is the time in months that the new product has been on the market. Find the total sales after 9 months.

32. The rate of change of a population of prairie dogs, in terms of the number of coyotes, x, which prey on them, is given by

$$P(x) = 25 - .1x.$$

Find the total number of prairie dogs as the coyote population grows from 100 to 200.

33. Find the area between the graphs of $f(x) = 5 - x^2$ and $g(x) = x^2 - 3$.

34. A company has installed new machinery which will produce a savings rate (in thousands of dollars) of

$$S(x) = 225 - x^2,$$

where x is the number of years the machinery is to be used. The rate of additional costs to the company due to the new machinery is expected to be

$$C(x) = x^2 + 25x + 150.$$

For how many years should the company use the new machinery? What will the net savings in thousands of dollars be over this period?

35. The rate of change of the population of a rare species of Australian spiders is given by

$$\dfrac{dy}{dx} = -.2x,$$

where y is the number of spiders present at time x, measured in months. Find y when $x = 10$, if there were 100 spiders at $x = 0$.

36. A manufacturer of electronic equipment requires a certain rare metal. He has a reserve supply of 4,000,000 units which he will not be able to replace. If the rate at which the metal is used is given by

$$f(t) = 100,000e^{.03t},$$

where t is the time in years, how long will it be before he uses up the supply? (Hint: Find an expression for the total amount used in t years and set it equal to the known reserve supply.)

Use substitution to find the following.

37. $\int 2x\sqrt{x^2 - 3}\ dx$

38. $\int x\sqrt{5x^2 + 6}\ dx$

39. $\int \dfrac{x^2\ dx}{(x^3 + 5)^4}$

40. $\int (x^2 - 5x)^4(2x - 5)\ dx$

41. $\int \dfrac{4x - 5}{2x^2 - 5x}\ dx$

42. $\int \dfrac{12(2x + 9)}{x^2 + 9x + 1}\ dx$

43. $\int \dfrac{x^3}{e^{3x^4}}\ dx$

44. $\int_2^3 (3x^2 - 6x)^2(6x - 6)\ dx$

45. $\int -2e^{-5x}\ dx$

46. $\int e^{-4x}\ dx$

Use the table of integrals to find the following.

47. $\int \dfrac{1}{\sqrt{x^2 - 64}}\ dx$

48. $\int \dfrac{5}{x\sqrt{25 + x^2}}\ dx$

49. $\int \dfrac{-2}{x(3x + 4)}\ dx$

50. $\int \sqrt{x^2 + 49}\ dx$

51. $\int 3x^2 \ln |x|\ dx$

52. $\int xe^x\ dx$

53. $\int \dfrac{12}{x^2 - 9}\ dx$

54. $\int \dfrac{15x}{2x - 5}\ dx$

Find the general solution for each differential equation.

55. $\dfrac{dy}{dx} = 2x^2 - 5x$

56. $\dfrac{dy}{dx} = 7x + x^2$

57. $\dfrac{dy}{dx} = -3 + 3x^2 + 5x^4$

58. $\dfrac{dy}{dx} = 6x^5 - x^4 + 9x^3$

59. $\dfrac{dy}{dx} = x^{3/2} - 2x$

60. $\dfrac{dy}{dx} = 5x^{2/3} + 4x^{1/3}$

61. $\dfrac{dy}{dx} = 4e^{-2x}$

62. $\dfrac{dy}{dx} = -e^{-8x}$

63. $\dfrac{d^2y}{dx^2} = x^3 + 5x^2 - 1$

64. $\dfrac{d^2y}{dx^2} = -x^4 + 2x + 6$

65. $\dfrac{d^2y}{dx^2} - e^x = 3x^2$

66. $\dfrac{d^2y}{dx^2} = 5e^{2x} - 7x$

Find particular solutions for the following.

67. $\dfrac{dy}{dx} = 5x^2 - 6x;\quad y = 4$ when $x = 0$

68. $\sqrt{x} + \dfrac{dy}{dx} = x;\quad y = 2$ when $x = 0$

69. $\dfrac{dy}{dx} = \dfrac{1}{x};\quad y = -2$ when $x = 1$

70. $\dfrac{dy}{dx} = 6e^x;\quad y = 1$ when $x = 0$

Case 17

Estimating Depletion Dates for Minerals

It is becoming more and more obvious that the earth contains only a finite quantity of minerals. The "easy and cheap" sources of minerals are being used up, forcing an ever more expensive search for new sources. For example, oil from the North Slope of Alaska would never have been used in the United States during the 1930s since there was so much Texas and California oil readily available.

Mineral usage tends to follow an exponential growth curve. Thus, if q represents the rate of consumption of a certain mineral at time t, while q_0 represents consumption when $t = 0$, then

$$q = q_0 e^{kt},$$

where k is the growth constant. For example, the world consumption of petroleum in a recent year was about 19,600 million barrels, with the value of k about 6%. Letting $t = 0$ correspond to this base year, then $q_0 = 19,600$, $k = .06$, and

$$q = 19,600 e^{.06t}$$

is the rate of consumption at time t, assuming that all present trends continue.

Based on estimates of the National Academy of Science, 2,000,000 million barrels of oil are now in provable reserves or are likely to be discovered in the future. At the present rate of consumption, how many years would be necessary to deplete these estimated reserves? Use the integral calculus of this chapter to find out.

To begin, find the total quantity of petroleum that would be used between time $t = 0$ and some future time $t = t_1$. Figure 1 shows a typical graph of the function $q = q_0 e^{kt}$.

Following the work in Section 13.2, divide the time interval from $t = 0$ to $t = t_1$ into n subintervals. Let the ith subinterval have width Δt_i. Let the rate of consumption for the ith subinterval be approximated by q_i^*. Thus, the approximate total consumption for the subinterval is given by

$$q_i^* \cdot \Delta t_i,$$

and the total consumption over the interval from time $t = 0$ to $t = t_1$ is approximated by

$$\sum_{i=1}^{n} q_i^* \cdot \Delta t_i.$$

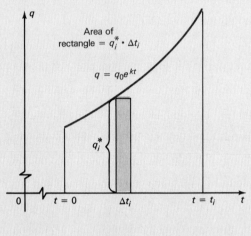

Figure 1

The limit of this sum as each of the Δt_i's approaches 0 gives the total consumption from time $t = 0$ to $t = t_1$. That is,

$$\text{Total consumption} = \lim_{\Delta t_i \to 0} \sum q_i^* \cdot \Delta t_i.$$

This limit is the definite integral of the function $\cdot \, q = q_0 e^{kt}$ from $t = 0$ to $t = t_1$, or

$$\text{Total consumption} = \int_0^{t_1} q_0 e^{kt} \, dt.$$

Evaluating this definite integral gives

$$\int_0^{t_1} q_0 e^{kt} \, dt = q_0 \int_0^{t_1} e^{kt} \, dt = q_0 \left(\frac{1}{k} e^{kt} \right) \Big]_0^{t_1}$$

$$= \frac{q_0}{k} e^{kt} \Big]_0^{t_1} = \frac{q_0}{k} e^{kt_1} - \frac{q_0}{k} e^0$$

$$= \frac{q_0}{k} e^{kt_1} - \frac{q_0}{k}(1)$$

$$= \frac{q^0}{k}(e^{kt_1} - 1). \tag{1}$$

Now return to the numbers given for petroleum: $q_0 = 19,600$ million barrels, where q_0 represents consumption in the base year; $k = .06$; and total petroleum reserves are estimated as 2,000,000 million barrels. Thus, using equation (1),

$$2,000,000 = \frac{19,600}{0.06}(e^{.06t_1} - 1).$$

Multiply both sides of the equation by .06:

$$120,000 = 19,600(e^{.06t_1} - 1).$$

Divide both sides of the equation by 19,600.

$$6.1 = e^{.06t_1} - 1$$

Add 1 to both sides.

$$7.1 = e^{.06t_1}$$

Take natural logarithms of both sides:

$$\ln 7.1 = \ln e^{.06t_1} = .06t_1 \ln e$$
$$= .06 \, t_1 \quad (\text{since } \ln e = 1).$$

Finally,

$$t_1 = \frac{\ln 7.1}{.06}.$$

From the table of natural logarithms or a calculator, estimate ln 7.1 as about 1.96. Thus,

$$t_1 = \frac{1.96}{.06} = 33.$$

By this result, petroleum reserves will last the world for 33 years.

The results of mathematical analyses such as this must be used with great caution. By the analysis above, the world would use all the petroleum that it wants in the thirty-second year after the base year, but there would be none at all in 34 years. This is not at all realistic. As petroleum reserves decline, the price will increase, causing demand to decline and supplies to increase.

Exercises

1. Find the number of years that the estimated petroleum reserves would last if used at the same rate as in the base year.

2. How long would the estimated petroleum reserves last if the growth constant was only 2% instead of 6%?

Estimate the length of time until depletion for each of the following minerals.

3. Bauxite (the ore from which aluminum is obtained), estimated reserves in base year 15,000,000 thousand tons, rate of consumption 63,000 thousand tons, growth constant 6%.

4. Bituminous coal, estimated world reserves 2,000,000 million tons, rate of consumption 2200 million tons, growth constant 4%.

Case 18

How Much Does a Warranty Cost?*

This case uses some of the ideas of probability. The probability of an event is a number p, where $0 \leq p \leq 1$, such that p is the ratio of the number of ways that the event can happen divided by the total number of possible outcomes. For example, the probability of drawing a red card from a deck of 52 cards (of which 26 are red) is given by

$$P \text{ (red card)} = \frac{26}{52} = \frac{1}{2}.$$

In the same way, the probability of drawing a black queen from a deck of 52 cards is

$$P \text{ (black queen)} = \frac{2}{52} = \frac{1}{26}.$$

In this case the cost of a warranty program to a manufacturer is found. This cost depends on the quality of the products made. These variables are used.

c = constant product price, per unit, including cost of warranty (We assume the price charged per unit is constant, since this price is likely to be fixed by competition.)

m = expected lifetime of the product

w = length of the warranty period

N = size of a production lot, such as a year's production

r = warranty cost per unit

$C(t)$ = pro rata customer rebate at time t

$P(t)$ = probability of product failure at any time t

$F(t)$ = number of failures occurring at time t.

Assume that the warranty is of the pro rata customer rebate type, in which the customer is paid for the proportion of the warranty left. Hence, if the product has a warranty period of w, and fails at time t, then the product worked for the fraction t/w of the warranty period. Hence, the customer is re-

*From "Determination of Warranty Reserves," by Warren W. Menke, from *Management Science*, Vol. 15, No. 10, June 1969. Copyright © 1969 The Institute of Management Sciences. Reprinted by permission.

imbursed for the unused portion of the warranty, or the fraction

$$1 - \frac{t}{w}.$$

Assuming the product cost c originally, and using $C(t)$ to represent the customer rebate at time t,

$$C(t) = c\left(1 - \frac{t}{w}\right).$$

For many different types of products, it has been shown by experience that

$$P(t) = 1 - e^{-t/m}$$

provides a good estimate of the probability of product failure at time t. The total number of failures at time t is given by the product of $P(t)$ and N, the total number of items per batch. If we use $F(t)$ to represent this total, we have

$$F(t) = N \cdot P(t) = N(1 - e^{-t/m}).$$

The total number of failures in some "tiny time interval" of width dt can be shown to be the derivative of $F(t)$,

$$F'(t) = \left(\frac{N}{m}\right)e^{-t/m},$$

while the cost for the failures in this "tiny time interval" is

$$C(t) \cdot F'(t) = c\left(1 - \frac{t}{w}\right)\left(\frac{N}{m}\right)e^{-t/m}.$$

The total cost for all failures during the warranty period is thus given by the definite integral

$$\int_0^w c\left(1 - \frac{t}{w}\right)\left(\frac{N}{m}\right)e^{-t/m} \, dt \ .$$

Using integration by parts (a method of integration not discussed) this definite integral can be shown to equal

$$Nc\left[-e^{-t/m} + \frac{t}{w} \cdot e^{-t/m} + \frac{m}{w}(e^{-t/m})\right]_0^w$$

or

$$Nc\left(1 - \frac{m}{w} + \frac{m}{w}e^{-w/m}\right).$$

This last quantity is the total warranty cost for all the units manufactured. Since there are N units per batch, the warranty cost per item is

$$r = \frac{1}{N}\left[Nc\left(1 - \frac{m}{w} + \frac{m}{w}e^{-w/m}\right)\right]$$

$$= c\left(1 - \frac{m}{w} + \frac{m}{w}e^{-w/m}\right).$$

For example, suppose a product which costs $100 has an expected life of 24 months, with a 12-month warranty. Then we have $c = \$100$, $m = 24$, $w = 12$, with r, the warranty cost per unit, given by

$$r = 100(1 - 2 + 2e^{-.5})$$
$$= 100[-1 + 2(.6065)]$$
$$= 100(0.2130) = 21.30,$$

where we found $e^{-.5}$ from Table 3.

Exercises

Find r for each of the following.

1. $c = \$50$, $m = 48$ months, $w = 24$ months

2. $c = \$1000$, $m = 60$ months, $w = 24$ months

3. $c = \$1200$, $m = 30$ months, $w = 30$ months

TABLE 1 SQUARES AND SQUARE ROOTS

n	n^2	$\sqrt{n}$	$\sqrt{10n}$	n	n^2	$\sqrt{n}$	$\sqrt{10n}$
1	1	1.000	3.162	51	2601	7.141	22.583
2	4	1.414	4.472	52	2704	7.211	22.804
3	9	1.732	5.477	53	2809	7.280	23.022
4	16	2.000	6.325	54	2916	7.348	23.238
5	25	2.236	7.071	55	3025	7.416	23.452
6	36	2.449	7.746	56	3136	7.483	23.664
7	49	2.646	8.367	57	3249	7.550	23.875
8	64	2.828	8.944	58	3364	7.616	24.083
9	81	3.000	9.487	59	3481	7.681	24.290
10	100	3.162	10.000	60	3600	7.746	24.495
11	121	3.317	10.488	61	3721	7.810	24.698
12	144	3.464	10.954	62	3844	7.874	24.900
13	169	3.606	11.402	63	3969	7.937	25.100
14	196	3.742	11.832	64	4096	8.000	25.298
15	225	3.873	12.247	65	4225	8.062	25.495
16	256	4.000	12.649	66	4356	8.124	25.690
17	289	4.123	13.038	67	4489	8.185	25.884
18	324	4.243	13.416	68	4624	8.246	26.077
19	361	4.359	13.784	69	4761	8.307	26.268
20	400	4.472	14.142	70	4900	8.367	26.458
21	441	4.583	14.491	71	5041	8.426	26.646
22	484	4.690	14.832	72	5184	8.485	26.833
23	529	4.796	15.166	73	5329	8.544	27.019
24	576	4.899	15.492	74	5476	8.602	27.203
25	625	5.000	15.811	75	5625	8.660	27.386
26	676	5.099	16.125	76	5776	8.718	27.568
27	729	5.196	16.432	77	5929	8.775	27.749
28	784	5.292	16.733	78	6084	8.832	27.928
29	841	5.385	17.029	79	6241	8.888	28.107
30	900	5.477	17.321	80	6400	8.944	28.284
31	961	5.568	17.607	81	6561	9.000	28.460
32	1024	5.657	17.889	82	6724	9.055	28.636
33	1089	5.745	18.166	83	6889	9.110	28.810
34	1156	5.831	18.439	84	7056	9.165	28.983
35	1225	5.916	18.708	85	7225	9.220	29.155
36	1296	6.000	18.974	86	7396	9.274	29.326
37	1369	6.083	19.235	87	7569	9.327	29.496
38	1444	6.164	19.494	88	7744	9.381	29.665
39	1521	6.245	19.748	89	7921	9.434	29.833
40	1600	6.325	20.000	90	8100	9.487	30.000
41	1681	6.403	20.248	91	8281	9.539	30.166
42	1764	6.481	20.494	92	8464	9.592	30.332
43	1849	6.557	20.736	93	8649	9.644	30.496
44	1936	6.633	20.976	94	8836	9.695	30.659
45	2025	6.708	21.213	95	9025	9.747	30.822
46	2116	6.782	21.448	96	9216	9.798	30.984
47	2209	6.856	21.679	97	9409	9.849	31.145
48	2304	6.928	21.909	98	9604	9.899	31.305
49	2401	7.000	22.136	99	9801	9.950	31.464
50	2500	7.071	22.361	100	10000	10.000	31.623

TABLE 2 SELECTED POWERS OF NUMBERS

n	n^2	n^3	n^4	n^5	n^6
2	4	8	16	32	64
3	9	27	81	243	729
4	16	64	256	1024	4096
5	25	125	625	3125	
6	36	216	1296		
7	49	343	2401		
8	64	512	4096		
9	81	729	6561		
10	100	1000	10,000		

TABLE 3 POWERS OF e

x	e^x	e^{-x}	x	e^x	e^{-x}
0.00	1.00000	1.00000	1.60	4.95302	0.20189
0.01	1.01005	0.99004	1.70	5.47394	0.18268
0.02	1.02020	0.98019	1.80	6.04964	0.16529
0.03	1.03045	0.97044	1.90	6.68589	0.14956
0.04	1.04081	0.96078	2.00	7.38905	0.13533
0.05	1.05127	0.95122			
0.06	1.06183	0.94176	2.10	8.16616	0.12245
0.07	1.07250	0.93239	2.20	9.02500	0.11080
0.08	1.08328	0.92311	2.30	9.97417	0.10025
0.09	1.09417	0.91393	2.40	11.02316	0.09071
0.10	1.10517	0.90483	2.50	12.18248	0.08208
			2.60	13.46372	0.07427
0.11	1.11628	0.89583	2.70	14.87971	0.06720
0.12	1.12750	0.88692	2.80	16.44463	0.06081
0.13	1.13883	0.87810	2.90	18.17412	0.05502
0.14	1.15027	0.86936	3.00	20.08551	0.04978
0.15	1.16183	0.86071			
0.16	1.17351	0.85214	3.50	33.11545	0.03020
0.17	1.18530	0.84366	4.00	54.59815	0.01832
0.18	1.19722	0.83527	4.50	90.01713	0.01111
0.19	1.20925	0.82696			
			5.00	148.41316	0.00674
0.20	1.22140	0.81873	5.50	224.69193	0.00409
0.30	1.34985	0.74081			
0.40	1.49182	0.67032			
0.50	1.64872	0.60653	6.00	403.42879	0.00248
0.60	1.82211	0.54881	6.50	665.14163	0.00150
0.70	2.01375	0.49658			
0.80	2.22554	0.44932	7.00	1096.63316	0.00091
0.90	2.45960	0.40656	7.50	1808.04241	0.00055
1.00	2.71828	0.36787	8.00	2980.95799	0.00034
			8.50	4914.76884	0.00020
1.10	3.00416	0.33287			
1.20	3.32011	0.30119	9.00	8103.08392	0.00012
1.30	3.66929	0.27253	9.50	13359.72683	0.00007
1.40	4.05519	0.24659			
1.50	4.48168	0.22313	10.00	22026.46579	0.00005

TABLE 4 NATURAL LOGARITHMS

x	ln x	x	ln x	x	ln x
0.0		4.5	1.5041	9.0	2.1972
0.1	−2.3026	4.6	1.5261	9.1	2.2083
0.2	−1.6094	4.7	1.5476	9.2	2.2192
0.3	−1.2040	4.8	1.5686	9.3	2.2300
0.4	−0.9163	4.9	1.5892	9.4	2.2407
0.5	−0.6931	5.0	1.6094	9.5	2.2513
0.6	−0.5108	5.1	1.6292	9.6	2.2618
0.7	−0.3567	5.2	1.6487	9.7	2.2721
0.8	−0.2231	5.3	1.6677	9.8	2.2824
0.9	−0.1054	5.4	1.6864	9.9	2.2925
1.0	0.0000	5.5	1.7047	10	2.3026
1.1	0.0953	5.6	1.7228	11	2.3979
1.2	0.1823	5.7	1.7405	12	2.4849
1.3	0.2624	5.8	1.7579	13	2.5649
1.4	0.3365	5.9	1.7750	14	2.6391
1.5	0.4055	6.0	1.7918	15	2.7081
1.6	0.4700	6.1	1.8083	16	2.7726
1.7	0.5306	6.2	1.8245	17	2.8332
1.8	0.5878	6.3	1.8405	18	2.8904
1.9	0.6419	6.4	1.8563	19	2.9444
2.0	0.6931	6.5	1.8718	20	2.9957
2.1	0.7419	6.6	1.8871	25	3.2189
2.2	0.7885	6.7	1.9021	30	3.4012
2.3	0.8329	6.8	1.9169	35	3.5553
2.4	0.8755	6.9	1.9315	40	3.6889
2.5	0.9163	7.0	1.9459	45	3.8067
2.6	0.9555	7.1	1.9601	50	3.9120
2.7	0.9933	7.2	1.9741	55	4.0073
2.8	1.0296	7.3	1.9879	60	4.0943
2.9	1.0647	7.4	2.0015	65	4.1744
3.0	1.0986	7.5	2.0149	70	4.2485
3.1	1.1314	7.6	2.0281	75	4.3175
3.2	1.1632	7.7	2.0412	80	4.3820
3.3	1.1939	7.8	2.0541	85	4.4427
3.4	1.2238	7.9	2.0669	90	4.4998
3.5	1.2528	8.0	2.0794	95	4.5539
3.6	1.2809	8.1	2.0919	100	4.6052
3.7	1.3083	8.2	2.1041		
3.8	1.3350	8.3	2.1163		
3.9	1.3610	8.4	2.1281		
4.0	1.3863	8.5	2.1401		
4.1	1.4110	8.6	2.1518		
4.2	1.4351	8.7	2.1633		
4.3	1.4586	8.8	2.1748		
4.4	1.4816	8.9	2.1861		

TABLE 5 COMMON LOGARITHMS

n	0	1	2	3	4	5	6	7	8	9
1.0	.0000	.0043	.0086	.0128	.0170	.0212	.0253	.0294	.0334	.0374
1.1	.0414	.0453	.0492	.0531	.0569	.0607	.0645	.0682	.0719	.0755
1.2	.0792	.0828	.0864	.0899	.0934	.0969	.1004	.1038	.1072	.1106
1.3	.1139	.1173	.1206	.1239	.1271	.1303	.1335	.1367	.1399	.1430
1.4	.1461	.1492	.1523	.1553	.1584	.1614	.1644	.1673	.1703	.1732
1.5	.1761	.1790	.1818	.1847	.1875	.1903	.1931	.1959	.1987	.2014
1.6	.2041	.2068	.2095	.2122	.2148	.2175	.2201	.2227	.2253	.2279
1.7	.2304	.2330	.2355	.2380	.2405	.2430	.2455	.2480	.2504	.2529
1.8	.2553	.2577	.2601	.2625	.2648	.2672	.2695	.2718	.2742	.2765
1.9	.2788	.2810	.2833	.2856	.2878	.2900	.2923	.2945	.2967	.2989
2.0	.3010	.3032	.3054	.3075	.3096	.3118	.3139	.3160	.3181	.3201
2.1	.3222	.3243	.3263	.3284	.3304	.3324	.3345	.3365	.3335	.3404
2.2	.3424	.3444	.3464	.3483	.3502	.3522	.3541	.3560	.3579	.3598
2.3	.3617	.3636	.3655	.3674	.3692	.3711	.3729	.3747	.3766	.3784
2.4	.3802	.3820	.3838	.3856	.3874	.3892	.3909	.3927	.3945	.3962
2.5	.3979	.3997	.4014	.4031	.4048	.4065	.4082	.4099	.4116	.4133
2.6	.4150	.4166	.4183	.4200	.4216	.4232	.4249	.4265	.4281	.4298
2.7	.4314	.4330	.4346	.4362	.4378	.4393	.4409	.4425	.4440	.4456
2.8	.4472	.4487	.4502	.4518	.4533	.4548	.4564	.4579	.4594	.4609
2.9	.4624	.4639	.4654	.4669	.4683	.4698	.4713	.4728	.4742	.4757
3.0	.4771	.4786	.4800	.4814	.4829	.4843	.4857	.4871	.4886	.4900
3.1	.4914	.4928	.4942	.4955	.4969	.4983	.4997	.5011	.5024	.5038
3.2	.5051	.5065	.5079	.5092	.5105	.5119	.5132	.5145	.5159	.5172
3.3	.5185	.5198	.5211	.5224	.5237	.5250	.5263	.5276	.5289	.5302
3.4	.5315	.5328	.5340	.5353	.5366	.5378	.5391	.5403	.5416	.5428
3.5	.5441	.5453	.5465	.5478	.5490	.5502	.5514	.5527	.5539	.5551
3.6	.5563	.5575	.5587	.5599	.5611	.5623	.5635	.5647	.5658	.5670
3.7	.5682	.5694	.5705	.5717	.5729	.5740	.5752	.5763	.5775	.5786
3.8	.5798	.5809	.5821	.5832	.5843	.5855	.5866	.5877	.5888	.5899
3.9	.5911	.5922	.5933	.5944	.5955	.5966	.5977	.5988	.5999	.6010
4.0	.6021	.6031	.6042	.6053	.6064	.6075	.6085	.6096	.6107	.6117
4.1	.6128	.6138	.6149	.6160	.6170	.6180	.6191	.6201	.6212	.6222
4.2	.6232	.6243	.6253	.6263	.6274	.6284	.6294	.6304	.6314	.6325
4.3	.6335	.6345	.6355	.6365	.6375	.6385	.6395	.6405	.6415	.6425
4.4	.6435	.6444	.6454	.6464	.6474	.6484	.6493	.6503	.6513	.6522
4.5	.6532	.6542	.6551	.6561	.6571	.6580	.6590	.6599	.6609	.6618
4.6	.6628	.6637	.6646	.6656	.6665	.6675	.6684	.6693	.6702	.6712
4.7	.6721	.6730	.6739	.6749	.6758	.6767	.6776	.6785	.6794	.6803
4.8	.6812	.6821	.6830	.6839	.6848	.6857	.6866	.6875	.6884	.6893
4.9	.6902	.6911	.6920	.6928	.6937	.6946	.6955	.6964	.6972	.6981
5.0	.6990	.6998	.7007	.7016	.7024	.7033	.7042	.7050	.7059	.7067
5.1	.7076	.7084	.7093	.7101	.7110	.7118	.7126	.7135	.7143	.7152
5.2	.7160	.7168	.7177	.7185	.7193	.7202	.7210	.7218	.7226	.7235
5.3	.7243	.7251	.7259	.7267	.7275	.7284	.7292	.7300	.7308	.7316
5.4	.7324	.7332	.7340	.7348	.7356	.7364	.7372	.7380	.7388	.7396
n	0	1	2	3	4	5	6	7	8	9

TABLE 5 (CONTINUED)

n	0	1	2	3	4	5	6	7	8	9
5.5	.7404	.7412	.7419	.7427	.7435	.7443	.7451	.7459	.7466	.7474
5.6	.7482	.7490	.7497	.7505	.7513	.7520	.7528	.7536	.7543	.7551
5.7	.7559	.7566	.7574	.7582	.7589	.7597	.7604	.7612	.7619	.7627
5.8	.7634	.7642	.7649	.7657	.7664	.7672	.7679	.7686	.7694	.7701
5.9	.7709	.7716	.7723	.7731	.7738	.7745	.7752	.7760	.7767	.7774
6.0	.7782	.7789	.7796	.7803	.7810	.7818	.7825	.7832	.7839	.7846
6.1	.7853	.7860	.7868	.7875	.7882	.7889	.7896	.7903	.7910	.7917
6.2	.7924	.7931	.7938	.7945	.7952	.7959	.7966	.7973	.7980	.7987
6.3	.7993	.8000	.8007	.8014	.8021	.8028	.8035	.8041	.8048	.8055
6.4	.8062	.8069	.8075	.8082	.8089	.8096	.8102	.8109	.8116	.8122
6.5	.8129	.8136	.8142	.8149	.8156	.8162	.8169	.8176	.8182	.8189
6.6	.8195	.8202	.8209	.8215	.8222	.8228	.8235	.8241	.8248	.8254
6.7	.8261	.8267	.8274	.8280	.8287	.8293	.8299	.8306	.8312	.8319
6.8	.8325	.8331	.8338	.8344	.8351	.8357	.8363	.8370	.8376	.8382
6.9	.8388	.8395	.8401	.8407	.8414	.8420	.8426	.8432	.8439	.8445
7.0	.8451	.8457	.8463	.8470	.8476	.8482	.8488	.8494	.8500	.8506
7.1	.8513	.8519	.8525	.8531	.8537	.8543	.8549	.8555	.8561	.8567
7.2	.8573	.8579	.8585	.8591	.8597	.8603	.8609	.8615	.8621	.8627
7.3	.8633	.8639	.8645	.8651	.8657	.8663	.8669	.8675	.8681	.8686
7.4	.8692	.8698	.8704	.8710	.8716	.8722	.8727	.8733	.8739	.8745
7.5	.8751	.8756	.8762	.8768	.8774	.8779	.8785	.8791	.8797	.8802
7.6	.8808	.8814	.8820	.8825	.8831	.8837	.8842	.8848	.8854	.8859
7.7	.8865	.8871	.8876	.8882	.8887	.8893	.8899	.8904	.8910	.8915
7.8	.8921	.8927	.8932	.8938	.8943	.8949	.8954	.8960	.8965	.8971
7.9	.8976	.8982	.8987	.8993	.8998	.9004	.9009	.9015	.9020	.9025
8.0	.9031	.9036	.9042	.9047	.9053	.9058	.9063	.9069	.9074	.9079
8.1	.9085	.9090	.9096	.9101	.9106	.9112	.9117	.9122	.9128	.9133
8.2	.9138	.9143	.9149	.9154	.9159	.9165	.9170	.9175	.9180	.9186
8.3	.9191	.9196	.9201	.9206	.9212	.9217	.9222	.9227	.9232	.9238
8.4	.9243	.9248	.9253	.9258	.9263	.9269	.9274	.9279	.9284	.9289
8.5	.9294	.9299	.9304	.9309	.9315	.9320	.9325	.9330	.9335	.9340
8.6	.9345	.9350	.9355	.9360	.9365	.9370	.9375	.9380	.9385	.9390
8.7	.9395	.9400	.9405	.9410	.9415	.9420	.9425	.9430	.9435	.9440
8.8	.9445	.9450	.9455	.9460	.9465	.9469	.9474	.9479	.9484	.9489
8.9	.9494	.9499	.9504	.9509	.9513	.9518	.9523	.9528	.9533	.9538
9.0	.9542	.9547	.9552	.9557	.9562	.9566	.9571	.9576	.9581	.9586
9.1	.9590	.9595	.9600	.9605	.9609	.9614	.9619	.9624	.9628	.9633
9.2	.9638	.9643	.9647	.9652	.9657	.9661	.9666	.9671	.9675	.9680
9.3	.9685	.9689	.9694	.9699	.9703	.9708	.9713	.9717	.9722	.9727
9.4	.9731	.9736	.9741	.9745	.9750	.9754	.9759	.9763	.9768	.9773
9.5	.9777	.9782	.9786	.9791	.9795	.9800	.9805	.9809	.9814	.9818
9.6	.9823	.9827	.9832	.9836	.9841	.9845	.9850	.9854	.9859	.9863
9.7	.9868	.9872	.9877	.9881	.9886	.9890	.9894	.9899	.9903	.9908
9.8	.9912	.9917	.9921	.9926	.9930	.9934	.9939	.9943	.9948	.9952
9.9	.9956	.9961	.9965	.9969	.9974	.9978	.9983	.9987	.9991	.9996
n	0	1	2	3	4	5	6	7	8	9

TABLE 6 COMPOUND INTEREST

$(1 + i)^n$

i n	2.00%	2.50%	3.00%	3.50%	4.00%	4.50%	5.00%	5.50%
1	1.02000	1.02500	1.03000	1.03500	1.04000	1.04500	1.05000	1.05500
2	1.04040	1.05063	1.06090	1.07123	1.08160	1.09203	1.10250	1.11303
3	1.06121	1.07689	1.09273	1.10872	1.12486	1.14117	1.15763	1.17424
4	1.08243	1.10381	1.12551	1.14752	1.16986	1.19252	1.21551	1.23882
5	1.10408	1.13141	1.15927	1.18769	1.21665	1.24618	1.27628	1.30696
6	1.12616	1.15969	1.19405	1.22926	1.26532	1.30226	1.34010	1.37884
7	1.14869	1.18869	1.22987	1.27228	1.31593	1.36086	1.40710	1.45468
8	1.17166	1.21840	1.26677	1.31681	1.36857	1.42210	1.47746	1.53469
9	1.19509	1.24886	1.30477	1.36290	1.42331	1.48610	1.55133	1.61909
10	1.21899	1.28008	1.34392	1.41060	1.48024	1.55297	1.62889	1.70814
11	1.24337	1.31209	1.38423	1.45997	1.53945	1.62285	1.71034	1.80209
12	1.26824	1.34489	1.42576	1.51107	1.60103	1.69588	1.79586	1.90121
13	1.29361	1.37851	1.46853	1.56396	1.66507	1.77220	1.88565	2.00577
14	1.31948	1.41297	1.51259	1.61869	1.73168	1.85194	1.97993	2.11609
15	1.34587	1.44830	1.55797	1.67535	1.80094	1.93528	2.07893	2.23248
16	1.37279	1.48451	1.60471	1.73399	1.87298	2.02237	2.18287	2.35526
17	1.40024	1.52162	1.65285	1.79468	1.94790	2.11338	2.29202	2.48480
18	1.42825	1.55966	1.70243	1.85749	2.02582	2.20848	2.40662	2.62147
19	1.45681	1.59865	1.75351	1.92250	2.10685	2.30786	2.52695	2.76565
20	1.48595	1.63862	1.80611	1.98979	2.19112	2.41171	2.65330	2.91776
21	1.51567	1.67958	1.86029	2.05943	2.27877	2.52024	2.78596	3.07823
22	1.54598	1.72157	1.91610	2.13151	2.36992	2.63365	2.92526	3.24754
23	1.57690	1.76461	1.97359	2.20611	2.46472	2.75217	3.07152	3.42615
24	1.60844	1.80873	2.03279	2.28333	2.56330	2.87601	3.22510	3.61459
25	1.64061	1.85394	2.09378	2.36324	2.66584	3.00543	3.38635	3.81339
26	1.67342	1.90029	2.15659	2.44596	2.77247	3.14068	3.55567	4.02313
27	1.70689	1.94780	2.22129	2.53157	2.88337	3.28201	3.73346	4.24440
28	1.74102	1.99650	2.28793	2.62017	2.99870	3.42970	3.92013	4.47784
29	1.77584	2.04641	2.35657	2.71188	3.11865	3.58404	4.11614	4.72412
30	1.81136	2.09757	2.42726	2.80679	3.24340	3.74532	4.32194	4.98395
31	1.84759	2.15001	2.50008	2.90503	3.37313	3.91386	4.53804	5.25807
32	1.88454	2.20376	2.57508	3.00671	3.50806	4.08998	4.76494	5.54726
33	1.92223	2.25885	2.65234	3.11194	3.64838	4.27403	5.00319	5.85236
34	1.96068	2.31532	2.73191	3.22086	3.79432	4.46636	5.25335	6.17424
35	1.99989	2.37321	2.81386	3.33359	3.94609	4.66735	5.51602	6.51383
36	2.03989	2.43254	2.89828	3.45027	4.10393	4.87738	5.79182	6.87209
37	2.08069	2.49335	2.98523	3.57103	4.26809	5.09686	6.08141	7.25005
38	2.12230	2.55568	3.07478	3.69601	4.43881	5.32622	6.38548	7.64880
39	2.16474	2.61957	3.16703	3.82537	4.61637	5.56590	6.70475	8.06949
40	2.20804	2.68506	3.26204	3.95926	4.80102	5.81636	7.03999	8.51331
41	2.25220	2.75219	3.35990	4.09783	4.99306	6.07810	7.39199	8.98154
42	2.29724	2.82100	3.46070	4.24126	5.19278	6.35162	7.76159	9.47553
43	2.34319	2.89152	3.56452	4.38970	5.40050	6.63744	8.14967	9.99668
44	2.39005	2.96381	3.67145	4.54334	5.61652	6.93612	8.55715	10.54650
45	2.43785	3.03790	3.78160	4.70236	5.84118	7.24825	8.98501	11.12655
46	2.48661	3.11385	3.89504	4.86694	6.07482	7.57442	9.43426	11.73851
47	2.53634	3.19170	4.01190	5.03728	6.31782	7.91527	9.90597	12.38413
48	2.58707	3.27149	4.13225	5.21359	6.57053	8.27146	10.40127	13.06526
49	2.63881	3.35328	4.25622	5.39606	6.83335	8.64367	10.92133	13.78385
50	2.69159	3.43711	4.38391	5.58493	7.10668	9.03264	11.46740	14.54196

TABLE 6 (CONTINUED)

n \ i	6.00%	8.00%	10.00%	12.00%	14.00%	16.00%	18.00%
1	1.06000	1.08000	1.10000	1.12000	1.14000	1.16000	1.18000
2	1.12360	1.16640	1.21000	1.25440	1.29960	1.34560	1.39240
3	1.19102	1.25971	1.33100	1.40493	1.48154	1.56090	1.64303
4	1.26248	1.36049	1.46410	1.57352	1.68896	1.81064	1.93878
5	1.33823	1.46933	1.61051	1.76234	1.92541	2.10034	2.28776
6	1.41852	1.58687	1.77156	1.97382	2.19497	2.43640	2.69955
7	1.50363	1.71382	1.94872	2.21068	2.50227	2.82622	3.18547
8	1.59385	1.85093	2.14359	2.47596	2.85259	3.27841	3.75886
9	1.68948	1.99900	2.35795	2.77308	3.25195	3.80296	4.43545
10	1.79085	2.15892	2.59374	3.10585	3.70722	4.41144	5.23384
11	1.89830	2.33164	2.85312	3.47855	4.22623	5.11726	6.17593
12	2.01220	2.51817	3.13843	3.89598	4.81790	5.93603	7.28759
13	2.13293	2.71962	3.45227	4.36349	5.49241	6.88579	8.59936
14	2.26090	2.93719	3.79750	4.88711	6.26135	7.98752	10.14724
15	2.39656	3.17217	4.17725	5.47357	7.13794	9.26552	11.97375
16	2.54035	3.42594	4.59497	6.13039	8.13725	10.74800	14.12902
17	2.69277	3.70002	5.05447	6.86604	9.27646	12.46768	16.67225
18	2.85434	3.99602	5.55992	7.68997	10.57517	14.46251	19.67325
19	3.02560	4.31570	6.11591	8.61276	12.05569	16.77652	23.21444
20	3.20714	4.66096	6.72750	9.64629	13.74349	19.46076	27.39303
21	3.39956	5.03383	7.40025	10.80385	15.66758	22.57448	32.32378
22	3.60354	5.43654	8.14027	12.10031	17.86104	26.18640	38.14206
23	3.81975	5.87146	8.95430	13.55235	20.36158	30.37622	45.00763
24	4.04893	6.34118	9.84973	15.17863	23.21221	35.23642	53.10901
25	4.29187	6.84848	10.83471	17.00006	26.46192	40.87424	62.66863
26	4.54938	7.39635	11.91818	19.04007	30.16658	47.41412	73.94898
27	4.82235	7.98806	13.10999	21.32488	34.38991	55.00038	87.25980
28	5.11169	8.62711	14.42099	23.88387	39.20449	63.80044	102.96656
29	5.41839	9.31727	15.86309	26.74993	44.69312	74.00851	121.50054
30	5.74349	10.06266	17.44940	29.95992	50.95016	85.84988	143.37064
31	6.08810	10.86767	19.19434	33.55511	58.08318	99.58586	169.17735
32	6.45339	11.73708	21.11378	37.58173	66.21483	115.51959	199.62928
33	6.84059	12.67605	23.22515	42.09153	75.48490	134.00273	235.56255
34	7.25103	13.69013	25.54767	47.14252	86.05279	155.44317	277.96381
35	7.68609	14.78534	28.10244	52.79962	98.10018	180.31407	327.99729
36	8.14725	15.96817	30.91268	59.13557	111.83420	209.16432	387.03680
37	8.63609	17.24563	34.00395	66.23184	127.49099	242.63062	456.70343
38	9.15425	18.62528	37.40434	74.17966	145.33973	281.45151	538.91004
39	9.70351	20.11530	41.14478	83.08122	165.68729	326.48376	635.91385
40	10.28572	21.72452	45.25926	93.05097	188.88351	378.72116	750.37834
41	10.90286	23.46248	49.78518	104.21709	215.32721	439.31654	885.44645
42	11.55703	25.33948	54.76370	116.72314	245.47301	509.60719	1044.82681
43	12.25045	27.36664	60.24007	130.72991	279.83924	591.14434	1232.89563
44	12.98548	29.55597	66.26408	146.41750	319.01673	685.72744	1454.81685
45	13.76461	31.92045	72.89048	163.98760	363.67907	795.44383	1716.68388
46	14.59049	34.47409	80.17953	183.66612	414.59414	922.71484	2025.68698
47	15.46592	37.23201	88.19749	205.70605	472.63732	1070.34921	2390.31063
48	16.39387	40.21057	97.01723	230.39078	538.80655	1241.60509	2820.56655
49	17.37750	43.42742	106.71896	258.03767	614.23946	1440.26190	3328.26853
50	18.42015	46.90161	117.39085	289.00219	700.23299	1670.70380	3927.35686

TABLE 7 AMOUNT OF AN ANNUITY

$$s_{\overline{n}|i} = \frac{(1 + i)^{n-1}}{i}$$

n \ i	1.00%	1.50%	2.00%	2.50%	3.00%	4.00%	5.00%
1	1.00000	1.00000	1.00000	1.00000	1.00000	1.00000	1.00000
2	2.01000	2.01500	2.02000	2.02500	2.03000	2.04000	2.05000
3	3.03010	3.04522	3.06040	3.07562	3.09090	3.12160	3.15250
4	4.06040	4.09090	4.12161	4.15252	4.18363	4.24646	4.31013
5	5.10101	5.15227	5.20404	5.25633	5.30914	5.41632	5.52563
6	6.15202	6.22955	6.30812	6.38774	6.46841	6.63298	6.80191
7	7.21354	7.32299	7.43428	7.54743	7.66246	7.89829	8.14201
8	8.28567	8.43284	8.58297	8.73612	8.89234	9.21423	9.54911
9	9.36853	9.55933	9.75463	9.95452	10.15911	10.58280	11.02656
10	10.46221	10.70272	10.94972	11.20338	11.46388	12.00611	12.57789
11	11.56683	11.86326	12.16872	12.48347	12.80780	13.48635	14.20679
12	12.68250	13.04121	13.41209	13.79555	14.19203	15.02581	15.91713
13	13.80933	14.23683	14.68033	15.14044	15.61779	16.62684	17.71298
14	14.94742	15.45038	15.97394	16.51895	17.08632	18.29191	19.59863
15	16.09690	16.68214	17.29342	17.93193	18.59891	20.02359	21.57856
16	17.25786	17.93237	18.63929	19.38022	20.15688	21.82453	23.65749
17	18.43044	19.20136	20.01207	20.86473	21.76159	23.69751	25.84037
18	19.61475	20.48938	21.41231	22.38635	23.41444	25.64541	28.13238
19	20.81090	21.79672	22.84056	23.94601	25.11687	27.67123	30.53900
20	22.01900	23.12367	24.29737	25.54466	26.87037	29.77808	33.06595
21	23.23919	24.47052	25.78332	27.18327	28.67649	31.96920	35.71925
22	24.47159	25.83758	27.29898	28.86286	30.53678	34.24797	38.50521
23	25.71630	27.22514	28.84496	30.58443	32.45288	36.61789	41.43048
24	26.97346	28.63352	30.42186	32.34904	34.42647	39.08260	44.50200
25	28.24320	30.06302	32.03030	34.15776	36.45926	41.64591	47.72710
26	29.52563	31.51397	33.67091	36.01171	38.55304	44.31174	51.11345
27	30.82089	32.98668	35.34432	37.91200	40.70963	47.08421	54.66913
28	32.12910	34.48148	37.05121	39.85980	42.93092	49.96758	58.40258
29	33.45039	35.99870	38.79223	41.85630	45.21885	52.96629	62.32271
30	34.78489	37.53868	40.56808	43.90270	47.57542	56.08494	66.43885
31	36.13274	39.10176	42.37944	46.00027	50.00268	59.32834	70.76079
32	37.49407	40.68829	44.22703	48.15028	52.50276	62.70147	75.29883
33	38.86901	42.29861	46.11157	50.35403	55.07784	66.20953	80.06377
34	40.25770	43.93309	48.03380	52.61289	57.73018	69.85791	85.06696
35	41.66028	45.59209	49.99448	54.92821	60.46208	73.65222	90.32031
36	43.07688	47.27597	51.99437	57.30141	63.27594	77.59831	95.83632
37	44.50765	48.98511	54.03425	59.73395	66.17422	81.70225	101.62814
38	45.95272	50.71989	56.11494	62.22730	69.15945	85.97034	107.70955
39	47.41225	52.48068	58.23724	64.78298	72.23423	90.40915	114.09502
40	48.88637	54.26789	60.40198	67.40255	75.40126	95.02552	120.79977
41	50.37524	56.08191	62.61002	70.08762	78.66330	99.82654	127.83976
42	51.87899	57.92314	64.86222	72.83981	82.02320	104.81960	135.23175
43	53.39778	59.79199	67.15947	75.66080	85.48389	110.01238	142.99334
44	54.93176	61.68887	69.50266	78.55232	89.04841	115.41288	151.14301
45	56.48107	63.61420	71.89271	81.51613	92.71986	121.02939	159.70016
46	58.04589	65.56841	74.33056	84.55403	96.50146	126.87057	168.68516
47	59.62634	67.55194	76.81718	87.66789	100.39650	132.94539	178.11942
48	61.22261	69.56522	79.35352	90.85958	104.40840	139.26321	188.02539
49	62.83483	71.60870	81.94059	94.13107	108.54065	145.83373	198.42666
50	64.46318	73.68283	84.57940	97.48435	112.79687	152.66708	209.34800

TABLE 7   (CONTINUED)

i / n	6.00%	8.00%	10.00%	12.00%	14.00%	16.00%	18.00%
1	1.00000	1.00000	1.00000	1.00000	1.00000	1.00000	1.00000
2	2.06000	2.08000	2.10000	2.12000	2.14000	2.16000	2.18000
3	3.18360	3.24640	3.31000	3.37440	3.43960	3.50560	3.57240
4	4.37462	4.50611	4.64100	4.77933	4.92114	5.06650	5.21543
5	5.63709	5.86660	6.10510	6.35285	6.61010	6.87714	7.15421
6	6.97532	7.33593	7.71561	8.11519	8.53552	8.97748	9.44197
7	8.39384	8.92280	9.48717	10.08901	10.73049	11.41387	12.14152
8	9.89747	10.63663	11.43589	12.29969	13.23276	14.24009	15.32700
9	11.49132	12.48756	13.57948	14.77566	16.08535	17.51851	19.08585
10	13.18079	14.48656	15.93742	17.54874	19.33730	21.32147	23.52131
11	14.97164	16.64549	18.53117	20.65458	23.04452	25.73290	28.75514
12	16.86994	18.97713	21.38428	24.13313	27.27075	30.85017	34.93107
13	18.88214	21.49530	24.52271	28.02911	32.08865	36.78620	42.21866
14	21.01507	24.21492	27.97498	32.39260	37.58107	43.67199	50.81802
15	23.27597	27.15211	31.77248	37.27971	43.84241	51.65951	60.96527
16	25.67253	30.32428	35.94973	42.75328	50.98035	60.92503	72.93901
17	28.21288	33.75023	40.54470	48.88367	59.11760	71.67303	87.06804
18	30.90565	37.45024	45.59917	55.74971	68.39407	84.14072	103.74028
19	33.75999	41.44626	51.15909	63.43968	78.96923	98.60323	123.41353
20	36.78559	45.76196	57.27500	72.05244	91.02493	115.37975	146.62797
21	39.99273	50.42292	64.00250	81.69874	104.76842	134.84051	174.02100
22	43.39229	55.45676	71.40275	92.50258	120.43600	157.41499	206.34479
23	46.99583	60.89330	79.54302	104.60289	138.29704	183.60138	244.48685
24	50.81558	66.76476	88.49733	118.15524	158.65862	213.97761	289.49448
25	54.86451	73.10594	98.34706	133.33387	181.87083	249.21402	342.60349
26	59.15638	79.95442	109.18177	150.33393	208.33274	290.08827	405.27211
27	63.70577	87.35077	121.09994	169.37401	238.49933	337.50239	479.22109
28	68.52811	95.33883	134.20994	190.69889	272.88923	392.50277	566.48089
29	73.63980	103.96594	148.63093	214.58275	312.09373	456.30322	669.44745
30	79.05819	113.28321	164.49402	241.33268	356.78685	530.31173	790.94799
31	84.80168	123.34587	181.94342	271.29261	407.73701	616.16161	934.31863
32	90.88978	134.21354	201.13777	304.84772	465.82019	715.74746	1103.49598
33	97.34316	145.95062	222.25154	342.42945	532.03501	831.26706	1303.12526
34	104.18375	158.62667	245.47670	384.52098	607.51991	965.26979	1538.68781
35	111.43478	172.31680	271.02437	431.66350	693.57270	1120.71295	1816.65161
36	119.12087	187.10215	299.12681	484.46312	791.67288	1301.02703	2144.64890
37	127.26812	203.07032	330.03949	543.59869	903.50708	1510.19135	2531.68570
38	135.90421	220.31595	364.04343	609.83053	1030.99808	1752.82197	2988.38913
39	145.05846	238.94122	401.44778	684.01020	1176.33781	2034.27348	3527.29918
40	154.76197	259.05652	442.59256	767.09142	1342.02510	2360.75724	4163.21303
41	165.04768	280.78104	487.85181	860.14239	1530.90861	2739.47840	4913.59137
42	175.95054	304.24352	537.63699	964.35948	1746.23582	3178.79494	5799.03782
43	187.50758	329.58301	592.40069	1081.08262	1991.70883	3688.40213	6843.86463
44	199.75803	356.94965	652.64076	1211.81253	2271.54807	4279.54648	8076.76026
45	212.74351	386.50562	718.90484	1358.23003	2590.56480	4965.27391	9531.57711
46	226.50812	418.42607	791.79532	1522.21764	2954.24387	5760.71774	11248.26098
47	241.09861	452.90015	871.97485	1705.88375	3368.83801	6683.43257	13273.94796
48	256.56453	490.13216	960.17234	1911.58980	3841.47534	7753.78179	15664.25859
49	272.95840	530.34274	1057.18957	2141.98058	4380.28188	8995.38687	18484.82514
50	290.33590	573.77016	1163.90853	2400.01825	4994.52135	10435.64877	21813.09367

TABLE 8   PRESENT VALUE OF AN ANNUITY

$$a_{\overline{n}|i} = \frac{1 - (1 + i)^{-n}}{i}$$

n	1.00%	1.50%	2.00%	2.50%	3.00%	4.00%	5.00%
1	0.99010	0.98522	0.98039	0.97561	0.97087	0.96154	0.95238
2	1.97040	1.95588	1.94156	1.92742	1.91347	1.88609	1.85941
3	2.94099	2.91220	2.88388	2.85602	2.82861	2.77509	2.72325
4	3.90197	3.85438	3.80773	3.76197	3.71710	3.62990	3.54595
5	4.85343	4.78264	4.71346	4.64583	4.57971	4.45182	4.32948
6	5.79548	5.69719	5.60143	5.50813	5.41719	5.24214	5.07569
7	6.72819	6.59821	6.47199	6.34939	6.23028	6.00205	5.78637
8	7.65168	7.48593	7.32548	7.17014	7.01969	6.73274	6.46321
9	8.56602	8.36052	8.16224	7.97087	7.78611	7.43533	7.10782
10	9.47130	9.22218	8.98259	8.75206	8.53020	8.11090	7.72173
11	10.36763	10.07112	9.78685	9.51421	9.25262	8.76048	8.30641
12	11.25508	10.90751	10.57534	10.25776	9.95400	9.38507	8.86325
13	12.13374	11.73153	11.34837	10.98318	10.63496	9.98565	9.39357
14	13.00370	12.54338	12.10625	11.69091	11.29607	10.56312	9.89864
15	13.86505	13.34323	12.84926	12.38138	11.93794	11.11839	10.37966
16	14.71787	14.13126	13.57771	13.05500	12.56110	11.65230	10.83777
17	15.56225	14.90765	14.29187	13.71220	13.16612	12.16567	11.27407
18	16.39827	15.67256	14.99203	14.35336	13.75351	12.65930	11.68959
19	17.22601	16.42617	15.67846	14.97889	14.32380	13.13394	12.08532
20	18.04555	17.16864	16.35143	15.58916	14.87747	13.59033	12.46221
21	18.85698	17.90014	17.01121	16.18455	15.41502	14.02916	12.82115
22	19.66038	18.62082	17.65805	16.76541	15.93692	14.45112	13.16300
23	20.45582	19.33086	18.29220	17.33211	16.44361	14.85684	13.48857
24	21.24339	20.03041	18.91393	17.88499	16.93554	15.24696	13.79864
25	22.02316	20.71961	19.52346	18.42438	17.41315	15.62208	14.09394
26	22.79520	21.39863	20.12104	18.95061	17.87684	15.98277	14.37519
27	23.55961	22.06762	20.70690	19.46401	18.32703	16.32959	14.64303
28	24.31644	22.72672	21.28127	19.96489	18.76411	16.66306	14.89813
29	25.06579	23.37608	21.84438	20.45355	19.18845	16.98371	15.14107
30	25.80771	24.01584	22.39646	20.93029	19.60044	17.29203	15.37245
31	26.54229	24.64615	22.93770	21.39541	20.00043	17.58849	15.59281
32	27.26959	25.26714	23.46833	21.84918	20.38877	17.87355	15.80268
33	27.98969	25.87895	23.98856	22.29188	20.76579	18.14765	16.00255
34	28.70267	26.48173	24.49859	22.72379	21.13184	18.41120	16.19290
35	29.40858	27.07559	24.99862	23.14516	21.48722	18.66461	16.37419
36	30.10751	27.66068	25.48884	23.55625	21.83225	18.90828	16.54685
37	30.79951	28.23713	25.96945	23.95732	22.16724	19.14258	16.71129
38	31.48466	28.80505	26.44064	24.34860	22.49246	19.36786	16.86789
39	32.16303	29.36458	26.90259	24.73034	22.80822	19.58448	17.01704
40	32.83469	29.91585	27.35548	25.10278	23.11477	19.79277	17.15909
41	33.49969	30.45896	27.79949	25.46612	23.41240	19.99305	17.29437
42	34.15811	30.99405	28.23479	25.82061	23.70136	20.18563	17.42321
43	34.81001	31.52123	28.66156	26.16645	23.98190	20.37079	17.54591
44	35.45545	32.04062	29.07996	26.50385	24.25427	20.54884	17.66277
45	36.09451	32.55234	29.49016	26.83302	24.51871	20.72004	17.77407
46	36.72724	33.05649	29.89231	27.15417	24.77545	20.88465	17.88007
47	37.35370	33.55319	30.28658	27.46748	25.02471	21.04294	17.98102
48	37.97396	34.04255	30.67312	27.77315	25.26671	21.19513	18.07716
49	38.58808	34.52468	31.05208	28.07137	25.50166	21.34147	18.16872
50	39.19612	34.99969	31.42361	28.36231	25.72976	21.48218	18.25593

TABLE 8 (CONTINUED)

i / n	6.00%	8.00%	10.00%	12.00%	14.00%	16.00%	18.00%
1	0.94340	0.92593	0.90909	0.89286	0.87719	0.86207	0.84746
2	1.83339	1.78326	1.73554	1.69005	1.64666	1.60523	1.56564
3	2.67301	2.57710	2.48685	2.40183	2.32163	2.24589	2.17427
4	3.46511	3.31213	3.16987	3.03735	2.91371	2.79818	2.69006
5	4.21236	3.99271	3.79079	3.60478	3.43308	3.27429	3.12717
6	4.91732	4.62288	4.35526	4.11141	3.88867	3.68474	3.49760
7	5.58238	5.20637	4.86842	4.56376	4.28830	4.03857	3.81153
8	6.20979	5.74664	5.33493	4.96764	4.63886	4.34359	4.07757
9	6.80169	6.24689	5.75902	5.32825	4.94637	4.60654	4.30302
10	7.36009	6.71008	6.14457	5.65022	5.21612	4.83323	4.49409
11	7.88687	7.13896	6.49506	5.93770	5.45273	5.02864	4.65601
12	8.38384	7.53608	6.81369	6.19437	5.66029	5.19711	4.79322
13	8.85268	7.90378	7.10336	6.42355	5.84236	5.34233	4.90951
14	9.29498	8.24424	7.36669	6.62817	6.00207	5.46753	5.00806
15	9.71225	8.55948	7.60608	6.81086	6.14217	5.57546	5.09158
16	10.10590	8.85137	7.82371	6.97399	6.26506	5.66850	5.16235
17	10.47726	9.12164	8.02155	7.11963	6.37286	5.74870	5.22233
18	10.82760	9.37189	8.20141	7.24967	6.46742	5.81785	5.27316
19	11.15812	9.60360	8.36492	7.36578	6.55037	5.87746	5.31624
20	11.46992	9.81815	8.51356	7.46944	6.62313	5.92884	5.35275
21	11.76408	10.01680	8.64869	7.56200	6.68696	5.97314	5.38368
22	12.04158	10.20074	8.77154	7.64465	6.74294	6.01133	5.40990
23	12.30338	10.37106	8.88322	7.71843	6.79206	6.04425	5.43212
24	12.55036	10.52876	8.98474	7.78432	6.83514	6.07263	5.45095
25	12.78336	10.67478	9.07704	7.84314	6.87293	6.09709	5.46691
26	13.00317	10.80998	9.16095	7.89566	6.90608	6.11818	5.48043
27	13.21053	10.93516	9.23722	7.94255	6.93515	6.13636	5.49189
28	13.40616	11.05108	9.30657	7.98442	6.96066	6.15204	5.50160
29	13.59072	11.15841	9.36961	8.02181	6.98304	6.16555	5.50983
30	13.76483	11.25778	9.42691	8.05518	7.00266	6.17720	5.51681
31	13.92909	11.34980	9.47901	8.08499	7.01988	6.18724	5.52272
32	14.08404	11.43500	9.52638	8.11159	7.03498	6.19590	5.52773
33	14.23023	11.51389	9.56943	8.13535	7.04823	6.20336	5.53197
34	14.36814	11.58693	9.60857	8.15656	7.05985	6.20979	5.53557
35	14.49825	11.65457	9.64416	8.17550	7.07005	6.21534	5.53862
36	14.62099	11.71719	9.67651	8.19241	7.07899	6.22012	5.54120
37	14.73678	11.77518	9.70592	8.20751	7.08683	6.22424	5.54339
38	14.84602	11.82887	9.73265	8.22099	7.09371	6.22779	5.54525
39	14.94907	11.87858	9.75696	8.23303	7.09975	6.23086	5.54682
40	15.04630	11.92461	9.77905	8.24378	7.10504	6.23350	5.54815
41	15.13802	11.96723	9.79914	8.25337	7.10969	6.23577	5.54928
42	15.22454	12.00670	9.81740	8.26194	7.11376	6.23774	5.55024
43	15.30617	12.04324	9.83400	8.26959	7.11733	6.23943	5.55105
44	15.38318	12.07707	9.84909	8.27642	7.12047	6.24089	5.55174
45	15.45583	12.10840	9.86281	8.28252	7.12322	6.24214	5.55232
46	15.52437	12.13741	9.87528	8.28796	7.12563	6.24323	5.55281
47	15.58903	12.16427	9.88662	8.29282	7.12774	6.24416	5.55323
48	15.65003	12.18914	9.89693	8.29716	7.12960	6.24497	5.55359
49	15.70757	12.21216	9.90630	8.30104	7.13123	6.24566	5.55389
50	15.76186	12.23348	9.91481	8.30450	7.13266	6.24626	5.55414

TABLE 9 COMBINATIONS

n	$\binom{n}{0}$	$\binom{n}{1}$	$\binom{n}{2}$	$\binom{n}{3}$	$\binom{n}{4}$	$\binom{n}{5}$	$\binom{n}{6}$	$\binom{n}{7}$	$\binom{n}{8}$	$\binom{n}{9}$	$\binom{n}{10}$
0	1										
1	1	1									
2	1	2	1								
3	1	3	3	1							
4	1	4	6	4	1						
5	1	5	10	10	5	1					
6	1	6	15	20	15	6	1				
7	1	7	21	35	35	21	7	1			
8	1	8	28	56	70	56	28	8	1		
9	1	9	36	84	126	126	84	36	9	1	
10	1	10	45	120	210	252	210	120	45	10	1
11	1	11	55	165	330	462	462	330	165	55	11
12	1	12	66	220	495	792	924	792	495	220	66
13	1	13	78	286	715	1287	1716	1716	1287	715	286
14	1	14	91	364	1001	2002	3003	3432	3003	2002	1001
15	1	15	105	455	1365	3003	5005	6435	6435	5005	3003
16	1	16	120	560	1820	4368	8008	11440	12870	11440	8008
17	1	17	136	680	2380	6188	12376	19448	24310	24310	19448
18	1	18	153	816	3060	8568	18564	31824	43758	48620	43758
19	1	19	171	969	3876	11628	27132	50388	75582	92378	92378
20	1	20	190	1140	4845	15504	38760	77520	125970	167960	184756

For $r > 10$, it may be necessary to use the identity

$$\binom{n}{r} = \binom{n}{n-r}.$$

TABLE 10 AREAS UNDER THE STANDARD NORMAL CURVE

The column under *A* gives the proportion of the area under the entire curve which is between $z = 0$ and a positive value of z.

z	A	z	A	z	A	z	A
.00	.0000	.49	.1879	.98	.3365	1.47	.4292
.01	.0040	.50	.1915	.99	.3389	1.48	.4306
.02	.0080	.51	.1950	1.00	.3413	1.49	.4319
.03	.0120	.52	.1985 ·	1.01	.3438	1.50	.4332
.04	.0160	.53	.2019	1.02	.3461	1.51	.4345
.05	.0199	.54	.2054	1.03	.3485	1.52	.4357
.06	.0239	.55	.2088	1.04	.3508	1.53	.4370
.07	.0279	.56	.2123	1.05	.3531	1.54	.4382
.08	.0319	.57	.2157	1.06	.3554	1.55	.4394
.09	.0359	.58	.2190	1.07	.3577	1.56	.4406
.10	.0398	.59	.2224	1.08	.3599	1.57	.4418
.11	.0438	.60	.2258	1.09	.3621	1.58	.4430
.12	.0478	.61	.2291	1.10	.3643	1.59	.4441
.13	.0517	.62	.2324	1.11	.3665	1.60	.4452
.14	.0557	.63	.2357	1.12	.3686	1.61	.4463
.15	.0596	.64	.2389	1.13	.3708	1.62	.4474
.16	.0636	.65	.2422	1.14	.3729	1.63	.4485
.17	.0675	.66	.2454	1.15	.3749	1.64	.4495
.18	.0714	.67	.2486	1.16	.3770	1.65	.4505
.19	.0754	.68	.2518	1.17	.3790	1.66	.4515
.20	.0793	.69	.2549	1.18	.3810	1.67	.4525
.21	.0832	.70	.2580	1.19	.3830	1.68	.4535
.22	.0871	.71	.2612	1.20	.3849	1.69	.4545
.23	.0910	.72	.2642	1.21	.3869	1.70	.4554
.24	.0948	.73	.2673	1.22	.3888	1.71	.4564
.25	.0987	.74	.2704	1.23	.3907	1.72	.4573
.26	.1026	.75	.2734	1.24	.3925	1.73	.4582
.27	.1064	.76	.2764	1.25	.3944	1.74	.4591
.28	.1103	.77	.2794	1.26	.3962	1.75	.4599
.29	.1141	.78	.2823	1.27	.3980	1.76	.4608
.30	.1179	.79	.2852	1.28	.3997	1.77	.4616
.31	.1217	.80	.2881	1.29	.4015	1.78	.4625
.32	.1255	.81	.2910	1.30	.4032	1.79	.4633
.33	.1293	.82	.2939	1.31	.4049	1.80	.4641
.34	.1331	.83	.2967	1.32	.4066	1.81	.4649
.35	.1368	.84	.2996	1.33	.4082	1.82	.4656
.36	.1406	.85	.3023	1.34	.4099	1.83	.4664
.37	.1443	.86	.3051	1.35	.4115	1.84	.4671
.38	.1480	.87	.3079	1.36	.4131	1.85	.4678
.39	.1517	.88	.3106	1.37	.4147	1.86	.4686
.40	.1554	.89	.3133	1.38	.4162	1.87	.4693
.41	.1591	.90	.3159	1.39	.4177	1.88	.4700
.42	.1628	.91	.3186	1.40	.4192	1.89	.4706
.43	.1664	.92	.3212	1.41	.4207	1.90	.4713
.44	.1700	.93	.3238	1.42	.4222	1.91	.4719
.45	.1736	.94	.3264	1.43	.4236	1.92	.4726
.46	.1772	.95	.3289	1.44	.4251	1.93	.4732
.47	.1808	.96	.3315	1.45	.4265	1.94	.4738
.48	.1844	.97	.3340	1.46	.4279	1.95	.4744

TABLE 10 (CONTINUED)

z	A	z	A	z	A	z	A
1.96	.4750	2.45	.4929	2.94	.4984	3.43	.4997
1.97	.4756	2.46	.4931	2.95	.4984	3.44	.4997
1.98	.4762	2.47	.4932	2.96	.4985	3.45	.4997
1.99	.4767	2.48	.4934	2.97	.4985	3.46	.4997
2.00	.4773	2.49	.4936	2.98	.4986	3.47	.4997
2.01	.4778	2.50	.4938	2.99	.4986	3.48	.4998
2.02	.4783	2.51	.4940	3.00	.4987	3.49	.4998
2.03	.4788	2.52	.4941	3.01	.4987	3.50	.4998
2.04	.4793	2.53	.4943	3.02	.4987	3.51	.4998
2.05	.4798	2.54	.4945	3.03	.4988	3.52	.4998
2.06	.4803	2.55	.4946	3.04	.4988	3.53	.4998
2.07	.4808	2.56	.4948	3.05	.4989	3.54	.4998
2.08	.4812	2.57	.4949	3.06	.4989	3.55	.4998
2.09	.4817	2.58	.4951	3.07	.4989	3.56	.4998
2.10	.4821	2.59	.4952	3.08	.4990	3.57	.4998
2.11	.4826	2.60	.4953	3.09	.4990	3.58	.4998
2.12	.4830	2.61	.4955	3.10	.4990	3.59	.4998
2.13	.4834	2.62	.4956	3.11	.4991	3.60	.4998
2.14	.4838	2.63	.4957	3.12	.4991	3.61	.4999
2.15	.4842	2.64	.4959	3.13	.4991	3.62	.4999
2.16	.4846	2.65	.4960	3.14	.4992	3.63	.4999
2.17	.4850	2.66	.4961	3.15	.4992	3.64	.4999
2.18	.4854	2.67	.4962	3.16	.4992	3.65	.4999
2.19	.4857	2.68	.4963	3.17	.4992	3.66	.4999
2.20	.4861	2.69	.4964	3.18	.4993	3.67	.4999
2.21	.4865	2.70	.4965	3.19	.4993	3.68	.4999
2.22	.4868	2.71	.4966	3.20	.4993	3.69	.4999
2.23	.4871	2.72	.4967	3.21	.4993	3.70	.4999
2.24	.4875	2.73	.4968	3.22	.4994	3.71	.4999
2.25	.4878	2.74	.4969	3.23	.4994	3.72	.4999
2.26	.4881	2.75	.4970	3.24	.4994	3.73	.4999
2.27	.4884	2.76	.4971	3.25	.4994	3.74	.4999
2.28	.4887	2.77	.4972	3.26	.4994	3.75	.4999
2.29	.4890	2.78	.4973	3.27	.4995	3.76	.4999
2.30	.4893	2.79	.4974	3.28	.4995	3.77	.4999
2.31	.4896	2.80	.4974	3.29	.4995	3.78	.4999
2.32	.4898	2.81	.4975	3.30	.4995	3.79	.4999
2.33	.4901	2.82	.4976	3.31	.4995	3.80	.4999
2.34	.4904	2.83	.4977	3.32	.4996	3.81	.4999
2.35	.4906	2.84	.4977	3.33	.4996	3.82	.4999
2.36	.4909	2.85	.4978	3.34	.4996	3.83	.4999
2.37	.4911	2.86	.4979	3.35	.4996	3.84	.4999
2.38	.4913	2.87	.4980	3.36	.4996	3.85	.4999
2.39	.4916	2.88	.4980	3.37	.4996	3.86	.4999
2.40	.4918	2.89	.4981	3.38	.4996	3.87	.5000
2.41	.4920	2.90	.4981	3.39	.4997	3.88	.5000
2.42	.4922	2.91	.4982	3.40	.4997	3.89	.5000
2.43	.4925	2.92	.4983	3.41	.4997		
2.44	.4927	2.93	.4983	3.42	.4997		

TABLE 11   INTEGRALS

(*C* is an arbitrary constant.)

1. $\int x^n \, dx = \dfrac{1}{n+1} x^{n+1} + C$ (if $n \neq -1$)

2. $\int e^{kx} \, dx = \dfrac{1}{k} e^{kx} + C$

3. $\int \dfrac{a}{x} \, dx = a \ln |x| + C$ ($a \neq 0$)

4. $\int \ln |ax| \, dx = x(\ln |ax| - 1) + C$

5. $\int \dfrac{1}{\sqrt{x^2 + a^2}} \, dx = \ln \left| \dfrac{x + \sqrt{x^2 + a^2}}{a} \right| + C$

6. $\int \dfrac{1}{\sqrt{x^2 - a^2}} \, dx = \ln \left| \dfrac{x + \sqrt{x^2 - a^2}}{a} \right| + C$

7. $\int \dfrac{1}{a^2 - x^2} \, dx = \dfrac{1}{2a} \cdot \ln \left| \dfrac{a + x}{a - x} \right| + C$ ($x^2 < a^2$)

8. $\int \dfrac{1}{x^2 - a^2} \, dx = \dfrac{1}{2a} \cdot \ln \left| \dfrac{x - a}{x + a} \right| + C$ ($x^2 > a^2$)

9. $\int \dfrac{1}{x\sqrt{a^2 - x^2}} \, dx = -\dfrac{1}{a} \cdot \ln \left| \dfrac{a + \sqrt{a^2 - x^2}}{x} \right| + C$ ($0 < x < a$)

10. $\int \dfrac{1}{x\sqrt{a^2 + x^2}} \, dx = -\dfrac{1}{a} \cdot \ln \left| \dfrac{a + \sqrt{a^2 + x^2}}{x} \right| + C$

11. $\int \dfrac{x}{ax + b} \, dx = \dfrac{x}{a} - \dfrac{b}{a^2} \cdot \ln |ax + b| + C$ ($a \neq 0$)

12. $\int \dfrac{x}{(ax + b)^2} \, dx = \dfrac{b}{a^2(ax + b)} + \dfrac{1}{a^2} \cdot \ln |ax + b| + C$ ($a \neq 0$)

13. $\int \dfrac{1}{x(ax + b)} \, dx = \dfrac{1}{b} \cdot \ln \left| \dfrac{x}{ax + b} \right| + C$ ($b \neq 0$)

14. $\int \dfrac{1}{x(ax + b)^2} \, dx = \dfrac{1}{b(ax + b)} + \dfrac{1}{b^2} \cdot \ln \left| \dfrac{x}{ax + b} \right| + C$ ($b \neq 0$)

15. $\int \sqrt{x^2 + a^2} \, dx = \dfrac{x}{2} \sqrt{x^2 + a^2} + \dfrac{a^2}{2} \cdot \ln |x + \sqrt{x^2 + a^2}| + C$

16. $\int x^n \cdot \ln x \, dx = x^{n+1} \left[\dfrac{\ln |x|}{n+1} - \dfrac{1}{(n+1)^2} \right] + C$ ($n \neq -1$)

17. $\int x^n e^{ax} \, dx = \dfrac{x^n e^{ax}}{a} - \dfrac{n}{a} \cdot \int x^{n-1} e^{ax} \, dx + C$ ($a \neq 0$)

Answers to Selected Exercises

Chapter 1

Section 1.1 (page 7)

1. Counting number, whole number, integer, rational number, real number   **3.** Integer, rational number, real number
5. Rational number, real number   **7.** Irrational number, real number   **9.** Irrational number, real number   **11.** True
13. True   **15.** False   **17.** True   **19.** Commutative   **21.** Commutative   **23.** Identity   **25.** Associative
27. Distributive   **29.** Commutative properties of addition and multiplication

31. 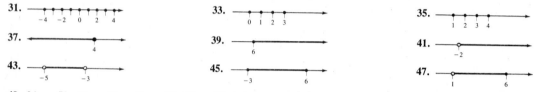 **33.** **35.**

37. **39.** **41.**

43. **45.** **47.**

49. 24   **51.** 17   **53.** -21 **55.** 37   **57.** $-6/7$ **59.** Meaningless   **61.** 2.22   **63.** -4 **65.** 2   **67.** -4
69. 17   **71.** 4   **73.** -19 **75.** $=$ **77.** $<$ **79.** $=$ **81.** $=$ **83.** $=$ **85.** $=$ **87.** $=$ **89.** Yes   **91.** Yes
93. Let x represent earnings in millions of dollars. Then $x \ge 26$.   **95.** Let x represent prices in dollars. Then $x \ge 3499$.
97. Let x represent cost of other IBM compatibles in dollars. Then $1000 < x < 1500$.   **99.** 100; 0   **101.** 67; 11

Section 1.2 (page 18)

1. 4   **3.** 12   **5.** $-2/7$ **7.** $-7/8$ **9.** -1 **11.** 3   **13.** 12   **15.** 3/4   **17.** $-12/5$ **19.** No solution
21. $-59/6$ **23.** $-9/4$ **25.** $x = -3a + b$ **27.** $x = (3a + b)/(3 - a)$ **29.** $x = (3 - 3a)/(a^2 - a - 1)$
31. $x = (2a^2)/(a^2 + 3)$ **33.** $V = k/P$ **35.** $g = (V - V_0)/t$ **37.** $B = (2A/h) - b$ or $(2A - bh)/h$ **39.** $R = (r_1 r_2)/(r_1 + r_2)$
41. .72   **43.** 6.53   **45.** -13.26 **47.** 68°F   **49.** 15°C   **51.** 37.8°C   **53.** 104°F   **55.** 13.0%   **57.** $432
59. $1500   **61.** $205.41   **63.** $66.50   **65.** 5 cm   **67.** $12,000 at 13%, $8000 at 16%   **69.** $54,000 at 8 1/2%, $6000
at 16%   **71.** $70,000 for land that made a profit, $50,000 for land that produced a loss   **73.** 38/3 kg   **75.** 400/3 liters
77. About 840 miles   **79.** 35 kph   **81.** 18/5 hr   **83.** 78 hrs   **85.** 40/3 hr

Section 1.3 (page 23)

1. $x \le -3$ **3.** $p > -6$ **5.** $a > 0$ **7.** $x \le 4$

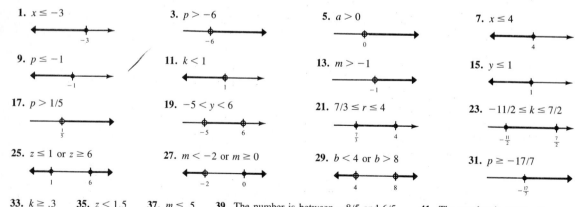

9. $p \le -1$ **11.** $k < 1$ **13.** $m > -1$ **15.** $y \le 1$

17. $p > 1/5$ **19.** $-5 < y < 6$ **21.** $7/3 \le r \le 4$ **23.** $-11/2 \le k \le 7/2$

25. $z \le 1$ or $z \ge 6$ **27.** $m < -2$ or $m \ge 0$ **29.** $b < 4$ or $b > 8$ **31.** $p \ge -17/7$

33. $k \ge .3$ **35.** $z < 1.5$ **37.** $m \le .5$ **39.** The number is between $-8/5$ and $6/5$.   **41.** The number is greater than or
equal to 2.   **43.** The number is at least 18.   **45.** 83 points   **47.** 4 summers   **49.** $x \ge 500$ **51.** $x \ge 45$ **53.** For
positive values of x, C is always greater than R, so it is impossible to make a profit.

Section 1.4   (page 27)

1. 1, 3   **3.** $-1/3$, 1   **5.** 2/3, 8/3   **7.** -6, 14   **9.** 5/2, 7/2   **11.** $-4/3$, 2/9   **13.** $-7/3$, $-1/7$ **15.** $-3/5$, 11
17. $-3 \le x \le 3$ **19.** $m < -1$ or $m > 1$ **21.** No solution   **23.** $-10 \le x \le 10$

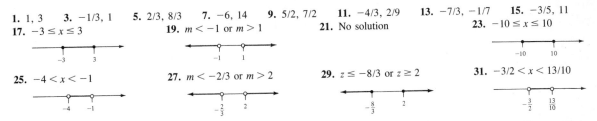

25. $-4 < x < -1$ **27.** $m < -2/3$ or $m > 2$ **29.** $z \le -8/3$ or $z \ge 2$ **31.** $-3/2 < x < 13/10$

33. The number is between -6 and 6, inclusive.   **35.** The number is between $-7/4$ and $-5/4$, inclusive.   **37.** The number is less than or equal to -14 or greater than or equal to 10.   **39.** $|x - 2| \le 4$ **41.** $|z - 12| \ge 2$ **43.** $|k - 9| = 6$ **45.** If $|x - 2| \le .0004$, then $|y - 7| \le .00001$.

Section 1.5   (page 33)

1. $14m + 6$ **3.** $-9k + 5$ **5.** $-x^2 + x + 9$ **7.** $-6y^2 + 3y + 10$ **9.** $-10x^2 + 4x - 2$ **11.** $-.327x^2 - 2.805x - 1.458$
13. $6p^2 - 15p$ **15.** $-18m^3 - 27m^2 + 9m$ **17.** $12k^2 - 20k + 3$ **19.** $6y^2 + 7y - 5$ **21.** $25r^2 + 5rs - 12s^2$
23. $.0036x^2 - .04452x - .0918$ **25.** $5(5k + 6)$ **27.** $4(z + 1)$ **29.** $2(4x + 3y + 2z)$ **31.** $2(3r^2 + 2r + 4)$
33. $m(m^2 - 9m + 6)$ **35.** $8a(a^2 - 2a + 3)$ **37.** $5p^2(5p^2 - 4pq + 20q^2)$ **39.** $(m + 7)(m + 2)$ **41.** $(x + 5)(x - 1)$
43. $(z + 5)(z + 4)$ **45.** $(b - 7)(b - 1)$ **47.** Prime   **49.** $(s - 5t)(s + 7t)$ **51.** $(y - 7z)(y + 3z)$ **53.** $6(a - 10)(a + 2)$
55. $3m(m + 3)(m + 1)$ **57.** $(2x + 1)(x - 3)$ **59.** $(3a + 7)(a + 1)$ **61.** $(2a - 5)(a - 6)$ **63.** $(5y + 2)(3y - 1)$
65. Prime   **67.** $(5a + 3b)(a - 2b)$ **69.** $(7m + 2n)(3m + n)$ **71.** $2a^2(4a - b)(3a + 2b)$ **73.** $4z^3(8z + 3a)(z - a)$
75. $(x + 8)(x - 8)$ **77.** $(3m + 5)(3m - 5)$ **79.** $(11a + 10)(11a - 10)$ **81.** Prime   **83.** $(z + 7y)^2$ **85.** $(m - 3n)^2$
87. $(3p - 4)^2$ **89.** $(a - 6)(a^2 + 6a + 36)$ **91.** $(2r - 3s)(4r^2 + 6rs + 9s^2)$ **93.** $(4m + 5)(16m^2 - 20m + 25)$
95. $(10y - z)(100y^2 + 10yz + z^2)$

Section 1.6   (page 38)

1. $m/4$ **3.** $z/2$ **5.** $5p/2$ **7.** 8/9   **9.** $3/(t - 3)$ **11.** $2(x + 2)/x$ **13.** $(m - 2)/(m + 3)$ **15.** $(x + 4)/(x + 1)$
17. $(2m + 3)/(4m + 3)$ **19.** $3k/5$ **21.** $25p^2/9$ **23.** $6/(5p)$ **25.** 2   **27.** 2/9   **29.** 3/10   **31.** $2(a + 4)/(a - 3)$
33. $(k + 2)/(k + 3)$ **35.** $(m + 6)/(m + 3)$ **37.** $(m - 3)/(2m - 3)$ **39.** $14/r$ **41.** $19/(6k)$ **43.** $5/(12y)$ **45.** 1
47. $(6 + p)/(2p)$ **49.** $(8 - y)/(4y)$ **51.** $137/(30m)$ **53.** $(3m - 2)/[m(m - 1)]$ **55.** $(r - 12)/[r(r - 2)]$
57. $14/[3(a - 1)]$ **59.** $23/[20(k - 2)]$ **61.** $(7x + 9)/[(x - 3)(x + 1)(x + 2)]$ **63.** $y^2/[(y + 4)(y + 3)(y + 2)]$
65. $k(k - 13)/[(2k - 1)(k + 2)(k - 3)]$ **67.** $(x + 1)/(x - 1)$ **69.** $-1/(x + 1)$ **71.** $-1/[x(x + h)]$ **73.** $(2 + b - b^2)/(b - b^2)$

Section 1.7   (page 43)

1. 343   **3.** 1/8   **5.** 1/8   **7.** 1/5   **9.** 1/512   **11.** 27/64   **13.** 25/64   **15.** 8   **17.** 49/4   **19.** 3^6 **21.** $1/7^4$
23. $1/3^6$ **25.** $1/2^3$ **27.** $1/6^2$ **29.** 4^3 **31.** $1/7^7$ **33.** 8^5 **35.** $1/10^8$ **37.** 5   **39.** x^2 **41.** $8k^3$
43. $r/7^2$ or $r/49$ **45.** $1/12k^4$ **47.** m^7/p^2 **49.** $9a^2/(5b)$ **51.** 5   **53.** 6   **55.** 0   **57.** 73/8   **59.** 1/6   **61.** 9
63. 1/6   **65.** 3/2   **67.** $-5/9$ **69.** $b + a$ **71.** $1/(ab)$

Section 1.8   (page 49)

1. 9   **3.** 3   **5.** 4   **7.** 100   **9.** 4   **11.** -25 **13.** 2/3   **15.** 4/3   **17.** 1/32   **19.** 4/3   **21.** 3/4   **23.** $2^2 = 4$ **25.** $27^{1/3} = 3$ **27.** $4^2 = 16$ **29.** $1/6^{11/5}$ **31.** $1/4^{32/15}$ **33.** $2^{5/6}p^{3/2}$ **35.** $1/(m^{14/15}n^{6/5})$ **37.** $x^{11/4} - 3x^{15/4}$
39. $6 + 2z$ **41.** 5   **43.** 6   **45.** -5 **47.** $5\sqrt{2}$ **49.** $20\sqrt{5}$ **51.** $-2\sqrt[4]{2}$ **53.** $-3\sqrt{5}/5$ **55.** $-\sqrt[3]{12}/2$

57. $\sqrt[4]{24}/2$ **59.** $24\sqrt{6}$ **61.** $2x^4z^2\sqrt{2z}$ **63.** $mn^3p^4\sqrt{n}$ **65.** $\sqrt{6x}/(3x)$ **67.** $(x^2y\sqrt{xy})/3$ **69.** $\sqrt[3]{2}$ **71.** $\sqrt[18]{x}$
73. $9\sqrt{3}$ **75.** $\sqrt{2}$ **77.** $-2\sqrt{7p}$ **79.** $2\sqrt{2}$ **81.** $7\sqrt[3]{3}$ **83.** -7 **85.** 10 **87.** $5\sqrt{6}$ **89.** $-3 - 3\sqrt{2}$
91. $(4\sqrt{3} - 3)/13$ **93.** $3 + \sqrt{2}$ **95.** $[p(\sqrt{p} - 2)]/(p - 4)$ **97.** \$64 **99.** \$64,000,000 **101.** About 86 miles
103. About 211 miles **105.** 29 **107.** 177

Section 1.9 (page 58)

1. $5, -4$ **3.** $-2, -3$ **5.** $6, -1$ **7.** $-8, 3$ **9.** $3, -1$ **11.** 4 **13.** $5/2, -2$ **15.** $4/3, -1/2$ **17.** $5, 2$
19. $4/3, -4/3$ **21.** $0, 1$ **23.** $\sqrt{29}, -\sqrt{29}$ **25.** $3 + \sqrt{5}, 3 - \sqrt{5}$ **27.** $(1 + \sqrt{19})/3, (1 - \sqrt{19})/3$ **29.** $2, -4$
31. $(1 + \sqrt{2})/2, (1 - \sqrt{2})/2$ **33.** $(5 + \sqrt{13})/6 \approx 1.434; (5 - \sqrt{13})/6 \approx .232$ **35.** $(1 + \sqrt{33})/4 \approx 1.686;$
$(1 - \sqrt{33})/4 \approx -1.186$ **37.** $5 + \sqrt{5} \approx 7.236; 5 - \sqrt{5} \approx 2.764$ **39.** $(-6 + \sqrt{26})/2 \approx -.450; (-6 - \sqrt{26})/2 \approx -5.550$
41. $5/2, 1$ **43.** $4/3, 1/2$ **45.** $-5, 2$ **47.** No real number solutions **49.** $0, -1$ **51.** $-\sqrt{5}, \sqrt{5}$
53. $\sqrt{15}/3, -\sqrt{15}/3$ **55.** $\pm\sqrt{6 + 2\sqrt{29}}/2$ **57.** $11/3, 3/2$ **59.** $-1, -15/2$ **61.** $-5, 3/2$ **63.** $-4 + 3\sqrt{2},$
$-4 - 3\sqrt{2}$ **65.** 16, 18 or $-18, -16$ **67.** $12, -2$ **69.** 50 m by 100 m **71.** 9 ft by 12 ft **73.** 50 mph
75. $\pm\sqrt{2Sg}/g$ **77.** $\pm(d^2\sqrt{kL})/L$ **79.** $\pm\sqrt{(S - S_0 - k)g}/g$ **81.** $(-\pi h \pm \sqrt{\pi^2h^2 + 2\pi S})/2\pi$

Section 1.10 (page 63)

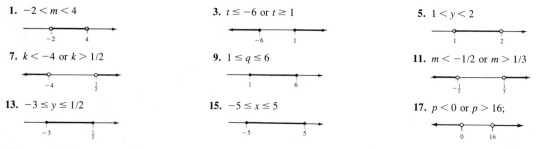

1. $-2 < m < 4$

3. $t \le -6$ or $t \ge 1$

5. $1 < y < 2$

7. $k < -4$ or $k > 1/2$

9. $1 \le q \le 6$

11. $m < -1/2$ or $m > 1/3$

13. $-3 \le y \le 1/2$

15. $-5 \le x \le 5$

17. $p < 0$ or $p > 16$;

19. $t \le -4$ or $0 \le t \le 4$ **21.** $-7 < r < 0$ or $r > 7$ **23.** $-2 < m < 0$ or $m > 1/4$ **25.** $-5 < m \le 3$ **27.** $k < -2$
29. $x < 6$ or $x \ge 15/2$ **31.** $-8 \le y < 5$ **33.** $k > -2$ **35.** \$54 **37.** 5/4 months and 6 months **39.** \$30

Chapter 1 Review Exercises (page 64)

1. 6 **3.** $-12, -6, -\sqrt{4}$ (or -2), $0, 6$ **5.** $-\sqrt{7}, \pi/4, \sqrt{11}$ **7.** Natural, whole, integer, rational, real **9.** ($\sqrt{36} = 6$)
natural, whole, integer, rational, real **11.** Integer, rational, real **13.** Irrational, real **15.** Irrational, real **17.** $-7, -3,$
$-2, 0, \pi, 8$ **19.** $-|3 - (-2)|, -|-2|, |6 - 4|, |8 + 1|$ **21.** -3 **23.** -1

25. ──────────────
        -3

27. ◄──────────────
           -2

29. -8 **31.** $-3/4$ **33.** 2 **35.** 6 **37.** $-11/3$ **39.** $-96/7$ **41.** -2 **43.** No solution **45.** $x = 2/(9p - 1)$
47. $x = -6b - a - 6$ **49.** $x = (6 - 3m)/(1 + 2k - km)$ **51.** \$500 **53.** 12 hours **55.** $x > -7/13$ **57.** $z \le 1$
59. $r > 1$ **61.** $4 \le x \le 5$ **63.** $3, -11$ **65.** $23, -13$ **67.** $15/8, -13/8$ **69.** $4/3, -2/7$ **71.** $-7 \le m \le 7$
73. $p < -3$ or $p > 3$ **75.** No solution **77.** $-2/7 < k < 8/7$ **79.** $r < -4$ or $r > -2/3$ **81.** $-7m^2 - m - 3$
83. $8r^4 - 8r^2 + 11r$ **85.** $-r^5 + r^4 + 19r^2$ **87.** $16y^2 + 42y - 49$ **89.** $21z^2 - 5zy - 50y^2$ **91.** $9k^2 - 30km + 25m^2$
93. $125x^3 - 150x^2 + 60x - 8$ **95.** $15w^3 - 22w^2 + 11w - 2$ **97.** $z(7z - 9z^2 + 1)$ **99.** $4p^3(3p^2 - 2p + 5)$
101. $(r + 7p)(r - 6p)$ **103.** $(3m + 1)(2m - 5)$ **105.** $(3m + 7)(m - 5)$ **107.** $(12p + 13q)(12p - 13q)$
109. $(2y - 1)(4y^2 + 2y + 1)$ **111.** $8p^2/5$ **113.** 4 **115.** $1/[2k^2(k - 1)]$ **117.** $17/(2r)$ **119.** $(5r + 3)/[(r - 1)r]$
121. $(q + p)/(pq - 1)$ **123.** $1/16$ **125.** $1/125$ **127.** $64/27$ **129.** 6 **131.** $1/8^2$ or $1/64$ **133.** 9^3 or 729
135. $3/4$ **137.** 25 **139.** $1/3^5$ **141.** $1/5^{2/3}$ **143.** $3^{7/2}a^{5/2}$ **145.** 3 **147.** -2 **149.** $2\sqrt{6}$ **151.** $3pq\sqrt[3]{2q^2}$

153. $n\sqrt{30m}/6m$ **155.** $\sqrt[6]{5}$ **157.** $-8\sqrt{3}$ **159.** $7s\sqrt[3]{2r^2}$ **161.** 4   **163.** $4 + 3\sqrt{15}$ **165.** $(-\sqrt{2} + \sqrt{6})/2$
167. $\sqrt{7}, -\sqrt{7}$ **169.** $-7 + \sqrt{5}, -7 - \sqrt{5}$ **171.** 1, 3   **173.** 1/2, -2 **175.** $1 + \sqrt{3}, 1 - \sqrt{3}$ **177.** $(6 + \sqrt{58})/2$,
$(6 - \sqrt{58})/2$ **179.** 5/2, -3 **181.** 7, $-3/2$ **183.** $\sqrt{3}/3, -\sqrt{3}/3$ **185.** $\sqrt{5}, -\sqrt{5}$ **187.** 5, $-2/3$ **189.** -5,
$-1/2$ **191.** $(-RP \pm E\sqrt{RP})/P$ **193.** $(a \pm \sqrt{a^2 + 4K})/2$ **195.** $-3 < r < 2$ **197.** $z \leq -5$ or $z \geq 3/2$ **199.** $a < -3/2$
or $a > 1/4$ **201.** $0 \leq r \leq 5/3$ **203.** $-2 \leq m < 0$ **205.** $-1 < p < 3/2$ **207.** $-19 \leq r < -5$ or $r > 2$
209. 50 m by 225 m; or 112.5 m by 100 m

Case 1   (page 68)

1. $.1A + 200$ **2.** $.0001A - .3$ **3.** 8000   **4.** 8000   **5.** 800   **6.** $800(10,000) = \$8,000,000$

Chapter 2

Section 2.1   (page 73)

1. $(-2, 17), (-1, 13), (0, 9),$
$(1, 5), (2, 1), (3, -3);$ range:
$\{17, 13, 9, 5, 1, -3\}$

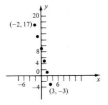

3. $(-2, -3), (-1, -4),$
$(0, -5), (1, -6), (2, -7),$
$(3, -8);$ range:
$\{-3, -4, -5, -6, -7, -8\}$

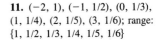

5. $(-2, 13), (-1, 11), (0, 9),$
$(1, 7), (2, 5), (3, 3);$ range:
$\{13, 11, 9, 7, 5, 3\}$

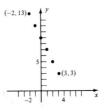

7. $(-2, 2), (-1, 0), (0, 0),$
$(1, 2), (2, 6), (3, 12);$ range:
$\{0, 2, 6, 12\}$

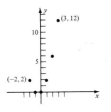

9. $(-2, -13), (-1, -1), (0, 3),$
$(1, -1), (2, -13), (3, -33);$
range: $\{-13, -1, 3, -33\}$

11. $(-2, 1), (-1, 1/2), (0, 1/3),$
$(1, 1/4), (2, 1/5), (3, 1/6);$ range:
$\{1, 1/2, 1/3, 1/4, 1/5, 1/6\}$

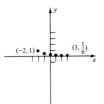

13. $(-2, -3), (-1, -3/2),$
$(0, -3/5), (1, 0), (2, 3/7), (3, 3/4);$
range: $\{-3, -3/2, -3/5, 0, 3/7, 3/4\}$

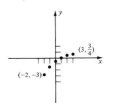

15. $(-2, 4), (-1, 4), (0, 4), (1, 4),$
$(2, 4), (3, 4);$ range: $\{4\}$

17. Function   **19.** Function   **21.** Not a function   **23.** Function   **25.** Not a function   **27.** Function   **29. (a)** 14
(b) -7 **(c)** 2 **(d)** $3a + 2$ **31. (a)** -12 **(b)** 2 **(c)** -4 **(d)** $-2a - 4$ **33. (a)** 6 **(b)** 6 **(c)** 6 **(d)** 6   **35. (a)** 48 **(b)** 6 **(c)** 0
(d) $2a^2 + 4a$ **37. (a)** 5 **(b)** -23 **(c)** 1 **(d)** $-a^2 + 5a + 1$ **39. (a)** 30 **(b)** 2 **(c)** 2 **(d)** $(a + 1)(a + 2)$ **41.** -3
43. -15 **45.** $2a - 3$ **47.** $2m + 3$ **49.** -1 **51.** 5.927   **53.** 459.449   **55. (a)** $11 **(b)** $11 **(c)** $18 **(d)** $32
(e) $32 **(f)** $39 **(g)** $39 **(h)** Continue the horizontal bars up and to the right.   **57. (a)** 1050 thousand dollars **(b)** 1100 thousand
dollars **(c)** 1150 thousand dollars **(d)** 1200 thousand dollars

Section 2.2 (page 82)

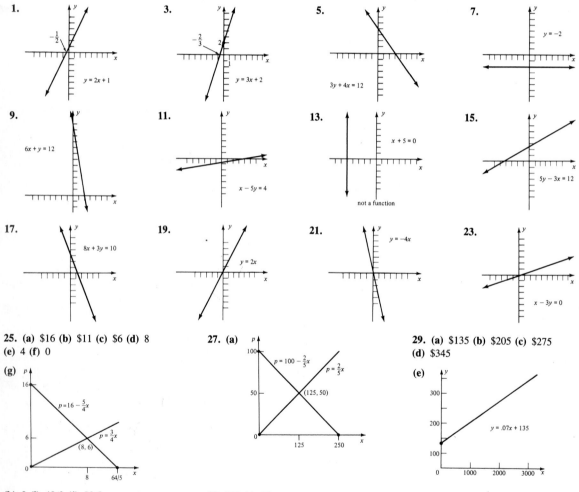

25. (a) $16 **(b)** $11 **(c)** $6 **(d)** 8
(e) 4 **(f)** 0

(g)

(h) 0 **(i)** 40/3 **(j)** 80/3
(k) See part (g) **(l)** 8 **(m)** $6

27. (a)

(b) 125 **(c)** 50

29. (a) $135 **(b)** $205 **(c)** $275
(d) $345

(e)

Section 2.3   (page 90)

1. −1/5   **3.** 2/3   **5.** −3/2   **7.** Undefined slope   **9.** 0   **11.** Perpendicular   **13.** Parallel   **15.** Neither
17. Neither   **19.** 3; 4   **21.** −4; 8   **23.** −3/4; 5/4   **25.** −3; 0   **27.** −2/5; 0   **29.** 0; 8   **31.** 0; −2
33. Undefined slope; no y-intercept

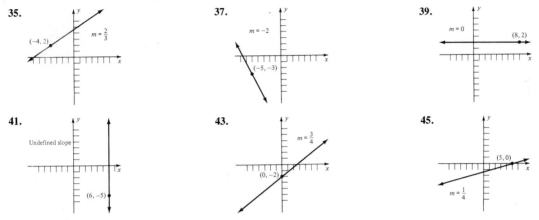

35. $m = \frac{2}{3}$ $(-4, 2)$

37. $m = -2$ $(-5, -3)$

39. $m = 0$ $(8, 2)$

41.   Undefined slope $(6, -5)$

43. $m = \frac{3}{4}$ $(0, -2)$

45. $(5, 0)$ $m = \frac{1}{4}$

47. $4y = -3x + 16$ **49.** $2y = -x - 4$ **51.** $4y = 6x + 5$ **53.** $y = 2x + 9$ **55.** $y = -3x + 3$ **57.** $4y = x + 5$
59. $3y = 4x + 7$ **61.** $3y = -2x$ **63.** $x = -8$ **65.** $y = 3$ **67.** $y = 640x + 1100$; $m = 640$ **69.** $y = 2.5x - 70$; $m =$
2.5   **71.** $y = -1000x + 40,000$; $m = -1000$ **73. (a)** $h = 3.5r + 83$ **(b)** About 163.5 cm; about 177.5 cm **(c)** About 25 cm

Section 2.4   (page 96)

1. (a) 2600 **(b)** 2900 **(c)** 3200 **(d)** 2000 **(e)** 300   **3. (a)** 100 thousand **(b)** 70 thousand **(c)** 0 **(d)** −5 thousand; the number is
decreasing   **5. (a)** $y = 82,500x + 850,000$ **(b)** $1,097,500 **(c)** About $1,510,000   **7. (a)** 480 **(b)** 360 **(c)** 120 **(d)** June 29
(e) −20   **9.** If $C(x)$ is the cost of renting a saw for x hours, then $C(x) = 12 + x$.   **11.** If $P(x)$ is the cost (in cents) of parking
for x half-hours, then $P(x) = 30x + 35$.   **13.** $C(x) = 30x + 100$ **15.** $C(x) = 25x + 1000$ **17.** $C(x) = 50x + 500$
19. $C(x) = 90x + 2500$ **21. (a)** $97 **(b)** $97.097 **(c)** $.097, or 9.7¢ **(d)** $.097, or 9.7¢   **23. (a)** $C(x) = 10x + 500$;
(b) $R(x) = 35x$ **(c)** (20, 700)   **25. (a)** $C(x) = 18x + 250$ **(b)** $R(x) = 28x$ **(c)** (25, 700)   **27. (a)** $C(x) = 100x + 2700$
(b) $R(x) = 125x$ **(c)** (108, 13,500)   **29. (a)** $100 **(b)** $36 **(c)** $24   **31. (a)** 1 **(b)** 2/3 **(c)** 6/10 or 3/5 **(d)** 4/10 or 2/5
33. $2000; $32,000   **35.** $2000; $52,000   **37.** $150,000; $600,000   **39. (a)** $20,000 per year **(b)** $80,000
41. (a) $4000; $3000; $2000; $1000 **(b)** $2500   **43.** $5090.91; $3563.64   **45.** $4171.43; $2085.71   **47.** 500 units; $30,000
49. Break-even point is about 41 units; produce the item   **51.** Break-even point is about 467 units; do not produce the item
53. Break-even is −200 units; product will never make a profit

55. (a)   **(c)** $y = 8x + 50$ **(d)**

Year	Sales (actual)	Sales (predicted from equation of (c))	Difference, actual minus predicted
0	48	50	−2
1	59	58	1
2	66	66	0
3	75	74	1
4	80	82	−2
5	90	90	0

(e) 106 **(f)** 122

Section 2.5 (page 103)

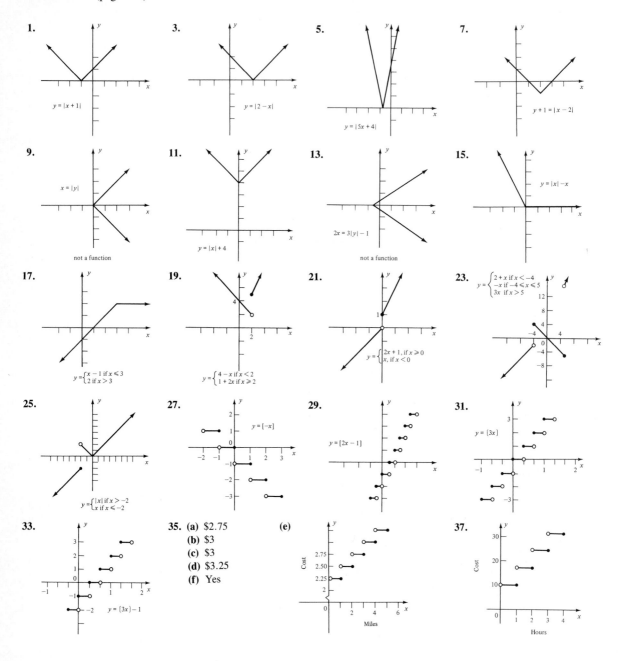

1. $y = |x + 1|$

3. $y = |2 - x|$

5. $y = |5x + 4|$

7. $y + 1 = |x - 2|$

9. $x = |y|$ — not a function

11. $y = |x| + 4$

13. $2x = 3|y| - 1$ — not a function

15. $y = |x| - x$

17. $y = \begin{cases} x - 1 \text{ if } x \leqslant 3 \\ 2 \text{ if } x > 3 \end{cases}$

19. $y = \begin{cases} 4 - x \text{ if } x < 2 \\ 1 + 2x \text{ if } x \geqslant 2 \end{cases}$

21. $y = \begin{cases} 2x + 1, \text{ if } x \geqslant 0 \\ x, \text{ if } x < 0 \end{cases}$

23. $y = \begin{cases} 2 + x \text{ if } x < -4 \\ -x \text{ if } -4 \leqslant x \leqslant 5 \\ 3x \text{ if } x > 5 \end{cases}$

25. $y = \begin{cases} |x| \text{ if } x > -2 \\ x \text{ if } x \leqslant -2 \end{cases}$

27. $y = [-x]$

29. $y = [2x - 1]$

31. $y = [3x]$

33. $y = [3x] - 1$

35. (a) $2.75 (b) $3 (c) $3 (d) $3.25 (f) Yes

(e)

37.

39.

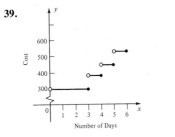

41. (a) 140
(b) 220
(c) 220
(d) 220
(e) 220
(f) 60
(g) 60

(h)

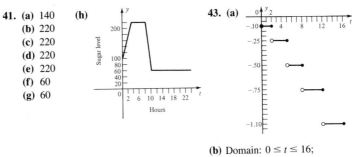

43. (a)

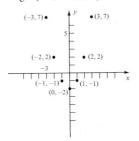

(b) Domain: $0 \le t \le 16$;
range: $\{-.10, -.25, -.50, -.75, -1.10\}$

Chapter 2 Review Exercises (page 105)

1. $(-3, -16/5)$, $(-2, -14/5)$, $(-1, -12/5)$, $(0, -2)$, $(1, -8/5)$, $(2, -6/5)$, $(3, -4/5)$;
range: $\{-16/5, -14/5, -12/5, -2, -8/5, -6/5, -4/5\}$

3. $(-3, 20)$, $(-2, 9)$, $(-1, 2)$, $(0, -1)$, $(1, 0)$, $(2, 5)$, $(3, 14)$;
range: $\{-1, 0, 2, 5, 9, 14, 20\}$

5. $(-3, 7)$, $(-2, 2)$, $(-1, -1)$, $(0, -2)$, $(1, -1)$, $(2, 2)$, $(3, 7)$;
range: $\{-2, -1, 2, 7\}$

7. $(-3, 1/5)$, $(-2, 2/5)$, $(-1, 1)$, $(0, 2)$, $(1, 1)$, $(2, 2/5)$, $(3, 1/5)$;
range: $\{1/5, 2/5, 1, 2\}$

9. $(-3, -1)$, $(-2, -1)$, $(-1, -1)$, $(0, -1)$, $(1, -1)$, $(2, -1)$, $(3, -1)$;
range: $\{-1\}$

11. (a) 23 **(b)** -9 **(c)** -17 **(d)** $4r + 3$
(d) 22 **(e)** $-k^2 - 4k$ **(f)** $-9m^2 + 12m$ **(g)** $-k^2 + 14k - 45$ **(h)** $12 - 5p$ **(i)** -28 **(j)** -21

13. (a) -28 **(b)** -12 **(c)** -28 **(d)** $-r^2 + 2r - 4$

15. (a) -13 **(b)** 3 **(c)** -32

17. $m = 4$ **19.** $m = 3/5$ **21.** Undefined slope   **23.** $m = 2$

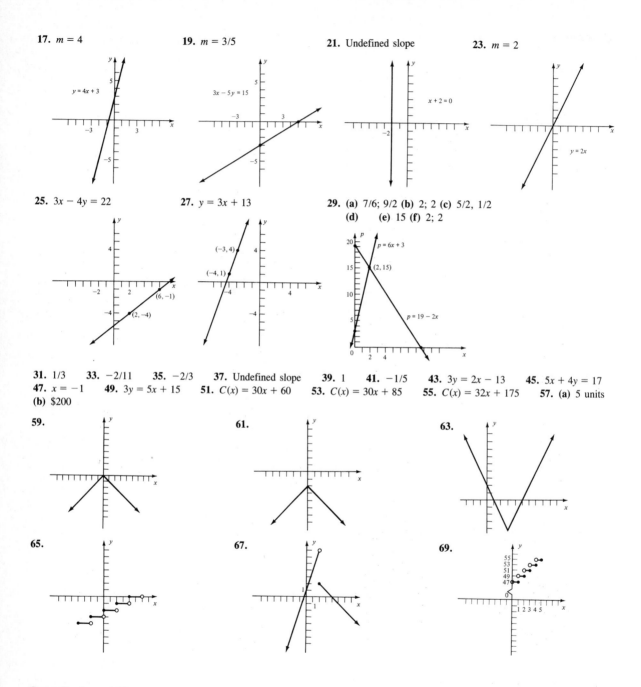

25. $3x - 4y = 22$ **27.** $y = 3x + 13$ **29. (a)** 7/6; 9/2 **(b)** 2; 2 **(c)** 5/2, 1/2 **(d)**   **(e)** 15 **(f)** 2; 2

31. 1/3   **33.** $-2/11$ **35.** $-2/3$ **37.** Undefined slope   **39.** 1   **41.** $-1/5$ **43.** $3y = 2x - 13$ **45.** $5x + 4y = 17$
47. $x = -1$ **49.** $3y = 5x + 15$ **51.** $C(x) = 30x + 60$ **53.** $C(x) = 30x + 85$ **55.** $C(x) = 32x + 175$ **57. (a)** 5 units
(b) $200

59.   **61.**   **63.**

65.   **67.**   **69.**

Case 2   (page 109)

1. 4.8 million units   **2.** Portion of a straight line going through (3.1, 10.50) and (5.7, 10.67)   **3.** In the interval under discussion (3.1 to 5.7 million units), the marginal cost always exceeds the selling price.   **4. (a)** 9.87; 10.22 **(b)** portion of a straight line through (3.1, 9.87) and (5.7, 10.22) **(c)** .83 million units, which is not in the interval under discussion

Chapter 3

Section 3.1 (page 117)

1.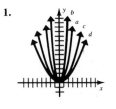

3.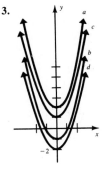

5.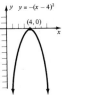

7.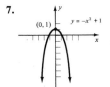

9.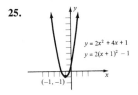

11.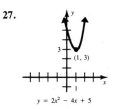

13.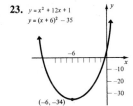

15.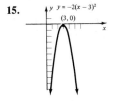

17.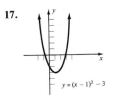

19.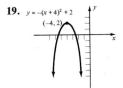

21.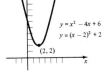

23.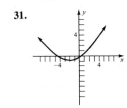

25.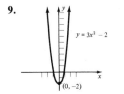

27.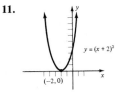

29.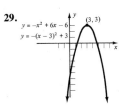

31.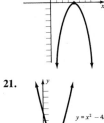

33.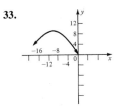

35. (a) $C(x) = (x - 5)^2 + 15$
(c) 5 units
(d) Minimum cost is $15

(b)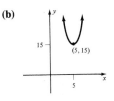

37. (a) $x = 3$ dimes or 30¢ **(b)** $p = 8$, or $800

39. (a) 640 **(b)** 515 **(c)** 140
(d) and (e)
(f) 8
(g) 320

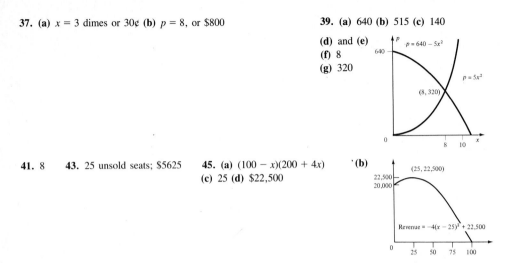

41. 8   **43.** 25 unsold seats; $5625   **45. (a)** $(100 - x)(200 + 4x)$
(c) 25 **(d)** $22,500

(b)

47. (a) 60 **(b)** 70 **(c)** 90 **(d)** 100 **(e)** 80 **(f)** 20   **49.** $6\sqrt{3}$ cm

Section 3.2   (page 125)

1.

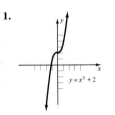

$y = x^3 + 2$

3.

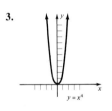

$y = x^4$

5.

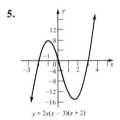

$y = 2x(x - 3)(x + 2)$

7.

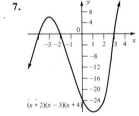

$(x + 2)(x - 3)(x + 4)$

9. $y = x^2(x - 2)(x + 3)$

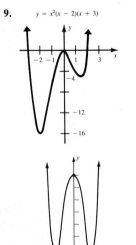

11.

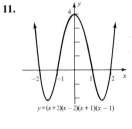

$y = (x + 2)(x - 2)(x + 1)(x - 1)$

13.

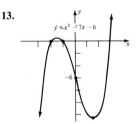

$y = x^3 - 7x - 6$

15.

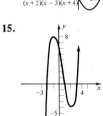

$y = x^3 - 2x^2 - 5x + 6$

19.

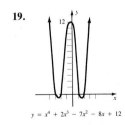

$y = x^4 + 2x^3 - 7x^2 - 8x + 12$

21.

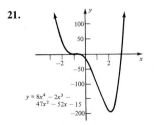

$y = 8x^4 - 2x^3 - 47x^2 - 52x - 15$

23.

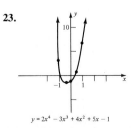

$y = 2x^4 - 3x^3 + 4x^2 + 5x - 1$

$y = x^4 - 5x^2 + 6$

25. (a) 1.0 tenths of a percent, or .1% **(b)** 2.0 tenths of a percent, or .2% **(c)** 3.3 tenths of a percent, or .33% **(d)** 3.1 tenths of a percent, or .31% **(e)** .8 tenths of a percent, or .08%

(f)

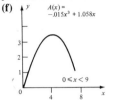

27. (a)

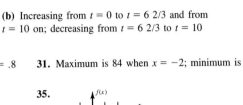

(b) Increasing from $t = 0$ to $t = 6$ 2/3 and from $t = 10$ on; decreasing from $t = 6$ 2/3 to $t = 10$

(g) Between 4 and 5 hours, closer to 5 hours
(h) From about 1 1/2 hour to about 7 1/2 hours

29. Maximum of -10.013 when $x = .3$; minimum of -11.328 when $x = .8$; -13 when $x = -1$

31. Maximum is 84 when $x = -2$; minimum is

33.

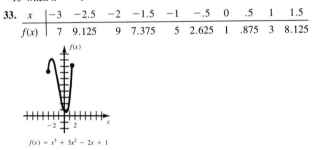

x	-3	-2.5	-2	-1.5	-1	$-.5$	0	$.5$	1	1.5
$f(x)$	7	9.125	9	7.375	5	2.625	1	.875	3	8.125

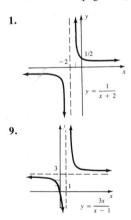

$f(x) = x^3 + 3x^2 - 2x + 1$

35.

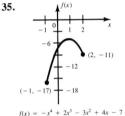

$f(x) = -x^4 + 2x^3 - 3x^2 + 4x - 7$

Section 3.3   (page 132)

1.

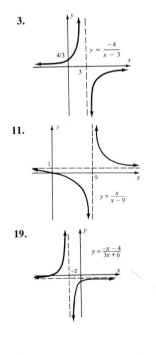

3.

5.

7.

9.

11.

13.

15.

17.

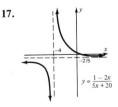

19.

21.

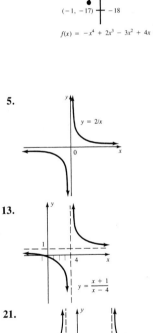

23.

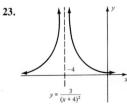

25.

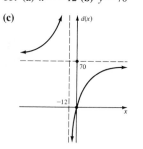

$$y = \frac{3x}{(x+1)(x-2)}$$

27.

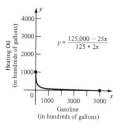

$$y = \frac{5x}{x^2 - 1}$$

29. **(a)** $12.50 **(b)** $10
(c) $6.25 **(d)** $5 **(e)** $3.85

(f)

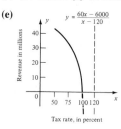

$$C(x) = \frac{500}{x + 30}$$

31. **(a)** $0 **(b)** $6250
(c) About $24,000
(d) About $48,800
(e) About $88,000
(f) $214,500 **(g)** $325,000

(h)

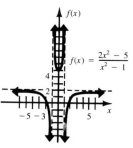

33. **(a)** About 60% **(b)** About 85%
35. **(a)** $x = -12$ **(b)** $y = 70$

(c)

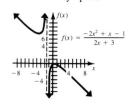

37. 500,000 gallons of gasoline,
100,000 gallons of oil

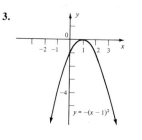

$$y = \frac{125,000 - 25x}{125 + 2x}$$

39. **(a)** $42.9 million **(b)** $40 million
(c) $30 million **(d)** $0 million

(e)

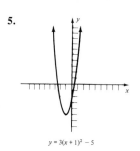

$$y = \frac{60x - 6000}{x - 120}$$

41. Vertical asymptote $x = -3/2$; no
horizontal asymptote

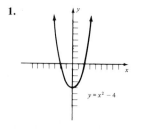

$$f(x) = \frac{-2x^2 + x - 1}{2x + 3}$$

43. Vertical asymptotes $x = 1$, $x = -1$;
horizontal asymptote $y = 2$

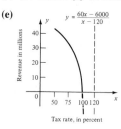

$$f(x) = \frac{2x^2 - 5}{x^2 - 1}$$

Chapter 3   Review Exercises   (page 134)

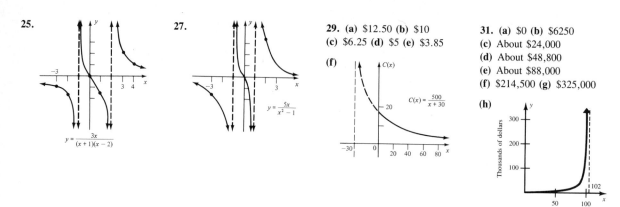

1.

$$y = x^2 - 4$$

3.

$$y = -(x - 1)^2$$

5.

$$y = 3(x + 1)^2 - 5$$

7.

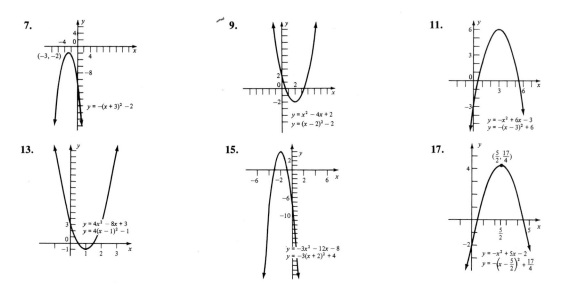

$y = -(x + 3)^2 - 2$
$(-3, -2)$

9.

$y = x^2 - 4x + 2$
$y = (x - 2)^2 - 2$

11.

$y = -x^2 + 6x - 3$
$y = -(x - 3)^2 + 6$

13.

$y = 4x^2 - 8x + 3$
$y = 4(x - 1)^2 - 1$

15.

$y = -3x^2 - 12x - 8$
$y = -3(x + 2)^2 + 4$

17.

$\left(\frac{5}{2}, \frac{17}{4}\right)$
$\frac{5}{2}$
$y = -x^2 + 5x - 2$
$y = -\left(x - \frac{5}{2}\right)^2 + \frac{17}{4}$

19. Minimum of -3 when $x = 2$ **21.** Maximum of 11 when $x = -2$ **23.** Minimum of -1 when $x = 1$ **25.** $(0, 5/4)$, $(6, +\infty)$ **27.** For any x in $(0, 5/3)$ or $(10, +\infty)$ **29.** 50 m × 50 m; 2500 sq m

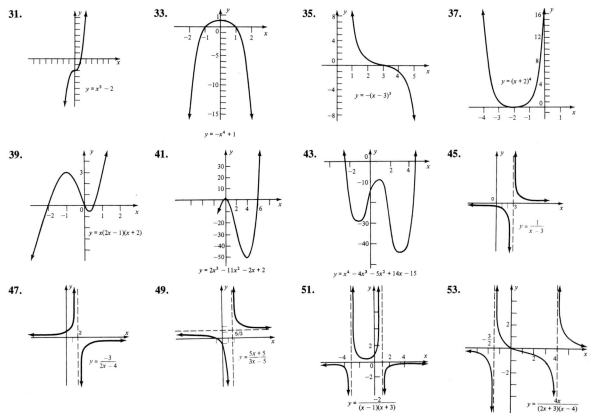

31.

$y = x^3 - 2$

33.

$y = -x^4 + 1$

35.

$y = -(x - 3)^3$

37.

$y = (x + 2)^4$

39.

$y = x(2x - 1)(x + 2)$

41.

$y = 2x^3 - 11x^2 - 2x + 2$

43.

$y = x^4 - 4x^3 - 5x^2 + 14x - 15$

45.

$y = \frac{1}{x - 3}$

47.

$y = \frac{-3}{2x - 4}$

49.

$y = \frac{5x + 5}{3x - 5}$

51.

$y = \frac{-2}{(x - 1)(x + 3)}$

53.

$y = \frac{4x}{(2x + 3)(x - 4)}$

55. **(a)** About \$8200 **(b)** About \$113,000 **(c)** About 75%

Chapter 4

Section 4.1 (page 140)

1. Exponential **3.** Not exponential **5.** 4/9 **7.** $\sqrt{6}/3$

9.

$y = 3^x$
$y = (\frac{1}{3})^{-x}$

11.

$y = 3^{-x}$

13.

$y = (\frac{1}{4})^x$

15. Same as answer 9

17.

$y = 3^{2x}$

19.

$y = 2^{-x/2}$

21.

$y = 2^{x^2}$

23.

$y = 3^{-x^2}$

25.

$y = 2^x + 1$

27.

$y = x \cdot 2^x$

29.

$y = 2^{|x|}$

31. 2 **33.** -3 **35.** 3/2 **37.** 3/2 **39.** -2 **41.** 3 **43.** $-5/4$ **45.** 1/9 **47.** 4, -4 **49.** 0, 8 **51.** 2
53. 5/3

55. (a)

t	0	1	2	3	4	5	6	7	8	9	10
y	1	1.12	1.25	1.40	1.57	1.76	1.97	2.21	2.48	2.77	3.11

57. (a) About \$151,000 **(b)** About \$39

(b)

$y = (1.12)^t$

59. (a) 1,000,000 **(b)** About 1,410,000 **(c)** 2,000,000 **(d)** 4,000,000 **(e)**

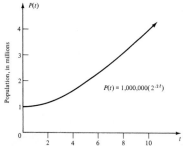

$P(t) = 1,000,000(2^{.2t})$

61. (a) About 1,120,000 **(b)** About 1,260,000 **(c)** About 1,410,000 **(d)** About 1,590,000 **(e)** 2,000,000 **(f)** 4,000,000
(g) 8,000,000 **(h)** 16,000,000   **63.** A series of points—one for each rational number   **65. (a)** 2.8×10^{16}, 3.7×10^8 cm^3
(b) 5.5×10^{12}, 3.25×10^4 cm^3 **(c)** 5.5×10^9, 32.5 cm^3

Section 4.2   (page 147)

1. (a) 1,000,000 **(b)** About 1,040,000 **(c)** About 1,080,000 **(d)** About 1,220,000   **3. (a)** 500 g **(b)** 409 g **(c)** 335 g **(d)** 184 g
5. (a) 25,000 **(b)** About 30,500 **(c)** About 37,300 **(d)** About 68,000   **7.** These values were found using the table. **(a)** $21,665.60
(b) $29,836.40 **(c)** $44,510.80   **9. (a)** 30 **(b)** About 42 **(c)** About 56 **(d)** About 59
11. (a) 0 **(b)** About 316 **(c)** About 432 **(d)** About 497 **(e)** Almost 500 **(f)** $y = 500$ **(g)**

$P(x) = 500 - 500e^{-x}$

13. (a) About 72,400 **(b)** About 48,500   **15. (a)** 40 **(b)** 21 **(c)** 11 **(d)** 6 **(e)** 3 **(f)** 3   **17. (a)** $.3 billion, or $300,000,000;
graph gives about the same **(b)** $1.61 billion, or $1,610,000,000; graph gives a little less **(c)** $6.15 billion, or $6,150,000,000; graph
gives about the same **(d)** $8.60 billion, or $8,600,000,000; graph gives about the same   **19.** 125.36°   **21.** 81.25°C
23. 117.5°C   **25.** 11.6 pounds per square inch

Section 4.3   (page 156)

1. $\log_2 8 = 3$ **3.** $\log_3 81 = 4$ **5.** $\log_{1/3} 9 = -2$ **7.** $2^7 = 128$ **9.** $5^{-2} = 1/25$ **11.** $10^4 = 10,000$ **13.** 3
15. -2 **17.** 2   **19.** 3   **21.** -2 **23.** -2 **25.** $-2/3$ **27.** 8   **29.** 1/2

31. 1/9; 1/3; 1; 3; 27   **33.**

$y = \log_3 x$

$y = \log_4 x$

35.

$y = \log_3(x - 1)$

37.

$y = \log_2 x^2$

39. .7781   **41.** .9542   **43.** 1.4771   **45.** 1.79176   **47.** .40546   **49.** 4.39444   **51.** .069315   **53.** $-.219722$
55. 2.89037   **57.** $\log_3 2 - \log_3 5$ **59.** $\log_9 7 + \log_9 m$ **61.** $1 + \log_3 x - \log_3 5 - \log_3 k$ **63.** $\log_k p + 2\log_k q - \log_k m$
65. Cannot be simplified with the properties of logarithms   **67.** $\log_3 5 + (1/2)\log_3 2 - (1/4)\log_3 7$ **69.** 2.9957   **71.** 4.0943
73. 6.6846   **75.** 6.2766   **77.** 6.6438   **79.** 10.9768   **81.** 4.0848   **83.** $-.8486$ **85.** $3a$ **87.** $a + 3c$ **89.** $2c +$
$3a + 1$ **91.** 1/5   **93.** 3/2   **95.** 16   **97.** 8/5   **99.** 1   **101.** 10   **103.** No solution
105. (a) $125 thousand **(b)** About $211 thousand **(c)** About $235 thousand **(d)** About $307 thousand **(e)**
107. About $60,000; about $450,000

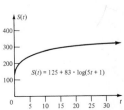

$S(t) = 125 + 83 \cdot \log(5t + 1)$

Section 4.4 (page 164)

1. 1.631 **3.** 1.069 **5.** −.535 **7.** −.080 **9.** 3.797 **11.** 2.386 **13.** −.123 **15.** No solution **17.** 1.104
19. 17.475 **21.** −2.141, 2.141 **23.** 1.386 **25.** 11 **27.** 5 **29.** 10 **31.** 5 **33.** 4 **35.** No solution
37. 11 **39.** 3 **41.** No solution **43.** −2, 2 **45.** 1, 10 **47.** 2.423 **49.** −.204 **51.** (a) 5000 (b) About 4520
(c) About 3350 (d) −(ln .5)/.02 ≈ 34.7 seconds **53.** (a) About 23 days (b) About 8.7 days (c) About 3466 days **55.** About
1800 years **57.** About 18,600 years **59.** (a) About 67% (b) About 37% (c) About 23 days (d) About 46 days **61.** (a) 20
(b) 30 (c) 50 (d) 60 **63.** (a) 3 (b) 6 (c) 8 **65.** .629 **67.** 4.3 ml/min; 7.8 ml/min **69.** (a) 11% (b) 36% (c) 84%

Appendix (page 170)

1. 2.8745 **3.** .9872 **5.** .8531 **7.** .9085 − 1 **9.** .5416 − 3 **11.** 2.24 **13.** 888 **15.** 30.8 **17.** .0862
19. .000850 **21.** 35.0 **23.** 273 **25.** 186 **27.** 8.83 **29.** 558 **31.** $19,400 **33.** (a) 0 (b) $389,000
(c) $699,000 (d) $849,000 **35.** 7.60 **37.** 3.2 **39.** 1.8

Chapter 4 Review Exercises (page 171)

1. −1 **3.** 2
5.

9. (a) 0 (b) About 55 (c) About 98 (d) About 100 **11.** $\log_2 64 = 6$ **13.** $\ln 1.09417 = .09$ **15.** $2^5 = 32$
17. $e^{4.41763} = 82.9$ **19.** 4 **21.** 4/5 **23.** 3/2 **25.** 1.79 **27.** .645 **29.** 1.8245 **31.** 6.1800 **33.** $\log_5 21k^4$
35. $\log_2 (x^2/m^3)$ **37.** 1.416 **39.** −2.807 **41.** −3.305 **43.** 2.115 **45.** .747 **47.** 28.463 **49.** ±1.097
51. 8 **53.** No solution **55.** 3/2 **57.** (a) .8 m (b) 1.2 m (c) 1.5 m (d) 1.8 m **59.** 2.3909 **61.** .5011 − 4
63. 3150 **65.** .0446 **67.** .0359 **69.** 5.44

Case 3 (page 174)

1. .0315 **2.** λy_0 is about 98,000; painting is a forgery **3.** About 157,000; forgery **4.** About 14,000; cannot be modern
forgery **5.** About 58,000; forgery **6.** About 127,000; forgery

Chapter 5

Section 5.1 (page 179)

1. $120 **3.** $3937.50 **5.** $186.54 **7.** $336 **9.** $143.46 **11.** $438.90 **13.** $376.11 **15.** $201.46
17. $13,554.22 **19.** $5056.06 **21.** $14,434.68 **23.** $6101.33 **25.** $318.69 **27.** $30,268.47 **29.** $1732.90
31. $6053.59 **33.** $3598; 20.1% **35.** $7547.13

Answers in the remainder of this chapter may differ by a few cents depending on whether a table or a calculator is used.

Section 5.2 (page 187)

1. $1593.85 **3.** $20,974.32 **5.** $1903.00 **7.** $82,379.10 **9.** $30,779.48 **11.** $13,213.14 **13.** $5105.58
15. $60,484.66 **17.** $2746.51 **19.** $2551.13 **21.** $4490.29 **23.** $583.78 **25.** $4016.21 **27.** $1000 now
29. $21,665.60 **31.** $44,510.80 **33.** $25,424.98 **35.** 4.04% **37.** 8.16% **39.** 12.36% **41.** $13,693.34
43. $7,743.93 **45.** About 18 years **47.** About 12 years **49.** $142,886.40 **51.** $123,506.50 **53.** $23,182.70
55. (a) $14,288.98 (b) $3711.02 (c) $2883.46 **57.** $63,685.27 **59.** $904.02

Section 5.3 (page 193)

1. 11, 17, 23, 29, 35; arithmetic **3.** $-4, -1, 2, 5, 8$; arithmetic **5.** $-2, -8, -14, -20, -26$; arithmetic **7.** 2, 4, 8, 16,
32; geometric **9.** $-2, 4, -8, 16, -32$; geometric **11.** 6, 12, 24, 48, 96; geometric **13.** 1/3, 3/7, 1/2, 5/9, 3/5; neither
15. 1/2, 1/3, 1/4, 1/5, 1/6; neither **17.** Arithmetic, $d = 8$ **19.** Arithmetic, $d = 3$ **21.** Geometric, $r = 3$ **23.** Neither
25. Arithmetic, $d = -3$ **27.** Arithmetic, $d = 3$ **29.** Geometric, $r = -2$ **31.** Neither **33.** 70 **35.** 65 **37.** 78
39. -65 **41.** 63 **43.** 678 **45.** 183 **47.** -18 **49.** 48 **51.** -648 **53.** 81 **55.** 64 **57.** 15
59. 156/25 **61.** -208 **63.** -15 **65.** 80 **67.** 125 **69.** 134 **71.** 70 **73.** -56 **75.** 125,250 **77.** 90
79. 170 **81.** 160 **83.** $15,600 **85.** (a) $1681 (b) $576

Section 5.4 (page 199)

1. 15.91713 **3.** 21.82453 **5.** 22.01900 **7.** 20.02359 **9.** $437.46 **11.** $305,390.00 **13.** $671,994.62
15. $4,180,929.88 **17.** $222,777.26 **19.** $40,652.46 **21.** $516,397.05 **23.** $6294.79 **25.** $158,456.00
27. $13,486.56 **29.** $8070.23 **31.** $526.95 **33.** $952.33 **35.** (a) $149,850.72 (b) $137,895.84 (c) $11,954.88
37. $118,667.74 **39.** (a) $159.49 (b) $190.47 **41.** $522.85 **43.** $112,796.87 **45.** $209,348.00 **47.** $354.79
49. $579.65 **51.** $4164.55 **53.** $126.91 **55.** $23,023.98
57. (a) $1200 (b) $3511.58 (c)

Payment number	Amount of deposit	Interest earned	Total
1	$3511.58	$ 0	$ 3511.58
2	3511.58	105.35	7128.51
3	3511.58	213.86	10,853.95
4	3511.58	325.62	14,691.15
5	3511.58	440.73	18,643.46
6	3511.58	559.30	22,714.34
7	3511.58	681.43	26,907.35
8	3511.58	807.22	31,226.15
9	3511.58	936.78	35,674.51
10	3511.58	1070.24	40,256.33
11	3511.58	1207.69	44,975.60
12	3511.58	1349.27	49,836.45
13	3511.58	1495.09	54,843.12
14	3511.59	1645.29	60,000.00

Section 5.5 (page 207)

1. 9.71225 **3.** 12.65930 **5.** 14.71787 **7.** 5.69719 **9.** $6246.89 **11.** $7877.72 **13.** $153,724.50
15. $148,771.76 **17.** $111,183.90 **19.** $97,122.50 **21.** $68,108.64 **23.** $12,493.78 **25.** (a) $158.00 (b) $1584.00
27. $6698.98 **29.** $160.08 **31.** $5570.58 **33.** $11,727.32 **35.** $274.58 **37.** $663.75 **39.** $742.51

41.

Payment number	Amount of payment	Interest for period	Portion to principal	Principal at end of period
0	—	—	—	$4000
1	$1207.68	$320.00	$ 887.68	3112.32
2	1207.68	248.99	958.69	2153.63
3	1207.68	172.29	1035.39	1118.24
4	1207.70	89.46	1118.24	0

43.

Payment number	Amount of payment	Interest for period	Portion to principal	Principal at end of period
0	—	—	—	$7184
1	$211.03	$107.76	$103.27	7080.73
2	211.03	106.21	104.82	6975.91
3	211.03	104.64	106.39	6869.52
4	211.03	103.04	107.99	6761.53
5	211.03	101.42	109.61	6651.92
6	211.03	99.78	111.25	6540.67

45. (a) $4025.90 **(b)** $2981.93

47.

Payment number	Amount of payment	Interest for period	Portion to principal	Principal at end of period
0	$5783.49	$ 0.00	$ 0.00	$37947.50
1	5783.49	3225.54	2557.95	35389.55
2	5783.49	3008.11	2775.38	32614.17
3	5783.49	2772.21	3011.29	29602.88
4	5783.49	2516.25	3267.25	26335.63
5	5783.49	2238.53	3544.96	22790.67
6	5783.49	1937.21	3846.28	18944.39
7	5783.49	1610.27	4173.22	14771.17
8	5783.49	1255.55	4527.94	10243.23
9	5783.49	870.67	4912.82	5330.41
10	5783.49	453.08	5330.41	0.00

Supplementary Exercises (page 209)

1. $57,513.15   **3.** $35,840.29   **5.** $12,125.91   **7.** $3903.93   **9.** $5054.47   **11.** $9097.32   **13.** $175,363.02
15. $574.71   **17.** $425.00   **19.** $1813.15   **21.** $19,356.63   **23.** $40,000.94; $7250.94   **25.** $69,331.75; $36,581.75
27. $44,313.15   **29.** $117,028.83   **31.** $62,197.46   **33.** $235,492.28   **35.** $12,500.00   **37.** 5 months   **39.** 15%
41. $5653.33   **43.** $5596.62   **45.** $21,384.28   **47.** $11,304.71   **49.** $586.66   **51.** $14,943.67; $2891.62
53. $2054.05   **55.** $14,702.23   **57.** $5842.10   **59.** $23,122.53   **61.** $9859.46

63.

Payment number	Amount of payment	Interest for period	Portion to principal	Principal at end of period
0	$1522.66	—	—	
1	1522.66	$510.00	$1012.66	$7487.34
2	1522.66	449.24	1073.42	6413.92
3	1522.66	384.84	1137.82	5276.10
4	1522.66	316.57	1206.09	4070.01
5	1522.66	244.20	1278.46	2791.55
6	1522.66	167.49	1355.17	1436.38
7	1522.66	86.18	1436.38	0

Chapter 5 Review Exercises (page 211)

1. $1908.36   **3.** $2686.84   **5.** $105.30   **7.** $21,897.81   **9.** $76,075.85   **11.** $665.54   **13.** $6194.13
15. 16.3%   **17.** $1999.00   **19.** $43,988.32   **21.** $7797.47   **23.** $4272.85   **25.** $6002.84   **27.** $6289.04
29. $18,306.34   **31.** $2501.24   **33.** $12,025.46   **35.** $845.74   **37.** $3137.06   **39.** $-2, -6, -10, -14, -18$;
arithmetic   **41.** 1/2, 4/7, 5/8, 2/3, 7/10   **43.** 61   **45.** -24 **47.** $15,162.14   **49.** $199,870.32   **51.** $41,208.86
53. $45,569.65   **55.** $886.05   **57.** $3339.86   **59.** $2815.31   **61.** $34,357.52   **63.** $4788.17   **65.** $12,806.37
67. $2806.66   **69.** $717.21   **71. (a)** $822.89; $216,240.43 **(b)** $842.58; $172,773.79 **(c)** $960.13; 92,824.20

73.

Payment number	Amount of payment	Interest for period	Portion to principal	Principal at end of period
0	—	—	—	$5000
1	$985.09	$250.00	$735.09	4264.91
2	985.09	213.25	771.84	3493.07
3	985.09	174.65	810.44	2682.63
4	985.09	134.13	850.96	1831.67
5	985.09	91.58	893.51	938.16
6	985.07	46.91	938.16	0

Case 5 (page 217)

1. $14,038   **2.** $9511   **3.** $8837   **4.** $3968

Chapter 6

Section 6.1 (page 228)

1. (3, 6)   **3.** $(-1, 4)$ **5.** $(-2, 0)$ **7.** (1, 3)   **9.** $(4, -2)$ **11.** $(2, -2)$ **13.** No solution   **15.** Same line
17. (12, 6)   **19.** $(7, -2)$ **21.** $(1, 2, -1)$ **23.** (2, 0, 3)   **25.** No solution   **27.** (0, 2, 4)   **29.** (1, 2, 3)
31. $(-1, 2, 1)$ **33.** (4, 1, 2)   **35.** $(x, 2 - x)$ **37.** $(x, -3x/4 + 3)$ **39.** $(x, x + 5, -2x + 1)$ **41.** $(x, x + 1, -x + 3)$
43. $(x, 4x - 7, 3x + 7, -x - 3)$ The answer is given in the form (x, y, z, w).   **45.** $(3, -4)$ **47.** No solution   **49.** No
solution   **51.** $a = -3, b = -1$ **53.** $a = 3/4, b = 1/4, c = -1/2$ **55.** Wife, 40 days; husband, 32 days   **57.** 5 model 201
and 8 model 301   **59.** $10,000 at 16%, $7000 at 20%, $8000 at 18%   **61.** $k = 3$; solution is $(-1, 0, 2)$

Section 6.2 (page 238)

1. $\begin{bmatrix} 2 & 3 & | & 11 \\ 1 & 2 & | & 8 \end{bmatrix}$ **3.** $\begin{bmatrix} 1 & 5 & | & 6 \\ 0 & 1 & | & 1 \end{bmatrix}$ **5.** $\begin{bmatrix} 2 & 1 & 1 & | & 3 \\ 3 & -4 & 2 & | & -7 \\ 1 & 1 & 1 & | & 2 \end{bmatrix}$ **7.** $\begin{bmatrix} 1 & 1 & 0 & | & 2 \\ 0 & 2 & 1 & | & -4 \\ 0 & 0 & 1 & | & 2 \end{bmatrix}$ **9.** $\begin{bmatrix} 1 & 0 & 0 & | & 5 \\ 0 & 1 & 0 & | & -2 \\ 0 & 0 & 1 & | & 3 \end{bmatrix}$

11. $x = 2$ **13.** $2x + y = 1$ **15.** $x = 2$
    $y = 3$ $3x - 2y = -9$ $y = 3$
                                         $z = -2$
17. (2, 3)   **19.** $(7/2, -1)$ **21.** No solution   **23.** $(-2, 1, 3)$ **25.** $(3, 2, -4)$ **27.** $(x, -x + 2, x + 3)$
29. $(0, 2, -2, 1)$ The answer is given in the form (x, y, z, w).   **31.** $(0, -2)$ **33.** (11/8, 5/4)   **35.** $(x, x - 1)$
37. $(1, 0, -1)$ **39.** $(x, (1 - 10x)/9, (5 - 23x)/9)$ **41.** $(-3, -2, 5)$

43. (a) $\begin{bmatrix} 1 & 0 & 0 & 1 & | & 1000 \\ 1 & 1 & 0 & 0 & | & 1100 \\ 0 & 1 & 1 & 0 & | & 700 \\ 0 & 0 & 1 & 1 & | & 600 \end{bmatrix}$; $\begin{bmatrix} 1 & 0 & 0 & 1 & | & 1000 \\ 0 & 1 & 0 & -1 & | & 100 \\ 0 & 0 & 1 & 1 & | & 600 \\ 0 & 0 & 0 & 0 & | & 0 \end{bmatrix}$ **(b)** $x_1 + x_4 = 1000$; $x_2 - x_4 = 100$; $x_3 + x_4 = 600$

(c) $x_4 = 1000 - x_1$; $x_4 = x_2 - 100$; $x_4 = 600 - x_3$ **(d)** 1000; 1000 **(e)** 100 **(f)** 600; 600 **(g)** $x_4 = 600$; $x_3 = 600$; $x_2 = 700$; $x_1 = 1000$ **45.** (4.12062, 1.66866, .117699) **47.** (30.7209, 39.6513, 31.386, 50.3966) **49.** 81 kg of the first chemical, 382.286 kg of the second, 286.714 kg of the third **51.** 243 of A, 38 of B, 101 of C (rounded)

Section 6.3 (page 246)

1. False; corresponding elements not all equal **3.** True **5.** True **7.** 2×2, square $\begin{bmatrix} 4 & -8 \\ -2 & -3 \end{bmatrix}$

9. 3×4 $\begin{bmatrix} 6 & -8 & 0 & 0 \\ -4 & -1 & -9 & -2 \\ -3 & 5 & -7 & -1 \end{bmatrix}$ **11.** 2×1, column $\begin{bmatrix} -2 \\ -4 \end{bmatrix}$ **13.** $x = 2$, $y = 4$, $z = 8$ **15.** $x = -15$, $y = 5$, $k = 3$

17. $z = 18$, $r = 3$, $s = 3$, $p = 3$, $a = 3/4$ **19.** $\begin{bmatrix} 9 & 12 & 0 & 2 \\ 1 & -1 & 2 & -4 \end{bmatrix}$ **21.** $\begin{bmatrix} 5 & 13 & 0 \\ 3 & 1 & 8 \end{bmatrix}$ **23.** Not possible

25. $\begin{bmatrix} 1 & 5 & 6 & -9 \\ 5 & 7 & 2 & 1 \\ -7 & 2 & 2 & -7 \end{bmatrix}$ **27.** $\begin{bmatrix} -12x + 8y & -x + y \\ x & 8x - y \end{bmatrix}$ **29.** $\begin{bmatrix} x & y \\ z & w \end{bmatrix} + \begin{bmatrix} r & s \\ t & u \end{bmatrix} = \begin{bmatrix} x+r & y+s \\ z+t & w+u \end{bmatrix}$ (a 2×2 matrix)

31. $\begin{bmatrix} x + (r+m) & y + (s+n) \\ z + (t+p) & w + (u+q) \end{bmatrix} = \begin{bmatrix} (x+r)+m & (y+s)+n \\ (z+t)+p & (w+u)+q \end{bmatrix}$ **33.** $\begin{bmatrix} m+0 & n+0 \\ p+0 & q+0 \end{bmatrix} = \begin{bmatrix} m & n \\ p & q \end{bmatrix}$

35. $\begin{bmatrix} 18 & 2.7 \\ 10 & 1.5 \\ 8 & 1.0 \\ 10 & 2.0 \\ 10 & 1.7 \end{bmatrix}$ $\begin{bmatrix} 18 & 10 & 8 & 10 & 10 \\ 2.7 & 1.5 & 1.0 & 2.0 & 1.7 \end{bmatrix}$ **37. (a)** $\begin{bmatrix} 2 & 1 & 2 & 1 \\ 3 & 2 & 2 & 1 \\ 4 & 3 & 2 & 1 \end{bmatrix}$ **(b)** $\begin{bmatrix} 5 & 0 & 7 \\ 0 & 10 & 1 \\ 0 & 15 & 2 \\ 10 & 12 & 8 \end{bmatrix}$ **(c)** $\begin{bmatrix} 8 \\ 4 \\ 5 \end{bmatrix}$

Section 6.4 (page 254)

1. 2×2; 2×2 **3.** 4×4; 2×2 **5.** 3×2; BA does not exist **7.** AB does not exist; 3×2

9. $\begin{bmatrix} -4 & 8 \\ 0 & 6 \end{bmatrix}$ **11.** $\begin{bmatrix} 24 & -8 \\ -16 & 0 \end{bmatrix}$ **13.** $\begin{bmatrix} -22 & -6 \\ 20 & -12 \end{bmatrix}$ **15.** $\begin{bmatrix} 13 \\ 25 \end{bmatrix}$ **17.** $\begin{bmatrix} -2 & 10 \\ 0 & 8 \end{bmatrix}$ **19.** $\begin{bmatrix} -4 & 1 \\ 2 & -3 \end{bmatrix}$

21. $\begin{bmatrix} 3 & -5 & 7 \\ -2 & 1 & 6 \\ 0 & -3 & 4 \end{bmatrix}$ **23.** $\begin{bmatrix} 13 & 5 \\ 25 & 15 \end{bmatrix}$ **25.** $\begin{bmatrix} 13 \\ 29 \end{bmatrix}$ **27.** $\begin{bmatrix} 110 \\ 40 \\ -50 \end{bmatrix}$ **29.** $\begin{bmatrix} 22 & -8 \\ 11 & -4 \end{bmatrix}$ **31. (a)** $\begin{bmatrix} 16 & 22 \\ 7 & 19 \end{bmatrix}$ **(b)** $\begin{bmatrix} 5 & -5 \\ 0 & 30 \end{bmatrix}$

(c) No; no **(b)** No   **37. (a)** $\begin{bmatrix} 47.5 & 57.75 \\ 27 & 33.75 \\ 81 & 95 \\ 12 & 15 \end{bmatrix}$ **(b)** [20 200 50 60]; [220 890 105 125 70] **(c)** [11,120 13,555]

39. (a) $C = \begin{bmatrix} 90,000 \\ 60,000 \\ 120,000 \end{bmatrix}$ **(b)** $PC = \begin{bmatrix} 72,000 \\ 48,000 \\ 96,000 \end{bmatrix}$ **(c)** $R = [1/2 \quad 1/3 \quad 1/6]$ **(d)** $\begin{bmatrix} 36,000 & 24,000 & 12,000 \\ 24,000 & 16,000 & 8,000 \\ 48,000 & 32,000 & 16,000 \end{bmatrix}$

41. $(5, -2)$ **43.** $\begin{bmatrix} 44 & 75 & -60 & -33 & 11 \\ 20 & 169 & -164 & 18 & 105 \\ 113 & -82 & 239 & 218 & -55 \\ 119 & 83 & 7 & 82 & 106 \\ 162 & 20 & 175 & 143 & 74 \end{bmatrix}$ **45.** Cannot be found   **47.** No

49. $\begin{bmatrix} -1 & 5 & 9 & 13 & -1 \\ 7 & 17 & 2 & -10 & 6 \\ 18 & 9 & -12 & 12 & 22 \\ 9 & 4 & 18 & 10 & -3 \\ 1 & 6 & 10 & 28 & 5 \end{bmatrix}$; $\begin{bmatrix} -2 & -9 & 90 & 77 \\ -42 & -63 & 127 & 62 \\ 413 & 76 & 180 & -56 \\ -29 & -44 & 198 & 85 \\ 137 & 20 & 162 & 103 \end{bmatrix}$; $\begin{bmatrix} -56 & -1 & 1 & 45 \\ -156 & -119 & 76 & 122 \\ 315 & 86 & 118 & -91 \\ -17 & -17 & 116 & 51 \\ 118 & 19 & 125 & 77 \end{bmatrix}$

$\begin{bmatrix} 54 & -8 & 89 & 32 \\ 114 & 56 & 51 & -60 \\ 98 & -10 & 62 & 35 \\ -12 & -27 & 82 & 34 \\ 19 & 1 & 37 & 26 \end{bmatrix}$; $\begin{bmatrix} -2 & -9 & 90 & 77 \\ -42 & -63 & 127 & 62 \\ 413 & 76 & 180 & -56 \\ -29 & -44 & 198 & 85 \\ 137 & 20 & 162 & 103 \end{bmatrix}$; yes

Section 6.5   (page 262)

1. $\begin{bmatrix} 2 & 4 \\ 0 & -1 \end{bmatrix}$ **3.** $\begin{bmatrix} 1 & 1/5 \\ 0 & 1 \end{bmatrix}$ **5.** $\begin{bmatrix} 1 & 5 & 6 \\ 0 & 13 & 11 \\ 4 & 7 & 0 \end{bmatrix}$ **7.** $\begin{bmatrix} -3 & 1 & -4 \\ 2 & 1 & 3 \\ -17 & 0 & -13 \end{bmatrix}$ **9.** Yes   **11.** No   **13.** No   **15.** Yes

17. $\begin{bmatrix} 0 & 1/2 \\ -1 & 1/2 \end{bmatrix}$ **19.** $\begin{bmatrix} 2 & 1 \\ 5 & 3 \end{bmatrix}$ **21.** None   **23.** $\begin{bmatrix} 1 & 0 & 0 \\ 0 & -1 & 0 \\ -1 & 0 & 1 \end{bmatrix}$ **25.** $\begin{bmatrix} 15 & 4 & -5 \\ -12 & -3 & 4 \\ -4 & -1 & 1 \end{bmatrix}$ **27.** None

29. $\begin{bmatrix} 7/4 & 5/2 & 3 \\ -1/4 & -1/2 & 0 \\ -1/4 & -1/2 & -1 \end{bmatrix}$ **31.** $\begin{bmatrix} 1/2 & 1/2 & -1/4 & 1/2 \\ -1 & 4 & -1/2 & -2 \\ -1/2 & 5/2 & -1/4 & -3/2 \\ 1/2 & -1/2 & 1/4 & 1/2 \end{bmatrix}$ **33.** $\begin{bmatrix} 1 & 0 \\ 0 & 1 \end{bmatrix} \cdot \begin{bmatrix} a & b \\ c & d \end{bmatrix} = \begin{bmatrix} a & b \\ c & d \end{bmatrix}$

35. $\begin{bmatrix} a & b \\ c & d \end{bmatrix} \cdot \begin{bmatrix} 0 & 0 \\ 0 & 0 \end{bmatrix} = \begin{bmatrix} 0 & 0 \\ 0 & 0 \end{bmatrix}$ **37.** $A^{-1} = \begin{bmatrix} d/(ad-bc) & -b/(ad-bc) \\ -c/(ad-bc) & a/(ad-bc) \end{bmatrix}$

$A^{-1}A = \begin{bmatrix} d/(ad-bc) & -b/(ad-bc) \\ -c/(ad-bc) & a/(ad-bc) \end{bmatrix} \cdot \begin{bmatrix} a & b \\ c & d \end{bmatrix} = \begin{bmatrix} \dfrac{ad-bc}{ad-bc} & \dfrac{db-bd}{ad-bc} \\ \dfrac{-ca+ac}{ad-bc} & \dfrac{-cb+ad}{ad-bc} \end{bmatrix} = \begin{bmatrix} 1 & 0 \\ 0 & 1 \end{bmatrix} = I$

39. $\begin{bmatrix} -.0447 & -.0230 & .0292 & .0895 & -.0402 \\ .0921 & .0150 & .0321 & .0209 & -.0276 \\ -.0678 & .0315 & -.0404 & .0326 & .0373 \\ .0171 & -.0248 & .0069 & -.0003 & .0246 \\ -.0208 & .0740 & .0096 & -.1018 & .0646 \end{bmatrix}$ **41.** $\begin{bmatrix} .0394 & .0880 & .0033 & .0530 & -.1499 \\ -.1492 & .0289 & .0187 & .1033 & .1668 \\ -.1330 & -.0543 & .0356 & .1768 & .1055 \\ .1407 & .0175 & -.0453 & -.1344 & .0655 \\ .0102 & -.0653 & .0993 & .0085 & -.0388 \end{bmatrix}$

43.

$C^{-1}C = \begin{bmatrix} 1 & 1.596046 \times 10^{-7} & .89407 \times 10^{-7} & 0 & -.149012 \times 10^{-7} \\ .540167 \times 10^{-7} & 1 & .596046 \times 10^{-7} & 0 & 0 \\ -.484288 \times 10^{-7} & .372529 \times 10^{-7} & 1 & .596046 \times 10^{-7} & 0 \\ -.372529 \times 10^{-8} & 0 & -.149012 \times 10^{-7} & 1 & .745058 \times 10^{-8} \\ .149012 \times 10^{-7} & -.447035 \times 10^{-7} & 0 & 0 & 1 \end{bmatrix}$

Section 6.6 (page 270)

1. $\begin{bmatrix} 3 \\ 4 \end{bmatrix}$ **3.** $\begin{bmatrix} 36 & 6 \\ 28 & -2 \end{bmatrix}$ **5.** $\begin{bmatrix} -22 \\ -18 \\ 15 \end{bmatrix}$ **7.** $\begin{bmatrix} 4 & -11 \\ -10 & 11 \\ 4 & -5 \end{bmatrix}$ **9.** $\begin{bmatrix} -6 \\ -14 \end{bmatrix}$ **11.** $(-8, 6, 1)$ **13.** $(15, -5, -1)$

15. $(-31, 24, -4)$ **17.** $(-4, -3, 1)$ **19.** No inverse, no solution for system   **21.** $(-7, -34, -19, 7)$

23. $\begin{bmatrix} 32/3 \\ 25/3 \end{bmatrix}$ **25.** $\begin{bmatrix} 6.43 \\ 26.12 \end{bmatrix}$ **27.** $\begin{bmatrix} 6.67 \\ 20 \\ 10 \end{bmatrix}$

29. About 1079 metric tons of wheat; about 1428 metric tons of oil   **31.** About 3077 units of agriculture; about 2564 units of manufacturing; about 3179 units of transportation   **33.** **(a)** 7/4 bushels of yams; 15/8 pigs **(b)** 167.5 bushels of yams; 153.75 pigs

35. $\begin{bmatrix} 38 \\ 102 \end{bmatrix}, \begin{bmatrix} -18 \\ 45 \end{bmatrix}, \begin{bmatrix} 19 \\ 61 \end{bmatrix}, \begin{bmatrix} -25 \\ 59 \end{bmatrix}, \begin{bmatrix} 34 \\ 175 \end{bmatrix}, \begin{bmatrix} -2 \\ 13 \end{bmatrix}, \begin{bmatrix} 5 \\ 116 \end{bmatrix}, \begin{bmatrix} 15 \\ 159 \end{bmatrix}, \begin{bmatrix} 10 \\ 88 \end{bmatrix}, \begin{bmatrix} 33 \\ 105 \end{bmatrix}, \begin{bmatrix} 35 \\ 173 \end{bmatrix}, \begin{bmatrix} 10 \\ 115 \end{bmatrix}, \begin{bmatrix} 39 \\ 165 \end{bmatrix}, \begin{bmatrix} 27 \\ 81 \end{bmatrix}, \begin{bmatrix} -4 \\ 44 \end{bmatrix}$

37. (a) 3 **(b)** 3 **(c)** 5 **(d)** 3   **39. (a)** $B = \begin{bmatrix} 0 & 2 & 3 \\ 2 & 0 & 4 \\ 3 & 4 & 0 \end{bmatrix}$ **(b)** $\begin{bmatrix} 13 & 12 & 8 \\ 12 & 20 & 6 \\ 8 & 6 & 25 \end{bmatrix}$ **(c)** 12 **(d)** 14

41. (a)

	dogs	rats	cats	mice
dogs	0	1	1	1
rats	0	0	0	1
cats	0	1	0	1
mice	0	0	0	0

$= C$

(b) $C^2 = \begin{bmatrix} 0 & 1 & 0 & 2 \\ 0 & 0 & 0 & 0 \\ 0 & 0 & 0 & 1 \\ 0 & 0 & 0 & 0 \end{bmatrix}$

C^2 gives the number of food sources once removed from the feeder. Thus, since dogs eat rats and rats eat mice, mice are an indirect as well as a direct food source for dogs.

Chapter 6 Review Exercises (page 273)

1. $(-1, 3)$ **3.** $(6, -12)$ **5.** $(x, -x/2 + 15/8)$, dependent system   **7.** $(-1, 2, 3)$ **9.** 10 at 25¢, 12 at 50¢

11. \$10,000 at 6%, \$20,000 at 7%, \$20,000 at 9%   **13.** $\left(\dfrac{15}{11}z + \dfrac{6}{11}, \dfrac{14}{11}z - \dfrac{1}{11}, z\right)$ **15.** $(-4, 6)$ **17.** $(-3, 2)$

19. $(0, 1, 0)$

21. 2×2; $a = 2$, $b = 3$, $c = 5$, $q = 9$; square   **23.** 1×4; $m = 12$, $k = 4$, $z = -8$, $r = -1$; row
25. 2×2; $r = 2$, $m = 7/2$, $k = 11/2$, $x = -7$; square

27. $\begin{bmatrix} 8 & 8 & 8 \\ 10 & 5 & 9 \\ 7 & 10 & 7 \\ 8 & 9 & 7 \end{bmatrix}$ **29.** $\begin{bmatrix} -4 & -10 \\ 2 & 3 \\ -6 & -9 \end{bmatrix}$ **31.** $\begin{bmatrix} 9 & 10 \\ -3 & 0 \\ 10 & 16 \end{bmatrix}$ **33.** $\begin{bmatrix} 8 & -6 \\ -10 & -16 \end{bmatrix}$ **35.** $\begin{bmatrix} -17 & 20 \\ 1 & -21 \\ -8 & -17 \end{bmatrix}$

37. $\begin{bmatrix} 4842 & -1/2 \\ 2722 & -1/8 \\ 10035 & -1 \\ 3566 & 1 \end{bmatrix}$ **39.** $\begin{bmatrix} 26 & 86 \\ -7 & -29 \\ 21 & 87 \end{bmatrix}$ **41.** $\begin{bmatrix} 6 & 18 & -24 \\ 1 & 3 & -4 \\ 0 & 0 & 0 \end{bmatrix}$ **43.** $\begin{bmatrix} 15 \\ 16 \\ 1 \end{bmatrix}$ **45.** $\begin{bmatrix} 138 & 646 \\ -43 & -209 \\ 129 & 627 \end{bmatrix}$

47. (a) $\begin{bmatrix} 1/4 & 1/3 \\ 1/2 & 1/3 \end{bmatrix}$ **(b)** 6 units of each size   **49.** $\begin{bmatrix} -1/4 & 1/6 \\ 0 & 1/3 \end{bmatrix}$ **51.** No inverse   **53.** $\begin{bmatrix} 1/4 & 1/2 & 1/2 \\ 1/4 & -1/2 & 1/2 \\ 1/8 & -1/4 & -1/4 \end{bmatrix}$

55. No inverse   **57.** $\begin{bmatrix} 6/7 & -5/7 \\ -1/7 & 2/7 \end{bmatrix}$ **59.** No inverse   **61.** $\begin{bmatrix} -1 & -1 \\ 2 & 3 \end{bmatrix}$ **63.** $\begin{bmatrix} 18 \\ -7 \end{bmatrix}$ **65.** $\begin{bmatrix} -22 \\ -18 \\ 15 \end{bmatrix}$ **67.** $(2, 2)$

69. $(2, 1)$ **71.** $(-1, 0, 2)$ **73.** No inverse; no solution for the system   **75.** 17.5 g of 12 carat, 7.5 g of 22 carat

77. 4 lbs   **79.** 164 fives; 86 tens; 40 twenties   **81.** $\begin{bmatrix} 218.09 \\ 318.27 \end{bmatrix}$ **83. (a)** $\begin{bmatrix} 0 & 1/2 \\ 2/3 & 0 \end{bmatrix}$ **(b)** 1200 units of cheese; 1600 units of goats

85. (a) $\begin{bmatrix} 1 & -1/4 \\ -1/2 & 1 \end{bmatrix}$ **(b)** $\begin{bmatrix} 8/7 & 2/7 \\ 4/7 & 8/7 \end{bmatrix}$ **(c)** $\begin{bmatrix} 2800 \\ 2800 \end{bmatrix}$

Case 6 (page 276)

1. $PQ = \begin{bmatrix} 1 & 2 & 0 & 2 & 1 & 1 \\ 0 & 1 & 0 & 1 & 0 & 0 \\ 1 & 1 & 0 & 1 & 2 & 1 \end{bmatrix}$ **2.** None   **3.** Yes, the third person

4. The second and fourth persons in the third group each had four contacts in all.

5. $PQ = \begin{bmatrix} 1 & 2 & 0 & 2 & 1 & 1 \\ 0 & 1 & 0 & 1 & 0 & 0 \\ 1 & 1 & 0 & 1 & 2 & 1 \end{bmatrix}$

Case 7 (page 279)

1. (a)

$$A = \begin{bmatrix} .245 & .102 & .051 \\ .099 & .291 & .279 \\ .433 & .372 & .011 \end{bmatrix} \quad D = \begin{bmatrix} 2.88 \\ 31.45 \\ 30.91 \end{bmatrix} \quad X = \begin{bmatrix} x_1 \\ x_2 \\ x_3 \end{bmatrix}$$

(b)

$$I - A = \begin{bmatrix} .755 & -.102 & -.051 \\ -.099 & .709 & -.279 \\ -.433 & -.372 & .989 \end{bmatrix}$$

(d)

$$\begin{bmatrix} 18.2 \\ 73.2 \\ 66.8 \end{bmatrix}$$

(e) $18.2 billion of agriculture, $73.2 billion of manufacturing, and $66.8 billion of household would be required (rounded to three significant digits).

2. (a)

$$A = \begin{bmatrix} .293 & 0 & 0 \\ .014 & .207 & .017 \\ .044 & .010 & .216 \end{bmatrix} \quad D = \begin{bmatrix} 138,213 \\ 17,597 \\ 1,786 \end{bmatrix}$$

(b)

$$I - A = \begin{bmatrix} .707 & 0 & 0 \\ -.014 & .793 & -.017 \\ -.044 & -.010 & .784 \end{bmatrix}$$

(d) Agriculture 195,000 million pounds; manufacture 26,000 million pounds; energy 13,600 million pounds

Chapter 7

Section 7.1 (page 287)

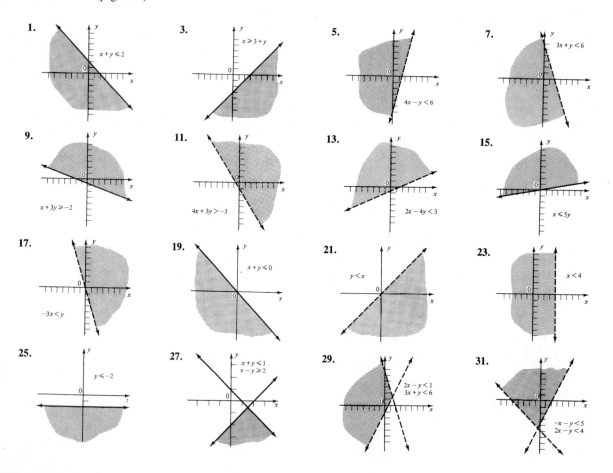

33.

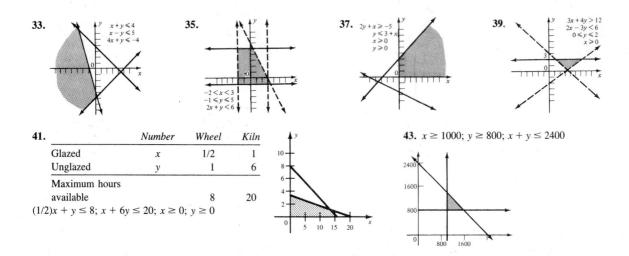

$x + y \leq 4$
$x - y \leq 5$
$4x + y \leq -4$

35.

$-2 < x < 3$
$-1 \leq y \leq 5$
$2x + y < 6$

37. $2y + x \geq -5$
$y \leq 3 + x$
$x \geq 0$
$y \geq 0$

39. $3x + 4y > 12$
$2x - 3y < 6$
$0 \leq y \leq 2$
$x \geq 0$

41.

	Number	Wheel	Kiln
Glazed	x	1/2	1
Unglazed	y	1	6
Maximum hours available		8	20

$(1/2)x + y \leq 8$; $x + 6y \leq 20$; $x \geq 0$; $y \geq 0$

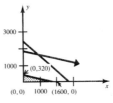

43. $x \geq 1000$; $y \geq 800$; $x + y \leq 2400$

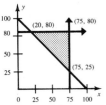

Section 7.2   (page 293)

1. Let x be the number of product A made, and y the number of B. Then $2x + 3y \leq 45$.   **3.** Let x be the number of green pills, and y be the number of red. Then $4x + y \geq 25$.   **5.** Let x be the number of pounds of \$6 coffee, and y the number of \$5 coffee. Then $x + y \geq 50$.
7. Let x be the number shipped to warehouse A and y the number shipped to warehouse B. Then $x + y \geq 100$, $x \leq 100$, $y \leq 100$, $x \leq 75$, $y \leq 80$ (these last two constraints make $x \leq 100$ and $y \leq 100$ redundant); also $x \geq 0$ and $y \geq 0$; minimize $12x + 10y$.

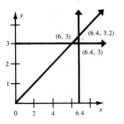

9. Let x be the number of type 1 bolts and y the number of type 2 bolts. Then $.1x + .1y \leq 240$, $.1x + .4y \leq 720$, $.1x + .5y \leq 160$, $x \geq 0$, $y \geq 0$; maximize $.10x + .12y$. The first two inequalities are redundant; only $.1x + .5y \leq 160$ affects the solution.

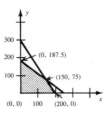

11. Let x be the number of gallons of gasoline (in millions), and let y be the number of gallons of fuel oil (in millions). Then $x \geq 2y$, $y \geq 3$, $x \leq 6.4$; maximize $1.25x + y$ (to find the total maximum revenue, this result must be multiplied by 1,000,000).

13. Let x be the number of kg of half and half mix, and y be the number of kg of the other mix. Then $x/2 + y/3 \leq 100$, $x/2 + (2y/3) \leq 125$, $x \geq 0$, $y \geq 0$; maximize $6x + 4.80y$.

15. Let x be the number of servings of A and y the number of servings of B. Then $3x + 2y \geq 15$, $2x + 4y \geq 15$, $x \geq 0$, $y \geq 0$; minimize $.25x + .40y$.

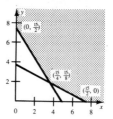

Section 7.3 (page 301)

1. Maximum of 65 at (5, 10); minimum of 8 at (1, 1) **3.** Maximum of 9 at (0, 12); minimum of 0 at (0, 0) **5.** No maximum; minimum of 18 at (3, 4) **7.** Maximum of 42/5 at $x = 6/5$; $y = 6/5$ **9.** Maximum of 25 at $x = 10$; $y = 5$ **11.** Maximum of 235/4 at $x = 105/8$; $y = 25/8$ **13.** (a) Maximum of 204; $x = 18$, $y = 2$ (b) Maximum of 117 3/5; $x = 12/5$, $y = 39/5$ (c) Maximum of 102; $x = 0$, $y = 17/2$ **15.** Ship 20 to A and 80 to B for a minimum cost of $1040 **17.** 1600 Type 1 and 0 Type 2 for maximum revenue of $160 **19.** 6.4 million gallons of gasoline and 3.2 million gallons of fuel oil, for maximum revenue of $11,200,000 **21.** 150 kg half-and-half mix, 75 kg other mix; $1260 **23.** 3 3/4 servings of A, 1 7/8 servings of B; $1.6875 **25.** (b) **27.** (c)

Section 7.4 (page 309)

1. $x_1 + 2x_2 + x_3 = 6$ **3.** $2x_1 + 4x_2 + 3x_3 + x_4 = 100$ **5.** (a) 3 (b) x_3, x_4, x_5 (c) $4x_1 + 2x_2 + x_3 = 20$
$$5x_1 + x_2 + x_4 = 50$$
$$2x_1 + 3x_2 + x_5 = 25$$

7. (a) 2 (b) x_4, x_5 (c) $7x_1 + 6x_2 + 8x_3 + x_4 = 118$ **9.** $x_1 = 0$, $x_2 = 0$, $x_3 = 20$, $x_4 = 0$, $x_5 = 15$ **11.** $x_1 = 0$,
$$4x_1 + 5x_2 + 10x_3 + x_5 = 220$$
$x_2 = 0$, $x_3 = 8$, $x_4 = 0$, $x_5 = 6$, $x_6 = 7$ **13.** $x_1 = 0$, $x_2 = 20$, $x_3 = 0$, $x_4 = 16$, $x_5 = 0$ **15.** $x_1 = 0$, $x_2 = 0$,
$x_3 = 12$, $x_4 = 0$, $x_5 = 9$, $x_6 = 8$ **17.** $x_1 = 0$, $x_2 = 0$, $x_3 = 50$, $x_4 = 10$, $x_5 = 0$, $x_6 = 50$
19.

$$\begin{array}{cccc} x_1 & x_2 & x_3 & x_4 \\ \begin{bmatrix} 2 & 3 & 1 & 0 & | & 6 \\ 4 & 1 & 0 & 1 & | & 6 \end{bmatrix} \end{array}$$

21.

$$\begin{array}{ccccc} x_1 & x_2 & x_3 & x_4 & x_5 \\ \begin{bmatrix} 1 & 1 & 1 & 0 & 0 & | & 10 \\ 5 & 2 & 0 & 1 & 0 & | & 20 \\ 1 & 2 & 0 & 0 & 1 & | & 36 \end{bmatrix} \end{array}$$

23.

$$\begin{array}{cccc} x_1 & x_2 & x_3 & x_4 \\ \begin{bmatrix} 3 & 1 & 1 & 0 & | & 12 \\ 1 & 1 & 0 & 1 & | & 15 \end{bmatrix} \end{array}$$

25. If x_1 is the number of kg of half-and-half mix and x_2 is the number of kg of the other mix, find $x_1 \geq 0$, $x_2 \geq 0$, $x_3 \geq 0$, $x_4 \geq 0$ so that $(1/2)x_1 + (1/3)x_2 + x_3 = 100$, $(1/2)x_1 + (2/3)x_2 + x_4 = 125$, and $z = 6x_1 + 4.8x_2$ is maximized.

$$\begin{array}{cccc} x_1 & x_2 & x_3 & x_4 \\ \begin{bmatrix} 1/2 & 1/3 & 1 & 0 & | & 100 \\ 1/2 & 2/3 & 0 & 1 & | & 125 \end{bmatrix} \end{array}$$

27. If x_1 is the number of prams, x_2 is the number of runabouts, and x_3 is the number of trimarans, find $x_1 \geq 0$, $x_2 \geq 0$, $x_3 \geq 0$, $x_4 \geq 0$, $x_5 \geq 0$, $x_6 \geq 0$ so that $x_1 + 2x_2 + 3x_3 + x_4 = 6240$, $2x_1 + 5x_2 + 4x_3 + x_5 = 10,800$, $x_1 + x_2 + x_3 + x_6 = 3000$, and $z = 75x_1 + 90x_2 + 100x_3$ is maximized.

$$\begin{array}{cccccc} x_1 & x_2 & x_3 & x_4 & x_5 & x_6 \\ \begin{bmatrix} 1 & 2 & 3 & 1 & 0 & 0 & | & 6240 \\ 2 & 5 & 4 & 0 & 1 & 0 & | & 10,800 \\ 1 & 1 & 1 & 0 & 0 & 1 & | & 3000 \end{bmatrix} \end{array}$$

29. If x_1 is the number of Siamese cats and x_2 is the number of Persian cats, find $x_1 \geq 0$, $x_2 \geq 0$, $x_3 \geq 0$, $x_4 \geq 0$, $x_5 \geq 0$ so that $2x_1 + x_2 + x_3 = 90$, $x_1 + 2x_2 + x_4 = 80$, $x_1 + x_2 + x_5 = 50$, and $z = 12x_1 + 10x_2$ is maximized.

$$\begin{array}{ccccc} x_1 & x_2 & x_3 & x_4 & x_5 \end{array}$$
$$\left[\begin{array}{ccccc|c} 2 & 1 & 1 & 0 & 0 & 90 \\ 1 & 2 & 0 & 1 & 0 & 80 \\ 1 & 1 & 0 & 0 & 1 & 50 \end{array} \right]$$

Section 7.5   (page 318)

1. Maximum is 20 when $x_1 = 0$, $x_2 = 4$, $x_3 = 0$, $x_4 = 0$, $x_5 = 2$ **3.** Maximum is 8 when $x_1 = 4$, $x_2 = 0$, $x_3 = 8$, $x_4 = 2$, $x_5 = 0$
5. Maximum is 264 when $x_1 = 16$, $x_2 = 4$, $x_3 = 0$, $x_4 = 0$, $x_5 = 16$, $x_6 = 0$ **7.** Maximum is 22 when $x_1 = 5.5$, $x_2 = 0$, $x_3 = 0$ and
$x_4 = .5$ **9.** Maximum is 120 when $x_1 = 0$, $x_2 = 10$, $x_3 = 0$, $x_4 = 40$, $x_5 = 4$ **11.** Maximum is 944 when $x_1 = 118$, $x_2 = 0$,
$x_3 = 0$, $x_4 = 0$, $x_5 = 102$ **13.** Maximum is 250 when $x_1 = 0$, $x_2 = 0$, $x_3 = 0$, $x_4 = 50$, $x_5 = 0$, $x_6 = 50$ **15.** 163.6 kg of food
P; none of Q; 1090.9 kg of R; 145.5 kg of S; maximum is 87,454.5   **17.** 150 kg of the half-and-half mix; 75 kg of the other;
maximum revenue is $1260   **19.** Make no 1-speed or 3-speed bicycles; make 2700 10-speed bicycles; maximum profit is $59,400
21. (a) 3 (b) 4 (c) 3   **23.** 6700 trucks and 4467 fire engines for a maximum profit of $110,997

Section 7.6   (page 326)

1. $2x_1 + 3x_2 + x_3 = 8$ **3.** $x_1 + x_2 + x_3 + x_4 = 100$ **5.** Change the objective function to maximize $z = -4x_1 - 3x_2 - 2x_3$.
    $x_1 + 4x_2 - x_4 = 7$ $x_1 + x_2 + x_3 - x_5 = 75$
                                 $x_1 + x_2 - x_6 = 27$
7. Change the objective function to maximize $z = -x_1 - 2x_2 - x_3 - 5x_4$.   **9.** Maximum is 480 when $x_1 = 40$ and $x_3 = 16$
11. Maximum is 750 when $x_2 = 150$ and $x_5 = 50$ **13.** Maximum is 300 when $x_2 = 100$, $x_3 = 50$, and $x_5 = 10$ **15.** Minimum
is 40 when $x_1 = 10$ and $x_4 = 50$ **17.** Minimum is 100 when $x_2 = 100$ and $x_5 = 50$ **19.** 3 of pill #1 and 2 of pill #2 for a
minimum cost of 70¢   **21.** 800,000 for whole tomatoes and 80,000 for sauce for a minimum cost of $3,460,900   **23.** Buy
1000 small and 500 large for a minimum cost of $210   **25.** 2 2/3 units of I, none of II, and 4 of III for a minimum cost of
$30.67   **27.** 1 2/3 ounces of I, 6 2/3 ounces of II, 1 2/3 ounces of III, for a minimum cost of $1.55 per gallon

Section 7.7   (page 335)

1. $\begin{bmatrix} 1 & 3 & 1 \\ 2 & 2 & 10 \\ 3 & 1 & 0 \end{bmatrix}$ **3.** $\begin{bmatrix} -1 & 13 & -2 \\ 4 & 25 & -1 \\ 6 & 0 & 11 \\ 12 & 4 & 3 \end{bmatrix}$ **5.** Minimize $5y_1 + 4y_2 + 15y_3$ subject to $y_1 + y_2 + 2y_3 \geq 4$; $y_1 + y_2 + y_3 \geq 3$; $y_1 + 3y_3 \geq 2$;

$y_1 \geq 0$, $y_2 \geq 0$, $y_3 \geq 0$.   **7.** Maximize $50x_1 + 100x_2$ subject to $x_1 + 3x_2 \leq 1$; $x_1 + x_2 \leq 2$; $x_1 + 2x_2 \leq 1$; $x_1 + x_2 \leq 5$; $x_1 \geq 0$;
$x_2 \geq 0$.   **9.** $y_1 = 0$, $y_2 = 7$; minimum is 14   **11.** $y_1 = 10$, $y_2 = 0$; minimum is 40   **13.** $y_1 = 0$, $y_2 = 100$, $y_3 = 0$; minimum
is 100   **15.** 3 of pill #1 and 2 of pill #2 for a minimum cost of 70¢   **17.** 800,000 for whole tomatoes and 80,000 for sauce
for a minimum cost of $3,460,000   **19.** (a) Minimize $200y_1 + 600y_2 + 90y_3 = w$ subject to $y_1 + 4y_2 \geq 1$; $2y_1 + 3y_2 + y_3 \geq 1.5$;
$y_1 \geq 0$, $y_2 \geq 0$, $y_3 \geq 0$. (b) $y_1 = .6$, $y_2 = .1$, $y_3 = 0$, $w = 180$ (c) $186 ($x_1 = 114$, $x_2 = 48$) (d) $179 ($x_1 = 116$, $x_2 = 42$)
21. Use 81 kg of the first, 382.3 kg of the second, 286.7 kg of the third, for minimum cost of $3652.25

Chapter 7 **Review Exercises** (page 337)

1.

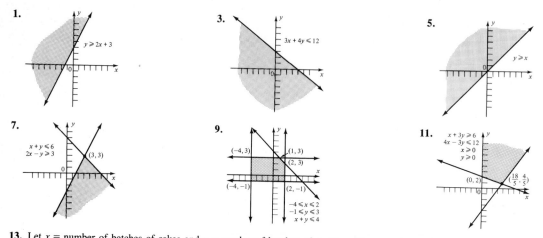

$y \geq 2x + 3$

3. $3x + 4y \leq 12$

5. $y \geq x$

7. $x + y \leq 6$ $2x - y \geq 3$ (3, 3)

9. $(-4, 3)$ $(1, 3)$ $(2, 3)$ $(-4, -1)$ $(2, -1)$ $-4 \leq x \leq 2$ $-1 \leq y \leq 3$ $x + y \leq 4$

11. $x + 3y \geq 6$ $4x - 3y \leq 12$ $x \geq 0$ $y \geq 0$ $(0, 2)$ $\left(\frac{18}{5}, \frac{4}{5}\right)$

13. Let x = number of batches of cakes and y = number of batches of cookies. Then $x \geq 0$, $y \geq 0$, and $2x + (3/2)y \leq 15$
$$3x + (2/3)y \leq 13.$$

(0, 10) (3, 6) (0, 0) $\left(\frac{13}{3}, 0\right)$ 8

15. Minimum of 8 at (2, 1); maximum of 40 at (6, 7)   **17.** Maximum of 24 at (0, 6)   **19.** Minimum of 40 at any point on the line segment connecting (0, 20) and (10/3, 40/3)   **21.** Make 3 batches of cakes and 6 of cookies for a maximum profit of $210
23. (a) Let x_1 = number of item A, x_2 = number of item B, and x_3 = number of item C she should buy.
(b) $z = 4x_1 + 3x_2 + 3x_3$
(c) $5x_1 + 3x_2 + 6x_3 \leq 1200$
$\quad x_1 + 2x_2 + 2x_3 \leq 800$
$\quad 2x_1 + \ x_2 + 5x_3 \leq 500$
25. (a) Let x_1 = number of gallons of fruity wine and x_2 = number of gallons of crystal wine to be made. (b) $z = 12x_1 + 15x_2$ (c) $2x_1 + \ x_2 \leq 110$
$$2x_1 + 3x_2 \leq 125$$
$$2x_1 + \ x_2 \leq 90$$
27. (a) $2x_1 + 5x_2 + x_3 = 50$; (b)
$\quad x_1 + 3x_2 + x_4 = 25$;
$\quad 4x_1 + x_2 + x_5 = 18$;
$\quad x_1 + x_2 + x_6 = 12$;

x_1	x_2	x_3	x_4	x_5	x_6	
2	5	1	0	0	0	50
1	3	0	1	0	0	25
4	1	0	0	1	0	18
1	1	0	0	0	1	12
−5	−3	0	0	0	0	0

29. (a) $x_1 + x_2 + x_3 + x_4 = 90$;   (b)
$\quad 2x_1 + 5x_2 + x_3 + x_5 = 120$;
$\quad x_1 + 3x_2 + x_6 = 80$

x_1	x_2	x_3	x_4	x_5	x_6	
1	1	1	1	0	0	90
2	5	1	0	1	0	120
1	3	0	0	0	1	80
−5	−8	−6	0	0	0	0

31. (68/5, 0, 24/5, 0, 0); maximum is 412/5; or (13.6, 0, 4.8, 0, 0); maximum is 82.4   **33.** (20/3, 0, 65/3, 35, 0, 0); maximum is 76 2/3   **35.** Change the objective function to maximize $z = -10x_1 - 15x_2$.   **37.** Change the objective function to maximize $z = -7x_1 - 2x_2 - 3x_3$.   **39.** Minimum of 40 at any point on the segment connecting (0, 20) and (10/3, 40/3)   **41.** Minimum of 1957 at (47, 68, 0, 92, 35, 0, 0)   **43.** (9, 5, 8, 0, 0, 0); minimum is 62   **45.** None of A; 400 of B; none of C; maximum profit is $1200   **47.** Produce 36.25 gallons of fruity and 17.5 gallons of crystal for a maximum profit of $697.50

Case 8   (page 341)

1. 1: .95; 2: .83; 3: .75; 4: 1.00; 5: .87; 6: .94   **2.** $x_1 = 100$; $x_2 = 0$; $x_3 = 0$; $x_4 = 90$; $x_5 = 0$; $x_6 = 210$

Case 9   (page 343)

1. (a) $x_2 = 58.6957$, $x_8 = 2.01031$, $x_9 = 22.4895$, $x_{10} = 1$, $x_{12} = 1$, $x_{13} = 1$ for a minimum cost of $12.55 (b) $x_2 = 71.7391$, $x_8 = 3.14433$, $x_9 = 23.1966$, $x_{10} = 1$, $x_{12} = 1$, $x_{13} = 1$ for a minimum cost of $15.05

Chapter 8

Section 8.1   (page 350)

1. False   **3.** True   **5.** True   **7.** True   **9.** False   **11.** False   **13.** False   **15.** True   **17.** False   **19.** True   **21.** False   **23.** True   **25.** False   **27.** True   **29.** True   **31.** True   **33.** True   **35.** False   **37.** True   **39.** True   **41.** False   **43.** False   **45.** False   **47.** 8   **49.** 32   **51.** 1   **53.** 32   **55.** {3, 5}   **57.** U, or {2, 3, 4, 5, 7, 9}   **59.** U, or {2, 3, 4, 5, 7, 9}   **61.** {7, 9}   **63.** Ø   **65.** Ø   **67.** U, or {2, 3, 4, 5, 7, 9}   **69.** All students in this school not taking this course   **71.** All students in this school taking accounting and zoology   **73.** All students in this school taking this course or zoology   **75.** (a) {s, d, c, g, i, m, h} (b) {s, d, c} (c) {i, m, h} (d) {g} (e) {s, d, c, g, i, m, h} (f) {s, d, c}

Section 8.2   (page 356)

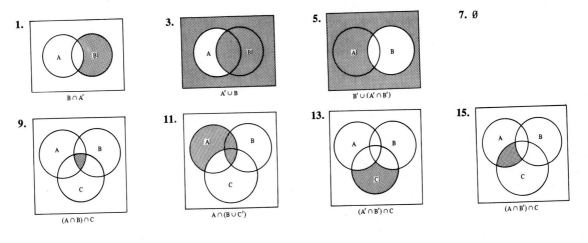

7. Ø

17. Yes; his data add up to 142 people.   **19. (a)** 89% **(b)** 66% **(c)** 4% **(d)** 1% **(e)** 4%   **21. (a)** 50 **(b)** 2 **(c)** 14   **23. (a)** 54 **(b)** 17 **(c)** 10 **(d)** 7 **(e)** 15 **(f)** 3 **(g)** 12 **(h)** 1   **25. (a)** 40 **(b)** 30 **(c)** 95 **(d)** 110 **(e)** 160 **(f)** 65   **27.** 9   **29.** 16

Section 8.3   (page 366)

1. 12   **3.** 56   **5.** 8   **7.** 24   **9.** 792   **11.** 156   **13.** 2.490952×10^{15} **15.** 240,240   **17.** 352,716   **19.** 2,042,975   **21.** $5 \cdot 3 \cdot 2 = 30$ **23. (a)** $2 \cdot 25 \cdot 24 \cdot 23 = 27,600$ **(b)** $2 \cdot 26 \cdot 26 \cdot 26 = 35,152$ **(c)** $2 \cdot 24 \cdot 23 \cdot 1 = 1104$ **25.** $3 \cdot 5 = 15$ **27.** $26^3 \cdot 10^3 = 17,576,000$ **29.** $2 \cdot 26 \cdot 25 \cdot 24 \cdot 10 \cdot 9 \cdot 8 = 22,464,000$ **31.** $6! = 720$ **33.** $6 \cdot 5 \cdot 4 = 120$ **35.** $15 \cdot 14 \cdot 13 = 2730$ **37.** $\binom{52}{2} = 1326$ **39.** 10   **41.** $\binom{5}{2}\binom{4}{1} + \binom{5}{3}\binom{4}{0} = 40 + 10 = 50$ **43.** Combinations; 6   **45.** Combinations; **(a)** 56 **(b)** 462 **(c)** 3080   **47.** Combinations; **(a)** 220 **(b)** 220   **49.** Permutations; $P(12,11) = 479,001,600$ **51.** Combinations; **(a)** $\binom{9}{3} = 84$ **(b)** $\binom{5}{3} = 10$ **(c)** $\binom{5}{2}\binom{4}{1} = 40$ **(d)** $1 \cdot \binom{8}{2} = 28$ **53.** Combinations; **(a)** $\binom{7}{2} = 21$ **(b)** $1 \cdot \binom{6}{1} = 6$ **(c)** $\binom{2}{1}\binom{5}{1} + \binom{2}{2}\binom{5}{0} = 11$ **55.** Permutations; $P(10, 4) = 5040$ **57.** Combinations; **(a)** $\binom{5}{3} = 10$ **(b)** 0 **(c)** $\binom{3}{3} = 1$ **(d)** $\binom{5}{2}\binom{1}{1} = 10$ **(e)** $\binom{5}{2}\binom{1}{1} = 30$ **(f)** $\binom{5}{1}\binom{3}{2} = 15$ **(g)** 0   **59. (a)** 840 **(b)** 180 **(c)** 420   **61. (a)** 362,880 **(b)** 6 **(c)** 1260   **63.** 2,598,960   **65.** 6.3501356×10^{11} **67.** 55,440

Section 8.4   (page 377)

1. $S = \{h\}$ **3.** $S = \{hhh, hht, hth, htt, thh, tht, tth, ttt\}$ **5.** Let c = correct, w = wrong. Then, $S = \{ccc, ccw, cwc, wcc, wwc, wcw, cww, www\}$.   **7. (a)** $\{hh, tt\}$; 1/2 **(b)** $\{hh, ht, th\}$; 3/4   **9. (a)** {2 and 4}; 1/10 **(b)** {1 and 3, 1 and 5, 3 and 5}; 3/10 **(c)** Ø; 0 **(d)** {1 and 2, 1 and 4, 2 and 3, 2 and 5, 3 and 4, 4 and 5}; 3/5   **11. (a)** 1/6 **(b)** 1/2 **(c)** 2/3 **(d)** 2/3   **13. (a)** 1/5 **(b)** 8/15 **(c)** 4/15   **15.** Feasible   **17.** Not feasible; sum of probabilities exceeds 1   **19.** Not feasible; one probability is negative   **21.** Subjective   **23.** Not subjective   **25.** Subjective   **27.** Subjective   **29.** Subjective   **31.** .24   **33.** .06   **35.** .27   **37.** 1 to 1   **39.** .35   **41.** .95   **43.** 0   **45. (a)** 1 to 5 **(b)** 1 to 1 **(c)** 2 to 1 **(d)** 1 to 5   **47.** 1 to 4   **49.** 2/5   **51.** The theoretical probability is 1/32.   **53.** The theoretical probability is 1/221 or .004525.

Section 8.5   (page 385)

1. 1/36   **3.** 1/9   **5.** 5/36   **7.** 1/12   **9.** 5/18   **11.** 11/36   **13.** 5/12   **15.** 1/2   **17.** 2/5   **19.** 2/13   **21.** 3/26   **23.** 3/13   **25.** 7/13   **27.** 1/10   **29.** 2/5   **31.** 7/20   **33.** 1/6   **35.** 3/10   **37. (a)** .5 **(b)** .24 **(c)** .09 **(d)** .83   **39. (a)** 4/25 **(b)** 47/50 **(c)** 39/50   **41. (a)** .4 **(b)** .1 **(c)** .6 **(d)** .9   **43.** 84%   **45.** .951   **47.** .473   **49.** .007   **51.** .25   **53.** .38   **55.** .89   **57. (a)** .51 **(b)** .67 **(c)** .39 **(d)** .12   **59. (a)** .45 **(b)** .75 **(c)** .35 **(d)** .46   **61.** 1/4   **63.** 1/2   **65.** .90   **67.** The theoretical answer is 7/8. Using the Monte Carlo method, answers will vary.

Section 8.6   (page 395)

1. 0   **3.** 1   **5.** 1/6   **7.** 4/17   **9.** 11/51   **11.** .012   **13.** .245   **15.** .052   **17.** .06   **19.** .7456   **21.** 7/10   **23.** 2/3   **25.** 3/10   **27.** 1/10   **29.** 1/7   **31.** 1/2   **33.** 4/7   **35.** The probability of a customer not making a deposit given that the customer cashed a check is 3/8.   **37.** The probability of a customer not cashing a check given that the customer made a deposit is 2/7.   **39.** 3/10   **41.** 1/3   **43.** 2/3   **45.** .02   **47.** 2/3   **49.** 1/4   **51.** 1/4   **53.** .527   **55.** .042   **57.** 6/7 or .857   **59.** Yes   **61.** .05   **63.** .25   **65.** Not very realistic   **67.** .000015; fairly realistic

Chapter 8 **Review Exercises** (page 398)

1. False **3.** False **5.** True **7.** False **9.** False **11.** Ø **13.** {1, 2, 3, 4} **15.** $2^4 = 16$ **17.** {a, b, e}
19. {c, d, g} **21.** {a, b, e, f} **23.** U **25.** All employees in the accounting department who also have at least 10 years with
the company **27.** All employees who are in the accounting department or who have MBA degrees **29.** All employees who are
not in the sales department and who have worked less than 10 years with the company

31.

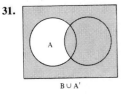

$B \cup A'$

33.

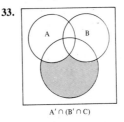

$A' \cap (B' \cap C)$

35. 52 **37.** 12 **39.** 6! = 720 **41.** $5 \cdot 4 = 20$ **43.** $\binom{8}{3} \cdot \binom{6}{2} = 840$ **45.** {1, 2, 3, 4, 5, 6}
47. {0, .5, 1, 1.5, 2, . . . , 299.5, 300} **49.** {(3, R), (3, G), (5, R), (5, G), (7, R), (7, G), (9, R), (9, G), (11, R), (11, G)}
51. {(3, G), (5, G), (7, G), (9, G), (11, G)} **53.** $E' \cap F'$ **55.** 1/4 **57.** 3/13 **59.** 1/2 **61.** 1 **63.** 1 to 25
65. .86 **67.**

	N_2	T_2
N_1	N_1N_2	N_1T_2
T_1	T_1N_2	T_1T_2

69. 1/2 **71.** 5/36 **73.** 1/6 **75.** 1/6 **77.** 2/11 **79.** .66

81. .71 **83.** .3 **85.** 2/5 or .4 **87.** 3/4 or .75

Case 10 (page 401)

1. 1/200 = .0005 **2.** 1999/2000 = .9995 **3.** $(1999/2000)^a$ **4.** $1 - (1999/2000)^a$ **5.** $(1999/2000)^{Nc}$
6. $1 - (1999/2000)^{Nc}$ **7.** $1 - .741 = .259$ **8.** $1 - .882 = .118$

Case 11 (page 402)

1. .715; .569; .410; .321; .271

Chapter 9

Section 9.1 (page 407)

1. 1/3 **3.** 2/41 **5.** 21/41 **7.** 8/17 **9.** 119/131 ≈ .908 **11.** 14/17 ≈ .824 **13.** 1/176 ≈ .006 **15.** 1/11
17. 5/9 **19.** 5/26 **21.** 178/343 ≈ .519 **23.** 72/73 ≈ .986 **25.** .146 **27.** .082

Section 9.2 (page 413)

1. $\binom{7}{1}/\binom{9}{1} = 7/9$ **3.** $\binom{7}{3}/\binom{9}{3} = 35/84$ **5.** .424 **7.** $\binom{6}{3}/\binom{10}{3} = 1/6$ **9.** $\binom{4}{2}\binom{6}{1}/\binom{10}{3} = 3/10$ **11.** $\binom{52}{2} = 1326$
13. $(4 \cdot 48 + 6)/\binom{52}{2} = 33/221$ **15.** $4 \cdot \binom{13}{2}/\binom{52}{2} = 52/221$ **17.** $\binom{40}{2}/\binom{52}{2} = 130/221$ **19.** $(1/26)^5$
21. $1 \cdot 25 \cdot 24 \cdot 23 \cdot 22/(26)^4 = 18{,}975/28{,}561$ **23.** $4/\binom{52}{5} = 1/649{,}740$ **25.** $13 \cdot \binom{4}{4}\binom{48}{1}/\binom{52}{5} = 1/4165$ **27.** $1/\binom{52}{13}$
29. $\binom{4}{3}\binom{4}{3}\binom{44}{7}/\binom{52}{13}$ **31.** $1 - P(365, 39)/365^{39}$ **33.** 1 **35.** 3/8; 1/4 **37.** 120/343 ≈ .3498 **39.** 6! = 720
41. 2/7 **43.** 2/7

Section 9.3   (page 418)

1. $\binom{5}{2}(1/2)^2(1/2)^3 = 5/16$ **3.** $\binom{5}{0}(1/2)^0(1/2)^5 = 1/32$ **5.** $\binom{5}{4}(1/2)^4(1/2)^1 + \binom{5}{5}(1/2)^5(1/2)^0 = 3/16$
7. $\binom{5}{0}(1/2)^0(1/2)^5 + \binom{5}{1}(1/2)^1(1/2)^4 + \binom{5}{2}(1/2)^2(1/2)^3 + \binom{5}{3}(1/2)^3(1/2)^2 = 13/16$ **9.** $\binom{12}{12}(1/6)^{12}(5/6)^0 \approx .0000000005$
11. $\binom{12}{1}(1/6)(5/6)^{11} \approx .269$ **13.** $\binom{12}{0}(1/6)^0(5/6)^{12} + \binom{12}{1}(1/6)^1(5/6)^{11} + \binom{12}{2}(1/6)^2(5/6)^{10} + \binom{12}{3}(1/6)^3(5/6)^9 \approx .875$
15. $\binom{5}{5}(1/2)^5(1/2)^0 = 1/32$ **17.** $\binom{5}{0}(1/2)^0(1/2)^5 + \binom{5}{1}(1/2)^1(1/2)^4 + \binom{5}{2}(1/2)^2(1/2)^3 + \binom{5}{3}(1/2)^3(1/2)^2 = 13/16$
19. $\binom{20}{0}(.05)^0(.95)^{20} \approx .358$ **21.** $\binom{6}{2}(1/5)^2(4/5)^4 \approx .246$ **23.** $\binom{6}{4}(1/5)^4(4/5)^2 + \binom{6}{5}(1/5)^5(4/5)^1 + \binom{6}{6}(1/5)^6(4/5)^0 \approx .017$
25. $\binom{3}{1}(5/50)^1(45/50)^2 \approx .243$ **27.** $\binom{10}{5}(.20)^5(.80)^5 \approx .026$ **29.** .999   **31.** $\binom{20}{17}(.70)^{17}(.30)^3 \approx .072$ **33.** .035
35. $\binom{10}{7}(1/5)^7(4/5)^3 \approx .00079$ **37.** $\approx .999922$ **39.** $\binom{3}{1}(.80)^1(.20)^2 = .096$ **41.** $\binom{3}{0}(.80)^0(.20)^3 + \binom{3}{1}(.80)^1(.20)^2 = .104$
43. .185   **45.** .256   **47.** $1 - (.99999975)^{10,000} \approx .0025$ **49. (a)** .000036 **(b)** .000088 **(c)** 0
51. (a) .148774 **(b)** .021344 **(c)** .978656

Section 9.4   (page 425)

1. Yes   **3.** Yes   **5.** No   **7.** No   **9.** Yes   **11.** Yes   **13.** No   **15.** Regular   **17.** Not regular   **19.** Regular
21. [2/5   3/5]   **23.** [4/11   7/11]   **25.** [14/83   19/83   50/83]   **27.** [170/563   197/563   196/563]
29. (a)   small   large   **(b)** [.10   .90] **(c)** [.74896   .25104] **(d)** [3/4   1/4]

$$\begin{array}{c}\text{small} \\ \text{large}\end{array}\begin{bmatrix} .80 & .20 \\ .60 & .40 \end{bmatrix}$$

31. (a) G_0 G_1 G_2 **(b)** 42,500, 5000, 2500 **(c)** 36,125, 8250, 5625

$$\begin{array}{c} G_0 \\ G_1 \\ G_2 \end{array}\begin{bmatrix} .85 & .10 & .05 \\ 0 & .80 & .20 \\ 0 & 0 & 1 \end{bmatrix}$$

(d) 30,706, 10,213, 9081 **(e)** 26,100, 11,241, 12,659 **(f)** [0   0   1]; in the long run, everybody has two or more accidents
33. (a) [.257   .597   .146] **(b)** [.255   .594   .151] **(c)** [.254   .590   .156] **(d)** [.244   .529   .227]
35.   A   B   C

$$\begin{array}{c} A \\ B \\ C \end{array}\begin{bmatrix} .3 & .3 & .4 \\ .15 & .3 & .55 \\ .3 & .6 & .1 \end{bmatrix}$$; Long range prediction is [60/251   102/251   89/251] or [.239   .406   .355]

37. (a)   Liberal   Conservative   Independent **(b)** [.40   .45   .15] **(c)** [44%   40.5%   15.5%]

$$\begin{array}{c} \text{Liberal} \\ \text{Conservative} \\ \text{Independent} \end{array}\begin{bmatrix} .80 & .15 & .05 \\ .20 & .70 & .10 \\ .20 & .20 & .60 \end{bmatrix}$$

(d) [46.4%   38.05%   15.55%] **(e)** [47.84%   36.705%   15.455%] **(f)** [50%   35%   15%]

39. [1/3   1/3   1/3]   **41.** The first power is the given transition matrix; $\begin{bmatrix} .2 & .15 & .17 & .19 & .29 \\ .16 & .2 & .15 & .18 & .31 \\ .19 & .14 & .24 & .21 & .22 \\ .16 & .19 & .16 & .2 & .29 \\ .16 & .19 & .14 & .17 & .34 \end{bmatrix}$;

$\begin{bmatrix} .17 & .178 & .171 & .191 & .29 \\ .171 & .178 & .161 & .185 & .305 \\ .18 & .163 & .191 & .197 & .269 \\ .175 & .174 & .164 & .187 & .3 \\ .167 & .184 & .158 & .182 & .309 \end{bmatrix}$; $\begin{bmatrix} .1731 & .175 & .1683 & .188 & .2956 \\ .1709 & .1781 & .1654 & .1866 & .299 \\ .1748 & .1718 & .1753 & .1911 & .287 \\ .1712 & .1775 & .1667 & .1875 & .2971 \\ .1706 & .1785 & .1641 & .1858 & .301 \end{bmatrix}$; $\begin{bmatrix} .17193 & .17643 & .1678 & .18775 & .29609 \\ .17167 & .17689 & .16671 & .18719 & .29754 \\ .17293 & .17488 & .17007 & .18878 & .29334 \\ .17192 & .17654 & .16713 & .18741 & .297 \\ .17142 & .17726 & .16629 & .18696 & .29807 \end{bmatrix}$; .18719

43. **(a)** .847423 or about 85% **(b)** [.032 .0998125 .0895625 .778625]
45. [.219086 .191532 .252352 .178091 .158938]

Section 9.5   (page 431)

1. 3.6   **3.** 14.64   **5.** −$.64   **7.** 9/7 or 1.3   **9.** **(a)** 5/3 or 1.67 **(b)** 4/3 or 1.33   **11.** 1   **13.** No; the expected value
is about −21¢.   **15.** −2.7¢   **17.** −20¢   **19.** **(a)** Yes; the probability of a match is still 1/2. **(b)** 40¢ **(c)** −40¢   **21.** 2.5
23. $4500   **25.** 118   **27.** 3.51   **29.** **(a)** $50,000 **(b)** $65,000   **31.** −87.8¢

Section 9.6   (page 436)

1. **(a)** Buy speculative **(b)** Buy blue-chip **(c)** Buy speculative; $24,300 **(d)** Buy blue-chip
3. **(a)** Set up in the stadium **(b)** Set up in the gym **(c)** Set up both; $1010
5. **(a)**

	New product better	Not better
Market new product	50,000	−25,000
Don't	−40,000	−10,000

(b) $5000 if they market new product, −$22,000 if they don't; market the new product

7. **(a)**

	Strike	None
Bid $30,000	−5500	4500
Bid $40,000	4500	0

(b) bid $40,000   **9.** Environment; 14.25

Chapter 9   Review Exercises   (page 438)

1. **(a)** 3/11 **(b)** 8/11   **3.** 3/22   **5.** 2/3   **7.** 4/9   **9.** $\binom{4}{3}/\binom{11}{3} = 4/165$ **11.** $\binom{4}{2}\binom{5}{1}/\binom{11}{3} = 2/11$ **13.** $\binom{5}{2}\binom{2}{1}/\binom{11}{3} = 4/33$
15. 25/102   **17.** 15/34   **19.** 5/16   **21.** 11/32   **23.** $\binom{20}{4}(.01)^4(.99)^{16} \approx .00004$
25. .81791 + .16523 + .01586 + .00096 + .00004 ≈ 1.0000   **27.** Yes   **29.** Yes   **31.** **(a)** [.54 .46] **(b)** [.6464 .3536]
(c) 2/3 of the market   **33.** **(a)** [.195 .555 .250] **(b)** [.1945 .5555 .2500] **(c)** [7/36 5/9 1/4]   **35.** −83.3¢; no
37. 2.5   **39.** −7¢   **41.** $1.29   **43.** **(a)** Oppose **(b)** Oppose **(c)** Oppose; 2700 **(d)** Oppose; 1400

Case 12   (page 442)

1. .076   **2.** .542   **3.** .051

Case 13   (page 444)

1. **(a)** $69.01 **(b)** $79.64 **(c)** $58.96 **(d)** $82.54 **(e)** $62.88 **(f)** $64.00   **2.** Stock only part 3 on the truck   **3.** p_1, p_2, p_3 are
not the only events in the sample space.   **4.** $\binom{n}{0} + \binom{n}{1} + \binom{n}{2} + \cdots + \binom{n}{n}$

Chapter 10

Section 10.1 (page 450)

1. (a) About 17.5% **(b)** About 8% **(c)** 20–29

3. (a)

0–24	4
25–49	3
50–74	6
75–99	3
100–124	5
125–149	9

(b)

Interval	Frequency	Cumulative frequency
0–24	4	4
25–49	3	7
50–74	6	13
75–99	3	16
100–124	5	21
125–149	9	30

(c)–(d)

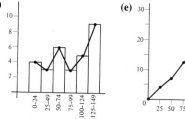

(e)

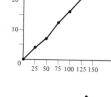

5. (a)–(b)

Interval	Frequency	Cumulative frequency
70–74	2	2
75–79	1	3
80–84	3	6
85–89	2	8
90–94	6	14
95–99	5	19
100–104	6	25
105–109	4	29
110–114	2	31

(c)–(d)

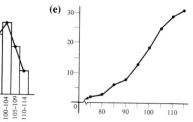

(e)

7.

Letter	Frequency	%	Letter	Frequency	%
E	41	16.9	O	10	4.1
T	22	9.1	U, F, G	9	3.7
N, R	17	7.0	D, Y	5	2.1
I	16	6.6	M, P, W	4	1.7
A	15	6.2	Q	3	1.2
S, H	14	5.8	V	1	.4
L	12	5.0	B, K, X,	0	0
C	11	4.5	J, Z		

9. Answers vary   **11.**

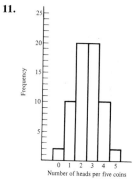

13. Answers vary

Section 10.2   (page 456)

1. 16   **3.** 150.3   **5.** 27,955   **7.** 7.7   **9.** .1   **11.** 6.7   **13.** 17.2   **15.** 118.8   **17.** 51   **19.** 130   **21.** 49
23. 562   **25.** 29.1   **27.** 9   **29.** 64   **31.** 68 and 74   **33.** No mode   **35.** 6.1   **37.** Mean = 86.2, modal
class 125–149   **39.** Mean = 69.4, modal class 60–69   **41.** 405.6 thousand or 405,600   **43.** 55.5°F   **45.** $3.29
47. Mean = 1.9, median = 1, mode = 1   **49.** Mean = 17.2, median = 17.5, no mode

Section 10.3   (page 462)

1. 6; $\sqrt{4} = 2$ **3.** 12; $\sqrt{18.3} = 4.3$ **5.** 53; $\sqrt{407.4} = 20.2$ **7.** 46; $\sqrt{215.3} = 14.7$ **9.** 24; $\sqrt{59.4} = 7.7$ **11.** 30;
$\sqrt{111.8} = 10.6$ **13.** About 44   **15.** $s = \sqrt{114.2} = 10.7$ **17.** 3/4   **19.** 24/25   **21.** At least 88.9%   **23.** At least
96%   **25.** No more than 11.1%   **27.** Forever Power, $\bar{x} = 26.2$, $s = 4.1$; Brand X, $\bar{x} = 25.5$, $s = 6.9$ **(a)** Forever Power
(b) Forever Power   **29.** 12.5   **31.** 4.3   **33.** 15.5   **35.** Mean = 5.0876, standard deviation = .1065
37. Mean = 46.6807, standard deviation = .9223

Section 10.4   (page 470)

1. 49.38%   **3.** 17.36%   **5.** 45.64%   **7.** 49.91%   **9.** 7.7%   **11.** 47.35%   **13.** 92.42%   **15.** 32.56%   **17.** 1.64
or 1.65   **19.** −1.04   **21.** 5000   **23.** 4332   **25.** 642   **27.** 9918   **29.** 19   **31.** 15.87%   **33.** .62%
35. 84.13%   **37.** 37.79%   **39.** 2.27%   **41.** 99.38%   **43.** 189   **45.** .0062   **47.** .4325   **49.** $62.64 and $41.86
51. .0823   **53.** 82   **55.** 70   **57.** .079614   **59.** .993333   **61.** .317246   **63.** .152730   **65.** .005107

Section 10.5   (page 477)

1. (a)

x	0	1	2	3	4	5	6
$P(x)$	.335	.402	.201	.054	.008	.001	.000

(b) 1.00 **(c)** .91   **3. (a)**

x	0	1	2	3
$P(x)$	.941	.058	.001	.000

(b) .06 **(c)** .24

5. (a)

x	0	1	2	3	4
$P(x)$	.0081	.0756	.2646	.4116	.2401

(b) 2.8 **(c)** .92   **7.** 12.5; 3.49   **9.** 51.2; 3.2   **11.** .1974   **13.** .1210
15. .0240   **17.** .9032   **19.** .8665   **21.** .0956   **23.** .0760   **25.** .6443   **27.** .0146   **29.** .1974   **31.** .0092
33. .0001   **35.** For Exercise 49: **(a)** .0001 **(b)** .0002 **(c)** 0; for Exercise 50: **(a)** .0472 **(b)** .9875 **(c)** .8814; for Exercise 51:
(a) .1686 **(b)** .0304 **(c)** .9575; for Exercise 52: **(a)** 0 **(b)** .0018

Chapter 10   Review Exercises   (page 478)

1. (a)

Interval	Frequency
450–474	5
475–499	6
500–524	5
525–549	2
550–574	2

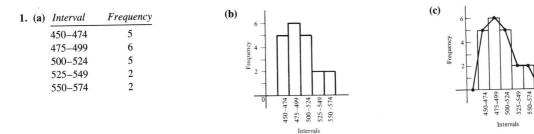

3. 73   **5.** 34.9   **7.** Median = 44, mode = 46   **9.** 30–39   **11.** 18; $\sqrt{38.8} = 6.2$ **13.** 12.5
15. (a) Stock 1: 8%, 6.5%; Stock 2: 8%, 2.2% **(b)** Stock 2   **17. (a)** At least 75% **(b)** No more than 69.4%
19. .3980   **21.** .3409   **23.** 2.27%   **25.** 68.25%

27. (a)

x	0	1	2	3	4
Probability	.9801	.0197	.0001485	.0000004975	6×10^{-10}

(b) Mean = .02, σ = .141

29. (a) .0959 **(b)** .0028   **31.** 28.10%   **33.** 56.25%   **35.** .0718   **37.** $\binom{100}{95}(.98)^{95}(.02)^5$
39. $\binom{100}{90}(.98)^{90}(.02)^{10} + \binom{100}{91}(.98)^{91}(.02)^9 + \cdots + \binom{100}{100}(.98)^{100}(.02)^0$

Case 14   (page 482)

1. No; yes   **2.** No; yes

Chapter 11

Section 11.1   (page 493)

1. 3   **3.** Does not exist   **5.** 2   **7.** 0   **9.** −2   **11.** 3   **13.** 9   **15.** No limit   **17.** 10   **19.** No limit
21. 41   **23.** 9/7   **25.** −1   **27.** 6   **29.** −5   **31.** $\sqrt{5}$ **33.** Does not exist   **35.** 4   **37.** −1/9   **39.** 1/10
41. $1/(2\sqrt{5})$ **43.** 2x   **45. (a)** $500 **(b)** Does not exist **(c)** $1500 **(d)** 15 units   **47.** 3/5   **49.** 1/2   **51.** 1/2   **53.** 0
55. No limit   **57. (a)** .57 **(b)** .53 **(c)** .50 **(d)** .5   **59.** −.24302   **61.** 1.000

Section 11.2   (page 501)

1. −1   **3.** −1   **5.** 0; 2   **7.** 1   **9.** Yes; no   **11.** No; no; yes   **13.** Yes; no; no   **15.** Yes; no; yes   **17.** Yes; no;
yes   **19.** Yes; yes; yes   **21.** No; yes; yes
23. (a) $96   **(b)** $150
(c)   **(d)** At $x = 100$

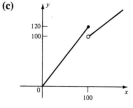

25. (a) $30 **(b)** $30 **(c)** $25 **(d)** $21.43 **(e)** $22.50 **(f)** $30 **(g)** $25   **27.** $x = 6, x = 8, x = 12, x = 16$ **29.** (−4, 6)
31. (−∞, −2]   **33.** [−11, 9]   **35.** (−6, −2)   **37.** (−4, ∞)   **39.** (−∞, 0)   **41.** Yes; yes; no   **43.** No; yes; yes

Section 11.3   (page 509)

1. (a) 2 **(b)** .6 **(c)** −.8 **(d)** −4/3 **(e)** −2.2 **(f)** −1   **3. (a)** 3 **(b)** 7/2 **(c)** 1 **(d)** 0 **(e)** −2 **(f)** −2 **(g)** −3 **(h)** −1   **5. (a)** 3°
per thousand feet **(b)** 1.25° per thousand feet **(c)** −7/6° per thousand feet **(d)** About −1/8° per thousand feet   **7. (a)** About .6
(b) About −2.4 **(c)** About −1 **(d)** About 2   **9.** 5   **11.** 8   **13.** 1/3   **15.** −1/3   **17. (a)** 7 **(b)** 5 **(c)** 3.02; 3.002;
3.0002 **(d)** 3   **19.** 3

Section 11.4 (page 521)

1. 27 **3.** 1/8 **5.** 1/8 **7.** $y = 8x - 9$ **9.** $5x + 4y = 20$ **11.** $3y = 2x + 18$ **13.** 2 **15.** 1/5 **17.** 0
19. $f'(x) = -8x + 11$; -5; 11; 35 **21.** $f'(x) = 8$; 8; 8; 8 **23.** $f'(x) = 3x^2 + 3$; 15; 3; 30 **25.** $f'(x) = 2/x^2$; 1/2; does not
exist; 2/9 **27.** $f'(x) = -4/(x-1)^2$; -4; -4; $-1/4$ **29.** $f'(x) = 1/(2\sqrt{x})$; $1/(2\sqrt{2})$; does not exist; does not exist **31.** 0
33. -6; 6 **35.** -3; 0; 2; 3; 5 **37.** $x = 0$ **39.** $x = -2$ **41.** $x = -1$; $x = 2$ **43.** $x = -4$; $x = 0$ **45.** Derivative is
never 0 **47. (a)** -8 **(b)** -20 **49. (a)** 30 **(b)** 20 **(c)** 10 **(d)** 0 **(e)** -10 **(f)** -30 **51.** From the left, our estimates are
$-.005$; .008; $-.00125$. **53.** $-.0435$ **55.** .994 **57.** -1.05

Section 11.5 (page 529)

1. $f'(x) = 18x - 8$ **3.** $y' = 30x^2 - 18x + 6$ **5.** $y' = 4x^3 - 15x^2 + 18x$ **7.** $f'(x) = 9x^{.5} - 2x^{-.5}$ or $9x^{.5} - 2/x^{.5}$
9. $y' = -48x^{2.2} + 3.8x^{.9}$ **11.** $y' = 36t^{1/2} + 2t^{-1/2}$ or $36t^{1/2} + 2/t^{1/2}$ **13.** $y' = 4x^{-1/2} + (9/2)x^{-1/4}$ or $4/x^{1/2} + 9/(2x^{1/4})$
15. $g'(x) = -30x^{-6} + x^{-2}$ or $-30/x^6 + 1/x^2$ **17.** $y' = 8x^{-3} - 9x^{-4}$ or $8/x^3 - 9/x^4$
19. $y' = -20x^{-3} - 12x^{-5} - 6$ or $-20/x^3 - 12/x^5 - 6$ **21.** $f'(t) = -6t^{-2} + 16t^{-3}$ or $-6/t^2 + 16/t^3$
23. $y' = -36x^{-5} + 24x^{-4} - 2x^{-2}$ or $-36/x^5 + 24/x^4 - 2/x^2$ **25.** $f'(x) = -6x^{-3/2} - (3/2)x^{-1/2}$ or $-6/x^{3/2} - 3/(2x^{1/2})$
27. $p'(x) = 5x^{-3/2} - 12x^{-5/2}$ or $5/x^{3/2} - 12/x^{5/2}$ **29.** $y' = (-3/2)x^{-5/4}$ or $-3/(2x^{5/4})$
31. $y' = (-5/3)t^{-2/3}$ or $-5/(3t^{2/3})$ **33.** $dy/dx = -40x^{-6} + 36x^{-5}$ or $-40/x^6 + 36/x^5$ **35.** $(-9/2)x^{-3/2} - 3x^{-5/2}$ or
$-9/(2x^{3/2}) - 3/x^{5/2}$ **37.** -28 **39.** 11/16 **41.** When $x = 5/3$ **43.** When $x = 2/3$ or 1 **45.** When $x = 5/2$ or -1
47. When $x = \sqrt[3]{4}$ **49.** Tangent line is never horizontal **51.** -28; $28x + y = 34$ **53.** 79/6 **55.** $-11/27$ **57.** 10
59. (a) 100 **(b)** 1 **(c)** $-.01$; the percent of acid is decreasing at the rate of .01 per day after 100 days **61. (a)** The blood sugar
level is decreasing at a rate of 4 points per unit of insulin. **(b)** The blood sugar level is decreasing at a rate of 10 points per unit
of insulin. **63. (a)** 264 **(b)** 510 **(c)** 97/4 or 24.25 **65. (a)** \$8 **(b)** \$192 **(c)** \$428 **(d)** \$760 **67. (a)** $v(t) = 6$ **(b)** 6; 6; 6
69. (a) $v(t) = 22t + 4$ **(b)** 4; 114; 224 **71. (a)** $v(t) = 12t^2 + 16t$ **(b)** 0; 380; 1360

Section 11.6 (page 536)

1. $y' = 12x - 15$ **3.** $y' = 4x + 3$ **5.** $y' = 48x + 66$ **7.** $y' = 18x^2 - 6x + 4$ **9.** $y' = 9x^2 - 4x - 5$
11. $y' = 36x^3 + 21x^2 - 18x - 7$ **13.** $y' = 4(2x - 5)$ or $8x - 20$ **15.** $y' = 4x(x^2 - 1)$ or $4x^3 - 4x$
17. $y' = 3x^{1/2}/2 + x^{-1/2}/2 + 2$ or $3\sqrt{x}/2 + 1/(2\sqrt{x}) + 2$ **19.** $y' = 10 + 3x^{-1/2}/2$ or $10 + 3/(2\sqrt{x})$ **21.** $y' = -3/(2x - 1)^2$
23. $y' = 53/(3x + 8)^2$ **25.** $y' = -6/(3x - 5)^2$ **27.** $y' = -17/(4 + x)^2$ **29.** $y' = (x^2 - 2x - 1)/(x - 1)^2$
31. $y' = (-x^2 + 4x + 1)/(x^2 + 1)^2$ **33.** $y' = (-2x^2 - 6x - 1)/(2x^2 - 1)^2$ **35.** $y' = (x^2 + 6x - 14)/(x + 3)^2$
37. $2(4x^2 + 3x + 1)/(1 - 4x^2)^2$ **39.** $y' = [-\sqrt{x}/2 - 1/(2\sqrt{x})]/(x - 1)^2$ or $(-x - 1)/[2\sqrt{x}(x - 1)^2]$
41. $y' = (5\sqrt{x}/2 - 3/\sqrt{x})/x$ or $(5x - 6)/(2x\sqrt{x})$ **43. (a)** 86.8 **(b)** 176.4 **(c)** $A(x) = 9x - 4 + 8/x$ **(d)** $A'(x) = 9 - 8/x^2$
45. (a) $G'(20) = -1/200$; go faster **(b)** $G'(40) = 1/400$; go slower **47. (a)** $s'(x) = m/(m + nx)^2$
(b) $s'(50) = 1/25600 \approx .0000391$

Section 11.7 (page 542)

1. $(2x + 5)^{1/3}$; $2x^{1/3} + 5$ **3.** $2/\sqrt{4 - 3x}$; $4 - 6/\sqrt{x}$ **5.** $125x^2 + 145x + 42$; $25x^2 - 5x + 3$ **7.** $\sqrt{x^2 + 12}$; $x + 12$
9. If $f(x) = x^{1/3}$ and $g(x) = 3x - 7$, then $y = f[g(x)]$. **11.** If $f(x) = x^{1/2}$ and $g(x) = 9 - 4x$, then $y = f[g(x)]$.
13. If $f(x) = (x + 3)/(x - 3)$ and $g(x) = \sqrt{x}$, then $y = f[g(x)]$. **15.** If $f(x) = x^2 + x + 5$ and $g(x) = x^{1/2} - 3$, then $y = f[g(x)]$.
17. $y' = 4(2x + 9)$ **19.** $y' = 90(5x - 1)^2$ **21.** $y' = -144x(12x^2 + 4)^2$ **23.** $y' = 36(2x + 5)(x^2 + 5x)^3$
25. $y' = 36(2x + 5)^{1/2}$ **27.** $y' = (-21/2)(8x + 9)(4x^2 + 9x)^{1/2}$ **29.** $y' = 16(4x + 7)^{-1/2}$ or $16/\sqrt{4x + 7}$
31. $y' = -(2x + 4)(x^2 + 4x)^{-1/2}$ or $-(2x + 4)/\sqrt{x^2 + 4x}$ **33.** $y' = 16x(2x + 3) + 4(2x + 3)^2$ or $12(2x + 3)(2x + 1)$
35. $y' = 2(x + 2)(x - 1) + (x - 1)^2 = 3(x - 1)(x + 1)$ **37.** $y' = 10(x + 3)^2(x - 1) + 10(x - 1)^2(x + 3)$ or
$10(x + 3)(x - 1)(2x + 2)$ or $20(x + 3)(x - 1)(x + 1)$ **39.** $y' = (x + 1)^2 \cdot x^{-1/2}/2 + 2x^{1/2}(x + 1)$ or $x^{-1/2}(x + 1)(5x + 1)/2$
41. $y' = -2(x - 4)^{-3}$ or $-2/(x - 4)^3$ **43.** $y' = (x + 3)(x - 5)/(x - 1)^2$ **45.** $y' = 8/(x + 2)^3$
47. $(3x^{1/2} - 1)/[4x^{1/2}(x^{1/2} - 1)^{1/2}]$ **49. (a)** -3.2 **(b)** -2.8 **(c)** -2.4 **(d)** $-.8$ **(e)** $A(x) = 1000/x - 4 + x/250$
(f) $A'(x) = -1000/x^2 + 1/250$ **51. (a)** $-1/2 = -.5$ **(b)** $-.02$ **(c)** $-.011$ **(d)** $-.008$

Section 11.8   (page 549)

1. $y' = 4e^{4x}$ **3.** $y' = 12e^{-2x}$ **5.** $y' = -16e^{2x}$ **7.** $y' = -16e^{x+1}$ **9.** $y' = 2xe^{x^2}$ **11.** $y' = 12xe^{2x^2}$
13. $y' = 16xe^{2x^2-4}$ **15.** $y' = xe^x + e^x = e^x(x + 1)$ **17.** $y' = 2(x - 3)(x - 2)e^{2x}$ **19.** $y' = -1/(3 - x)$ or $1/(x - 3)$
21. $y' = (4x - 7)/(2x^2 - 7x)$ **23.** $y' = 1/[2(x + 5)]$ **25.** $y' = (12x + 17)/[(2x + 7)(3x - 2)]$
27. $y' = 22/[(2x + 4)(5x - 1)]$ **29.** $y' = 3(2x^2 + 5)/[x(x^2 + 5)]$ **31.** $y' = -3x/(x + 2) - 3 \ln |x + 2|$ **33.** $y' = x + 2x \ln |x|$
35. $y' = (2xe^x - x^2e^x)/e^{2x} = x(2 - x)/e^x$ **37.** $y' = (2x^3 - 1 + 6x^3 \ln |x|)/x$ **39.** $y' = (4x + 7 - 4x \ln |x|)/[x(4x + 7)^2]$
41. $y' = (1 - x \ln |x|)/(xe^x)$ **43.** $y' = (6x \ln |x| - 3x)/(\ln |x|)^2$ **45.** $y' = [4(\ln |x + 1|)^3]/(x + 1)$
47. $y' = [x(\ln |x|)e^x - e^x]/[x(\ln |x|)^2]$ or $e^x[x \ln |x| - 1]/[x(\ln |x|)^2]$ **49.** $y' = [x(e^x - e^{-x}) - (e^x + e^{-x})]/x^2$
51. $y' = e^x/x + 2xe^{x^2} \ln |x|$ **53.** $y' = -20,000e^{.4x}/(1 + 10e^{.4x})^2$ **55.** $y' = 8000e^{-.2x}/(9 + 4e^{-.2x})^2$ **57.** $y' = 1/(x \ln |x|)$
59. (a) 20 (b) 1.34   **61.** (a) .005 (b) .0007 (c) .000013 (d) −.022 (e) −.0029 (f) −.000054   **63.** (a) 610.70
(b) 745.91 (c) 911.06 (d) 1112.77 **65.** $200e^{.4} \approx 298$; $200e^{1.6} \approx 991$ **67.** (a) 7.42 (b) 21.2 (c) .8

69.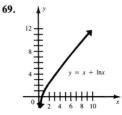

71.

Chapter 11   Review Exercises   (page 550)

1. 4   **3.** −3   **5.** 4   **7.** 17/3   **9.** 8   **11.** −13   **13.** 1/6   **15.** 1/5   **17.** 3/4   **19.** None   **21.** No; yes; yes
23. Yes; no; yes; no   **25.** Yes; no; yes   **27.** Yes; yes; yes
29. (a) $150   (b) $187.50   (c) $189
(e) When $x = 125$
(d)

31. $(-3, \infty)$ **33.** $[-4, 2]$ **35.** $(-\infty, -1]$ **37.** 1/4   **39.** −3/2   **41.** 30   **43.** 9/77   **45.** $y' = 4$
47. $y' = -3x^2 + 7$ **49.** −2; $y + 2x = -4$ **51.** −3/4; $3x + 4y = -9$ **53.** 38; $y = 38x + 22$ **55.** 3/4; $4y = 3x + 7$
57. $x = 0$ **59.** $x = 1/6$ **61.** None   **63.** (a) 315 (b) 1015 (c) 1515   **65.** (a) 55/3 = 18 1/3 (b) 65/4 = 16 1/4 (c) 15
67. (a) −9.5 (b) −2.375   **69.** $y' = 10x - 7$ **71.** $y' = 14x^{4/3}$ **73.** $f'(x) = -3x^{-4} + (1/2)x^{-1/2}$ or $-3/x^4 + 1/(2x^{1/2})$
75. $y' = 15t^4 + 12t^2 - 7$ **77.** $y' = 12x^{1/2} - 6x^{-1/2}$ or $12x^{1/2} - 6/x^{1/2}$ **79.** $g'(t) = -20t^{1/3} - 7t^{-2/3}$ or $-20t^{1/3} - 7/t^{2/3}$
81. $y' = 9x^{-3/4} - 45x^{-7/4}$ or $9/x^{3/4} - 45/x^{7/4}$ **83.** $k'(x) = 15/(x + 5)^2$ **85.** $y' = (2 - x + 2x^{1/2})/[2x^{1/2}(x + 2)^2]$
87. $y' = (x^2 - 2x)/(x - 1)^2$ **89.** $f'(x) = 12(3x - 2)^3$ **91.** $y' = 1/(2t - 5)^{1/2}$ **93.** $y' = 3(2x + 1)^2(8x + 1)$
95. $r'(t) = (-15t^2 + 52t - 7)(3t + 1)^4$ **97.** $y' = (x^2 - 4x + 4)(x - 2)^2 = 1$ **99.** $y' = -12e^{2x}$ **101.** $y' = -6x^2e^{-2x^3}$
103. $y' = 10xe^{2x} + 5e^{2x} = 5e^{2x}(2x + 1)$ **105.** $y' = 2x/(2 + x^2)$ **107.** $y' = (x - 3 - x \ln |3x|)/[x(x - 3)^2]$
109. $-1/[x^{1/2}(x^{1/2} - 1)^2]$ **111.** $dy/dt = (1 + 2t^{1/2})/[4t^{1/2}(t^{1/2} + t)^{1/2}]$ **113.** −2/3

Chapter 12

Section 12.1   (page 564)

1. Relative minimum of -4 at $x = 1$ **3.** Relative maximum of 3 at $x = -2$ **5.** Relative maximum of 3 at $x = -4$; relative minimum of 1 at $x = -2$ **7.** Relative maximum of 3 at $x = -4$; relative minimum of -2 at $x = -7$ and $x = -2$ **9.** Relative minimum of -44 at -6 **11.** Relative maximum of 17 at -3 **13.** Relative maximum of -8 at -3; relative minimum of -12 at -1 **15.** Relative maximum of 827/96 at $-1/4$; relative minimum of $-377/6$ at -5 **17.** Relative maximum of 57 at 2; relative minimum of 30 at 5   **19.** Relative maximum of -4 at 0; relative minimum of -85 at 3 and -3 **21.** Relative maximum of 0 at 8/5   **23.** Relative maximum of 1 at -1; relative minimum of 0 at 0   **25.** No relative extrema   **27.** Relative minimum of 0 at 0   **29.** Relative maximum of 0 at 1; relative minimum of 8 at 5   **31.** A maximum of $1/e$ at $x = -1$ **33.** A minimum of 0 at $x = 0$; a maximum of $4/e^2$ at $x = 2$ **35.** A minimum of 2 at $x = 0$ **37.** Relative minimum of $-1/e$ when $x = 1/e$; by symmetry and absolute value there is a relative maximum of $1/e$ when $x = -1/e$ **39.** No absolute maximum; absolute minimum at x_1 **41.** Absolute maximum at x_1; absolute minimum at x_2 **43.** Absolute maximum at x_2; absolute minimum at x_3 **45.** Absolute maximum at 0; absolute minimum at -3 **47.** Absolute maximum at -1; absolute minimum at -5 **49.** Absolute maximum at -2; absolute minimum at 4   **51.** Absolute maximum at -2; absolute minimum at 3   **53.** Absolute maximum at 6; absolute minimum at -4 and 4   **55.** Absolute maximum at 0; absolute minimum at 2   **57.** Absolute maximum at 6; absolute minimum at 4   **59.** Absolute maximum at $\sqrt{2}$; absolute minimum at 0   **61.** Absolute maxima at -3 and 3, absolute minimum at 0   **63.** Absolute maximum at 0; absolute minimum at -4 **65.** Absolute maximum at 0; absolute minima at -1 and 1   **67.** 10; 180   **69.** 10 hundred thousand tires; 700 thousand dollars   **71.** 25; 16.1   **73.** 12°C   **75.** 119/2 items   **77.** Absolute maximum is about 3.4, at about $x = 1.3$; absolute minimum is -7 at $x = 0$ **79.** Absolute maximum on $[-5, -4]$ is about 60.4 at -4.5 or -4.4; absolute minimum is 58, at -5; absolute maximum on [0, 1] is 4 at 1; absolute minimum is about 1.6 at about .4 or .5

Section 12.2   (page 569)

1. $f''(x) = 18x$; 0; 36; -54 **3.** $f''(x) = 36x^2 - 30x + 4$; 4; 88; 418   **5.** $f''(x) = 6(x + 4)$; 24; 36; 6   **7.** $f''(x) = 10/(x - 2)^3$; $-5/4$; $f''(2)$ does not exist; $-2/25$ **9.** $f''(x) = 2/(1 + x)^3$; 2; 2/27; $-1/4$ **11.** $f''(x) = -1/[4(x + 4)^{3/2}]$; $-1/32$; $-1/(4 \cdot 6^{3/2}) \approx -.0170$; $-1/4$ **13.** $f''(x) = (-6/5)x^{-7/5}$ or $-6/(5x^{7/5})$; $f''(0)$ does not exist; $-6/(5 \cdot 2^{7/5}) \approx -.4547$; $-6/[5 \cdot (-3)^{7/5}] \approx .2578$ **15.** $f''(x) = 2e^x$; 2; $2e^2$; $2e^{-3} = 2/e^3$ **17.** $f'''(x) = -24x$; $f^{(4)} = -24$ **19.** $f'''(x) = 240x^2 + 144x$; $f^{(4)} = 480x + 144$ **21.** $f'''(x) = 18(x + 2)^{-4}$ or $18/(x + 2)^4$; $f^{(4)}(x) = -72(x + 2)^{-5}$ or $-72/(x + 2)^5$ **23.** $f'''(x) = 40e^{2x}$; $f^{(4)}(x) = 80e^{2x}$ **25.** $f'''(x) = 2/x^3$; $f^{(4)}(x) = -6/x^4$ **27.** $f'''(x) = -1/x^2$; $f^{(4)}(x) = 2/x^3$ **29.** Relative minimum at -3 **31.** Relative minimum at 1   **33.** Relative minimum at 0; relative maximum at $-4/3$ **35.** Relative maximum at 1/2; relative minimum at -3 **37.** Relative maximum at -1; relative minimum at 1/2   **39.** Critical value at 0, but neither a relative maximum nor minimum there   **41.** Relative maximum at 0; relative minimum at -2 and 2   **43.** Relative maximum at $-\sqrt{3}$; relative minimum at $\sqrt{3}$ **45.** Relative maximum at -3; relative minimum at 3   **47.** No critical values; no maximum or minima   **49.** 6 units; $1072   **51.** 8 inches   **53. (a)** After 2 hours **(b)** 3/4%   **55. (a)** At 4 hours **(b)** 1160 million

Section 12.3   (page 578)

1. (a) $y = 100 - x$ **(b)** $P = x(100 - x)$ **(c)** $P' = 100 - 2x$; $x = 50$ **(d)** 50 and 50 **(e)** $50 \cdot 50 = 2500$ **3.** 100; 100   **5.** 100; 50   **7.** 10; 0   **9. (a)** $R(x) = 100,000x - 100x^2$ **(b)** 500 **(c)** 25,000,000 cents   **11. (a)** $\sqrt{3200} \approx 56.6$ mph **(b)** $45.25   **13. (a)** $1200 - 2x$ **(b)** $A(x) = 1200x - 2x^2$ **(c)** 300 m **(d)** 180,000 sq m   **15.** 405,000 sq m   **17.** 20,000 sq ft   **19. (a)** $40 - 2x$ **(b)** $100 + 5x$ **(c)** $R(x) = 4000 - 10x^2$ **(d)** Pick now **(e)** $40 per tree   **21. (a)** 90 **(b)** $405   **23.** 4 in by 4 in by 2 in   **25.** 3 ft by 6 ft by 2 ft   **27.** 40 mph gives the minimum cost of $800.   **29.** 1 mile from A   **31.** 10   **33.** 60   **35.** 5

Section 12.4 (page 591)

1. Increasing on $(1, \infty)$; decreasing on $(-\infty, 1)$ **3.** Increasing on $(-\infty, -2)$; decreasing on $(-2, \infty)$ **5.** Increasing on $(-\infty, -4)$ and $(-2, \infty)$; decreasing on $(-4, -2)$ **7.** Increasing on $(-7, -4)$ and $(-2, \infty)$; decreasing on $(-\infty, -7)$ and $(-4, -2)$
9. $(-6, \infty)$; $(-\infty, -6)$ **11.** $(-\infty, 3/2)$; $(3/2, \infty)$ **13.** $(-\infty, -3)$, $(4, \infty)$; $(-3, 4)$

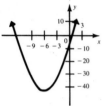

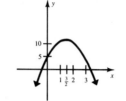

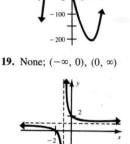

15. $(-\infty, -3/2)$, $(4, \infty)$; $(-3/2, 4)$ **17.** None; $(-\infty, \infty)$ **19.** None; $(-\infty, 0)$, $(0, \infty)$

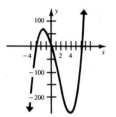

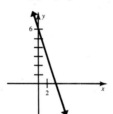

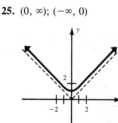

21. $(-4, \infty)$; $(-\infty, -4)$ **23.** None; $(1, \infty)$ **25.** $(0, \infty)$; $(-\infty, 0)$

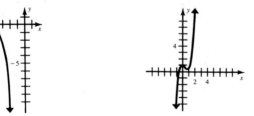

27. Concave upward on $(2, \infty)$; concave downward on $(-\infty, 2)$; point of inflection at $(2, 3)$ **29.** Concave upward on $(-\infty, -1)$ and $(8, \infty)$; concave downward on $(-1, 8)$; points of inflection at $(-1, 7)$ and $(8, 6)$ **31.** Concave upward on $(2, \infty)$; concave downward on $(-\infty, 2)$; no points of inflection   **33.** Always concave upward; no points of inflection   **35.** Always concave downward; no points of inflection   **37.** Concave upward on $(-1, \infty)$; concave downward on $(-\infty, -1)$; point of inflection at $(-1, 44)$ **39.** Concave upward on $(-\infty, 3/2)$; concave downward on $(3/2, \infty)$; point of inflection at $(3/2, 525/2)$ **41.** Concave upward on $(5, \infty)$; concave downward on $(-\infty, 5)$; no points of inflection   **43.** Concave upward on $(-10/3, \infty)$; concave downward on $(-\infty, -10/3)$; point of inflection at $(-10/3, -250/27)$
45. No points of inflection   **47.** Point of inflection at $x = 1/3$ **49.** Point of inflection at $x = -3/2$

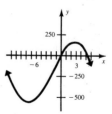

51. Point of inflection at $x = -7/12$

53. No points of inflection

55. Points of inflection at $x = -\sqrt{3}$ and $x = \sqrt{3}$

57. No points of inflection

59. No points of inflection

61. No points of inflection

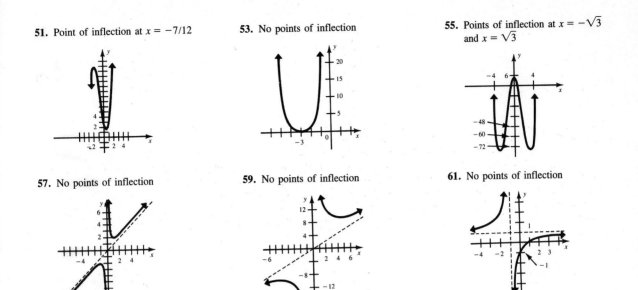

63. **(a)** Near trash compactors **(b)** Near wringer washers **(c)** Color tv's, room air conditioners: rate of growth of sales will now decline **(d)** Black and white tv's and wringer washers: the rate of decline is starting to slow

Section 12.5   (page 597)

1. **(a)** 6 **(b)** -8 **(c)** -20 **(d)** 43   **3.** **(a)** $\sqrt{43}$ **(b)** 6 **(c)** $\sqrt{19}$ **(d)** $\sqrt{11}$ **5.** **(a)** 446 **(b)** 286 **(c)** 816 **(d)** 91   **7.** **(a)** 7.85
(b) 4.02

9.

11.

$2x + 3y + 4z = 12$

13.

$3x + 2y + z = 18$

15.

$x + y = 4$

17.

$x = 2$

Section 12.6 (page 603)

1. (a) $24x - 8y$ **(b)** $-8x + 6y$ **(c)** 24 **(d)** -20 **3.** $f_x = -2y$; $f_y = -2x + 18y^2$; 2; 170 **5.** $f_x = 9x^2y^2$; $f_y = 6x^3y$; 36; -1152
7. $f_x = e^{x+y}$; $f_y = e^{x+y}$; e^1 or e; e^{-1} or $1/e$ **9.** $f_x = -15e^{3x-4y}$; $f_y = 20e^{3x-4y}$; $-15e^{10}$; $20e^{-24}$
11. $f_x = (-x^4 - 2xy^2 - 3x^2y^3)/(x^3 - y^2)^2$; $f_y = (3x^3y^2 - y^4 + 2x^2y)/(x^3 - y^2)^2$; $-8/49$; $-1713/5329$
13. $f_x = 6xy^3/(1 + 3x^2y^3)$; $f_y = 9x^2y^2/(1 + 3x^2y^3)$; 12/11; 1296/1297 **15.** $f_x = e^{x^2y}(2x^2y + 1)$; $f_y = x^3e^{x^2y}$; $-7e^{-4}$; $-64e^{48}$
17. $f_{xx} = 36xy$; $f_{yy} = -18$; $f_{xy} = f_{yx} = 18x^2$ **19.** $R_{xx} = 8 + 24y^2$; $R_{yy} = -30xy + 24x^2$; $R_{xy} = R_{yx} = -15y^2 + 48yx$
21. $r_{xx} = -8y/(x + y)^3$; $r_{yy} = 8x/(x + y)^3$; $r_{xy} = r_{yx} = (4x - 4y)/(x + y)^3$ **23.** $z_{xx} = 0$; $z_{yy} = 4xe^y$; $z_{xy} = z_{yx} = 4e^y$
25. $r_{xx} = -1/(x + y)^2$; $r_{yy} = -1/(x + y)^2$; $r_{xy} = r_{yx} = -1/(x + y)^2$ **27.** $z_{xx} = 1/x$; $z_{yy} = -x/y^2$; $z_{xy} = z_{yx} = 1/y$
29. $f_x = 2x$; $f_y = z$; $f_z = y + 4z^3$; $f_{yz} = 1$ **31.** $f_x = 6/(4z + 5)$; $f_y = -5/(4z + 5)$; $f_z = -4(6x - 5y)/(4z + 5)^2$; $f_{yz} = 20/(4z + 5)^2$
33. $f_x = (2x - 5z^2)/(x^2 - 5xz^2 + y^4)$; $f_y = 4y^3/(x^2 - 5xz^2 + y^4)$; $f_z = -10xz/(x^2 - 5xz^2 + y^4)$; $f_{yz} = 40xy^3z/(x^2 - 5xz^2 + y^4)^2$
35. (a) 80 **(b)** 180 **(c)** 110 **(d)** 360 **37.** $.7x^{-.3}y^{.3}$ or $.7y^{.3}/x^{.3}$; $.3x^{.7}y^{-.7}$ or $.3x^{.7}/y^{.7}$ **39. (a)** 168 **(b)** 5448 **(c)** 96 **(d)** 3882
41. (a) $.08585\, W^{-.575}H^{.725}$ **(b)** $.0112$ **(c)** $.14645\, W^{.425}H^{-.275}$ **(d)** $.783$

Section 12.7 (page 611)

1. Saddle point at $(1, -1)$ **3.** Relative minimum at $(-1, -1/2)$ **5.** Relative minimum at $(-2, -2)$ **7.** Relative minimum at $(15, -8)$ **9.** Relative maximum at $(2/3, 4/3)$ **11.** Saddle point at $(2, -2)$ **13.** Saddle point at $(1, 2)$ **15.** Saddle point at $(0, 0)$; relative minimum at $(4, 8)$ **17.** Saddle point at $(0, 0)$; relative minimum at $(9/2, 3/2)$ **19.** Saddle point at $(0, 0)$ **21.** Saddle point at $(0, 0)$; relative maximum at $(1, 1)$ **23.** Saddle point at $(0, 0)$; relative minimum at $(1, 0)$; relative maximum at $(-1, 0)$ **25.** Saddle points at $(0, 0)$, $(1, 1)$, $(1, -1)$, $(-1, 1)$, and $(-1, -1)$; relative maxima at $(0, 1)$ and $(0, -1)$; relative minima at $(1, 0)$ and $(-1, 0)$ **27.** $P(12, 40) = 2744$ **29.** $x = 12$, $y = 72$ **31.** $C(12, 25) = 2237$

Chapter 12 Review Exercises (page 612)

1. Relative maximum of -4 when $x = 2$ **3.** Relative minimum of -7 when $x = 2$ **5.** Relative maximum of 101 when $x = -3$; relative minimum of -24 when $x = 2$ **7.** Relative maximum of $-151/54$ at $1/3$; relative minimum of $-27/8$ at $-1/2$ **9.** Relative maximum of 1 when $x = 0$; relative minimum of $-2/3$ when $x = -1$ and $-29/3$ when $x = 2$ **11.** None **13.** Relative minimum of $-e^{-1}$ when $x = -1$ **15.** Relative minimum of e^2 when $x = 2$ **17.** Absolute maximum of 29/4 at 5/2; absolute minimum of 5 at 1 and 4 **19.** Absolute maximum of 39 at -3; absolute minimum of $-319/27$ at 5/3 **21.** $f''(x) = 36x^2 - 10$; 26; 314 **23.** $f''(x) = 14/(x + 1)^3$; 7/4; $-7/4$ **25.** $f''(x) = 2e^{-x^2}(2x^2 - 1)$; $2e^{-1}$; $34e^{-9}$ **27.** Increasing on $(5/2, \infty)$; decreasing on $(-\infty, 5/2)$ **29.** Increasing on $(-4, 2/3)$; decreasing on $(-\infty, -4)$ and $(2/3, \infty)$ **31.** Always decreasing **33.** Concave downward on $(-1/12, \infty)$; concave upward on $(-\infty, -1/12)$; point of inflection when $x = -1/12$ **35.** Always concave upward; no point of inflection **37.** Concave upward on $(-\infty, 0)$; concave downward on $(0, \infty)$; no point of inflection **39.** Concave upward on $(-\infty, 3)$; concave downward on $(3, \infty)$; no point of inflection **41.** 12.5 and 12.5 **43. (a)** $x = 2$ **(b)** \$20 **45.** 225 m by 450 m **47.** 30 **49.** $f_x = 16x + 2y$; $f_y = 2x + 3$; $f_{xx} = 16$; $f_{yy} = 0$; $f_{xy} = 2$ **51.** $f_x = 2xy^3 + 8y$; $f_y = 3x^2y^2 + 8x$; $f_{xx} = 2y^3$; $f_{yy} = 6x^2y$; $f_{xy} = 6xy^2 + 8$ **53.** Saddle point at $(0, 2)$ **55.** Saddle point at $(3, 1)$ **57.** Saddle point at $(4, 2)$ **59. (a)** $x = 11$; $y = 12$ **(b)** $C(11, 12) = 431$

Case 15 (page 615)

1. 20 knots (for a maximum profit of \$6535.71)

Case 16 (page 617)

1. $-C_1/m^2 + DC_3/2$ **2.** $m = \sqrt{2C_1/(DC_3)}$ **3.** About 3.33   **4.** $m^+ = 4$ and $m^- = 3$ **5.** $Z(m^+) = Z(4) = \$11,400$; $Z(m^-) = Z(3) = \$11,300$ **6.** 3 months; 9 trainees per batch

Chapter 13

Section 13.1 (page 624)

1. $2x^2 + C$ **3.** $5t^3/3 + C$ **5.** $6k + C$ **7.** $z^2 + 3z + C$ **9.** $x^3/3 + 3x^2 + C$ **11.** $t^3/3 - 2t^2 + 5t + C$
13. $z^4 + z^3 + z^2 - 6z + C$ **15.** $10z^{3/2}/3 + C$ **17.** $2u^{3/2}/3 + 2u^{5/2}/5 + C$ **19.** $6x^{5/2} + 4x^{3/2}/3 + C$
21. $4u^{5/2} - 4u^{7/2} + C$ **23.** $-1/z + C$ **25.** $-1/(2y^2) - 2y^{1/2} + C$ **27.** $9/t - 2\ln|t| + C$ **29.** $e^{2t}/2 + C$
31. $-15e^{-.2x} + C$ **33.** $3\ln|x| - 8e^{-.5x} + C$ **35.** $\ln|t| + 2t^3/3 + C$ **37.** $e^{2u}/2 + 2u^2 + C$ **39.** $x^3/3 + x^2 + x + C$
41. $6x^{7/6}/7 + 3x^{2/3}/2 + C$ **43.** $C(x) = 2x^2 - 5x + 8$ **45.** $C(x) = .2x^3/3 + 5x^2/2 + 10$ **47.** $C(x) = 2x^{3/2}/3 - 8/3$
49. $C(x) = x^3/3 - x^2 + 3x + 6$ **51.** $C(x) = \ln|x| + x^2 + 7.45$ **53.** $P(x) = -x^2 + 20x - 50$ **55.** $5t^2 - 30t^{1/2}$ tons
57. $f(x) = (3/5)x^{5/3}$

Section 13.2 (page 632)

1. 18   **3.** 65   **5.** 20   **7.** 8   **9.** 56   **11.** 32; 38   **13.** 15; 31/2   **15.** 20; 30   **17.** 16; 14   **19.** 12.8; 27.2
21. 2.67; 2.08   **23.** (a) 3 (b) 3.5 (c) 4   **25.** About 10,000,000 cars   **27.** A concentration of about 19 units   **29.** About
1900 feet   **31.** About 4   **33.** 13.9572   **35.** 1.28857

Section 13.3 (page 643)

1. -6 **3.** 3/2   **5.** 28/3   **7.** 38   **9.** 1/3   **11.** 76   **13.** 4/3   **15.** $5\ln 5 - (12/25) \approx 7.567$
17. $20e^{-.2} - 20e^{-.3} + 3\ln 3 - 3\ln 2 \approx 1.676$ **19.** $e^{10}/5 - e^5/5 - 1/2 \approx 4375.1$ **21.** 49/3   **23.** $447/7 \approx 63.857$
25. 42   **27.** 76   **29.** 24   **31.** 41/2   **33.** $e^2 - 3 - 1/e \approx 4.757$ **35.** 1   **37.** 23/3   **39.** $e^3 - 2e^2 + e \approx 8.026$

41. 19,000 units   **43.** (a) \$22,000 (b) \$62,000 (c) 49.5 days   **45.** About 2240   **47.** (a) $c(t) = 1.2e^{.04t}$ (b) $\displaystyle\int_0^{10} 1.2e^{.04t}\, dt$

(c) $30e^{.04} - 30 \approx 14.75$ billion (d) About 12.8 years (e) About 14.4 years   **49.** $(72/.014)(e^{.014T} - 1)$

Section 13.4 (page 653)

1. \$280; \$11.67   **3.** \$2500; \$30,000   **5.** No   **7.** (a) 12 years (b) About \$293 (c) \$2340   **9.** (a) 19.6 days (b) \$176,792
(c) \$80,752 (d) \$96,040   **11.** 31.5   **13.** 54   **15.** 5733.33   **17.** 1024/3; 1288/3   **19.** \$5.76 million dollars
21. (a) \$2000 (b) \$64,000   **23.** No   **25.** \$8700   **27.** 6960   **29.** (a) 75 (b) 100   **31.** .32 ft   **33.** (a) $I(.1) = .019$;
the lower 10% of the income producers earn 1.9% of the total income of the population (b) .184; the lower 40% of the income
producers earn 18.4% of the total income of the population (c) $I(.6) = .384$; the lower 60% of the income producers earn 38.4% of
the total income of the population (d) .819; the lower 90% of the income producers earn 81.9% of the total income of the population

Section 13.5 (page 662)

1. $2(2x + 3)^5/5 + C$ **3.** $-2(y - 2)^{-2} + C$ **5.** $-(2m + 1)^{-2}/2 + C$ **7.** $-(x^2 + 2x - 4)^{-3}/3 + C$ **9.** $(z^2 - 5)^{3/2}/3 + C$
11. $-2e^{2p} + C$ **13.** $e^{2x^3}/2 + C$ **15.** $e^{2t - t^2}/2 + C$ **17.** $-e^{1/z} + C$ **19.** $-8 \ln |1 + 3x|/3 + C$ **21.** $(\ln |2t + 1|)/2 + C$
23. $-(3v^2 + 2)^{-3}/18 + C$ **25.** $-(2x^2 - 4x)^{-1}/4 + C$ **27.** $[(1/r) + r]^2/2 + C$ **29.** $(x^3 + 3x)^{1/3} + C$
31. $(p + 1)^7/7 - (p + 1)^6/6 + C$ **33.** $2(5t - 1)^{5/2}/125 + 2(5t - 1)^{3/2}/75 + C$ **35.** $2(u - 1)^{3/2}/3 + 2(u - 1)^{1/2} + C$ **37.** 15/4
39. 512/3   **41.** $[\ln(e^2 - 4e + 6) - \ln 3]/2$ or $\ln[(e^2 - 4e + 6)/3]^{1/2} \approx -.0880$ **43.** $(e^2 - 1)/4 \approx 1.597$
45. $x \ln 4 + x(\ln |x| - 1) + C$ **47.** $-4 \ln |(x + \sqrt{x^2 + 36})/6| + C$ **49.** $\ln |(x - 3)/(x + 3)| + C, (x^2 > 9)$
51. $(4/3) \ln [(3 + \sqrt{9 - x^2})/x] + C, (0 < x < 3)$ **53.** $-2x/3 + 2 \ln |3x + 1|/9 + C$ **55.** $(-2/15) \ln |x/(3x - 5)| + C$
57. $\ln |(2x - 1)/(2x + 1)| + C$ **59.** $-3 \ln |(1 + \sqrt{1 - 9x^2})/(3x)| + C$ **61.** $x^5[(\ln |x|)/5 - 1/25] + C$
63. $(1/x)(-\ln |x| - 1) + C$ **65.** $-xe^{-2x}/2 - e^{-2x}/4 + C$ **67.** \$1,466,840   **69.** .5 million dollars

Section 13.6 (page 667)

1. $y = 4x^2 - 7x + C$ **3.** $y = -x^2 + x^3 + C$ **5.** $y = -4x + 3x^4/4 + C$ **7.** $e^{4x}/4 + C$ **9.** $-e^{-6x}/3 + C$
11. $-(1 - x^2)^{3/2}/3 + C$ **13.** $-e^{-x^3}/3 + C$ **15.** $y = 2x^3 - 2x^2 + 4x + C$ **17.** $y = -4x^3/3 + C_1x + C_2$
19. $y = 5x^2/2 - 2x^3/3 + C_1x + C_2$ **21.** $y = x^4/3 - x^3/3 + C_1x + C_2$ **23.** $y = -5x^3/6 + 3x^2/2 + C_1x + C_2$
25. $y = e^x + C_1x + C_2$ **27.** $y = x^3 - x^2 + 2$ **29.** $y = 5x^2/2 + 2x - 3$ **31.** $y = x^3 - 2x^2 + 2x + 8$
33. $y = x^3/3 + x^2/2 - 5x/6 + 2$ **35.** $y = e^x + x^2/2 - ex + 1$ **37.** $y = x^4/4 - x^2 + x + 4$ **39. (a)** $y = 5t^2 + 250$
(b) 12 days   **41. (a)** 156 **(b)** 198 **(c)** 206 **(d)** 6   **43. (a)** 1080 thousand **(b)** 1125 thousand **(c)** 1000 thousand
(d) 625 thousand   **45.** About 13.4 grams   **47.** About 332,000

Chapter 13   Review Exercises (page 669)

1. $x^4/2 - x^2/2 + C$ **3.** $2x^3 - x^2 + 5x + C$ **5.** $2x^{5/2}/5 - x^2/2 + 5x^{7/5}/7 + C$ **7.** $2x^{5/2}/5 + C$ **9.** $-x^{-2}/2 + C$
11. $-4x^{1/2} - 3 \ln |x| + C$ **13.** $-e^{-5x}/5 + C$ **15.** $-5 \ln |x| + C$ **17.** $C(x) = 3x^2 + 2x + 75$
19. $C(x) = 4x^{7/4}/7 + x^2 + 321/7$ **21.** About 34   **23.** 12   **25.** 72/25   **27.** 2 ln 3, or ln 9   **29.** $2(e^4 - 1)$
31. 36,000   **33.** 64/3   **35.** About 90   **37.** $2(x^2 - 3)^{3/2}/3 + C$ **39.** $-1/[9(x^3 + 5)^3] + C$ **41.** $\ln |2x^2 - 5x| + C$
43. $-e^{-3x^4}/12 + C$ **45.** $2e^{-5x}/5 + C$ **47.** $\ln |(x + \sqrt{x^2 - 64})/8| + C$ **49.** $-(1/2) \ln |x/(3x + 4)| + C$
51. $3x^3[(\ln |x|)/3 - 1/9] + C$ or $x^3[3 \ln |x| - 1]/3 + C$ **53.** $2 \ln [(x - 3)/(x + 3)] + C, x^2 > 9$ **55.** $y = 2x^3/3 - 5x^2/2 + C$
57. $y = -3x + x^3 + x^5 + C$ **59.** $y = 2x^{5/2}/5 - x^2 + C$ **61.** $y = -2e^{-2x} + C$
63. $y = x^5/20 + 5x^4/12 - x^2/2 + C_1x + C_2$ **65.** $y = x^4/4 + e^x + C_1x + C_2$ **67.** $y = 5x^3/3 - 3x^2 + 4$ **69.** $y = \ln |x| - 2$

Case 17 (page 673)

1. About 102 years   **2.** About 55.6 years   **3.** About 45.5 years   **4.** About 90 years

Case 18 (page 675)

1. 10.65   **2.** 175.8   **3.** 441.46

Index

Definition of the Derivative The derivative of the function $y = f(x)$ is written $f'(x)$ and is defined as

$$f'(x) = \lim_{h \to 0} \frac{f(x + h) - f(x)}{h},$$

provided this limit exists. Alternative notations for the derivative include dy/dx, $D_x y$, $D_x[f(x)]$, and y'.

Rules for Derivatives (Assume all indicated derivatives exist.)

Constant Function If $f(x) = k$, where k is any real number, then

$$f'(x) = 0.$$

Power Rule If $f(x) = x^n$, for any real number n, then

$$f'(x) = n \cdot x^{n-1}.$$

Constant Times a Function Let k be a real number. Then the derivative of $y = k \cdot f(x)$ is

$$y' = k \cdot f'(x).$$

Sum or Difference Rule If $y = f(x) \pm g(x)$, then

$$y' = f'(x) \pm g'(x).$$

Product Rule If $f(x) = g(x) \cdot k(x)$, then

$$f'(x) = g(x) \cdot k'(x) + k(x) \cdot g'(x).$$

Quotient Rule If $f(x) = \dfrac{g(x)}{k(x)}$, and $k(x) \neq 0$, then

$$f'(x) = \frac{k(x) \cdot g'(x) - g(x) \cdot k'(x)}{[k(x)]^2}.$$

Chain Rule If y is a function of u, say $y = f(u)$, and if u is a function of x, say $u = g(x)$, then $y = f(u) = f[g(x)]$, and

$$\frac{dy}{dx} = \frac{dy}{du} \cdot \frac{du}{dx}.$$

Chain Rule (alternate form) Let $y = f[g(x)]$. Then

$$y' = f'[g(x)] \cdot g'(x).$$

Generalized Power Rule Let u be a function of x, and let $y = u^n$ for any real number n. Then

$$y' = n \cdot u^{n-1} \cdot u'.$$

Natural Logarithmic Function If $y = \ln|g(x)|$, then

$$y' = \frac{g'(x)}{g(x)}.$$

Exponential Function If $y = e^{g(x)}$, then

$$y' = g'(x) \cdot e^{g(x)}.$$